Advances in GIS Research II

⌐ Advances in GIS Research II ⌐

Proceedings of the Seventh International Symposium on Spatial Data Handling

Edited by

Menno-Jan Kraak (Director SDH'96)
Martien Molenaar (Chair, Programme Committee)
Elfriede M. Fendel

Faculty of Geodetic Engineering
Delft University of Technology

International Geographical Union Commission
on GIS

Supported by
Aula Congress Office of the Delft University of Technology

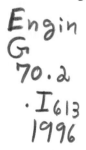

Taylor & Francis
Publishers since 1798

UK   Taylor & Francis Ltd, 1 Gunpowder Square, London EC4A 3DE
USA   Taylor & Francis Inc., 1900 Frost Road, Suite 101, Bristol, PA 19007

British Library Cataloguing in Publication Data

A catalogue record for this book is available from the British Library
ISBN 0-7484-0591-7

Library of Congress Cataloging Publication Data are available

Cover design by Hybert Design & Type

Printed in Great Britain by T.J. International Ltd, Padstow

Martien Molenaar studied Geodesy at Delft Technical University from 1967 to 1972. After finishing this study he joined the International Institute for Aerospace Survey and Earth Science (ITC) in Enschede, The Netherlands, as a lecturer in photogrammetry, geodesy and mathematics. His research was on quality control of geodetic networks and photogrammetric blocks. He wrote a PhD thesis on this topic and received his degree in Delft in 1981. In 1983 he got the Chair on GIS and remote sensing at Wageningen Agricultural University, where he became head of the Department of Geo-Information Processing and RS. His research was there on the integration of GIS and RS and on datamodelling for GIS, for application fields like soil science, physical planning, land use planning etc. He returned to ITC in October 1996 where he now occupies the Chair on Spatial Information Production from Photogrammetry and RS of the Department of Geoinformatics.

Menno-Jan Kraak studied Physical Geography and Cartography at the Utrecht University between 1976 and 1981. In 1983 he started as a lecturer in Cartography and GIS at the Faculty of Geodetic Engineering at the Delft University of Technology. Here he got a PhD degree on the topic of Three-dimensional Cartography in 1988. In 1996 he moved to the ITC in Enschede where he holds the Chair in Cartography of the Department of Geoinformatics. He is co-Chair of the International Cartographic Association's Commission on visualization and co-author of the book *Cartography: visualization of spatial data*.

Contents

Contents

Contents

Contents

Programme Committee
Chair: M. Molenaar

H.J.G.L. Aalders (NL)
P. Boursier (F)
K.E. Brassel (CH)
A. Bregt (NL)
P. Burrough (NL)
A.G. Cohn (UK)
H. Couclelis (USA)
D.J. Cowen (USA)
B. David (F)
L. De Floriani (I)
G. Edwards (CDN)
M.J. Egenhofer (USA)
P. Fisher (UK)
A.U. Frank (A)
K.P. Gapp (D)
C.M. Gold (CDN)
M. Goodchild (USA)
D. Green (UK)
D. Grünreich (D)
S.C. Guptill (USA)
C. Heipke (D)
S.C. Hirtle (USA)
S. de Hoop (NL)
C. Jones (UK
W. Kainz (NL
Z. Kemp (UK
M. Konecny (CZ)

M.-J. Kraak (NL)
R.E. Kuunders (NL)
J.-P. Lagrange (F)
W. Mackaness (UK)
D.F. Marble (USA)
D. Mark (USA)
S. Morehouse (USA)
J.C. Müller (D)
P.J.M. van Oosterom (NL)
F.-J. Ormeling (NL)
D.J. Peuquet (USA)
S. Pigot (AUS)
E. Puppo (I)
J.F. Raper (UK)
D. Richardson (CDN)
S.A. Robers (UK)
J. van Roessel (USA)
A. Ruas (F)
T. Sarjakoski (SF)
I.D.H. Shepherd (UK)
A. Stein (NL)
T. Vijlbrief (NL)
M. Visvalingam (UK)
T.C. Waugh (UK)
R. Weibel (CH)
M.F. Worboys (UK)

Sponsors

Main
ESRI Europe
Taylor & Francis
Delft University of Technology, Faculty of Geodetic Engineering

Subsidiary
Bridgis
Grontmij Geogroep
International Cartographic Association
Netherlands Cadastre
Netherlands Cartographic Society
Netherlands Geodetic Commission
Netherlands Photogrammetric Society
Ravi - Netherlands Council for Geographic Information
Survey Department of the Ministry of Transport, Public Works and Water Management
University of Utrecht/Department of Spatial Sciences

Author Index

Author Index

Foreword

Advances in GIS Research II

With this edition of the Proceedings of the Seventh International Symposium on Spatial Data Handling (SDH) a 12-year tradition in research for Geographical Information Systems continues. It started in 1984 when a Commission of the International Geographical Union (IGU) organized the first meeting in Zürich, Switzerland. Following this initial symposium, meetings were organized every two years. From Zürich SDH moved to Seattle (USA) in 1986, followed by Sydney (Australia) in 1988. The symposium returned to Zürich in 1990, after which it went to Charleston (USA) in 1992 and Edinburgh (UK) in 1994.

The 1996 SDH took place in Delft, The Netherlands, hosted by the Delft University of Technology. As started by the Edinburgh meeting, the proceedings contain papers which were fully reviewed by three international referees who are recognized as specialists in their field. More than 100 full papers were submitted from 26 different countries. After the selection procedure the programme features 63 high-quality papers from participants from 18 different countries, spread over 4 continents. Most of the papers rejected were of a very high quality as well, but because of the standards set by those who submitted a paper rejections were inevitable.

The symposia of the IGU Commission on GIS have become the bi-annual focal point for advances in GIS research. Unlike most other GIS meetings, SDH has been established to bring together in an intimate setting a relatively small, international group of interdisciplinary researchers who are working at the cutting edge of new approaches of handling geographic data. It offers a unique opportunity to exchange ideas, and present research progress and results. Next to the papers, which are generally considered to represent the state of the art in this field, the proceedings include an opening and a closing keynote paper. The results of the Seventh International Symposium on Spatial Data Handling are collected together and published in this post-symposium edition of the proceedings.

At the end of a foreword it is a good tradition to look back at the activities and efforts that happened before an event such as SDH took place and the proceedings were published. First, words of thanks should go to Elfriede M. Fendel, Managing Assistant of the Faculty of Geodetic Engineering's Geo-information Department, who played a major role in compiling the proceedings and assisted me with every aspect of the symposium. Axel Smits designed the SDH-logo, and Theo Tijssen and Robert Kuunders created the SDH WWW-site. Thanks also goes to Martien Molenaar and the Programme Committee who are responsible for the quality of the programme.

Menno-Jan Kraak
Delft, The Netherlands, 1996

TWELVE YEARS IMPACT OF SPATIAL DATA HANDLING

RESEARCH ON THE GIS COMMUNITY

François Salgé[1]

Institut Géographique National/GIE MEGRIN
Paris, France

1. Introduction

- Spatial or spacial, adj, of or relating to space ; existing or happening in space
- Space, n., the unlimited three-dimensional expanse in which all material objects are located
- Data, n., a series of observations, measurements or facts ; information ; computer technology : the information operated on by a computer programme.
- Handling, n., the act or an instance of picking up turning over or touching something ; the process by which a commodity is packaged, transported, etc.

From time to time it is fruitful to look again at the semantics of the words we use. Spatial Data Handling as a buzzword could be defined though as :

"the act of picking up, turning over, touching, packaging, transporting series of observations, measurements or facts, eventually operated on by computers, which are related or which exist or happen in the unlimited three dimensional expanse in which all material objects are located".

Which such a wide definition a lot of activities of humanity and society are relevant to "Spatial Data Handling". The counterpart of such a wide definition is that a lot of actors just simply do not realize that some of their problems are simplified or easier to handle with the use of Spatial Data.

---

[1]François Salgé is "Ingénieur en Chef Géographe" at Institut Géographique National in France and is currently Executive Director of the GIE MEGRIN, Chairman of CEN/TC 287 and Co Director of the ESF/GISDATA Programme. The opinion expressed in this paper reflects the Author's opinion and must not be taken as ideas from the organisations the Author works with.

This reminds me of a joke that will underpin the paper.

One man is crossing the Sahara and due to a lack of caution he ran out of drinkable water. He then slightly starts to get thirsty. He is almost to dying when he meet a merchant offering some goods for sale : "please I am out of water, he said, could you sell me something to drink ?". "No, was the answer, I just sell ties". The merchant disappears. More and more thirsty the man continue his way. Twice he meet other merchants. They have no drinks, just selling ties. Finally, he arrives close to an Oasis and just in front, a luxurious Pub with gloomy lights. Almost creeping he gets to the door. A gigantic Groom was controlling the entrance. "Get away, said the Groom, no entry for people without ties".

Who are thirsty people in our society ? Who are the ties merchants ? What is the pub, the desert ?

In order to try to answer these questions I will firstly illustrate the way, in the desert, NMAs have been going for 12 years. Then I will look with you at the past 12 years of the SDH Conferences and finally I would like to introduce a description of the desert (or the Jungle) which we are going to step into.

2. From maps to EGII

2.1 Creating DLM

Back to the early eighties, Digital Cartography was mainly a large scale issue (i.e. very local maps from 1:1250, (in UK) to 1:5000, as in France). Small scale mapping was digital only for thematic mapping. The keyword was not geographic information but digital mapping. The demand for providing information useful to the interpretation of Remote Sensing images was emerging. so was the ideas that NMAs should provide the basic, topographic, fuel for GIS.

Gradually the new challenges for the NMA were to create geographic data base instead of simply mimicking the analog maps in a computer environment. Using the German terminology it was thought more useful for the society to produce Digital Landscape Model (DLM) rather than Digital Cartographic Model (DCM). Model in that context should be understood as a "representation" of the world/geographic reality according to a given level of abstraction.

Being in charge of devising what is now known as the French BD Carto, a 10 to 20 m accuracy DLM, IGN-France started to investigate within four main areas
- content definition
- data base modeling
- data capture technique
- data storage

Content definition could have been an easy task if IGN was to mimic the content of the existing maps (1:100 000, 1: 250 000). The analysis of the history of these two map series showed a lot of problems difficult to circumvent : some of the sources used to create the map were from the XIX century which geometry was uncertain, some of the maps have been badly updated. It came rapidly evident that the only reliable source was the 1: 50 000 series. Using it as a source was not meaning digitizing the appearance of the maps : creating a DLM consist in revisiting them in order to identify the topographic objects that are cartographically represented instead of focusing on how it is represented. This definition of suitable content for DLMs was in 1986 the main challenge some NMAs were facing.

Content definition is a first step while defining a DLM. Immediately follows, the database modeling. Geographic Information is now seen as a standard information which is special as it has a Spatial component . In the 80's it was quite revolutionary to think GI as an information. At that time the "graphic" part of "geographic" was the key word. Designing DLM, with the aim of informing on the Landscape, needed to overpass the graphical aspects of GI and to think in term of information : there is geographic objects of the real world having their own characteristics and which interact with other geographic objects. Mid 80's was the "apogee" of the relational model and though NMAs wanted to model topography using object, attributes and relationships concepts.

Conceptualization is first in the DLM creation, data acquisition is second. Mid 80's tools for mass data acquisition were not widely distributed. NMAs were forced to develop their own data capture software . It was the early days of computer graphics and PC-based solution was not able to cope with graphics. Interactive data capture was in its infancy, most of the processes were batch : scanning, vectorizing, attributing.

Having digitized the DLM, one was thinking of allowing the data to the viewed as seamless over territories (such as a country). Mid 80's seamless was a dream. No DB systems was able to cope with such huge amount of data (Mega bytes was still an unusual unit for counting mass storage). It was a time were indexing (tiling, quadtries, ...) was a research topic.

2.2 Integrating DLMs from various levels of abstraction

The end of the 80's was a time when the term scaleless was much discussed ; another dream was starting. NMAs were facing three main issues as researchable questions
- how consistent databases could be derived regardless their level of abstraction
- how National Geographic Information systems could be organized
- how traditional cartography could be derived from the DB

"Scaleless" as a keyword was invented to reflect that in the Digital arena the concept of scale should be irrelevant. Discussion on the meaning of "Scale" was central.

Nowadays it is accepted that the word scale covers two aspects :
* the graphical aspect which allows to fix parameters such as legibility (what is the minimum distance between graphical primitives permitting to still differentiate features in the map).
* the level of abstraction aspects which allows to determine what are the key features and what are the details (relevant to the level of depicting the reality).

"Scaleless" was a term to escape from the graphical limits of the analog paper maps.

Linked to that concept, NMAs started to consider how to make consistent the databases assembled at different levels of abstraction. It was thought difficult to explain to users that one database at a given level of abstraction, was not consistent with the information provided at an other level of abstraction. These issues led, for example in Germany, to the definition of ATKIS as a system allowing to describe consistently the German topographical databases regardless their level of abstraction.

Scaleless was also a leading concept in the creation of National Geographic Information systems. Thoughts were developed towards using the same tools whatever the "scale" of the National database to be created. For example, IGN-France started to develop its concept of SIGNIGN "Système d'Informations Géographiques Nationales de l'Institut Géographique National". The principle was to decentralize, i.e. organize in different production centers, the creation of the national databases (BDTopo, BDCarto, BDAlti, Géoroute) but centralize the storage and exploitation for internal and external use.

End of the 80's, object oriented DB systems was so promising that it was thought possible to set up procedures allowing to extent from any geographic databases subset fulfilling geographic criteria (geographical extent) and semantic criteria (thematic extent) in a consistent way.

National Geographic databases started to be assembled in order to fulfill expected user requirements for digital data. A second objective was to enable the creation of the analog maps that was required for example by the general public from the databases. NMAs still have within their remits the provision of the National paper-based mapping which still represent, for those having a cost recovery constraint, a significant part of their revenue. For example one of the French BD Topo objectives, was to enable the production of several cartographic series : technical drawings at 1 : 5000 and 1: 10 000 or basic cartography at 1 : 25 000 and 1 : 50 000. Creating drawings from database is relatively easy, although illegible ones are also easy to produce ! Creating maps which respect the cartographic rules is far more difficult. NMAs at the end of the 80s started to raise the issue of generalization in a way different to the old topic of line simplification. Changing the level of abstraction was the question research had to solve.

Production lines of the NMAs was also increasingly using different systems for their different phases. Each step of the production was performed on a system specialized for the tasks to be undertaken. That lead to methods enabling the transfer of digital data from systems based on different software and hardware inside each production line. Consequently NMAs devised internal exchange formats which purpose was to allow communication from systems. FEIV in IGN France (Format d'échange interne en mode vecteur) or DLG-E in USGS (Digital Line Graph enhanced) exemplify such formats.

In parallel, delivery of data sets to users started to raise the issue of National Transfer Standards such NTF in the United Kingdom or EDIGéO in France. With a pragmatic approach, these standards were based more on practical usage than on a solid information theory. Starting to emerge were researchable questions such as the conceptualization of the exchange of geographic data from systems which was not limited to the simple transfer of bytes but could encompasses the question of transferring some level of knowledge : conceptual data schema, object, attribute and relationship catalogues.

The provision of geographical data sets which fit the user requirements and the ability to combine over a same area data from various sources lead to define new researchable questions such as the definition and assessment of the quality of geographical data sets. From history, maps produced by NMAs were always considered as the "best" ones although often viewed as out of date, and the confidence to the professionalism of the NMAs served as assessment of the quality of cartography. Within the digital view of geographic data, this statement started to become insufficient to the users. Concepts such as "fitness for use" or "suitability to applications" lead also to researchable questions .

2.3 Europeanisation of the Geographic Information Market

Beginning of the 90's, that is yesterday, it was clear for NMAs that they can no longer be considered as simple map-makers but need to behave as information providers in a more and more open market. Cost recovery concerns, Europeanisation of the market was key in the context the NMAs were operating. Evidence of that change lead the GI community to organize themselves at a European level : beginning of the 90's was the creation of EUROGI, MEGRIN, CEN/TC 287, GISDATA and other initiatives which major aim was to create synergies at a European level. In that context what was the new researchable questions the NMAs had to face ?

Knowing which data are on the market and available to the users lead to new developments in the area of meta data. The availability of information on geographic data sets was considered to be central to the development of a GI market. The role of such GI was thought to become prominent in the development of computer science and also in the improvement of the way we use our living space. Initiatives such as the GDDD (Geographical Data Description Directory) of MEGRIN or the SINES System in UK or the "Catalogue des sources en information géographique" in France exemplify this new concern.

Other key issues of Europeanisation was the creation of European wide geographical data sets. To initiate studies in that area MEGRIN undertook at the beginning of the 90's the creation of the Seamless Administrative Boundaries of Europe data set (SABE). Lack of such harmonized data sets was explicated late 80's by the results of the CORINE Project. Merging geographic data sets from various countries implies to solve difficult problems in four main areas : geodetic, geometrical and topological, semantic and temporal. Each area provided topics for research. NMAs within CERCO took over the responsibility to establish a new geodetic network (ETRF) covering the entire Europe in order to enable the computation of the parameters necessary to transform coordinates from national systems (horizontal datum, ellipsoid, height system) to the ETRF system and back. The use of techniques based on GPS was key to the issue.

Harmonizing geometry and topology is conceptually a non-researchable topic although practically it needs the development of efficient methods allowing to reduce the costs of harmonization. Nonetheless with the view of designing generalized methods allowing to assemble data sets having a different geometric & topographic structure, research was necessary.

The semantic side of merging data sets from different countries is far to be the most critical issue. Each country has its own tradition of structuring the semantic of geographic information. Here "semantic" means "of or relating to meaning or arising from distinctions between the meanings of different words or symbols". Semantic covers the conceptual modeling of the reality with regard to the nature of objects and their characteristics, as opposed to their position. Research was needed to find ways of going from one structure to the harmonized one.

Finally the temporal dimension of merging data sets from various countries still was an open problem. For each country, the manner in which geographic phenomena are observed and the very nature of the phenomena affect the way in which time information is handled. Attention of the research community was brought to the maintenance of temporal consistency across heterogeneous databases and on the aspects of temporal data that influence data quality.

Linked to that issue of time, NMAs in the beginning of the 90's started to be confronted with the problem of maintenance and update of digital geographic databases, including the difficult problem of history where geographic information on past situation of the Earth needed to be combined with current situation in order to study evaluation of phenomena.

Beginning of the 90's the issue of standardization started to be real concern at the European level. CEN (Comité Européen de Normalisation), the European Committee on standards, set up a new technical committee (CEN/TC 287) tasked with the creation of standards in the field of Geographic Information.

Lot of National initiatives and discipline oriented international work pre-existed, including NTF in UK, EDIGÉO in France, SDTS in USA, DIGEST for military purposes, SP57 for hydrography, GDF for road transport & traffic telematics. But there was a consensus to say that transfer formats such as previously listed, were the emerging part of the iceberg. More needed to be archived such as a common geometric/spatial conceptual model, meta data standards, quality models.

The fundaments for the standardization effort was clearly laid down in the scope of CEN/TC 287: "Standardization in the field of digital geographic information : this comprises a structured set of standards which specifies a methodology to define describe and transfer representations of the real world". In term of research topic, it was necessary to understand all the mechanisms which underpin the definition, description and transfer of representations of the real world. A reference model was needed to focus the concepts necessary to undertake standardization. Also it was necessary to evaluate existing or emerging standards such as EDIFACT, STEP, SQL or IRDS, against their availability or adaptability to geographic information. Defining standardization in the GI field while the Information and Communication Technology (ICT) was rapidly mutating needed careful attention towards the emerging de-facto standards. Early 90's Internet was far to be the unavoidable tool it is nowadays. Interoperability was a buzzword with a variety of meaning.

3. Twelve years of spatial data handling research

It would be soporific to undertake a comprehensive survey of the research on Spatial Data Handling matters undertaken during the last twelve years.

As we are in 1996 at the beginning of a new Olympiad, the following sections discuss the matters through the proceedings of SDH opening each past Olympiad. Browsing into the key note and the programme of the conferences provide interesting hints into the hot topics of the previous and coming years. One may think that the questions addressed in the past are still relevant where the way of addressing them simply changes.

3.1 SDH 84 in Zürich - Olympic games 84 in Los Angeles

In Zürich, Roger Tomlinson immediately pointed out that the " eighties would seem to be poised for significant technical break through in the speed, ease and flexibility with which geographical data can be handled and yet at the same time important methodological and institutional problems are emerging that the very existence of GIS is causing us to address". He was pointing at eight key questions related to this statement opening doors for new research.

1. "What is the nature of the queries we want to address to the data". A lot of manipulation functions have been developed since then : buffering, network analysis (shortest path, connectivity), layer combination. Few real spatial analysis functions are in fact available. Basic queries such as "what phenomenon exists at a given point", "what are the information I have there" or "where is located a given phenomenon" are routine in nowadays GIS. Still pending are satisfying answers to questions such as "what are the future consequences of such scenario" or " what will be the impact on the living space, at short, medium or long term, of a proposed decision ".

2. "Should there be a proliferation of data structures ?". Nowadays each GIS is based on differing data structures. Commercial interests have for long prevented the knowledge of internal data structure implying great difficulties in importing / exporting data from heterogeneous systems. Open GIS initiative may nonetheless overcome the issue.

3. "Should we be seeking for a new generation of data structures and algorithms ?". Research provided quite a large amount of proposals in that area where mutating technology created dinosaurs.

4. "Can existing data structure cope with very large data volumes". Answers to that are quite largely available. The question may be rapidly no longer relevant as the concept of distributed data resources spread on the net may limit the size of what is locally handled to strictly what is really necessary.

5. "Should relationships in databases be explicit or implicit ?". Although algorithms allow nowadays to explicit on the fly many implicit information , the question is still relevant as the complexity of the information structure increases with the requirements for maintaining the consistency of databases.

6. "How do we handle problems of data with different degrees of accuracy and precision". Due to the globalisation of the information as well as the requirements for global and regional monitoring of our living space, the advances permitted by past research are not sufficient to correctly provide solutions.

7. "Who get access to the data, how is it administered ?". Considered to be a hot topic in 84 by the research community, it took several years before meta data services and documenting geographical data sets start to become reality.

8. "What are the problems of interchange of data between databases". More and more experts are mastering the issue. Nonetheless as the number of users of Geographic Information is more rapidly increasing, the question is still valid and fall into education, training and awareness but also simplification of the user interaction with the system.

Tomlinson was concluding his keynote with a broader question "does spatial digital databases represent spatial reality or should they be copies of conventional views such as map images or both". With the advent of multimedia and the progress of personal computers and Internet, this question lead to a sharper segmentation of the usage of digital geographic information : from contextual display for human information to geographic information treated as key element of decision support systems through virtual reality, the representation of the spatial reality in computer environment need to be thought again in order to select the representation which best fits the intended usage. To the question of "how do we describe the world ?" we need to answer by another question "what purpose do we serve when we describe the world ?.

Browsing into the table of content of the SDH'84 volumes, the place of spatial handling in policy contexts was addressed from national standpoints : remote sensing, digital mapping, GIS on the one hand, large scale on the other hand. Spatial Data Models were discussed from the stand point of topography and large databases. Relational model was evaluated for spatial data handling together with indexing methods applied to GIS. Assistance to manual digitizing operations was evaluated through Conceptual Data Models as well as acoustic channel (voice recognition & synthetic voice). Analytic techniques were suggested. Application of the GIS technology ranged from hydrographic research and planning to land use and municipal databases, through environmental maps and atlases.

SDH'84 conclusion was given by Tomlinson's key note : "Given our new tools, how do we describe the world ? The answers to these questions will have a fundamental impact on the nature of geography as a discipline, and on the rational development of the world's resources. These present a new frontier to be approached and crossed in the development of GIS".

3.2 SDH' 88 in Sidney - Olympic games 88 in Seoul

At the time when a lot of people complains in Europe on an overweighed role of NMAs into the development of the Geographic Information Economic sector, it is worthwhile to recall Michael Dobson's keynote at SDH'88 : "Map publishers have been slow to adopt digital cartography as a production method due to the following factors : quality of output, cost of entry, speed of production, stability of systems, appropriateness of either general purpose drafting systems or GIS for map publishing".

In term of opportunities for printed map publishers a strategic choice was needed for the future in deciding whether the company will be a publisher of print products, a purveyor of digital data or perhaps examine the bottom line that might result from offering both services.

Many publishers however mainly in the private sector considered adventurous to turn into data provider for many reasons including protection of the investment made in the collection of their information as a result of many years of hard and costly work in assembling "facts" from many sources that competitors could use to market products into the same niches. Dobson in 88 was convinced that the "digital market place was somewhat further away than most suspect". In addition, he considered that the emerging technology in 88 would allow one to market digital data while maintaining its market position of maps only if one can successfully develop an effective and economical digital production system.

Dobson, in the late eighties, considered that the justification for turning production lines into digital methods could only be made by comparing the production costs : investment and running costs. The benefits of versatility of digital methods was of low weight as you cannot obtain decisions based on hopes instead of on tangible figures. Thus due to the economic constraints, most publishers considered automating those product lines that could be leveraged from a "DLM" database which could provide the greatest number of "DCM" products that significantly contribute to the publisher's revenue stream. A key research topic was identified in this area : how database systems could be made able to efficiently handle large, complicated, highly attributed and structured geographical databases ?

Deriving maps from so called "leveraged database" are implied in 88 large and repetitive processes to spin data into product databases that are maintained by time marked changes that are propagated from the master DB to each of the product DB for symbolization. In itself that statement induced a major research topic.

Dobson recognized that digital tools were faster image generator than manual scribing. Nonetheless the necessary (in order to meet public's requirements for crisp and clear cartographic design) interactive editing, required by digital methods, needed to be taken into account when making relevant comparisons. That lead other key research issues : how to decrease the time response of the systems for interactive editing ? How to improve the man-machine interface taking account of the operator's real behavior while in day to day action ? How to organize a better split and integration between the automated steps and the interactive editing ? These concerns lead to the problem of customization of commercial tools.

Tools for creating cartographic design, i.e. legend generators, was also a key concern as desktop solutions was emerging allowing "non cartographers" to easily design maps ignoring the basic principles of efficient communication using the cartographic language. In 88 was discussed the topic of experts systems for cartographic design.

Dobson in its conclusion introduced two topics rarely discussed which are still valid nowadays. The first one was called the "psoriasis of the data handling world" : the cost of data collection. It costs more to "get" the data from the reality than it does to digitize it. Dobson considered that "the economics of data collection will clearly decide whether our spatial data handling achievements ever come to fruition". Turned to nowadays environment of creating a growing market for digital GI in the information society this issue could be turned into "how can the large investment needed to collect data be protected so that private funding could complement public money to enable digital data to be widely available ?".

The second issue is quite straight forward and is still outstanding : in 88 SDH systems were simply not capable of dealing with large databases. How that could be improved ?

Echoing Dobson's keynote, Helen Mounsey and David Briggs stressed important key issues as a result of the creation of the environmental database for the European community "CORINE". A wide range of problems have been faced as challenges to databases development :
- how to locate suitable data sets and develop an inventory of sources. That echoes the Dobson's "psoriasis" of the SDH world.
- how to optimize the tiling structure so that the seamless characteristic of CORINE database is kept while the performance for retrieving and use the data remains efficient.
- how such a large data base can be updated given the sensitivity of many decisions which might be taken on the basis of information held in the database.
- how to assess the quality of the large database and how the users of the systems can be made aware of possible inconsistencies or can be constrained from misusing the information.
- David Rhind also provided an R & D agenda as part of his discussion on the Chorley committee including :
- symbol & feature recognition within digitized map and satellite images
- generalization from detailed maps and voluminous data sets to simpler results
- strategic research in large databases
- SHD'88 conclusion can be drawn from David Rhind's abstract "though the application areas of GIS are exceedingly diverse and apparently very different, there is much underlying commonality of concepts and tools utilized. Moreover, the supply of geographical data in a form, which preserves confidentiality and also provides significant revenue to the 'data owner's is clearly a major difficulty : nonetheless, the importance of signposts to data sources is critical".

3.3 SDH 92 in Charleston Olympic games 92 in Barcelona

Setting the scene, David Cowen in its introduction to SDH92 pointed out that the main goal of the meeting was to provide a forum about how we use computer technology to handle geographic information. The papers covered the best way to store, retrieve, analyze and display this type of information. Surprisingly no mention of the best way to acquire, transfer and maintain it. The way users interact with computers and how we can improve the performance of computer were also discussed with the hope that the new ideas, concepts and algorithms presented will ultimately improve the way society utilizes such systems to cope with the geographical problems that affect our social, economic and natural world.

David Rhind, as a keynote speaker, addressed the important and emerging issue of the information infrastructure of GIS. Nowadays he probably would say the "geographic information dimension of the information society". As a matter of principle he stated that no position on the earth's surface is absolute as the way location is specified always refer to a global or local frame. To enable proper use of GIS, he argues that the topography over which other spatial distributions are draped form the basic information infrastructure of any GISs.

In 92 clear specification of the basic template was country dependent and the way it can be made available was evaluated as many NMAs were subject of searching inquiries for various good reasons including the perception that national mapping is an expensive operation, the large delays in producing the complete map coverage, and its "out of date nature". In 92 an increasing number of governments sought to treat government held information as a commodity and to recover at least a fraction of the costs of its creation. In Europe considering that the tax payers have already paid the data creation had little weight considering the trend to limit public expenses.

Government can be seen as having two parts : that part which deals with policy and administration of policy and that which provides a service. The latter may stay within government if, through market testing rather than assertion, this is shown to be the most efficient mechanism. Nowadays we would qualify it as the "universal service". As a complement David Rhind pointed out that unless services are paid for, no meaningful assessment of priorities is possible. Following are two major questions : can the necessary quality be provided, i.e. what are the real requirements of the Market (as all the users) ? How can we avoid the focus on short term market needs accumulates long term problems and expenses ? The latter question is quite relevant for topographic mapping given the delay necessary to develop a National coverage : when the coverage is complete the requirement may have significantly changed.

Trading off known short term needs against unknown longer term ones is however a major policy and management dilemma.

In 92, and it is still the case, NMAs were facing several challenges including the expected rapid growth of the market for geographic information but with an immature technology, the contradiction between the requirements to improve cost recovery and the obligation to maintain records of little value for the users, the opening of competition both inside the country and at a European and international level, the vacuum in which current legislation in Europe on the information market drives information providers, the difficulty in preventing piracy. In 92 David Rhind considered that the measurement of user satisfaction with products was the only indicator of NMAs success and in parallel the specification of products for filling user's requirements could no longer be the concern of Expert committees but should be addressed via marker research.

What was then the key research topics David Rhind listed in his SDH 92 keynote ?

1. how can a NMA handle and thus market a nationally complete "seamless" database of information derived from its mapping, which is topologically structured and having consistent geometry ?
2. how topographical data could be made compatible and integrated with geographic identifiers such as post codes, addresses, administrative units identifiers census units ?
3. how to improve the updating of databases in order to be close to real time (in a cartographic scale !) and similarly how to operate faster and more customized delivery of data and better after sale services ?
4. how to routinely allow for direct access to data via-computer networks and similarly how to enable linking and combining data from various origin ?

To these questions, Davis Rhind added other topics to be added into the research agenda such as the question of multiple representation of geographical objects and the linkage of objects through scale, the provision of change only data and the maintenance of customer records, the economy of the provision of geographical data including setting prices for information, and last but not least how to circumvent the differences in terminology and culture not only across discipline but also among the European countries.

Browsing into the table of contents of SDH 92, the range of issues addressed is quite wide but it is difficult to evaluate how much of the new ideas presented in 92 are presently about to be integrated into commercial GIS. In line with that is the highly sophisticated solutions that research was investigating in 92 and the relatively basic use of GIS in the day to day life of administration and production ! Perhaps is it due to the fact that in average 10 years are necessary for a new idea to mature into commercial system for the day to day life !

4. And now key question for 1996: SDH 96 in Delft - Olympic games 96 in Atlanta

1996, SDH is back in Europe, the Olympic games are back in the states. The cycle is complete what are the key questions we are facing in 1996. I see three main contextual elements which may affect our future research :

1. the advent of the information society and its related Global Information Infrastructure (GII) and Geographical Information Infrastructure (GeoII)
2. the discussion on international standards and the advent of initiatives such as Open GIS
3. the intrusion of social science concerns exemplified by the GISDATA programme

All of these lines aim at the same objective : a better shared knowledge of our living space in order to improve the decision making process related to it. Sustainable development has long been a buzz word, perhaps shadowed by new fashionable worlds such as Information Society or unemployment, but it is still a goal to achieve.

4.1 GII and GeoII

For many years there has been increasing talk about the information society, often in terms of a second industrial revolution. This talk has received added impetus from US vice-president Al Gore, the G7 Group, Delor's white paper on "Croissance, Compétitivité et Emploi", the Bangemann report and the European Commission itself. This has resulted in the translation of words into concrete action to create what is widely recognized as the Global Information Society "GIS". GIS is not just the latest buzzword. It is a focal point for exploiting within society new developments in the information and communication technology (ICT) industries and source of political support.

A distinction is made between the two terms "GIS" and the Global Information Infrastructure "GII". An article in the October 1995 issue of the ISO bulletin gives the definition that the GII "provides the vital services and capabilities required for the information society to prosper". Another definition often used is that the GII is the enabler for achieving the objective of the Global Information Society.

The consequence of this large tidal wave on our Geographic Information economic sector is yet unknown and implies for us to undertake socio-economic research into that area. There is an immediate consequence for we will have to abandon our old term GIS, standing for Geographical Information System. That is fortunate because in my opinion GIS was one of the most wholly terms we invented in our area. While defining the concept of an European umbrella organization, now known as EUROGI, the team of experts early understood that the issue needed to be widely understood as related to Geographic Information in general, rather than limited to GIS in its strict meaning. Remember Mike Goodchild keynote in Zurich when he was advocating the S of GIS as Sciences instead of System. Remember also the very name of this conference series "Spatial Data Handling".

Recognizing that GI was an important part of the information society, some Ministers in Europe urged the European Commission to take a political initiative to further study the Geographic component of the GIS, G for Global, S for Society, and the related concept of GeoII. They also suggested that the EU should provide a stronger political impetus to GI at the European level. DGXIII, with the help of the professionals of the GI economic sector including European bodies such as EUROGI and MEGRIN, undertook to justify an EU involvement through a document now known as GI 2000 "Towards a European Policy Framework for GI".

The latest version of GI 2000 dated 15 May 1996 showed a significant change in the wording of the title where the word infrastructure disappeared for the profit of "Policy framework". Looking at the word "Infrastructure" in the GII abbreviation it seams that the common understanding limit the concept to very basic elements such as the network (i.e. the media for transporting information either wires , cables, radio waves) and the basic services (i.e. electronic services, telephone, TV etc.). In the GI economic sector it was soon recognized that more was required in the "content" field, i.e. the data side, because of the importance of the costs of basic data acquisition as well as education for using GI. Therefore "Infrastructure" was seen improper and "framework" preferred.

Exemplified not only by the GI 2000 document but also NSDI initiatives in US, Japan and other countries, actions plan are developed encompassing various areas such as :
- stimulating the creation of base data
- stimulating the creation of meta data services
- lowering legal barriers and reducing potential risks
- stimulating Public/Private synergy
- coordinating at national , regional and global levels

Research agendas are starting to be developed in that area. GI 2000 for example is suggesting "that the existing efforts regarding GI and GIS in the commission's R&D programmes should be clustered together under the 5th Framework Programme ensuring better exploitation of synergies and reflection of user's and industry's needs".

GI 2000 also list possible research for the creation of new applications :
- the development of new algorithms and modelisation techniques involving advanced spatial and temporal analysis techniques for solving complex technical problems with a good control of scale and error propagation
- the integration of new visualization methodologies coming out of research in virtual reality

4.2 Standards and open systems

Standardization has a specific role to play in achieving the potential of the GIS, G for global, S for society, and its supporting GII.

Standardization bodies have already responsibility for providing the necessary standards in key IT application areas which have been identified as necessary for the realization of the GIS. Other partners have equal concerns for areas such as telecommunications infrastructure and applications, digital video and broadcasting, and last but not least multimedia content provision.

- Standardization bodies identify their responsibility as falling into the area of sector applications and this in itself can be separated into two groups :
- intersectorial : i.e. definition of "intermediate tools" such as smart cards, character sets and bar coding
- sectorial : i.e. application of ICT by specific industrial sectors such as road transport, health care or geographic information

STEP (standard for the exchange of product model data) also is a typical intermediate tool that will be used by other applications or programmes and hence will be quite often transparent to the end user. Special attention must be given to the coherence of various data elements representation techniques within STEP or elsewhere (EDI), for they might be incompatible.

There is here an area in which the Spatial data handling community lacked to get involved. Many emerging standards are more and more including in their agenda issues related to geographic information. STEP is one of them, SQL is another. So far I have no information on real investigation into the appropriateness of such emerging standards for application of geographic information to any type of discipline. A lot of standardization experts are boosting the use of STEP for the exchange of Geographic Information between systems without providing convincing report on the appropriateness of the proposal.

The aim of standardization in the field of GI is to define a family of standards which enables the definition, the description, the structure, the query, the encoding and the transfer of geographic information and related information. The basic purpose is also to enable GI to be "delivered" to different users, applications and systems. In order to draw a clear context in which standardization takes place, a "Reference Model" is needed. Its purpose is to guide the structuring of standards to enable the universal usage of digital data including to allow integration of GI with other digital information and applications.

Beyond the needs within traditional applications of GI, there is a growing recognition among users of information technology to index data by location for use in a diversity of applications. Very little attention has been given by the Spatial Data Handling community to the conceptual issues addressed by the writing of a Reference Model for the standardization in the field of Geographic Information. At a time where the reusability of Geographic Information in an open System environment seams to be a major concern, it is urgent to start research into that field. I see personally two main areas in which research is of up most importance for the definition of the Reference Model for Geographic Information:

- understanding the relations between the processes involved in conceptualizing and creating geographical data aiming at depicting the reality (see figure 1)
- understanding the processes involved when data is delivered from one environment (Modeling of information and modeling of data together with the computer environment) to another.

The latter issue is key not only for the future in <u>Internet</u> environment where any supplier may be delivering data to any user but also for the future in <u>Intranet</u> (network internal to an organization) environment where both the suppliers and the users are belonging to the same organization and have both objectives clearly definable and ideally known to each other.

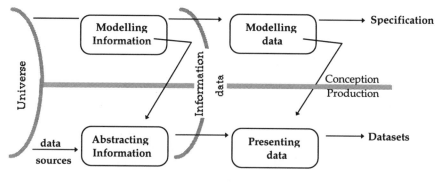

Figure 1.

Those are key questions for the future of GI if we accept the axiom of "Geographic Information as part of the GIS, G for Global and S for society".

4.3 GI and social sciences

The GISDATA Programme is funded by the European Science foundation and is instrumental to create a European Community of researchers having interest in Geographic information. Among the specialist meeting organized in this programme, which each time involved some twenty experts from various origin each time being largely different, a stream of meetings was focusing on the integration of data at the European level.

The first one prepared the ground for a study of GIS diffusion in local government in nine European countries. It confirmed that the diffusion of GIS and the overall development of the market depends to a large extent on the availability of digital data. The review of the current policies on geographic information in Europe and the assessment of the opportunities and limitations created for social science by the increased availability of accurate remotely sensed data were concerns of two other meetings. The last of the series investigated the European dimension of geographic information and the need for European wide co-ordination of GI.

A number of obstacles to the implementation of the concept of a GeoII (a framework of agreements, standards and procedures to collect and exchange GI in Europe, a core set of data, a framework within which meta data services can be extended) were identified including the significant lack of awareness, among decision-makers, of how important is GI vis à vis broader issues of the information society, the immaturity of the market specifically with regard to European wide or transborder data sets, the wide diversity of the actors having interest in GI.

GI, in a wider-sense market, is key to many applications which can be classified as follows :
• the mass market, i.e. the consumption of GI by the general public,

- the operational functions, i.e. the use of GI to sustain the core business of organizations both public and private,
- the strategic functions, i.e. all the activities which intend to hamper the future of our living place often the duties of government or private sector vested with a public service mission
- and last but not least GI for pure and applied research.

As research dimension is identified by the GISDATA programme with an overall aim towards a better use of GI within the broader social sciences sector. From an economic point of view, the analysis of the economy of the GI sector needs to be undertaken. It should include the definition of parameters which could enable a better understanding of the GI sector including the identification of the market niches that exist in each country and the understanding of the reasons explaining why in certain country specific niches are more developed than others or in other countries. It may include modeling the consequences of political choices, such as simplified access to publicly held GI or cost recovery constraints on public bodies, on the development of the GI sector and the use of GI in the information society.

Linked to that economic issue, it is needed to study the intellectual property rights (IPR) and related rights that exist in each country together with the new directives at the European level, such as the Data Base directive, and their consequences for the development of the GI use.

Following are the study of the appropriate safeguards that have to be put in place so that privacy is protected, warranty and liability clearly defined, without unduly restricting the market.

There are also numerous research aspects which need to be undertaken from a more Spatial Data Handling viewpoint such as the development of analytical and exploring functions for very large data sets or the integration of GI and multi media for data visualization.

5. Conclusion

Now comes the time for me to answer the questions at the start of the paper. Who are the thirsty men ? Somewhere it seems to me that any governmental agency or administration or private sector body and also perhaps the general public are those men. They are fiddling around the desert (or the Jungle) looking for water or drinks to help them to accomplish their businesses. They are searching information, strategic information, to help them to take the right decision at the right time. Information is the drinks they need.

Here at this stage of the analogy I don't know whether geographic information is "ties" for these thirsty men or is the available GI the "ties". Two ways to look at it. Effectively what we, i.e. the SDH community, offer to the thirsty people are ties, where what they need is drink. In fact, may be, the thirsty people are in the situation that they need ties and they don't know yet they need them not for an immediate profit to them but as an investment of the future allowing them to enter the pub !

More seriously and given the title of this paper "a twelve years impact of SDH research on the GIS community" I am wondering if we, SDH people, should rather concentrate on how we can impact on the development of the Information Society in such a way that GI technology starts to be considered as an enabling technology for many human activities.

Contact address
François Salgé
Institut Géographique National
2 Avenue Pasteur
94160 St. Mandé
France
Tel.: +33-143988440
Fax: +33-143988443
E-mail: salge@megrin.ign.fr or francois.salge@ign.fr

AN INTEGRATED DATABASE DESIGN

FOR TEMPORAL GIS

Donna Peuquet and Liujin Qian

Department of Geography
The Pennsylvania State University
University Park, USA

Abstract
Current data representation techniques for Geographic Information Systems (GIS) are geared toward representation of static situations. Cartography has traditionally focused on the visual presentation of a portion of the world at a specific point in time. Although GIS are intended to provide a flexible and integrated tool for analyzing large volumes of data, representations historically used in GIS are derived from these cartographic traditions and are geared toward a similar static view.

In this paper we will describe the **TRIAD** model for representing spatiotemporal data in a GIS context. This database model represents a completely integrated approach for handling such multidimensional data. The **TRIAD** model is also the basis of TEMPEST, a prototype GIS designed to support spatiotemporal queries and analysis. We first briefly review the rationale behind the approach. Then we describe the three views composing the **TRIAD** model in detail. Finally, the method used for dealing with queries is briefly outlined.

Key words
Temporal GIS, space-time data models, object-oriented representation.

1. Introduction

The development of temporal capabilities for GIS has been receiving a significant amount of attention over the past few years. The ability to examine the *dynamics* of geographic phenomena is urgently needed as an essential tool for examining and increasing our understanding of man-environment interactions at local, regional, and global scales. Problems requiring the analysis of *change* through time and of *patterns* of change range from urban growth and agricultural impacts to global warming.

A key element in the development of such capabilities is how to effectively represent geographic phenomena in time, as well as in space. Most of the research on this topic, however, has focused upon the extension of individual models, including traditional raster and vector models to incorporate the temporal dimension (Hazelton 1991; Kelmelis 1991; Langran 1992; Pigot and Hazelton 1992). The only data model available within existing GIS that can be used to represent space-time data is a series of 'snapshot' images. Instead of a single grid representing a complete thematic map layer as in a static (i.e., non-temporal) spatial database, this type of representation simply employs a sequence of spatially-registered grids. Each gridded image represents a 'world state' relative to a given thematic domain by storing a complete image, or 'snapshot' (S_i) at a known point in time (t_j) (Peuquet and Duan 1995).

If future GISs are to provide sophisticated temporal analysis capabilities as well as the ability to answer a wide range of queries effectively, it is necessary to utilize an integrated, multi-representational scheme which facilitates time-based as well as other types of queries.

It is the purpose of this paper to describe a unified database model for handling spatiotemporal data, called the TRIAD model. The TRIAD model also serves as the basis for a Temporal GIS, called TEMPEST, which is currently under development as a demonstration prototype. In the remainder of the paper, the concept of the TRIAD approach, as originally proposed in Peuquet (1994), is reviewed briefly before describing the model itself in detail. Then, query strategies based upon this model will be described. The paper will conclude with some comments on implementation issues and future directions.

2. An integrative approach

The fundamental commonality of all traditional vector data models is that the basic organizational unit is the (cartographic) line, or vector. Locational coordinates and all other information are stored in relation to these vectors. Vectors may be further organized hierarchically into specific types of geographic/cartographic entities, or features (roads, lakes, political units, etc.). Such models can thereby said to be feature-based. The raster or grid data model, conversely is location-based in the sense that all other information is stored relative to specific locations. Because of these two fundamentally different orderings of stored information, the feature-based model is more effective for query and retrieval of information about spatial features or objects; the raster model more effective for query and retrieval of information about a specific location or set of locations. Similarly, a time-based representation, where all other information is stored relative to specific locations in time, is more effective than either feature-based or location-based for retrieving information regarding specific times or changes through time. Therefore, just as feature-based and location-based data models are considered both necessary and complementary in a current GIS, so should feature-based, location-based and time-based representations be considered complementary in any Temporal GIS.

It is these three types of representation that serve as the basis of the TRIAD framework. Unlike traditional spatial data models that are designed to be self-contained, however, all three components of the TRIAD model are designed to be cooperative and interdependent. This need for representational integration of the multiple dimensions inherent in a Temporal GIS has also been recognized by Langran (1993) and by Worboys (1994) as necessary before the full potential capabilities of Temporal GIS can be realized. This amalgamation is essential, not only to overcome the obvious data redundancy issue, but more importantly, so that the multiple dimensions required by complex geographical analysis applications can be handled in an integrated manner. A parallel from the art world provides insight into the general idea: Cubism as developed by Picasso and Braque means to display several different aspects of one object simultaneously. It is useful to bear the notion of Cubism in mind to understand the very high level of dimensional integration desired in the current work. Our goal is to integrate all spatial, temporal and feature dimensions of geographic data in a unified and mutually-supporting fashion. Such a dimensional integration not only makes all dimensions of the data accessible to the user, it also enables a user to observe and analyze data from varying dimensional perspectives within a *single* representation.

3. Conceptual overview of the TRIAD model

The object-oriented approach was used as the technique for building the TRIAD model. Using this approach, any conceptual representation consists of objects, attributes and relationships. An object in this context is any type of data element that serves as the basis of the representation. Entities have attributes, or properties, associated with them. Thus, an object is defined on the basis of the data it encapsulates, which includes attributes and specification of the operations which can be performed upon it. These operations may include inheritance, which allows the definition of a new class of object in terms of existing objects. Each object is a unit of information and may or may not belong to some class of objects. Encapsulation in the object-oriented approach helps to enforce a scheme where data access and data manipulation are uniform for all objects of the same class.

In the TRIAD model as shown in Figure 1, there are three interdependent representations; the feature-based, the location-based and the time-based views. The feature-based view appears at the top of the diagram. A feature is a thing that exists and can be seen (a road, a mountain, a lake) or is purely conceptual (a historic district). In the TRIAD model, a feature serves as the basic "object" (and thereby the organizational basis) of the feature-based organization, and is defined in terms of its associated attributes.

The location-based representation appears on the lower left-hand portion of Figure 1. In this case, the basic object is a location; an elemental unit of discretized, two- or three-dimensional space, represented as an areal cell or pixel in a (locationally ordered) gridded data structure. Locations possess attributes that may include identification of the objects therein (e.g., "New York"), as well as attributes for that specific location which vary over time (e.g., population density or land use).

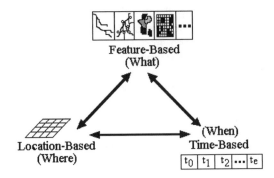

Figure 1 The TRIAD Model.

The time-based view appears on the lower right-hand portion of Figure 1. In this view, the basic object is an event, which occurs at a particular moment in time. The attributes associated with a particular event denote changes which occurred at that specific moment in time. Attributes defining a specific change can also be of two types; (1) changing at that time and relating to a specific feature or set of features as a whole, or (2) changing at that time and relating to a specific location or set of locations. The first type of dynamic, feature-related attribute for "New York," for example might include total population.

The time-based view organized in this fashion constitutes an open-ended "time-line" in which events are chronologically ordered. Although the temporal location is the organizational basis of this view, not every consecutive temporal location, or "tick of the clock," is recorded, in contrast to the location-based view. Another difference is that the time-based view continues to have events added to the end of it as new events occur. Previously unknown past events may also be inserted into the time-line. The idea of a time-based representation had been discussed earlier as a stand-alone data model for recording only locational changes in Peuquet and Duan (1995). In the current time-based view, however, an event represents some change relating to a feature as well as to a location.

Features, locations and events thus form the three interrelated "views" of the spatiotemporal TRIAD scheme. This scheme is not entirely new. Precedents range from the interrelationship known as the "Kronecker Delta" in tensor calculus to journalism's old axiom on "what, when, and where" as the most fundamental rule of reporting.

4. The three views of the TRIAD model

As stated above, we address the multidimensional complexity of spatiotemporal data by encapsulating information according to the dominant dimension relating to it. This results in the three views of the TRIAD scheme, and results in three interrelated object templates for defining these views.

When all three views are used in concert, only generalized locational indicators, such as the bounding rectangle, need to be stored as object attributes in the feature-based representational view. These generalized locational indicators are used to retrieve the detailed information within the location-based view. Similarly, only generalized temporal indicators, such as when a feature began and ended its existence, need to be stored in the feature-based view. The details of what changes occur over time are stored in the time-based view, and similarly, changes occurring over space are detailed in the location-based view. This allows a fairly simple, yet effective generic object template to be defined for objects within each view of the TRIAD model, which we will now describe in more detail.

4.1 Feature-based view

As said previously, the feature-based view encapsulates information which uniquely defines the feature, including general indicators as to where and when the feature occurs. Information within the feature template includes attributes which are inherent to the feature itself and are not directly associated with location or time of existence. Examples of such nonspatial, nontemporal attributes include feature name (e.g., New York), feature class, and other static attributes, such as type of governance (mayor or city manager), etc. Spatial attributes defined within the feature template consists only of a bounding spatial delimiter for the entire feature. Similarly, temporal attributes are limited to a bounding temporal delimiter.

Figure 2 shows the structure of the feature template, used to define all features and feature types. The components of the feature template are defined as follows:
FID: Every feature is assigned a unique Feature ID which is assigned automatically upon entry into the database. Encoded within the FID is identification of the geometric type of that feature; point, line or polygon

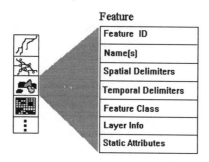

Feature

| Feature ID |
| Name[s] |
| Spatial Delimiters |
| Temporal Delimiters |
| Feature Class |
| Layer Info |
| Static Attributes |

Figure 2 The structure of the feature template.

Name: This is the user-assigned name for the given geographic feature, such as the name of a city ("Pleasantville"). This name is not necessarily unique, and there can potentially be more than one name stored for the same feature (e.g., Main Street / Route 32).

Spatial Delimiters: Two Spatial Delimiters are stored within this data element - one denoting the feature's location at the beginning of its existence (or the beginning of the recorded time period), and a second denoting the feature's most recent location. This second location can represent either the feature's current location or its final location at the end of its existence.

The Spatial Delimiter for the first and latest location can be stored in one of three possible forms, determined by the geometric type of the feature. For all three forms, *each* of these delimiters consists of two x-y coordinate pairs to denote the bounding rectangle of an areal feature (i.e., a polygon), two x-y coordinate pairs to denote the endpoints of a line, and a single x-y coordinate to denote the location of a point-type feature at a specific point in time.

The Spatial Delimiters are intended to serve two functions. The first function is as the primary link to the Location-based view. The spatial limits are used as pointers into the location-based view for retrieval of the detailed locational information, if specifically needed. Of course, the Spatial Delimiters can also be used directly as a general indicator of location, areal extent, shape or length, depending upon the specific geometric type. Similarly, comparing the spatial coordinate values for the first and latest delimiters provides a general indicator of spatial change. For example, comparison of first and last spatial delimiters for "Pleasantville" would give an indication of how much urban expansion has occurred over the entire time period.

Temporal Delimiters: In correspondence with the Spatial Delimiters, two times can be stored within the Temporal Delimiter data element - one specifying the time the specific feature came into existence, and another which specifies when the feature ceased to exist. If the feature is still extant, there is no second time stored. Each of the (possibly) two times stored consists of a "time stamp," or unique temporal location, such as year/month/day.

Similar in purpose to the Spatial Delimiters, the Temporal Delimiters serve as the primary link to the Time-based view. The specific times are used as pointers into the Event Bank for retrieval of some or all events associated with a specific feature. The Temporal Delimiters can also be used directly to determine total duration of the feature's existence.

Feature Class: This element identifies the class or classes to which the given feature is a member. This can be thought of as an "is-a" relationship. For example, a feature whose Name is "Pleasantville" would have an "is-a" relationship of "city" specified within this element as the class to which it belongs. This class identification, besides being a descriptive attribute in itself, serves to link the specific feature to a feature hierarchy or set of feature hierarchies. Both the member class and the geometric type have added importance in that they define which specific procedures can and cannot be applied to specific features.

Feature class definitions, which comprise feature class hierarchies, are defined separately and do not contain observational data. Feature class hierarchies, as well as all characteristic attributes and valid operations pertaining to those classes, can be defined through inheritance. Exactly how these feature class hierarchies are defined is itself a complex topic and beyond the scope of the current paper.

Layer Information: Information stored within the location-based and time-based views are separated into a series of thematic layers. These thematic layers are the same for both the location- and time-based views. Because spatial and temporal information relating to features are stored within these location-based and time-based views, each geographic feature must be associated with one or more of these thematic layers. This data element thus contains the name of the thematic layers within those views which contain that information.

4.2 Time-based view

The time-based view encapsulates information which uniquely defines the event. Types of attributes include all changes specific to one or more features which occur at that time, and similarly all changes specific to one of more locations which occur at that same time. The event template, as shown graphically in Figure 3, is defined as follows:

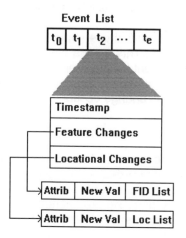

Figure 3 The event template.

Timestamp: This is the time at which the event occurred. This Timestamp may be of a greater temporal resolution than those used as Temporal Delimiters within the Feature-based view. The Timestamp, combined with the layer name associated with the entire event list, constitutes a unique event identifier.

Feature Changes: This data element records the specific changes which occurred to one or more features at that given time. It consists of a list, wherein each member of the list consists of three things: the FID for the feature or features that changed, the name of the feature-based attribute that changed, and the new value for that attribute.

Location Changes: This data element records the specific changes which occurred to one or more locations at a given time. Similar to the Feature Changes element above, it consists of a list, wherein each member of the list contains the grid cell location (i.e., unique location identifier) of the location or locations that changed in 2-d run-length encoded form, the name of the attribute which changed, and the new value for that attribute.

Thus, the Feature Changes and Location Changes elements of the Event Template can show the same change occurring to multiple features or locations. It can, at the same time, show a complex set of differing changes occurring at the same time. This is very necessary as it is simply a reflection of the complexity of changes which normally occur in the real world, and the multi dimensionality of those changes. A complex of changes can be caused by some specific disturbance (e.g., a major factory closing), or simply by natural evolution (e.g., urban growth), and affect whole classes of features over a spatial extent. Some disturbance factors can themselves be represented as features (e.g., forest fires, earthquakes) with their own set of stored characteristic attributes, spatial and temporal delimiters, associated class ("disturbance"?) and associated operators. This also means that such "disturbance" features can be stored as components of events, with specific attributes for that occurrence, such as intensity, at the times when they occur.

It should also be noted that *only* those times when some change occurs are recorded within the time-line. This means that the temporal interval between recorded times within the event list is variable. The degree of variability, as is the temporal resolution needed, phenomenon-dependent. Indeed, analysis of the temporal pattern of when change occurs can itself yield valuable insights.

4.3 Location-based view

As shown in Figure 4, the Location-based view defined within the TRIAD model uses a gridded array of pixels as its organizational basis. Each pixel is represented by a list of changes that have occurred at that location. Any time a change occurs, an entry denoting the change is added to the beginning of the list for the affected location(s). These change entries can be of two types. The first type records a change in the presence or absence of a feature at a specific location. What feature or features occupy a given location as features come into and go out of existence and move through space. The second records a change in an attribute value. The structure of the template to denote a single change is shown in Figure 4. Each change can be in one of two forms, corresponding to the two types of change defined within this view: The template for a feature change has three components: Feature ID, whether that feature begins or ends its presence at that location, the time of the change (Timestamp). This presence/absence can represent either movement of a feature over space, or the actual coming-into or going-out of existence of a feature. The second type of change template also contains three components; attribute name, the new value for that attribute, and the

Timestamp. Of course, it is possible to have both types of change noted as happening at the same time at a given location. This would occur when a new feature, with associated attributes, comes into existence at that location.

Instead of recording only one state for all pixels over the entire spatial area for a single point in time as in the snapshot approach, each pixel contains a history for that location over the entire recorded time span as a chronologically-ordered sequence of events. Each pixel initially contains the Feature Identifier(s) for all features relevant to that layer which are present at the beginning of the recorded time span. The initial values over the entire grid thus represents the initial spatial feature distribution. This general idea of using a gridded array of chronologically-ordered changes specific to each location was originally used in electronic circuitry design analysis (Fujimoto, 1990). The advantage of this approach is that *only* changes are noted. This means that if no changes occur at a given location, that location only contains the initial, single entry, no matter how long the time span. The length of the chronological list is variable; the more changes at a given location, the longer the associated list.

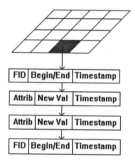

Figure 4 The structure of the template to denote a single change.

5. Handling queries

Of course, there are many queries that can be satisfied by accessing only one of the three views in the **TRIAD** model. Nevertheless, much of the real power of the model derives from two query-related features: First, using the **TRIAD** model, queries can be launched from any combination of the three views; when, where, and/or what. Second, the specific ordering of view access in satisfying a query is determined very simply. This is accomplished by evaluating what has been called the "dimensional dominance" of the query (Langran, 1993). The "dominant" dimension of a multidimensional query is defined here as the most constrained dimension within the query. It is the specific information regarding these dimension-related constraints which are used subsequently as search constraints to limit search space within the remaining dimensions, or views. In the following list of sample queries, the dimension where only one of "what," "where" or "when" is requested becomes the dominant dimension (or dimensions) of the query. Here, we have shown only "all" or "one" for the sake of clarity, but these can be further expanded to include ranges of times, locations, and features.

WHAT	WHERE	WHEN	
1	1	1	Did this feature exist at this location at this time?
1	1	all	When did this feature exist at this location?
1	all	1	Where was this feature at this time?
1	all	all	Where has this feature ever existed?
all	1	1	What features are at this location at this time?
all	1	all	What features have ever been at this location?
all	all	1	What features are present at this time over the entire area?

6. Conclusions

The description above provides details of the **TRIAD** model for representing the dynamics of geographic phenomena over space and time. As stated at the beginning of the paper, the primary goal is to develop a generic, spatiotemporal representation for use in a Temporal GIS which combines the multiple dimensions of features, space and time in a completely integrated and unified manner. We feel that the representation presented provides a strikingly balanced set of components among the three interrelated "views," yet each such view represents a dimension with fundamentally different conceptual characteristics.

Nevertheless, the development of spatiotemporal representations remains a complex problem. There are many other issues which could not be addressed in the current paper given constraints of length while maintaining clarity of key concepts. One such issue is world time vs. database time. We have chosen to deal only with world time in the current implementation of the **TRIAD** model as a conscious choice and leave this extension for further work.

References
Fujimoto, R. M. (1990). "Parallel Discrete Event Simulation." *Communications of the ACM* 33(10): 30-53.
Hazelton, N. W. J. (1991). *Integrating Time, Dynamic Modeling and Geographical Information Systems: Development of Four-Dimensional GIS*. Dept. of Surveying and Land Information, The University of Melbourne.
Kelmelis, J. (1991). *Time and Space in Geographic Information: Toward a Four-Dimensional Spatio-Temporal Data Model*. Department of Geography, The Pennsylvania State University.
Langran, G. (1992). *Time in Geographic Information Systems*. London, Taylor & Francis.
Langran, G. (1993). "Manipulation and Analysis of Temporal Geographic Information." Canadian Conference on GIS, Ottawa.

Peuquet, D. J. (1994). "It's About Time: A Conceptual Framework for the Representation of Temporal Dynamics in Geographic Information Systems." *Annals of the Association of American Geographers,* 84(3): 441-461.

Peuquet, D. J. and N. Duan (1995). "An Event-based Spatio-temporal Data Model (ESTDM) or Temporal Analysis of Geographic Data." *International Journal of Geographical Information Systems,* 9(1): 2-24.

Pigot, S. and B. Hazelton (1992). "The Fundamentals of a Topological Model for a Four-Dimensional GIS. A *Proceedings of the 5th International Symposium on Spatial Data Handling,* Charleston, South Carolina.

Worboys, M.F. (1994). "Unifying the Spatial and Temporal Components of Geographical Information." *Sixth International Symposium on Spatial Data Handling,* Edinburgh, Scotland, International Geographical Union.

Contact address
Donna Peuquet and Liujin Qian
Department of Geography
The Pennsylvania State University
University Park, Pa. 16802
U.S.A.
Phone: 814/863-0390
Fax: 814/863-7943
E-mail: peuquet@geog.psu.edu
E-mail: qian@geog.psu.edu

CONCEPTUAL, SPATIAL AND TEMPORAL REFERENCING

OF MULTIMEDIA OBJECTS

Christopher B. Jones, Carl Taylor, Douglas Tudhope and Paul Beynon-Davies

Department of Computer Studies
University of Glamorgan
Pontypridd
Mid Glamorgan, UK

Abstract
The increasing usage in GIS of multimedia objects, such as scanned photographs and video recordings, has introduced the need to provide flexible indexing methods to help in accessing objects that may be of relevance to a range of possible applications. An experimental system is described in which media objects may be referenced to several classifications concepts, as well as qualitative descriptions of geographic location and time period. Semantic modelling techniques are used to create semantic networks that represent the hierarchical and lattice structures of conceptual, temporal and geographic space. The associated query methods enable inexact matching between user interests and the media object descriptors. Several metrics of semantic closeness, in concept, time and geography have been developed, based on scaled measures of path lengths within the semantic networks that represent the three dimensions of indexing.

Key words
Multimedia; qualitative reasoning; semantic modelling; spatio-temporal data; semantic closeness.

1. Introduction

The effectiveness of geographical information systems has been enhanced in recent years by the now commonplace facility to access media objects, primarily in the form of scanned images, though also as digital video and sound recordings. As the number of such media items increases within any one system, problems arise in providing access to the specific items that are relevant to a particular query on the system. A single photograph may contain a multiplicity of features that could be relevant to different types of query. A natural landscape scene may include features of interest, for different reasons from the perspectives of, for example, geomorphology, geology, soil science, ecology and archaeology. An urban scene might be of interest with regard to studies of housing condition, landscape architecture, road safety and transport systems. Furthermore, photographic material, whether still or video, provides a potentially very rich archive of historical data. Thus any one media object might record a specific event or it may illustrate environmental and cultural phenomena each of which belong to, or are characteristic of, particular historical or cultural periods.

There are numerous examples of GIS application areas in which access to media objects is useful. In local and regional planning, photographs can assist in evaluating the likely impact of developments associated with planning applications, and in evaluating the need for and the most appropriate design of new urban or rural development. In police work, photographs assist in understanding events leading to, and consequent upon, road accidents. In environmental conservation work, photographs may form an essential part of the documentation of a site of 'special scientific interest'. They may also be used in the process of monitoring geomorphological and ecological aspects of environmental change. Some survey organisations store very large collections of vertical and oblique aerial photographs. Because the images from such collections often date back over the last fifty years, they may serve as invaluable historical records to be used in understanding the nature of change in environmental phenomena.

Despite the importance of media objects, GIS technology is not at present very well adapted to take full advantage of such material. GIS database development has to date tended to focus on spatial and, to a lesser extent, temporal indexing methods intended to access geometric representations of real world phenomena. This has led to a proliferation of multidimensional indexing schemes that depend upon quantitative geometric descriptions of specific phenomena. More recently the need for facilities to handle qualitative descriptions of spatio-temporal objects has been recognised and the various efforts to formalise qualitative spatial and temporal reasoning reflect this trend (Egenhofer and Herring, 1990; Cui et al, 1993; Frank, 1994). Much of the work on spatial reasoning has however been concerned with representations which, in making a precise distinction between interiors and boundaries of spatial objects, may be more appropriate to computer-aided design for engineering, than to the representation of geographic space. The need to address the complex and often inexact variation that characterises geographical phenomena has led in turn to study of the concepts of regions with undetermined boundaries (Burrough and Frank, 1995). Clearly such work is of considerable benefit in improving the manipulation of spatial representations in GIS. However, in attempting to develop access methods for media objects, other types of facility are required, that take account of the multiple interpretations and the often imprecise terminology associated with descriptions of natural and man-made scenes. The importance of these issues was emphasised by Walker et al (1992) in their design for a system to access geographical datasets.

In this paper we describe an experimental hypermedia information system that specifically addresses problems of access to historical photographs with regard to the need to reference the information content of media objects in terms of concepts and of geographic space and time. Section 2 summarises the problems in the context of access to historical records in museums. Section 3 provides an overview of the Semantic Hypermedia Architecture (SHA) developed with the objective of modelling and reasoning with the qualitative semantic relationships inherent in the mediabase. Section 4 describes user interface tools that provide access via the three dimensions of concept, space and time. This includes a summary of alternative approaches to determining similarity between the encoded description of a media object and the user's specification of their interests. Section 5 concludes with a brief discussion of further work.

2. Spatio-temporal access to historical media objects

GIS technology is now being seen increasingly as relevant to assisting in spatio-temporal referencing of historical data in museums. Conventional GIS functions are certainly of use in mapping the distribution of historical and archaeological sites and in performing various spatial modelling operations to understand associations and interactions between phenomena (Allen, 1990). Much historical information, of which photographic records may be an important part, is however referenced spatially only in terms of place names and historical periods (MDA, 1989). When creating archives of media objects, each object may typically be stored with a set of descriptive terms. These may include place names, such as a building, a street, a parish, a town or a topographic term such as the name of a valley or hill. Temporal referencing may be provided by exact dates if known, otherwise by historical periods such as Victorian, World War 2, and Iron Age. Conceptual referencing might include the names of people or specific artifacts (such as a work of art), but more generally would include classification terms which describe for example a particular implement, such as a type of plough and its higher level classification, such as agricultural tools.

In the museum domain, standard classification systems exist for purposes of archiving. In the UK the Social History and Industrial Classification (SHIC, 1983) is one such system. It is hierarchical and consists of five levels of classification. Some museums also make use of place name hierarchies, which might for example relate parish names to their counties and in turn to their containing nations (MDA, 1989). Using conventional database access methods, based on exact matching of terms, the use of standard classification systems and place names introduces major problems in answering queries. In the case of classification concepts these may be summarised as follows:
* the user of the system must be familiar with the particular classification system, thus differences in use of terminology between users and archivists could result in failure to find relevant objects;
* there may be uncertainty in the classification, leaving the possibility that a particular object may not be found due to a perhaps small degree of misclassification;
* the archivist must anticipate all possible perspectives on the media object when allocating classification terms.

In the case of geographical place names, problems include:
* uncertainty in the exact geographical location associated with a media object;
* a photograph of an urban or rural landscape could include many places, not all of which were registered against the media object when stored;

- place names often change over time, so that a historical name may not match the current name, either due to its having changed entirely or due to a change in its usage;
- for a given point in time, certain place names may have slightly different interpretations by different people, reflecting the fact that topographic place names often have imprecise boundaries or boundaries that are not agreed;
- users may not remember, or be aware of, certain names of places in the geographic area in which they are interested.

Problems with temporal referencing are of a very similar nature to those of place names. Thus
- there may be uncertainty in dating;
- a scene could include man-made or physical phenomena relating to different times of origin;
- temporal terminology differs between countries and cultures (e.g. different geological classification systems between Europe and America);
- some named time periods have different meaning to different people (e.g. 'medieval', 'renaissance');
- the user may not be familiar with certain temporal terminology.

In a museum context, and indeed in various public access information systems, some of these problems have been addressed through the use of thesauri which provide some capacity to translate between similar terms (Aitchison and Gilchrist, 1990; Molholt and Petersen, 1993). Furthermore, the presence of hierarchical classification systems and place name hierarchies is sometimes exploited to allow some flexibility in referring to concepts and places at different levels of generalisation (Harpring, 1992). Access to text has also been the subject of considerable research in the context of the computer science field of information retrieval (Salton, 1989), with particular emphasis upon vector processing methods in which text items are located in a multidimensional space defined by the contained words. Such facilities are however not normally to be found in commercial GIS systems or in museum access systems. Carlotto (1995) is notable however for describing their use in some GIS-related projects.

3. The semantic hypermedia architecture

Although facilities for accessing multimedia objects in GIS are quite restricted, this is in distinct contrast to the multitude of multimedia and hypermedia systems in which a variety of access techniques is to be found. Despite the widespread use of such systems however, it is not the case that commercial hypermedia systems provide the levels of sophistication in spatio-temporal or conceptual information access required by applications such as those in the museum domain or in many of the potential application areas familiar to GIS users.

Hypermedia systems are characterised by providing the facility to link together media objects, or *nodes*, in a manner that enables the user to traverse in a non-linear manner between them. In doing so, hypermedia systems place greater importance, for purposes of querying, on the logical relationships and associations between individual entities. The most basic approach allows the author of the system to specify what are in effect *fixed links* between specified media objects. Typically each media object is accompanied by several buttons or hot spots that function as the *anchors* of the ends of individual links. Typically these buttons provide links to previous and next logically connected nodes as well as a few other nodes that have been deemed to be related in some way. In principle, the presence of such links provides great flexibility for the user to move around the information space. However, with fixed, or explicit links, requirements for direct links must have been anticipated at the time of authoring, and all traversals through the information space can only be performed by moving along the predefined links. As a consequence, the total link distance, or path length, between some nodes may be very great and relatively complex to navigate, though it will depend upon the structure of linkage. It may also be noted that the insertion of explicit links can itself constitute a major overhead in building a hypermedia system.

A more versatile approach to hypermedia development entails the use of intensional links (DeRose, 1989) that are not embedded within the individual media objects, or necessarily stored explicitly at all, but are generated as required using references to the unique identifiers of the relevant nodes (Hall, 1994). In the Dexter hypertext reference model (Halasz, 1993) links are maintained within the Storage Layer and as such are quite distinct from the contents of individual media objects which are stored in the Within Component Layer. In this way the link to a logically associated node may be computed on demand. The major benefit of computed links is that it is not necessary to predict in advance, and store the links for, all likely routes through the information space. It therefore leaves the possibility of incorporating a procedural or deductive process to identify nodes that may be related to a particular node on the basis of the user's interests.

The semantic hypermedia architecture (SHA) adopts the approach of computed links and applies the concepts of semantic modelling and semantic databases (Hull and King, 1987; Schnase et al, 1993) to represent classification structures and the qualitative structure of space and time, which, together with references to individual instances of media objects, constitute the index space of the information system. This stored knowledge of semantic, spatial and temporal relationships is then used by the query mechanism to compute links between media objects. The approach differs significantly from the vector processing of text referred to above in that the emphasis is upon modelling the meaning and structure of concepts in time and space by means of semantic nets. An analogy may be made with the work of Kuhn (1994) who has described an algebraic approach to representing the semantics of terminology in geographical data, for purposes of spatial data transfer.

In the SHA, semantic, spatial and temporal relationships are all treated as first class categories and are modelled by binary relations, thus facilitating deductive processing in the conceptual domain, as well as qualitative spatial reasoning and temporal reasoning. The implementation of the binary relations of the SHA consists of triples which consitute a Binary Relational Store or BRS (Frost, 1982). The system has been programmed in LISP in combination with HyperCard (Taylor et al, 1994). The BRS provides a flexible basis for subsequent reasoning across all three aspects of indexing.

3.1 Semantic relationships

The SHA employs four fundamental relationships from the domain of semantic modelling. These are *Has-a*, *Is-a*, *A-Kind-Of (AKO)* and *Part-Of*.

The *Has_a* relationship is used to define the attributes or properties of a class. For example the class *Bridge* has attributes that include *Span* and *Builder*.

(Bridge **Has-a** Span)   (Bridge **Has-a** Builder)

The *A-Kind-Of* relationship implements generalisation and specialisation relationships between classes and subclasses. Examples include:

(Foot_Bridge **A-Kind-Of** Bridge)   (Bridge **A-Kind-Of** Construction) (Public-House **A-Kind-Of** Building)
(Railway **A-Kind-Of** Transport)   (Building **A-Kind-Of** Construction)

The *A-Kind-Of* relationship has been used to represent the classification hierarchy of the SHIC system. Examples of triples used to represent SHIC, starting at the highest level of the hierarchy, are:

(Community-Life **A-Kind-Of** SHIC)   (Domestic-&-Family-Life **A-Kind-Of** SHIC)
(Personal-Life **A-Kind-Of** SHIC)   (General-Personal-Life **A-Kind-Of** Personal-Life)
(Tobacco-Food-&-Drink **A-Kind-Of** Personal-Life)   (Costume **A-Kind-Of** Personal-Life)
(General-Costume **A-Kind-Of** Costume)   (Mens **A-Kind-Of** Costume)
(Womens **A-Kind-Of** Costume)   (Childrens-Unisex **A-Kind-Of** Costume)

Classes that are sub classes of other classes inherit the attributes of the parent class. A class such as *Foot-Bridge* is a subclass of the class *Bridge* and as such inherits the attribute *Span*.

The *Is-a* relationship associates an instance of a phenomenon with its classification. For example:

(Maltsters **Is-a** Public-House)   (Pontypridd-Bridge **Is-a** Foot-Bridge)   (Taff-Vale-Railway **Is-a** Railway)

The *Part-Of* relationship implements the semantic modelling construct of aggregation in the sense of part-whole relationships. This may be distinguished from the *Has-a* relationship which implements aggregation of attributes to constitute a class. Examples of *Part-Of* in the context of semantic relationships are:

(Railway-Station **Part-Of** Railway)   (Railway-Line **Part-Of** Railway)

3.2 Temporal relationships

Classes referring to real-world phenomena such as bridges, streets and buildings are associated with the class Temporal Interval, which has attributes of a *name*, a *start* and, for certain types of interval, an *end*. There are several specialisations of Temporal-Interval represented by the *Is-a* relationship. BRS tuples that represent a schema for Temporal-Intervals are as follows:

(Temporal-Interval **Part-of** Temporal Model)
(Temporal-Interval **Has-a** Name)
(Temporal-Interval **Has-a** Start)
(Year **A-Kind-Of** Temporal-Interval)
(Period **A-Kind-Of** Temporal-Interval)
(Period **Has-a** End)

Examples of instances of temporal intervals are as follows:

(TI161 **Is-a** Year)	(TI162 **Is-a** Year)	(TI7 **Is-a** Period)
(TI161 **Name** 1910)	(TI162 **Name** 1911)	(TI7 **Name** 1910s)
(TI161 **Start** 1910)	(TI162 **Start** 1911)	(TI7 **Start** 1910)
		(TI7 **End** 1919)
(TI8 **Name** C1900)	(TI255 **Name** VictoranEra)	
(TI8 **Start** 1900)	(TI255 **Start** 1837)	
(TI8 **End** 1999)	(TI255 **End** 1901)	

Individual phenomena such as a particular bridge or building will then have temporal intervals defined for them as necessary.

3.3 Spatial relationships

The present version of the experimental SHA implementation uses the two spatial relationships of containment and adjacency. These relationships apply to objects classified directly or indirectly (through generalisation relationships) as *Geographic-Area*. The class *Geographic-Area* is subject only to qualitative spatial relationships. Its class attributes include its *Name* along with semantic and temporal instantiation via *Is-a* and *Temporal-Class* relationships. They do not include geometric data. Examples of BRS triples designating *Geographic-Area* directly and through generalisation include:

(Town **A-Kind-Of** Geographic-Area)   (District **A-Kind-Of** Geographic-Area)
(River **A-Kind-Of** Geographic-Area)   (Religious-Building **A-Kind-Of** Geographic-Area)
(Market-Place **A-Kind-Of** Geographic-Area) (Chapel **A-Kind-Of** Religious-Building)
(Church **A-Kind-Of** Religious-Building)

Instances of the geographic area class are allocated unique identifers which are then used for defining semantic relationships. For example:

(GA4 **Name** Pontypridd)   (GA216 **Name** Pontypridd-Town-Centre) (GA13 **Name** Taff-Street)
(GA4 **Is-a** Town)   (GA16 **Is-a** District)   (GA13 **Is-a** Street)

(GA11 **Name** Bridge-Street) (GA12 **Name** Chapel-Street)
(GA11 **Is-a** Street)   (GA12 **Is-a** Street)

The hierarchical structure of space can then be recorded through strict containment relationships, designated here by *Inside*. For example:

(GA216 **Inside** GA4)   (GA11 **Inside** GA216)
(GA12 **Inside** GA216)   (GA13 **Inside** GA216)

Containment relationships enable the definition of composite, hierarchically structured geographic areas (Jones and Luo, 1994), such as Pontypridd-Town-Centre. In the present implementation, the *Inside* relationship is, for purposes of processing queries, one of strict containment. Thus phenomena such as rivers which may pass through several geographic areas cannot participate in an *Inside* relationship with the overlapped regions except by means of qualitative subdivision into uniquely identified constituent parts, which may be modelled by the *Part-Of* relationship. The definition of the subdivisions is the subject of ongoing work which will lead to the implementation of a richer set of qualitative spatial relationships.

The adjacency (or meets) relationship *Next-to* is used to refer to topological connectivity between geographic areas, of whatever sort. It has been used particularly to represent connectivity between streets, for example:

(GA12 **Next-to** GA11)
(GA13 **Next-to** GA11)

3.4 Spatio-temporal relationships

The SHA employs several relationship types to represent the changes that geographical areas may undergo over time. The relationship *Redefines* is used to designate a change in either the name, the function, or the composition of a geographic area. The changes refer here to real-world or event time, as opposed to transaction or database time (Snodgrass, 1992). If the change involves the loss of a component part of the previous geographic area, then the *Removal* relationship is used to indicate the contained part that is be removed from the composition. The *Inside* relationship can be used to add areas. A geographic area is associated with a Temporal-Interval through the relationship *Temporal-Class*. The geographic area that redefines another area will therefore have a different associated Temporal-Interval. The following set of triples provide an example of spatio-temporal change (note that the names of geographic areas have been substituted for their unique identifiers in order to aid clarity):

(Penuel-Burial-Ground **Inside** Penuel-Square)   (Penuel-Fountain **Inside** Penuel-Burial-Ground)
(Penuel-Chapel **Inside** Penuel-Square)   (Penuel-Road **Inside** Penuel Square)
(Market-Square **Redefines** Penuel-Square)   (Market-Square **Removal** Penuel-Burial-Ground)
(Pontypridd-Market **Inside** Market-Square)
(Penuel-Square **Temporal-Class** TI20)   (Market-Square **Temporal-Class** TI21)

In this example, Penuel Square changes its name to, and hence is redefined by, Market Square. The change coincides with the disappearance of Penuel-Burial-Ground, indicated by the *Removal* relationship, and the appearance of Pontypridd-Market. Insertion of the *Redefines* tuple triggers an update of the BRS to provide an explicit record of the composition of the new geographic area of Market square. This results in the insertion of the triples:

(Penuel-Chapel **Inside** Market-Square)
(Penuel-Road **Inside** Market-Square)

Note that because Market-Square and Penuel-Square have different associated temporal intervals there is no logical conflict between the multiple *Inside* relationships for Penuel-Chapel and Penuel-Road.

3.5 Media objects

The relationships referred to so far serve the purpose of modelling the structure of classifications, space and time. Clearly links must be provided between the media objects and this structural knowledge. The schema for media objects includes the following triples:

(Media-Object **Has-a** Name)
(Photograph **A-Kind-Of** Media-Object)
(Audio **A-Kind-Of** Media-Object)
(Text **A-Kind-Of** Media-Object)

Individual media objects are then named and classified, in terms of type of media, with triples such as:

(MI1 **Is-a** Photograph)   (MI1 **Name** A4.Lewis-Merthyr-Colliery)
(MI86 **Is-a** Text)   (MI86 **Name** History-Of-BrownLenox)
(MI104 **Is-a** Audio)   (MI104 **Name** Traveller-Tale-1)

Individual media objects are linked to specific geographic areas using relationships such as *photo*, meaning 'is photographed by', examples of which are:

(Taff-Street **Photo** MI100)   (Taff-Street **Photo** MI101)
(Bridge-Street **Photo** MI100)   (Bridge-Street **Photo** MI102)   (Bridge-Street **Photo** MI103)

Note that in this way individual photographs (such as MI100) may be referenced initially to several geographic areas. The same type of relationship is used to associate media objects with classification terms and events, for example

(retail-distribution **Photo** MI100).

4. Query and navigation methods

A wide variety of query and navigational tools has been implemented with the SHA. Here we will describe a few of these methods, focusing on search procedures that attempt to find media objects that are similar in some sense to a specified term in any or all of the three index spaces of classification, geography and time.

4.1 Classification, generalisation and specialisation

Media objects can be referenced to classification terms at any level of the classification hierarchy. A query may specify a single classification term which can be used by the BRS to find all media items that are a direct match on that term. Whether or not there is a direct match, the term can also be used to allow the user to browse through media objects that are classified in content as specialisations and generalisations of the given term. This is done easily by following the transitive *A-Kind-Of* relationships between classification terms, and selecting media objects that are classified by the term. This type of classification hierarchy browsing can be expanded by exploiting the presence of *Part-Of* relationships between classification terms that constitute aggregated objects (such as a Railway). Thus if for example a query started by specifying the term Transport, the classification hierarchy could be descended via the *A-Kind-Of* relationship to find Railway, which in turn leads, via the *Part-Of* relationships, to media objects classified as Railway-Station and Railway-Line.

4.2 Semantic closeness

A problem with the hierarchy traversal as just described is that it may result in significant differences of meaning and hence level of user interest, as traversal progresses. To provide some measure of the difference in meaning between classification terms it is possible to adopt an approach similar to the concept of semantic distance (Rada, 1989; Brooks, 1995), which was intended to measure the difference in meaning between hypermedia nodes. It is the minimum number of edges in the class generalisation graph between the two nodes of interest. The generic formula for semantic closeness (Tudhope et al, 1995) uses a diminishment factor which affects the rate at which semantic closeness decreases with path distance through the graph. Thus

SemanticCloseness(Term1, Term2) = 1 - (|Path(Term1, Term2)|*DiminishmentFactor)

The diminishment factor chosen here is 0.3 which means that up to path lengths of three the SemanticCloseness metric returns positive values, which are to be regarded as close to some extent. Beyond this the values will become negative. The use of 0.3 appears appropriate to the structure of the SHIC hierarchy, since it means that terms within each of the four primary branches of the hierarchical classification scheme will appear similar as will siblings at any one level. Terms in different major branches are less likely to appear similar, as most media objects are attached to terms low down the hierarchy. Having defined such a metric, it can be used in a navigation or query tool to find all media objects that are semantically close in the classification system, as determined by a specified threshold value. Typically, this threshold value is set to be non-negative. Semantically close objects are then found by performing a search through the hierarchy, in all directions from the initial classification term, until the threshold is exceeded.

Various experiments have been performed to examine the effects of modifying the value and usage of the diminishment factor. For example, it may be given a lower value for graph edges defined by *Is-a* than for *A-Kind-Of* or *Part-Of*. An approach which appears to improve performance of the metric with regard to perceived sense of closeness is to reduce the dimishment factor in proportion to the depth within the hierarchy, since as terms become progressively more specialised their semantic difference may be regarded as decreasing whereas, higher up, there are greater generic differences.

4.3 Semantic closeness in time

As the temporal intervals are largely quantitative, measures of semantic closeness in the temporal domain should exploit such numerical data when present. A simple metric for temporal SemanticCloseness would be

$$\text{SemanticCloseness(Year1, Year2)} = 1 - \frac{|\text{Year1} - \text{Year2}|}{\text{Range}}$$

where Range refers to the maximum temporal extent of the database or project. Since many phenomena are defined by temporal intervals that could extend over several years, this metric has been modified to take account of the size of the respective intervals as well as the distance between them. The resulting formula (Tudhope et al, 1995) is:

$$\text{SemanticCloseness(TI1, TI2)} = W1(MP/IU) + W2(IU/(NMP+IU)) + W3(IU/D+IU))$$

where TI1 and TI2 are the temporal intervals to be compared; D is the difference between the end of one non-overlapping interval and the beginning of the next (otherwise zero); MP is the length of overlap in years between overlapping intervals; NMP is the number of years (of the combined intervals) that intervals do not have in common; IU is the interval being used as the basis of comparison. The values W1, W2 and W3 are weights to be attached to each of the terms of the formula.

4.4 Semantic closeness in geographic space

A qualitative measure of distance in geographic space can be obtained as a function of shortest path length between locations connected by the *Next-to* relationships. This path length can be modified, as in the case of distances in classification space, by means of a diminishment factor. This factor has been set somewhat arbitrarily in our experimental system at 0.2, which means that, in the case of streets, distances equal to or less than four blocks can be considered close and result in non-negative values. An attraction of the qualitative measure of geographic closeness is that, provided it operates on *Next-to* relationships between geographic areas (including streets) that are at the same level of geographic hierarchy, then it is adaptive to scale and context.

A very similar process to that of following *A-Kind-Of* and *Part-Of* relationships in the classification space can be used to expand seaches in the vicinity of a particular geographical location. Thus having specified a geographic area at some level of the geographical hierarchy, the *Inside* relationships can be used to search downwards through the hierarchy to find all places that are contained within the current geographic area.

4.5 Search in the three dimensions of concept, space and time

User interface tools have been developed which allow the user to control each of the three dimensions simultaneously. Thus the user can specify a classification term, a geographic area and a temporal interval, that are used for an initial search. Note that the hierarchies of concept, time and space will all be used to find any media objects that have been classified as specialisations of the specified terms. If no media objects are found, then knowledge of the density of media objects in each of the index spaces can be used to extend the search by generalising its extent in the most sparsely populated of the three index spaces. Suppose, for example, the user specified the concept *retail distribution*, and the geographic area of *High Street*, and nothing was found because, although many media objects specified *retail distribution* (or its specialisation), none was also associated with *High Street*. In this case the search is extended automatically beyond *High Street* by means of the *Next-to* relationship. In fact *High Street* is a continuation of the street *Taff Street*, for which there are several media objects that also reference retail distribution.

5. Concluding remarks

The semantic hypermedia architecture summarised here has provided a basis for implementing a set of access methods for finding media objects referenced to conceptual, temporal and geographic space, each of which is represented by semantic networks that display hierarchical structure and, as in the case of spatial adjacency, lattice structure. The structure of the index space has been exploited to enable indirect matching of user queries and to apply metrics of semantic closeness to assist in finding media objects that may be indexed only indirectly with respect to the user's specified interest.

Future work will increase the number of types of spatial relationship and will provide an integration between the qualitative reasoning methods used here and more conventional GIS access techniques that exploit the geometric coordinate-based representations of geographic space. It is also intended to extend the capacity for imprecise match between user items and media classification terms by using a structured thesaurus. Thus, in addition to the aggregation and generalisation relationships currently encoded, this will include linguistic relationships of for example synonym and agency that can assist in translation and association between semantically related terminology.

References

Aitchison J. and A. Gilchrist 1990. 'Thesaurus construction: a practical manual', ASLIB, London.

Allen M.S. et al 1990. Interpreting Space: GIS and Archaeology , Taylor and Francis, London.

Brooks T. 1995. 'People, words, and perceptions: A phenomenological investigation of textuality', Journal of the American Society for Information Science 46(2), 103-115.

Burrough P.A. and Frank A.U. 1995. 'Concepts and paradigms in spatial information: are current geographical information systems truly generic?'. International Journal of Geographical Information Systems 9(2), 101-116.

Carlotto M.J. 1995. 'Text attributes and processing techniques in geographical information systems' International Journal of Geographical Information Systems 9(6), 621-635.

Cui Z., A.G. Cohn and D.A. Randall, 1993. 'Qualitative and topological relationships in spatial databases'. Advances in Spatial Databases. LNCS 692, Springer-Verlag. 296-315.

DeRose S. 1989. 'Expanding the notion of links'. Proceedings Hypertext '89. Pittsburg USA, ACM. 249-250.

Egenhofer M. J. and J. R. Herring, 1990. 'A mathematical framework for the definition of topological relationships'. Fourth International Symposium on Spatial Data Handling, Zurich, International Geographical Union, 803-813.

Frank A. U., 1994. 'Qualitative temporal reasoning in GIS-ordered time scales'. SDH'94 Sixth International Symposium on Spatial Data Handling, Edinburgh, Taylor and Francis, 410-430.

Frost R.A. 1982. 'Binary-relational storage structures'. The Computer Journal 25(3), 358-367.

Halasz F. and M. Schwartz 1994, 'The Dexter Hypertext Reference Model', Communications of the ACM 37(2), 30-39.

Hall W. 1994. 'Ending the tyranny of the button', IEEE Multimedia 1(1), 60-68.

Harpring P. 1992. 'The Thesaurus of Art-Historical Place Names (TAP)', Visual Resources Association Bulletin, 19(3), 26-32.

Hull R. and R. King 1987. 'Semantic database modelling: survey, applications and research issues', ACM Computing Surveys 19(3), 201-260.

Jones C. B. and L. Q. Luo, 1994. 'Hierarchies and objects in a deductive spatial database'. SDH'94 Sixth International Symposium on Spatial Data Handling, Edinburgh, 588-603.

Kuhn W. 1994. Defining semantics for spatial data transfers. SDH'94 Sixth International Symposium on Spatial Data Handling, Edinburgh. Taylor and Francis, London, 973-985.

MDA. 1989. Occasional Paper 16, Place Names Recording Guidelines. Museum Documentation Association, 347 Cherry Hinton Road, Cambridge, UK.

Molholt P. and T.Petersen 1993. The role of the 'Art and Architecture Thesaurus' in communicating about visual art', Knowledge Organisation 20(1), 30-34.

Rada R., H. Mili, E. Bicknell and M. Blettner 1989. 'Development and application of a metric on semantic nets', IEEE Transactions on Systems, Man and Cybernetics 19(1), 17-30.

Salton G. 1989. Automatic Text Processing. Addison-Wesley. Reading:MA.

Schnase J., J. Leggett, D. Hicks and R. Szabo 1993. Semantic data modelling of hypermedia associations', ACM Transactions on Information Systems 11(1), 27-50.

SHIC Working Party. 1983. Social History and Industrial Classification: A Subject Classification for Museum Collections. (2 vols.) Centre for English Cultural Tradition and Language, University of Sheffield, UK.

Snodgrass R.T. 1992. 'Temporal databases'. In Frank A.U., Campari I., Formentini U. (eds) Theories and Methods of Spatio-Temporal Reasoning in Geographic Space'. LNCS 639, Springer Verlag, 22-64.

Taylor C., D. Tudhope, P. Beynon-Davies 1994. 'Representation and manipulation of conceptual, temporal and geographical knowledge in a museum hypermedia system'. Proceedings ACM European Conference on Hypermedia Technology. Edinburgh. 239-244.

Tudhope D., P. Beynon-Davies and C. Taylor C., 1995. 'Navigation via similarity in hypermedia and information retrieval', Proceedings of Conference on Hypertext, Information Retrieval, Multimedia. Konstanz, April, 203-218.

Walker D.R.F., I.A. Newman, D.J. Medyckyj-Scott and C.L.N. Ruggles 1992. 'A system for identifying datasets for GIS users'. International Journal of Geographical Information Systems 6(6), 511-527.

Contact address
Christopher B. Jones, Carl Taylor, Douglas Tudhope and Paul Beynon-Davies
Department of Computer Studies
University of Glamorgan
Pontypridd
Mid Glamorgan, CF37 1DL
United Kingdom
Phone: 44 1443 482722
Fax: 44 1443 482715
E-mail: cbjones@glamorgan.ac.uk

TOWARD SEMANTICS FOR MODELLING SPATIO-TEMPORAL

PROCESSES WITHIN GIS

Christophe Claramunt[1] and Marius Thériault[2]

[1]Swiss Federal Institute of Technology
Department of Rural Engineering, Spatial Information Systems
Lausanne, Switzerland
[2]Laval University
Department of Geography, Planning and Development Research Centre
Quebec, Canada

Abstract

This paper presents a taxonomy of processes and semantics for describing and modelling spatio-temporal evolution within GIS. The approach is based on an event-oriented representation of spatial dynamics. Events are described by sets of basic and composite processes. We show how this semantic formalism allows for comprehensive decomposition and representation of complex spatio-temporal phenomena. The explicit representation of processes is designed to record temporal relationships between geographic entities and links describing their joint evolution.

Key words

Temporal GIS, events, processes, modelling, spatial dynamics.

1. Introduction

Time is a fundamental dimension for understanding and modelling evolution of geographic phenomena. Previous papers propose temporal models and query languages for GIS applications [Hazelton 1990, Langran 1992 and 1993, Cheylan 1993, Frank 1994, Peuquet 1994 and 1995, Worboys 1994a]. Many of them use a linear scale to model time. However, time is too complex to be reduced to a simple collection of time-stamps [Peuquet 1994, Frank 1994]. A temporal GIS (TGIS) must be able to monitor and analyse successive states of spatial entities, and also be equipped to study dependencies between linked entities [Claramunt 1995]. This implies formal representation of spatio-temporal processes in order to fully describe their topological networks, which involve both temporal and spatial dimensions.

In a previous paper, we have proposed a model to describe basic spatio-temporal processes using an event-oriented approach [Claramunt 1995]. It provides a topological framework for representing entity-based events and processes. It describes basic processes of evolution, distinguishes between the metric and topological dimensions of time, and suggests temporal query operators. The purpose of this paper is to develop a taxonomy of spatio-temporal processes that retain the original meaning of events and processes while they are represented within a GIS database. The approach allows for complete representation of processes involving both space and time, and the encoding of known relationships between entities that produces complex spatio-temporal evolution. It is based on the Event Pattern Language (EPL) described by Gehani (1992) and Motakis (1995). EPL is already used for active database applications and provides logical constructors for complex event specifications. The proposed model takes advantage of EPL's expressiveness to summarise complex spatio-temporal processes.

In section 2, we present the temporal ontology used to define the framework. Section 3 concerns modelling of spatio-temporal objects. Identification and classification of basic processes follow in section 4. A summary of the necessary EPL constructs is presented in section 5. Section 6 addresses the semantic specification of composite processes. Section 7 concludes with a summary about capabilities and the usefulness of this approach and outlines further research issues.

2. Temporal ontology

The proposed process-oriented TGIS model combines the quantitative concept of time (measured dates and durations) [Langran 1992] with a qualitative temporal reasoning involving topologically ordered events [Frank 1994, Peuquet 1994]. An event is defined as a set of processes (emphasis on global result) while a process is an action which modifies entities (emphasis on the mechanisms) [Bestougeff 1992]. The model represents local and discrete spatial dynamism and retains the Kowalski's principle of local event representation [Kowalski 1986]. It is not oriented toward global changes which are rather described by space and time series (E.g. statistical or numerical analysis of spatio-temporal behaviour) [Peuquet 1994, Thériault 1995] or by temporal logics dealing with global event descriptions [McCarthy 1969].

To respond adequately to scientific needs, a TGIS should explicitly preserve known or possible links (the temporal topology) between events and their consequences, the facts [Beller 1991]. Observed relationships should be noted (E.g., entities A and B generate entity C) to help scientists develop models that reproduce the dynamics of spatio-temporal processes. Researchers will thus be able to study complex relationships, draw conclusions and search for causal links that associate entities through influence and transformation processes. Recording known relationships between entities is an essential requirement for scientific TGIS because these facts are the basic constructs used in the research.

We define an entity as a real-world abstraction of an existing feature while 'object' means its database representation. Object versions correspond to successive states of a specific entity. Additionally, we will use the term 'instance' to designate versions that are linked by a spatio-temporal process and that do not concern the same entity.

Various metaphors represent complementary dimensions of time: linear time, cyclical time, branching time (inter-entity time dependencies) and multidimensional time (valid-time, transaction-time, user-time) [Jensen, 1992]. The model is built using valid-time to describe real-world events occurrence and duration, as opposed to transaction-time and user-time that are used for management and user purposes [Jensen 1992]. The time range of any process or entity is represented by its lifespan [Segev 1994]. The quantitative attributes of time may be represented as an instant or a period. An instant is generally modelled as a time-stamp [Navathe 1988]. A period is defined by a composite attribute involving two time-stamps with an order constraint [Snodgrass 1995]. The chronon describes the shortest duration of time that can be represented in the computer system [Jensen 1992], and the time granularity represents the time line partitioning related to application needs (E.g., geologic, historic or annual scales) [Snodgrass 1995].

3. Spatio-temporal entities modelling

Recent thinking about integrating time into GIS has shown that it is a complementary facet of spatial and thematic domains [Peuquet 1994 and 1995, Claramunt 1995]. This paper retains the triad representational framework proposed by Peuquet (1994) and uses an object-oriented model to define the data structure. Object-oriented models provide a natural method for describing real-world spatial entities [Worboys 1994a and 1994b], avoids data fragmentation, and enables useful capabilities for managing time (E.g., inheritance of time properties) [Snodgrass 1994].

We represent an entity class as a set of objects (A) of the same type. These objects share common attribute sets. The object type (A_i) of a spatial entity is defined by three attribute sets:

- The temporal domain (A_iTemp) represents entities' temporal data {Valid time, Transaction time, Last Process, Next Process}. Last Process and Next Process attributes link each version of an entity to the processes that generate and replace it. The Valid time attribute describes the total duration of each version; thus the lifespan of an entity is the total duration of its successive versions.
- The thematic domain (A_iThem) describes the thematic attributes {$Them_1$, ..., $Them_j$}.
- The spatial domain (A_iSpat) conveys the geographical representation of entities {$Spat_1$, ..., $Spat_j$}.

Object versions use this triad structure to describe evolution manageable at the domain level. Successive object versions can refer to the same spatial attribute set if the mutation is thematic, or to the same thematic attribute set if the mutation is only spatial. This behaviour specialisation allows for asynchronous value changes of a same object within the thematic or the spatial domain without affecting the other one. When needed, time-stamps (Valid time and Lifespan) may be used to take a snapshot of the contents of the database at a specific instant or during a period. Specifying time granularity enables simultaneous use of many temporal scales within the same database.

Rules for inheritance of temporal attributes depend on the characteristics and semantics of abstraction mechanisms. Generalisation and aggregation are the main abstractions generally identified for spatial applications. Generalisation assembles objects with common behaviours and properties while aggregation groups different objects into a composite object. We define temporal inheritance as follows:

- Each object member of a generalisation hierarchy has its set of temporal attributes (A¡Temp) declared at the highest supertype object level.

- Each object member of an aggregation hierarchy has its set of temporal attributes (A¡Temp) defined at its own type level. An aggregated object has a set of temporal attributes that is a composite of the constituent subtype objects. For example, the lifespan of a flooded parcel has to be an intersection of the parcel's and the flood area's respective lifespans. These consistency rules may change according to the entities' characteristics and to the composition type.

Moreover, object hierarchies introduce temporal constraints. The existence of a subtype object depends on the previous existence of the appropriate supertype object [Wachowicz 1994]. Conversely, there is no fixed rule for transaction-time composition since it is independent of any order constraints (E.g., the second process of a sequence may be registered in the database before the first one).

4. Basic spatio-temporal processes

Events are defined through processes that transform entities. We distinguish between evolving and mutating entities [Langran 1992] to define three main classes of basic spatio-temporal processes (Figure 1):

- Evolution of a single entity represents basic changes (appearance, disappearance, etc.), transformations and movements of that entity.

- Functional relationships involve spatio-temporal processes between several entities (replacement and diffusion processes). They convey known dependence links that need to be modelled explicitly in the temporal domain. The range of phenomena that can be processed in a TGIS is probably inexhaustible. Therefore, a rule must be provided to distinguish between relationships with predominant temporal incidence (which must be modelled within the temporal structure) and relationships in which time plays a secondary role. Certain events are ordered in an unchanging sequence (E.g., an ancestor must live before his descendants) that dates alone cannot always reproduce. A review of the principal types of temporal relationships studied in geography leads us to set forth the following rule: functional processes that involve a dependence link between entities, while they impose a precedence constraint, must be modelled on the temporal structure in order to clarify the topological relationships they convey. Furthermore, processes with only one time-period overlap may be described in the thematic domain without creating confusion [Claramunt 1995].

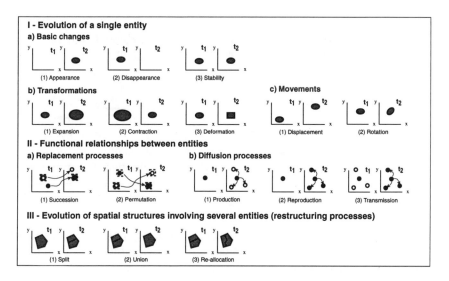

Figure 1 Typology of spatio-temporal processes.

- Evolution of spatial structures describes spatio-temporal processes involving several land-based entities. Restructuring processes are used to model changes that affect the way a given piece of land is partitioned according to a system of mutually exclusive zones (E.g., counties, municipalities, census tracts, land parcels). This involves a mix of simultaneous transformation, appearance and disappearance of interrelated entities with an obvious precedence constraint. However, evolving spatial patterns of independent entities in a population conveys overall tendencies that can be modelled using aggregation mechanisms.

This typology identifies three main classes of spatio-temporal processes that allows for the hierarchic decomposition of real-world phenomena (E.g., a displacement is a movement and a movement is an evolution). Although we postulate that these basic processes are mutually exclusive and their number is finite, the model can further integrate other basic processes defined with the same rules.

We also model processes as objects. This provides a homogeneous representation structure for both entities and processes. Process objects have to be kept distinct from those describing entity characteristics and behaviour. Thus, entity objects may use process objects (through methods) to describe their evolution and link their respective versions or instances (Figure 2).

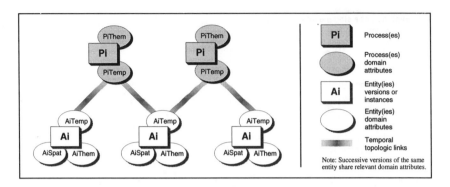

Figure 2 Temporal topology.

A basic process type (P) represents a class of spatio-temporal processes with the same attribute set. Each process class (P_i) has a temporal (P_iTemp) and thematic (P_iThem) attribute set:

- The temporal domain (P_iTemp) is a set of temporal attributes {Valid time, Transaction time, –Entities lists}. The entities lists are used to reference active entities generating the process, passive entities being modified by the process and other entities directly involved during the process. These lists contain versions related to only one object when they describe the evolution of a single entity. For multiple-entity relationships, they refer to the appropriate instances of each entity involved. Links between the active, passive and involved entities define a relation, taking the form: Process_Type{set(s) of related instances or versions}.

- The thematic domain (P_iThem) is a set of thematic attributes {Process type, Attribute$_1$, ..., Attribute$_j$} which describes and qualifies the process. The Process type represents the basic processes shown in Figure 1.

Spatial properties of processes are inferred from the entities included in the process, thereby avoiding the need for a spatial domain to locate it. On the other hand, temporal topologic relationships between successive entities (versions or instances) are inferred through the process that produces the modifications.

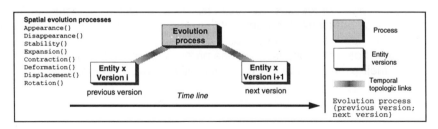

Figure 3 Syntactic graph of single entity spatial evolution processes.

Spatio-temporal processes describing the evolution of a single entity (Figure 3) are:

- Appearance (Appearance{0;X}) occurs when an object version of X has no preceding version (0 is the null set).
- Disappearance (Disappearance{X;0}) occurs when an object version of X has no subsequent version.
- Stability (Stability{X;X}) occurs when an object version of X has the same state for a certain period. This process is needed to handle the case of an entity that changes in the thematic domain without corresponding evolution of its spatial properties.
- Transformation processes involve changes in entity's size or shape. They are Expansion (Expansion{X;X}), Contraction (Contraction{X;X}) and Deformation (Deformation{X;X}).
- Movement processes convey location changes related to Displacement (Displacement{X;X}) and Rotation (Rotation{X;X}).

The {X;X} operands link any given version of X with the immediately succeeding version of the same entity after the event occurs.

Functional spatio-temporal processes are:

- Succession (Succession{X;Y;A}) occurs when an instance of entity Y is the immediate successor of an instance of X at the position A. Both instances may survive the replacement (Figure 4), instance of X being permuted to another location tracked with the appropriate evolution process. We represent the succession process as a relationship of one-to-one entities over one position. This process built networks showing the succession of individuals or groups holding a given position. The position is a real (E.g., a parking place) or a virtual (E.g., a set of coordinates) entity with a spatial domain that is used to define a spatial link between a set of succeeding owner entities. In Figure 4, the ownership link expresses the relationship between one owner entity and the position during a given period of time, whilst the temporal link is used to model the succession of owners for a same position.

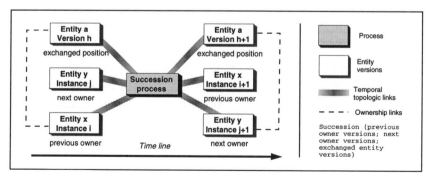

Figure 4 Syntactic graph of a succession functional process.

- Permutation (Permutation{X;Y;A;B}) occurs when an instance of Y is the immediate successor of an instance of X at position A, and when this same instance of X is the immediate successor of the instance of Y at position B,during the same instant or period. This is a relationship of one-to-one entities over one-to-one positions (Figure 5). The Permutation{X;Y;A;B} expression is not strictly equivalent to the combination of Succession{X;Y;A} and Succession{Y;X;B} because the simultaneity of the process is an essential property of the changes occurring. For example, this process is needed to model two hockey teams (X and Y) exchanging their goals (A and B) at the end of a period, because this change must be simultaneous when it occurs and there is a complementary relationship between entities A and B.

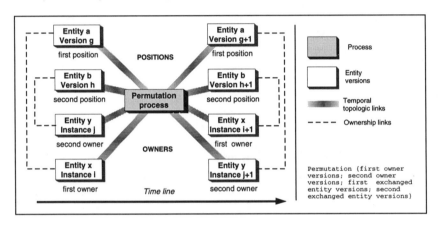

Figure 5 Syntactic graph of a permutation functional process.

- Production (Production{$X_1...X_n;B_1...B_p;A_1...A_q$}) occurs when a set of entities (X) produce a set of entities (B) while consuming a third set of involved entities (A). The production process is modelled as a relationship of many-to-many entities using many component entities (Figure 6). The producers, the involved components and the products must be entities of different classes. The production process implies the disappearance or tranformation of involved components and the appearance of produced entities. However, it is more meaningful than Disappearance{$A_1;0$} and Appearance{$0;B_1$} because it describes the linkages between entities and the mechanisms involved (via the process thematic attributes). The Disappearance{$A_1;0$} and Appearance{$0;B_1$} are used only when the mechanism involved is missing (E.g., disappearance and appearance in unknown circumstances). For example, this process can model a plant (producer) using pieces (components) to manufacture cars (product). The relationship between the involved components is needed to model the exchange of products between firms and describe complex industrial networks.

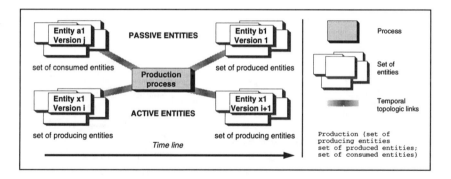

Figure 6 Syntactic graph of a production functional process.

- Reproduction (Reproduction$\{X_1...X_n;Y_1...Y_p\}$) occurs when a first set of entities (X) creates a new set of entities (Y) of the same class. The reproduction process is modelled as a many-to-many entities relationship (Figure 7). It shares the linkage capabilities of the production process while restricting the parent and the child entities to the same class. It is valid only for life phenomena, and the need to specify the contributed material is not obvious. Successive reproduction involving grandparents, parents and children may be used to model complex reproduction networks needed in biology or genealogy.

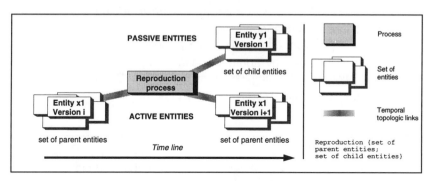

Figure 7 Syntactic graph of a reproduction functional process.

- Transmission (Transmission$\{X_1...X_n;Y_1...Y_p;A_1...A_q\}$) occurs when a set of entities (Y) has its attributes modified by some contact with a set of entities (X) with the collaboration of a third set of involved entities (A). The third set of entities (A) acts as a vector of transmission (Figure 8). The main difference with the production and reproduction processes is that the Y entities existed prior to transmission. Their previous histories are tracked through evolution processes. We represent the transmission process as a network of many-to-many entities using many vector entities. This type is needed to model diffusion mechanisms for telecommunications and innovations, as well as contagion processes in epidemiology. In both fields there is an obvious need to make the relation with the transmission medium explicit (the communication network or the virus).

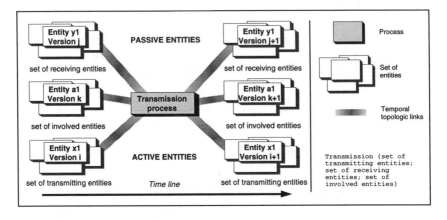

Figure 8 Syntactic graph of a transmission functional process.

Spatial restructuring processes (Figure 9) are:

- Land subdivision (Split$\{X;Y_1...Y_n\}$) occurs when a spatial entity (X) is replaced by a set of distinct entities (Y) of the same class. It is a one-to-many entities relationship that implies a simultaneous Disappearance$\{X;0\}$ and Appearance$\{0;Y_1\}$ ... Appearance$\{0;Y_n\}$ while adding a functional link between all those simultaneous events.
- Land union or fusion (Union$\{X_1...X_n;Y\}$) occur when a set of spatial entities (X) is replaced by another spatial entity (Y). It is a many-to-one entities relationship that implies a simultaneous Disappearance$\{X_1;0\}$ ... Disappearance$\{X_n;0\}$ and Appearance$\{0;Y\}$ plus the simultaneous functional meaning.

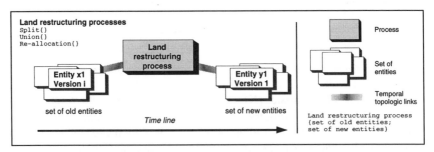

Figure 9 Syntactic graph of land restructuring processes.

- Territory re-allocation (Re-allocation$\{X_1...X_n;Y_1...Y_p\}$) occurs when a set of spatial entities (Y) is the immediate successor of a distinct set of entities (X), overlapping the same common territory. It is a many-to-many entities relationship (Figure 9). Figure 10 illustrates a re-allocation of land parcels: Re-allocation$\{(A,B,R1,C1);(D,E,F,G,H,R2,C2)\}$. This re-allocation cannot be modelled as a land union followed by a land subdivision for many reasons.

Firstly, entities C and R survived the process. Secondly, original parcels are split among many new parcels which may themselves be formed by the union of parts of various original ones. Thirdly, one original parcel (R) extends its territory over contemporaneous parcels (A, B and C) while giving some part of its previous area to form a new parcel (H). Fourthly, parcel C has shrunk during the operation. We get a complex mix of simultaneous related (one process involved) appearance, disappearance, expansion, contraction and land exchange. Therefore, representing this process as many distinct processes is inadequate because it hides the underlying common cause.

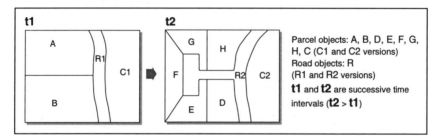

Figure 10 Example of land redistribution among parcels.

Land subdivision and land union processes are related to administrative procedures, like day-to-day surveying operations. Territory re-allocation is needed to model a comprehensive change involving many entities caused, for example, by a law. It is also the appropriate process for modelling borderline change between two or more existing spatial entities.

Evolution processes are generally sufficient to operate with the adjacent versions of an object, even if time-stamps are missing or corrupted. The other basic processes are mandatory for time-dependent applications in many fields, including industry (production), biology or genealogy (reproduction), communications or epidemiology (transmission). Temporal integrity constraints are enforced by appropriate controls on versions' or instances' lifespans (E.g., the begin-time of a produced instance must be compatible with the respective lifespan of its parent instances) and transitive rules to maintain consistent temporal networks (E.g., an ancestor cannot be the child of one of its descendants).

5. Event pattern language constructs

Composite processes enable description of complex natural phenomena by reproducing their dynamics (epidemiology, mobile entities, fire evolution, innovation diffusion, etc.). A knowledge representation of composite processes requires a formal language to understand and describe their semantics. Although temporal logics propose different theories to describe evolving systems, the issue of formally specifying the semantics of complex events is only partially solved [Long 1989].

The proposed approach uses event-oriented constructs [Gehani 1992, Motakis 1995] to represent such complex processes. The expressive power of the Event Pattern Language (EPL) is efficient for modelling complex spatio-temporal processes. A composite process is defined as an expression formed through processes and operators. The thematic attributes of a composite process are embodied in its constituents.

Let us introduce some basic notions and operations of the Event Pattern Language:

- P is a basic or a composite process.
- *:P is a sequence of zero or more consecutive occurrences of P.
- !P is the exclusion and indicates that there is no occurrence of P.
- $(P_1,P_2,...,P_n)$ is a sequence of processes consisting of an occurrence of P_1, immediately followed by an occurrence of P_2, ..., immediately followed by an occurrence of P_n.
- $(P_1\&P_2\&...\&P_n)$ is a conjunction of processes occurring simultaneously (at a same instant or period).
- $\{P_1,P_2,...,P_n\}$ is a disjunction of processes. It happens when at least one process among P_1, P_2, ..., P_n occurs.
- $[P_1,P_2,...,P_n]$ is a relaxed sequence consisting of an occurrence of P_1, followed later by an occurrence of P_2, ..., followed later by an occurrence of P_n. It is logically equivalent to $(P_1,*:any,P_2,...,*:any,P_n)$, where *:any means a sequence of zero or more other processes.

We introduce a cycle operator which is formed by the recursive repetition of the same sequence of processes using the ⊕: operator. It can be combined with sequences $(⊕:(P_1,P_2,...,P_n))$, conjunctions $(⊕:(P_1\&P_2\&...\&P_n))$, disjunctions $(⊕:\{P_1,P_2,...,P_n\})$, and relaxed sequences $(⊕:[P_1,P_2,...,P_n])$. This operator is requested to model recurrent natural and human phenomena (E.g., seasons in climatology or cultural rotations in agriculture). The cycle operator is significantly different than the sequence operator (*:): there are at least two occurrences of the sequence; the recurrence is regular on the time line (I.e., equally spaced events); and the successive instances may be separated by any other events.

6. Composite processes

Dynamic phenomena involving real-world entities may be represented by composite process expressions. Let us develop a few representative examples:

- A boat trip described successively by a stability position, a casting off and a sailing: (Stability{A;A}, Displacement{A;A}, *:{Displacement{A;A}, Rotation{A;A}}, Stability{A;A}).

- A forest fire as it evolves: (Appearance{0;A}, *:(Expansion{A;A} & Displacement{A;A}), Contraction{A;A}, Disappearance{A;0}).
- An epidemic phenomenon involving a virus (V) and many individuals (A, B, C_1...C_5, D_1...D_{45}): [Transmission{A;B;V}, Transmission{B;C_1...C_5;V}, Transmission{C_1...C_5;D_1...D_{45};V}].
- An industrial plant (A) producing cars (B) and industrial rejects (C) using M_1...M_{35} components:
  *:(Production{A;B;M_1...M_{35}} & Production{A;C;M_1...M_{35}}).
- Two boats (A and B) exchanging their dockyard positions (C and D) while transmitting radio information (I and J): (Permutation{A;B;C;D} & Transmission{A;B;I} & Transmission{B;A;J}).
- The genealogical tree of an individual (X): [Reproduction{(A,B);(C,D)}, Reproduction{(M,N);O}, Reproduction{(D,O);X}].
  In this example, there is a pair of twins in the first generation.
- A cyclic three-year rotation of three crops (W: Wheat, G: Grass, V: Vegetables) on a farm divided into three parcels (A, B, C):
  ⊕:(((Succession{W;G;A} & Succession{G;V;B} & Succession{V;W;C}),   *Year*   *1*
     (Succession{G;V;A} & Succession{V;W;B} & Succession{W;G;C}),   *Year*   *2*
     (Succession{V;W;A} & Succession{W;G;B} & Succession{G;V;C}))   *Year*   *3*

Transition graphs [Gertz 1995] express the structure of composite processes and may improve users' comprehension of the underlying system with explicit mapping of relevant constraints. Figure 11 shows a proposal for presenting composite processes in graphical form. This example presents a composite process (P=[P1,P2,P3]) involving two cars (entities A and B) arriving successively (relaxed sequence P1) at a ferryboat (entity C), a ferryboat trip to another port (process P2), and a simultaneous landing of cars (processes conjunction P3):

- P1=[*:{Displacement{A;A}, Rotation{A;A}}, (Stability{A;A} & Succession{0;A;C}),
      *:{Displacement{B;B}, Rotation{B;B}}, (Stability{B;B} & Succession{0;B;C})]
- P2=(Stability{C;C}, Displacement{C;C},
      *:{Displacement{C;C}, Rotation{C;C}}, Stability{C;C})
- P3=(((Succession{A;0;C} & Displacement{A;A}),
      *:{Displacement{A;A}, Rotation{A;A}})
      & ((Succession{B;0;C} & Displacement{B;B}),
      *:{Displacement{B;B}, Rotation{B;B}}))

The ferry trip is implicitly transmitted to the cars using the positional link defined by the succession processes. The ferry's storage capacity limits the number of cars it can move.

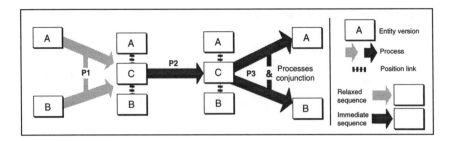

Figure 11 Diagram of composite processes.

This diagram uses few constructors, which enhances understanding of composite process sequences and structure.

Such semantics provides a language for describing processes with an object-oriented database model. The process decomposition principle offers a way to represent different time granularities or user point of views. The model allows for representation of ordered, linear and cyclic temporal processes. A regular composite process is defined by a constant granularity (i.e., regular data points). The lifespan of each composite process depends on the set of involved processes. Composite processes enable snapshot generation of historic states (using time-stamps).

The described processes connect entities and define the necessary lineages to discover the causal relationships that produce the observed events. Although temporal query languages provide operators that allows to compare time intervals [Snodgrass 1995], there is a need for a new class of topological operators to efficiently access the explicit links between evolving or mutating entities [Wuu 1992, Claramunt 1995].

The method outlined in this paper may be applied in many fields. It provides basic constructors for defining concepts specific to each discipline. In a paper dealing with temporal and spatial patterns of vegetation, Miles (1992) defined many concepts related to ecology that can be modelled through formal semantics. They can distinguish between the local and the global extinction of a species. The first is related to an entity disappearance while the latter involves the disappearance of all the members of a class (defined through the generalisation mechanism). The permanence of a vegetative species (A) may be defined as !Disappearance{A;0}; its absolute regression as [Contraction{A;A}, !Expansion{A;A}]; its global progression as [Expansion{A;A}, !Contraction{A;A}]; its replacement by species B in locality L with Succession{A;B;L}; etc. Using the concept of time granularity to study the rhythms of change, it is also possible to distinguish between fluctuations (rapid and cyclical temporary changes) and trends (slow and permanent modifications).

Moreover, in many scientific fields, it is only possible to measure time through indirect methods like C14. These measures include a factor of inaccuracy (E.g., 3200 years before present ± 75 years) that must be stored in the database to define the time granularity. In many fields of natural sciences there is a clear need to handle fuzzy time-stamps and to define the time line as a sequence of ordered and related events.

7. Conclusion

Studies of spatial dynamics need formal semantics to decompose and represent the characteristics of spatio-temporal processes. Scientists seeking causal links need qualitative models that describe functional links between entities and improve analysis of event sequences. These models must represent spatial entities as well as processes involved in their evolution.

The semantics outlined in this paper enable homogeneous and flexible representation of spatio-temporal processes and related spatial entities. The framework is based on a classification of basic processes combined with an object-oriented formalism and the EPL language. Representation of processes and spatial entities enable specialised temporal, spatial and thematic behaviours, thereby allowing object versions to be managed by domains in an asynchronous mode. The language developped can model complex spatial processes. It includes an order syntax, coordinating rules and a graphical presentation proposal.

The model provides formal support for representing evolving phenomena via database and for developing simulation models. Since the proposed semantics are oriented toward spatio-temporal applications, they could be used for TGIS and other temporal applications involving qualitative reasoning.

Further research perspectives include specifying spatio-temporal operators to interact with the database. Spatio-temporal process analysis may also be extended by integrating non-spatial processes related to geographic phenomena (e.g., non-spatial successions or permutations). Implementation in real case studies is an essential validation perspective. Applications to study soil dynamics and ecological impacts in Canada and Switzerland are currently in development.

Acknowledgements
We are indebted to anonymous reviewers for their useful comments and suggestions. We also acknowledge Christine Parent for stimulating discussions about several issues of this paper. We thank Yves Brousseau and Sylvie Saint-Jacques for designing the graphics and Patricia Dyksterhuis for carefully revising preliminary versions of this paper.

This research has been developed with the support of the Spatial Information System Institute of the Swiss Federal Institute of Technology in Lausanne, and the Planning and Development Research Centre, Laval University, Quebec, Canada. It is partly funded by the Social Sciences and Humanities Research Council of Canada (410-93-1097) and the Tri-Council Secretariat (Social Sciences and Humanities Research Council, Natural Sciences and Engineering Research Council and Medical Research Council) of Canada (922-94-0015).

References
[Beller 1991] Beller, A., 1991, Spatial/temporal Events in a GIS. In Proceedings of GIS/LIS '91, Bethesda, Maryland, ASPRS/ACSM, 2: 766-775.

[Bestougeff 1992] Bestougeff, H., and Ligozat, G., 1992, Logical Tools for Temporal Knowledge Representation, Ellis Horwood, United-Kingdom, 311p.

[Cheylan 1993] Cheylan J. P. and Lardon, S., 1993, Toward a Conceptual Model for the Analysis of Spatio-temporal Processes. In Spatial Information Theory, Frank, A. and Campari, I. Eds, Springer-Verlag, Berlin, 158-176.

[Claramunt 1995] Claramunt, C. and Thériault, M., 1995, Managing Time in GIS: An Event-oriented Approach. In Recent Advances in Temporal Databases, Clifford, J. and Tuzhilin, A. Eds., Berlin, Springer-Verlag, 23-42.

[Frank 1994] Frank, A. U., 1994, Qualitative Temporal Reasoning in GIS - Ordered Time Scales. In Proceedings of the Sixth International Symposium on Spatial Data Handling Conference, Waugh, T. C. and Healey, R. C. Eds., London, Taylor and Francis, 410-430.

[Gehani 1992] Gehani, N. H., Jagadish, H. V. and Shmueli, O., 1992, Composite Event Specification in Active Databases: Model and Implementation. In Proceedings of the 18th VLDB Conference, Vancouver, Canada, 327-338.

[Gertz 1995] Gertz, M. and Lipeck, U. W., 1995, "Temporal" Integrity Constraints in Temporal Databases. In Recent Advances in Temporal Databases, Clifford J. and Tuzhilin A. Eds., Berlin, Springer-Verlag, 77-92.

[Hazelton 1990] Hazelton, N. W. J., Leahy, F. J. and Williamson, I. P., 1990, On the Design of Temporally-referenced, 3-D GIS: Development of Four Dimensional GIS. In Proceedings of GIS/LIS '90, Anaheim, ASPRS/ACSM, 357-372.

[Jensen 1992] Jensen, C. S., Clifford, J., Gadia, S. K., Segev, A., and Snodgrass, R. T., 1992, A Glossary of Temporal Database Concepts. Sigmod Record, 21(3): 35-43.

[Kowalski 1986] Kowalski, R. and Sergot, M., 1986, A Logic-based Calculus of Events. In New Generation Computing, 4:67-95.

[Langran 1992] Langran, G., 1992, Time in Geographic Information Systems. London, Taylor & Francis, 189 p.

[Langran 1993] Langran, G., 1993, Issues of Implementing a Spatio-temporal System. International Journal of Geographical Information System, 7(4): 305-314.

[Long 1989] Long, D., 1989, A Review of Temporal Logics. In The Knowledge Engineering Review, 4(2): 141-162.

[McCarthy 1969] Mc Carthy, J. and Hayes, P. J., 1969, Some Philosophical Problems from the Standpoint of Artificial Intelligence, Machine Intelligence, Meltzer, B. and Michie, D. Eds., (4), Edinburgh University Press.

[Miles 1992] Miles, J. , Schmidt, W. and Van der Maarel, E., 1992, Temporal and Spatial Patterns of Vegetation Dynamics, The Netherlands, Kluwer Academic Publishers.

[Motakis 1995] Motakis, I. and Zaniolo, C., 1995, Composite Temporal Events in Active Databases: A Formal Semantics. In Recent Advances in Temporal Databases, Clifford, J. and Tuzhilin, A. Eds., Berlin, Springer-Verlag, 332-354.

[Navathe 1988] Navathe, S. B., and Ahmed, R., 1988, TSQL: A Language Interface for History Databases. In Temporal Aspects in Information Systems, Rolland, D., Bodart, F. and Leonard, M. Eds., Paris, North Holland, 109-121.

[Peuquet 1994] Peuquet, D. J., 1994, It's About Time: A Conceptual Framework for the Representation of Temporal Dynamics in Geographic Information Systems. Annals of the Association of the American Geographers, 84(3): 441-461.

[Peuquet 1995] Peuquet, D. J., and Duan N., 1995, An Event-based Spatiotemporal Data Model (ESTDM) for Temporal Analysis of Geographical Data. International Journal of Geographical Information Systems, 9(1): 7-24.

[Segev 1994] Segev, A. and Shoshani, A., 1994, Modeling Temporal Semantics. In Temporal Aspects in Information Systems, Rolland, D., Bodart, F. and Leonard, M. Eds., Paris, North Holland, 47-57.

[Snodgrass 1994] Snodgrass, R. T. 1994, Temporal Object-oriented Databases. A Critical Comparison. In Modern database systems: The Object Model, Interoperability and Beyond, Kim, W. Ed., 386-408.

[Snodgrass 1995] Snodgrass, R. T. et al., 1995, The TSQL2 Query Language. The TSQL2 Language Design Committee, Kluwer Academic Publishers, 674 p.

[Thériault 1995] Thériault, M., and Des Rosiers, F., 1995, Combining hedonic modeling, GIS and spatial statistics to analyze the residential markets in the Quebec Urban Community. In Proceedings of the Joint European Conference and Exhibition on Geographical Information, The Hague, 2: 131-136.

[Wachowicz 1994] Wachowicz, M., and Healey, G., 1994, Towards Temporality in GIS. In Innovations in GIS, Worboys, M. F., Ed., London, Taylor and Francis, 105-115.

[Worboys 1994a] Worboys, M. F., 1994, A Unified Model of Spatial and Temporal Information. Computer Journal, 37(1): 26-34.

[Worboys 1994b] Worboys, M. F., 1994, Object-oriented Approaches to Geo-referenced Information. International Journal of Geographical Information Systems, 8(4): 385-400.

[Wuu 1992] Wuu, G. and Dayal, U., 1992, A Uniform Model for Temporal Object-Oriented Databases, In Proceedings of the International Conference on Data Engineering, IEEE Computing Society, 584-593.

Contact address
Christophe Claramunt[1] and Marius Thériault[2]
[1]Swiss Federal Institute of Technology
Department of Rural Engineering, Spatial Information Systems
Lausanne, CH-1015, Switzerland
Phone: (41) (21) 693 57 83
Fax :(41) (21) 693 57 90
E-mail:Christophe.Claramunt@dgr.epfl.ch
[2]Laval University
Department of Geography, Planning and Development Research Centre
Quebec, G1K 7P4, Canada
Phone: (418) 656-2131 ext. 5899
Fax :(418) 524-6701
E-mail: Marius.Theriault@ggr.ulaval.ca

A DECISION SUPPORT SYSTEM FOR DESIGNING SAMPLING

STRATEGIES FOR POTENTIALLY CONTAMINATED SITES

Peter Tucker[1], Colin C. Ferguson[2], Ammar Abbachi[2] and Paul Nathanail[2]

[1]University of Paisley
Paisley , UK
[2]Centre for Research into the Built Environment
The Nottingham Trent University
Nottingham, UK

Abstract
Designing efficient and cost effective spatial sampling strategies for investigating suspected contaminated sites requires an adequate use of all available prior information derived from a desk study review of site history and a walkover survey.

This paper describes an expert support system for helping site assessors to compile this prior information and to develop an initial hypothesis on the likely locations of hotspots on a contaminated site. Using a knowledge-base approach the system converts desk study information into a score of indicators in order to build an *a prior* probability map of hotspot locations. The total number of sample locations is then computed and distributed over the site to reflect the prior information and hotspot specification.

A case study is presented in which a desk study report and site walkover survey were the sources of prior information. Using the system, a series of spatial sampling strategies were worked out for different hotspot sizes to assist in designing subsequent site investigation phases.

Key words
Expert support system, desk study, spatial sampling , soil contamination.

1. Introduction

Pressure to redevelop former industrial land and to avoid developing greenfield sites has increased the need for better site investigation tools. An important component of environmental risk assessment is determining the nature, extent and significance of soil contamination, and how to manage it. Sufficient information of acceptable quality is required to arrive at defensible decisions based on the likely positions of contaminant hotspots and the estimated spatial distribution of contaminants in soil.

The *Site-ASSESS* decision support system, developed under contract for the UK Department of the Environment is designed to streamline site investigation planning and increase the likelihood of collecting appropriate data cost-effectively.

The system aggregates information relevant to possible soil contamination and thus indicates the areas most likely to contain hotspots. The indicators are then used to calculate prior probabilities and hence to design spatial sampling strategies.

A case study illustrating the use of *Site-ASSESS* to design a first stage sampling strategy is presented. A desk study report (ICC 1995) and site walkover survey were used to compile a list of weighted indicators of likely contamination. The user enters this preliminary information into *Site-ASSESS* in order to help develop an initial hypothesis on the location of suspected hotspots. Then a cost effective spatial sampling strategy is designed using this initial hypothesis. The number of samples required is expressed as a function of hotspot size, and sampling locations reflect the spatial distribution of the weighted indicators.

2. Using prior information

Regular sampling designs, which assume that all parts of a site have an equal probability of containing a hotspot, can lead to large numbers of sampling points (Ferguson 1992). More cost-effective sampling strategies can be devised when there are grounds for relaxing the equiprobability assumption and concentrating investigative efforts in areas suspected of being contaminated. The two main approaches for improving the efficiency of sampling designs are:

*   variable density sampling using weight-of-evidence scores
*   multi-stage sampling

Combinations of both approaches may be used when appropriate. On some sites on-site analytical methods can be used to advantage in a multi-stage strategy.

Experienced assessors will often suspect that some parts of a site are more likely to contain a hotspot than others. They may then wish to design a sampling strategy that reflects their degree of suspicion or strength of belief about the target's likely location. One approach to this type of sampling design is to partition the site into subareas and then to score subareas (say, on a 1 to 10 scale) to reflect strength of belief as to where the target is likely to be.

Figure 1 shows an example of this type of scoring, and the corresponding sampling plan. The scoring scale is arbitrary, the highest score being used as a normalising factor so that strength of belief is expressed relative to that in the most favoured subarea. Sampling density for the most favoured subarea is calculated to give 0.95 probability of hitting a target if it exists in that subarea (Ferguson & Abbachi 1993; Department of Environment 1994a). Other subareas carry lower sampling point densities to reflect the assessor's lower expectation that the target is located in these areas.

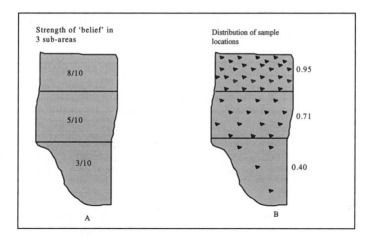

Figure 1 Variable density sampling for different strengths of belief.

If the assessor's judgment is correct the 0.95 probability of success is thus achieved with fewer sampling points and lower cost. But if the assessor is wrong the penalty is a reduced probability of success in locating the target (see probability values adjacent to Figure 1B).

The problem with this approach is that it requires a site investigator to convert a variety of disparate information into a score reflecting his or her strength of belief about hotspot location. To overcome this difficulty a computer-based decision support system has been designed so that specific items of information derived from a review of site history and from a preliminary walkover survey can be used directly to optimise sampling designs. *Site-ASSESS*, described below, has been developed to help professionals make informed judgments about sampling designs.

3. The *Site-ASSESS* Decision Support System

Site-ASSESS (Assessment of Sampling Strategies Expert Support System) is a decision support system for the design of sampling strategies for contaminated sites. Eventually the system will comprise the following linked modules:

* Likely contaminants
* Location of contaminant hotspots
* Soil gas survey
* Groundwater sampling
* Data analysis of first-stage sampling results
* Design of second-stage sampling

In version 1.0, which is currently being field-tested, only the hotspot location module (*Site-ASSESS* Hotspot) is fully developed.

The *Site-ASSESS* Hotspot module is intended to help site investigators design the initial, or first stage, sampling of a site. *Site-ASSESS* Hotspot develops an optimised sampling pattern for detecting (though not delineating) hotspots, and provides a statistical justification for the chosen strategy.

How much of a site is covered by sampling will depend on the density of sample locations. A compromise between total coverage and cost must therefore be achieved, with sampling becoming a statistical exercise to maximise useful information at a cost as low as possible consistent with the objectives of site investigation. When the objective is to locate hotspots, it is sufficient to place just one sampling point on the area covered by a hotspot. Therefore if a circle equal to the radius of the putative hotspot is drawn around a sampling point, the circle can be thought of as the zone of coverage for that sampling point.

The total sampling coverage is the sum of all zones of coverage excluding overlaps. The ratio of the total coverage to the total site area can also be thought of as the probability of hitting a hotspot if it exists. *Site ASSESS* is developed around this concept of coverage and hit probability. The computational part calculates an optimum sampling coverage, covering areas with the strongest prior evidence for the existence of a hotspot.

Before running *Site-ASSESS* Hotspot the user should have already completed a preliminary survey of the site and have compiled:

(i)   data on the historic use of the site
(ii)   a record of any visual, or other, indications of potential contamination gained during a site walkover (Department of Environment 1994b)

The user may additionally have some preliminary chemical analysis data from previous investigations or from *ad hoc* sampling of the site.

Archive data on previous site use (chemicals and processes) should, when possible, relate to the **length of time in use, time since last use, chemicals present, quantities of chemicals handled** and the properties of these chemicals, specifically **toxicity, biodegradation potential and leaching potential**. It is recognised that these data are often very difficult to obtain precisely. In *Site-ASSESS* high precision is not required. The user may simply make a best judgment according to a 3-point scale "High", "Low" or "Unknown".

It is helpful when conducting a walkover survey to subdivide the site into a number of square cells to compile the visible or other indicators for each cell. These cells represent an information grid (see Figures 3 and 4) into which all prior information is aggregated. Users must specify the size of the information cells, balancing spatial resolution with the time required to input all the information cell by cell. Experience has taught us to avoid too fine a grid as this rarely leads to a significantly improved solution but substantially increases data entry and computation time.

The presence or absence of an indicator (or attribute) in a given information cell is registered in terms of a score allocated to that cell; scores are then summed for each of the individual cells.

These total scores can be viewed on the screen (Figure 6), their values being rounded to the nearest integer for display. A high score indicates a high *a-priori* probability (i.e. strong evidence that a hotspot exists) and therefore the need for a relatively high sampling density in order to locate it with confidence. Attribute scores are converted to *a priori* probabilities by the user specifying the probabilities he or she thinks most appropriate for the highest and lowest scoring information cells; intermediate scoring cells are scored proportionately. Default values are provided in *Site-ASSESS* as initial guesses.

3.1 Estimating local sampling densities

Analysis of the prior information provides a spatially distributed set of *a priori* probabilities that a hotspot exists within specified subareas of the site defined by the information grid cells. The following approach is used to convert the *a priori* probability assigned to each information cell (grid square) into a target number of samples for that grid square.

The primary motivation for a sampling scheme is to ask; "What is the probability of locating a hotspot *if it exists*?" If, however, the sampling scheme fails to locate a hotspot, the question then becomes: "What is now the probability that a hotspot exists given that the sampling scheme has failed to find one?" This (after the event) probability is termed the *a posteriori* probability. The probability of locating the hotspot, if it exists, can be considered as the hit probability which is equal to the sampling coverage as discussed above.

A Bayesian probability approach has been advocated to relate the above probabilities (Gilbert 1987, Ferguson & Abbachi 1993):

$$PrH_i = 1 - PrA_i.(1 - P_i)/(P_i.(1- PrA_i))$$

where PrH_i is the hit probability for grid square i, P_i the *a priori* probability and PrA_i the *a posteriori* probability. By setting one of the sampling objectives (i.e. PrH_i or PrA_i) to a fixed target value, the above equation can be used to compute the other probability. We generally set all PrA_i to 0.05, i.e. if, after sampling, a hotspot has not been located in any given square, there is 95% confidence that a hotspot *does not exist* within that square. The value 0.05 is not prescriptive although it provides a confidence level with which most users seem to feel comfortable.

The hit probabilities PrH_i for each grid square are used to calculate the target number of samples $N_T = \Sigma N_i$ for the whole site, where N_i is the nominal number of samples allocated for grid square i. The problem is how to distribute the total number of samples N_T over the site such that each individual requirement on N_i is satisfied.

3.2 Optimisation of sample locations

The given number of samples, N_T, needs to be placed such that their aggregate weighted coverage is maximised. Equivalently, the problem is to minimise the aggregate a priori probability score of all parts of the site fall outside the zones of coverage of the sampling points. This is a well-defined optimisation problem that can be solved using the Quasi-Newton method (Fletcher 1987). However, this is a local optimisation procedure and in practice, a good initial estimate was found to be essential.

The initial estimate is found using an approximate sequential placement algorithm. For computational convenience each information cell is subdivided into a fine grid, typically comprising 3x3 or 4x4 smaller grid squares. The centres of the fine grid squares define the set of possible sample locations in the approximation. The *a priori* probability scores of all the fine grid squares are first placed in rank order. When fine grid squares have the same score, a possible sample location whose zone of coverage lies wholly a high probability subarea is ranked higher than one whose zone of coverage overlaps into an adjacent lower probability subarea. More generally, the rank order is based on the average a priori probability over the whole zone of coverage rather than the probability at the sample placement point. Any remaining ties in rank order are broken arbitrarily. In practice the sequential solution usually performs almost as well as the optimised solution, which typically improves the overall weighted sampling coverage by less than 5%.

4. Case study

The initial objective in the case study was to design a preliminary sampling strategy to locate suspected hotspots and to provide an overall picture of the spatial distribution of soil contaminants within the former industrial site.

The site used to illustrate *Site-ASSESS* extends over approximately 4ha and has been the location of several past industrial activities. Figure 2 shows survey maps of the site at different periods. The presence of a gas works in the central area (from 1890 onwards) is regarded as a subarea with high potential for containing contaminant hotspots. Loading and off loading areas are not accurately recorded and could be anywhere next to the railway tracks to the west, or the road to the south of the site. The canal junction to the east of the site may also have been used as a loading and off-loading area, as waterways were in use until the middle of this century.
A full search of historical data for this site would be time consuming and reporting it would be beyond the scope of this paper. However, a list of possible attributes has been compiled.

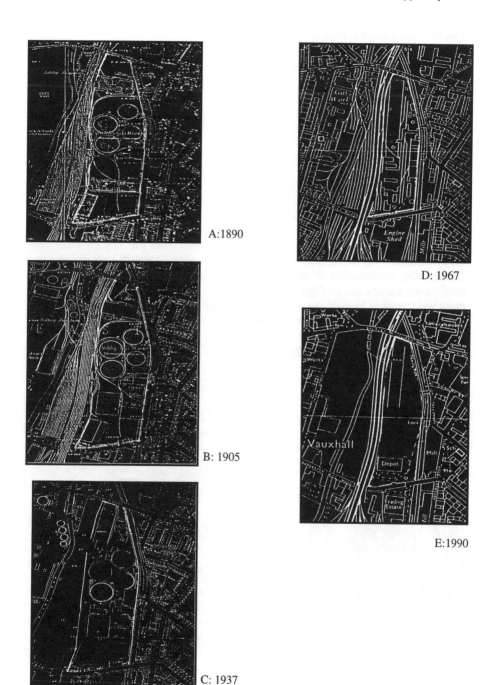

A:1890

B: 1905

C: 1937

D: 1967

E:1990

Figure 2 Different Ordnance Survey maps of the site with gas works in central area (A: 1980, B: 1905 and gas works replaced with new infrastructure (D: 1967) the infrastructure cleared (E: 1990).

The following are the historical attributes used:
- Process areas
- Storage areas of raw materials
- Waste disposal areas
- Loading and off-loading areas
- Filled areas

However, a site walkover indicated the existence of the following attributes:
- Irregular surface
- Poor drainage
- Anomalous soil type
- Oily patches
- Bare areas with sparse vegetation
- Remains of site infrastructure
- Waste tips

The scanned map of the site was imported into *Site-ASSESS* and divided into 66 25*25m information cells (Figure 3). Cells outside the site have been eliminated as shown in Figure 4. Although a greater number of information cells will result in higher resolution it will considerably increase the amount of time required to input all the information.

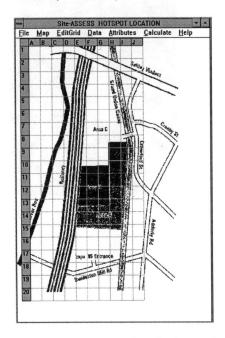 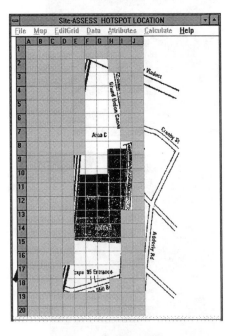

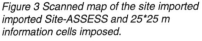

*Figure 3 Scanned map of the site imported imported Site-ASSESS and 25*25 m information cells imposed.*

Figure 4 Information cells outside the site eliminated.

Figure 5 shows two ways of allocating attributes to information cells and working out their influence on the final score. Figure 5A shows that user judgement on the importance of an attribute can be set using a sliding scale to specify a score in the nominal range 0 - 5. Alternatively, Figure 5B shows a menu which allows site assessors to respond to simple questions; the answers are then used to work out a score for the designated information cell. The user can overrule the knowledge base if specific data on items such as quantity handled, years in use, years since last use and leaching potential, are available.

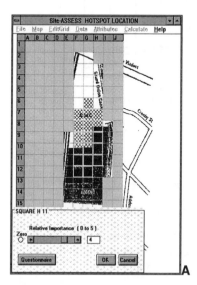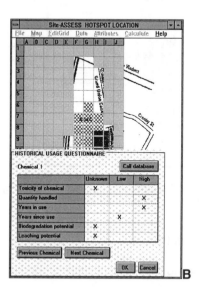

Figure 5 Attributes allocation and different scoring systems
(A: expert judgement and B: using knowledge base).

The hotspot size is made variable to study its impact on the number of sample locations required. The hotspot size, assumed of circular shape, is expressed as a percentage of the total site area.The scores shown in Figure 6 are then reviewed and cross-checked with the assessor's strength of belief on level of contamination in different part of the site. The central area is most suspected of being contaminated, and was the location of a gas works for many years (see Figure 2). The far north-east part of the site is also of high *a priori* probability; it is suspected to be a site where waste material was deposited.

The sampling strategies worked out using *Site-ASSESS* allow the number of sample locations to be expressed as a function of hotspot size as shown in Figure 7. An optimum sampling strategy could be selected if the target hotspot size and other economic parameters of the site investigation were known.

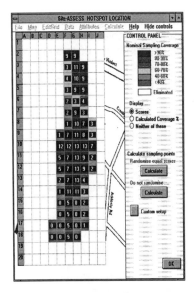

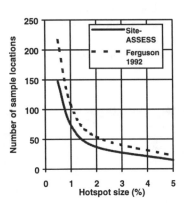

*Figure 6 Scoring results
(high scores in central area).*

*Figure 7 Number of sample locations
as a function of hotspot size, expressed
as a percentage of the total site area.*

The spatial distribution of the samples is characterised by higher sampling density in the central areas and the north-east part of the site (Figure 8) reflecting the higher *a priori* probabilities as described above.

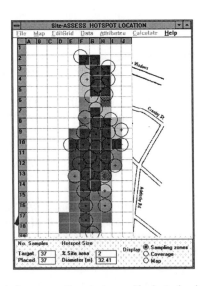

Figure 8 First stage sampling strategy with clustering in central area.

Using *Site-ASSESS* the assessor is able to compare sampling strategies designed on the basis of prior information with strategies based on the equiprobable assumption (Ferguson 1992). The reduction in total number of sample locations is apparent especially at smaller hotspot sizes (Figure 7); for this site roughly a 30% reduction in total number of sample locations is achieved. Of course this reduction is obtained at the expense of relaxing the hit probability in some areas of the site. Site investigators should make their own judgement as to whether prior information gathered on a site is robust enough to give confidence in the sampling strategy adopted.

5. Conclusion

The system developed here provides a standardised framework for sampling design. It uses readily available prior information acquired during a desk study and walkover survey. It should be noted that many practitioners fail to use these data in sampling design (although the data have been collected at some expense) and recourse to a regular grid pattern is the norm for site sampling. Sampling and analysis costs may amount to hundred of pounds for each sample taken, so minimisation of sample numbers is important. The subsequent costs of missing a contaminant hotspot through insufficient sampling could, however, be orders of magnitude higher. It is necessary, therefore, to find a balance between minimising sample numbers and maintaining an acceptable statistical confidence. The system provides a pioneering methodology for helping to achieve that balance.

The system is centred on expert knowledge. The knowledge has been distilled and structured as a series of numerical coefficients. These coefficients are a 'snapshot' of current knowledge. As knowledge increases through a better understanding of contamination indicators, it is expected that these coefficients can be improved.

An overall evaluation of system performance is difficult to undertake. One criterion would be to compare system predictions with those made by experts. An objective criterion could be the relative success in the long term at hitting hotspot targets found on investigated sites. It must be borne in mind however that there is no definitive answer to the sampling design problem.

Acknowledgment
This work was funded by the UK Department of the Environment but the views expressed are those of the authors and do not necessarily represent those of the Department. We are grateful to British Gas Properties for permission to publish the case study.

References
Department of the Environment (1994a) Sampling Strategies for Contaminated Land. CLR Report No 4., Department of the Environment, London.
Department of the Environment (1994b) Guidance on Preliminary Site Inspection of Contaminated Land. CLR Report No 2., Department of the Environment, London.
Ferguson,C.C. (1992) The statistical basis for spatial sampling of contaminated land. Ground Engineering 25/5, 34-38.
Ferguson, C.C. and Abbachi, A (1993) Incorporating expert judgement into statistical sampling designs for contaminated sites. Land Contamination & Reclamation 1, 135-142.
Fletcher, R.O. (1987) Practical methods of optimisation. Academic Press, London.

Gilbert, R.O. (1987) Statistical Methods for Environmental Pollution Monitoring. Van Nostrand Reinhold, New York.

ICC [ICC Information Group Ltd] (1995) Search for Contaminative Uses Report at EXPO'95 Site. Available from ICC Information Group Ltd., 16-26 Banner Street, London EC1Y 8QE.

Contact address
Peter Tucker[1], Colin C. Ferguson[2], Ammar Abbachi[2] and Paul Nathanail[2]
[1]University of Paisley
High Street
Paisley PA1 2BE
UK
Phone: (44) 141 8483205
Fax: (44) 141 8483204
[2]Centre for Research into the Built Environment
The Nottingham Trent University
Nottingham NG1 4BU
UK
Phone: (44) 115 9418418
Fax: (44) 115 9486510
E-mail: cbe3abbaca@ntu.ac.uk

AN ARCHITECTURE FOR KNOWLEDGE-BASED SPATIAL

DECISION SUPPORT SYSTEMS

Xuan Zhu

Department of Geography
University College Cork
Cork, Ireland

Abstract
This paper presents an architecture for knowledge-based spatial decision support systems. This architecture applies knowledge-based techniques and mechanisms based on a graphical knowledge representation scheme called spatial influence diagrams to provide assistance for the formulation of a spatial problem, and the design and execution of a solution process, by automatically integrating models and data existing within the system. Spatial influence diagrams provide a means for the representation and formulation of spatial problems, together with automation of the solution process. From the system design perspective, this architecture is an integration and use of GIS and knowledge-based techniques for spatial modelling and knowledge processing.

Key words
Geographical information systems, expert systems, knowledge-based diagrams.

1. Introduction

Geographical information systems (GIS) and spatial decision support systems (SDSS) are becoming important tools for supporting managers and planners in making decisions for resource and environmental management. They place the immense spatial data storage, computation and analysis capabilities of modern computers at the finger tips of decision makers. Since the mid-1960s, artificial intelligence has achieved considerable success in the development of expert systems. Expert systems are also known as knowledge-based systems. They are designed to represent and apply human knowledge in specific areas of expertise to solve real-world problems (Jackson 1990). Expert systems or knowledge-based techniques have proven to be very attractive for a variety of problems, which can be solved through the use of heuristic methods or rules of thumb.

In recent years, attention has turned to the integration of existing GIS, expert systems and other problem-solving techniques to develop more powerful SDSS. Zhu and Healey (1992) argue that while GIS and expert systems can each be used to support spatial decision making as stand-alone systems, they have some limitations. A number of these difficulties may be avoided by integrating the two types of system, while also allowing advantage to be taken of their respective strengths. Many efforts have been made towards such integration in the last decade (Morse 1987; Diamond and Wright 1988; Miller and Xiang 1992; Williams 1992; Loh and Rykiel 1992), given the significant competitive advantages which could result, such as cost savings, the ability to couple analytical modelling with heuristic reasoning, and automated explanation facilities for interpreting and justifying the results of modelling studies. However, early attempts have also demonstrated several drawbacks, such as user unfriendliness, lack of flexible model management capabilities and poor adaptation to users' needs (Zhu 1995).

SDSS need to be user-friendly so that they can be used by planners and managers who are knowledgeable in their domains of expertise, but may have little knowledge of the necessary computer system commands. Generally, SDSS have been designed for specific domains. The knowledge embedded in them mainly involves theories and concepts relating to a particular problem for the domain which the SDSS is to be used. Such knowledge is termed "domain-specific". The focus of most SDSS has been on the ability to perform analyses and modelling with data and models. However, the spatial decision process involves more than data interpretation. Before data and results are obtained, planners or managers are faced with a series of tasks: firstly, building the database relations and models; then, deciding modelling strategies; selecting appropriate data sets; choosing sequences of commands for analyses; and finally, displaying the results of the analyses or offering solutions to the problems. Few of these tasks involve domain-specific knowledge of a particular resource and environmental application. Instead, they involve knowledge of how to perform spatial modelling and how to run and use a set of tools for particular analytical purposes. This type of knowledge is called "tools knowledge", and it might relate, for instance, to GIS functions, decision models and other spatial modelling techniques, needed in problem solving. Traditional SDSS lack tools knowledge. The user has to act as both domain expert and tools expert. Incorporation of tools knowledge into SDSS would eliminate the latter requirement, and thereby improve user-friendliness.

On the other hand, if an SDSS is to become a useful tool in the task of spatial problem analysis and decision support, it should provide the user with effective support in selecting appropriate models and solution strategies to solve his or her problems. That is, an SDSS should have flexible model management capabilities, such as model selection, integration and formulation. However, the problem may only be solved after it has been structured and a suitable model has been formulated. A decision support system that can structure and simplify the problem and present the complex interactions of the problem parameters in a comprehensible manner, is of great potential value (Banerjee and Basu 1993). Thus, an appropriate technique for spatial problem structuring and representation is needed to facilitate the machine representation and manipulation of spatial problem models. In this light, the research develops a spatial influence diagram-based representation scheme and mechanisms for spatial problem representation, formulation and evaluation. Based on this, an architecture for knowledge-based spatial decision support systems (KBSDSS) is proposed. It aims to provide an interactive and flexible spatial decision making environment which allows the decision maker to evaluate alternative solutions based on his or her objective and subjective judgement.

A KBSDSS can be seen as an interactive tool that uses knowledge-based or expert systems techniques and integrates models and data to make it a model-based expert system for resource and environmental decision support. A KBSDSS is able to provide the assistance for formulation of the spatial problem according to users' preferences over its solutions, and for the design and execution of a solution process by automatically integrating different types of models and data available in the system.

This paper is organised as follows. The next section will discuss representation and formulation of spatial problems using spatial influence diagrams. The mechanism for the automation of the solution process based on spatial influence diagrams is detailed in Section 3. Section 4 describes a KBSDSS architecture. The concluding section will summarise the paper.

2. Spatial influence diagrams - a graphical knowledge representation scheme for spatial problems

A spatial influence diagram is a graphical knowledge representation of a spatial problem, consisting of nodes and directed arcs with no cycles (Figure 1). The nodes in the diagram represent the resource and environmental variables or factors relevant to a particular spatial problem. The directed arcs between nodes represent influences or dependencies between the variables.

There are three types of node: chance, value and border nodes. The value node, drawn as a rounded rectangle, represents the goal or expected value in solving the spatial problem. There is at most one value node. Chance nodes, drawn as circles, represent resource and environmental variables, which influence (directly or indirectly) the value node and enable an outcome to be computed for it. There are zero or more chance nodes in a spatial influence diagram. Border nodes, drawn as ellipses, represent resource and environmental variables that correspond to available or acquirable data. That is, the variables denoted as border nodes already have values, or their values can be acquired from the decision makers. The values of nodes in a spatial influence diagram may be represented as attribute data only or maps with different thematic values.

Figure 1 shows a spatial influence daigram constructed to structure a land use problem for depicting the land use potential for development in a certain area. The decision-maker's expected value in this land use problem is the potential areas for development, which are directly influenced by the proximity to roads, proximity to nature conservation areas, slope, aspect and visual exposure to water. The proximity to roads is determined by the existing road network and weighted by intervening slopes. However, the proximity to nature conservation areas is determined only by the locations of the designated areas for nature conservation. Slope and aspect are dependent on variations of elevation on the ground. Visual exposure to water is determined by elevation and positions of water bodies.

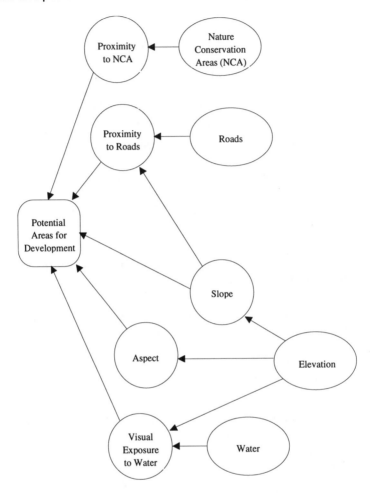

Figure 1 An example of a spatial influence diagram.

We define the predecessors of a node as the set of nodes having an arc directly connected and pointed to the node. For example, in Figure 1, the value node "Potential Areas for Development" has five predecessors: "Proximity to Roads", "Proximity to Nature Conservation Areas", "Slope", "Aspect" and "Visual Exposure to Water". The successors of a node are the set of nodes having an arc directly connected and pointed from the node. As can be seen from Figure 1, "Elevation" has three successors "Slope", "Aspect" and "Visual Exposure to Water".

The relations between the variables and their predecessors may be not only functional (e.g. logical and probabilistic), but also spatial. Here, we differentiate between attribute relations and spatial relations. Attribute relations are represented by analytical models (procedures composed of mathematical equations, such as arithmetic equations, probabilistic formula, regression equations and linear programming functions) or rule-based models (sets of knowledge base rules that perform reasoning to infer a solution to a particular problem). Spatial relations are represented by GIS models (constructed using GIS analysis functions, which operate on spatial data or digital maps). Every chance or value node may have an associated functional model and/or a GIS model specified for representing the relations with its predecessors, and through which the values of the node will be derived.

The spatial influence diagram brings problem considerations together and shows how they are related. The process by which a real resource and environmental problem is formulated is indeed that for the formulation of a spatial influence diagram. The spatial influence diagram-based representation scheme provides an interface through which heuristics and algorithms may be developed to build the capabilities for integrating data and models logically, and driving the solution process for spatial problems.

3. A solution procedure for spatial problems

A spatial influence diagram represents a spatial problem model wherein the major factors influencing the decision maker's objective are explicitly shown. The process of calculating the values of all nodes in their natural topological order is called evaluation.

Algorithm. Solving a spatial problem through evaluation of a spatial influence
            diagram.
Input. A spatial influence diagram constructed for a particular spatial problem.
Output. The outcome of the value node v.

Procedure SOLUTION OF A SPATIAL PROBLEM
begin
   let N be the set of all nodes in the input spatial influence diagram;
   let N' be the nodes to be removed, N'=N\{v}; / all nodes but the value node for*
                                           the spatial problem./*
   let K be the set of the nodes which can be evaluated currently; $K = \varnothing$;
   while *there exist unassessed nodes $\in$ N* **do**
     add all the border nodes into K;
   begin
     while *$N' \neq \varnothing$* **do**
       begin
         if *there exists a node $i \in$ K*
         then if *i is an unassessed node*
             then *evaluate i; set i assessed;*
             propagate the value of i to its successor set S(i);
             remove i from K and N';
         else *update K;*
       end
     evaluate the value node v;
     set v assessed;
     return the value of v;
   end
end

Figure 2 A solution procedure for a spatial problem.

The set of directed arcs in a spatial influence diagram corresponds to an evaluation order. For example, a value of a variable i could be obtained by first evaluating a variable j, and then evaluating i based on the value of j. This evaluation order would correspond to a two-node spatial influence diagram with an arc from i to j. Thus, a spatial influence diagram implies an evaluation order for its solution. Such an order always exists by virtue of the acyclicity assumption.

A node is said to be assessed if it has a value, or unassessed if otherwise. Border nodes are inputs to the spatial problem model. Most of them are assessed. In the case of an unassessed border node, the user will be asked to supply its value. A solution procedure for a spatial problem is shown in Figure 2.

4. An architecture for KBSDSS

An expert system in combination with powerful facilities for managing and handling spatial data, spatial modelling, formulating and evaluating spatial problem models, constitutes a KBSDSS. In contrast to a conventional SDSS system, a KBSDSS aims to provide the following capabilities (Zhu 1995):
- knowledge-based support for spatial problem solving;
- structuring and representation of spatial problems;
- integration of GIS, analytical and rule-based models.

4.1 Knowledge requirements for KBSDSS

A KBSDSS provides the user with a substantial amount of domain-specific and tools knowledge. The knowledge within a KBSDSS can be used to guide the formulation and evaluation of a spatial problem model for a specific resource and environmental problem, and provide guidance in employing different types of models, including GIS, analytical and rule-based models, to solve the problem. Thus, a KBSDSS should enable the user to exploit the power of spatial problem analysis and modelling, GIS modelling, analytical modelling and rule-based modelling in a relatively simple and fast way. It can also provide explanations during the spatial problem solving process, and guide the user in the task of capturing the essential elements of a spatial problem, and incorporating these into a spatial problem model, which can then be automatically evaluated by access to the database and knowledge base; and by integrating different types of models.

The design of a KBSDSS using knowledge-based techniques requires, as a first step, the development of a knowledge base. We may divide the knowledge in the knowledge base of a KBSDSS into at least five broad categories: domain knowledge, model knowledge, utility program knowledge, meta-data and process knowledge.

Domain knowledge
Domain knowledge contains the knowledge and expertise for finding solutions to spatial problems in a particular problem domain. Most domain-specific

knowledge concerns the evaluation of each domain variable relevant to the problem domain. A domain variable is depicted as a node in a spatial influence diagram. The domain knowledge for each of these variables or nodes will contain details of its predecessors (the variables that directly influence it), and its attribute relation and spatial relation with these predecessors.

In addition, a KBSDSS also contains domain knowledge regarding preferences and circumstances that decision makers may have or encounter within a certain decision-making context. These are expressed in the form of objective models, which represent the decision-maker's preferences for a particular problem solving approach. Each objective model accounts for a set of attributes, which directly influence the outcome of the problem solving. For example, one user may specify the five attributes to be considered in selecting development areas (the goal of the decision maker) as shown in Figure 1. Then, an objective model will be formed accounting for the proximity to roads, proximity to nature conservation areas, slope, aspect and visual exposure to water. Another user may only be concerned about the proximity to roads and slope. In this case, another objective model can be formed accounting for only two attributes. After an objective model is formed, a particular spatial influence diagram can be formulated by choosing a domain variable as the value node and searching the predecessors of the node until the border nodes are added, and then evaluation of a spatial influence diagram starts. The domain knowledge regarding preferences and circumstances is used interactively with the decision maker to form an objective model for a specific spatial problem, and incorporate the decision-maker's knowledge and preferences for problem solving into the overall analysis.

Model knowledge
The model knowledge serves to assist the user with the different types of model available to solve a particular spatial problem. It involves descriptions of the models, selection of those appropriate for a particular problem, and assignment of the relevant parameter values to the selected models. The description of a model may include its name (unique), the tool on which it is executed, its required data model and data structure and a list of parameters that are in the required calling sequences. The knowledge about selection and parameterisation of models is encoded in rules.

Utility program knowledge
During the evaluation process of a spatial problem, some utility programs may be needed. These programs are used to convert the types of data structures, display spatial data in maps and so on. The knowledge relating to them is described in a similar way to that for models. Utility programs are selected and called automatically when needed. The knowledge about selection and execution of utility programs is mostly coded in rules, which dictate when and where to call an utility program during the spatial modelling process. Just like models, heuristic rules are used to assign parameter values to selected utility programs for their implementation.

Meta-data

Meta-data is knowledge about data. It may include name, data model and data structure, geographical feature type (point, line, area and surface) and scale if it is a map data layer, definitions of attributes, and data source.

Process knowledge

Process knowledge refers to knowledge which provides support during the spatial problem solving process. Within a KBSDSS, this process can be typically described as one consisting of specifying a problem, determining preferences (an objective model), formulating a spatial problem model, evaluating the formulated model, and generating a solution to the problem. Process knowledge is used to guide the successful execution of these steps and provide help messages in the course of consultation. The help messages provide information about the current modelling process, inform the next modelling operations, and give explanations of spatial modelling terminology, etc.

4.2 An architecture for KBSDSS

KBSDSS can be considered as a class of expert systems with an inference engine based on the spatial influence diagram-based mechanisms. Figure 3 shows a KBSDSS architecture, which is composed of a user interface, query processing subsystem, modelling subsystem, problem processor, knowledge base, and back-end subsystem.

The query processing subsystem is developed to accept user queries, display modelling results and provide on-line help. It allows the user to retrieve existing data from the database together with data derived during the modelling process, and display them in the forms of maps, images or tables. It also allows the user to retrieve the meta-information about data and models.

The modelling subsystem is designed to help the user in developing useful models by supporting knowledge acquisition, user-assisted modelling and automatic modelling. The system acquires relevant knowledge through the knowledge acquisition module, such as meta-data for new data sets, meta-information about new models, or structures for spatial problems defined by the user. A spatial influence diagram editor can be used to guide the user to capture, elicit and represent the problem structure and develop the problem model.

Within the scope permitted by the classes of problems inherent to its design, the system can elicit important features of the given situation from the decision maker and can then incorporate these features in the formulation of a spatial problem model (a spatial influence diagram) using the domain knowledge in the system to address the problem at hand. This process is called automatic modelling, and it is handled through the automatic modelling module in the modelling subsystem.

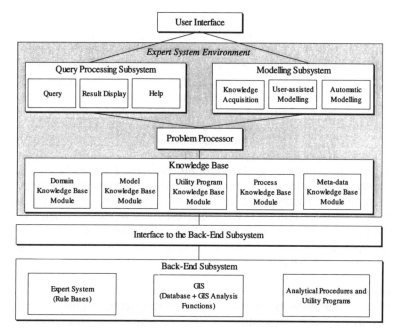

Figure 3 An architecture for KBSDSS.

After a spatial problem model is formulated, the system allows the user to modify the model. This process is termed user-assisted modelling. It also includes the creation of new spatial problem models using spatial influence diagrams by the user, addition of new analytical, rule-based and GIS models, and addition of new data sets or data layers in order to meet the requirements of particular analyses.

The problem processor plays a central role in a KBSDSS. It performs two main functions: inference and control. It is used to drive the evaluation process for spatial problems, and to integrate the knowledge base, database and different types of models for spatial problem solving. The problem processor accepts a spatial problem model from the modelling subsystem and then evaluates it. The evaluation is based on the algorithm shown in Figure 2. This involves the evaluation of each node within the spatial influence diagram through invocation of database calls and pertinent knowledge, and assignment of parameters values to the models and utility programs (Zhu 1995). The successful completion of these processes will trigger execution of the selected models and utility programs, and each node gets an outcome. When the value node denoting the decision maker's objective is assigned an outcome, a solution to the problem is obtained. The problem parameters in the problem model may be modified and the process repeated, allowing the user to make a "best" decision by comparing the outcomes of different scenarios.

The problem processor can also accept commands translated from the request and actions issued by the user through the query processing subsystem and modelling subsystem, execute these commands, control access to the data base and knowledge base, execute models, retrieve knowledge from the knowledge base and make inferences.

The knowledge base is structured into different parts, called knowledge base modules. There are at least five knowledge base modules in the knowledge base. These include the domain knowledge base module, the meta-data knowledge base module, the model knowledge base module, the utility program knowledge base module and the process knowledge base module. The modular structure of the knowledge base makes the system adaptive to different back-end software tools and different problem domains. It can easily be seen that the inference and control mechanism of the problem processor is not only independent of the specific back-end software tools used for implementation of models, but also of the problem domain. To adapt the system to new back-end software tools, only the model and utility program knowledge base modules have to be changed. To adapt it to a new problem domain, appropriate adjustments are required to only the domain and the process knowledge base modules.

The back-end subsystem consists of the software tools employed in the system for implementing different types of models and utility programs. It contains three basic separate modules: an expert system, a GIS and a collection of analytical procedures and utility programs. The expert system here is a tool for implementing the rule-based models. The subsystem provides appropriate tools for implementing GIS, analytical and rule-based models and necessary utility programs defined in the system. The architecture places no restrictions upon the number of software tools in the back-end subsystem. Different software systems, such as statistical packages, can be added, if required, in order to meet the needs of the problem solving task. When new software tools are added, the meta-information about the models and the utility programs they support should be added to the system through the knowledge acquisition module in the modelling subsystem, or by modifying the model and utility program knowledge in the knowledge base.

The query processing subsystem, modelling subsystem, problem processor and the knowledge base are built within an expert system environment. The interface to the back-end subsystem provides the link between the expert system environment and the back-end subsystem. It handles inter-process communication between the expert system environment and the back-end subsystem.

The user interface interacts directly with the user. Through the user interface, the user can access the query processing and modelling subsystems.

5. Conclusion

The paper has presented an architecture for KBSDSS, based on a spatial influence diagram-based representation scheme and mechanisms. This architecture has been applied to develop ILUDSS (the Islay Land Use Decision Support System) — a prototype KBSDSS designed for strategic planning of land use in a rural area (Zhu 1995). With this architecture, a KBSDSS can overcome some of the problems of current SDSS systems mentioned in the introduction: namely, their user unfriendliness, inflexibility in model management (model formulation and integration) and inability to adapt to users' needs.

Densham and Goodchild (1989) point out that an SDSS "should incorporate knowledge used by expert analysts to guide the formulation of the problem, the articulation of the desired characteristics of the solution and the design and execution of a solution process". KBSDSS can achieve this goal. A KBSDSS captures domain knowledge that can assist users in formulating a problem model (a spatial influence diagram) appropriate to their particular situation. The system then provides an automated solution procedure for evaluating the problem through the spatial influence diagram-based mechanism.

The spatial influence diagram-based approach provides flexibility in model integration and problem formulation in response to users' problems. As the demand for "what-if" analysis increases, the user may require solutions for different sets of the problem parameters. The system in turn is needed to identify alternative models to represent and solve the problem as modified by each set of parameters. The spatial influence diagram-based approach has mechanisms to provide such assistance. With each set of parameters, a new spatial influence diagram can be automatically formulated to represent the modified problem. Thus, the system can be easily adapted to the user's needs and desires.

A KBSDSS contains not only domain knowledge for problem formulation, but also model knowledge, utility program knowledge, meta-data, process knowledge and other kinds of knowledge. The extensive knowledge base in a KBSDSS can help a range of users in the appropriate use of GIS and other modelling and analysis techniques to solve their problems. As Cowen and Shirley (1991) suggest, a good SDSS should be able to make GIS software accessible to users with different levels of technical expertise. A KBSDSS can help users who have limited experience in GIS and spatial modelling to formulate and solve their problems without the need for expertise in the use of the relevant models and data. It can also help users who have experience in GIS and spatial modelling to formulate and solve their problems by using the existing models and data without the need to know the implementation details of the models. Thus, a KBSDSS enables the user to exploit the power of spatial problem analysis and modelling, in particular GIS modelling, analytical modelling and rule-based modelling in a simple and fast way. The spatial influence diagram-based representation scheme and mechanisms, as well as the extensive knowledge in the knowledge base, make a KBSDSS effective in supporting the formulation of the problem, the design and execution of a solution process, and the proper use of modelling and analysis techniques.

Furthermore, since they contain much of the functionality of expert systems, KBSDSS can be used to distribute the expertise of domain experts. In addition, a KBSDSS can be developed incrementally. As with all expert systems, a KBSDSS can be started with a small scope. Once experience is gained with the system, more and more areas can be supported.

KBSDSS technology can enhance the capabilities of SDSS. However, much remains to be done before an SDSS can provide fully automated, intelligent, reliable and user-friendly decision support in resource and environmental planning and management. The current research aims to facilitate the process of problem solving by solving the more structured components without needing extensive interaction from the decision makers and guiding them by providing relevant information at appropriate stages.

References

Banerjee, S. and A. Basu, 1993, Model type selection in an integrated DSS environment. Decision Support Systems, Vol.9, pp.75-89.

Jackson, P., 1990, Introduction to Expert Systems, Addison-Wesley Publishing Company.

Cowen, D.J. and W.L. Shirley, 1991, Integrated Planning Information Systems. In: Maguire, D.J., Goodchild, M.F. and Rhind, D.W. (eds.), Geographical Information Systems: Principles and Applications, Vol.1, Longman, London, pp.297-310.

Densham, P.J. and M.F. Goodchild, 1989, Spatial decision support systems: a research agenda. Proceedings of GIS/LIS'89, Vol.2, ACSM/ASPRS/AAG, Virginia, pp.707-716.

Diamond, J.T. and J.R. Wright, 1988, Design of an integrated spatial information system for multiobjective land-use planning. Environment and planning B: Planning and Design, Vol.15, pp.205-214.

Loh, D.K. and E.J. Rykiel, 1992, Integrated resource management systems: coupling expert systems with data-base management and geographical information systems. Environmental Management, Vol.16, No.2, pp.167-177.

Miller, Ralph and Wei-Ning Xiang, 1992, A knowledge-based GIS method for visual impact assessment in transmission line siting. Proceedings of GIS/LIS'92, Vol.2, San Jose, California, pp.577-584.

Morse, B., 1987, Expert interface to a geographic information system. Auto Carto-8, Maryland, pp.535-541.

Williams, S.B., 1992, INFORMS-TX overview. Paper Presented at the Seventh Forest Service AI Workshop, Blacksburg, Virginia.

Zhu, Xuan, 1995, A Knowledge-Based Approach to the Design and Implementation of Spatial Decision Support Systems. Unpublished PhD dissertation, the University of Edinburgh, Edinburgh.

Zhu, Xuan and Richard Healey, 1992, Towards intelligent spatial decision support: integrating geographical information systems and expert systems. Proceedings of GIS/LIS'92, San Jose, California, Vol.2, pp.877-886.

Contact address
Xuan Zhu
Department of Geography
University College Cork
Cork
Ireland
Phone: +353-21-902891
Fax: +353-21-271980
E-mail: zhu@ucc.ie

A PROBLEM MODEL FOR SPATIAL DECISION SUPPORT

SYSTEMS

M. A. Cameron[1] and D. J. Abel[2]

[1]Department of Computer Science
Australian National University
[2]CSIRO Division of Information Technology
Canberra, Australia

Abstract
This paper presents a model for Spatial Decision Support System interface design based on a 5-component decomposition of the structure of problems. The model offers an alternative to process based methods for structuring a decision-maker's interaction with ill-defined problems. The model does not impose a problem-solving or decision-making methodology on the decision-maker. We report a case study using the model to design a forest zoning decision support system interface.

1. Introduction

Spatial Decision Support Systems (SDSS) present many challenges in design and implementation. SDSS are often defined in terms of dealing with ill-structured or ill-defined problems which vary significantly case-by-case (e.g. Simon 1973). The solution process is then typified by a growing and changing understanding of the nature and structure of the problem in terms of discovery of causal relationships between process variables, recognition of process variables that can be manipulated, and the decision-maker's preference structures. SDSS then require a design approach differing from most other types of information systems for which data requirements, constraints and objectives are assumed to be identifiable by a structured systems analysis.

The most common approach to design of SDSS has been to define architectures which assemble a set of data, visualisation, and modelling components that a decision-maker can draw upon freely in the course of an investigation. The early Decision Support System (DSS) architecture of Sprague (1980) with its database, model management and dialogue management subsystems contrasts with other researchers who propose the more general components of language system, knowledge management and a generic problem processing component (e.g. Bonczek et al., 1981; Dos Santos and Holsapple, 1989; Chang et al., 1993). Essentially these approaches aim at providing the decision-maker with a set of tools which support an investigation while not being prescriptive about the problem-solving methodology. The content and structure of the dialogue between the end-user and the system (equivalently, the design of the user interface) remains a creative undertaking.

There is, however, a considerable body of theory and experience with techniques for solving ill-defined problems (e.g. Van Gundy, 1988). Collectively labelled as 'creative problem-solving' methodologies, these are based on process models which decompose problem-solving to a number of phases or procedures in which problem formulation is pre-eminent. It is attractive to investigate whether the phases identified in these models can be applied to the design of SDSS. An obvious motivation is that the steps involved might be a useful basis for the design of the user interface. Rather than presenting the user with an unstructured collection of tools, the SDSS designer might seek to organise the models, database and visualisation to conform to the key components of the problem-solving process. More subtly, knowledge of the components might allow design of the various subsystems of the SDSS to better conform to likely usage patterns for those components.

This paper advances a 5-component model of problem-solving for ill-defined problems. The model adapts previous research on decision-making and problem-solving, specifically to provide a basis for design of SDSS. The approach is not prescriptive. Rather it seeks to isolate aspects of problems which would be separately manipulated by the decision-maker and which then form a reasonable approach to organisation of operations accessible through the user interface. The approach is demonstrated by a pilot SDSS for forest zoning.

2. Ill-defined problems

Ill-defined problems, after Simon (1973), can be defined as a problem where the decision-maker must develop an understanding of the problem as part of the investigation, typically by changing elements of the problem definition as the investigation progresses. Additional data, information and knowledge might be required and there is no obvious stopping rule for the investigation. Understanding the problem involves such aspects as identifying observable and controllable process variables, discovering causal relationships, and striking a preference structure to rank outcomes.

'Decision-making' and 'problem-solving' are closely related, although problem-solving has an additional stage of adopting a preferred solution. We use the terms interchangeably in this paper, without lessening the significance of implementing the chosen solution (Turban, 1990).

Investigation of decision-making has focused on two key areas: the models of choice and methodologies for problem formulation.

Models of choice can be classed as behavioural, normative and satisficing. Behavioural models are based on observation of human decision-making to discover common processes used to solve a given problem (e.g. Newell and Simon, 1972). The normative (or rational) models assert that a decision-maker is able to discover a measure describing the worth of outcomes and so can select the 'best' solution by generating and ranking alternative solutions. These measures can be multi-criteria (e.g. Nijkamp et al., 1990; Carver, 1991) or fuzzy (e.g. Banai, 1993; Seo and Sakawa, 1988). Selection of the optimal solution might be by a rational evaluation of alternatives or an exhaustive enumeration of possible solutions. The satisficing approaches (Simon, 1977) are motivated by the observation that some decision-making is made in the light of imperfect information and detailed analysis of possible outcomes is not practical. In this situation, a decision-maker introduces simplifications or bounds to the problem in terms of the information gathered and analysed and of the complexity of modelling of situations. There is limited evaluation of alternative solutions and the choice of solution might be biased. The emphasis in satisficing is finding a 'near enough' decision rather than the optimal.

Problem solving methodologies have concentrated on developing a set of procedures as a framework within which to deal with a problem. Simon (1977) proposed a three phase model of decision-making. The phases of Simon's model, intelligence, design and choice, classify processes or activities undertaken within a decision-making cycle. The exact nature, sequence of investigation and timing of activities within each phase is not specified as these are highly dependent on the decision context. Brightman (1978) and Van Gundy (1988) extended Simon's decision-making model into a problem-solving process model. The decision model's three phases of intelligence, design and choice are repeated within each phase and augmented with an additional phase of solution implementation. Explicit control flow between phases provides a firm statement of the input and output requirements for the outer phases. These extensions to Simon's model opened the structure of activities within each of the original phases. The result of the intelligence trio is a statement of the problem, which feeds into a design process resulting in solution generation, leading to solution selection and implementation.

Elaborations of these models of unstructured decision-making have been proposed. Mintzberg et al., (1976), for example, propose seven steps within three main phases of identification, development and selection while retaining the view that the sequencing of these phases is not linear.

All the methodologies recognise the critical importance of problem formulation. Simon's model covers problem formulation implicitly within the intelligence phase. Brightman's extension identifies problem statement as the bridge between intelligence and design phases. Mintzberg et al. have recognition and diagnosis routines within the identification phase and they assert that diagnosis is the single most important routine.

3. SDSS architectures

Previous reports on SDSS design and implementation have focused on architectural approaches to assemble tools allowing flexibility in solving ill-defined problems. Broadly, SDSS designs adopt the same constraints and objectives of mainstream DSS while incorporating specialist spatial components such as map-like presentations of data and spatial optimisation solution techniques (e.g. Armstrong et al. 1986; Densham and Rushton, 1990). The seminal architecture of Sprague and Carlson (1982) has been the basis for many subsequent architectures. Sprague and Carlson identified subsystems for dialogue management, data management and model management (for connecting modules for prediction, optimisation and so on). Recognition of the value of knowledge-based techniques have led other workers to propose that DSS architectures include knowledgebase or hyperknowledge subsystems (e.g. Bonczek et al. 1981; Chang et al. 1993). Greater emphasis on interface flexibility leads Chang, Holsapple and Whinston (1993) to view the dialogue management system of Sprague and Carlson as a specialisation of the language and knowledge subsystems of Bonczek et al. Similarly, the language system of Bonczek et al., has been suggested as better subdivided into presentation and language systems (Dos Santos and Holsapple 1989).

Designing systems that incorporate a variety of such tools remains a complex task. At one level, the traditional structured analysis-and-design methodologies are clearly unsuitable for applications where requirements and objectives cannot be established a priori. This has encouraged design of SDSS as tool sets of database, visualisation and modelling facilities. The tool set is essentially unstructured, to provide the user with the greatest flexibility in exploring alternative outcomes. Integrating the extensive set of facilities required to solve problems of realistic complexity is itself non-trivial, particularly where pre-existing facilities such as databases and modelling systems are to be used (Abel et al., 1994a, 1994b; Blair, 1994). Design of subsystems specifically intended to be components of SDSS is then attractive (Fedra, 1993).

These design approaches focus on assembling a set of tools as a collection of subsystems and enabling effectively ad hoc use of those tools. The design of the end-user interface has apparently not been addressed in any depth. In SDSS with a narrow context (i.e. a small set of variables able to be manipulated), simple visual formalisms can provide the basis for the end-user interface (e.g. Abel et al., 1994b). In general, however, the requirement to treat the problem formulation as changing during an investigation limits the value of the usual Human-Computer Interaction (HCI) design approaches: basing the interface design on the end-user's conceptual model, metaphors and visual formalisms. The difficulty of ensuring ease-of-use and flexibility clearly becomes greater as the tool set grows in size and functionality.

The key concept of the problem-solving models—that a set of separable activities (or problem-solving process components) are present—offers an alternative approach. Essentially, we propose that the components provide a logical basis for organisation of functions in the system.

4. The 5-component model

Using a set of problem-solving process components to design an SDSS is subtly different from using components as a basis for models intended to develop an understanding of problem-solving. From our perspective, a 'good' set of components will lead to identification of a set of functions and organisation of these functions which will allow the user to progress smoothly through the phases of problem-solving. While it appears feasible to adopt a model of problem-solving processes directly as a work-flow model and to structure the interface directly, we consider here the alternative approach of seeking core functions which the user can apply flexibly and which are more obviously related to the applications domain of the SDSS. Similarly, in identifying the problem-solving process components, we do not require a direct mapping between the components and the functions of the software subsystems of the SDSS.

Importantly, we distinguish between structural and value manipulations within problem formulation and problem solution. A structural manipulation influences the set of controllable and uncontrollable variables or the specification of functions computed from controllable or uncontrollable variables. A value manipulation alters the values of controllable variables

We propose a model with five components (Figure 1): the Current World, the Solution, the Desired World, the Evaluation Function and the Gap Function. We informally present the definitions of these components and then provide a description and commentary on using the 5-component model to design an SDSS interface. To explain the model, we will use a simple problem of purchase of a house by the Smith family.

The Current World models the context of the decision-making event. It represents the current reality relevant to the problem. For example, for the house purchase, the Current World will include information on the Smith family (their current residential location and rent, the number of members, their savings, their income, their financial commitments, their places of work, schools attended, and so on), the houses for sale (location, size, prices), facilities in the region (schools, churches, recreational facilities) and available sources of finance (lending organisations, interest rates, repayment periods).

A Solution is a manipulation to the value of one or more decision variables. For example, a solution for the Smith family would be to buy a specific house with their current savings. Another solution would be to buy another house using some savings and some borrowed funds from a bank with specific repayments over a specific period. Importantly, the Solution is not necessarily optimal or generated by some algorithmic process. Rather the set of Solutions represent all possible assignments of values to decision variables. The derived state is a representation of the result of implementing a solution to the Current World and is implicitly linked with the Solution. For example, purchase of a specific house determines the Smith's new residential location, their financial commitments and so on.

The Desired World records an idealised representation of the decision-maker's goals. Importantly, the representation of the Desired World cannot include elements not present in the model of the Current World, nor values for elements unable to be derived from applying a solution to the Current World. The Desired World records the range of financial commitment acceptable to the Smiths, for example.

The Evaluation Function provides a comparison of the relative value (or utility) to the decision-maker of alternative derived states with respect to the Desired World of the decision-maker. The Smith Family's Evaluation Function might include income, financial commitments, and travel times to work, schools, shops and recreational facilities. The Evaluation Function does not necessarily yield a numerical result, neither is it necessarily an algorithmic procedure. We simply require that it yield a reproducible ranking of alternative derived states by reference only to those alternative states.

The Gap Function provides a comparison of the Current World with a specific derived state. The Smith Family's Gap Function might measure differences in travel times to work, the quality of schools and the amenity value of the new house versus the value of the current apartment.

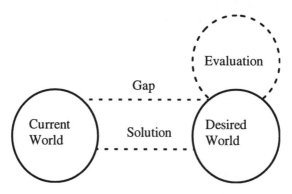

Figure 1 The 5-Component Model.

A complete (normative) problem model has elements within each of the five components. In developing a complete model of a problem, the decision-maker is free to interact with any of the five components at all times. Normative, satisficing and behavioural decision models are supported by the style of evaluation function used. Problem solving can be goal driven, or data driven.

5. The Forest Zoning SDSS

To demonstrate the application of the 5-component model to design of an SDSS, we consider the Forest Zoning System, a SDSS aimed at policy formulation for preservation, conservation and use of Australian forests. We note that the Forest Zoning System is an experimental system built to explore SDSS end-user interfaces and architectures. The description presented here does not follow the practices followed by any particular governmental agency and must be recognised as dealing with a small subset of the issues considered in actual policy formulation.

Planning by government for conservation and exploitation of forests is an archetypal ill-defined problem. The physical system—the areas covered by forests and the uses of those areas—is complex and some interactions such as the effects of logging and the ecological associations between flora and fauna are incompletely understood. It is undoubtedly infeasible for a planner to assemble a full collection of applicable data, information and knowledge. Planning must also recognise the divergent preferences of such stakeholders as the three layers of Australian government, the forestry, tourism and other industries, special interest groups such as conservationists, and the wider community.

A fuller statement of the problem in terms of the 5-component model demonstrates its general strength in decomposing an ill-defined problem and provides a framework for presentation and explanation of the architecture of the Forests Zoning System. Broadly, the problem can be defined as assignment of permitted uses for parcels of forest in a specified geographic region. These permitted uses range from strict preservation to replacement by plantations.

The Current World comprises both the current state of the region and knowledge of the physical and economic processes. The current state includes the current uses of the region (including existing reserves and logging licences), socio-economic activity and so on. Knowledge encompasses material accessed by the planner in the course of the investigation but not applied directly in computational processes. For example, it might include reference material describing in plain text the habitats required by bird species or the texts of agreements which must be respected in the uses assigned.

Solutions necessarily involve the assignment of uses to parcels. These assignments might be made effectively arbitrarily; in this case they are simply presented by the planning officer. Alternatively they arise from application of some numerical procedure such as (for example) a mathematical programming technique. In some cases, the solution will also include complementary actions such as compensation to holders of logging licences for withdrawal of licences. The derived state represents the state of the region after changes to the usage assignments to some or all parcels. Derivation of the new state would consider the areas under different permitted use categories and the status of the parcels under the usage regimes after certain periods. The new status of a parcel (for example, changes in the species composition and the forest communities) might be derivable by application of rules based on expert knowledge or by use of predictive models. Also relevant are the status of other biological and biophysical characteristics of the parcels, such as suitability for particular species of animals and the likelihood of soil erosion. Socio-economic values predicted might be the size of the logging and associated industries dependent on harvesting in the region.

The Desired World establishes the goals used by an evaluation of derived states. Draft intergovernmental guidelines include key criteria dealing with biological and environmental aspects of the Desired World. The criteria are comprehensiveness, adequacy and representativeness (JANIS, 1995a). Further draft guidelines have suggested representative ranges for monitoring the criteria (JANIS, 1995b). Broadly, comprehensiveness guards against extinction of ecosystems and individual species; adequacy covers continued viability of ecosystems and species; and representativeness deals with retaining the current aerial distributions of ecosystems and species. Individual criteria within these groups are assignable to the Evaluation Function and the Gap Function depending on whether they are absolute or relative.

The Evaluation Function assesses the merit of the derived state after reassignment of parcels in terms of a Desired World alone (i.e. it is not concerned with comparisons with the Current World). For example, adequacy includes the requirement that '10% of all pre-European forests should be preserved' so that this is tested within the Evaluation Function. Comprehensiveness can be treated as part of the Evaluation Function, as it can be tested by examining a derived state alone.

The Gap Function tests a derived state by comparison with the Current World. Representativeness deals with a comparison between the current state and a derived state and so is part of the Gap Function.

5.1 The end-user interface

The end-user interface of the Forest Zoning System is a direct expression of the 5-component model. The design follows the EDSS (Abel et al., 1992) in applying the concepts of views (i.e. windows), linked views and object-oriented interaction with features within views. There are five major views. The Current World, Gap, Desired World and Evaluation represent the corresponding elements of the 5-component model. The Working World combines the Solution component and the derived state.

The Current World view is map-like and is the point of access to supporting databases. These databases include the current zoning, other data supporting assessment of alternative zonings and background information (such as the text of relevant legislation and of submissions from industry and community groups). Importantly the Current World view gives the user access to sets of predictive functions or rules which describe the interaction between key attributes. To extend or adapt the formulation of forest processes, the user can choose from available functions or rules. For example, he can choose between alternative functions to predict forest re-growth under certain logging regimes.

The Gap view and Desired World view have similar features. The Gap view has two sections. The first section enables specification of value functions to assess a derived state in relation to the Current World. The second section (when activated by the user) shows the values of these measures. The

Desired World view allows the user to specify checking of hard constraints (i.e. violation of necessary conditions for a valid outcome). A second Desired World section shows the set of constraints not satisfied by the current zonings (i.e. by the current derived state) and the values of the measures. Constraints and measures are composed from menus of variables and functions or by direct text entry. A number of value functions can be specified, so that multi-criteria objectives can be specified. These facilities allow the user to specify target values for decision variables, effectively specifying the preferred outcome. Sets of constraints or measures can be stored for reference in future runs.

The Evaluation view is very similar though constraints and measures can be composed only from variables from the Desired World description. The Gap, Desired World and Evaluation views are linked to the Working World view. As a solution is built by edit changes, or after an automatic solution technique is applied, the constraints not yet satisfied and the measures are automatically updated.

The Working World view is essentially a working surface and is again map-like. It does not support interrogation of the background information. The user can specify a solution by directly editing the working copy or by editing tabular displays of alterable variables. Alternatively, he can pull down a menu to choose from a set of solution techniques to derive a solution, or solution component, automatically. Solutions and the resultant derived state can be captured and stored persistently for future reference, and can then be chosen as the starting point for a new run.

6. Conclusion

Investigations of decision-making and problem-solving methodologies have highlighted problem formulation as the critical component and have suggested strongly that problem formulation proceeds through an evolutionary process in which the formulation is continually revised. The 5-component model is essentially an approach to designing SDSS interfaces which seeks to map the elements of problem formulation processes to the structure of the interface. Fundamentally, our approach can be seen as addressing SDSS interface design by direct manipulation of the statement of the problem, rather than by the tools solving the problem once formulated.

Application of the 5-component model has been demonstrated in terms of an experimental system for a complex planning task. While it remains to test the 5-component model through usability studies of the system, the simple establishment of the structure of the interface and the apparent flexibility of the use provide tentative confirmation of the validity of the approach. Importantly, the experimental system is not prescriptive in terms of the precise methodology applied by a planner to develop a policy for forest use. The planner is free to adopt a goal-oriented approach, in which an initial investigation establishes decision variables and a statement of preferences and later stages find the 'best' solution. The system also supports proposition of alternative statements of preferences, so that the outcomes of a given policy can be evaluated from the viewpoints of different sector groups such as the logging industry, conservationists and so on.

As for SDSS design and implementation, the 5-component model is not prescriptive as far as the sets of tools to be provided nor in the systems architecture. It does offer some guidelines for establishing tool requirements. Clearly, deficiencies in the available tools, such as the absence of desirable data and information in databases or in restricted forms of predictive simulation or optimisation, will be reflected in the support offered in problem formulation and in the solutions able to be posed and explored. The 5-component model then provides a basis for assessing just how a tool might contribute to the utility of the SDSS and for identifying gaps in a current set of tools.

The 5-component model does provide some guidelines for dealing with the top-level design of the end-user interface. It is not prescriptive in the detailed design of the interface, and indeed the HCI aspects of representing and manipulating the five components of a problem will continue to require creativity, expert knowledge and skill in HCI.

Strategically, the 5-component model addresses a basic problem in SDSS. To date, the solution to dealing with ill-defined problems has essentially been to provide the user with a workbench of tools which might be required. For SDSS dealing with problems of real-world complexity, it is likely that the collection of tools will be extensive. Usability of the SDSS will then be impaired by the complexity of navigation within the tool set and of choice of tools at different stages of the investigation. Fundamentally, the 5-component model is a methodology for structuring and presenting the tool set to overcome these problems.

Acknowledgements

The authors wish to acknowledge Jeffrey X. Yu and Kerry Taylor for their comments on an earlier draft of this paper.

References

Abel, D. J., Yap, S. K., Ackland, R., Cameron, M. A., Smith, D. F., and Walker, G., 1992, Environmental Decision Support System Project: An Exploration Of Alternative Architectures For Geographical Information Systems, International Journal of Geographical Information Systems, Vol. 6, No. 3, pp. 193-204.

Abel, D. J., Kilby, P. J., and Davis, J. R., 1994a, The Systems Integration Problem, International Journal of Geographical Information Systems, Vol. 8, No. 1, pp. 1-12.

Abel, D. J., Kilby, P. J., and Cameron, M. A., 1994b, A Federated Approach To Design Of Spatial Decision Support Systems, Proceedings of the Sixth International Symposium on Spatial Data Handling, SDH6, T. C. Waugh and R. G. Healey (Eds.) Edinburgh.

Armstrong, M. P., Densham, P. J., and Rushton, G., 1986, Architecture For A Microcomputer Based Spatial Decision Support System, Proceedings of the Second International Symposium on Spatial Data Handling, Seattle.

Banai, R., 1993, Fuzziness In Geographical Information Systems: Contributions From The Analytic Hierarchy Process, International Journal of Geographical Information Systems, Vol. 7, No. 4, pp. 315-329.

Blair, A., 1994, Inter-Connection Of Software Components In Intelligent Decision Support Systems, The Australian Computer Journal, Vol. 26, No. 1, pp. 11-19.

Bonczek, R. H., Holsapple, C. W., and Whinston, A. B., 1981 Foundations Of Decision Support Systems, Academic Press, New York.

Brightman, H., 1978, Differences In Ill-Structured Problem Solving Along The Organizational Hierarchy, Decision Sciences, Vol. 9, No. 1, pp. 1-18.

Carver, S. J., 1991, Integrating Multi-Criteria Evaluation With Geographical Information Systems, International Journal Geographical Information Systems, Vol. 5, No. 3, pp. 321-339.

Chang, A., Holsapple, C. W., and Whinston, A. B., 1993, Model Management Issues And Directions, Decision Support Systems, Vol. 9, No. 1, pp. 19-37.

Dos Santos, B. L., and Holsapple, C. W., 1989, A Framework For Designing Adaptive DSS Interfaces, Decision Support Systems, Vol. 5, pp. 1-11.

Densham, P. and Rushton, G., 1988, Decision Support Systems For Locational Planning, in Behavioural Modelling in Geography and Planning, R. G. Golledge and H. Timmermans (Eds.), Croom Helm.

Fedra, K., 1993, GIS And Environmental Modeling, in Environmental Modeling With GIS, M. Goodchild, B. O. Parks and L. T. Steyaert (Eds.), Oxford University Press, New York.

JANIS, 1995a, Joint ANZECC-MCFFA National Forest Policy Statement Implementation Sub-committee, Broad Criteria For The Establishment Of A Comprehensive, Adequate And Representative Forest Reserve System In Australia, Draft Report July 1995, Prepared by the Technical Working Group on Reserve Criteria.

JANIS, 1995b, Joint ANZECC-MCFFA National Forest Policy Statement Implementation Sub-committee, The Development Of Consistent Nationwide Baseline Environmental Standards For Native Forests, Draft Report July 1995, Prepared by the Technical Working Group on Forest Use and Management.

Mintzberg, H., Raisinghani, D. and Theoret, A., 1976, The Structure Of The Unstructured Decision Process, Administration Science Quarterly, Vol. 21, No. 2, June 1976.

Newell, A., and Simon, H. A., 1972, Human Problem Solving. Prentice-Hall, Englewood Cliffs, New Jersey.

Nijkamp, P., Rietveld, P., and Voogd, H., 1990, Multicriteria Evaluation In Physical Planning, North Holland.

Seo, F., and Sakawa, M., 1988, Multiple Criteria Decision Analysis In Regional Planning: Concepts, Methods, And Applications, D. Reidel Publishing Co., Dordrecht.

Simon, H. A., 1973, The Structure Of Ill Structured Problems, Artificial Intelligence 4, pp. 181-201.

Simon, H. A., 1977, The New Science Of Management Decision, Revised Edition, Prentice-Hall, Englewood Cliffs, New Jersey.

Sprague, R. H., 1980, A Framework For The Development Of Decision Support Systems, MIS Quarterly, December 1980, pp. 1-26.

Sprague, R. H. and Carlson, E. D., 1982, Building Effective Decision Support Systems, Prentice-Hall, Englewood Cliffs, New Jersey.

Turban, E., 1990, Decision Support And Expert Systems: Management Support Systems, Second Edition, Macmillan Publishing, New York.

Van Gundy, A. B. Jr., 1988, Techniques Of Structured Problem Solving, Second Edition, Van Nostrand Reinhold, New York.

Contact address
M. A. Cameron[1] and D. J. Abel[2]
[1]Department of Computer Science,
Australian National University
[2]CSIRO Division of Information Technology,
GPO Box 664,
Canberra, ACT 2601
Australia
E-mail: {mark.cameron david.abel}@cbr.dit.csiro.au

THE SPATIAL LOCATION CODE

Peter van Oosterom[*1] and Tom Vijlbrief[2]

[1]Cadastre Netherlands, Company Staff
Apeldoorn, the Netherlands
[2]TNO Human Factors Research Institute
Soesterberg, the Netherlands

Abstract

In this paper we present the Spatial Location Code (SLC) which is used for indexing and clustering geographic objects in a database. It combines the strong aspects of several known spatial access methods (Quadtree, Field-tree, and Morton code) into one SLC value per object. The *unique* aspect of the SLC is that both location and extent of possibly non-zero-sized objects are approximated by this *single* value. These SLC values can then be used in combination with traditional access methods, such as the b-tree, available in every database. It is expected that the typical query response time for spatial objects is reduced by orders of magnitude for a reasonably sized data set. The examples in this paper are all given in two-dimensional space, but the SLC is quite general and can be applied in higher dimensions.

Key words

Spatial searching, DBMS, field-tree, Morton code, benchmark.

1. Introduction

The Spatial Location Code (SLC) is designed to enable efficient storage and retrieval of spatial data in a standard (relational) DBMS. The requirements for the SLC are summarized below:

- be suitable for two-dimensional data;
- solve range queries efficiently (two-dimensional search rectangles), which implies spatial clustering and indexing;
- use one code per object and not sets of codes as these might be more difficult to manage within the DBMS;
- be (as) transparent (as possible) to the data model and queries;
- do not require DBMS kernel modifications, that is, use standard data types and access methods.

Because the solution must be implementable within any DBMS environment, the SLC extension has to be based on an existing type (e.g. `integer4`) in combination with an existing access method, such as the b-tree (Bayer & McCreight, 1973). A rough outline of this solution:

---

[*] A part of this research was performed while the author was still affiliated with the TNO Physics and Electronics Laboratory, P.O. Box 96864, 2509 JG The Hague, The Netherlands.

1. Add one 'spatial location code' SLC attribute to every table in the database which has a spatial (point, line, polygon, or box) attribute. The SLC is a one-dimensional code and every object gets exactly one code. This code is an approximation of the `location` and `extent` of possibly non-zero-sized objects in the two-dimensional space. It is possible that two different objects are approximated by the same SLC, but these objects will have about the same size and location.

2. Modify the table structure to b-tree according to the SLC attribute, that is, the objects are more or less stored on disk in the order defined by the b-tree. This primary index is responsible for the spatial clustering.

3. Define two functions:
   a) `Compute_SLC`: computes the SLC of a two-dimensional box. First, the bounding box (bbox) of an object is computed, then the SLC for this box is computed. So, only one `Compute_SLC` is needed;
   b) `Overlap_SLC`: determines the ranges of SLCs that do overlap the given two-dimensional search rectangle (query box).

4. Finally, one could define rules/procedures within the DBMS to make the SLC transparent to application:
   a) fill the SLC-attribute in case of insert or update of tuples (possible in most DBMSs);
   b) 'rewrite' box-overlap queries into SLC-queries (only possible in advanced extensible DBMSs, always possible in GIS front-end).

A combination of the Field-tree and Morton code (Quadtree) is used as a basis for the SLC. These structures will be described in Section 2. Abel and Smith (1983) describe a method based on the combination of the Quadtree and Morton code. However, a small object crossing a top level boundary will get a code corresponding to a large Quadtree region. Later on, during querying, this small object will be retrieved whenever the search area overlaps this large Quadtree region. This problem can be reduced by allowing more than one code per object; e.g. 4 codes (Abel & Smith, 1984). Drawbacks of this method are the increased storage use and the query where-clause will become more complex and therefore slower. The SLC method does not have these disadvantages. More details with respect to the `Compute_SLC` and `Overlap_SLC` functions will be given in Section 3. The benchmarks are presented in the subsequent section. Finally, conclusions can be found in Section 5.

2. Basic structures

In this section brief descriptions of the Region Quadtree, the Morton code, and the Field-tree are given. More details can be found in the given references.

2.1 The Region Quadtree

The *Quadtree* is a generic name for all kinds of trees that are built by recursive division of space into four quadrants. Samet (1984, 1989) gives an excellent overview. The best known Quadtree is the *region Quadtree*, which is used to store a rasterized approximation of a polygon. First, the area of interest is enclosed by a square. A square is repeatedly split into four squares of equal size until it is completely inside (a black leaf) or outside (a white leaf) the polygon or until the maximum depth of the tree is reached (dominant color is assigned to the leaf); see Figure 1. Each leaf can be assigned an unique label, the *quadcode*. During every split, the quads are assigned a number (e.g. SW=0, NW=1, SE=2, and NE=3). From the root to the leaf these number form a unique string; short strings (only a few splits) correspond to large regions, long strings correspond to small regions (a lot of splits).

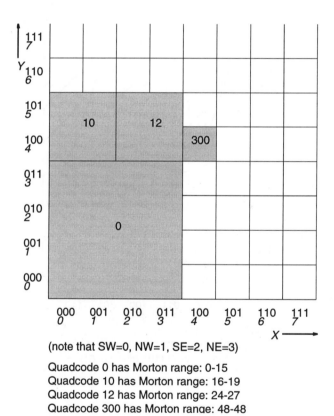

(note that SW=0, NW=1, SE=2, NE=3)

Quadcode 0 has Morton range: 0-15
Quadcode 10 has Morton range: 16-19
Quadcode 12 has Morton range: 24-27
Quadcode 300 has Morton range: 48-48

Figure 1 The Quadtree and quadcodes.

2.2 The Morton code

The ordering of point locations with integer coordinates (or the raster cells of a two-dimensional grid or two-dimensional nodes of a Quadtree) can be used for *tile indexing*. It transforms a two-dimensional problem into a one-dimensional one. Several orderings have been described in the literature

(Abel & Mark, 1990; Goodchild & Grandfield, 1983; Jagadish, 1990; Nulty & Barholdi, III, 1994): row, row prime, Morton, Hilbert, Sierpinski, Gray code, Cantor-diagonal, and spiral.

During the storage of spatial data, the transformation to one dimension has to be made as the two-dimensional data has to be stored in the one-dimensional computer memory structure. The purpose of these orderings is clustering the spatial data: objects that are close together in the two-dimensional space should also be stored close together in memory (on the disk), as it is likely that they will be retrieved by the same query.

The bitwise interleaving of the two coordinates results in a one-dimensional key, called the *Morton* key (Orenstein & Manola, 1988). The Morton key is also known as Peano key, or N-order, or Z-order. For example, row $2 = 010_{bin}$ column $1 = 001_{bin}$ has Morton key $6 = 000110_{bin}$ (see Figure 2).

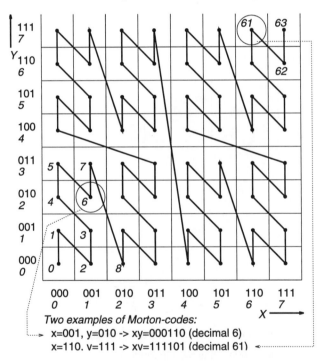

Figure 2 The Morton-codes (shown on a 88 grid).

The Morton key has several relationships with the Quadtree. As can be seen in Figure 1, each quadcode can directly be translated into a range of Morton odes. Short quadcode strings relate to large Morton code ranges and long quadcode strings relate to small Morton code ranges. (Abel & Mark, 1990) have identified three desirable properties of spatial orderings:

1. An ordering is *continuous* if the cells in every pair with consecutive keys are 4-neighbors.
2. An ordering is *quadrant-recursive* if the cells in any valid Quadtree subquadrant of the matrix are assigned a set of consecutive integers as keys.
3. An ordering is *monotonic* if for every fixed x, the keys vary monotonically with y in some particular way, and vice versa.

Abel and Mark (Abel & Mark, 1990) conclude from their practical comparative analysis of five orderings (they do not consider the Cantor-diagonal, the Spiral and, the Sierpinski orderings) that the Morton ordering and the Hilbert ordering are, in general, the best alternatives. Goodchild (Goodchild, 1989) proved that the expected difference of 4-neighbor keys of an n by n matrix is $(n + 1) / 2$, which is exactly the same as in the case of row and row-prime ordering. However, the average farthest distance of the neighbors in the Morton ordering is higher than in the Hilbert ordering (Faloutsos & Roseman, 1989).

2.3 The Field-tree

The *Field-tree* can store polylines and polygons in a non-fragmented manner. The Field-tree is one of the few structures which takes account of the fact that there may exist geometric objects of varying importance. During the last decade several variants of the Field-tree have been published by Frank *et al.* (1983, 1986, 1989). In this subsection attention will be focused on the *Partition Field-tree*. Conceptually, the Field-tree consists of several levels of grids, each with a different resolution and a different displacement; see Figure 3. A grid cell is called a *field*. Actually, the Field-tree is not a hierarchical tree, but a directed acyclic graph, as each field can have one, two, or four ancestors. At one level the fields form a partition and therefore never overlap.

A newly inserted object is stored in the smallest field in which it completely fits. As a result of the different displacements and grid resolutions, an object never has to be stored more than three levels above the field size that corresponds to the object size. Note that this is not the case in a Quadtree-like structure, because the edges at different levels are collinear. A drawback of the Field-tree is that an overflow page is sometimes required, as it is not possible to move relatively large or important objects of an overfull field to a lower level field.

3. Spatial location code

Why should the SLC values not directly use a quadcode (or range of Morton codes)? There are two possible options, but each one has a drawback:
1. Assign one quadcode to every object. However, a small object may cross one of the major split lines and therefore get a short string quadcode. It is obvious that this large region does not approximate the object very well.
2. Alternatively, several quadcodes could be assigned to a single object. The corresponding regions now better approximate the object, but they

are too complex to be managed efficiently in the DBMS (a set of values is associated with every tuple).

max domain

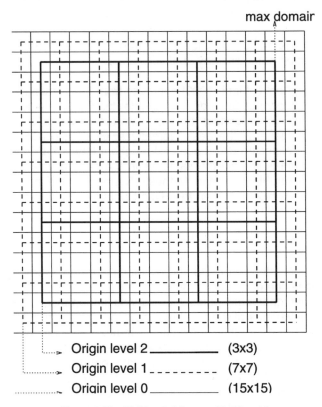

→ Origin level 2 —————— (3x3)
→ Origin level 1 _ _ _ _ _ _ _ _ (7x7)
→ Origin level 0 —————— (15x15)

Figure 3 The Field-tree (shown with 3 levels).

The Field-tree was used to get a better approximation of an object by a single value (the actual field in which the object is stored). Due to the 'shifting grids' of the Field-tree, the small objects will never be assigned to very large regions. The SLC encodes both the grid level and the grid cell at this level (e.g. by a Morton code) in one value. The SLC, as it has roughly been described in the introduction, could be used as shown in the SQL example of Figure 4 Query 1 defines a table with a SLC-attribute. Then two records are inserted in this table in queries 2 and 3. Somehow, the SLC-attribute is also filled; e.g. by using a database rule which derives this value from the place, a box type, attribute. In query 4 a spatial selection is defined by using the overlap function with a given box. Again, somehow this query has to be translated into terms of SLC-values.

In subsection 3.1, the details of the SLC bit encoding are described. The pseudo code for the Compute_SLC and Overlap_SLC functions is given in subsection 3.2.

```
1. create table spatial (name char(100), place box, SLC integer4)
2. insert values into table spatial (name="jan", place="(0,0,1,2)")
     /* SLC filled with rule; e.g. SLC=115679 */
3. insert values into table spatial (name="klaas", place="(1,1,2,2)")
     /* SLC filled with rule; e.g. SLC=113456 */
4. select * from spatial where place overlaps "(0,2,2,3)"
     /* where-clause translated into where clause that uses SLC:
        where (SLC>100000 and SLC<150000) or (SLC>200000 and SLC<225000)
           or...*/
```

Figure 4 Example use of the SLC within the database.

3.1 SLC bit encoding

For encoding objects with an SLC only their bounding box (bbox) is needed, because if an object has to fit in a cell of the Field-tree, just its bounding box has to fit. The following SLC encoding schema is based on `integer4`:

- Use variable length bit patterns for addressing grid-cells (finer grids require more bits for addressing);
- Start with a coarsest level of '255*255 grid'. The next finer level contains 2*255+1 = 511 cells in one direction. The number of cells for the next finer levels are computed in the same way (1023, 2047, etc.); see columns 'level' and 'actual #cell 1 dim' in Figure 5 (the other columns will be explained later);
- Allow flexible encoding of the number of levels (e.g. 5, 8, 11) and domain, so the SLC can be fine-tuned for a given application.

The basis is `integer4` values because they are more efficient in combination with the b-tree in many databases, compared to using a simple string encoding for the SLC (char).

The second column in Figure 5 'SLC-code' describes the bit-pattern on a given level. The higher/coarser levels have more bits for the level indication and less 'xy'-bits, but this is no problem as there are also less grid cells to address. Note that the first bit is not used (to avoid possible problems with negative values of `integer4`). The column '#bit grid code' gives the number of bits available for addressing grid cells; 'x' and 'y' bits are added together. The column 'max #cell 1 dim' gives the maximum number of cells in one dimension with the available number of bits. The next column ('actual #cell 1 dim') gives the actual number of grid cells in one dimension for this level in the Field-tree. It can be observed that these are always less or equal than the available number. The total number of grid cells per level in the Field-tree is given in column '#cells per level'.

In Figure 5 an example is given in which the finest level is 'level 0' with 25m*25m cells (this can be adapted if one needs a larger domain; see '*25' in the column 'len cell in m'), the coarsest level is 'level 7' (with 3.2km*3.2km cells). Everything too large to fit into the coarsest cells (3.2km), gets a special SLC code (e.g. -1): the *overflow bucket*, which should not become 'too full'[1]. If

---

[1] If the overflow bucket becomes too full, a more coarse level should be added to the Field-tree (SLC value).

one assumes the origin or lower-left point of the domain at (0,0), then the upper-right point of the domain is at (816,816) in km. In the column 'origin of domain (x) in m' it can be seen that the finer levels get a little offset in the Field-tree; see also Figure 3.

level	SLC-code: - 01's are for level - xy's are for Morton code	#bit grid code	max #cell 1 dim	actual #cell 1 dim	len cell in m	#cells per level	origin domain (x) in m
-	11xyxyxy xyxyxyxy xyxyxyxyxyxyxyxy	NOT	USED				
-	10xyxyxy xyxyxyxy xyxyxyxyxyxyxyxy	NOT	USED				
0	01xyxyxy xyxyxyxy xyxyxyxyxyxyxyxy	30	32768	32767	*25	1073676289	-1587.
1	0011xyxy xyxyxyxy xyxyxyxyxyxyxyxy	28	16384	16383	50	268402689	-1575.
2	0010xyxy xyxyxyxy xyxyxyxyxyxyxyxy	28	16384	8191	100	67092481	-1550.
3	0001xyxy xyxyxyxy xyxyxyxyxyxyxyxy	28	16384	4095	200	16769025	-1500.
4	000011xy xyxyxyxy yxyxyxyxyxyxyxy	26	8192	2047	400	4190209	-1400
5	000010xy xyxyxyxy xyxyxyxy yxyxyxy	26	8192	1023	800	1046529	-1200.
6	000001xy xyxyxyxy xyxyxyxy yxyxyxy	26	8192	511	1600	261121	-800.
7	00000011 xyxyxyxy xyxyxyxy yxyxyxy	24	4096	255	3200	65025	0

Figure 5 The bits in the SLC coding; ORIGIN=(0,0), FINEGRID=25m.

The x- and y-coordinates of each grid cell are bitwise interleaved, which has the effect of the Morton encoding per level (which is very much related to the Quadtree; see Figure 1 The level indication (bit-pattern; Figure 5) together with this Morton code, forms the complete SLC of each possible object in the given domain.

Note that this schema is general and can be extended to less or more levels (e.g. 5, 8, 11 or 15) or even different dimensions (e.g. three-dimensional). It might turn out that 8 levels are not required, given a certain data set, for stating the queries efficiently (ranges at 8 levels). When only 5 levels are used, the size of the smallest cell should probably be larger than 25m in order to avoid too many objects in the overflow bucket for a given application, e.g. 100m (and 1.6km on level 4).

3.2 SLC pseudo code

The algorithms in this section are given in C-style pseudo code. First, remember the definition of the different Field-grids; see Figure 5. In the pseudo code the levels are numbered 0 (finest grid) to 7 (coarse). The coarsest Field-tree grid has no (additional) displacement, the finer levels get more and more displacement; see Figure 5, column 'origin of domain (x) in m'. In Figure 6 some constants used in the pseudo code are initialized. Note that ORIGX[NRLEVELS-1] is initialized without additional displacement, and that the other levels get their additional displacement starting with ORIGX[NRLEVELS-2] down to finest level ORIGX[0]. The function Compute_SLC (bbox) computes the SLC for a given object (bbox); Figure 7. First, it determines the finest grid in which the object might fit (lines 1 and 2). Then the exact grid level and SLC value is computed in line 3, by testing if the minimum and maximum point of the bbox have the same SLC value. The SLC value of these points is computed with the function Compute_SLC_Point (level,x,y); Figure 8.

```
NRLEVELS=8   // 5 levels might be better
FINEGRID=25 // 100m might be better
THEORIGX=0
THEORIGY=0

for (i=0; i<NRLEVELS; i++) do
  NR[i] = 2^(15-i);   // - 1;
for (i=0; i<NRLEVELS; i++) do
  SIZE[i] = FINEGRID * 2^i;

ORIGX[NRLEVELS-1] = THEORIGX;
ORIGY[NRLEVELS-1] = THEORIGY;
for (i=NRLEVELS-2; i>=0; i--) do
  ORIGX[i]= ORIGX[i+1] - SIZE[i+1]/4;
  ORIGY[i]= ORIGY[i+1] - SIZE[i+1]/4;

// bitpatterns for encoding level
BITS[MAXLEVEL]={
  01000000 00000000 00000000 00000000,
  00110000 00000000 00000000 00000000,
  00100000 00000000 00000000 00000000,
  00010000 00000000 00000000 00000000,
  00001100 00000000 00000000 00000000,
  00001000 00000000 00000000 00000000,
  00000100 00000000 00000000 00000000,
...}
```

Figure 6 Some constants (generic solution).

```
0. SLC = -1 // does not fit value!
1. get bbox max extent (in x or y dimension) -> max_ext
2. go to finest level with size>max_ext -> level[i]
3. for (j=i; j<i+3 && j<NRLEVELS && SLC== -1; j++)
                                    // worst case on level i+3
     if (Compute_SLC_Point(j,x_min(bbox),y_min(bbox)) ==
         Compute_SLC_Point(j,x_max(bbox),y_max(bbox)))
       SLC=Compute_SLC_Point(j,x_min(bbox),y_min(bbox))
```

Figure 7 Compute_SLC(bbox).

```
1. x = (short)trunc((x-ORIGX[level])/SIZE[level]);
   y = (short)trunc((y-ORIGY[level])/SIZE[level]);
2. bitwise interleave xy -> SLC;
3. add level bitpattern (bitwise OR) -> SLC = SLC | BITS[level];
```

Figure 8 Compute_SLC_Point(level,x,y).

The `Overlap_SLC (bbox)` computes the set of SLC ranges that belong to the specified search box `bbox`; see Figure 9. For each level, the quadcodes (or Morton ranges) corresponding to the bbox are computed (step 1). This is done with the function `Compute_quadcodes (level, bbox)` (Figure 10), which in turn uses the recursive function `Add_quad_level (quaddomain, minSLC, maxSLC, bbox)`. This recursive function adds the new ranges for the specified domain to the set of ranges; Figure 11.

Note that when the number of ranges becomes large (e.g. too large for the database where-clause), then a few 'tricks' may be applied to reduce the number of ranges (these will include all required SLC-codes, but also include some unneeded SLC-codes, outside the search box):

- do not descend to the lowest level of the Quadtree, or
- join ranges even if there is a small gap between them, or
- reduce the number of levels to begin with (e.g. only 5).

Especially at the fine-level grids (e.g. 0, 1, and 2), the problem of the large number of ranges may occur. So, the 'tricks' mentioned above should first be applied at these levels. Experimental results and data distributions should be used to fine-tune the SLC encoding and functions.

```
1. for each level do
     Compute_quadcodes(level, bbox)      // see function below
2. ranges of SLC codes can be joined to larger ranges
```

Figure 9 Overlap_SLC(bbox).

```
Initialize quaddomain                  // depends on level
set_of_ranges={}                       // start with empty set
minSLC=0; maxSLC=(nr[level])^2 - 1 // the SLC ranges

Add_quad_level(quaddomain, minSLC, maxSLC, bbox)
```

Figure 10 Compute_quadcodes(level, bbox).

```
if (minSLC==maxSLC) do
  1. add (minSLC, maxSLC) to set_of_ranges                 // finished
else
  2. split quaddomain in quad[0], quad[1], quad[2], quad[3]
  3. for i=0 to 3 do
     3a. new_minSLC = minSLC + i*((maxSLC+1-minSLC)/4))
         new_maxSLC = minSLC + (i+1)*((maxSLC+1-minSLC)/4)) -1
     3b. if bbox FULL_COVERS quad[i]
            add (new_minSLC,new_maxSLC) to set_of_ranges   // finished
     3c. else if bbox PARTIAL_COVERS quad[i]               // recursion
            add_quad_level(quad[i], new_minSLC, new_maxSLC, bbox)
```

Figure 11 Add_quad_level(quaddomain, minSLC, maxSLC, bbox).

4. Benchmarks

Two types of benchmarks are presented. In the first subsection, measurements with respect to the number of ranges (and the reduction thereof) are presented. In the second subsection, some results with actual (cadastral) data are given.

4.1 The number of ranges

As mentioned in the previous section, the number of ranges produced by the Overlap_SLC may become critical. Therefore, a function was created which closes the smaller gaps between two successive ranges. This function continues until the required number of ranges is reached. Every time two successive ranges are joined, other (unwanted) cells are also included; e.g. assume range 1 is (10,17) and range 2 is (20,23), then the joined range (10,23) contains unwanted cells 18 and 19.

| max   # | .25   *   .25 | .5 * .5 km² | 1 * 1 km² | 2 * 2 km² | 4 * 4 km² | 8 * 8 km² |
ranges	km²					
6	36.6	36.5	18.1	23.8	10.8	19.1
10	1.5	1.8	2.3	4.2	3.0	2.8
15	1.0	1.2	1.4	1.7	1.8	1.8
20	1.0	1.1	1.3	1.4	1.5	1.5
30	1.0	1.0	1.1	1.2	1.3	1.3

Table 1 Ratio retrieved/required cells for the different query sizes (with a 5 level grid SLC).

For each test, 100 random queries of a certain size are generated in the following way: the location of the query is random over the whole domain (uniformly distributed), the size of the edge of a square query region is random over the range (0.75*X, 1.25*X), where X is the specified average size. The SLC values are based on a 5 level Field-tree.

Table 1 shows the measurements for six different query sizes (X is 0.25, 0.5, 1, 2, 4, and 8 km) and for five different upper bounds of the allowed number of ranges (at most 6, 10, 15, 20, and 30 ranges). Over the 100 random queries, the average initial number of corresponding SLC ranges per query (without closing range gaps > 0) are 13.1 (X=0.25), 22.5 (X=0.5), 40.3 (X=1), 77.9 (X=2), 155.6 (X=4), and 308.6 (X=8) respectively. These large number of ranges in the where-clause may be of problem for some RDBMSs. The effect of reducing the number of ranges is captured by the ratio between the number of cells addressed in this way (including the unnecessary cells) and the number of required cells. Except for the upper bound of at most 6 ranges, the same measurements are visualized in Figure 12: on the x-axis, the average size of the query region; on the y-axis, the ratio between the number of retrieved and required cells. The different lines in the graph represent the different upper bounds of the allowed number of ranges. For interactive applications a ratio of at most 2 is acceptable. From Table 1 and Figure 12, it can be concluded that about 15 ranges are required (for SLC values based on 5 levels).

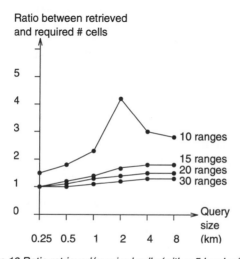

Figure 12 Ratio retrieved/required cells (with a 5 level grid SLC).

4.2 The 'Flevoland' test

The test data set was provided by the Dutch Cadastral and Public Registers Agency. The data set Flevoland `grens` contains all (land owner) parcels in the 'new' province Flevoland created by man on the floor of the former sea; see Figure 13 which shows the SLC implementation in our GIS GEO++ (Vijlbrief & van Oosterom, 1992; van Oosterom & Vijlbrief, 1994). For spatial indexing, the SLC method has been applied with the following specifications: 5 levels, finest grid 100m, maximum number of ranges in a query where clause is 20. The total area of the bounding box of this province is about 3000 km^2 of which about 2000 km^2 are actually filled with data.

The data density varies a lot: very high density in the center of towns, medium density in the other urban areas, low density in the natural/agricultural areas, and very low density in the large lakes 'Markerwaard' and 'IJsselmeer'. Note that in Figure 13 all objects with an SLC value in Field-tree 'level 0' are gray, and all others are black. The overview window (upper-right) is a selection of all features with SLC values above Field-tree 'level 3'. The representation is quite good and it can be retrieved and displayed very efficiently. This indicates that the SLC values can also be used for *map generalization*.

4.3 The timings

The hardware configuration consists of a Sun SparcStation 10 (2 processors, 64Mb Main memory). The most important software components are Solaris 2.4 and CA OpenIngres 1.1/03 (beta) with OME/SOL (Object Management Extension/Spatial Object Library). In the benchmarks 10 representative queries are used (see Table 2: the 10 different queries) and every bbox range query is translated into a similar SLC query. Besides the `grens` table with 156.998 records, also the `lynstring` table with 633.397 records has been used in the benchmarks. The `lynstring` table contains the topographic features. Both tables use the OME/SOL spatial data types line(33) and box (and compression). The table `grens` takes 66.8Mb when using a btree storage structure. For the table `lynstring` this Figure is 264.8Mb. The benchmarks are based on counting objects and not on actually retrieving them. This gives the best impression of the effect of (spatial) indexing.

When using no SLC value the DBMS has to do a sequential scan, which takes always about the same time. For the SLC index scan benchmarks, the response times (column 4 in Table 3 are proportional to the number of counted objects (column 5 in Table 3): about 200 counted objects per second, unless the number of counted objects is very low. This indicates that the SLC method with btree is functioning well and reduces the query times from several hundreds of seconds to several seconds for restrictive queries. When comparing the number of objects with overlap (column 3) with the number of objects reported by the SLC method (column 5), it can be seen their ratio (column 6) is quite close the expected ratio; see previous subsection.

The effect of using the SLC method becomes even more important when the table becomes larger as can be seen in Table 4, which present the measurements for the large `lynstring` table.

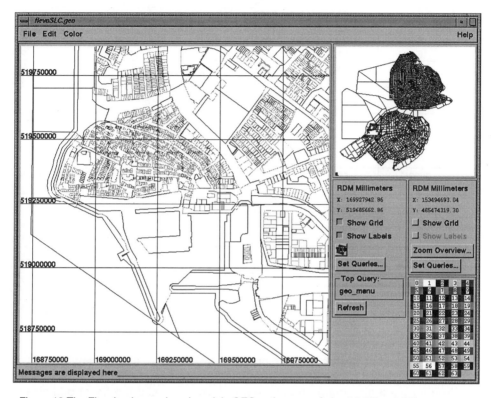

Figure 13 The Flevoland grens benchmark in GEO++ (an area of about 1.25km 1.25km near the harbor of Urk): 'level 0' SLC objects displayed in gray, others in black.

query nr.	bounding box of query region (coordinates in RD km)	area (km^2)
1	(150.0,490.0,160.0,500.0)	100
2	(155.0,495.0,160.0,500.0)	25
3	(170.0,525.0,175.0,530.0)	25
4	(172.0,526.0,173.0,527.0)	1
5	(172.0,526.0,172.5,526.5)	0.25
6	(180.0,520.0,190.0,525.0)	50
7	(180.0,520.0,181.0,521.0)	1
8	(180.5,520.5,181.0,521.0)	0.25
9	(175.0,500.0,180.0,505.0)	25
10	(179.0,500.0,180.0,501.0)	1

Table 2 The 10 different queries.

query	no slc		slc		
	sec	#obj	sec	#slc_obj	#slc_obj/#obj
1	691	652	15	3650	5.6
2	718	454	7	1212	2.7
3	724	2609	14	3267	1.3
4	922	1000	5	1182	1.2
5	734	682	5	952	1.4
6	712	10434	82	23458	2.2
7	655	91	4	253	2.8
8	647	3	2	72	24.0
9	647	16045	60	17924	1.1
10	660	31	3	76	2.5

Table 3 Querying the table grens.

query	no slc		slc		
	sec	#obj	sec	#slc_obj	#slc_obj/#obj
1	2861	32623	193	51243	1.6
2	3023	12528	73	21371	1.7
3	2871	5992	39	7403	1.2
4	2973	1818	11	2247	1.2
5	3222	1132	9	1610	1.4
6	2843	21477	262	53938	2.5
7	3457	169	14	441	2.6
8	2862	11	3	80	7.3
9	2707	51497	229	62397	1.2
10	2773	304	15	575	1.9

Table 4 Querying the table lynstring.

Using 5 levels seems a very practical solution. In the current Cadastral software (based on a network DBMS and an own implementation of a Field-tree for geometric data), the number of levels is 5 and the finest grid-size is 100m. This Field-tree has been satisfactory for the large amounts of data stored in the database (called LKI). Now the LKI database is moving from a network DBMS to a relational environment, and the Field-tree is maintained by means of the SLC values. In July 1994, the LKI database contained about 20.000.000 lynstring features and 10.000.000 grens features. So in the heavily used part of the domain, the number of SLC values is about equal to the number of lynstring features. Not all SLC values will be used, and some will be used more than once.

5. Conclusion

The pros and cons of the SLC values are summarized in the next subsection, followed by a description of possible future work.

5.1 Advantages and disadvantages

The disadvantages of the current implementation of the SLC values:
1.   The SLC approximation of an object can be quite coarse, especially in case of a 'long' (non-square) object. This will result in unnecessary selection of some objects based on their SLC.
2.   The structure has to be tuned for a given application: the domain and the number of levels have to be chosen, but this can be done by means of automatic data analysis.
3.   The number of SLC ranges in a query (as result of the `Overlap_SLC (bbox)` function) can be large. Several solutions for this problem have been described in Sections 3 and 4. One other alternative is to have multiple queries; e.g. one per level.

The most important advantages of the SLC value can be summarized as:
1.   The SLC is very compact, only 4 bytes per object when `integer4` is used. Compared to a `bbox` based on 8 byte floating point coordinates as used in the R-tree (and many other structures), this is only 1/8.
2.   The SLC can be used in many DBMS's as long as `integer4` and an index structure is supported (b-tree).
3.   The SLC can be transparent to the user if the DBMS supports triggers and procedures.
4.   The SLC will be even more efficient if it is implemented inside a DBMS.
5.   The SLC provides good clustering and indexing, and will therefore enable efficient spatial DBMS queries.
6.   The SLC can also be used to support 'map generalization' and multi-scale queries by retrieving only the coarse levels (which contain only large or important objects).
7.   Empty SLC grid cells do cost nothing. Further, overfull cells are not possible, because many objects can have the same SLC value.

5.2 Future work

Future work will include performing more benchmarks with other real data sets and evaluating other promising orderings instead of the Morton code; e.g. Hilbert or Sierpinski orderings.

One important use of the SLC-values, other than spatial searching, is selection for interactive generalization: for an overview map (small scale) only the coarse grids have to be used; see Figure 13. However, small but important objects may be missed in this way, because they are stored in one of the finer grids. Therefore, the `Compute_SLC` function could be modified in a way that a user may specify a lower bound for the finest grid. In this way, important objects may be stored at coarser ('more important') grid levels even if they would fit at finer levels.

Another improvement is to have at each level two grids of the same resolution: the *shadow grid technique*. The second grid is just translated in x and y-direction by halve the size of a grid cell. In this approach an object is never stored more than 1 level higher (coarser) than its own size (instead of 3 levels in the original method). This will reduce the number of overflow objects and is also better for generalization. The drawback is that twice as much ranges of SLC-values are required in the where-clause of a spatial overlap query.

Acknowledgments
Many valuable comments and suggestions were made in preliminary discussions with Chrit Lemmen and Tapio Keisteri. Paul Strooper made important additional comments with respect to both the contents and the presentation of the paper. Without their contributions we could not have achieved the current results. Finally, three anonymous reviewers of SSD'95 inspired us to rewrite the paper and to add the timings of queries to the benchmark section. The Dutch Cadastral and Public Registers Agency sponsored the described research and also provided the cadastral maps, which were used in our benchmarks.

References
Abel, D. J., & Mark, D. M. (1990). A Comparative Analysis of Some Two-Dimensional Orderings. International Journal of Geographical Information Systems, 4 (1), 21-31.

Abel, D. J., & Smith, J. L. (1983). A Data Structure and Algorithm Based on a Linear Key for a Rectangle Retrieval Problem. Computer Vision, Graphics and Image Processing, 24, 1-13.

Abel, D. J., & Smith, J. L. (1984). A Data Structure and Query Algorithm for a Database of Areal Entities. The Australian Computer Journal, 16 (4), 147-154.

Bayer, R., & McCreight, E. (1973). Organization and Maintenance of Large Ordered Indexes. Acta Informatica, 1, 173-189.

Faloutsos, C., & Roseman, S. (1989). Fractals for secondary key retrieval. in Eighth ACM SIGACT-SIGMOD-SIGART Symposium on Principles of Database Systems (PODS), pp. 247-252.

Frank, A. U., & Barrera, R. (1989). The Field-tree: A Data Structure for Geographic Information System. in Symposium on the Design and Implementation of Large Spatial Databases, Santa Barbara, California, pp. 29-44 Berlin. Springer-Verlag.

Frank, A. (1983). Storage Methods for Space Related Data: The Field-tree. Tech. rep. Bericht no. 71, Eidgen"ossische Technische Hochschule Z"urich.

Goodchild, M. F., & Grandfield, A. W. (1983). Optimizing Raster Storage: An Examination of Four Alternatives. in Auto-Carto 6, pp. 400-407.

Goodchild, M. F. (1989). Tiling Large Geographical Databases. in Symposium on the Design and Implementation of Large Spatial Databases, Santa Barbara, California, pp. 137-146 Berlin. Springer-Verlag.

Jagadish, H. V. (1990). Linear Clustering of Objects with Multiple Attributes. in ACM/SIGMOD, Atlantic City, pp. 332-342 New York. ACM.

Kleiner, A., & Brassel, K. E. (1986). Hierarchical Grid Structures for Static Geographic Data Bases. in Auto-Carto London, pp. 485-496 London. Auto Carto.

Nulty, W. G., & Barholdi, III, J. J. (1994). Robust Multidimensional searching with spacefilling curves. in Proceedings of the 6th International Symposium on Spatial Data Handling, Edinburgh, Scotland, pp. 805-818.

Orenstein, J. A., & Manola, F. A. (1988). PROBE Spatial Data Modeling and Query Processing in an Image Database Application. IEEE Transactions on Software Engineering, 14 (5), 611-629.

Samet, H. (1984). The Quadtree and Related Hierarchical Data Structures. Computing Surveys, 16 (2), 187-260.

Samet, H. (1989). The Design and Analysis of Spatial Data Structures. Addison-Wesley, Reading, Mass.

van Oosterom, P., & Vijlbrief, T. (1994). Integrating Complex Spatial Analysis Functions in an Extensible GIS. in Proceedings of the 6th International Symposium on Spatial Data Handling, Edinburgh, Scotland, pp. 277-296.

Vijlbrief, T., & van Oosterom, P. (1992). The GEO++ System: An Extensible GIS. in Proceedings of the 5th International Symposium on Spatial Data Handling, Charleston, South Carolina, pp. 40-50 Columbus, OH. International Geographical Union IGU.

Contact address
Peter van Oosterom[1] and Tom Vijlbrief[2]
[1]Cadastre Netherlands
Company Staff
P.O. Box 9046
7300 GH Apeldoorn
The Netherlands
Phone: +31 55 528 5163
Fax: +31 55 355 7931
E-mail: oosterom@kadaster.nl
[2]TNO Human Factors Research Institute
P.O. Box 23
3769 ZG Soesterberg
The Netherlands
Phone: +31 3463 56309
Fax: +31 3463 53977
E-mail: vijlbrief@tm.tno.nl

A DBMS REPOSITORY FOR THE APPLICATION DOMAIN OF

GEOGRAPHIC INFORMATION SYSTEMS

Emmanuel Stefanakis and Timos Sellis

Department of Electrical and Computer Engineering
National Technical University of Athens
Athens, Greece

Abstract
Much attention has been devoted in the past to support classes of applications which are not well served by conventional database systems. Focusing on the application domain of geographic information systems (GISs), several architectural approaches have been proposed to implement commercial or prototype systems and satisfy the urgent needs for operational spatial data handling. All these approaches adopt an application layer, which lies on the top of the database management system (DBMS) repositories and aims at supplementing the set of capabilities offered by the underlying repositories. The execution of operations within the application layer is frequently accompanied by up-/down-loading of voluminous data from/to the database and consequently it should be avoided. What is usually suggested as an alternative solution is pushing the operations of the application layer into the DBMS repositories. After a classification of operations available in GIS packages, the paper examines the features of a DBMS repository for the application domain of GISs.

Key words
GIS architectures, DBMS repositories, GIS operations.

1. Introduction

Traditional database management systems (DBMSs) have only dealt with alphanumeric domains. These systems have proved not to be suitable for non-standard applications, such as computer-aided design (CAD), geographic information systems (GISs) and image databases, which are characterized by the existence of more complex domains.

Much attention has been devoted in the past to support classes of applications that are not well served by conventional relational systems. Focusing on the application domain of GISs, four architectural approaches have been adopted generally, to implement commercial or prototype systems and satisfy the urgent needs for operational spatial data handling. These approaches are [3,12,16]:

- *Single conventional DBMS:* In this approach, both spatial and non-spatial data are represented in tabular form in the pure relational model. Operators needed to define and manipulate spatial entities are contained in an application layer which is built on the top of the conventional DBMS.

- *Partial conventional DBMS:* In this approach, a conventional DBMS is used to represent thematic information associated to spatial entities, while a separate subsystem is used to handle spatial data. Operators needed to define and manipulate spatial entities are contained in an application layer which provides a uniform interface to both subsystems.
- *Extended conventional DBMS:* In this approach, a conventional DBMS is modified to also support GIS application domains. This is accomplished by adding new constructs to a conventional DBMS so as to enhance its modeling power and provide better support for spatial applications. These new constructs include support for abstract data types (ADTs), procedural fields, complex objects, composite attributes, and so on. Additional operators needed to define and manipulate spatial entities are contained in an application layer which lies on top of the extended DBMS.
- *Object-Oriented DBMS:* In this approach, the object-oriented model is used as a basis for the application domain of GIS. The concepts of the underlying model (e.g., classes, inheritance, encapsulation, types, methods) are very convenient. Any additional operators needed to define and manipulate spatial entities are contained in an application layer which lies on top of the object-oriented DBMS.

The role of the *application layer* in all approaches above is to supplement the set of capabilities offered by the underlying system architectures, so that the operational needs for spatial data handling are satisfied. In other words, the application layer provides the set of GIS operations which are not available in the underlying DBMSs. Obviously, this set varies from one software package to another and heavily depends upon the system architecture. The execution of operations within the application layer is frequently accompanied by up-/down-loading of voluminous data[1] from/to the database. This is an expensive task which should be avoided in a production environment. What is usually suggested as an alternative solution is pushing the operations of the application layer into the DBMS repository.

This study examines the shrinkage of the application layer and the generation of a DBMS repository for the application domain of GISs. The discussion is organized as follows. Section 2 presents a classification of operations available in GISs, while Section 3 shows a simplified example of a sequence of GIS operations to support a real-world problem. Section 4 presents the design aspects for the DBMS repository. These involve the design of both the structure of the database and operations available in the DBMS. Sections 5 and 6 demonstrate the features of an extensible relational DBMS repository and an object-oriented DBMS repository for the application domain of GISs. Finally, Section 7 concludes the discussion.

---

[1] Spatial data are usually voluminous.

2. Classification of GIS operations

The operations available in geographic information systems (GIS) vary from one system to another. However, their fundamental capabilities can be expressed in terms of four types of operations [15]:

- *Programming operations:* They consist of a number of routines in the operating system level, such as supervise and direct the system operations and control the communication with peripheral devices connected to the computer.
- *Data preparation operations:* They encompass a variety of methods for capturing data from different sources (e.g., digital or paper maps, land measurements), processing and storing them appropriately in the database.
- *Data presentation operations:* They encompass a variety of methods for presentation of data, such as drawing maps, drafting charts, generating reports, and so on.
- *Data interpretation operations:* These operations transform data into information and as such they comprise the heart of any geographic information system. Consequently, the discussion that follows focuses on them.

Operations for data interpretation can be viewed as dealing with a hierarchy of data [12,15]. At the highest level, there is a library of *maps* (more commonly referred to as *layers*), all of which are in registration (i.e., they have a common coordinate system). Each layer is partitioned into *zones* (regions), where the zones are sets of locations with a common *attribute value*. Examples of layers are the land-use layer, which is divided into land-use zones (e.g., wetland, river, desert, city, park and agricultural zones) and the road network layer, which contains the roads that pass through the portion of space that is covered by the layer.

Data interpretation operations available in GISs characterize [3,4,12,15]:

- *individual locations,*
- *locations within neighborhoods,* and
- *locations within zones,*

and constitute respectively the three classes of operations, i.e., *local, focal* and *zonal* operations. Notice that all data interpretation is done in a layer-by-layer basis. That is, each operation accepts one or more existing layers as input (the operants) and generates a new layer as output (the product).

The first class of data-interpreting operations (local operations) includes those that compute a new value for each location on a layer as a function of existing data explicitly associated with that location. The data to be processed by these operations may include the zonal values associated with each location on one or more layers. *Local operations* include:

- *Search operations*, i.e., retrieval of information associated with individual locations on a layer.
- *Classification and recoding operations*, i.e., assignment of new attribute values to individual locations on a layer.
- *Generalization operations*, i.e., reduction of detail associated with individual locations on a layer.
- *Overlay operations*, i.e., assignment of new attribute values to individual locations resulting from the combination of two or more layers.

Focal operations compute new values for every location as a function of its *neighborhood*. A neighborhood is defined as any set of one or more locations that bear a specified distance and/or directional relationship to a particular location, the *neighborhood focus*. *Focal operations* include:

- *Search operations*, i.e., retrieval of information characterizing the immediate or extended vicinity (the region of interest) of individual locations on a layer.
- *Proximity operations*, i.e., assignment of new attribute values to individual locations on a layer, which depict their distance or direction in a neighborhood with respect to the neighborhood focus.
- *Interpolation operations*, i.e., assignment of new attribute values to individual locations on a layer derived by averaging sets of two or more target values associated to selected locations in their immediate or extended vicinity.
- *Surfacial operations*, i.e., assignment of new attribute values to individual locations on a layer indicating their surfacial characteristics (slope, aspect, volume, etc.).
- *Connectivity operations*, i.e., assignment of new attribute values to individual locations on a layer derived from a running total of the results being retained in a quantitative or qualitative step-by-step fashion and considering the values associated to locations in the immediate or extended vicinity (optimum path finding, etc.).

The third and final class of data-interpreting operations (zonal operations) includes those that compute a new value for each location as a function of existing values associated with a zone containing that location. *Zonal operations* include:

- *Search operations*, i.e., retrieval of information characterizing individual locations on a layer that coincide with the zones of another layer.
- *Measurement operations*, i.e., assignment of new attribute values to individual locations on a layer that correspond to a measurement (e.g., area, length) characterizing their zones.

3. Site selection based on a sequence of GIS operations

The purpose of this section is to present a sequence of data-interpretation operations which may compose one or more procedures[2] to accomplish the task of *site selection* for a *residential housing development*. The basic approach to this is to create a set of *constraints*, which restrict the planned activity, and a set of *opportunities*, which are conducive to the activity. The combination of the two is considered in order to find the best locations.
In the simplified situation that follows the set of constraints and opportunities consists of[3]:

- vacant area (i.e., no development),
- dry land,
- level and smooth site (e.g., slope < 10%),
- nearness to the existing road network, and
- south-facing slope.

In addition all candidate sites should have an adequate size to satisfy the needs of the planning activity (e.g., between 1 and 1.5 sq km).

The whole task requires as input three layers of the region under examination:

- *hypsography* layer: the three-dimensional surface of the region (altitude values),
- *development* layer: it depicts the existing infrastructure of the region (e.g., roads, buildings, etc.), and
- *moisture* layer: it depicts the soil moisture of the region (e.g., lakes, wet-lands, dry-lands, etc.).

The procedure of site selection, based on the sets of constraints and opportunities determined above, may consist of the following sequence of operations[4]:

1. *Vacant areas:* A new layer of vacant areas is produced from the layer of development by classifying, generalizing and finally performing a selective search on the result.
    - *development-classes* = Local(classification) of *development*
    - *vacant-developed* = Local(generalization) of *development-classes*
    - *vacant* = Local(search) of *vacant-developed*
2. *Dry lands:* A new layer of dry lands is produced from the layer of moisture by classifying, reducing detail and performing a selective search on the result.
    - *moisture-classes* = Local(classification) of *moisture*

---

[2] A *procedure* is any finite sequence of one or more operations that are applied to meaningful data with deliberate intent [15].
[3] A wider set could be taken into account, but this subset is enough to illustrate some basic data-interpreting operations available in GISs.
[4] The syntax adopted for the operations is:
   new-layer = Operation-class(operation-subclass) of *existing-layer* and ...

- *dry-wet* = Local(generalization) of *moisture-classes*
- *dry* = Local(search) of *dry-wet*

3. *Level sites:* A new layer of level and smooth sites is produced from the layer of hypsography by computing, classifying, generalizing and finally performing a selective search on the result.
   - *slope* = Focal(surfacial) of *hypsography*
   - *slope-classes* = Local(classification) of *slope*
   - *level-steep* = Local(generalization) of *slope-classes*
   - *level* = Local(search) of *level-steep*

4. *Accessible areas:* A new layer of accessible sites by the existing road network is produced implicitly from the layer of development by highlighting the road network, computing, classifying, generalizing and finally performing a selective search on the proximities.
   - *roads* = Local(search) of *development*
   - *road-proximity* = Focal(Proximity) of *roads*
   - *road-proximity-classes* = Local(classification) of *road-proximity*
   - *accessible-inaccessible* = Local(generalization) of *road-proximity-classes*
   - *accessible* = Local(search) of *accessible-inaccessible*

5. *South-facing areas:* A new layer of south-facing areas is produced from the layer of hypsography by computing, classifying, generalizing and finally performing a selective search on the aspects.
   - *aspect* = Focal(surfacial) of *hypsography*
   - *aspect-classes* = Local(classification) of *aspect*
   - *south-north* = Local(generalization) of *aspect-classes*
   - *south* = Local(search) of *south-north*

6. *Good-sites:* A new layer of sites that satisfy the set of constraints and opportunities is produced by the successive overlay of layers produced in the previous steps. Finally, good sites are highlighted by performing a selective search on the result.
   - *vacant-dry* = Local(overlay) of *vacant* and *dry*
   - *vacant-dry-level* = Local(overlay) of *level* and *vacant-dry*
   - *vacant-dry-level-accessible* = Local(overlay) of *accessible* and *vacant-dry-level*
   - *vacant-dry-level-accessible-south* = Local(overlay) of *south* and *vacant-dry-level-accessible*
   - *good-sites* = Local(search) of *vacant-dry-level-accessible-south*

7. *Candidate sites:* A new layer of sites that satisfy the set of constraints and opportunities and have adequate size is produced from the layer of good sites by measuring the sizes of zones and highlighting those that are within the predefined size interval.
   - *good-sites-size* = Zonal(measurement) of *good-sites*
   - *candidate-sites* = Local(search) of *good-sites-size*

4. Design of a DBMS repository for GISs

Depending on the underlying architecture (Section 1) commercial and prototype GISs support the execution of data-interpreting operations either entirely in the application layer or partially in the DBMS repositories and partially in the application layer. In general, the execution of operations within the application layer is accompanied by up-/down-loading of massive data from/to the database and should be avoided in order to ameliorate the response time of the system. The shrinkage of the application layer can be accomplished by enriching the DBMS repository with GIS operations.

In fact, this is not applicable in the approaches of *single conventional DBMS* and *partial conventional DBMS*. As for the former, the relational model is not rich or powerful enough to model the structural complexities of most types of spatial data, and relational algebra has not the expressive power to support data-interpreting operations. As for the latter, provided that spatial and non-spatial data reside in separate databases, processing is mostly performed in the application layer. However, the shrinkage of the application layer can be accomplished when the approaches of *extended conventional DBMS* (Section 5) and *object-oriented DBMS* (Section 6) are adopted.

This Section aims at presenting the design aspects of a DBMS repository for the application domain of GISs. This task involves the design of both the database and DBMS components.

4.1 The database component

A database is a collection of *entities*. Each entity is associated with a number of *attributes* describing its characteristics. Entities with common attributes compose a *class* of entities. The links among entities are described through *relationships*.

For the application domain of GISs, the *zones* (e.g., a city, a woodland area, a road network, etc.) constitute the entities in the database. A zone is a set of individual locations with common attribute values. Zone attributes have a *spatial* and a *thematic* component. The spatial component specifies the zone location with respect to a given coordinate system (e.g., [long., lat.] of Athens), while the thematic component describes the qualitative or quantitative properties associated with the zone (e.g., archaeological interest and population of Athens).

From the literature two main concepts for representing the spatial component of a zone are known: the *raster* and the *vector* models. In the raster model each zone is represented by the set of individual locations which constitute vertices of a regular grid in space of a fixed resolution and fall within the zone. In the vector model the zone is represented by the set of individual locations which describe its outline[5]. The distance between successive locations defines the resolution in representation. The thematic component of a zone is represented by a set of alphanumeric data types.

Each collection of zones with common attributes composes a class of entities, such as a layer (e.g., land-use layer). The links among zones describe their relationships. The relationships may be spatial (e.g., distance between two cities or intersection of two highways) or non-spatial (e.g., the capital city of a province).

4.2 The DBMS component

A Database Management System (DBMS) is a computerized system for managing a database. Conventional DBMSs provide some analytical operations such as simple statistics.

A DBMS repository for the application domain of GISs should provide additional operations, such as those presented in Section 2. Specifically, the following categories of operations for data-interpretation should be available:

- *Selection operations:* They perform a selective search on zones based on their spatial and non-spatial attribute values (i.e., local search).
- *Combination operations:* They compute the relationships (intersection, union, relative position, etc.) between a pair of zones or layers (i.e., overlay, focal and zonal search).
- *Analytical operations:* They process further selections and combinations of existing sets of zones (i.e. classification, generalization, proximity, interpolation, surfacial, connectivity, measurement).

The presence of *indices* over both the spatial and non-spatial attributes is required in order to facilitate and speed up the execution of operations.

5. An extensible relational DBMS repository for GISs

This Section briefly presents the features of an extensible relational DBMS repository for the application domain of GISs. The relational model [6] represents the data in a database as a collection of relations (tables), which consist of rows and columns. Each row (tuple) in a relation is a collection of related data values. These values can be interpreted as a fact describing an entity or a relationship instance. A class of entities can be viewed as a relation in a relational DBMS. The records of a relation represent the entities of a class, while the columns accommodate the attribute values characterizing those entities. All attribute values are defined over a set of domains, each of which has one or another form of an alphanumeric data type. The relational algebra provides a collection of operations to manipulate

---

[5] In 2-D space a point feature (e.g., a village) is represented by a single individual location; a linear feature (e.g., a highway) is represented by a sequence of individual locations which trace its axis; and a polygonal feature (e.g., a forest) is represented by a sequence of individual locations which trace its boundary.

entire relations. These operations are used to select tuples from individual relations and to combine related tuples from several relations. The result of each operation is a new relation, which can be further manipulated by the relational algebra operations. The relational algebra operations are divided into two groups. One group includes set of operations from the mathematical set theory, (e.g., union, intersection, difference and cartesian product), while the other group consists of operations developed specifically for relational databases (e.g., select, project, join). Several index structures for files (e.g., hashing technique, B-tree variations) are adopted in order to speed up the execution of these operations. In an extensible relational DBMS, like Postgres [14] and Starburst [8], new data types (Abstract Data Types), operations and specialized indices can be incorporated into the system.

In an extensible relational DBMS a zone can be viewed as a tuple in one or more relations, which compose classes of zones (e.g., layers). Attribute values are of alphanumeric data types for thematic data (e.g., city population or vegetation type) or of user defined ADTs (e.g., point, line, polygon boundary) for geometric data (e.g., road or forest geometry). The minimum set of operations provided by relational algebra is extended to support GIS operations. Spatial index structures, such as grid-files, quadtrees and R-trees [11], are employed, along with standard alphanumeric indices, to speed up the execution of operations.

Several prototype systems have been built on top of extensible relational DBMS to support spatial applications. Haas and Cody [7] describe how Starburst can be extended with user defined ADTs and operations to support the needs of two real-world applications and show how some sample queries may be processed efficiently. The Geo++ system [10] is built on top of Postgres. The system adopts the geometric capabilities of Postgres (R-tree index structure; primitive data types: point, line segment, path and box) and makes use of additional data types, as well as operators to incorporate the functionalities of commercial GISs. Aref and Samet [2] describe the architecture of SAND system. In SAND spatial and non-spatial data are linked bidirectionally. Spatial attributes are stored in spatial data structures. A spatial data structure is chosen based on the attribute's spatial data type (e.g., point, line, polygon, etc.), and serves as an index for spatial objects and an environment for the execution of spatially related operations. Non-spatial attributes are maintained in database relations. Two logical links are maintained between spatial and non-spatial attributes of an entity: forward and backward links. The linked instances and the links form what is termed a spatial relation.

6. An object-oriented DBMS repository for GISs

This Section briefly presents the features of an object-oriented DBMS repository for the application domain of GISs. The object-oriented model [6] represents the data in a database as a collection of objects. An object represents a physical entity, a concept, an event or some aspect of interest to the database application and is characterized by its identity, state and behavior. The object identity is maintained via an object identifier. The state is stored in data attributes, which can also be objects. The behavior is encoded in subroutines and functions, called methods, which are encapsulated into objects. Objects with common state and behavior form classes of objects, which are organized into type hierarchies. A subclass inherits both data attributes and methods from its superclass (inheritance). The execution of operations is supported by appropriate index structures for objects.

In an object-oriented DBMS repository for GISs, a zone can be viewed as an object, with a unique identifier, a set of spatial and non-spatial attribute values, and a collection of operations describing its behavior. Zones with common data attributes and behavior (e.g., a layer) can be viewed as a class of objects.

A few attempts to build prototype systems on top of object-oriented DBMS to support spatial applications have been made in the past. Milne et al. [9] examine the construction of a system on top of the object-oriented DBMS, ONTOS. ONTOS provides a persistent class library, written in C++, and allows the construction of new persistent classes using C++. The authors extent ONTOS with spatial object classes, a graphical user interface, graphic display module and geographical database module. They also report an impressive performance gain over a proprietary relational DBMS (Oracle) and an extended relational DBMS with special GIS capabilities (SIRO-DBMS [1]). A couple of attempts have been made to build an object-oriented GIS on top of the object-oriented DBMS O_2. David et al. [5] have implemented GeO$_2$ system. The conceptual data model of GeO$_2$ is made of a semantic data model, which is an extension of the Entity-Relationship data model [6], and a localization data model, which is defined by an ADT. Three levels of data structures (spaghetti, network and map structure) are provided to support the efficient and appropriate storage of geographic coordinates and relationships between entities. Scholl and Voisard [13] implemented a system in which the database consists of a set of maps, where a map is a relation that has at least one spatial attribute. Spatial attributes are represented by ADTs (e.g., points, lines, polygons, etc.). As for map operations, they are implemented as methods on map objects.

7. Conclusion

The design of a DBMS repository for the application domain of GISs is a difficult task due to peculiarities of spatial data and operations involved. The contribution of the paper can be summarized as follows:

- After a classification of operations available in current GIS packages, it is shown how a sequence of them may compose one or more procedures to support the spatial decision-making process.
- The fundamental design aspects of a DBMS repository for the application domain of GISs are presented.

- The features of both an extensible relational and an object-oriented DBMS repositories for spatial applications are given briefly and some representative prototype systems developed in the past are summarized.

Future research in the area includes:

- The design and implementation of an object-oriented DBMS repository for the application domain of GISs.
- The incorporation of operations available in commercial GIS packages into the DBMS repository and the shrinkage of the application layer.
- The development of efficient optimization strategies for query processing and the evaluation of the system performance versus other commercial and prototype packages.

References
[1] D.J. Abel. A database toolkit for Geographical Information Systems. International Journal of Geographical Information Systems, 3(2):103–115, 1989.
[2] W.G. Aref and H. Samet. Extending a DBMS with spatial operations. In O. Guenther and H. J. Schek, editors, Advances in Spatial Databases (Proceedings of the 2nd Symposium on Large Spatial Databases - SSD'91), pages 299–318. Spinger-Verlag, LNCS-525, Zurich, Switzerland, 1991.
[3] S. Aronoff. Geographic Information Systems: A Management Perspective. WDL Publications, 1989.
[4] J.K. Berry. Fundamental operations in computer-assisted map analysis. International Journal of Geographical Information Systems, 1(2):119–136, 1987.
[5] R. David, I. Raynal, G. Schorter, and V. Mansart. GeO$_2$: Why objects in a geographical DBMS. In D. Abel and B.C. Ooi, editors, Advances in Spatial Databases (Proceedings of the 3rd Symposium on Large Spatial Databases - SSD'93), pages 264–276. Spinger-Verlag, LNCS-692, Singapore, 1993.
[6] R. Elmasri and S.B. Navathe. Fundamentals of Database Systems. Benjamin-Cummings, 1989.
[7] L.M. Haas and W.F. Cody. Exploiting extensible DBMS in integrated Geographic Information Systems. In O. Guenther and H.J. Schek, editors, Advances in Spatial Databases (Proceedings of the 2nd Symposium on Large Spatial Databases - SSD'91), pages 423–449. Spinger-Verlag, LNCS-525, Zurich, Switzerland, 1991.
[8] L.M. Haas et al. Starburst mid-flight: As the dust clears. IEEE Transactions on Knowledge and Data Engineering, 2(1):143–160, 1990.
[9] P. Milne, S. Milton, and J.L. Smith. Geographical object-oriented databases: A case study. International Journal of Geographical Information Systems, 7(1):39–55, 1993.
[10] P. van Oosterom and T. Vijlbrief. Building a GIS on top of the open DBMS Postgres. In Proceedings of the 2nd European Conference on Geographical Information Systems (EGIS'91), 1991.
[11] H. Samet. The Design and Analysis of Spatial Data Structures. Addison-Wesley, 1989.
[12] H. Samet and W.G. Aref. Spatial data models and query processing. In W. Kim, editor, Modern Database Systems, pages 339–360. ACM Press, 1995.
[13] M. Scholl and A. Voisard. Object-oriented database systems for geographic applications: An example with O$_2$. In G. Gambosi, M. Scholl, and H. W. Six, editors, Geographic Database Management Systems, pages 103–137. Springer-Verlag, 1992.
[14] M. Stonebraker and L. Rowe. The design of Postgres. In Proceedings of the ACM-SIGMOD Conference, pages 340–355, Washington, D. C., 1986.
[15] C.D. Tomlin. Geographic Information Systems and Cartographic Modeling. Prentice Hall, 1990.
[16] T. Vijlbrief and P. van Oosterom. The GEO++ System: An extensible GIS. In Proceedings of the 5th International Symposium on Spatial Data Handling, pages 40–50, Charleston, South Carolina, 1992.

Contact address
Emmanuel Stefanakis and Timos Sellis
Department of Electrical and Computer Engineering
National Technical University of Athens
Zographou
Athens
Greece 15773
Phone: +30-1-7721402,
Fax: +30-1-7722459
E-mail: {stefanak,timos}@cs.ntua.gr

BULK INSERTION IN DYNAMIC R-TREES

Ibrahim Kamel[1], Mohamed Khalil[2] and Vram Kouramajian[3]

[1]Matsushita Information Technology Laboratory Panasonic Technologies Inc.
Princeton, USA
[2]Computer Science Deptartment, Wichita State University Kansas
Kansas, USA
[3]Knight Ridder Information Inc., El Camino Real
Mountain View, USA

Abstract
The R-tree variants that allow the update e.g., insertion and deletion operations are called Dynamic R-trees. On the other hand, if the R-tree does not allow update it is called Static R-trees. In this paper we introduce two variants of the insertion algorithms for Dynamic R-trees. Our proposed algorithms reduce the insertion cost by making use of the fact that new object arrives in large groups. In one proposal objects are sorted according to their locational proximity and then are inserted. In the second proposal the objects are grouped into partially full nodes. The new nodes are then inserted in the tree using an algorithm that is similar to the traditional insertion algorithm.

We implemented the new algorithms and compared them with the traditional insertion algorithms. We achieve more than two orders of magnitude savings in the insertion cost. At the same time the new insertion algorithms produce R-trees that have search performance similar to the one produced by the traditional insertion algorithms.

Key words
Spatial indexing, R-trees.

1. Introduction

One of the requirements for the database management systems (DBMSs) of the near future is the ability to handle spatial data. Spatial data arise in many applications including Cartography [Whi81]; ComputerAided Design CAD [OHM+84]; computer vision and robotics [BB82]; traditional databases where a record with k attributes corresponds to a point in a k-d space temporal databases where time can be considered as one more dimension; scientific databases with spatial-temporal data such as the ones in the Grand Challenge applications [Gra92], etc. We focus on the R-tree [Gut84] family of methods, which contains some of the most efficient methods that support range queries. The advantage of the R-tree-based methods over the methods that use linear quadtrees and z-ordering is that R-trees treat the data objects as a whole while quad-tree based methods typically divide objects into quadtree blocks, increasing the number of items to be stored.

All the above applications require the ability to update the database from time to time. One of the important operations is the insertion operation where new objects need to be added to the existing index (as opposed to the packed R-trees [RL85, KF93] where update is not allowed.) New objects usually arrive in large sets, rather than one object at a time. For example, the TIGER files from the US Bureau of Census, which store the road map of the U.S. To add data objects that represent only one county, e.g. MGCounty, MD, you need to insert 40,000 line objects. These objects will be available together to the database administrator. All the insertion algorithms that are defined for the R-tree data structures insert data objects one at a time. The goal of all insertion algorithms is to insert the new object in the best location that minimizes the search cost.

To add 40,000 new objects to an existing R-tree index requires executing the insertion algorithm 40,000 time (one for each object). The same node might be retrieved updated and stored back to the disk several (even hundred) times during this process. This is both inefficient and impractical. In this paper, we propose variants of the insertion algorithm that insert the new objects in groups to decrease the insertion cost. The challenge is to produce after the insertion an R-tree that is as efficient as the one produced by the traditional insertion algorithm.

The paper is organized as follows. Section 2 gives a brief description of the R-tree and its variants. Section 3 describes the proposed insertion algorithms. Section 4 presents our experimental results. Section 5 gives the conclusions and directions for future research.

2. R-trees

One of the most promising approaches in the last class is the R-tree [Gut84]: compared to the linear quadtrees the R-trees do not need to divide the spatial objects into several pieces (quadtree blocks.) The R-tree is an extension of the B-tree for multidimensional objects. A geometric object is represented by its minimum bounding rectangle MBR. Nonleaf nodes contain entries of the form (*R, ptr*) where *ptr* is a pointer to a child node in the R-tree. R is the MBR that covers all rectangles in the child node. Leaf nodes contain entries of the form (objid, R) where objid is a pointer to the object description and R is the MBR of the object. The main innovation in the R-tree is that father nodes are allowed to overlap. This way the R-tree can guarantee at least 50% space utilization and remain balanced.

Figure 1 illustrates data rectangles (in black) organized in an R-tree with fanout 3. Figure 2 shows the file structure for the same R-tree where nodes correspond to disk pages. In the rest of this paper the term node and the term page will be used interchangeably.

R-tree variants can be divided into two groups. The first group contains the structures which allow update insertion and deletion in real time. Insertion and deletion can take place at any time. We call them *Dynamic R-trees*. The second group contains the structures which assume that the whole data set is known before creating the R-tree. Once the R-tree is built no insertion nor deletion is allowed. We call these structure *Static R-trees*.

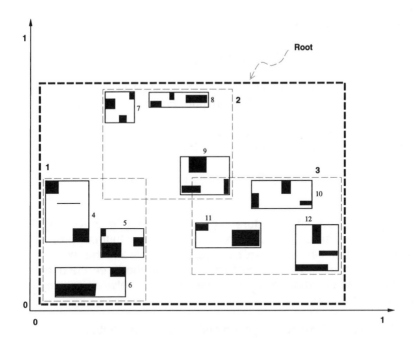

Figure 1 Data (dark rectangles) organized in an R-tree with fanout=3.

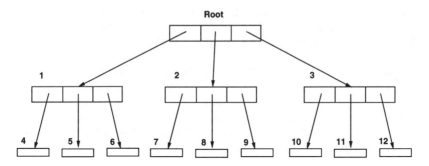

Figure 2 The file structure for the R-tree of the previous figure (fanout=3).

2.1 Dynamic R-trees

Dynamic R-trees grow by insertion and shrink by deletion in the same manner the B-tree does. When a node become full it splits into 2 nodes each of them about half full. This way dynamic R-trees guarantee 50% minimum space utilization. It has been shown in [KF93, PSTW93, FK94] that the search response time increases with increasing the area and perimeter of the R-tree nodes. Different variants propose different insertion and/or split algorithms to minimize the area and perimeter of the resulting nodes. The following is an outline for the insertion algorithm. The new object is represented by its minimum bounding rectangle *w*.

Algorithm Insert-node (node *Root*, rect *w*):
I1. For each entry in the node:
       find the cost *Fn* to accommodate the rectangle *w*
I2. Find the (*R*, *ptr*) that gives the minimum *Fn* and
       invokes Insert(node *R*, rect *w*)
I3. Repeat I1 and I2 until a leaf is reached.
       Insert *w* in the appropriate order inside the leaf.

The value of *Fn* increases with increasing the area and/or the perimeter of the node.

The original R-tree data structure is proposed by Guttman [Gut84]. The cost function *Fn* minimizes only the area of the node. Guttman proposed three splitting algorithms the *linear split*, the *quadratic split* and the *exponential split*. Their names come from their complexity; among the three the quadratic split algorithm is the one that achieves the best trade-of between splitting time and search performance.

The *R**-tree [BKSS90] of Beckmann etal. differs from Guttman R-tree in that it takes both area and perimeter into account. The *R**-tree also employs the concept of *forced re-insert*. When a node overflows some of its children are carefully chosen; they are deleted and reinserted, usually resulting in a R-tree with better structure. The *R**-tree produces smaller nodes and gives better performance than Guttman R-tree "quadratic split".

Kamel and Faloutsos proposed the Hilbert R-tree [KF94] which behaves like a B-tree during the insertion and R-tree during the search. Objects, represented by their MBR, are inserted according to the Hilbert value of their centers. This makes the tree construction simple and fast. Another advantage of the Hilbert R-tree is that it can increase the space utilization as desired at the expense of the insertion algorithm.

2.2 Static R-trees

Other structures are proposed for static environment where there is no or rare update [RL85, KF93]. Packed R-trees achieve almost 100% space utilization (last node might be partially full.) The volume of data in these databases is expected to be enormous, in which case it is crucial to minimize the space overhead of the index. Static data appear in several applications. For example, in cartographic databases, insertions and deletions are rare; databases with census results are static; the same is true for databases that are published on CD-ROMs; databases with spatio-temporal meteorological and environmental data are seldom modified too. The search algorithm is similar to the original algorithm by Guttman.

3. The proposed insertion algorithms

In this section we propose different variants of the R-tree insertion algorithm that make use of the fact that several records are inserted at the same time. The cost of inserting a rectangle, in terms of the number of disk accesses, equal to the number of nodes (i.e., pages) retrieved during the insertion[1]. Inserting two rectangles that are nearby in the Euclidean space might retrieve the same set of nodes during the insertion. If these two rectangles are inserted separately the touched nodes are retrieved twice. We present variants of the insertion algorithm that insert nearby rectangles together. Mainly we make use of the Euclidean proximity in two different ways. In one scheme the data objects are sorted according to their spatial proximity and then inserted in the R-tree one at a time according to the sorting order. In the second scheme, which is useful when we have a large set of objects, the objects are grouped to form nodes that are similar to the R-tree leaf nodes and then the nodes are inserted one at a time. It has been shown in [KF93] that the Hilbert order is one of the best space filling curves that preserves the spatial proximity among the different objects. We use space filling curves, and specially, the Hilbert curve [Bia69] to order the objects.

Next we provide a brief introduction to the Hilbert curve. The basic Hilbert curve on a 2x2 grid denoted by H_1, is shown in Figure 3. To derive a curve of order i each vertex of the basic curve is replaced by the curve of order $i - 1$, which may be appropriately rotated and/or reflected. Figure 3 also shows the Hilbert curves of order 2 and 3. When the order of the curve tends to infinity, the resulting curve is a *fractal*, with a fractal dimension of 2 [Man77]. The Hilbert curve can be generalized for higher dimensional space. Algorithms to draw the two-dimensional curve of a given order can be found in [Gri86], [Jag90]. An algorithm for higher dimensional space is in [Bia69].

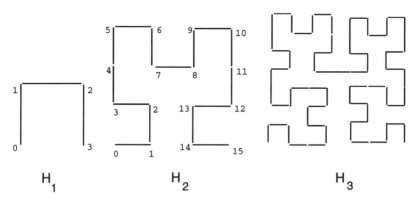

Figure 3 Hilbert Curves of order 1, 2 and 3.

The path of a space filling curve imposes a linear ordering on the grid points, which may be calculated by starting at one end of the curve and following the path to the other end. Figure 3 shows one such ordering for a 4x4 grid (see

---

[1] If no split is required.

curve H_2). For example, the point $(0,0)$ on the H_2 curve has a Hilbert value of 0, while the point $(1,1)$ has a Hilbert value of 2.

In the following we describe two insertion schemes.

Sort and insert Individual object ("SI"): Our first proposal is to sort the data rectangles according to the Hilbert order [KF93].

We define the function $H(o)$ as follow:

$$H(o) = \text{The Hilbert value of the center of the MBR of the object } o$$

The new objects are sorted according the value of the function $H(o_i)$. Then insert the rectangles, one at a time, in order, in the R-tree. The original R-tree insertion algorithm is used. We assume that there are some memory buffers available to keep the nodes that are retrieved from the disk. When the following rectangle is inserted the nodes from the previous insertion would be in the main memory buffer. Since objects that are nearby in the insertion order are expected to be nearby in the Euclidean space, some or all of the nodes required to insert the new rectangle might be already in the buffer.

Sort and insert Nodes ("SN"): More saving in the insertion cost can be achieved if we organize the rectangles into nodes. As in SI, the data objects are sorted according to the Hilbert order of the center of their MBR. Then each set of objects are grouped to form a new leaf node. The resulting nodes are inserted, one at a time, in the R-tree and added in the appropriate place in the leave level. Each node is treated as a big rectangle represented by the Minimum Bounding Rectangle (MBR) of all the objects in the node and inserted in the R-tree. The outline of the insertion (SN) insertion algorithm is as follow:

Algorithm Bulk-Insertion:
B1. For each object o calculate $H(o)$.
B2. Sort the new objects on $H(o)$.
B3. Using the sorted object construct
       leaf nodes that are partially full.
B4. For each new leaf node n:
       calculate the MBR M of the node.
B5. Starting from the root node:
       find the (R, ptr) that gives the minimum Fn and
       invoke Insert(node R, rect M)
B6. Repeat B1 and B2 until the last nonleaf node is reached.
       Let R points to n.

It is clear that the saving in the insertion cost is significant in this case. Because we execute the insertion algorithm once to insert a set of objects. This also avoids several split operations. However, the resulting R-tree might not be as good as the one which is built by inserting the objects one at a time. The better the tree is packed the lower the query response time is. The quality of the resulting R-tree structure depends mainly on the heuristic that is used to pack the objects into nodes. In Section 4 we show that our methods result in an R-tree that is as good as the one which is built by inserting the objects one at a time.

In many cases the new objects are spatially clustered, e.g. in the TIGER system, that stores the map of all U.S. roads produced by the U.S. Bureau of Census. When a set of objects that constitutes a township, a county or a city is added the objects are expected to be close to each other. But this is not always the case the new objects might be scattered through the space. The worst case for the distribution of the new objects is when the new objects are uniformly distributed across the whole space. Notice that this is the distribution used in our experiments in Section 4.

The algorithm (SN) is more appropriate for applications where new objects arrive in groups of large number, e.g., at least an order of magnitude of the fanout (= node capacity). But if the new objects arrives in small sets, SN might adversely affect, after several insertions, the performance of the search operation. This is because we force a set of objects that are far from each other to be in the same node. In this case, the algorithm (SI) is more appropriate. In fact (SI) and (SN) can be used interchangeably depending on the size of the new dataset.

4. Experimental results

To assess the merit of our proposed insertion algorithms we ran many simulation experiments. Recall that the proposed algorithms can be used with any Dynamic R-tree variant, the quadratic R-tree, R^*-tree or the Hilbert R-tree. In our experiments we used the R^*-tree code because it employs the most sophisticated insertion algorithms among the other variants. We compared the performance of the following insertion algorithms: (SI) which sorts the data then inserts individual rectangles (SN) which sorts the objects then inserts nodes and the original algorithm which inserts **individual** objects of the **unsorted** list of objects (UI).

In all the experiments we assume that there is an existing R-tree that contains rectangles uniformly distributed in the 2D space. The R-tree contains 345 nodes. Then an additional set of objects is inserted in the existing R-tree using the different algorithms described above. Each node occupies a 4KBytes page. The maximum and minimum fanout of the tree are 204 and 102 respectively.

The experiments are divided into two groups. The first set of experiments compares the performance of the resulting R-tree when each of the above algorithms is used. The second set of experiments compare the insertion cost of each algorithm.

4.1 The performance of the resulting R-trees

Before we claim that our insertion algorithms cost less we should show that the performance of the R-tree produced by the new insertion algorithms is satisfactory and the saving in the insertion cost is not at the expense of performance of the resulting R-tree. We performed several experiments to measure the query response time for the R-trees resulting from the three insertion algorithms described above. In each experiment we ask 100 square queries and calculate the average number of nodes retrieved. The query size is 0.01 of the search space. We assume that there is no main memory buffer, i.e., each node retrieval costs one disk access.

Recall that the R-tree, originally, contains 50,000 objects. The size of the datasets to be inserted varies from 10,000 to 200,000 objects. Table 1 shows that the performance of the final R-tree, after using algorithms (SI) and (SN), are as good as the one resulted from using the original insertion algorithm (UI). The SN algorithms results in an R-tree which is slightly inferior to the one produced by the (UI) algorithm. The reason is that (UI) algorithm optimizes the placement of every individual object to be inserted in the tree. On the other hand, (SN) applies similar optimization procedure to each group of objects.

size of the	Response time nodes/query		
inserted dataset	SN	SI	UI
10,000	2.41	2.39	2.37
50,000	3.29	3.20	3.08
100,000	4.36	4.42	4.25
200,000	6.38	6.44	6.43

Table 1 Performance comparison of the final R-tree after insertion.

4.2 Comparison of insertion cost

In this section we calculate the savings in the insertion cost that is achieved by algorithms (SI) and (SN) over the algorithms (UI). In each experiment, we insert a new set of objects into an existing R-tree, which contains 50,000 rectangles, and calculate the average insertion cost per object, in terms of the number of node touched. When a node is requested by the insertion algorithm, the main memory buffer is checked first. If the node is in the buffer the cost is considered zero, otherwise the node is read from the disk to the buffer, which costs one disk read. The FIFO buffer replacement strategy is used. When the buffer is full and a new node is read from the disk, the new node replaces the oldest node that exist in the memory buffer.

In Figure 4 we show how the insertion cost is affected with increasing the size of the buffer when the (SI) algorithm is used. We used real data from the TIGER system of the US Bureau of Census. This file consists of 39717 line segments representing the roads of Montgomery county in Maryland. Using the minimum bounding rectangles of the segments, we obtained 39717 rectangles. We refer to this dataset as the *'MGCounty'* dataset.

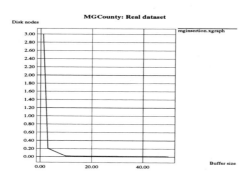

Figure 4 Effect of the buffer size on the insertion cost for the (SI) algorithm.

Next we present two sets of experiments to show the savings in the insertion cost achieved by our proposed algorithms. In one set of experiments, we inserted 10,000 new objects in an R-tree which already have 50,000 objects. Figure 5.a shows the insertion cost, in terms of the number of nodes that are read from the disk, for the (SI) and the (UI) algorithms. The buffer size is varied from 1 to 100 pages (4k bytes each.) Figure 5.b shows a similar experiment but the results are averaged over 50,000 objects. The insertion cost of the (SI) algorithms is significantly lower than that of the (UI) algorithm. For example, the (SI) algorithm achieves at least an order of magnitude saving in the insertion cost when 10 buffer pages or more are used. It is also interesting to notice that using only 3 buffer pages reduces the insertion cost to about half page/insertion.

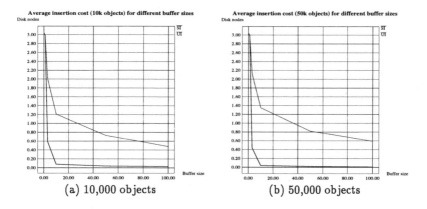

(a) 10,000 objects   (b) 50,000 objects

Figure 5 Average insertion cost in terms of nodes retrieved from the disk.

In another experiment, we calculated the insertion cost of (SN) algorithms. The buffer size is assumed to be zero in this experiment. All the nodes requested by the insertion algorithm are retrieved from the disk. Table 2 shows the insertion cost/object of the three algorithms (SN), (SI) and (UI). Notice that (SN) achieves more than two order of magnitude saving in the insertion cost over (UI) and (SI). Because we assume no buffer available in this experiment, the insertion cost of the (SI) algorithm and the (UI) algorithms are the same.

size of the	Insertion cost nodes/query		
inserted dataset	SN	SI	UI
10,000	0.015	3.06	3.04

Table 2 Insertion cost in terms of the number of nodes read from the disk per insertion, no buffer.

5. Conclusions and future works

We introduced two variants of the R-trees insertion algorithms to handle bulk insertion in Dynamic R-trees. In the first algorithm (SI) data objects are sorted according to the Hilbert value of the center of the objects MBR. Then objects are inserted in order. Since objects that are close in the Hilbert order are expected to be close in the native space, the nodes requested by the insertion algorithm are likely to be in the buffer (i.e., no disk access is required.)

The second algorithm (SN), organizes the new data objects into nodes that are 70% full and insert the nodes into the leaf level. Our experimental results showed that (SI) achieves up to 70% improvement in the insertion cost over the traditional algorithms when 10 pages of buffer or more are used, while (SN) achieves more than two orders of magnitude savings in the insertion cost.

Moreover, the new algorithms produces R-trees that are as efficient as the ones produced by the traditional insertion algorithm.

References
[BB82] D. Ballard and C. Brown. Computer Vision. Prentice Hall, 1982.
[Bia69] T. Bially. Space-fillling curves: Their generation and their application to bandwidth reduction. IEEE Trans. on Information Theory, IT-15(6):658-664, November 1969.
[BKSS90] N. Beckmann, H. Kriegel, R. Schneider and B. Seeger. The R*-tree an efficient and robust access method for points and rectangles. In Proc. of ACM SIGMOD, pages 322-331, Atlantic City, NJ, May 1990.
[FK94] C. Faloutsos and I. Kamel. Beyond uniformity and independence: Analysis of R-trees using the concept of fractal dimension. In Proc. of ACM on Principles of Databases (PODS), pages 4-13, Minneapolis, MN, May 1994.
[GAV94] D. Gavrila. R-tree Index Optimization. Advances in GIS Research, Taylor & Francis, 1994.
[Gra92] Grand challenges: High performance computing and communications, 1992. The FY 1992 U.S. Research and Development Program.

[Gri86] J. Griffiths. An algorithm for displaying a class of space filling curves. Software-Practice and Experience, 16(5):403-411, May 1986.

[Gut84] A. Guttman. R-trees: a dynamic index structure for spatial searching. In Proc. of ACM SIGMOD, pages 47-57, Boston, MA, June 1984.

[Jag90] H. Jagadish. Linear clustering of objects with multiple attributes. In Proc. of ACM SIGMOD Conf., pages 332-342, Atlantic City, NJ, May 1990.

[KF93] I. Kamel and C. Faloutsos. On packing R-trees. In Proc. of 2nd International Conference on Information and Knowledge Management(CIKM), pages 490-499, Arlington, VA, November 1993.

[KF94] I. Kamel and C. Faloutsos. Hilbert R-tree an improved R-tree using fractals In 20th International Conference on Very Large Data Bases (VLDB), pages 500-509, SANTIAGO, CHILE, September 1994.

[Man77] B. Mandelbrot. Fractal Geometry of Nature. W.H. Freeman, NY, 1977.

[OHM+84] J. K. Ousterhout, G. T. Hamachi, R. N. Mayo, W. S. Scott, and G. S. Taylor. Magic a vlsi layout system. In 21st Design Automation Conference, pages 152-159, Alburquerque, NM, June 1984.

[PSTW93] B. Pagel H. Six H. Toben and P. Widmayer. Towards an analysis of range query performance. In Proc. of ACM on Principles of Database Systems (PODS), pages 214-221, Washington, D.C. May 1993.

[RL85] N. Roussopoulos and D. Leifker. Direct spatial search on pictorial databases using packed R-trees. In Proc. of ACM SIGMOD, pages 17-31, Austin, TX, May 1985.

[SRF87] T. Sellis N. Roussopoulos and C. Faloutsos. The R+-tree: a dynamic index for multi-dimensional objects. In Proc. of VLDB Conf., pages 507-518, Brighton, England, September 1987.

[Whi81] M. White. N-Trees: Large Ordered Indexes for Multi-Dimensional Space. Application Mathematics Research Staff, Statistical Research Division, US Bureau of the Census, December 1981.

[YY94] J. Yu and V. Yong. Initial Loading of spatial objects into R*-tree. unpublished report, June 1994.

Contact address
Ibrahim Kamel[1], Mohamed Khalil[2] and Vram Kouramajian[3]
[1]Matsushita Information Technology Laboratory Panasonic Technologies Inc.
2 Research Way
Princeton NJ 08540-6628
USA
E-mail: ibrahim@research.panasonic.com
[2]Computer Science Deptartment
Wichita State University Kansas
Kansas
USA
E-mail:oprojs@cs.twsu.edu
[3]Knight Ridder Information Inc.
El Camino Real
Mountain View CA 94040
USA
E.mail: Vram@dnt.dialog.com

DEVELOPMENT OF A SHAPE FITTING TOOL

FOR SITE EVALUATION

Geoffrey B. Ehler[1], David J. Cowen[1] and Halkard E. Mackey, Jr.[2]

[1]Department of Geography and
Liberal Arts Computing Laboratory
University of South Carolina
Columbia, USA
[2]Savannah River Technology Center
Westinghouse Savannah River Company
Aiken, USA

Abstract
The objective of this research was to develop and implement a shape fitting measure within a geographic information system, specifically one that incorporates analytical modeling for site location planning. The procedure was incorporated into a graphical user interface that facilitates linkage between a vector based data browsing systems and full featured GIS. The procedure created binary matrices, which approximate the object's geometrical form. The system allows for different orientations of the shape. The system is fully implemented and its performance has been verified.

Key words
Shape analysis, system design, system evaluation, cartographic modeling, land use planning.

1. Introduction

Methods of designing spatial models for site planning were first discussed over 30 years. In his seminal paper McHarg (1969) mapped thematic site criteria onto Mylar transparencies and, when superimposed, was able to differentiate between acceptable and unacceptable zones. These same principles can be applied to the task of finding a suitable location for businesses, goods and services, utilities, and other facilities. Poorly made location decisions can often be attributed to several factors, most common of which is uninformed land use planning (O'Hare, et al., 1983). This location decision making process may be organized and structured utilizing recent advances in geographic information systems (GIS) theory, in particular that of spatial decision support systems (Densham, 1991).

The research reported describes the design and implementation of a user-friendly site selection system for use by research scientists and land-use planners at the Department of Energy's Savannah River Site in South Carolina. This system includes a function to allow the user to automatically search for potential locations that require specific shape criteria (e.g., a facility's footprint).

Over the past two years, the Environmental Impact Data Analysis and Retrieval System was developed as a cooperative effort between the University of South Carolina and Westinghouse Savannah River Company

(Cowen, et al., 1995). The system was developed in an effort to provide SRS researchers with the necessary tools to increase efficiency in their everyday geographic decision making. An important part of this system was the design and implementation of a user-friendly site selection system. The site selection process involves the implementation of several exclusionary and inclusionary criteria that specifically eliminate or retain certain types of geographic features based on regulations or mandates (WSRC, 1992a; WSRC, 1992b). A site's individual score or rating is calculated by summing criteria ratings found at each potential site location. A critical aspect of some site selection projects is the need to find specific acceptable areas within the site boundary that can accommodate a site of a specific size and shape. Therefore, it was proposed that a generic shape fitting analysis functions be incorporated into the system.

2. Review of shape analysis research

Davis (1986) maintained that the determination of an object's shape is an extremely difficult task to measure. Perhaps this is the reason that a great abundance of shape measures have been developed and have been met with varying degrees of success. While it is certain that different objects yield unique shape measurements, it has been shown that no single shape measurement can accurately distinguish one shape from another (Davis, 1986; Lee and Sallee, 1970).

2.1 Single-value measures of shape

Many of the earliest measures of shape include single valued calculations or some index comprised of two or more calculations (Davis, 1986; Moellering and Rayner, 1979). Some of the more successful measurements based on single value measures are those that measure an object's circularity, compactness, thinness, form, elongation, and curvature. As an example, FRAGSTATS, landscape ecology software for analyzing spatial patterns, is able to compute two types of shape indices based on area to perimeter metrics and fractal analysis (McGarigal and Marks, 1994). More recently, researchers in computer vision have developed a series of general purpose shape representation methods. For example Mokhtarian and Mackworth (1992) define a set of procedures for describing a closed planar curve at increasing levels of abstraction. They demonstrate that through an evolutionary process a geographical shape such as the boundary of Africa can systematically evolve into a simple curve such as an oval. Rosen (1993) extends the notion of codons to accomplish the same objective. The codons that he describes or based on a syntax for defining shapes based on local minima and maxima. In many ways this work is a modernization of the early line generalization work of Douglas and Pecker (1973).

2.2 Fourier measures of shape

Another approach for analyzing the shape of an object involves Fourier analysis. In this method, a series of measurements from the object's centroid to its perimeter are taken. Given the length of these line segments (radii) and to its perimeter are taken. Given the length of these line segments (radii) and their orientation, a series of polar coordinate pairs may be extrapolated for analysis involving Fourier methods (Davis, 1986; Moellering and Rayner, 1982). One drawback of Fourier shape analysis is that a line segment from the centroid to the perimeter may only intersect the perimeter once, therefore excluding accurate measurements of extremely convoluted shapes. Therefore, Fourier measurements are more appropriate for sinuous lines rather than objects such as buildings.

2.3 Image content queries

As reported by Niblack and Flickner (1993), a system has been developed recently by IBM that is able to perform a visual search of a digital image based on a graphic query defined by the user. An example of a graphic query is "find all images that contain a graphic object that looks like this" (and the user would sketch or digitize the objects shape). Methods by which a computer may search for visual content of an image include neural network technology (where the system learns patterns in an image) and feature-based methods that compute the image's visual properties (including shape). The shape measures that are calculated in the Query By Image Content (QBIC) system include area, circularity, eccentricity, major-axis direction, and a set of tangent angles around the object's perimeter (Niblack and Flickner, 1993).

2.4 Binary shape matrices

Shape metrics for feature recognition based on binary shape matrices were proposed by Flusser (1992). An object's shape matrix was based on the following algorithm:

1. Find the center of gravity $T = (x_t, y_t)$ of object G.
2. Find such point $M = (x_m, y_m)$ that $M \in G$ and

$$d(M,T) = \begin{array}{c} Max\ d(A,T) \\ \\ A \in G \end{array} \quad (A,T),\text{where: } d \text{ is Euclidean distance in } R2.$$

3. Construct the square with the center in T and with the size of the side $2 * d (M, T)$. Point M lies in the center of one side.
4. Divide the square into $n \times n$ subsquares.
5. Denote S_{kj} the subsquares of the constructed grid; k, j = 1, ... , n.
6. Define the $n \times n$ binary matrix B:

$$B_{kj} = \begin{cases} 1 \Leftrightarrow m(S_{kj} \cap G) \\ 0, otherwise \end{cases} \geq m(S_k)/2$$

where: $m(F)$ is the area of the planar region F.

(From Flusser, 1992)

As reported by Flusser, shape matrices can be utilized as the feature for recognizing objects, with high degrees of success. In addition, there is no limit to the number of shapes these matrices can identify, and the methodology can even describe shapes that contain holes. An object's rotation and scaling can also be resolved by using binary shape matrices. Lastly, the degree of similarity between the object and the target can be measured by comparing their shape matrices (Flusser, 1992).

3. Implementation

It was very important that the system be designed around standard commercially supported software that utilized a simple graphical user interface. Customization of the graphical user interface (GUI) and incorporation of the analytical site selection model were developed using ARC/INFO's Arc Macro Language (AML) and ArcView's Avenue scripting language. Through the customized interface the user is allowed to set locational criteria as it pertains to a particular siting task via graphical interfaces, run this analytical model against the raster-based GIS data, and produce a series of potential site locations.

A "Model Builder" tool was designed to allow the user to interactively build a site selection model in real time (Figure 1). As the user builds the model, the appropriate ARC/INFO commands are written to an executable batch file. Categorical weights are assigned to broad categories of themes (e.g., geology, ecology, or engineering). Criteria weightings are assigned to each individual theme and may vary within a category based on importance. The final weight that is applied to each layer is based on the product of the categorical weight and the criteria weight. The user selects each criteria, applies the appropriate weightings and spatial operations, and adds this criteria to the overall site location model. Immediately following grid processing of the model, the grid is converted into a polygon format that is compatible with the data browsing environment. The new theme is automatically added to the active ArcView "View". Each polygonal area has its own individual numerical score based on the summed criteria weights.

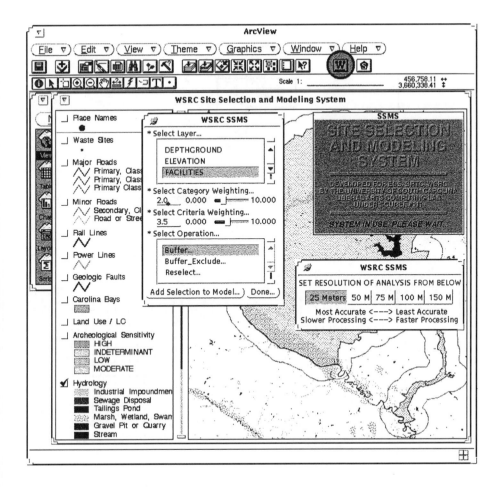

Figure 1 User Interface for Site Selection System.

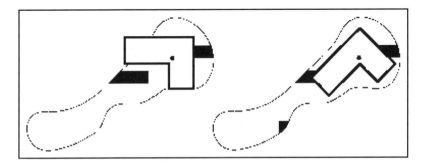

Figure 2 Need to rotate shape to make it fit.

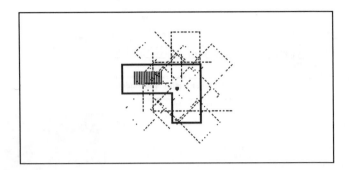

Figure 3 Spinning shape filter.

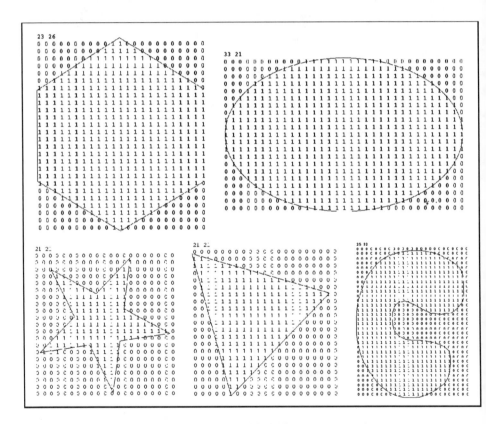

Figure 4 Geometric objects and their binary representations.

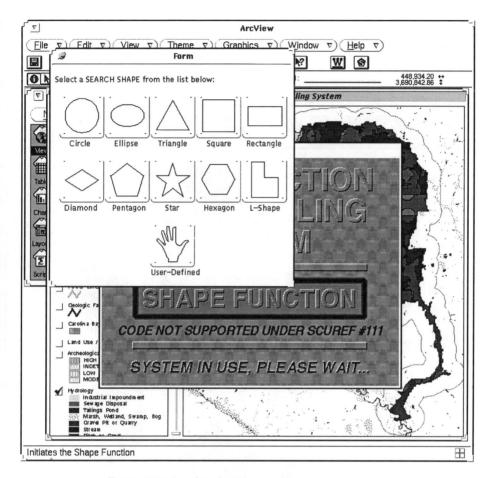

Figure 5 User interface for the shape selection procedure.

3.1 Development of shape fitting function

The shape fitting funciton represented an interesting challenge. In addition to specification of the size and shape of the desired object the system had to check for a variety of orientations. (Figure 2). In effect, the process could be conceptualized as fitting roving, spinning filter that examines every cell and some combination of neighbors (Figure 3). The implementation of the procedure followed Flusser's (1992) binary shape matrix methodology. The binary matrices are constructed by geometrically constructing vector-based shapes (Figure 4) within ARC/EDIT, converting them to GRID (raster) format, and exporting these shape matrices (ESRI, 1994). The function produces binary shape matrices as proposed by Flusser's research.

An object's rotation is accounted for by rotating each shape in the eight cardinal directions. This is the same methodology utilized by Podolsky within FullPixelSearch (Podolsky, 1996; Norr, 1995; Seiter, 1995). Given these techniques, a maximum of eight matrices (or kernels) may be produced for each individual shape-based target.

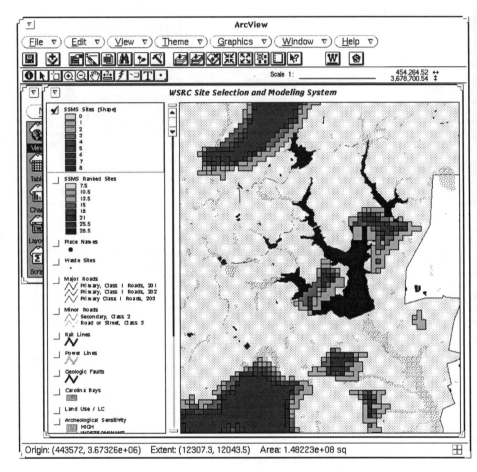

Figure 6 Enlargement of area of interest for shape location.

The user interacts with the shape analysis GUI in much the same manner as the model builder tool (Figure 5). A customized ArcView icon links the appropriate scripts and menus to perform the shape searches. After the user selects the appropriate search shape and enters the appropriate dimensions of the shape, the binary kernels are constructed and applied to the final site ranking layer. Each kernel is passed over the layer on a cell-by-cell basis, producing an output layer denoting target shape identification. Up to eight output shape layers may be produced, which are combined using map algebra to produce a final shape identification layer. The unique numerical values for each cell location denote the successful angle of rotation of the object's shape (Figure 6).

4. System evaluation

While the most serious evaluation of the system will be conducted by the actual users at the Savannah River Site a controlled experiment was performed to determine whether the procedure can find the shapes that it should and not find those that it should not. The method employed was an indirect proof as discussed by Goodenough and Gerhardt (1975). A series of specifically sized shapes were placed within a feature space at varying angles of orientation. Some shapes were oriented along the eight cardinal directions while the others were placed off the these 45 degree increments. The binary shape detection procedure was run on these synthetic sample spaces for ten different shapes. These shapes included eight regular geometric shapes and two irregular shapes. In every case the procedure "found" all the shapes that were oriented in cardinal directions and was unable to fit the shape into the feature space in which the shape was oriented in a non-cardinal direction (Figure 7). These findings demonstrate that the procedure is reliable.

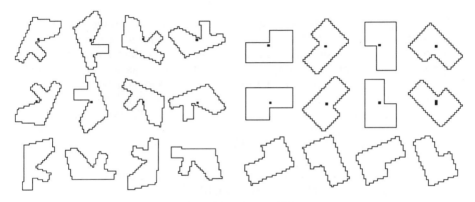

Figure 7 L and user defined shapes - those oriented in cardinal directions (top 8) corrected found. Off cardinal (bottom 4) not found.

Another test was conducted to determine how the procedure performed when cell size varied. The series of time trials were run for a siting model that located a rectangular shaped facility at cell resolutions that ranged from 25 to 200 meters (Figure 8). It is interesting to note that an increase in cell size from 25 to 50 meters reduced the processing time from 4498 seconds to 849 seconds while the increase from 50 to 100 meters reduced the processing time to 283 seconds. Therefore, there is more than a five time improvement in performance by doubling the cell size when going from 25 to 50 meters but only 3.5 time increase when the cell size was doubled again. Of course, the actual number of cells in the study area decreased geometrically in a ratio of 64, 16 to 4 as cell size doubled. These finding suggest that there can be sizable performance gains by using a 50 meter cell size but the gains from moving to the 100 meters only provides marginal increases in performance.

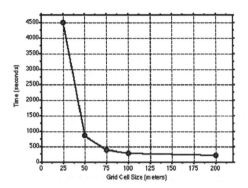

Figure 8 Relationship between cell size and processing time for finding square shape.

5. Conclusions

The objectives of the research were to incorporate a generic shape fitting procedure into a system for site location. While other spatial data handling systems have analyzed the specific shapes of objects, no existing system appears to allow the user to specify a specific shape as a part of suitability analysis. Current research from the document imagining field allowed the design and prototyping of the necessary "shape fitting" techniques. This involved designing irregularly shaped binary kernels (filters) as proposed by Flusser (1992 that approximate the form of the target in question. In particular the system meets the following objectives:

1. The GUI provides a linkage between a vector based data browsing system and the grid module of a full featured GIS;
2. The inter-application linkage can also be supported across operating systems;
3. Supports selection and weighting functions required for suitability analysis;
4. Conversion between vector and raster data structures are efficiently performed;
5. The shape fitting function is integrated with the other functions of the system;
6. The GUI allows the user to describe the desired shape and size of the object;
7. The user can select the level of spatial resolution;
8. The shape function evaluates up to eight different orientations;
9. Major performance gains are obtained by increasing cell size from 25 to 50 meters but is only marginally improved beyond 50 meters.

Acknowledgements
The authors gratefull acknowledge the support provided by the South Carolina Universities Research and Educational Foundation grant number 111.

References

Cowen, D.J., J.R. Jensen, P.J. Bresnahan, G.B. Ehler, D. Graves, X. Huang, C. Wiesner, and H.E. Mackey, 1995a. "The Design and Implementation of an Integrated Geographical Information System for Environmental Applications", Photogrammetric Engineering and Remote Sensing, November, Vol. 61, No. 11, pp. 1393-1404

Davis, J.C., 1986. Statistics and Data Analysis in Geology. New York: John Wiley and Sons.

Densham, P.J., 1991. "Spatial Decision Support Systems". In Geographical Information Systems: Principles and Applications, edited by Maguire, Goodchild, and Rhind. Essex, England: Longmans, 1, pp. 403-412.

Douglas, D and T. Peucker, 1973. " Algorithms for the Reduction of the Number of Points Required to Represent a Digitized Line or its Caricature" Canadian Cartographer Vol. 10, No. 4, pp.110-122.

Environmental Systems Research Institute (ESRI), 1994. GRID functions: FocalRange, FocalSum. ArcDoc Ver. 7.1. Redlands, CA: ESRI.

Flusser, J., 1992. "Invariant Shape Description and Measure of Object Similarity". Proceedings of the International Conference on Image Processing and its Applications, Maastricht, The Netherlands, pp. 139-142.

Goodenough, J. and S. Gerhert, 1975. "Toward a Theory of Test Data Selection", IEEE Transactions in Software Engineering, Vol. SE 1, No. 2 pp. 156-173.

Lee, D.R., and G.T. Sallee, 1970. "A method of Measuring Shape." Geographical Review, 60, No. 4, pp. 555-563.

McGarigal, K., and B.J. Marks, 1994. FRAGSTATS: Spatial Pattern Analysis Program for Quantifying Landscape Structure, Version 2.0. Corvallis, OR: Forest Science Department, Oregon State University.

McHarg, I.L., 1969. Design With Nature. Garden City, NY: Doubleday and Company, Inc..

Moellering, H., and J.N. Rayner, 1979. "Measurement of Shape in Geography and Cartography". Report of the Numerical Cartography Laboratory, Ohio State University, NSF Grant No. SOC77-11318, pp. 1-24.

Moellering, H., and J.N. Rayner, 1982. "The Dual Axis Fourier Shape Analysis of Closed Cartographic Forms". The Cartographic Journal, 19, No. 1, pp. 53-59.

Mokhtarian F. and A. K. Macworth, 1992. "A Theory of Multi-Scale Curvature-based Shape Representation for PlanarCurves", IEEE Trans.Pattern Anal. Mach. Intelligence, Vol. 14 .pp. 789-805.

Niblack, W. And M. Flickner, 1993. Find Me the Pictures that Look Like This: IBM's Image Query Project. Advanced Imaging, 8, No. 4, pp. 32-35.

O'Hare, M., L. Bacaw, and D. Sanderson, 1983. Facility Siting and Public Opposition . New York: Van Nostrand Reinhold Company.

Podolsky, 1996. Functional Overview of Avian Systems Inc.'s FullPixelSearch. Telephone interview by Geoffrey Ehler, January 5, Personal notes, Columbia, SC.

Rosin, P. L., 1993. " Multiscale Representation and Matching of Curves Using Codons" CVGIP: Graphical Models and Image Processing, Vol. 55 No. 4, July, pp.286-310.

Seiter, C., 1995. "Software Review: FullPixelSearch 1.5". MacWorld, July, p. 79.

Norr, H., 1995. "FullPixelSearch Locates Patterns in Image Data". MacWeek, Vol. 9, No. 8, p. 13.

Tomlin, C.D., 1990. Geographic Information Systems and Cartographic Modeling Englewood Cliffs, NJ: Prentice-Hall.

Westinghouse Savannah River Company (WSRC), 1992a. Replacement Power Facility Site Selection Report. WSRC-RP-92-672.

Westinghouse Savannah River Company (WSRC), 1992b. Preliminary Site Selection Report for the New Sanitary Landfill at the Savannah River Site. WSRC-RP-92-1397.

Contact address

Geoffrey B. Ehler[1], David J. Cowen[1] and Halkard E. Mackey, Jr.[2]
[1]Department of Geography and
Liberal Arts Computing Laboratory
University of South Carolina
Columbia, SC 29208
USA
Phone: (803) 777-6803
Fax: (803) 777-7489
E-mail: geoff@floyd.cla.sc.edu, cowend@sc.edu
[2]Savannah River Technology Center
Westinghouse Savannah River Company
Aiken, SC 29808
USA

GEOMETRIC MATCH PROCESSING:

APPLYING MULTIPLE TOLERANCES

Francis Harvey[1] and François Vauglin[2]

[1]Department of Geography
University of Washington
Seattle, USA
[2]IGN COGIT Laboratory
Saint-Mandé, France

Abstract
GIS overlay algorithms are very effective for general layer joining tasks. For applications requiring more control, fuzzy overlay may introduce positional error due to arbitrary node movement during processing. Further, the reduction of all accuracy and error issues to a single fuzzy tolerance has decided disadvantages. In this paper we formulate a general framework for extending fuzzy overlay as geometric match processing. The proposed heuristic framework presents a design for a "toolbox" that accounts for issues of scale and error by implementing multiple tolerances. The setting of tolerances is made based on application purpose and scale and on the statistical analysis of line form and error. We emphasize the complementary nature of the experiential and stochastic components in the heuristic. This framework extends previous work on multiple tolerances, called shift and match.

Key words
Vector overlay, cluster analysis, algorithms, accuracy.

1. Overlays of cartographic data

Overlay is the most ubiquitous GIS operations and fuzzy vector overlay, in particular, belongs to the essential components of GIS technology. This algorithm calculates inexact and exact intersections of two cartographic data sets by employing a tolerance around nodes to merge other nodes (Dougenik, 1980, Frank, 1987). In spite of its core role in GIS, the fuzzy vector overlay algorithm has the acknowledged limitation of "creep" (Harvey, 1994; Pullar, 1993) arbitrary node movement that adversely affects cartographic applications. It is, however, essential to vector GIS because it provides the most effective means to combine cartographic data sets and resolve the problems of spurious polygon creation. Conversely, the fuzzy overlay algorithm exacerbates the problems of positional and attribute accuracy associated with cartographic generalization.

Beginning with an evaluation of the uses of vector overlay and of statistical models for assessing data error, we present a new heuristic for the geometric combination of cartographic data. We base this modified approach to overlay, which we call geometric match processing, on the statistical evaluation of geometric line error and the purpose for which the cartographic data was prepared. This continues the vein of work developed around cartographic generalization and quality control (CNIG, 1993; Plazanet, 1995; Werschlein &

Weibel, 1994). Statistical models of error distribution provide the means to carry out overlay in a more rigorous fashion. Determining better epsilon filters allows us to resolve positional accuracy issues in overlay and deal with partial incompatibilities between source data.

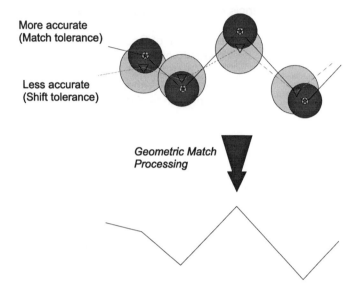

More accurate
(Match tolerance)

Less accurate
(Shift tolerance)

*Geometric Match
Processing*

The line element output retains the accuracy of
the more accurate (match) line elements

Figure 1 Overview of geometric matching process.

Specifically, this paper presents our initial preliminary work on a heuristic for vector overlay that accounts for issues of scale and application purpose by using multiple tolerances. While we find fuzzy overlay suited for many applications, geographic data processing presents many cases that demand more exacting control of overlay. In terms of scale, we refer to situations when multiple data sets, collected at different scales, must be combined, or when a data set is to be matched to a larger scale data set. Application purpose refers to the perspective and goals that guide the preparation of a cartographic data set. As described in an earlier paper (Harvey, 1994) the situation often arises that data sets are to be combined through overlay, but the boundaries of similar phenomena do not exactly match. For instance, when a coverage of vegetation types is combined with a sealed surfaces coverage, often boundaries will coincide, but overlay frequently will result in subtle differences due to different source materials, digitizing, etc. (Goodchild 1978). Without an exhaustive examination it is impossible to statistically assess the accuracy of the data sets. However, an experiential heuristic allows the user to evaluate the accuracy of an data set based on knowledge, both empirical and ancillary, and assign a tolerance that can be tested in an iterative fashion.

Geometric match processing brings more control over overlay to the user. The combination of stochastic, empirically evaluated, measures and experiential knowledge leads to a heuristic extending fuzzy overlay that capitalizes on the advantages of these otherwise distinct approaches and helps to overcome the disadvantages (complexity on one hand, and arbitrariness on the other). Current implementations of the fuzzy overlay algorithm just leave the user with too little control. Furthermore, ensuing spurious polygons of single tolerance fuzzy overlay make analysis and representation more difficult. Applying a larger fuzzy tolerance is a possible solution, but this brings new problems. The increased arbitrary node movement potential leads to even greater distortion of the cartographic elements and may in fact even increase the number of spurious polygons created in overlay.

Different data needs different tolerances. Pushing every data set through the same epsilon filter in overlay (Chrisman, Dougenik & White, 1992; Dougenik, 1980) reduces the accuracy and cartographic fidelity of the resulting overlay. Multiple tolerances, at minimum, one tolerance for each data set, enable the application of parameters that better represents a data sets' character. These tolerances can be based on stochastic models of line error or expert knowledge. Because it is easier to operationalize one tolerance for each data set, our discussion will focus on an approach for assessing the geometric data quality of data sets.

At present, our work applying multiple tolerances is guided by what we refer to as the fixed/moveable axiom. This axiom is motivated by our experience assessing multiple data sets for overlay. In many cases one data set is more accurate or simply cartographically "better" than the other. The lines of the higher quality data set represent the geographic phenomena more distinctly and precise. Because overlay processing should not alter the geometric representation of this cartographic data set, we refer to it as the fixed data set. The other data set is a less accurate cartographic representation and may be altered to a larger degree in overlay processing. It is the moveable data set.

In the next section of this paper (2), we review fuzzy vector overlay and the conceptual basis for geometric match processing. The following section (3) develops geometric match processing as a pragmatic approach for resolving critical issues in overlay related to scale and purpose. We illustrate its utility using an example that draws on our experiences. The next section (4) develops the statistical components of the geometric match processing heuristic. This is followed (section 5) by the presentation and discussion of determining and operationalizing multiple tolerances. We conclude (section 6) with a summary and directions for further research as we move beyond theoretical work and write the algorithm for geometric match processing.

2. Beyond the current overlay algorithm

Fuzzy vector overlay utilizes an algorithm designed to efficiently combine cartographic data (Frank 1987). This algorithm has established itself as the standard in the GIS industry after multiple other attempts to perform vector overlay were tried. They range from conversion of vector data to raster and then back to vector (Goodchild, 1978), utilizing only integers for coordinate storage (Cook, 1978), or using rational arithmetic (Wu & Franklin, 1990). Once the problem of inexact intersection detection was resolved (White, 1978), the key problem facing vector overlay is the creation of spurious polygons during processing (Goodchild, 1978; MacDougall, 1975). These "slivers", as they are also known, made up to 75% of all polygons resulting from overlay (using a non-fuzzy overlay algorithm) in a test Michael Goodchild performed with CGIS data (Goodchild, 1978). They increase the complexity of the data set and the attribute information used to describe the polygons many times over (Goodchild, 1978).

Although, as Goodchild wrote, spurious polygons can be removed based on their small size (Goodchild, 1978), this too brings problems of increased positional and attributal inaccuracy. Lester and Chrisman examined positional errors resulting in spurious polygons and found there was no correspondence between spurious polygon shape and the line shape until a large enough size was reached (Lester & Chrisman, 1991). Removing polygons according to size introduces more uncontrolled distortion.

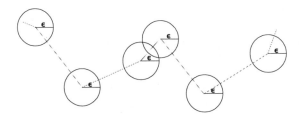

Figure 2 Epsilon tolerance.

Fuzzy vector overlay resolves many of the spurious polygon problems. Examining the area surrounding a node based on a set tolerance, called the epsilon distance or error distance (Dougenik 1980), uses clustering and a band-sweep approach to detect near intersections and exact intersections. In the case of near intersections, the node, or closest point on a segment, is extended to meet the nearest segment of the other coverage (Dougenik 1980). This is also called epsilon tolerance after the work of the Polish mathematician J. Perkal (Perkal, 1956).

However, current GIS software packages utilize clustering in an piece-meal fashion. A node in the cluster is dissolved and the affected line arbitrarily extended to connect to another node. Obviously this can lead to grave inaccuracies in the output data. Implementations of the fuzzy overlay algorithm can also create new spurious polygons through the arbitrary movement of nodes. With the goal of circumventing these problems, we set out to improve the processing of overlay by eliminating the arbitrary error produced by fuzzy vector overlay by providing more control over tolerances. Our emphasis is maintaining positional accuracy. Since attribute accuracy is entangled with positional accuracy (Beard, Chrisman, and Patterson, 1984), we maintain that an overlay algorithm that enables more control over positional accuracy benefits many GIS applications.

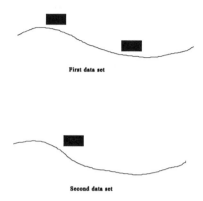

Figure 3 Positional or attribute accuracy? (see text).

Because it is frequently very difficult (or impossible) to extricate positional accuracy from attribute accuracy, the heuristic we propose helps account for both. If we consider the example in figure 3, involving two data sets, the first of which has better positional and attribute accuracy, and a second which is less accurate, how do we know—from the data sets—what is the main problem with the second data set. It could be a problem of positional accuracy (the ruins and the house have been shifted on the left) or a problem of attribute accuracy (the house has been misclassified in ruins, and the ruins have been forgotten)? Even with metadata, in many cases, the intricacies of positional accuracy and attribute accuracy cannot be readily decomposed.

The utilization of multiple tolerances provides a good means for controlling positional accuracy in GIS overlay. David Pullar presents an approach involving multiple tolerances (Pullar, 1993). Whereas Pullar's paper is concerned with resolving the specific problems of arbitrary node movement in fuzzy overlay, a paper from Zhang and Tulip applies a clustering algorithm to coalesce less accurate nodes to more accurate nodes (Zhang & Tulip, 1991). Their algorithm design attempts to avoid splinters by ranking possible intersection points in fuzzy overlay solely according to geometric criteria. Zhang and Tulip evaluate the coalescence of nodes inside the epsilon tolerance without considering the distribution of geometric error.

Our work aims to develop a new set of "tools" for performing geometric combinations based on understanding differences between the purposes for which data is collected (experiential component) and the statistical analysis of line error (stochastic component). We examine the accuracy of a cartographic data set in terms of its origin and errors created in processing. Using statistical analysis of line error (probability density functions) we ascertain quality parameters that describe the degree to which the line coordinates approximate the ground coordinates. These parameters can be used as a basis for setting shift and match tolerances. The experiential component can incorporate knowledge of the data automation process, categories, etc..

We can use the probability density function (pdf) to describe the shape of the geometric tolerance. This function is often complicated but can be easily simplified in one tolerance value for each data set. This way, epsilon-like tolerances can be objectively defined for each data set. The match tolerance is assigned the more accurate (fixed) data set, and the shift tolerance is assigned the less accurate (moveable) data set.

In summary, we conclude that fuzzy overlay is an effective technique for combining polygons from multiple coverages for generic purposes. However, it oversimplifies cartographic representations and creates additional errors. Cartographic boundaries will coincide, but they never align (Goodchild, 1978). Fuzzy overlay combines boundaries, but distorts spatial and attribute accuracy. Although this may be acceptable in many situations we recognize the need for an algorithm that retains positional accuracy, instead of irrevocably diluting it by the fuzzy tolerance. A unique tolerance for every element of the cartographic data sets would afford the ultimate in flexibility for vector overlay that retains spatial and attribute accuracy. Since the implementation of this ideal seems impractical, we apply the fixed/moveable axiom as the basis for a heuristic. Geometric match processing turns overlay into an extendable geometric processing tool that responds to specific application needs.

3. Scale and purpose

Geometric match processing considers scale and purpose, both intricately related in the application. For an overlay to provide an useful and accurate geometric structure for cartographic representation, retain the input data sets' cartographic fidelity, and, concurrently, consistently create an integrated composite of the attributes, the purpose and cartographic representation of both data sets must be reconciled. These difference correspond to distinctions made in generalization literature (McMaster & Shea, 1992). Specifically, our usage of scale and purpose as principles that guide the development of the geometric match heuristic, ties into work presented by W. Mackaness in 1994. His holistic approach to map generalization points to similar intricacies we find between locational and attributal accuracy.

In our consideration of the geometry of overlay, we refer to the fundamental issues scale and purpose. An explanation of each follows.

The purpose for which a cartographic data set is prepared defines many aspects of its representation. The examination of purpose ties into earlier work (Harvey 1994). This paper describes the overlay of a land use coverage that was digitized from a Mylar of a topographic map at a large scale (more accurate) with a vegetation map, prepared from blueprints of topographic maps (less accurate). The application for which each data set was prepared influences the delineation of boundaries. The boundaries that indicate road borders should coincide, but because of media distortion, digitizing error, coordinate inaccuracy, etc. they frequently will only approximate each other. A fuzzy overlay of these two data sets will either result in numerous spurious tolerances if the fuzzy tolerance is small. If the fuzzy tolerance is increased, many "slivers" will disappear, but arbitrary node movement can possibly reduce the accuracy needed for output to intolerable levels.

Likewise, scale can be assessed in terms of purpose, but it duly receives special attention due to its fundamental role in reducing data observed in a reality to a meaningful cartographic representation. In terms of overlay, scale changes, part of generalization, can lead to substantial geometric changes, but subtle cartographic changes.

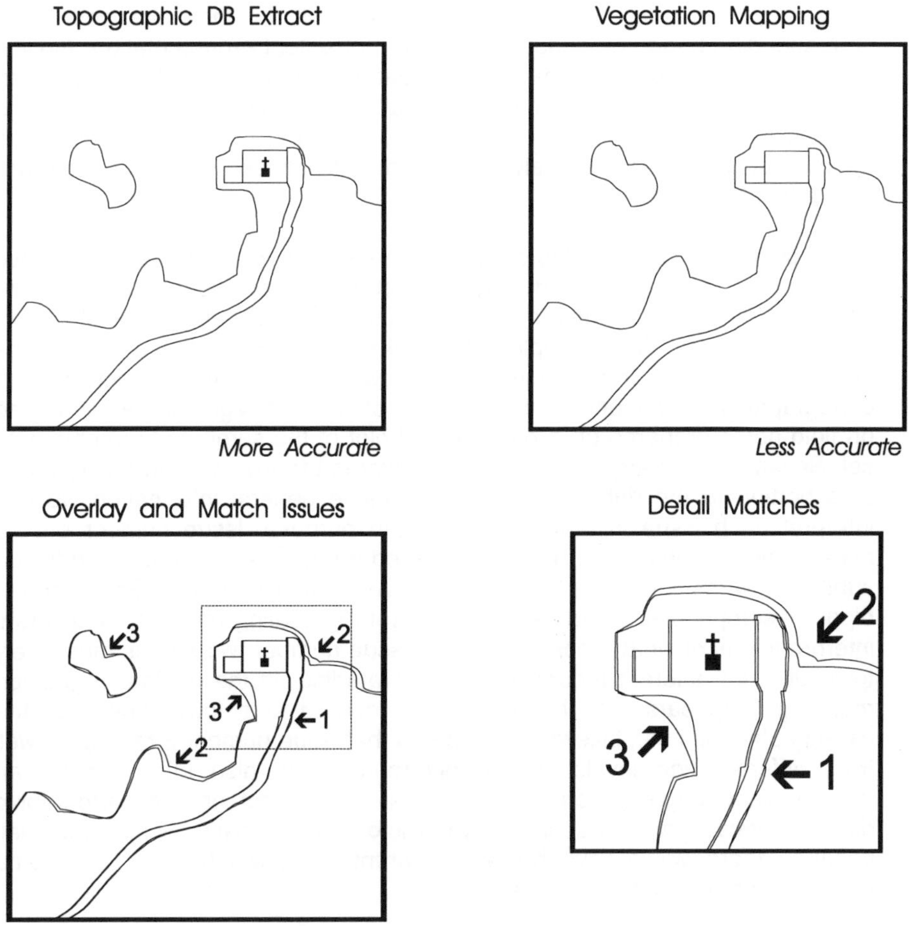

Topographic DB Extract

Vegetation Mapping

More Accurate

Less Accurate

Overlay and Match Issues

Detail Matches

Figure 4 Example of match situations.

We address purpose and scale issues by employing cartographic generalization concepts to assess positional accuracy issues of overlay. Evaluating the application of overlay in terms of work from Buttenfield (1991), McMaster (1991), McMaster and Shea (1992), and numerous others stress aspects of cartographic generalization that have great effect on the data sets used in overlay.

Generalization, following McMaster and Shea (1992) strives for a balance between spatial accuracy and attribute accuracy. When there are scale and purpose differences there is a conflict between spatial accuracy and attribute accuracy. For instance, where the real world distance between a lake and village is too small at a given scale to graphically display the railroad and the road that run between the edge of the lake and the village there is a clear conflict. Displacement is a solution, but always leads to a loss in positional accuracy. Displacing the railroad (perhaps abandoned) into the lake may do it's relative importance justice, but would introduce attribute and positional inaccuracy, because a railroad can't be built in the lake without a corresponding bridge or viaduct. The more consistent representation, moving the shoreline to allow both road and railroad to run on dry land seems better, but introduces positional inaccuracy and attribute inaccuracy. A slight shift in the shoreline, does seem more justifiable in terms of both positional and attribute accuracy. There is always a trade-off between positional accuracy and attribute accuracy.

In the overlay of cartographic data sets, users work with the results of these modifications and must find ways to retain the positional and attribute accuracy from the input data sets without unintentionally altering positional or attribute accuracy. The example in figure 4 shows some of these issues. The upper two frames from figure four show the accurate extract from a cartographic data set, and, to the right, the outlines of vegetation areas based on field notes prepared on a photocopied plot of the same cartographic data set. Several issues are identified in the lower left frame, with an enlargement of a portion to the right. Based on the above general information we can interpret each issue in terms of scale and purpose. Issue number one, a misalignment between the borders of paved areas, seems largely to reflect a subtle leftwards shift of all lines in the vegetation mapping. Some of the bigger discrepancies could result from errors in scaling the plot, or different interpretations of what areas are paved. Issue number two seems most likely to result from different recordings of the waterline. In the original vegetation mapping it appears that the map was prepared at a lower tide than the cartographic data set. Likewise, it appears that issue number 3 results as well from different tide levels. These differences can also all result (to an undetermined degree) from generalization carried out on the cartographic data set necessary to present the map information clearly. Scale is crucial, because there will always be two different representations of the same information at two different scales.

We can see in this example how positional accuracy and attribute accuracy are intertwined. Instead of exacerbating positional and attribute errors by arbitrarily moving nodes, geometric match processing aims to retain the highest level of coordinate accuracy in the input data set and help void some of the errors extenuated by fuzzy overlay.

4. Statistical measures

The production of cartographic data is a process that starts with measurements and produces an abstract representation of reality. In vector data bases, small areas are modeled as points. Linear features like roads or rivers are modeled as polylines (ordered lists of points) and areas are represented as polygons made of polylines. The rules that describe the procedure of modeling real world phenomena are often called "specifications of the data base". They define a sort of abstracted view of geographic phenomena.

The first problem, in creating a database, is the definition of specifications, for example of a forest: Where is the edge of a forest? Most cartographic representations contain fuzziness. Who defines and explains to the stereo plotter technician how they should draw the line that delimits the forest in a way that any technician can reproduce? Of course, we know that "exact replication" is impossible and that most cartographic feature specifications contain inherent inaccuracy.

Even the digitization of points itself is an uncertain process. We never can be sure that the digitized coordinates are correct, i.e. that the technician pointed at the exact location for each point. It is even impossible to get perfect coordinates. Of course, an approximation is found that is acceptable for making data usable, except in case of gross error. This approximation is comparable to the result of a dart player aiming at a bull's eye: the distribution of the shots looks like a bell, which fits a Gauss distribution (Normal law). And there are many more examples that explain why location is never perfect (generalization, changing of references system, data exchange via different standards, finite precision calculations, any other stage of the production or exploitation processes) (Vauglin, 1994).

It is possible to cope with the positional inaccuracy inherent to data by considering the coordinates as a random variable: As the errors are too complicated for being modeled step by step, a global approach is preferred. Second, we need to know what is the shape of the random variable "positional error". In other words, how is the error dispersed: what is the probability density function (pdf) for the error? This approach leads to statistical measures and probability-based modeling.

For vector overlay, it is especially important to know the error along lines (see Figure 1). Some studies examined the corresponding pdf's to assess the accuracy of the lines (like quality controls at the IGN[1]). It often was found to be Gaussian (Hord & Brooner, 1976; Rosenfield & Melley, 1980) or even more complex. A specific study at the IGN was launched for assessing the positional error along lines of the French BDCarto. (a vector cartographic data base at the scale 1:100,000). A comparison was made using a raster of the land use layer lines and corresponding SPOT images. It was found that the probability density function is the sum of the Laplace-Gauss law and a

---

[1] Institut Géographique National (France).

symetric exponential law (LeMen, 1994):

$$f(x) = \frac{\alpha}{\sigma\sqrt{2\pi}} e^{-\frac{x^2}{2\sigma^2}} + (1-\alpha)\frac{\lambda}{2} e^{-\lambda|x|}$$

The Gaussian part stems from the digitizing error, while the other part stems from some degradation of the geometry (e.g. generalization processes). The ratio α between these two parts usually varies between 60 % and 100 % (100 % corresponds to a pure Gaussian pdf). Because the land use layer contains much generalized data the lower values for α are reached when about 40 % of the pdf is exponential, but the exponential portion is often lower, between 5 and 15 %. We observe as well that the pdf f has relatively long branches. This is the plot for f:

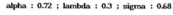

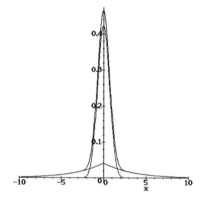

Figure 5 Pfd computed on Perpignan, compared with a Gaussian distribution.

The above analysis gives a good idea of the phenomenon's origin, but it is not totally exact: digitizing and the generalization errors are probably correlated. Further, a cumulated contribution of the two processes would not give a sum of the digitizing pdf (Gaussian) and the generalization pdf (exponential) as a result. That is why this interpretation gives an idea how the two part contribute to error, but it is not a convincing analysis.

This analysis also does not describe mistakes (gross errors). Considering the exponential part as the effect of a generalization processes hides the fact that there are mistakes. It seems more probable that the exponential part should be separated into two parts: a sum of two exponential laws with two different lambda and α.

Getting the pdf f from the quality controls is done by fitting the examined pdf to error measurements utilizing a Kolmogorov test. The Kolmogorov test gives the probability that a distribution is followed by the data set. The use of this test on "classical" functions gives interesting results: e.g. performed on the modeling by a Gaussian like formula, the Kolmogorov test proves it is definitely not a good approximation for the geometric error in this kind of data. Its value on a data set from the BDCarto land use layer (about 30,000 measurements) is 10^{-60} for the Gaussian pdf! It is sometimes hard to tune the parameters performed on the sum of a Gaussian and an exponential pdf, but the best fit corresponds to a Kolmogorov estimate of 99.8 % (LeMen, 1994). While none of these studies found an epsilon-like distribution, epsilon tolerances are very useful in GIS.

$$f_\varepsilon(x) = \frac{1}{2\varepsilon} \forall x \in [-\varepsilon; +\varepsilon], \text{ and}$$

$$f_\varepsilon(x) = 0 \text{ elsewhere}$$

A good way to explain the utility of epsilon tolerances is to conceive of them as confidence intervals: the pdf for geometric error may be a complicated function, but it is possible to use it through the definition of an adapted confidence interval. This can explain the positive results of epsilon tolerance techniques as well. For example, examining the matching of close points

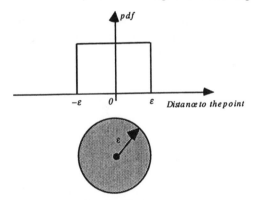

using an epsilon tolerance has no mathematical justification: the tests for matching the points are based on the calculation of the geometric intersection of the coalescence area. This purely geometric procedure is used in most GIS software packages, but it does not rely on the confidence interval nature of the epsilon tolerance.

The next section explains how these results are used for geometric match processing and how the match and the shift tolerances are set.

5. Multiple tolerances heuristic
(experiential or mathematical substantiation)

The statistical approach makes it possible to assess the pdf of the positional accuracy for each layer or data set. The localization information is augmented by a statistical function (the pdf) representing the possibility to be at a position not too different from the ground coordinates. The inherent information on geometrical inaccuracy can be used correctly through a probabilistic interpretation of these statistical laws: matching points is done in terms of pdf's and, as simplification is needed for the computer, in the terms and techniques of confidence intervals.

Confidence intervals connect the pdf's to match and shift tolerances. They are defined for each data set to be overlaid and its width is simply the match tolerance for the more accurate data set or the shift tolerance for the less accurate data set. A common value that is chosen for confidence intervals is 90%. Taking the example of the above pdf (Figure 5), the values for α, λ, Σ are respectively 0.72%, 0.30, and 0.68. The corresponding value for ϵ is the value that verifies the condition :

$$\int_{-\varepsilon}^{\varepsilon}\left(\frac{\alpha}{\sigma\sqrt{2\pi}}e^{-\frac{x^2}{2\sigma^2}} + (1-\alpha)\frac{\lambda}{2}e^{-\lambda|x|}\right)dx = 0.9$$

which corresponds to epsilon = 70 m.: For that set of data, the best choice for the tolerance (match or shift) is actually of 70 m. For comparison, here is the pdf and the corresponding tolerance set:

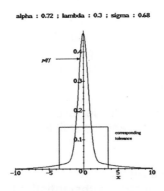

alpha : 0.72 ; lambda : 0.3 ; sigma : 0.68

Figure 6 Pfd and corresponding tolerance.

The choice for ε is made on a graph representing the confidence rate as a function of ε. The corresponding function is:

$$\text{confidence}(\varepsilon) = \int_{-\varepsilon}^{\varepsilon} \left(\frac{\alpha}{\sigma\sqrt{2\pi}} e^{-\frac{x^2}{2\sigma^2}} + (1-\alpha)\frac{\lambda}{2} e^{-\lambda|x|} \right) dx$$

In this figure, the unit for ε is 20 meters. The slope of the curve becomes slightly lower after graduation 2 (40 meters). The graph shows that it would be hazardous to choose a tolerance less than 40 meters. The corresponding confidence would quickly decrease from 84% to 0%.

These statistical measurements show it is possible to have a statistically validated rationale for geometric match processing. Using epsilon bands, the geometric matching of points is done in terms of confidence intervals. These procedures are coherent with the data that actually exist and they do not rely on non-relevant information like geometric consideration of epsilon bands without considering the actual distribution of errors.

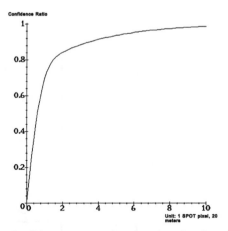

Figure 7 Confidence rate as a function of epsilon, i.e. tolerance.

The method could be improved by assessing the pdf for feature classes instead of for each data set. This would require more time devoted to comparing a sample of the set against a reference. Several tests are being prepared at the IGN with this goal in mind.

6. Conclusion—a geometric processing toolbox

Geometric match processing responds the intricacies of positional and attributal accuracy by jointly considering scale and purpose. Offering improved control over the process of combining cartographic data sets through overlay, this heuristic fills a gap in automated cartography. Statistical support for this approach ties into earlier work that also came to the conclusion that confidence intervals are suitable approximations of positional accuracy (Chrisman, 1982). Geometric match processing should be able to overcome the limitations of current implementations of fuzzy vector overlay that introduce errors and reduce positional accuracy. We find that the rationale behind heuristic may also be extended to other geometric operations in vector GIS. The multiple uses of overlay as an analytical and cartographic operation (Bailey, 1988; Chrisman, 1987; Hopkins, 1977; MacDougall, 1975; Pullar, 1993; Zaslavsky, 1995) that reach back to the late 1950's and earlier (Finch, 1933; Sherman & Tobler, 1957), succinctly point to the importance of having an extensive set of overlay capabilities for cartographic data. A toolbox of geometric processing functions would fit the bill.

Because of the complexity assessing the accuracy of cartographic data, geometric match processing couples experiential and statistical approaches. Given the importance of positional accuracy in overlay our emphasis is on the statistical evaluation of line error. Because of the number of unknown factors in cartographic data, our heuristic also relies on the experiential assessment of positional accuracy. Our rationale for extending overlay to facilitate a more exacting control over nodal movement combines these two aspects as an application orientated method.

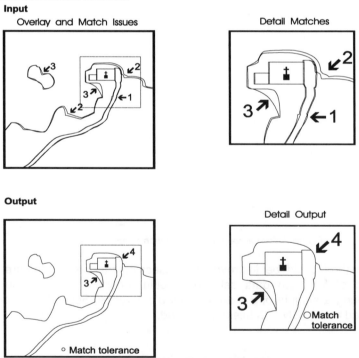

Figure 8 The result of geometric match processing.

We can illustrate the utility of geometric match processing by considering the example of realigning a vegetation mapping with a more accurate cartographic database extract (Figure 4) further. The match cases (shown in the upper two frames of Figure 8) are identical to those describe earlier. The processing resolves issues one and two because the chains lie within the specified match tolerance. Issue three is not resolved, because the chain from the vegetation coverage does not come close enough to the chain from the cartographic data set. It remains as it was digitized from the vegetation source map and requires further examination. Issue four is new, the result of two chains from the cartographic data set lying within the match tolerance. This alteration does in fact induce a new positional and attribultal inaccuracy. Is this an extreme case? Further research is necessary to determine how to process this situation.

Since in this paper we present our basic considerations such issues represent important aspects to consider as we develop this heuristic and algorithm.

An issue requiring more examination is the use of multiple tolerances in geometric match processing. We have thought of three approaches. First, a default tolerance could be set for all the nodes of one coverage and nodes could also be specifically selected and given other tolerance values. This would permit a more precise setting of tolerances. Another approach is to define tolerances for nodes in particular areas. This could be done automatically, based, for instance on meta-information, or interactively by the user. Third, tolerances could be set for each feature type. The automation of such settings can be done via statistical studies of error.

Undoubtably, developing geometric match processing will involve testing the algorithm on data. We plan to begin with this soon.

At this point, we would conclude by suggesting the concept of a geometric processing toolbox for continuing the development of advanced, stochastically grounded overlay processing. This "toolbox" would contain the algorithms and control capabilities for users to evaluate and apply overlay operations that best suit their requirements. From our research, the heuristic for any geometric processing must involve three issues:

- **scale**   -   combining different scales requires an examination of geometrical similarities and differences
- **purpose**   -   integration of different geographic data needs reviewing the boundary similarities and classification differences
- **error**   -   reducing arbitrary node movement involves assessing error; control of fuzzy tolerances reduces error potential

Our extension of fuzzy vector overlay ties together experiential and stochastic components. Geometric match processing resolves many spurious polygon problems, creates a new cartographic tool that accounts for positional and attribultal accuracy, and helps control "fuzzy creep". Consideration of scale, purpose, and error in GIS overlay extends the usefulness of the fuzzy overlay algorithm. Geometric match processing points to the usefulness of further research in this direction.

References

Bailey, R. G. (1988). Problems with using overlay mapping for planning and their implications for geographic information systems. (1), 11-17.

Beard, K. M., Chrisman, N. R., & Patterson, T. D. (1984). Paper presented at the URISA Meetings.

Buttenfield, B. (1991). A rule for describing line feature geometry. In B. P. Buttenfield & R. B. McMaster (Eds.), Map Generalization: Making Rules for Knowledge Representation. Essex: Longman Scientific and Technical.

Chrisman, N. R. (1982). Methods of Spatial Analysis Based on Error in Categorical Maps. Unpublished Ph.D, University of Bristol.

Chrisman, N. R. (1987). The accuracy of map overlays: A reassessment. Landscape and Urban Planning, 14(1987), 427-439.

Chrisman, N. R. (1989). Modeling error in overlaid categorical maps. In M. Goodchild & S. Gopal (Eds.), The Accuracy of Spatial Databases (pp. 21-34). London: Taylor & Francis.

Chrisman, N. R., Dougenik, J., & White, D. (1992). Lessons for the design of polygon overlay processing from the Odessey Whirlpool algorithm. Paper presented at the International Symposium on Spatial Data Handling, Charleston, NC.

CNIG. (1993). CNIG, Qualité des données géographiques echangées (Report). Paris: CNIG.

Cook, B. G. (1978). The Structural and Algorithmic Basis of a Geographic Data Base. In G. Dutton (Ed.), Harvard Papers on GIS, First International Advanced Study Symposium on Topological Data Structures for Geographical Information Systems (Vol. 4). Cambridge: Harvard University.

Dougenik, J. (1980). WHIRLPOOL: A geometric processor for polygon coverage data. Paper presented at the AutoCarto 4.

Finch, V. (1933). Montfort: A study in landscape types in southwestern Wisconsin. Geographic Society of Chicago, Bulletin 9.

Frank, A. U. (1987). Overlay processing in spatial information systems. Paper presented at the AutoCarto 8 Conference, Baltimore.

Goodchild, M. F. (1978). Statistical Aspects of the Polygon Overlay Problem, Harvard Papers on GIS, First International Advanced Study Symposium on Topological Data Structures for Geographical Information Systems. Cambridge: Harvard University.

Harvey, F. (1994). Defining unmoveable nodes/segments as part of vector overlay: The alignment overlay. In T. C. Waugh & R. C. Healey (Eds.), Advances in GIS Research (Vol. 1, pp. 159-176). London: Taylor and Francis.

Hopkins, L. (1977). Methods for generating land suitability maps: A comparative evaluation. AIP Journal, 386-400.

Hord, R. M., & Brooner, W. (1976). Land-use Map Accuracy Criteria. Photogrammetric Engineering and Remote Sensing, 42(5), 671-677.

LeMen, B. (1994). Géométrie de l'occupation du sol dans la BD Carto (Internal Report). Saint-Mande: IGN.

Lester, M. K., & Chrisman, N. R. (1991). Not all slivers are skinny: A comparison of two methods for detecting error. Paper presented at the GIS/LIS '91.

Mackaness, W. A. (1994, 5-9 September). Issues in resolving visual spatial conflicts in automated map design. Paper presented at the Sixth International Symposium on Spatial Data Handling, Edinburgh, UK.

MacDougall, E. B. (1975). The accuracy of map overlays. Landscape Planning, 2, 25-30.

McMaster, R. B. (1991). Conceptual frameworks for geographical knowledge. In B. P. Buttenfield & R. B. McMaster (Eds.), Map Generalization: Making Rules for Knowledge Representation (pp. 21-39). Essex: Longman Scientific and Technical.

McMaster, R., & Shea, K. R. (1992). Generalization in Digital Cartography. Washington D.C.: The American Association of Geographers.

Perkal, J. (1956) On epsilon length. Bulletin de l'Academie Polonaise des Sciences, 4:399-403.

Plazanet, C. (1995). Measurement, Characterization and classification for automated line feature generalization. Paper presented at the AutoCarto 12, Charlotte, NC.

Pullar, D. (1993). Consequences of using a tolerance paradigm in spatial overlay. Paper presented at the AutoCarto 11, Minneapolis, Minnesota.

Rosenfield, G. H., & Melley, M. (1980). Applications of statistics to thematic mapping. Photogrametric Engineering and Remote Sensing, 48(1), 1287-1294.

Sherman, J. C., & Tobler, W. R. (1957). Multiple use concept in cartography. The professional Geographer, 9(5), 5-7.

Vauglin, F. (1994). Modelisation de la precision géometrique dans les SIG. Paper presented at the CNRS, Avignon.

Werschlein, T., & Weibel, R. (1994). Use of neural networks in line generalization. Paper presented at the EGIS/MARDI, Paris.

White, D. (1978). A Design for Polygon Overlay, Harvard Papers on GIS, First International Advanced Study Symposium on Topological Data Structures for Geographical Information Systems: Harvard University.

Wu, P. Y. F., & Franklin, R. W. (1990). A logic programming approach to cartographic map overlay. Computational Intelligence, 6(2, May 1990), 61-70.

Zaslavsky, I. (1995). Local Inference about Categorical Coverages in Multi-Layer GIS. Unpublished Dissertation, University of Washington.

Zhang, G., & Tulip, J. (1991). An algorithm for the avoidance of sliver polygons and clusters of points in spatial overlay. Paper presented at the Fourth International Conference on Spatial Data Handling, Zurich.

Contact address

Francis Harvey[1] and François Vauglin[2]
[1]Department of Geography
University of Washington
Seattle, WA 98115
USA
[2]IGN COGIT Laboratory
2 avenue Pasteur
94160 Saint-Mandé
France

SIMILARITY OF SPATIAL SCENES

H. Tom Bruns[1] and Max J. Egenhofer[2]

[1]National Center for Geographic Information and Analysis
[2]Department of Spatial Information Science and Engineering
University of Maine
Orono, USA

Abstract
Similarity is the assessment of deviation from equivalence. Spatial similarity is complex due to the numerous constraining properties of geographic objects and their embedding in space. Among these properties, the spatial relations between geographic objects—topological, directional, and metrical—are critical, because they capture the essence of a scene's structure. These relations can be categorized as a basis for similarity assessment. This paper describes a computational method to formally assess the similarity of spatial scenes based on the ordering of spatial relations. One scene is transformed into another through a sequence of gradual changes of spatial relations. The number of changes required yields a measure that is compared against others, or against a pre-existing scale. Two scenes that require a large number of changes are less similar than scenes that require fewer changes.

1. Introduction

Similarity of spatial scenes is a casual judgment people make frequently in everyday life. It is intuitive, subjective, and displays no strict mathematical models. Geographers have for a long time investigated methods to describe similarity of point sets for spatial analyses, addressing such properties as pattern, density, and dispersion 1981). Recently, spatial similarity has found new champions with the advent of content-based image retrieval (del Bimbo *et al.* 1994; Faloutsos *et al.* 1994; Flickner *et al.* 1995) and sketch-based spatial query languages for geographic databases. While the principle remains the same—the quantification of deviations from equivalence for spatial configurations—the methods employed are different in order to accommodate cognitive considerations as well as database processing constraints. This paper focuses on similarity measures that are appropriate for the retrieval of similar spatial configurations, expressed in a spatial query language, from a geographic database. Such formalizations provide an integral of Naive Geography (Egenhofer and Mark 1995b).

An example of a spatial similarity query is a sketch, which approximates reality and often is sufficient for communicating meaning about salient properties of the objects and their geometries. The sketch is similar to reality and represents to the human observer a close approximation of reality. For a spatial database and spatial query processor, however, the sketch is not necessarily equivalent to any stored representation. There are too many spatial variations and differences between the geometries sketched and the spatial variations and differences between the geometries sketched and the geometries present in a geographic database. The mere retrieval of the exact match between the sketch and the database content would therefore be insufficient in most cases. The same holds true for the comparison of scenes stored in a database. Rarely will two scenes match exactly: the spatial relations between objects are slightly different; the directions and distances are not exactly the same; and shape, if considered, may vary considerably. Therefore, constraining exact matches leads to empty query results.

Spatial knowledge representation schemes absorb much of the differences among spatial scenes, allowing a focus on the important aspects and thereby facilitating comparison. These abstractions afford spatial reasoning to derive new knowledge from existing knowledge. In computer vision and image processing, the similarity of scenes is based on specific measures of objects. Their locations, shapes, and orientations have been addressed extensively, favoring metrical approaches to determine similarity (Chang and Lee 1991; Uhlmann 1991; Gudivada and Raghavan 1995). The problem addressed here is similarity of spatial scenes contained in geographic databases. A spatial scene is a set of geographic objects together with their spatial relations— topological relations, distance relations, and direction relations—and optionally other types of spatial characteristics, including such unary object descriptors as shape, ratio-type relations such as relative sizes (areas and lengths), or attributes specifying the semantics of the spatial objects. Assessing the similarity between different scenes involves comparing individual tuples or sets of tuples in the geographic database.

To retrieve the most similar configurations to a target scene, this paper employs the concept of *gradual change*, which imposes an order on sets of spatial relations. Gradual change originates from the gradual deformation of spatial objects until the spatial relation between them is changed. It applies to topological relations (Egenhofer and Al-Taha 1992) as well as to cardinal directions (Freksa 1992) and approximate distances (Sharma 1996). Two spatial relations that require little change to deform one into the other are more similar than relations that require more change. By extension, spatial scenes that require little change to deform the topological, distance, and direction relations of one into the other, are more similar than scenes that require more change.

The remainder of this paper discusses the design of the conceptual neighborhoods for the three types of spatial relations being considered: topology (Section 2), distance (Section 3), and direction (Section 4). The core is the process by which these concepts are integrated into an assessment of spatial similarity (Section 5). The paper concludes with comments on future work.

2. Similarity of topological relations

Topological relations often capture the essence of a spatial configuration—topology matters, metric refines (Egenhofer and Mark 1995b). Topological constraints are attractive for assessing similarity as they are largely immaterial to subtle geometric variations and when they get changed, usually significant alterations occur. If several of such significant changes occur, a chain reaction gets triggered. Initially two relations are slightly changed, or still just one, only a bit more dramatically. The new scene is still similar, only less so. As the number and extent of the changes increases, the new scene becomes less and less similar. The change is gradual, from equivalent, to highly similar, to less and less similar.

This concept of *gradual change* has been used to model conceptual neighborhoods of topological relations (Egenhofer and Al-Taha 1992; Egenhofer and Mark 1995a). Conceptual neighborhoods facilitate an ordering of topological relations, and support the determination of similar relations. Figure 1 shows the eight topological relations for simple areal objects (Egenhofer and Franzosa 1991). Relations connected by a line in the figure represent conceptual neighbors.

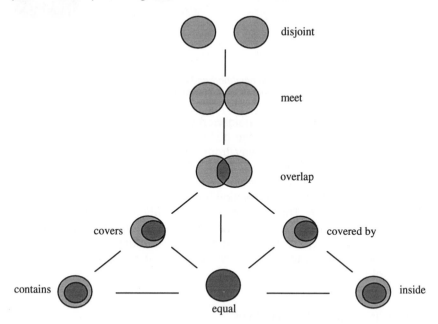

Figure 1 Conceptual neighborhood of topological relations.

The application of some gradual, spatially consistent process can change a relation into any of its conceptual neighbors. Using this concept, statements like "meet is similar to overlap" or "meet is more similar to overlap than to contains" are made.

Topological similarity may have little meaning beyond simple, two-object scenes. Its real utility comes with a combination of equivalent concepts for distance and direction relations, and scenes with more than two objects.

3. Similarity of distance relations

Qualitative distance relations are difficult to define for general spatial objects. The concepts and terms used are highly subjective, and sensitive to the scale of the spatial data being considered. Irregular shapes introduce special cases that confound most defined distance relations. Figure 2 shows one possible distance relation scheme based on increasing buffer distances. Objects range from having no distance between them, *zero*, to being *very close*, to *close*, and then *far*. The actual definitions of these distances have been investigated elsewhere (Hong 1994; Hernández *et al.* 1995).

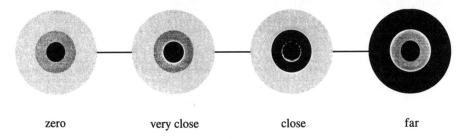

zero   very close   close   far

Figure 2 Conceptual neighborhoods of distance relations,
and the symbols that represent them.

For such distances, conceptual neighborhoods are derived by imposing an order relation < (less than) over the distance symbols, which corresponds to two objects gradually moving away from each other. Adjacent symbols are more similar than non adjacent symbols. For example, *very close* is more similar to *close* than to *far*, because *zero* < *very close* < *close* < *far*. Transitivity applies to this order relation, supporting such statements as *far* is greater than *very close*. This type of reasoning supports the determination of the difference in spatial relations between two scenes, which can guide the process of determining the number and type of gradual changes required to transform one scene into another, which forms the basis of the similarity assessment presented here.

4. Similarity of direction relations

Cardinal directions describe, qualitatively, the orientation between spatial objects (Frank 1991). Figure 3 shows a conceptual neighborhood diagram of a subset of the 169 possible spatial relations between rectangles (Chang and Lee 1991). Such schemes have been formally derived from Allen's interval relations applied to orthogonal projections (Sharma and Flewelling 1995). The subset shown depicts the relations between two equally-sized squares. Neighborhood diagrams for other shapes can be generated similarly, but they are more difficult to depict and decipher in the 2-dimensional plane. The gradual changes accommodated by these neighborhoods are translations of either object in any direction. The directions on the outside edge of the diagram are labeled with acronyms such as N for "north," SWS for "south-southwest," etc. Direction relation schemes are highly sensitive to the orientation of the objects in a scene, as well as the scene itself. Still, any direction scheme can be arranged into conceptual neighborhoods, which is the important concept here.

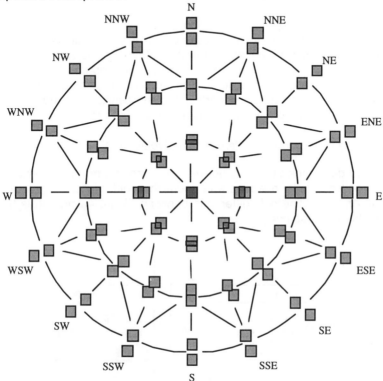

Figure 3 Conceptual neighborhoods of direction relations for same-size square objects.

Figure 3 illustrates how to produce, by gradual change, any direction relation given any other direction relation. It is assumed that each object can move in a continuous, smooth manner in any direction, but may not suddenly jump to a new location. (Note, however, that all relations can be represented by fixing one object, and moving the other.) The lines in the figure show the links between conceptual neighbors of direction relations. If two relations are not conceptual neighbors, then one cannot be derived from the other via gradual change without first producing one or more intermediate relations. The assumption made here is that the number of intermediate relations is proportional to the similarity of relations. The similarity of scenes of multiple relations is somehow proportional to the sum of the similarities of the individual relations.

The structure of the figure shows the interplay between direction and topological relations. There are four rings, each representing a topological relation. The center ring represents the topological relation equal. The next ring out shows all direction relations for the topological relation overlap. Further out are the rings for meet and disjoint, in that order. Additional rings would represent associations between direction and distance relations, while preserving topology (disjoint) and the corresponding direction relation.

5. Assessing spatial similarity

The three models of topological relations, cardinal direction, and approximate distances form the basis for the assessment of spatial similarity of scenes. Given two scenes with an equal number of spatial objects, and different spatial relations, there exists a minimum set of gradual changes that will transform one scene into the other. The number of gradual changes in this process is considered to be proportional to the similarity of the two scenes. The more steps required, the less similar the scenes are.

5.1 Counting the number of different spatial relations

In order to determine the steps between two scenes, one may evaluate both scenes' binary spatial relations and count their differences. Figure 4 shows such an analysis. The only difference between Figure 4a and 4b is that one object, object C, has been moved.

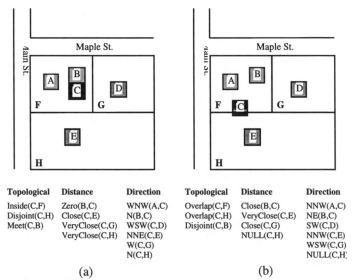

Topological	Distance	Direction	Topological	Distance	Direction
Inside(C,F)	Zero(B,C)	WNW(A,C)	Overlap(C,F)	Close(B,C)	NNW(A,C)
Disjoint(C,H)	Close(C,E)	N(B,C)	Overlap(C,H)	VeryClose(C,E)	NE(B,C)
Meet(C,B)	VeryClose(C,G)	WSW(C,D)	Disjoint(C,B)	Close(C,G)	SW(C,D)
	VeryClose(C,H)	NNE(C,E)		NULL(C,H)	NNW(C,E)
		W(C,G)			WSW(C,G)
		N(C,H)			NULL(C,H)

(a)   (b)

Figure 4 Two similar scenes and the differences in their databases.

For even this simple scene, such a slight change can have a dramatic impact on the set of spatial relations. The similarity of the two scenes could be assessed as the number of different relations, and the degree of difference between the relations, but it is unclear how the individual assessments would be combined, and what the priorities are for the different types of spatial relations. More important, since spatial relations are often highly correlated, it would be too simplistic to consider each of them in isolation.

5.2 Counting the gradual changes spatial relations

An alternative is to determine a set of gradual changes that affect the observed differences in the spatial scenes. For example, consider the scene in Figure 5a (a sub-scene of Figure 4a) and how it might be transformed into the scene in Figure 5b (a sub-scene of Figure 4b).

(a)   (b)

Figure 5 How is scene (a) transformed into scene (b)?

There are differences in distance, direction, and topological relations, and the goal is to determine the fewest number of gradual changes required to transform the start scene (Figure 5a) into the target scene (Figure 5b). In general, the set of gradual changes required, and the order in which they are applied, is unknown. The approach, then, is to systematically apply all possible gradual changes to a scene, producing a set of possible new scenes. Each new scene is the result of applying one gradual change. Each gradual change results from replacing one spatial relation, distance, direction, or topological, with one of its conceptual neighbors. For each conceptual neighbor for each type of relation, a new possible scene is produced. This process is recursively applied to all the derivative scenes, until the destination scene is among the new derivatives. The number of steps required in this process is the measure for similarity.

Figure 6 shows the results of applying this process once to the scene in Figure 5a. The start scene, Scene 0, is in the center of Figure 6. The scenes surrounding the start scene were produced by applying a single gradual change of some spatial relation. Scenes 1 through 6 result from a gradual change in a direction relation. Scenes 7 and 8 result from a gradual change in a distance relation. And scenes 9 through 12 result from a gradual change in a topological relation. Each scene results from a single gradual change, so each is said to be equally similar to the original scene.

This process continues by taking each of the derivative scenes from Figure 6 and again applying all possible gradual changes. It is limited so as not to reproduce any scene visited before. Eventually, a scene network is formed, which corresponds to a partially ordered set. Figure 7 shows such a network for a two object scene, starting with the objects meeting, and one object strictly north of the other. All scenes at each level of the partially ordered set are considered equally similar to the original scene.

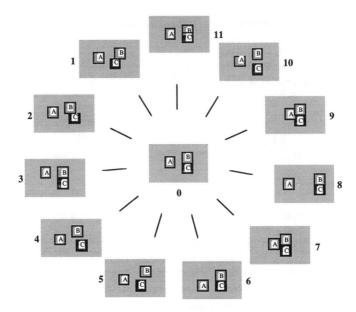

Figure 6 Derivative scenes resulting from applying a single gradual change to the center scene from Figure 5a.

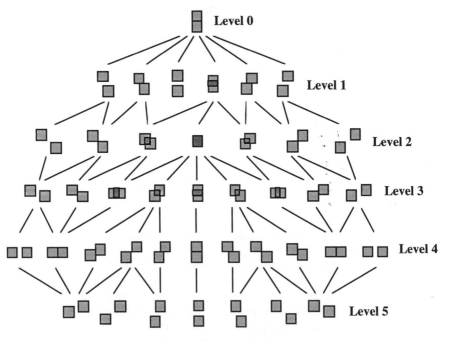

Figure 7 Similarity network.

This search process can be thought of as dividing the space of all possible scenes into regions of equal similarity. Figure 8 shows the conceptual neighborhood diagram of Figure 3 divided into 5 similarity levels. The lines between the levels are *similarity contours* in scene-space.

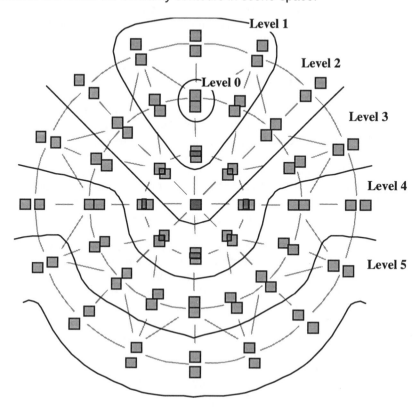

Figure 8 *Similarity contours in scene-space indicating levels of equal similarity.*

Levels of similarity can be used to directly assess the *relative* similarity of a set of scenes. One scene is compared against two or more others, which are ranked according to the similarity level at which they occur. In Figure 9, scene (a) is being compared against three other scenes (b-d). To produce the scene in Figure 9b from Figure 9a, one gradual change is required, so the scene is ranked at level 1. For the scene in Figure 9c, two gradual changes are needed, putting the scene at level 2. It takes four gradual changes to achieve Figure 9d, placing it at level 4. Given these level assessments, it can be said that scene 9b (Figure 9b) is more similar to scene 9a than scene 9c, and scene 9c is more similar to 9a than 9d.

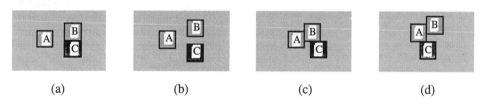

Figure 9 *Relative similarity.*

Within any similarity level there may be many scenes, and it would be useful to be able to establish a ranking within a level as well. Note that not all gradual changes will have an equal impact on a scene database. This impact, the number and type of changed spatial relations, could be used to rank the similarity of scenes within similarity level.

6. Conclusions and future work

This paper presents a method for assessing spatial similarity. It is based on the topological, directional, and distance relations in a spatial scene, and their conceptual neighborhoods. Similarity is determined by a process that gradually replaces spatial relations in a scene by their conceptual neighbors, in an attempt to construct one scene from another. Derivative scenes are ordered into levels by the number of required gradual changes. Scenes that require fewer changes, which are therefore placed at a lower level, are said to be more similar than scenes requiring more changes.

There are numerous aspects of scene similarity that have not been discussed in this paper. Most obvious is the need for testing whether the model chosen matches with human intuition, in the tradition of earlier evaluations and calibrations of natural-language spatial relations (Mark *et al.* 1995). Two formal aspects of this research are of immediate concern: the derivation of conceptual neighbors for the full set of 169 spatial relations, and a ranking scheme for scenes at a given level of similarity. The types of gradual changes need to be expanded to accommodate more realistic scenes. The ability to gradually change size, rotation, and add or remove objects, holes, and other features is needed. Eventually, this research will consider more complex scenes, with line-region relations, objects with holes, islands, and irregular shapes.

The similarity of the details of spatial relationships will also be investigated. For example, topological invariants beyond the emptiness/non-emptiness of the 9-intersection have been used to model the details of a complex topological relation (Egenhofer and Franzosa 1995). What are the conceptual neighbors of a set of topological invariants? Consider the *overlap* relation at the top of Figure 10. A possible set of conceptual neighbors could be those that introduce another basic topological invariant, such as a one-dimensional *meet*. The bottom of Figure 10 shows the different ways such a *meet* could be added to the simple *overlap*. Other deformations would have to be defined such that any set of topological invariants could be produced from any other set. A search similar to the one discussed here would then yield the smallest set of deformations required to transform one scene to another. Further refinements based on the amount of overlapping areas are complementary, taking over the role of distance relations for disjoint relations.

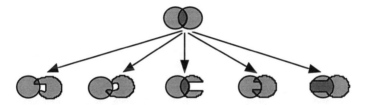

Figure 10 Derivative scenes resulting from adding a meet to a single-overlap scene.

A related topic is the generation of similarity queries, given some initial query. In Figure 11, a user has posed the query on the left side. This may or may not result in any database matches. The concepts of gradual change and conceptual neighborhoods can be used to modify this query. Each specification is relaxed to include a disjunction of its conceptual neighbors. Any combination of specifications describes, generally speaking, a similar scene. However, there may be some illegal combinations, and consistency constraints would have to be applied first.

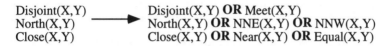

Figure 11 Fully relaxed similarity query.

Similarity assessments based on the relaxation of spatial relations is an promising alternative to the purely quantitative methods applied in image retrieval. Relation-based similarity stresses qualitative aspects, and allows for incrementally more detailed measures where necessary. Ultimately, such methods will need to be complemented by qualitative shape descriptions and analyses of the semantics of the objects involved in order to provide intelligent, computer-based spatial reasoning.

Acknowledgments
This work was partially supported by a Massive Digital Data Systems contract sponsored by the Advanced Research and Development Committee of the Community Management Staff and administered by the Office of Research and Development, by the National Science Foundation under grant IRI-9309230, and by Rome Laboratories under grant number F30602-95-1-0042. Max Egenhofer's work is further supported by NSF under grant number SBR-8810917 for the National Center for Geographic Information and Analysis, the Scientific and Environmental Affairs Divisions of the North Atlantic Treaty Organization, Intergraph Corporation, Environmental Systems Research Institute, and Space Imaging Inc.

References
C. Chang and S. Lee (1991) Retrieval of Similar Pictures on Pictorial Databases. Pattern Recognition 24(7): 675-680.

A. del Bimbo, E. Vicario, and D. Zingoni (1994) A Spatial Logic for Symbolic Description of Image Content. Journal of Visual Languages and Computing 5: 267-286.

M. Egenhofer and K. Al-Taha (1992) Reasoning About Gradual Changes of Topological Relationships. in: A. Frank, I. Campari, and U. Formentini (Eds.), Theories and Methods of Spatio-Temporal Reasoning in Geographic Space. Lecture Notes in Computer Science 639, pp. 196-219, Springer-Verlag, New York.

M. Egenhofer and R. Franzosa (1991) Point-Set Topological Spatial Relations. International Journal of Geographical Information Systems 5(2): 161-174.

M. Egenhofer and D. Mark (1995a) Modeling Conceptual Neighbourhoods of Topological Line-Region Relations. International Journal of Geographical Information Systems 9(5): 555-565.

M. Egenhofer and D. Mark (1995b) Naive Geography. in: A. Frank and W. Kuhn (Eds.), Spatial Information Theory—A Theoretical Basis for GIS, International Conference COSIT '95, Semmering, Austria. Lecture Notes in Computer Science 988, pp. 1-15, Springer-Verlag, Berlin.

M. Egenhofer and R. Franzosa (1995). On the Equivalence of Topological Relations. International Journal of Geographical Information Systems 9(2): 133-152.

C. Faloutsos, R. Barber, M. Flickner, J. Hafner, W. Niblack, D. Petrovic, and W. Equitz (1994) Efficient and Effective Querying by Image Content. Journal of Intelligent Information Systems 3: 231-262.

M. Flickner, H. Sawhney, W. Niblack, J. Ashley, Q. Huang, B. Dom, M. Gorkani, J. Hafner, D. Lee, D. Petkovic, D. Steele, and P. Yanker (1995) Query by Image and Video Content: The QBIC System. IEEE Computer 28(9): 23-32.

A. Frank (1991) Qualitative Spatial Reasoning about Cardinal Directions. in: D. Mark and D. White (Eds.), Autocarto 10, Baltimore, MD, pp. 148-167.

C. Freksa (1992) Using Orientation Information for Qualitative Spatial Reasoning. in: A. Frank, I. Campari, and U. Formentini (Eds.), Theories and Methods of Spatio-Temporal Reasoning in Geographic Space. Lecture Notes in Computer Science 639, pp. 162-178, Springer-Verlag, New York, NY.

V. Gudivada and V. Raghavan (1995) Design and Evaluation of Algorithms for Image Retrieval by Spatial Similarity. ACM Transactions on Information Systems 13(2): 115-144.

D. Hernández, E. Clementini, and P. di Felice (1995). Qualitative Distances. in: A. Frank and W. Kuhn (Eds.), Spatial Information Theory—A Theoretical Basis for GIS, International Conference COSIT '95, Semmering, Austria. Lecture Notes in Computer Science 988, pp. 45-57, Springer-Verlag, Berlin.

J.-H. Hong (1994). Qualitative Distance and Direction Reasoning in Geographic Space, Ph.D. Dissertation, Department of Surveying Engineering, University of Maine.

J. -H. Hong, M. Egenhofer, and A. Frank (1995). On the Robustness of Qualitative Distance- and Direction Reasoning. in: D. Peuquet (Ed.) Autocarto 12, Charlotte, NC, pp. 301-310.

S. Lee, and F. Hsu (1990). 2D C-String: A New Spatial Knowledge Representation for Image Database Systems. Pattern Recognition 23(10): 1077-1087.

D. Mark, D. Comas, M. Egenhofer, S. Freundschuh, M. Gould, and J. Nunes (1995) Evaluating and Refining Computational Models of Spatial Relations Through Cross-Linguistic Human-Subject Testing. in: A. Frank and W. Kuhn (Ed.), Spatial Information Theory—A Theoretical Basis for GIS, International Conference COSIT '95, Semmering, Austria. Lecture Notes in Computer Science 988, pp. 553-568, Springer-Verlag, Berlin.

J. Sharma, J. and D. Flewelling (1995). Inferences from Combined Knowledge about Topology and Directions. in: M. Egenhofer and J. Herring (Eds.) Advances in Spatial Databases—4th International Symposium, SSD '95, Portland, ME. Berlin, Springer-Verlag. 951: 279-291.

J. Sharma (1996) Heterogeneous Spatial Reasoning in Geographic Databases, Ph.D. Thesis, University of Maine (forthcoming).

J. Uhlmann (1991) Satisfying General Proximity/Similarity Queries with Metric Trees. Information Processing Letters 40(4): 175-179.

D. Unwin (1981) Introductory Spatial Analysis. Methuen & Co, New York, NY.

Contact address
H. Tom Bruns[1] and Max J. Egenhofer[2]
[1]National Center for Geographic Information and Analysis
[2]Department of Spatial Information Science and Engineering
University of Maine
Orono, ME 04469-5711
USA
E-mail: {tbruns, max}@spatial.maine.edu

RECOGNITION OF BUILDING CLUSTERS FOR

GENERALIZATION

Nicolas Regnauld

Institut Géographique National
Laboratoire Cogit
Saint-Mandé, France

Abstract

Generalizing buildings in dense areas aims at reducing the number of objects while preserving their relative positioning. For this task, operations such as aggregation or typification must be used. The first step to automate this task is the selection of a set of buildings and of the appropriate generalization operator. This firstly requires to automate the extraction of spatial relationships between buildings of a same area. The aim is to isolate the same building clusters as can be visually identified by looking at the map, so that they can be processed together. Analysing and comparing these clusters will provide the necessary information to decide which generalization operation is to be applied on each cluster.

This paper presents a method to isolate such clusters. It is based on a criterion of regularity on the distances between neighbour buildings of a same cluster. Because this is not the sole criterion to influence the recognition of a cluster, a method to take other structure types into account will be studied. Information previously computed, such as alignments of buildings along roads, will be used. Finally, a way to store these clusters in the data base will be presented.

Key words

Automated generalization, spatial analysis, automated cartography.

1. Introduction

In order to have a map which satisfies a set of specifications from a geographical database whose resolution is not compatible with the map scale, cartographic generalization has to be performed on the data. At this stage, each object in the data base is associated with its symbolization (defined in the specifications), while the legibility constraints imposed by the eye's perception capacity are satisfied. Reducing the scale as well as adding object symbolization can make the map unreadable. Three types of visual conflicts can be distinguished :

- local congestion : too many objects in too small an area make the map unreadable
- confusion of objects : two objects are overlapping or too close to be distinguished
- imperceptibility : an object is too small, too short, too narrow...to be visible

One of the projects of the Institut Géographique National is to achieve a graphic display at the 1:25000 scale from its one meter resolution topographic data base. That is why research on generalization problems is being undertaken at IGN's Cogit Laboratory. My own work is to generalize buildings in dense areas. I must take information about urban fabric into account. Extraction of this kind of information is being studied in the context of a second thesis [Hangouet 96]. The global process of how to make the choice of the fittest generalization strategy according to the context is the subject of a third thesis [Ruas 96].

To generalize buildings, there are two types of main, basic operators, common to different classifications [McMaster & Shea 89], [Beard & Mackaness 91] :
- Enhancement : Making the size of buildings compatible with the new scale.
- Simplification : Reducing the complexity of contours.

But induced by these operations, and because of the number of objects and visibility constraints, the three following operators must also be used :
- Displacement : Displacing objects to emininate problems of object confusion. [Jäger 91], [Ruas 95], [Mackaness 94].
- Aggregation : Unifying objects too close to be distinguished when they can not or must not be displaced [Ware, Jones & Bundy 95], [Regnauld 95].
- Typification : Eliminating some objects of a cluster while trying to preserve the characteristics of the group.

Choosing one or a sequence of operators in each case requires some knowledge on the objects involved and their environment. As far as possible, conflicts are solved by displacements, but in case of too dense areas, displacing may prove impossible. Then a choice between typification and aggregation has to be made. This choice depends on the knowledge of the spatial relationships between objects, an information that is not present in the original database. A primary stage of data base enrichment is then required [Lagrange & Ruas 94].

Processing an area with a high density of buildings first requires to recognize their spatial distribution. Visually, looking at a map, a cluster of buildings is identified when some kind of homogeneity is noticed. This can be inspired by the repetition of a same shape, of a regular spacing or of a typical disposition (orientation). Whatever the pattern, a cluster is a set of buildings that does not show any sudden breaking in the distances that separate each object from its nearest neighbour. Generally, a breaking makes the eye distinguish two distinct clusters. The same clusters must be identifiable before and after generalization.

As to aggregation and typification, both operations apply on clusters that show some kind of regularity in the spacing of their objects. Aggregation is based on a proximity criterion in addition to the regularity criterion.

So the regularity of distances between neighbouring buildings will be used as a first criterion to form a cluster of buildings. The first part of the paper will be devoted to the presentation of a method to detect clusters in a set of points that is based on techniques to construct and segment graphs. In the second part, the contruction of the initial graph will be adapted to the case of buildings. The third part will focus on the segmentation of this graph, taking a priori known structures of buildings into account. In fine, a data structure will be presented to store these clusters of buildings in an object-oriented data base, that allows easy management of proximity relationships between these clusters. It will be useful to study the characteristics of a cluster in comparison with its neighbours. The conclusion will introduce the characterization works necessary to take advantage of the clusterization of buildings.

2. Cluster detection techniques

The clusters to identify are composed of closely and regularly spaced-out objects. The first step consists in linking each object with its neighbours. Different techniques have been presented to represent the neighbourhood of points [Ahuja & Tuceryan 89].

One of the most widespread techniques currently used in spatial analysis is the Voronoi Diagram that for each point defines an area corresponding to its neighbourhood. All of these areas form a tessellation, and adjacency of two areas expresses a neighbourhood relationship between the two points from which they stem. By linking all the pairing points, the Delaunay triangulation is obtained, that is a neighbourhood graph on a set of points. In the present case, the triangulation produces a graph with too many cycles to be efficiently used for the segmentation step. It is more suitable for processing local areas, as was shown by its use for guiding object displacement [Ruas 95] or for guiding aggregation [Ware, Jones & Bundy 95].

A method for the representation of neighbourhood which is less complete but more suitable for our case has been chosen : the construction and the segmentation of a Minimal Spanning Tree (MST) [Zahn 71].

A MST is a graph without any cycle that allows representation of neighbourhood relationships in a set of points. The neighbourhood notion is very local. For each point, its nearest neighbour only is stored. This corresponds to the principal information we need, and especially for aggregation that usually involves one building and its nearest neighbour - otherwise, the building disposition would become skewed.

Definition:
A MST is a graph (without any cycle) that links all the points of a set, the sum of the weights of its edges being minimal. Each edge links two points and is weighted by its length. In a MST, the path that links two points is among all possible connections, the one whose longest edge is the shortest.

Principle of construction :
Each point is linked with its nearest neighbour. This makes a set of connected graphs. Then each graph is linked with its nearest neighbouring graph (the distance between two graphs is the minimal distance between two points not belonging to the same graph). This link is made by adding the edge

connecting these two points. This task is iterated until only one connected graph remains.

Now this graph can be hierarchically segmented. Neighbourhood search at each edge reveals if it is strikingly longer than its neighbours on either side. Figure 1 shows a situation where the length of edge AB is not homogeneous with the lengths of its neighbours in A. The neighbours search is limited to depth 2 (search depth is the max number of edges on the path between a point and its neighbours). This depth is a parameter of the method. The hierarchy is found by ordering the edges to be deleted, each segmentation generating two graphs.

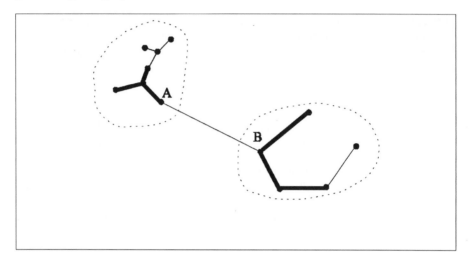

Figure 1 Segmentation of a MST.

This is the technique we will use to define the building clusters on which we will apply aggregation or typification operators. It has the advantage to produce clusters with a small number of links, and to preserve the hierarchical relationships between clusters. By representing the segmentation of the initial graph by a tree, the most important breakings happen by the root.

3. Application to buildings

Now points are replaced by buildings whose contours already are clusters of points. These points can not be used to make a MST. It could generate cases where points of a same building contour are separated by other points. The segmentation process would generate clusters integrating just portions of buildings.

Each building must thus be viewed as a forced cluster. The MST is composed with buildings as nodes and segments that represent the distance between two buildings as edges. In this context, the minimal distance between contours must be taken. The distance separating two centroids of buildings could have been choosen, but, because it is too sensitive to size and shape of buildings, it would not have reflected the information needed: the variation of *spacing* between buildings (see figure 2).

The main drawback with the minimal distances separating all pairs of buildings lies in the computational complexity. In a first step, the minimal distance between two buildings is computed as follows :

Let B1 and B2 be two buildings.
for each node of B1's contour
        for each edge of B2's contour
                compute the minimal distance between the point and the edge

Take the minimum of these values as the minimum distance "from B1 to B2". Compute in the same way the minimum distance "from B2 to B1" and take the minimum of these two results as the minimum distance between B1 and B2.

Then, the MST is built by linking each building with its closest neighbour and iterating in the same way as in the previous section. All the points of a same contour have to be represented by a same node in the graph. This is necessary to simulate connectivity between edges and thus to allow the detemining of neighbourhood. That is why a specific model for storing these connectivity relationships is needed. Such a model is presented in the fourth section.

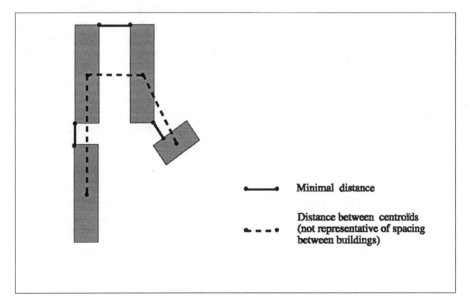

Figure 2 Distances between buildings.

4. Segmentation

The segmentation of a MST takes two criteria into account. The first one, presented in the first section, expresses the need to segment a cluster when a remarkably long distance among the edges of the tree is detected. The second one is used to modulate the first one to force buildings of a well-known structure to be part of a same cluster. In the following, the case of building alignment with regard to a road will be studied. This is the kind of information we consider to be at our disposal (some works are parallely in progress at the IGN to automate the detection of such structures [Hangouët 95]).

4.1 Segmentation process

The segmentation process of the initial tree starts by computing, for each edge, a maximum value depending on its neighbours (and taking possible alignment into account), and then the ratio between its length and this threshold. Each ratio greater than 1 (edge length greater than its threshold) denotes that the edge can be eliminated. The edge whose ratio is the highest (and superior to 1) is selected to be eliminated. It is considered to be the one whose length constitutes the most obvious breaking from its neighbours. Then the process iterates. At each step the edge to eliminate is selected. Two cases can occur :

- the elimination generates the division of an alignment that can be recovered (how is explained in part "*taking alignments into account*") : The eliminated edge is replaced by another one. There is no segmentation, but thresholds of the neighbouring edges of the the eliminated edge and the replacing edge have to be recomputed. Then the process iterates on this updated tree.
- the elimination generates the division of an alignment that can not be recovered, or generates no such division : the edge is eliminated, generating two subtrees in which thresholds have to be recomputed (in the neighbourhood of the eliminated edge). Alignments are passed on subtrees in which they appear, and the segmentation process iterates on the two subtrees.

The process stops when no edge satisfies the segmentation condition (ratio greater than 1). Then a post-processing is performed to eliminate edges that intersect objects of other types (as explained in part "*taking surrounding objects of other types into account*"). A graph hierarchy is obtained. Its lower level is constituted by building clusters on which aggregation or typification operations will be applied in case of too high densities.

4.2 Distance criterion to delete an edge

At first, for each edge, two thresholds are computed : one for each end. Either expresses the maximum length to be homogeneous with its neighbours at one endpoint. The most restrictive value (the minimum threshold) is assigned to the edge.

Detail of computation of these thresholds:
Let G be a MST and A the set of the edges of G.
For any edge *edge* of A, the sets of its "left" neighbours V₁(*edge*) and its "right" neighbours Vᵣ(*edge*) are computed. Which end of the edge is "left" and which end is "right" does not matter, this is just used here to distinguish the two extreme points. Searching for neighbours is made by following adjoining edges at either end of the edge in the graph. The depth of research is limited by the parameter *depth*.

We associate to *edge* one threshold for each of these two sets :
dist-limit₁(*edge*) = *p1**mean₁(*edge*) + *p2**st-dev₁(*edge*)
dist-limitᵣ(edge) = *p1**meanᵣ(*edge*) + *p2**st-devᵣ(*edge*)
with
        mean₁(*edge*), mean length of the neighbouring edges' lengths on the left side of edge :

$$\left(\sum_{e_i \in \, V_1(\text{edge})} \lg(e_i) \right) \bigg/ \text{card}(V_1(\text{edge}))$$

        with lg(x) the length of edgex and card(X) the number of elements in set X.
        st-devl(edge), the associated standard deviation :

$$\sqrt{ \left(\sum_{e_i \in \, V_1(\text{edge})} \lg(e_i)^2 \right) \bigg/ \text{card}(V_1(\text{edge})) \; - \; \text{mean}_1(\text{edge})^2 }$$

        p1 and p2, two parameters which we will subsequently describe.

With this segmentation technique, some border problems occur when an edge has only one neighbour at either of its two ends. The mean value loses parts of its meaning and the standard deviation is null. This may induce the inopportune elimination of some edges when the length of the neigbouring edge is smaller. This problem is illustrated in figure 3, where the value of parameter *p1* is less than 1.33. The edge of length 12 is eliminated because of its left neighbour. What is more, this problem passes onto the subtrees.

So the threshold of an edge's end is never computed if there is only one neighbour. It means that a cluster of three buildings or less can not be segmented.

191

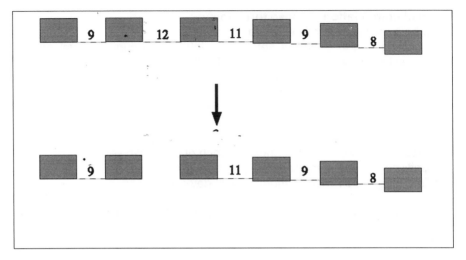

Figure 3 Side effect.

Setting the parameters:

Depth = 2 : If the examination level in the tree is deeper, clusters of three buildings interfer with their neighbours. In figure 4, the neighbours of edge *a* are edges *b*, *c* and *d*. Edge *d* increases the mean length of this group, and also the corresponding standard deviation, which may impede the elimination of edge *a*. In the same way, *a* if taken into account to compute the limit value of edge *d*, will increase it.

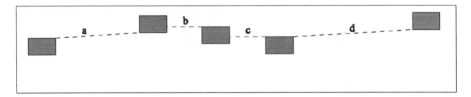

Figure 4 Neighbourhood depth set at 2.

The other two parameters have been chosen empirically. Their values depend on the difference of scale before and after generalization. The greater the difference, the more flexible the grouping, and the higher the values *p1* and *p2*. Here the parameters have been fixed to group buildings in a semi-urban zone at the 1:10000 scale to be generalized at the 1:25000 scale.

p1 = 1.2 : when the neighbours at an edge end have all nearly the same length, the standard deviation is close to 0. The only way to tolerate a small deviation in length for an edge adjacent to this group is to allow a length value slightly higher than the mean value. This ensures a minimum tolerance in case of a null standard deviation.

p2 = 2 : when lengths in a group of neighbouring edges are heterogeneous, they probably do not compose an entire building cluster. The length mean is not representative of a group's maximum lengths, so the breaking threshold is increased with regard to the standard deviation of the lengths of the group's edges.

Proximity constraints:
Once the edge to eliminate has been selected at any step of the process, we must verify that its length is greater than the perception threshold corresponding to the minimum separation distance possible between two buildings on the map. If two buildings are too close, they will not be preserved as they stand but on the other hand they must not be generalized independently. That is why they must stay in the same cluster. Definitions of these cartographic thresholds can be found in [Cuenin 72], [Lichtner 79].

4.3 Taking alignments into account

During the segmentation process, whenever possible, alignments of buildings with regard to space-structuring objects have to be preserved. The disposition of buildings is highly correlated to the roads that allow people to access them. For the moment, information on these alignments is given by the system user, but a study is being carried out at the IGN to automatically detect particular spatial relationships between buildings and roads [Hangouët 95].

Preserving alignments raises two problems :

1- avoiding the elimination of an edge that links two buildings of a same alignment (figure 5).

To avoid such eliminations, the limit value of all edges that link two buildings of a same alignment is increased. This value is multiplied by a coefficient ranging from 1 to 2 which is related to the alignment of the edge on the structuring object (based on the angular deviation between the line connecting the two building centroids and the line connecting their projections on the structuring object). This has for consequence the limitation (not exclusion) of edge elimination inside a known alignment.

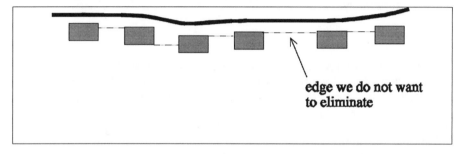

Figure 5 Known alignment of buildings with regard to a road.

2- Recovering a segmented group that contained an alignment (figure 6).

Such a segmentation occurs when the two groups were linked (before segmentation) by an edge that does not link two objects of the same alignment. We search the minimal distance separating the two groups (see section "application to buildings"). For this computation, only buildings that take part to the alignment are concerned. Then this edge in added to the MST. At the next step, it is less likely to be eliminated, because of its membership in an alignment (see 1-).

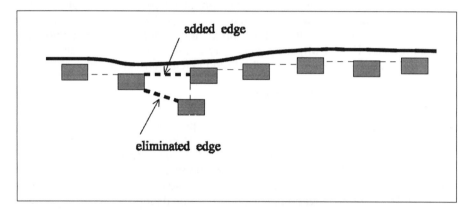

Figure 6 Linking a segmented alignment.

4.4 Taking surrounding objects of other types into account

Once the process has ended, a last segmentation criterion has to be taken into account : when an object stands between two buildings linked by an edge. This often occurs with a narrow path or track. When this obstructing object is preserved, the edge of the MST must be eliminated before recovering the alignement if necessary. Indeed, aggregation can not apply on these separated buildings. Nevertheless the information that the two clusters generated by this last segmentation can be typified together is memorised. The interfering object will be a constraint that the typification operation will have to satisfy. In case of impossibility the system will process the two clusters separately.

4.5 Example

Figures 7 and 8 show an example of cluster detection. Figure 7 shows the construction of the four MSTs within the four topological faces delimited by the roads in the zone (one on the lowest part spanning the whole width, the other three above). Figure 8 shows the result of the segmentation process on each of these four MSTs. In the lowest one, a known alignment has been taken into account (the 6 buildings on the right, aligned with a road above them). The vertical edge, linking one of the buildings of the alignment with a building above the road, has been eliminated because of the length criterion. The building on the left of the alignment is then isolated from the others. The alignment recovering process adds a new edge. At the finishing, the edge that still linked the alignment with the two buildings above is eliminated because of the crossing with the road.

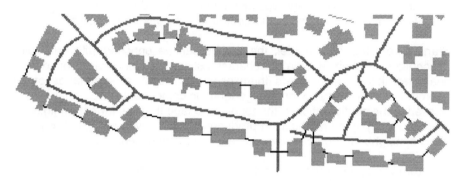

Figure 7 Initial MSTs.

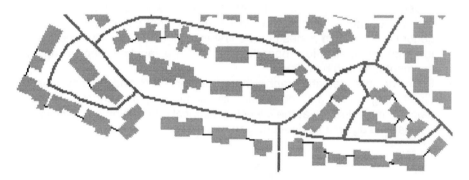

Figure 8 Segmented MST.

5. Cluster representation

In order to represent this new spatial information in the database, a data model that focuses on preserving the hierarchy between clusters has been chosen. This is essential information indeed to compare neighbouring clusters and to trace a cluster's characteristics down to its sub-clusters.
The implementation was made within *Stratège*, the object-oriented platform dedicated to research on contextual generalization developed at IGN's Cogit laboratory. Classes have been added to represent clusters by means of graphs

* The *Graphe* class : a graph consists of the list of its edges, the list of the geographic features at its nodes, the links to its two sub-graphs and the link to its father-graph. Edges are not duplicated up in the hierarchy: the graphs at the lowest level (corresponding to the final clusters after segmentation) contain all their edges while higher level graphs contain only the edge that pairs their two sub-graphs.

- The *A-graphe* class : regroups all the edges. Each edge is linked to its two end-nodes, its length, and its two indices for homogeneity control with its left and right neighbours.
- The *N-graphe* class : regroups all the nodes. A node is linked to the edges of which it is an end, and to the geographic feature it represents.

Starting from this data model, the graphs can be enriched with an explicit description of their geometric characteristics : either shape or density characteristics of the clusters themselves, or statistical summaries of the geometrical characteristics of their geographic components (size, shape ...).
Applied to the configuration in the topological face that lies lowest on fig.8 (re-sketched in fig.9), hierarchization produces the tree given in fig.9. There, two kinds of neighbourhood between clusters can be identified :

- Downwards, there first occur the major cuts, and then refinements in the segmentation. This emulates the physiological perception of clusters. The example shows indeed that cluster 2 is first separated from the rest (cluster 3). Cluster 4 then isolates itself from what eventually splits into clusters 6 and 7. This comes in handy for characterizing zones when generalizing : broad zones can be first investigated, and when no clue as to what generalization operation should be applied shows up, the segmentation can be refined.
- From the intermediate graphs (here, clusters 1, 3 and 5), any cluster's direct neighbours can be found. By comparing the characteristics of neighbouring clusters at the lowest level, local information on any cluster's pecularities against its direct neighbours can be identified.

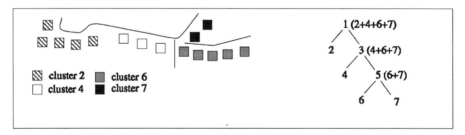

Figure 9.

6. Conclusion and outlook

Here has been described a tool for extracting proximity relationships between buildings. The result provides a set of building clusters. In dense areas, building groups fit for aggregation or typification are found in the resulting clusters. We obtain two types of clusters :

- those whose number of buildings is less than 4 : these clusters are too small to be segmented, and a complementary analysis is needed to know what kind of operator is suitable.
- those whose number of buildings is greater than 3 : these clusters contain buildings whose regularity of spacing is relevent with the scale reduction

under consideration. Depending on the mean spacing between buildings and the final scale, their number must be reduced (operation of aggregation or typification) or can be preserved.

This tool will be integrated in a system that makes automatic contextual generalization on urban zones. It partakes to the analysis stage, giving information on the distribution of buildings in an area. The information created at the end of the process is a tree of graphs, whose leaves are groups of buildings for which the most adequate generalization operation has to be decided on. It still remains to enrich the clusters with the information required for the selection and application of the adequate generalization operation. They must be characterized. There will be two aspects to the characterization:

- The shape, size and orientation characteristics to be found on the objects of a same group [Coster & Cherman 89]. Statistical processes can be carried out aiming at exhibiting recurring patterns within a cluster. What a recurring pattern is should first be defined however. Are exceptions allowed, and what is to be done of them ? Should they be considered as important landmarks in the group, or as defacing elements?
- The shape and density characteristics of a group. The density of objects in a group is something remarkable. If it is meaningless to fix a target density for any initial density of a group, preserving the balance and differences in densities between groups is of foremost importance.

The result of this characterization after any segmentation will determine the clusters that must be resegmented and those on which an aggregation operation can already be performed.

References

Ahuja, Narendra and Tuceryan, Mihran 1989 : Extraction of Early Perceptual Structure in Dot Patterns: Integrating Region, Boundary, and Component Gestalt. Extracted from Computer Vision, Graphics, and Image Processing 48 , Pages 304-356.

Coster, Michel and Chermant, J.L. 1989: Précis d'analyse d'images. Presses du CNRS, Chap 9, Pages 291-339

Cuenin R. : Cartographie Générale, Editions Eyrolles 1972, Tome 1, Chap 5.8, Pages 163-167

Jäger, Ernst 1991 : Investigations on Automated Feature Displacement for Small Scale Maps in Raster Format.Procedings of ICA 1991, pages 245-256, Hanover

Hangouët, Jean-François 1996 : City-Scape Generalization Based on Structural Principles. To be published

Lagrange, J.P. and Ruas, Anne 1994 : Geographic Information Modelling : GIS and Generalization. Proceedings of SDH 94, volume 2, Pages 1099-1117

Lichtner, Werner 1979 : Computer-assisted Processes of Cartographic Generalization in Topographic Maps.Proceedings of Geo-Processing, Volume 1, Pages 183-199.

Mackaness, William A. 1994 : An Algorithme for Conflict Identification and Feature Displacement in Automated Map Généralization. Procedings of Cartography and Geography Information Systems, volume 4 Pages 219-232

Mackaness, William A. and Beard, Kate 1991 : Generalization Operations and Supporting Structures Proceedings of Autocarto 10, Pages 29-45

McMaster, Robert B. and Shea, K. Stuart 1989 : Cartographic Generalization in a Digital Environment: When and How to Generalize. Proceedings of Autocarto 9

Regnauld, Nicolas 1995 : Agrégation de bâtiments par prolongation d'arêtes existantes. Actes de Cassini. Vol. 1

Ruas, Anne 1995 : Multiple Paradigms for Automating Map Generalization : Geometry, Topology, Hierarchical Partitioning and Local Triangulation. Proceedings of AutoCarto 12, volume 4.

Ruas, Anne 1996 : Strategies for Automated Generalization.To be published

Ware, J. Mark, and Jones, Christopher B. and Bundy, Geraint Li 1995 : A Triangulated Spatial Model for Cartographic Generalization of Areal Objects. Proceedings of COSIT 95.

Zahn, Charles T. 1971 : Graph-Theoretical Methods for Detecting and Describing Gestalt clusters. Extracted from IEEE Transactions on Computers , Vol. C-20, No. 1, janvier 1971.

Contact address

Nicolas Regnauld (PhD student at LIM, Laboratoire Informatique Marseille)
Institut Géographique National
Laboratoire Cogit
2, avenue Pasteur
94160 Saint-Mandé
France
Phone: (33-1) 43 98 84 38
Fax: (33-1) 43 98 81 71

AN OBJECT-ORIENTED DESIGN FOR

AUTOMATED DATABASE GENERALIZATION

W. Peng and K. Tempfli

Department of Geoinformatics
International Institute for Aerospace Survey and Earth Sciences
Enschede, the Netherlands

Abstract

This paper presents the interim results of an on going research project on automated generalization in GIS. It provides a logical concept for implementing database generalization in an object-oriented database environment. A scheme for a generalization rule-base and templates for constructing rules are proposed. Operations are defined at database, container, and object levels, according to the database structure proposed. Examples of some of the operations at different levels are given by means of C++ procedures. In conclusion, the paper indicates the feasibility of developing database generalization in a GIS, and summarizes the benefit and power of object-orientation in supporting the process.

1. Introduction

In developing a generic automated generalization in a GIS, one critical question is how can we define operations for problems which are unknown at the moment the system is constructed, as generalization solutions are usually application dependent and theme related? Moreover, users may wish to introduce their own rules and indicate what they expect from the new database. How can a system deal with such demands in a more flexible way than embedding generalization rules in procedures? If a rule-based approach is the solution, then what should be the principle for constructing rules so that they can be interpreted and translated into meaningful equivalent geometric/thematic descriptions and generalization operations (Peng and Molenaar, 1995)?

This paper presents an attempt to deal with these issues. It particularly looks into the solutions based on today's "data theory" and GIS technology. It first outlines the ten generalization processes defined by Peng (Peng et al., 1996). After having provided a brief description of the Formal Data Structure model (FDS), it proposes a logical concept for implementing database generalization in an object-oriented database environment. Operations are first defined at database level, then "propagated" to container level, and finally some of them are "propagated" to object level according to their natures and the database structure. The concept is illustrated using examples of some of the operations. A generalization rule-base scheme and templates for constructing rules are introduced, and a method for reasoning the rule-base is proposed. In conclusion, it gives a discussion on the feasibility of developing database generalization in a GIS, as well as how object-orientation can benefit the process.

2. The ten generalization processes

Since the main purpose of this study is to provide a logical concept for the implementation of the ten processes of database generalization, we first review them briefly.

- Extracting application-relevant feature classes: a (thematic) **selection** operation that selects a subset of feature classes that are relevant to an application, or need further processing. For example, for urban planning, road and river may be selected; telephone wire pole may not be selected; building may be selected to form built-up area. Note that to avoid confusing, the term *features* are used for objects in geography sense, and the term *objects* are used for objects in programming sense.
- Extracting target features: a **selection** operation that selects, from a selected class, a subset of features having particular geometric and/or thematic properties. Selecting buildings located on both sides of a road is an example.
- Extracting target feature components: different from selecting a subset of objects from a given class, this **selection** operation selects a subset of feature components of a (complex) feature that have particular geometric and/or thematic properties. Typical examples include selecting inter-city roads from a road network and selecting those rivers with a flow capacity more than a certain value from a river network, assuming both networks are represented as complex features, not two classes of features.
- Changing thematic resolution: an **universalization** operation, which is equivalent to the *generalization* operation in semantic data modelling (Nyerges, 1991, Molenaar, 1993). It can be applied to both classes and attributes (Figure 1).

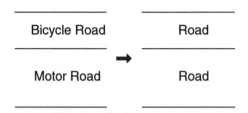

Figure 1 Change thematic resolution.

- Homogenization: this (**homogenization**) operation creates a homogeneous feature by merging a subset of adjacent features of same type, or a subset of adjacent features of same type that have the same values of some attributes. Changing thematic resolution normally requires this operation to follow (Figures 2 and 1).

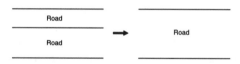

Figure 2 Homogenization.

- Changing scope: a thematic **simplification** operation that reduces the number of attributes of a class by taking out some attributes but remaining the theme and resolution unchanged. E.g., in an application the class road may have attributes *number-of-lanes* and *traffic-volume*, whereas in another application these may not be relevant.

- Changing theme: different from changing thematic resolution which moves along a classification hierarchy, this **reclassification** process aims at creating a new feature class by changing the theme of an existing class to *one of its attributes*. For instance, a cadastral parcel may include attribute *land-use*, it is then possible to create a new class *land-use homogeneous zone* by changing the theme *parcel* to land use, assuming that the land-use is homogeneous in each parcel (Figure 3a). As the boundary of an object is related to its theme (e.g., ownership defines the boundary of a cadastral parcel), geometric operation may need to follow afterwards (Figure 3b). This reclassification process, together with the universalization and homogenization processes, can provide a very useful and powerful operation for data acquisition and data sharing.

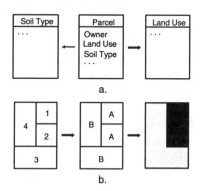

Figure 3 Change theme.

- Combining object classes: a **combination** operation, which in this case is similar to the *aggregation* operation in semantic data modelling. Different from the homogenization process this operation represents a specific subset of features that may belong to different classes by a simple feature, based on their geometric and semantic relationships. For example, representing several buildings and (recreation) grounds that belong to an university by a simple feature *university* Ä the container-feature.

- Changing geometric representation type: a **collapse** operation that changes the geometric representation of a feature from area type to line or point type, or from line type to point type.
- Changing spatial resolution: three types of operations are involved, namely, **aggregation**, **deletion**, and geometric **simplification**, each of which corresponding to a spatial resolution component (Peng et al., 1996): 1) among the selected features of a selected class, if the space between two close features is smaller than the required minimum space, *aggregate* them to form a new object without moving any of them; 2) among the selected area or line features of a selected class, *delete* features of which the sizes are smaller than the required minimum value; 3) plane away or ignore small details of a selected feature if their sizes are smaller than the required value.

Based on these identified processes, an operation-matrix and operation-network were presented in (Peng et al., 1996) for modelling basic operations for complex generalization problems. They will be used to guide generalization rule-bases (to be discussed in section 4.2.2).

3. The formal data structure model

The implementation of the generalization processes relies on a formal description of the thematic and geometric aspects of (terrain) features and the relationships between them. While the thematic aspect can be described as a set of properties in a class definition, the geometric aspect depends on the supporting data model. The Formal Data Structure model developed by Molenaar (Molenaar, 1989, 1991) is an object-oriented topological (conceptual) data model. It consists of 1) three feature types, namely *point feature, line feature,* and *area feature*, classified according to the geometric description of terrain features; 2) four geometric data types defined based on graph theory, including *coordinates, node, arc,* and *shape*; 3) a set of links among geometric data types (g-g links), and a set of links between geometric data types and feature types (g-f links). It supports a number of elementary topological relationships, including *area-area, line-line, point-point, area-line, point-line,* and *point-area* relationships.

This model is the basis based on which we define the fundamental geometric transformations which in turn will support complex generalization processes. The basic idea is the following: any geometry-related (complex) generalization process can be broken down into lower level processes, and no matter how different the thematic aspects are, eventually solutions are based on unambiguous and reliable transformations at the primitive level. At the primitive level, we can predefine a set of fundamental geometric transformations, and any geometric operation at a higher level will be a combination of some of these fundamental transformations with a proper modelling. The data model determines which transformation can be safely defined. In the FDS, for instance, adjacency is only defined for faces that have a common boundary. Therefore, the model does not support the operation of merging two disconnected faces. It will be extremely difficult to prevent topologic violation if such an operation would nevertheless be implemented.

4. An O-O conceptual design for database generalization

Having the FDS as a basis, we can now study how to introduce database generalization. We first introduce an o-o database structure and then elaborate the generalization operations. The C^{++} programming language will be used for procedure description (note that C^{++} is chosen for formal purpose. In practice, a more efficient approach is to use an existing OODBMS, such as O_2 or Ontos, and its associated database programming language). For the sake of simplicity, we will not pay attention to the problems of object storage strategy, persistent storage management, transaction, and other related issues, as these are the problems that any o-o database management system must solve.

4.1 A General FDS database

The proposed (2D) database structure based on the FDS is shown in Figure 4. Note that in this diagram, an additional data type *face* is introduced as an additional geometric data type based on the following considerations: 1) although *face* is not a necessary primitive component in defining the (2D) geometric descriptions of features, it helps to increase efficiency of spatial analysis and to define the fundamental geometric transformation set described in chapter 3. A face as a unit instead of a list of arcs has advantage in defining the geometric transformation of area features; 2) with *face* as a data type, the *left-feature* and *right-feature* properties of an arc become *left-face* and *right-face* respectively, which leads to the four basic properties of an arc (i.e., *begin-node, end-node, left-face*, and *right-face*) to stay in the same (primitive) level. However, this additional data type implies redundancy which must be taken care of by consistency check and concurrency management.

The diagram implies the following concepts: 1) a database is an object which contains and manipulates a list of object-containers; 2) an object-container is also an object that contains and manipulates database objects; 3) a database is therefore a super container; 4) an object always maintains a reference to its associated container, which represents the *part-of* relationship; 5) a thematic object (i.e., feature) holds a reference to its geometric description, which itself is a (geometric) object holding a link to the thematic object it belongs to, and having methods to access and collect its primitive geometric components; 6) a geometric primitive is an object maintaining a link to its associated geometric object; 7) the six geometric object containers are the basic components of a database and are always created by the database at the stage of building the database, whereas thematic object containers are normally created by the database upon the request of the user.

This database structure has important implications for implementing generalization operations.

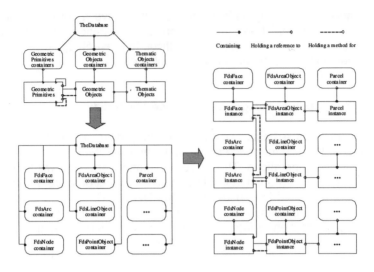

Figure 4 A general FDS structured database.

Figure 5 shows the basic classes and auxiliary classes as well as the inherence hierarchy. Their definitions are given below.

BaseObject: an abstract base class with an unique identity and a pointer pointing to its container as its attributes. All other classes except *Location2D* will be derived from this class to maximize the benefits of inheritance, dynamic binding, and type casting.

Location2D: a class that defines and manipulates the plane coordinates of points.

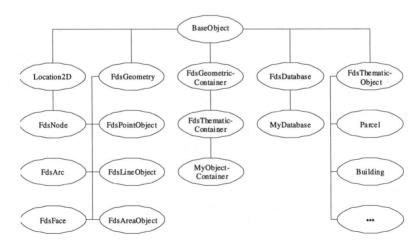

Figure 5 Object classes and class inheritance hierarchy.

FdsGeometry: a class derived from *BaseObject*. It holds a link, i.e., *part-of*, to the associated thematic object or geometric complex (to be explained later), and will serve as a base class for all the object classes related to geometry.

FdsNode, FdsArc, FdsFace: derived from *FdsGeometry*, these three classes represent the three geometric data types in the FDS, and their instances are referred to as geometric primitives. *FdsNode* is also derived from *Location2D*. *FdsArc* has additional attributes beginNode, endNode, leftFace, and rightFace, its *shape* is implicitly defined by the beginNode and endNode. It is also possible that *FdsArc* holds a list of sequential coordinates between the begin-node and end-node to allow "curve" arcs. *FdsFace* holds a method *GetBoundary* to get all the arcs/nodes that form the boundary of the face. Note that an alternative is that each face holds references to the arcs to increase efficiency, which is at the cost of increasing the work of consistency check and concurrency management due to the redundancy.

FdsPointObject, FdsLineObject, FdsAreaObject: derived from *FdsGeometry*, these three (geometric object) classes represent the geometric parts of the three feature types in the FDS and their instances are referred to as geometric complex (or geometric objects). Each of these classes has an operation to get their (primitive) geometric components, that is, a node for a point object, a list of arcs for a line object, and a face for an area object (assuming that an area object has only one face). Note that an alternative is that each geometric complex holds references to its components to increase efficiency, which requires extra work on consistency check and concurrency management.

FdsGeometricContainer: a container class derived from *BaseObject*. It is used to create, maintain, retrieve, detach/delete instances (or objects) of the above six geometry related classes.

FdsThematicContainer: derived from *FdsGeometricContainer*, this container class will serve as the base class for thematic object containers used to create, maintain, retrieve, detach/delete instances of the corresponding thematic object classes. It has an additional attribute *fdsObjectType* to indicate the associated object type (i.e., POINT, LINE, or AREA) of the thematic object class.

FdsDatabase: a (container) class derived from *BasedObject*. It is used to create, maintain, retrieve, detach/delete object-containers contained in a database.

FdsThematicObject: an object class derived from *BaseObject*. It maintains a link to its geometric component, and will serve as a base class for any thematic object class such as *Parcel* and *Building*.

4.2 Introducing generalization in a FDS structured database

The generalization model should be implemented according to the logical structure of a database. With the database structure described above, generalization operations should be defined at database level, and be "propagated" to object-container level and object level if necessary. In defining these operations, generalization rules will play a role since operations are in fact the "consequences" of rules that (implicitly or explicitly) specify which operations to be invoked and how they will be conducted in general.

4.2.1Generalization rule-base

To deal with (different) user demands, and be able to handle different themes and applications, a generalization rule-base should comprise a list of statements or rules that are constructed according to the following general principles:

- each statement should lead to actions on one or more object classes, therefore;
- each statement should comprise an action part and an argument part. The action part specifies the desired (generalization) operations. The argument part gives a list of classes on which the operations will apply, and a list of attributes of which the values require to be modified in a process, as well as conditions and spatial tolerances, as options.

Although there are numbers of schemes available for constructing a rule-base (Hughes, 1991), it is possible to design a template for each generalization operation type. The following is an example for the nine operations, in which the capitalized words are system defined *key-words*. Their names are self explanatory, however, some of the key-words such as COPY and SUM may need further explanation. This will be done during the discussion.

- OPERATION **selection** ON CLASS <class name> WHERE <condition>
     {AND/OR <condition> ...}
- OPERATION **reclassification** ON CLASS <class name> NEW CLASS <class name>
     COPY <attribute name> {, <attribute name>, ...}
- OPERATION **universalization** ON CLASS <class name> NEW CLASS <class name>
- OPERATION **universalization** ON ATTRIBUTE <attribute name> OF CLASS
     <class name> NEW LEVEL <level>
- OPERATION **combination** ON CLASS <class name>
- OPERATION **simplification** ON CLASS <class name> NEW CLASS <class name>
- OPERATION **homogenization** ON CLASS <class name> {BASED ON <attribute name>}
     {SUM <attribute name> {, <attribute name>, ...}}
- OPERATION **collapse** ON CLASS <class name> NEW TYPE <POINT/LINE/AREA>
- OPERATION **aggregation** ON CLASS <class name> TOLERANCE <tolerance>
     {SUM <attribute name> {, <attribute name>, ...}}
- OPERATION **deletion** ON CLASS <class name> TOLERANCE <tolerance>
- OPERATION **simplification** ON CLASS <class name> TOLERANCE <tolerance>

In fact, this scheme is very much similar to a batch file of database query processes (e.g. the SQL), which reflects the nature of database generalization as a database process. It provides the user the means to define the target database and the transformation, so that the user remains the control for what he/she wants, while the implementation is actually done by the system in an automated or batch manner. To illustrate the concept and usage of these templates, consider the following simple example:

(1) OPERATION **selection** ON CLASS Parcel AND Building WHERE *

(2) OPERATION **selection** ON CLASS Parcel WHERE landUse = "green" OR "water-body"

(3) OPERATION **reclassification** ON CLASS Parcel NEW CLASS LandUseZone COPY landUse

(4) OPERATION **homogenization** ON CLASS LandUseZone BASED ON landUse

(5) OPERATION **universalization** ON CLASS Building NEW CLASS BuiltUpArea

(6) OPERATION **homogenization** ON CLASS BuiltUpArea SUM ATTRIBUTE population

(7) OPERATION **aggregation** ON CLASS BuiltUpArea TOLERANCE 3.0

(8) OPERATION **deletion** ON CLASS BuiltUpArea TOLERANCE 6.0

In this example, rule (1) selects classes *Parcel* and *Building*; rule (2) selects parcels with land use "green" or "water-body"; rule (3) creates a new class *LandUseZone* by the use of attribute *landUse* of class *Parcel*; rule (4) creates land-use homogeneous zones according to the *LandUse* of class *LandUseZone*; rule (5) changes the thematic resolution of type *Building* to built-up-area; rule (6) creates larger built-up-area homogeneous zones and assigns a new value to the attribute *population* by taking the sum; rules (7) and (8) changes the spatial resolution of *BuiltUpArea* with a tolerances 3.0 and 6.0 for distance and size respectively.

4.2.2 Generalization operations at database level

Conceptually, when a generalization process is required, the user will first define the rule-base, and then "communicate" with the system through its interface, and send a message to the system by, for example, clicking on an icon or strike a function key. Upon receiving this message, the system may ask some further information (e.g., the database and rule-base names) by, for example, popping up a dialogue window. After confirmation, the system should take over the control and start a batch process by passing a message to an adequate database. Upon receiving the message, the database starts its *Generalization* operation defined in the following, to complete the task:

```
void FdsDatabase::Generalization( NameType* ruleBaseName )
{ NameType className[MaxClassName], newClassName[MaxClassName],
  condition[MaxCondition];
  NameType attributeName; NameList attributeList;
  int operation, level, newType; double tolerance;
  Restart(); /* set the current index to the top of the container array */
  for( IdType i = 1; i <= numberOfClasss; i++ )  /* no class is selected at the begging */
  { FdsThematicContainer* container = GetNextClass(); container->SetSelection( FALSE );
  }
  ...; /* open the rule-base indicated by 'ruleBaseName */
  while( ... )  /* while not every rule has been executed */
  { ...;    /* reason the rule-base and get next rule according to the sequence specified in the
            operation-network and divide it into operation part (e.g. COLLAPSE) and argument part
            (e.g. className) (see 4.2.1) */

    switch( operation )
    { case Selection:          Selection( className, condition ); break;
      case Reclassification:   Reclassification( className, newClassName, attributeList );
                               break;
      case UniversalizationC:  Universalization( className, newClassName ); break; // on
                               class
      case UniversalizationA:  Universalization( className, attributeName, level ); break; //
                               on attribute
```

```
        case Combination:      Combination( className ); break;

        case SimplificationT:   Simplification( className, newClassName ); break; //
                                thematic simplification
        case HomogenizationC:  Homogenization( className, attributeList ); break; // based
                                on theme
        case HomogenizationA:  Homogenization( className, attributeName, attributeList );
                                break; // attri.
        case Collapse:         Collapse( className, newType ); break;
        case Aggregation:      Aggregation( className, tolerance, attributeList ); break;
        case Deletion:         Deletion( className, tolerance ); break;
        case SimplificationG:   Simplification( className, tolerance ); break; // geometric
                                simplification
    }
  }
}
```

The task of this operation is not to perform any generalization activity, but acts as a control centre conducting decision-making at the highest level and determining what to do, in a class by class manner, based on the current and historical information. To be able to do so, a reasoning mechanism (or inference engine) must be introduced, that can interpret and search the rule-base, and keep track of the historical states according to the structure defined in the operation-network (Peng et al., 1996). Such an inference engine can be developed using C/C++ programming language without much effort. Rule interpretation will detect the operation name, class name(s), and other arguments/conditions included in a rule. The key-words used in the rule-base play an important role in the process.

According to the structure defined in section 4.1, a database does not directly manipulate database objects, but does so via their containers. For instance, no methods are defined in the class *FdsDatabase* to create, store, retrieve and detach/delete nodes, arcs, faces, parcels, and buildings. Instead, these objects (and of similar kinds) are contained and manipulated by their associated object-containers, such as *FdsNodeContainer, FdsArcContainer, FdsFaceContainer, parcelContainer, and buildingContainer*, which in turn are created and manipulated by the database. Therefore, generalization operations defined at this level will not apply to database objects, but object-containers, such as indicating adequate containers, creating new containers, and passing generalization messages to related containers for further processes (to be discussed later in "generalization operations at object-container level"). The following examples are given to illustrate this concept:

```
void FdsDatabase::Universalization( NameType* className, NameType* newClassName )
{ FdsThematicContainer* container = GetClass( className );
  if( container->IsSelected() == FALSE ) return;
  FdsThematicContainer* newContainer = CreateClass( newClassName,
  container->GetFdsObjectType());
  container->Universalization( newContainer );
  DetachClass( container, FALSE );  //should not detach geometry
}
```

```
void FdsDatabase::Homogenization( NameType* className, NameList& attributeList )
{ FdsThematictContainer* container = GetClass( className );
  if( container->IsSelected() == FALSE ) return;
  container->Homogenization( attributeList );
}
```

In the *Universalization* operation, the first line gets the container associated to a class indicated by argument *className*; the second line makes sure that only selected classes will be generalized; the third line creates a new container associated to a new class indicated by argument *newClassName*; the fourth line sends a message *Universalization(newContainer)* to *container*, which, upon receiving the message, will perform a further (universalization) process (to be discussed later). The *attributeList* in the second example specifies which attribute values to be accumulated.

4.2.3 Generalization operations at object-container level

From the programming point of view, object-containers are objects controlled by the database, and they in turn contain and manipulate the database objects. Their existence is hidden from the user in the sense that he/she does not need to be aware of their existence when defining the rule-base. Because of these properties, those operations that act at class level (e.g., reclassification and thematic simplification) will conduct substantive generalization activities at the container level, whereas those that act at object level (e.g., homogenization, aggregation, deletion, and geometric simplification) will "propagate" the processes to the object level. In this sense, an object container is a "message transition station", that receives messages from the database and transforms them to the objects it contains. The following examples show how an operation is implemented at this level:

```
void FdsThematicContainer::Universalization( FdsThematicContainer* newContainer )
{ Restart();
  for( IdType i = 1; i <= numberOfObjects; i++ )
    { FdsThematicObject* object = GetNextObject(); // get an object of the sub-class
      if( object->IsSelected() == FALSE ) continue; // only selected objects need to be
      generalized
      FdsThematicObject* newObject = newContainer->CreateObject(); // create a new object
      of the super-class
      newObject->SetGeometry( object->GetGeometry() ); // set the geometric component of the
      new object
      newObject->CopyAttributes( object ); // copy attribute value (to be discussed later in 4.2.4)
      FdsGeometry* geometry = newObject->GetGeometry());
      geometry->SetTheme( newObject ); // the geometric component maintains a link to the
      new object
    }
}
```

```
void FdsThematicContainer::Homogenization( NameList& theAttributeList )
{ if(( fdsObjectType == POINT ) || ( fdsDataType == LINE )) return; // this operation only applies
    to area object
    Restart();
    for( IdType i = 1; i <= numberOfObjects; i++ )
      { FdsThematicObject* object = GetNextObject();
        if( object->IsSelected() == FALSE ) continue; // only selected objects need to be
        generalized
        object->Homogenization( theAttributeList ); // send message to the object, and ask it to
        conduct further process
      }
}
```

It is important to note that in the *Universalization* operation, although the sub-class was derived from the super-class, we cannot simply take out some of the attributes and convert its objects into instances of the super-class. We have to explicitly create an instance of the super-class, and then copy the geometric component and attribute values to the new object. The same applies to thematic simplification operation.

4.2.4 Generalization operations at object level

Only some of the processes are "propagated" to this level, namely, selection, universalization, homogenization, collapse, aggregation, deletion, and (geometric) simplification, among which, the selection operation is very much similar to the (database) retrieval query process. Several additional attributes and functions are introduced to the class *FdsThematicObject*. These include: 1) a boolean type variable *selected* used to indicate if an object is selected; 2) a structure *organization* introduced to indicate if an object also belongs to a "container-feature" (e.g., a building belongs to an university); 3) four functions *GetSomething*, *CopyAttributes* (two versions), and *SumAttributeValues*. They are all defined as **virtual** to allow dynamic binding, and must be redefined in any further derived object class.

For two reasons these functions are introduced. First, some generalization operations may need to copy the values of some attributes of an object of one class to an object of another class. For instance, when creating an object of class *LandUseZone* from class *Parcel* through *reclassification*, one may require the value of attribute *postCode* of a parcel to be copied to the new object of class *LandUseZone*. Another example is related to the *universalization* operation, in which a new object of the super-class is created based on an object of the sub-class, and because the sub-class also has the attributes that the super-class has, we would like the values of these (common) attributes of the object of the sub-class to be copied to the corresponding object of the super-class. Second, some operations, such as *homogenization* and *aggregation*, may need to sum the attribute values of two or more objects of same type for a new (larger) object. For example, when aggregating two buildings into a larger one, one may ask to sum the population values of the two original buildings for the new object. The key-words COPY and SUM given in 4.2.1 were particularly introduced to specify these requests. The following examples show how these functions are implemented in class *Parcel*.

```
ErrorType Parcel::GetSomething( NameType* name, void* result )
{ if( !stricmp( name, "landUse" )) { strcpy( (NameType*)result, landUse ); return OK; }
  if( !stricmp( name, "address" )) { strcpy( (NameType*)result, address ); return OK; }
  ...;
  return FAIL;
}
void Parcel::CopyAttributes( FdsThematicObject* sourceObject) // copy values of all common
attributes
{ sourceObject->GetSomething( "landUse", (void*)landUse );
  sourceObject->GetSomething( "address", (void*)address );
  ...;
}
void Parcel::SumAttributeValues( FdsThematicObject* anotherObject, NameList& attributeList )
{ attributeList.Restart();
  while( !attributeList.End() )
    { NameType* attribute = attributeList.Peek();
      if( !stricmp( attribute, "population" ))
        SetPopulation( GetPopulation() + ((Parcel*)anotherObject)->GetPopulation() );
      ...; attributeList++;
    }
}
```

Note that if the object storage format of a system is transparent and can be accessed by the programming language, then the implementation of these functions will be different and need **not** be redefined in any derived object class.

Because an object of class *FdsThematicObject* (or its descendants) does not directly own and manipulate its geometric component, but holds the identifier number (can also be a pointer or reference) of that component, which itself is an object having its own properties and methods to perform necessary geometric operations, generalization operations defined at this level will only carry out decision-making processes and change the thematic properties of an object. The final geometric processes (such as merging two faces, simplify a line, and shrink an area to a point) will be conducted by the geometric components upon request. To be able to perform necessary geometric transformation, two methods, that is, *Simplification* and *Collapse*, need to be introduced to the class *FdsLineObject* for line simplification and shrinking respectively, and four functions should be introduced to the class *FdsAreaObject*, namely, *Aggregation, Simplification, Collapse, GetNeighbours*. Note that these functions can be predefined and implemented as standard functions. Thematic consideration can be taken into account in advance by the use of attribute *part-of* defined in the class *FdsGeometry*.

The following example shows how an operation is implemented in this level:

```
void FdsThematicObject::Homogenization( NameList& attributeList )
{ ObjectArray array;
  FdsAreaObject* myGeometry = (FdsAreaObject*)(GetGeometry());
  while(1)
    { Boolean done = FALSE;
      myGeometry->GetNeighbours( array );
      for( CountType i = 1; i <= array.GetNumberOfObjects(); i++ )
        { FdsAreaObject* geometryNeighbour = (FdsAreaObject*)(array[i]);
          FdsThematicObject* neighbour = geometryNeighbour->GetTheme();
```

```
        if( neighbour->IsSelected() == FALSE ) continue;
        if( stricmp( neighbour->GetClassName(), GetClassName())) continue;      // if not the
        same class
        myGeometry->Aggregation( geometryNeighbour );
        SumAttributeValues( neighbour, attributeList );
        ((FdsThematicContainer*)myContainer)->DetachObject(neighbour, FALSE);
        done = TRUE;
      }
    if( done == FALSE ) break;
    }
}
```

5. Discussion

The approach to database generalization introduced in this paper is illustrated in Figure 6. It is clear that (automated) database generalization can be implemented in a GIS given the support of a good data model and an adequate system development environment, and should be embedded in the database structure. In this study, a three-level structure (database/container/object) was proposed, which allows a complex generalization problem to be decomposed and solved at different levels according to its nature, which in turn leads to a more simple, clear, and structured implementation of generalization operations. The implementation could be more straightforward and simple if the DBMS maintains a "Class and Class Hierarchy Manager" that takes care of the definition of a class and its hierarchical relationships with other classes, and provides facilities to access this information. The benefit of object-orientation is obvious from the context. First the FDS was translated smoothly into a database structure without any loss of its semantic meaning; second, apart from other commonly recognized advantages, the inheritance and dynamic binding mechanisms in the sense of programming ensured generalization operations and other associated functions to be defined without knowing exactly the definition of a future (thematic) object class, which is critical for developing a generic generalization in a GIS. The facility of overloading was also beneficial, we could use the same operation name (e.g. simplification) for different tasks. Encapsulation enabled generalization operations to be bound to the object itself, rather than existing as separate procedures, so that the whole generalization mechanism can be embedded in the database structure. The proposed concept has the advantage that generalization operations can be redefined with different algorithms for a new (thematic) object type without changing the existing controlling structure.

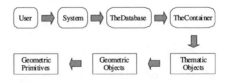

Figure 6 The generalization flow.

Acknowledgements
We would like to thank the anonymous reviewers for their useful comments.

References

Hughes J.G., 1991, Object-Oriented Databases, Cambridge: University Press, .

Molenaar M., 1989, "Single Valued Vector maps - A Concept in GIS", Geo-Informations-Systeme 2(1), pp. 18-26.

Molenaar M., 1991, "Terrain Objects, Data Structures and Query Spaces", in Schilcher M.(ed.), Geo-Informatik, Siemens-Nixdorf Informationssystemme A.G., Munchen, 1991, pp.53-70.

Molenaar M. 1993, "Object Hierarchies and Uncertainty in GIS or Why is Standardisation so Difficult", GeoInformations-Systeme, Vol. 6, No. 3, pp. 22-28.

Nyerges T.L. 1991, "Representing Geographical Meaning", in Buttenfield B.P., McMaster R.B. (ed), 'Map Generalization: Making Rules for Knowledge Representation', Longman House Essex, United Kingdom. p.59-85.

Peng W., Molenaar M., 1995, "An Object-Oriented Approach to Automated Generalization", Proceedings of GeoInformatics'95, Hong Kong. pp. 295-304.

Peng W., Tempfli K. and Molenaar M., 1996, "Automated Generalization in a GIS Context", International Symposium on GIS/RS, Research, Development and Application, Florida, USA, pp. 135-144.

Contact address
W. Peng and K. Tempfli
Department of Geoinformatics
International Institute for Aerospace Survey and Earth Sciences
P.O. Box 6
7500 AA Enschede
The Netherlands
Phone: +31-53-4874358
Fax: +31-53-4874335
E-mail: peng@itc.nl, tempfli@itc.nl

A HIERARCHIC RULE MODEL FOR GEOGRAPHIC

INFORMATION ABSTRACTION

John W.N. van Smaalen

Wageningen Agricultural University
Department of Surveying and Remote Sensing
Wageningen, The Netherlands

Abstract

The model covers the use of a hierarchically ordered set of rules and object classes for geographic information abstraction. It provides a generic approach for use with many types of geographic datasets. The model is based on the definition of database object classes to represent terrain features. Superclasses and composite object classes are derived from this basic object classification, using spatial or thematic 'nearness' of the classes.

Three types of data description are distinguished: the basic geometric level, a topological description of the dataset, and the already mentioned description of classes and their mutual relationships. Based on the defined classes a set of abstraction rules is created. These rules are related to the three description methods having, successively: geometric, topological, and thematic conditions. The class level is used to control the abstraction process, the end-user is expected to define the subject of the abstraction in terms of a selection of (composite) classes.

Key words

GIS, generalization, abstraction, classification, rule-based.

1. Introduction

1.1 Map generalization

Geographic information consists of data with a spatial aspect, mostly related to the earth's surface. A characteristic of this information is that it normally consists of a thematic and a spatial component. The information is usually gathered, structured, and stored with a certain purpose in mind. Optimizing data for a certain use often means that it is badly accessible, or not particularly applicable, for other purposes. One of the properties related to the use is the desired level of detail. A dataset can be derived to provide as much detail to its user as possible. For another user this amount of data can be overwhelming, making it impossible to get an overview. A solution to this problem is in generalizing the data. João [João et.al 1993] defines generalization as the process which creates a derived data set with more desirable and usually less complex properties than those of the original data set. But what is more desirable? Just like the structure of the original data this depends strongly on the purpose. Consequently the generalization strategy to be followed is also dependent on the use of the generalized dataset.

1.2 Information abstraction

When looking at the components of generalization distinguished by different authors, we find that most tend to divide the generalization process into two distinct parts. Geographical information abstraction mainly concerns managing geographical meaning in databases, and map generalization mainly concerns structuring map presentations. For these reasons, it is convenient and useful to separate geographical information abstraction and map generalization [Nyerges 1991]. Or, as [Kilpeläinen 1992] states: 'It is emphasized that the conceptual generalization should be done before the visualization phase. These parts were often put together, and this makes the problems of generalization even worse. We should, whenever possible, separate the actual generalization problems from the problems related to the graphic display.'

These distinctions roughly agree with the distinction in geographic information abstraction and graphic generalization (≈ map generalization) used in the research presented here (Figure 1). Geographic information abstraction specifies what must be expressed to fulfill the purpose defined by the user. Graphic generalization specifies what can be expressed, taking into account graphical limitations like pixel size and the human ability to see. Therefore graphic generalization is scale-dependent, geographic information abstraction is not.

Figure 1 Geographic information abstraction as it relates to graphic generalization.

Geographic information abstraction can be seen as a pre-stage for graphic generalization and to create datasets better suited to the requirements of spatial analyses. Although organizations involved in the production of geographic and cartographic material are highly interested in automation of the process it is still mostly rendered interactively.

2. Concepts of the research

2.1 Introduction

The topic of the research project is geographic information abstraction. The research concentrates primarily on thematic and topological issues. The method described is primarily being developed to generalize detailed cadastral and topographic data but is intended to provide a more generic approach.

2.2 Object-based data structure

The strategy of the research is based on the definition of a powerful data structure to describe the basic information. Elementary terrain features and their mutual relationships are described using geometric, topological and thematic components. The data will be stored in vector format, structured in an object-based1 way [Molenaar 1989][Richardson 1993] (for an example of terminology used see Table 1). By now the value of the object-based or object-oriented2 approach for many applications, including generalization, is widely recognized [Mark 1991].

'object class3'	roads
'object instance'	Route 66
'attribute'	length of roads
'attribute value'	4000 km (2448 mls.) length of Route 66
'relationship type'	roads connect cities
'relationship'	Route 66 connects Los Angeles and Chicago

Table 1 Object related terminology.

The most promising approach towards a powerful generalization concept seems to be based on a combination of classification hierarchies (top-down) [Robinson 1995], topological/spatial analysis (bottom-up) and rules.

In the generalization domain, little is known about the user and with that the purpose of the generalization. The approach taken here to overcome this problem is to provide a method to structure the available data and the rules to be applied to these data, so that an expert user (a user who is familiar with both generalization/classification theory and the data to be generalized) will be able to define the purpose of the generalization in terms of this structure.

2.3 Classification and aggregation hierarchies

In an object-based geographic database the object instances are member of a basic class, a 'flat' data structure (e.g. a certain object is a house, a member of the class 'house'). I order to structure the dataset these basic classes can be grouped into more general or aggregate classes. This is done by means of assigning relationships to the classes:
- 'is-a' relationships, referring to class generalization (Figure 4); the combination of several classes to a more general superclass;
- 'part-of' relationships, referring to aggregation (Figure 4) grouping multiple individuals to a new composite object [Frank & Egenhofer 1988];
- and other, more loosely defined, relationships; often referred to as associations. Selections are considered to be part of these.

---

[1] Object-based: real world objects are represented as units in the database
[2] Object-oriented: object is a 'package' that contains both data and related procedures [Taylor 1990]
[3] The terms 'class' and 'type' are equivalent

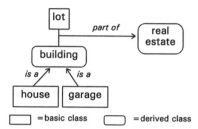

Figure 2 Example of classification generalization (is a) and aggregation (part of) of object classes.

Figure 2 shows an example of the classification of object classes into superclasses followed by aggregation to create composite object classes. The (in this case) basic object classes 'house' and 'garage' are generalized to create the (super)class 'building'. Instances of this (super)class can be aggregated with instances of the class 'lot' to create a composite object instance of the class 'real estate'. A composite object (aggregate) can be seen as a relationship between two (or more) objects seen as a new object [after: Hansen & Hansen 1992]. Similarly a composite object class can be seen as a relationship type between two classes viewed as a new object class. In the case of Figure 2 the relationship between 'building' and 'lot' ('building' is on 'lot') is seen as a composite class 'real estate'

Besides creating superclasses or supersets before aggregation takes place ('building' in Figure 2) it may also be necessary to create subsets within newly created classes in order to discern certain instances for aggregation or deletion. These subsets are created based on characteristics which can be either thematic, topological or (geo)metric in nature, for example being dead ends in a network of roads (Figure 3).

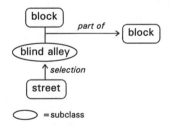

Figure 3 Selection of instances within an object class to enable deletion/aggregation.

2.4 Operations used for geographic information abstraction

The operations related to thematic classification of basic object classes are illustrated in Figure 4 Class generalization is based on 'is a' (or 'member of') relationships between object classes. Aggregation is performed through 'part of' relationships. Class generalization primarily leads to thematic abstraction but can also lead to spatial abstraction if subsequently adjacent members of a resulting class are aggregated. Aggregation leads to spatial abstraction and generally more complex thematic object description. Deletion primarily results in spatial abstraction (thematic abstraction if a complete class is deleted).

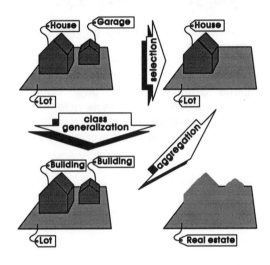

Figure 4 Operations used in geographic information abstraction.

2.5 Describing geographic datasets

Figure 5A shows a description of reality in a GIS by means of area features. In detailed datasets, such as cadastral data, this is the predominant description form. Topological relationships which are implicit in this description can be made more explicit by representing the object instances and their neighbor relationships as an adjacency graph Figure 5B) [Molenaar 1995]. In the graph description both thematics and topology of the dataset are described, the geometry (georeference) is missing. By adding the graph description to the normal area description (containing the geometry) of the dataset and the classification structure, we can distinguish three levels of description for the dataset. By identifying these three levels of description we are better able to analyze the structure of the dataset. It will also help in defining abstraction rules as these rules can be associated to the different levels, thus introducing structure in the set of rules.

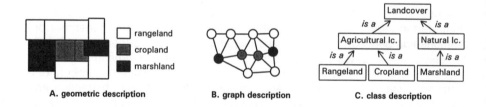

A. geometric description   B. graph description   C. class description

Figure 5 Various descriptions of a dataset.

Instead of moving directly from a detailed geometric description to a less detailed geometric description (lowest level in Figure 6), we first change to an abstract description of the dataset, the classification network (top level Figure 6). At the class level we decide what should be included and what rules and operations are required to move from the basic classes contained in the database to the classes needed. When this is determined we go into detail again and actually modify the geometry of the dataset. In this process the topological description is used to analyze spatial relationships between the objects in the dataset.

Thus, the class level is the level used to control the information abstraction process, the topological (graph) level is mostly used for analysis, and the final implementation takes place at the geometric level.

2.6 Abstraction rules

After the explanation of processes, relationships and properties involved, we come to the structure of the rule set. The general composition of rules is:
if <condition> then <action>

Based on the operations (Figure 4) we distinguish rules with actions referring to:
- generalization
- aggregation
- selection within a class
  We can distinguish selection rules with conditions referring to data-classification based on:
  - thematic properties
    example: if city.population < 100.000 then .....
  - topological relationships
    example: if street is a blind alley then .....
  - geometric properties
    example: if barn.area < 50m2 then .....
  - geometric relationships
    example: if distance(barn, road) > 100m then .....

Generalization and aggregation rules apply to complete classes (all instances of the class) and are mostly dependent on the basic classification of the dataset. Selection rules apply to subsets of instances within classes and are more dependent on the purpose of the abstraction. Members of the selected subset can subsequently be deleted or aggregated with instances of another subset or class. Aggregation rules are used to move from one column in Figure 6 to the next, generalization and selection rules to prepare sets of instances to be used for aggregation.

In Figure 6 the largest units are blocks and streets, using this method of stepwise aggregation it is possible to aggregate even further. City quarters and cities would probably be logical next steps. Proceeding like that it is possible to end up with only one object representing the whole dataset, therefore the user has to decide on the largest relevant aggregation level.

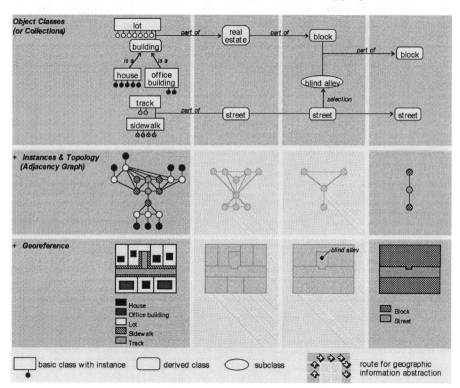

Figure 6 Information abstraction procedure.

3. Description of the system

Describing the implementation of the method we are able to distinguish two phases (Figure 7):
• Definition
• Use

3.1 Definition

In the definition phase (Figure 7) the system is set up for use with a certain dataset. This is done by an 'expert', a user who is familiar with both generalization/classification theory and the data to be generalized. The system will provide a graphic interface (similar to the class level description in Figure 6) to define, visualize and modify classification and aggregation hierarchies. Using this interface the 'expert' builds a hierarchical classification network for the dataset. Abstraction rules refer to the same object classes and can therefore be included in the same structure.

In order not to rely completely on expert knowledge to produce the classification network, spatial or thematic 'nearness' of classes can be employed. This will require spatial or thematic analysis of the complete dataset in order to detect these correlations.

Not only object classes with a spatial representation (building, parcel) but also non-spatial object classes (resident, value) can be included because these can also play a role in the generalization process. An example is the aggregation of residential areas based on the income of the residents.

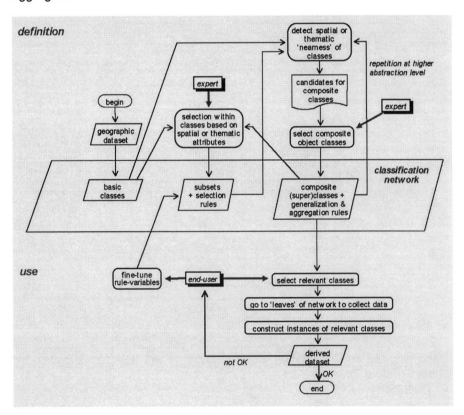

Figure 7 Functional description of the system.

3.2 Use of the system

In the second phase the actual generalization process is carried out. A particular spatial situation leads to the application of abstraction rules formulated in the definition phase. In this case not the class is addressed but the instances of that class, the rules are applied to individual object instances. Spatially related (e.g. neighboring) objects can be aggregated based on the relationships defined.

Generally speaking the steps in this phase are:

- end-user selects relevant classes (Figure 8),
- go to 'leaves' of the network, where the connection to the instances in the database is found,
- follow path back and reclassify and aggregate basic database object instances into the (composite) object instances of the selected classes according to the rules connected to the relationships between classes in the network.

Additional flexibility can be introduced by allowing the end-user to adjust spatial variables within the rule while using the rule set. If certain rule reads 'if width(road) > 2 meters then select road' it is possible to take a variable width instead of 2 meters.

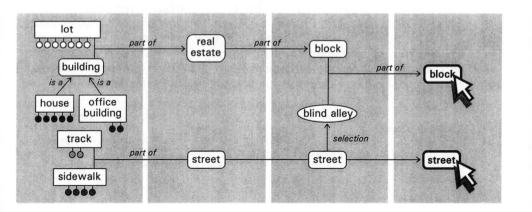

Figure 8 Selection of object classes.

In this second phase user interaction will be limited to the absolute minimum. After adjusting the required parameters applying to the rules, the abstraction process is executed without interruption. Often used sets of parameters could be stored as predefined 'profiles', like the user-defined settings of a word processor.

4. Conclusions

Introducing a hierarchically ordered structure to contain the abstraction rules will make it easier to define a set of rules. The ability to assign rules at the appropriate aggregation/generalization level is considered very important, this has also been recognized by [Robinson 1995]. It structures the decision making process of the 'expert' and directs him through the process of rule definition. In this process the 'expert' is guided by the results of spatial and thematic analyses of the dataset. Once the classification network and related rules are created the model allows 'on the fly' generalization, which is desirable in a GIS environment.

Using different descriptions of a dataset (classes, topology, geometry) enables to clearly identify inter-object relationships and to structure abstraction rules according to these relationships. Furthermore, the graphic portrayal of the classification structure of a dataset provides a high-level interface to the data. At class level it is possible to define rules and relationships that apply to the instances of these classes. Because of the high level of abstraction it provides a clear perception of the structure of the dataset.

If composite instances are completely - both the geometry and the thematics - derived through rules they can only have derived attributes like area, perimeter and, at higher abstraction levels, attributes like 'road density', 'fragmentation index' etc. [Turner 1989]. The behavior of the attributes under aggregation depends on their domain: the type, units and scaletype in which the attribute is described. Further investigation into the implications of attribute behavior during the information abstraction process is needed. Future research also includes the implementation of the model described.

Acknowledgments
The research project is taking place at Wageningen Agricultural University, Department of Geographic Information Processing and Remote Sensing (GIRS), funded by the Dutch Cadastre and the Survey Department of the Directorate General for Public Works and Water Management.

References
Frank, A. and M. Egenhofer, 1988. Object-Oriented Database Technology for GIS. Seminar workbook, San Antonio, Texas.
Hansen, G.W., J.V. Hansen, 1992. Database management and design. Prentice Hall, Englewood Cliffs.
João, E., G. Herbert, D. Rhind, S. Openshaw and J. Raper, 1993. Towards a generalisation machine to minimise generalisation effects within a GIS, in: Mather, P. (ed.) Geographical Information Handling - Research and Applications. John Wiley, Sussex.
Kilpelaïnen, T., 1992. Multiple representations and knowledge-based generalization of topographical data, Proceedings ISPRS commission III, Washington.
Mark, D.M., 1991. Object modelling and phenomenon-based generalization, in: Buttenfield B.P., and R.B. McMaster (ed.), Map Generalization: Making rules for knowledge representation. Longman, Harlow.
Molenaar, M., 1989. Single Valued Vector Maps; a concept in Geo-Information Systems. Geo-Informationssysteme 2(1).

Molenaar, M., 1995. The role of topologic and hierarchical spatial object model in database generalization, Publication in Geodesy, nr. , Netherlands Geodetic Commission, Delft.

Richardson, D.E., 1993. Automated Spatial and Thematic Generalization Using a Context Transformation Model. Ph.D. Dissertation, Wageningen Agricultural University, R&B Publications, Ottawa.

Robinson, G.J., 1995. A hierarchical top-down bottom-up approach to topographic map generalization. in: Muller, J-C, J-C Lagrange and R. Weibel, GIS and Generalization; Methodology and Practice. Taylor & Francis, London, 1995.

Taylor, D.A., 1990. Object-Oriented Technology, A Manager's Guide. Addison Wesley, Reading.

Turner, M.G., et. al., 1989. Effects of changing spatial scale on the analysis of landscape pattern. Landscape Ecology Vol. 3, No. 3/4.

Contact address
John W. N. van Smaalen
Wageningen Agricultural University
Dept. of Surveying and Remote Sensing
Hesselink van Suchtelenweg 6
PO Box 339
6700 AH Wageningen
The Netherlands
Phone: +31 317 482906
Fax: +31 317 484643
E-mail: john.vansmaalen@aio.lmk.wau.nl

DEVELOPING APPLICATIONS WITH THE OBJECT-ORIENTED

GIS-KERNEL GOODAC

Ludger Becker, Andreas Voigtmann and Klaus H. Hinrichs

FB15, Institut für Informatik, Westfälische Wilhelms-Universität
Münster, Germany

Abstract
We review OOGDM, an extensible, object-oriented data model for geographic information systems. This model is intended to be a general base for the development of geo-information systems. Currently we are implementing a prototype GIS-kernel GOODAC which realizes OOGDM. We sketch the overall system architecture of this prototype and describe how GOODAC can be used to develop GIS-applications.

Key words
Database, system architecture, data model.

1. Introduction

The study of extensible and object-oriented database systems started almost a decade ago. This research aimed at the support of advanced applications, e.g. CAD, office information systems, and geo-information systems (GIS). Several extensible and object-oriented systems have been built (e.g. [AHS 91], [Deux 90], [Güti 88], [LLOW 91]), and there has been a significant progress on the engineering side (e.g. [GrMc 93], [Haas 89], [Haas 90], [SRH 90]) of these systems. However, the new technology has been applied to GIS in part only.

Many GIS use a classical relational database system to store thematic information based on standard data types. In addition a file management system provides access to geometric and topological data. As a consequence, extensions and adaptations can hardly be performed, and information describing a single object of the real world is distributed in the system. Moreover, typical database features are not offered, and there is a mismatch between the two modules and the two query languages. ARC/INFO is a prominent example for such a GIS [More 89]. Another generation of GIS is based on extended relational database systems supporting additional data types and index structures. Examples of such systems are System 9 [EcUf 89], Smallworld [CNT 90], GEO++ [ViOo 92], and GeoSabrina [LPV 93]. Since extensible relational database systems (e.g. [Güti 89], [SRH 90]) can be extended by data types and index structures, such systems are a good base for GIS. For example, GEO++ [ViOo 92] has been implemented on top of POSTGRES [SRH 90]. A third class of GIS is based on object-oriented database systems which avoid the structural deficiencies of relational systems (e.g. [AHS 91], [Deux 90], [LLOW 91], [AGGL 95]). Such GIS are for example presented in [DRSM 93] and [ScVo 92].

We have developed an Object-Oriented Geo Data Model (OOGDM) [VBH 95]. For querying databases based on OOGDM we have modified the query language OQL presented in [ODMG 93] to meet our requirements. This variant of OQL is called Object-Oriented Geo Query Language (OOGQL), it is presented in detail in [VBH 96a]. Similar to SAIF [KuSo 92] OOGDM provides a class hierarchy which may be used to describe data models for arbitrary GIS applications. In contrast to SAIF which has been designed to support interoperability of existing GIS the focus of our research is on the development of an implementation environment for GIS. A discussion of the modelling concepts of OOGDM and SAIF is beyond the scope of this paper. Compared with other data models and query languages proposed for GIS (e.g. [Alve 90], [BDQV 90], [CNT 90], [ChZu 95], [EcUf 89], [Egen 94], [Feuc 93], [Herr 87], [More 89], [ODD 89], [ScVo 92], [ViOo 92]) the benefits of OOGDM and OOGQL are extensibility **and** support for raster- **and** vector-based data in 2-D and 3-D space.

Currently, we are implementing a Geo Object-Oriented DAtabase Core (GOODAC) which realizes the basic class hierarchy of OOGDM. We expect GIS for concrete applications to be implemented on top of GOODAC. In this paper we present the basic design of GOODAC and describe how to implement a GIS application on top of GOODAC. The requirements of GIS-applications may differ [GüLa 92]. Hence, GOODAC provides only the basic environment for application development. To adapt GOODAC to the requirements of special GIS-applications there are extension mechanisms for adding (1) representations of types, (2) methods and functions for operations, (3) index structures, (4) query processing methods, and (5) extensions of the optimizer. Some of these extensions are supported by database systems (e.g. [Güti 89], [SRH 90]). However, to our knowledge there is no GIS which supports such extensions. By offering an object-oriented data model in an extensible database core, GOODAC provides the benefits of recent research in database technology for the development of GIS.

The paper is organized as follows. Section 2 reviews the data model OOGDM. Section 3 gives an overview of the overall architecture of GOODAC. In section 4 we discuss the development of GIS applications based on GOODAC, and we describe how to customize GOODAC for special GIS-applications. Section 5 concludes this paper and gives an overview of future work.

2. The structure of the OOGDM

2.1 Types, classes, and objects

As usual, objects have an identity, a state, and a behavior. The possible states and the behavior of an object are defined by the attributes and methods of the object's class. Attributes are described by names and types. The definition of attribute types is based on either data types or class names. In addition, they may be structured by the collection type constructors **Set**, **List**, **Bag**, and **Array** (cf. [ODMG 93]). Attribute definitions based on data types are marked by the keyword attribute, and definitions based on class names are marked by the keyword **relationship**. A relationship between two classes c_1 and c_2 which is established by an attribute of c_1 and an attribute of c_2 can be maintained by the system if one attribute is declared to be the **inverse** of the other attribute. We further support generic classes having formal type parameters. Each type parameter T can be constrained by the definition of a supertype S for the actual type parameter. This constraint is denoted by T -> S. Furthermore, we define the use of a generic class c without a formal type parameter as a short notation for c <*Void*>.

2.2 Basic structure

The core of OOGDM is a hierarchy of classes. The most abstract class of our data model is the *spatial_object*. It represents a set of attributes and operations common to all instances which are stored in the GIS-database. The class *spatial_object* is specialized into two subclasses – *feature* and *geo_object*. Geo-objects are used to describe complex objects of the real world. They are composed of *features*, other *geo_objects*, and atomic values. A *feature* represents a single geo-abstraction. It is either elementary or composed of several features (*feature_set*). Figure 1 gives a representation of the class hierarchy described so far.

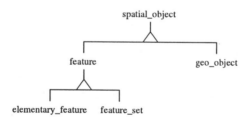

Figure 1 Top-levels of the inheritance tree.

2.3 Elementary features

Elementary features are atomic objects of the database describing a single geo-abstraction. An elementary feature contains a description of the geometry and thematic information related to this abstraction. The thematic information can be customized by providing appropriate type parameters. Based on the geometry of the objects methods for basic operations can be provided which are useful in all GIS application areas. We distinguish the elementary features by their spatial dimensionality.

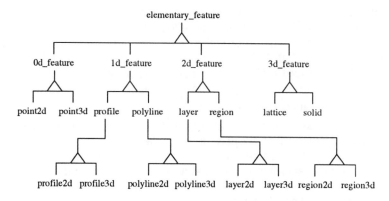

Figure 2 Child-classes of elementary features.

Except for the zero-dimensional elementary features which are described by the class *0d_feature* we distinguish between vector-based and raster-based representations. One-dimensional elementary features are either raster-based *profile* objects or vector-based *polyline* objects. Two-dimensional elementary features are specialized into *layer* objects which are used for storing raster-based data or into *region* objects which are used to store vector-based data. A three-dimensional elementary feature is either a raster-based *lattice* object or a vector-based *solid* object. We assume that all vector-based representations describe a contiguous part *p* of the data space, i.e. the thematic information associated with these features is known for all points in *p* whereas the raster-based representation describes a discrete part of the data space.

To support both 2-D and 3-D GIS-applications we have to distinguish whether an elementary feature is used in a two- or in a three-dimensional data space. Hence, we introduce corresponding subclasses of *0d_feature*, *profile*, *polyline*, *layer*, and *region*. Figure 2 introduces the names of these classes and presents the complete hierarchy of classes for elementary features.

The geometry of the elementary features is described by data types for 2-D and 3-D points, lines, rectangles, polygons, regions, polyhedrons, and solids. A polygon is described by a list of points. A region is a simple polygon with holes. Each hole is again described by a simple polygon. Similarly, a polyhedron is represented by a set of 3-D polygons describing its boundary. A solid consists of a polyhedron with holes. Each concrete class of elementary features has an attribute geometry describing the geometry underlying the objects of the class. This geometry is defined based on the corresponding data types mentioned above. In addition:

- The geometry of a *polyline* is a list of points.
- The geometry of a *profile* is a line in 2-D or 3-D space.
- The geometry of a *layer2d* and a layer3d is a rectangle.
- The geometry of a *lattice* is a box.

All concrete classes of elementary features are generic classes having a formal type parameter *T* which may be used to associate thematic data with the objects. All vector-based classes have an attribute *data* of type *T*, and all raster-based classes have an attribute *data* which is an array of *T* having the appropriate dimension, i.e. each array element corresponds to a raster point.

In addition all raster-based classes have
- an appropriate number of *frequency* attributes describing the raster
- an array *validPoints* of boolean values, which is used to mark each cell in the *data* array of a raster-based object as valid or invalid, and
- a default value which is assigned to non-occupied cells of the data array.

2.4 Geometric and topological operations

The basic classes of OOGDM introduced above represent spatial data. To support processing of this data, a set of geometric and topological operations is required. Geometric operations include a *distance* operation, *length*, *area*, or *volume* operations for one-, two-, or three-dimensional spatial objects, a *center* operation returning the center of an object, and a *bounding_box* operation returning the minimal bounding box of an object.

Topological predicates have been discussed intensively, e.g. [EgFr 91], [EgTh 92], [CFO 93]. For our purposes, we use the following set of five basic topological predicates: *disjoint, touch, in, cross, overlap*. Originally, these relations have been introduced in [CFO 93]. [CFO 93] show that these predicates are sufficient for representing all possible relationships. The predicates can easily be adapted to support 3-dimensional data. Based on these predicates corresponding topological functions *touch, cross*, and *overlap* can be defined which construct for each pair of objects satisfying the topological relationship the resulting intersection object.

All operations and predicates which are not dimension-dependent and the *bounding_box* are introduced in *spatial_object*. It is obvious that some operations have to be redefined in the subclasses to return appropriate values or to be implemented efficiently.

2.5 Feature sets

Elementary features are not sufficient to describe the objects used by a GIS. There is often the need to model compound features, e.g. a river consisting of many segments or a network of rivers. The class *feature_set* is used to represent these types of features. Similar to the class *elementary_feature*, we perform a subdivision of *feature_set* according to the spatial dimensionality, the kind of the resulting object, and the kind of the elementary features being composed. Since objects consisting of components of different types are described by geo-objects, feature sets are homogeneous.

In figure 3 we show the definitions for the class *feature_set* and its subclasses. It is obvious that 0-D elementary features can only be used to define raster-based feature set objects. A set of 0-D features or a set of profiles can describe a profile in 2-D or 3-D space, a layer in 2-D or 3-D space, or a lattice. The corresponding classes are given in parts (1) and (3) of figure 3.

```
class feature_set <T -> feature> : feature { relationship Set <T> members; };
```
(1) Sets of 0-D features:
   class profile2d_0d_set <T -> point2d> : feature_set <T>, profile2d;
   class profile3d_0d_set <T -> point3d> : feature_set <T>, profile3d;
   class layer2d_0d_set <T -> point2d> : feature_set <T>, layer2d;
   class layer3d_0d_set <T -> point3d> : feature_set <T>, layer3d;
   class lattice_0d_set <T -> point3d> : feature_set <T>, lattice;
(2) Sets of polylines
   class network2d <T -> polyline2d> : feature_set <T>;
   class network3d <T -> polyline3d> : feature_set <T>;
(3) Sets of profiles
   class profile2d_1d_set <T -> profile2d> : feature_set <T>, profile2d;
   class profile3d_1d_set <T -> profile3d> : feature_set <T>, profile3d;
   class layer2d_1d_set <T -> profile2d> : feature_set <T>, layer2d;
   class layer3d_1d_set <T -> profile3d> : feature_set <T>, layer3d;
   class lattice_1d_set <T -> profile3d> : feature_set <T>, lattice;
(4) Sets of regions
   class region2d_set <T -> region2d> : feature_set <T>;
   class region3d_set <T -> region3d> : feature_set <T>;
(5) Sets of layers
   class layer2d_2d_set <T -> layer2d> : feature_set <T>, layer2d;
   class layer3d_2d_set <T -> layer3d> : feature_set <T>, layer3d;
   class lattice_2d_set <T -> layer3d> : feature_set <T>, lattice;
(6) Sets of solids
   class solid_set <T -> solid> : feature_set <T>;
(7) Sets of lattices
   class lattice_3d_set <T -> lattice> : feature_set <T>, lattice;

Figure 3 Definition of feature_set and its subclasses (methods omitted).

A set of polylines is a network in 2-D or 3-D space (part (2)). The classes of part (4) describe sets of regions. A set of layers can define a layer in 2-D or 3-D space or a lattice (part (5)). A set of solids is described by the class *solid_set* of part (6). A set of lattices can define another lattice. Such feature sets are denoted by the class *lattice_3d_set* in part (7).

All classes of feature sets which combine raster-based objects of class x basically inherit from *feature_set<x>* and from the class of elementary feature constructed by the set. The first class provides the methods for processing sets and the second class describes the methods and the geometry for the resulting feature.

3. Overview of the system architecture

The system architecture of GOODAC is shown in figure 4. OOGDM is realized by the top layer of GOODAC which is called the *descriptive layer*. There are two interfaces to this layer: the query language OOGQL and a C++-application programming interface (API). The application-dependent classes which can be accessed by these two interfaces are defined using the data definition language of OOGDM called OOGDM-ODL [VBH 96a], a variant of ODL [ODMG 93].

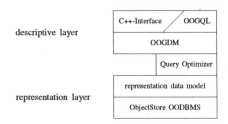

descriptive layer

representation layer

Figure 4 Architecture of GOODAC.

The first step to develop a GIS-application on top of GOODAC is the design of application dependent classes. Features are described by the OOGDM feature classes or by classes which are derived from these feature classes. All geo-objects are modeled by classes which inherit from class *geo_object*. A precompiler translates the corresponding OOGDM-ODL class definitions into C++-code usable in connection with the API. Based on this code the application programmer implements the methods of the new classes. Figure 5 shows the architecture of the C++-programming interface.

In object-oriented programming languages attributes are encapsulated. However, in object-oriented database systems these attributes should be visible at least in the query language [BCD 92]. Hence, attributes of OOGDM classes are visible to the user in the query language and in the data definition language while the API has to ensure privacy for all attributes and therefore has to provide methods for accessing the attributes. Furthermore, relationships and collections must be provided in a C++ suitable way. These demands are satisfied by representing each class derived from class *geo_object* by an interface class and an internal class. The interface classes are the base for the development of GIS-applications. The internal classes which are part of the so called *representation layer* are superclasses of the corresponding interface classes. For classes derived from feature classes the distinction between interface class and internal class is not required. Details are discussed below.

The application programmer has to provide implementations for the methods declared by the OOGDM-ODL definitions. To realize these implementations further classes with non-persistent objects may be provided by the OOGDM-user. The interface classes as well as the internal classes are then compiled and linked with the OOGDM-library which provides implementations for the standard classes to obtain the final OOGDM-based application.

In addition to the basic C++-classes the OOGDM-ODL precompiler also generates information for the OOGDM data dictionary. The OOGDM data dictionary is used by the OOGDM-library linked to the application program to access the stored data. The query language OOGQL can be used to compose ad hoc queries to the database. Furthermore, OOGQL is also accessible from the C++-application programming interface to integrate complex queries into application programs. These queries will be optimized by the query optimizer of GOODAC.

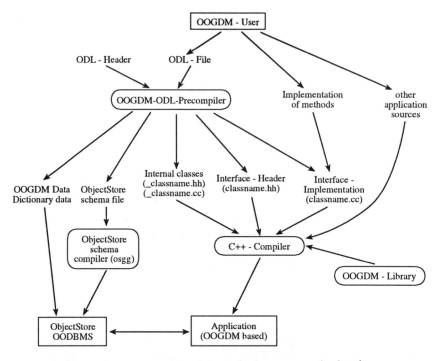

Figure 5 Architecture of the C++ application programming interface.

The bottom layer of the GOODAC architecture is the representation layer. This layer consists of the representation data model and the underlying object-oriented database system ObjectStore [LLOW 91]. The schema for the *ObjectStore* database is generated by the OOGDM-ODL precompiler. Since the internal representation must support efficient access to the objects, this representation may differ from the user's view of a class. For example, an OOGDM region is described by a polygon and a set of holes. For the representation of OOGDM regions we might choose to use a decomposition into trapezoids [ScKr 91]. To support efficient query processing on spatial data, spatial index structures (e.g. [Same 89], [ScKr 91]) may be provided at the representation layer. These index structures may be used as primary or secondary indices for geometric objects. GOODAC can be customized by adding index structures, internal representations, query processing strategies, OOGQL-predicates and functions, and implementations for OOGQL-predicates and functions. This issue is discussed below.

The layer between the descriptive and the representation layer consists of the query optimizer which translates OOGQL-queries into access plans for the representation level. Since GOODAC is extensible, the query optimizer must be adaptable like those optimizers presented in [BeGü 92], [GrMc 93], or [KMP 93]. We plan to use an optimization technique which is based on *Second-Order Signatures* described in [Güti 93] and [BeGü 95]. This optimization technique is currently developed independent of GOODAC.

4. An example application

In this section we describe the development of GIS-applications based on GOODAC. We show how to
- define appropriate application dependent classes in the OOGDM-ODL
- use the C++-API to realize a GIS-application
- customize GOODAC by adding new index structures and new representations for the basic types describing the geometry of the features and geographic objects.

4.1 Class definitions

The first step in application development is the definition of the application dependent classes in the OOGDM-ODL [VBH 96a]. Our running example in this section is a simple GIS-application involving cities and states. Since cities and states are objects of the real world, we describe them by classes derived from *geo_object*. The corresponding class definitions are shown in figure 6.

```
class vegetation_map : layer2d<String> {
        Short coverage (String vegetation_type, region2d query_region) const; };
class elevation : point2d<Short> { } ;
class elevation_map : layer2d_0d_set <elevation> {
        elevation min_elevation () const;
        elevation max_elevation () const;
        Double avg_elevation () const; };
class city : geo_object {
        Relationship region2d geometry;
        attribute String name;
        attribute Long population;
        Relationship state belongs_to inverse state::cities;
        Relationship vegetation_map vegetation;
        Short avg_elevation() const;};
class state : geo_object {
        Relationship region2d geometry;
        attribute String name;
        attribute Long population;
        Relationship Set<city> cities inverse city::belongs_to;
        Relationship city capital;
        Relationship Set<vegetation_map> land_use;
        Relationship Set<elevation_map> topography;};
```

Figure 6 Class definitions for the example application.

A city has the attributes name and population and belongs to a state. In addition, a method *avg_elevation* is provided, which determines the average elevation for the region corresponding to a city. Furthermore, a vegetation map is associated with each city. Since a vegetation map is an object having a geometry and thematic information, such a map is described by the subclass *vegetation_map* of the elementary feature class *layer2d*. For this class we have defined a method coverage which returns the percentage of a certain *vegetation_type* in a given *query_region*. A state is described by its name, its population, the set of cities belonging to the state, and by its capital. Each state has an associated set of elevation maps describing its topography and an associated set of vegetation maps describing the land use. Since vegetation and elevation maps are objects, these relationships are realized by sets of references. Hence, a single vegetation or elevation map may be assiciated with several states. An elevation map consists of a geometry and corresponding thematic information, i.e. elevation maps are features. We represent elevation maps by feature sets composing a layer object from a set of points representing single elevation points. Note that we could describe this application by other class definitions. For example, it might be reasonable to describe the elevation information by a layer instead of a set of points.

The syntax of OOGDM-ODL is very similar to the data definition language ODL proposed in [ODMG 93]. However, note the use of the generic classes. The class *vegetation_map* inherits from class *layer2d<String>* which is an elementary raster-based feature having an array of strings as thematic attributes, i.e. a string is associated with each point of the raster underlying the layer. This string denotes the vegetation at the corresponding raster point.

4.2 The C++-application programming interface

As mentioned in the previous section the OOGDM-ODL class definitions are processed by a precompiler. Each OOGDM-class *c* which is derived from class *geo_object* is translated into an internal C++ class *_c* and a C++ interface class *c* which inherits from *_c*. *c* is the interface which is used for application programming whereas *_c* describes the representation of the objects and the methods required to maintain the data in the underlying ObjectStore database. As we will see, the distinction between an interface class and an internal class is not required for subclasses of *feature*. Hence, only an interface class is generated for such classes. As usual, each C++ class is described by a header and a corresponding implementation file. In figures 7 and 8 we show the generated headers for the OOGDM classes *city* and *vegetation_map*.

For each interface class the precompiler generates method definitions for constructors and destructors and method declarations for each method declared in the ODL (e.g. for method *avg_elevation* of class *city*). The constructors and destructors call the appropriate constructors and destructors of the underlying internal class. The internal class defines methods for reading and updating all attributes of the class. In addition, the internal class provides protected data members for each attribute of the corresponding OOGDM-class (e.g. attribute *population* of class *_city*). These data members of the C++-class are only visible in classes derived from the internal class, i.e. they can be used in the interface class to implement methods which may read and update these attributes. The methods for updating attributes are responsible for maintaining the underlying database. E. g. these methods maintain index structures, additional representations for geometric types, meta data, or time stamps[1].

---

[1] We are currently extending OOGDM to support time varying data [VBH 96b].

```
class _city : geo_object {
public:
  _city() {};
  _city(Ref<region2d<void> > geometry, String name, long population,
         Ref<state> belongs_to, Ref<vegetation_map> vegetation) {};
  virtual ~_city();
  Ref<region2d<void> > get_geometry() const;
  String get_name() const;
  long get_population() const;
  Ref<state> get_belongs_to() const;
  Ref<vegetation_map> get_vegetation() const;
  void set_geometry(Ref<region2d<void> > geom);
  void set_name (String name);
  void set_population(long population);
  void set_belongs_to(Ref<state> belongs_to);
  void set_vegetation(Ref<vegetation_map> vegetation);
  void update();
protected:
  Ref< region2d<void> > geometry;
  String name;
  long population;
  Ref_1_m<city, state> belongs_to;
  Ref< vegetation_map > vegetation;
private:
  Ref< region2d<void> > _geometry;
  String _name;
  Ref _1_m< city, state > _belongs_to;
  long _population;
  Ref< vegetation_map > _vegetation;
};

class city : public _city {
public:
  city() {};
  city(Ref<region2d<void> > geometry, String name, long population,
        Ref<state> belongs_to, Ref<vegetation_map> vegetation) :
             _city(geometry, name, population, belongs_to, vegetation) {};
  ~citiy() {};
  short avg_elevation() const;
};
```

Figure 7 Header for interface and implementation of class city.

By supporting both, the access via data members and the access via methods, two different views on the objects are realized. If the attributes are accessed and updated by the methods only, the attributes appear to be completely encapsulated. This is the view object-oriented programming methodology prefers. However, if the attributes are accessed via the protected data members of the internal class, the attributes may be used like private data members without any additional method calls. This is the intuitive view of the application classes which an application programmer has. Since the update methods for the attributes maintain all required data structures, the stored information is always consistent when these methods are used. If the protected data members are used to update objects, the method *update* must be applied. This method is provided by each internal class and performs all changes which are required due to updates on any attribute of the corresponding object.

GOODAC offers the application programmer the possibility to perform index updates immediately by using the update methods for the individual attributes or to perform index updates after several changes by calling method *update*. Although the underlying data structure may be inconsistent using the second alternative, this policy is useful for some applications. Consider for example a data structure which classifies the temperatures stored in a lattice. If an application wants to predict the temperatures for some point in the future it is not useful to update this data structure after each step of this computation, i.e. for each lattice point - it is more efficient to finish the computation first and to update the data structure afterwards.

For each attribute which is defined as a relationship we generate a reference to the related objects. If the relationship has no inverse, a data member of type *Ref<R>* is generated where R is the type of the object to which the relationship exists (e.g. a *vegetation_map* is referenced by a data member of type *Ref<vegetation_map>* in class *_city*). If there is an inverse for the relationship, a data member of type *Ref_1_1*, *Ref_1_m*, *Ref_m_1*, or *Ref_m_m* is generated. E.g. a state is referenced by each city object via the data member *belongs_to* of type *Ref_1_m<city, state>*, and the set of cities is referenced by each state object via the data member *cities* of type Ref_m_1<state, city>. The *Ref_i_j<T,R>* classes are subclasses of *Ref<R>*. T denotes the class which specifies the relationship, R the type of the object to which the relationship exists. The type of the data member depends on the kind of the relationship which can be recognized by the OOGDM-ODL precompiler. The reference classes realize logical pointers to objects, i.e. in contrast to ObjectStore which uses physical pointers to access objects immediately we use an additional indirection to access objects. Since updates of a primary index may change the physical address of objects, logical pointers are required to keep secondary indices and relationships consistent (cf. section 4.3).

In contrast to classes derived from class *geo_object* it is not allowed to add attributes to classes derived from a feature class. One may only specify the thematic information of a feature by providing a parameter type. Hence, there is no need to provide the update mechanism described above, and we need not distinguish an interface class and an internal C++-class for such classes (Figure 8 shows the C++-class corresponding to the *vegetation_map* OOGDM-class).

```
class vegetation_map : public layer2d<String> {
public:
  vegetation_map() : layer2d<String>() {};
  vegetation_map(RectangleDesc rect, Array< Array<String> > _data)
    : layer2d<String>(rect, _data) {};
  virtual ~vegetation_map() {};
  short vegetation_map::coverage(String type, region_2d<void> query_region) const
};
```

Figure 8 Header of C++ class vegetation_map.

The classes generated by the OOGDM-ODL precompiler are used to implement the methods declared by OOGDM-ODL commands. We show the implementation of method *coverage* in figure 9.

```
short vegetation_map::coverage(String type, region_2d<void> query_region) const {
        layer2d<struct{String data1; void data2;}> part;
        int i, j, allPoints, matchingPoints;

        part = overlap(query_region);
        for (i = 0; i <= part.data.upper_bound(); i++)
                for (j = 0; j <= part.data[i].upper_bound(); j++)
                        if (part.validPoints[i][j]) {
                                if (part.data[i][j].data1 == type) matchingPoints ++;
                                allPoints ++;
                        }
        return (short) 100*matchingPoints/allPoints;
}
```

Figure 9 Definition of method coverage.

First the relevant part of the vegetation layer is determined. This is done by the *overlap* operation which is inherited from class *spatial_object*. The operation overlap is redefined in class *layer2d<T₁>* to return a *layer2d<Struct {T₁ data1; T₂ data2}>* when called with an object of class *region2d<T₂>*. Since the region parameter need not denote a rectangular region and the overlap of the layer and the region can consist of several regions (figure 10), *overlap* will most likely create a *layer2d* object having unoccupied cells. These cells are marked by assigning „false" to the corresponding entries in the *validPoints* array of the resulting *layer2d* object. This attribute is used to determine all points lying inside the *query_region*.

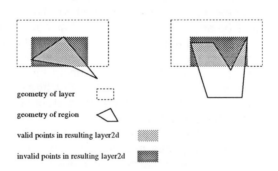

geometry of layer

geometry of region

valid points in resulting layer2d

invalid points in resulting layer2d

Figure 10 Overlap of a layer 2d and a region2d.

The implementation of method *avg_elevation* (figure 11) in the C++ interface class *city* uses the relationship between a city and the corresponding state and the relationship between a state and its topography. The topography of a state is defined as a set with elements of class *elevation_map*. Since a geometry is defined for the elements of the set but not for the complete set, we have to iterate over all elements to access the relevant parts of the topography. This is done by creating an iterator. Iterators can be used to access all elements of a collection. An iterator is constructed by providing the collection which shall be accessed. The collection may be specified to the constructor by immediately passing a collection or by passing a query which is evaluated first. A method *advance()* moves the iterator to the next element in the collection. Method *not_done()* determines whether the end of the collection has been reached, and method *get_element()* returns a reference to the current element inside the collection. Details about the iterator class can be found in [ODMG 93]. In method *avg_elevation* we use the *overlap* operation to restrict an element of the elevation map corresponding to the topography of a state to the region corresponding to the city. For this part of the topography layer we can easily compute the average elevation by accessing all data elements.

```
short city::avg_elevation () const {
        layer2d<struct{elevation data1; void data2;}> part;
        int i, j, allPoints, elev;

        for(Iterator iter(belongs_to->topography); iter.not_done();iter.advance()) {
                part = iter.get_element().overlap(geometry);
                for (i = 0; i <= part.data.upper_bound(); i++)
                        for (j = 0; j <= part.data[i].upper_bound(); j++)
                                if (part.validPoints[i][j]) {
                                        elev = elev+part.data[i][j].data1.data;
                                        allPoints ++;
                                }
        }
        return (short) elev/allPoints;}
```

Figure 11 Definition of method avg_elevation.

4.3 Customization - adding index structures and representations

To support special GIS applications it must be possible to integrate further index structures and new representations for the basic types [GüLa 92]. Index structures which shall be added to GOODAC must at least support the operations defined in the predefined abstract class *OOGDM_Index*. *OOGDM_Index* defines virtual methods for creating an index, adding and removing individual objects from an index, and operations for scanning the index. Specialized operations which are not supported by all index structures (e.g. special geometric query operations) are defined for the concrete index structures only. *OOGDM_Index* is the base class for all primary and all secondary index structures which are available in the system. As usual, the objects can be stored in a primary index whereas a secondary index contains only references to the objects. The implementation of an index structure should use the cluster mechanism offered by ObjectStore to ensure a good performance. Based on existing index structures appropriate indices may be defined using the OOGDM-ODL. E.g. assume there is a class *RTree* implementing a R^*-tree [BKSS 90]. In figure 12 we show the relevant part of a class definition for class *state* if a R^*-tree on the bounding boxes of the state objects is used to organize the data. The OOGDM-ODL precompiler translates this definition into two class definitions for C++-classes *_state* and *state* (figure 13 shows class *_state*). However, there is an additional protected attribute in the internal class *_state* referencing the index which is also created. The index is maintained by the generated update methods *set_geometry* and update of class *_state*. Each time one of these methods is called the old entry is removed from the index and the new entry is added. Since this may change the physical address of the object, logical references are required to keep relationships and secondary index structures consistent.

class state : geo_object (primary index RTree on geometry->BoundingBox()){
     ...
}

Figure 12 Class definition for class state having an R-tree index.

class _state : geo_object {
public:
  _state() {};
  _state(region2d<void> geometry, String name, Set<city> cities, city capital) {};
   virtual ~_state();
  Ref<region2d<void> > get_geometry() **const**;
  String get_name() **const**;
  Set<city> get_cities() **const**;
  Ref<city> get_capital() **const**;
  Ref<_state> set_geometry(Ref<region2d<void> > geom);
  void set_name(String new_name);
  void set_cities(Set<city> new_cities);
  void add_to_cities(Ref<city> new_city);
  void remove_from_cities(Ref<city> old_city);
  void set_capital(Ref<city> new_capital);
  Ref<_state> update();
protected:
  Ref< region2d<void> > geometry;
  String name;
  Ref_m_1 <state, city> cities;
  Ref< city > capital;
  Ref<RTree<_state> > primary_index;
private:
  Ref< region2d<void> > _geometry;
  String _name;
  Ref_m_1<state, city > > _cities;
  Ref< city > _capital;
};

Figure 13 Internal class for OOGDM class state.

Each feature class is based on a special geometric type. These types have quite simple definitions for the descriptive layer defined by OOGDM. This descriptive view is also the default representation for the representation layer. However, applications may require more sophisticated representations and corresponding implementations of the geometric types. Since there is no set of representations which meets the requirements of all GIS-applications, there must be a corresponding extension mechanism. This is achieved by an inheritance hierarchy. The basic representations are provided by GOODAC. Further subtypes in which we add new representations as additional views of the geometry may be derived from these basic representations.

If we want to use a special representation for the geometry of an elementary feature class or a class derived from such a class the corresponding geometry attribute is described by a subtype of the corresponding basic geometry type. This must be specified in the OOGDM-ODL definition. Assume there is a representation of regions as a decomposition into trapezoids as described in [ScKr 91] which is realized by class *Tz_RegionDesc2d*, a subtype of *RegionDesc2d* which describes the geometry of a 2-dimensional region. If we want to use the class *Tz_RegionDesc2d* to represent the geometry of a state, we have to change the attribute definition for attribute geometry of class state (figure 6) to

relationship region2d (**representation** Tz_RegionDesc2d) geometry;

Due to inheritance, all properties of *RegionDesc2d* are visible in *Tz_RegionDesc2d*. Hence, all predefined functions, predicates, and methods may be applied to this region. However, we may override implementations by using more efficient algorithms or by providing additional functions, predicates, or methods.[2]

5. Conclusions and future work

We have reviewed the object-oriented data model OOGDM for geo-information systems. This model basically consists of a hierarchy of abstract classes. By using an object-oriented data model as a base for OOGDM we may benefit from current database technology and avoid the drawbacks of current GIS architectures and models. We have shown how to use the prototype GOODAC implementing OOGDM to realize GIS-applications.

Clearly, much work has still to be done to finish this research. We have to extend our work to query processing. During the implementation appropriate algorithms and data structures must be chosen to realize the proposed predicates and functions. In addition, the extensible optimizer based on the second order signature technique must be designed.

Another important extension to OOGDM is to incorporate time-varying data. Basic work has already been done in this context. We propose to use an attribute-timestamped approach. Timestamps may be either discrete points in time or intervals. A major issue is the handling of timestamps in the implementation of GOODAC. The API has to hide the implementation of timestamped attributes from the user. Suitable methods have to be implemented to retrieve timestamped attributes. Additional work has to be done to integrate time into OOGQL. The temporal extensions of OOGDM and OOGQL are subject of a companion paper [VBH 96b].

---

[2] In the OOGDM-ODL we may specify alternative representations for each use of a feature class.

References

[AHS 91] T. Andrews, C. Harris, K. Sinkel: ONTOS: A persistent database for C++, in: R. Gupta and E. Horowitz (eds.): Object-Oriented Databases with Applications to CASE, Networks and VLSI Design, Series in Data and Knowledge Base Systems, Prentice Hall, 1991.

[Alve 90] D. Alves: A Data Model for Geographic Information Systems, Proc. Spatial Data Handling 1990, 879 - 887.

[AGGL 95] R. Arlein, J. Gava, N. Gehani, D. Lieuwen: Ode 4.0 User Manual, AT&T Bell Laboratories, Murray Hill, 1995.

[BCD 92] F. Bancilhon, S. Cluet, C. Delobel: A Query Language for O_2, in: F. Bancilhon, C. Delobel, P. Kanellakis: Building an Object-Oriented Database System - The Story of O_2, Morgan Kaufmann Publishers, 1992, 234 - 255.

[BeGü 92] L. Becker and R.H. Güting: Rule-Based Optimization and Query Processing in an Extensible Geometric Database System, ACM TODS 17 (1992), 247 - 303.

[BeGü 95] L. Becker and R.H. Güting: The GraphDB Algebra: Specification of Advanced Data Models with Second-Order Signature, Informatik-Berichte, No. 183-5/1995, Fernuniversität Hagen, Germany, 1995.

[BDQV 90] K. Bennis, B. David, I. Quilio, Y. Viémont: GéoTropics Database Support Alternatives for Geographic Applications, Proc. Spatial Data Handling 1990, 599 - 610.

[BKSS 90] T. Brinkhoff, H. P. Kriegel, R. Schneider, B. Seeger: The R^*-tree: An Efficient and Robust Access Mehod for Points and Rectangles. Proc. ACM SIGMOD, 1990, 322 - 331.

[CFO 93] E. Clementini, P. Di Felice, P. van Oosterom: A Small Set of Formal Topological Relationships Suitable for End-User Interaction, Proc. Large Spatial Databases (SSD'93), Springer, LNCS 692, 277 - 295.

[ChZu 95] E. P. F. Chan, R. Zhu: QL / G - A Query Language for Geometric Data Bases, Technical Report, Department of Computer Science, University of Waterloo, 1995.

[CNT 90] A. Chance, R. Newel, D. Theriault: An object-oriented GIS - Issues and solutions, Proc. EGIS'90, 179-188.

[DRSM 93] B. David, L. Raynal, G. Schroter, V. Mansart: GeO$_2$: Why objects in a geographical DBMS, Proc. Advances in Spatial Databases 1993, LNCS 692, 264 - 276.

[Deux 90] O. Deux: The Story of O_2, IEEE Transactions on Knowledge and Data Engineering, 2(1), 91 - 108, 1990.

[EcUf 89] J. W. van Eck, M. Uffer: A Presentation of System 9, Proc. Photogrammetry and Land Information Systems (1989), 139 - 178.

[Egen 94] M. J. Egenhofer: Spatial SQL: A Query and Presentation Language, IEEE Trans. on Knowledge and Data Engineering, 6 (1), 1994, 86 - 95.

[EgFr 91] M. J. Egenhofer, R. D. Franzosa: Point-set topological spatial relations, Int. Journal of Geographic Information Systems, 5 (2), 1991, 161 - 174.

[EgTh 92] M. J. Egenhofer, K. K. Al-Thaha: Reasoning about Gradual Changes of Topological Relationships, Int. Conference GIS, 1992, Springer, LNCS 639, 196 - 219.

[ElCa 83] A. A. Elassal, V. M. Caruso: Digital Elevation Models, USGS Digital Cartographic Data Standards, Geological Survey Circular 895-B, 1983.

[Feuc 93] M. Feuchtwanger: Towards a Geographic Semantic Database Model, Ph. D. Thesis, Simon Fraser University, Vancouver, 1993.

[GrMc 93] G. Graefe, W.J. McKenna: The Volcano Optimizer generator, Proc. 9th Intern. Conf. on Data Engineering (1993), 209 - 218.

[Güti 88] R. H. Güting: Geo-Relational Algebra: A Model and Query Language for Geometric Database Systems, Proc. Conf. on Extending Database Technology, 1988.

[Güti 89] R. H. Güting: Gral: An Extensible Relational Database System for Geometric Applications, Proc. VLDB 1989, 33 - 44.

[Güti 93] R. H. Güting: Second-Order Signature: A Tool for Specifying Data Models, Query Processing and Optimization, Proc. ACM SIGMOD 1993, 277 - 286.

[GüLa 92] O. Günther, J. Lamberts: Objektorientierte Techniken zur Verwaltung von Geodaten. In: O. Günther, W. F. Riekert: Wissensbasierte Methoden zur Fernerkundung der Umwelt. Herbert Wichmann Verlag, Karlsruhe, 1992 (in German)

[Haas 89] L. M. Haas, J. C. Freytag, G. M. Lohman H. Pirahesh: Extensible Query Processing in Starburst, Proc. ACM SIGMOD 1989.

[Haas 90] L. M. Haas et al.: Starburst Mid-Flight: As the Dust Clears, IEEE Transactions on Knowledge and Data Engineering, 2 (1), 1990.

[Herr 87] J. Herring: TIGRIS: Topologically Integrated Geographic Information Systems, Proc. Auto-Carto 8 (1987), 282 -291.

[KMP 93] A. Kemper, G. Moerkotte, K. Pleithner: A Blackboard Architecture for Query Optimization in Object Bases, Proc. VLDB 1993.

[KuSo 92] H. Kucera, M. Sondheim: SAIF - Conquering Space and Time, Proc. GIS'92 - Working Smarter, 1992.

[LLOW 91] C. Lamb, G. Landis, J. Orenstein, D. Weinreb: The ObjectStore Database System. Communications of the ACM, 34 (10), 50 - 63.

[LPV 93] T. Larue, D. Pastre, Y. Viémont: Strong Integration of Spatial Domains and Operators in Relational Database Systems, Proc. Advances in Spatial Databases 1993, LNCS 692, 53 - 72.

[More 89] S. Morehouse: The Architecture of ARC/INFO, Auto-Carto 9 Conf. (1989), 266-277.

[ODMG 93] R. G. G. Cattell (ed.): The Object Database Standard: ODMG - 93, Morgan-Kaufmann Publishers, San Francisco, 1994.

[ODD 89] B. C. Ooi, R. S. Davis, K. J. McDonell: Extending a DBMS for Geographic Applications, 5th Intern. Conf. on Data Engineering (1989), 590 - 597.

[Same 89] H. Samet: The Design and Analysis of Spatial Data Structures, Addison-Wesley, 1989.

[SRH 90] M. Stonebraker, L. A. Rowe, M. Hirohama: The Implementation of Postgres, IEEE Transactions on Knowledge and Data Engineering, Vol. 2 (1), 1990.

[ScKr 91] R. Schneider, H. P. Kriegel: The TR*-tree: A new Representation of Polygonal Objects Supporting Spatial Queries and Operations, Proc. Int. Workshop on Computational Geometry (CG' 91), Springer, LNCS 553, 249 - 263.

[ScVo 92] M. Scholl, A. Voisard: Geographic Applications: An Experience with O$_2$: In F. Bancilhon, C. Delobel, P. Kanellakis (eds): Building an Object-Oriented Data-base System: The Story Of O$_2$. Morgan Kaufman Pub. 585 - 618.

[ViOo 92] T. Vijlbrief, P. van Oosterom: The GEO++ system: An Extensible GIS, Proc. Spatial Data Handling, 1992, 40 - 50.

[VBH 95] A. Voigtmann, L. Becker, K. Hinrichs: An Object-Oriented Data Model and a Query Language for Geographic Information Systems, Internal Report 15/95-I, Institut für Informatik, Westf. Wilhelms-Universität Münster, 1995.

[VBH 96a] A. Voigtmann, L. Becker, K. Hinrichs: A Query Language for Geo-Applications, Internal Report, Institut für Informatik, Westf. Wilhelms-Universität Münster, 1996.

[VBH 96b] A. Voigtmann, L. Becker, K. Hinrichs: Temporal Extensions for an Object-Oriented Geo-Data-Model, Internal Report 6/96-I, Institut für Informatik, Westf. Wilhelms-Universität Münster, 1996.

Contact address
Ludger Becker, Andreas Voigtmann and Klaus H. Hinrichs
FB15, Institut für Informatik
Westfälische Wilhelms-Universität
Einsteinstr. 62,
D-48149 Münster
Germany
Phone & Fax: (++49) 251 / 83 - 3755
E-mail: {beckelu,avoigt,khh}@math.uni-muenster.de

AN OBJECT-ORIENTED, FORMAL APPROACH

TO THE DESIGN OF CADASTRAL SYSTEMS

Andrew U. Frank

Department of Geoinformation E127.1
Technical University Vienna
Vienna, Austria

Abstract
Modeling GIS or applications of GIS with object-oriented tools is difficult for the practitioners. In this case study the deed registration system is studied using formal, object-oriented tools.

The formal modeling approach also shows the differences and the similarities in the deed registration system used in the USA and the property registration system in Europe. A formal modeling approach will demonstrate the functional similarities, caused by the similar social demands, and the differences owing to different historic developments.

It is found that the models are comparable and the rules are fundamentally similar. The checks for a valid transfer of ownership are quite similar, but the time of checking is different. A Continental title registration system checks documents when they are received, whereas the US registers first without checking and only later the validity is checked by a prospective buyer's lawyer.

Several results can be generalized to other administrative GIS applications:
- The question of identity plays a primary role and several types of *identities* may be hierarchically nested.
- Administrative rules are less strict than database logic on what constitutes an entity or an identifier (the *unique name assumption* is maintained only within a context).
- Administrative time for a registry is a simple ordered scale.
These results are of practical importance for the design of GIS software which includes a *land ownership layer* and should be applicable in countries with different property registration systems. It shows the designers of property registration software what can be built into a general, widely usable system and when adaptation to local rules is necessary.

The study was motivated by efforts to design a property registration system for a reform country in Eastern Europe. In this situation, one has to understand how social demands and legal rules relate to the technical solution. To start with a formal model is the best way to achieve this.

1. Introduction

There is an international interest in the design or improvement of property registration systems. Several studies show a linkage between economic development and cadastral systems [Dale, 1988]. The common recommendation for the installation of a GIS points to the *ownership layer* as the desired base for a GIS [National Research Council, 1980; National Research Council, 1983]. This study was motivated by an effort to design a property registration system for Moldavia, a newly independent country of the former Soviet Union, but similar problems are posed in other countries. The goal was to understand the linkage between social demands, legal rules and technical solutions for property registration. The formal method presented here revealed dependencies and led to new insights useful for design.

To make progress with the study of cadastral systems - and with other related administrative GIS applications - the administrative and legal context must be included in the formal (computer) model. Application designers do this routinely, but the resulting code rarely demonstrates principles of the system - it just works. Better tools are necessary to bring about various solutions to overcome the linguistic barriers between the descriptions of different systems and to render the descriptions in a comparable format.

The deed registration system used in the USA and the property registration systems in Europe appear to be quite different solutions to a similar problem. How much do they have in common? The answer is important for the design of commercial GIS software which includes a *land ownership layer* and can be applied in many countries with different property registration systems. What can be built into a general, widely usable system? Where must a system leave scope for adaptation to local rules?

It is found that the models are comparable; the rules are fundamentally similar and many of the differences in real estate law do not affect the operation of the registry, properly speaking. The checks for a valid transfer of ownership are quite similar (but differ in detail even within the USA), but the check is carried out at a different time. A Continental property registration system checks documents when they are received, whereas the US counterpart registers every document without checking and only later prospective buyers check the validity.

The formal modeling approach demonstrates the functional similarities, which are caused by the similar social demands, and the differences, related to different historic developments and initial technical solutions. It can point out the options for change and what changes in legal rules or administrative process would be required. It becomes evident that many of the recommendations to improve the US deed registration system do not address the roots of the problem, but are only a superficial technical improvement [Dale, 1988].

Modeling GIS or applications of GIS with object-oriented tools is difficult for the practitioners. The tools used - database data description languages, programming languages and software design methods - pose technical difficulties and require much time to master. But it is even more difficult to achieve the conceptual clarity in understanding how the system works.

In this case study, property registration systems are studied using formal, object-oriented tools. The approach taken here starts from the most fundamental operations of a property registration system and shows how they can be expanded later. The tools used here lead to an extremely compact design document (3 pages of code for the core part of the registry and 3 pages to code a simplistic model of a database).

Several results achieved can be generalized to other administrative GIS applications:

- The question of identity plays a primary role and several types of *identities* may be hierarchically nested.
- Administrative rules are less strict than database logic on what constitutes an entity.
- Administrative time for a registry is a simple ordered scale.

It is necessary to point out that any discussion about the 'typical' legal rules must generalize broadly and specific countries may have adopted different ones. The discussion here is further complicated by the substantially incompatible legal theories and jargon describing Anglo-Saxon deed registration systems and European property registration.

2. The charge of a property registry

In very general terms, society needs a method to determine ownership of land. Ownership of small scale objects is determined by possession, the person holding an object is assumed to be the owner. A legal theory from Roman times protects good faith in the appearance of ownership: If you buy in good faith a watch from a thief, then you become the new owner.

Ownership of land is not apparent; today it is not possible any more to grant ownership only to persons physically living on a piece of land. 'Absentee landlords' are a necessity in our society, despite the fact that some legal systems strongly favor the present possessor and make it difficult for absentee landlords to maintain ownership (often in Latin America, e.g. Ecuador).

Acts of transfer of ownership were originally rituals. In medieval Germany, it was required to live on a piece of land for seven days and maintain a fire for seven nights. Transfer of ownership was later declared by the courts during their annual sittings (where everybody was assumed present). The court registration book documented all properly executed transfers and made them known to anybody with a need to know. This registration system replaces the rituals and publicizes (in the sense of to *make publicly known*) the transfer of ownership, to give everybody notice of the ownership rights of the new owner. Nobody can claim not knowing facts, which are properly registered in this public register. In this respect, a property register is similar to other public registers.

A charge to a property registration system might be described as:

> The property registry maintains a record of transfers of ownership of land with the date and time of their registration. It reports all transfers which affect a piece of land in order of their registration. These records constitute evidence for the ownership of land.

It is implied that one can sell a part of a parcel, which means spatial subdivision. It is also generally possible to transfer other rights than full ownership. Most important is the use of real estate for collateral to secure a loan - here called a mortgage -, often received to improve the property and build a building. Encumbrances are rights granted to others, which diminish the value of a property. Liens are laid against a property to secure debts arising by law.

The fact that ownership of land is not readily visible allows different types of fraud: selling a parcel one does not own, selling the same parcel to two different buyers etc. The registry must help a reasonably careful buyer to avoid such problems. It must be possible to establish if a seller owns the land he sells and a mechanism must be available to prevent selling the same piece more than once. In this case, the time of a transfer of ownership is crucial. The general rule is: Who buys first in good faith, owns the land and the other one can only try to get his money back from the seller. The legal effects are (often) bound to the time of registration, as only this time is easy to prove and only from this time on another buyer can check the registry and detect the potential fraud.

3. Legal history of a parcel

The registry must help the owner prove his right to the land. In theory this is very simple: the current owner receives his rights from the previous owner, reduced for all this one transferred away. The previous owner received from his previous owner what that one had received, again reduced for what he had transferred away, and so forth till one reaches the original grant from the emperor (or whoever in modern times replaces the emperor), whose ownership cannot be questioned, as it is assumed to be 'directly from God' (or some similar theory). The metaphor underlying is that ownership is like a bundle of sticks, each signifying a particular right. The owner can give away specific rights or transfer ownership (i.e. all that remains in the bundle). The owner can also divide the land and create new parcels, which each inherit their bundle of rights. The owner can then transfer these parts independently. Nobody can transfer more than he owns.

To prove ownership the owner has to produce the valid document with which he acquired ownership plus all the valid documents of his predecessors. The registry stores all these documents and makes them available on demand. For each document the time when it was registered must be noted. The organization of a registry of documents to prove ownership is a practical problem and countries have found different forms of organization of the registration.

A search in a registry for a document starts with the name of the current owner and the document with which he acquired the parcel (in this document, he is called the *grantee*). One has to find all events where this owner has transferred rights to others because he holds only the remainder *(search forward in time from the data of his document)*. From his document (naming the person transferring ownership to him: *the grantor* in this document) one must find the document with which the previous owner acquired ownership *(search backwards in time)*. It is often customary, to do all the backward search first (to establish a so-called *chain of titles*) and then search for all the intervening documents.

Documents are registered chronologically to assure their precedence. Precedence is assumed here to be strictly by registration time, but when a document was signed may also play a role; for an extensive discussion of temporal issues in the property registration system from a database point of view, see [Al-Taha, 1992]. Searching for a document in chronological order is very time consuming and a grantor/grantee index is generally provided, which lists under the name of a grantor or grantee all documents he appears in. With this index it is possible to find the previous document for the grantor of any successive documents, as the grantor of the successive document must be the grantee of the previous one. Thus using the grantor/grantee index the chain can be established.

Once all the relevant documents are found, they must be examined for their validity. A lawyer can then produce a *legal opinion* and declare the title to be merchantable, i.e. of sufficient quality to be bought or mortgaged.

4. Formalization

The formalization starts with the core of the application to achieve a prototype rapidly. This is an efficient way to understand the major charge of an administrative unit without getting lost in the additional functions which are added in the course of time.

4.1 Simplified ontology

The ontology, i.e. the entities and operations relevant can be deduced from the charge. Essentially the nouns form the object types and the verbs the operations. For the most typical case, it consists of the following classes of objects:

Persons transfer parcels (from and to). Persons are characterized by a name.

Parcels with an identifier are owned by persons. Parcels can be sold as a whole or in part. Their boundaries are described in the documents registered.

Documents (technically called instruments) are the evidence of the transfer of ownership or other rights to land. They must be signed and registered.

The registry, which received the contracts to transfer ownership (documents), gives them a register number (id) and a time stamp when received. All documents regarding a parcel can be retrieved and presented in the order they were received for registration.

4.2 Formalization and prototype code

A functional programming language [Bird, 1989], namely Gofer [Jones, 1995; Jones, 1994] used here, produces readable specifications and, at the same time, the code can be executed at least as a prototype to check out if it captures the intentions correctly [Frank, 1995][1].

A very simple view of the registry gives the following code, which is then refined, but gives an overview form.

```
type Parcel = Int           -- parcels are represented by their parcel Id
type Name = String          -- a name is simply a string
type Date = Int             -- the date is simplified to an integer

class Registers r where     -- the registry with the operations
    register :: Contract -> r -> r              -- to register
    history :: Parcel -> r -> [Contract]        -- to find all

class Contracts c where
    close :: Name -> Name -> Parcel ->  Date -> c
    refParcel :: Parcel -> c -> Bool

data Contract = Contract Name Name Parcel Date
                        -- a contract names grantee, grantor, parcel and data

type Register = [ Contract ] -- the register is a list of Contracts

instance Registers Register where
    register c r = c:r          -- to register a contract ad it to the list
    history pp r = filter (refParcel pp) r

instance Contracts Contract where
    close s b p t = Contract s b p t
    refParcel pp (Contract s b p t) = pp == p
```

4.2.1 Entities

For the entities *parcel* and *person* nothing else than an identifier is necessary. For the level of detail permitted in this paper, details of the boundary definition are not considered.

Time is modeled by the integer series, so that two events can be compared (total order). For the registry, it can be excluded that two events happen at the same time. There are no requirements for calculating time differences in this simple legal situation [Frank, 1994].

Transfer documents come in different types: to transfer ownership or ownership for part of a parcel, and to establish a mortgage (as an example for limited rights). In the class Transfers the operations that apply to all kinds of transfers are gathered: a document is signed and registered. Observer-operations allow to retrieve specific information of a document (the grantee, the grantor, when the document was signed, registered, etc.). There are finally operations to compare two documents: which one was signed or

---

[1] Classes list the operations and their arguments and result types. Objects are defined with data or type statements, based on predefined types for Int, String, Float and Bool; lists of objects are marked with [ ] alternatives separated by |.

Instances describe how classes apply to a particular object (Ôinstance Registers Register' is read as 'Register is from the class Registers (and has the corresponding operations)'. Executable operation definitions are written as mathematical functions (without parenthesis around the arguments). All text on a line following "--" is comment.

registered earlier, which one transfers less rights than the other. The often used test for earlier registration date is the order relation on documents.

```
type Parcel = Int
type Person = String          -- a person is represented by its name
type Money = Int

class Transfers d where
    signTransfer :: Person -> Person  -> Parcel -> Time ->  d
    register ::  Time -> d -> d
                                    -- observer operations:
    seller, buyer :: d -> Person
    parcelNew, parcelFrom  :: d -> Parcel
    signed, registered :: d -> Time
    sigBefore, regBefore :: d -> d -> Bool
    encumbrance, ownershipTransfer, isOriginalGrant :: d -> Bool
    lessEqRights :: d -> d ->Bool
```

The class SpecialDocs groups the operations that are particular for a special type of transfer.

```
class Sales d where
    makeGrant :: Person -> Parcel -> Time -> d
    closeSale :: Person -> Person -> Parcel  -> Time -> Money ->  d
    closePartialSale :: Person -> Person ->
                    Parcel -> Parcel -> Time -> Money -> Float ->  d
    establishMortgage :: Person -> Person -> Parcel  -> Time -> Money ->  d
    amount :: d ->Money                    -- the only additional observer op
```

The data type for Documents can be one of 4 kinds, each of them including the common elements found in the GeneralDoc, which groups the grantee, the grantor, the parcel, time of signature, time of registration and an identifier.

```
data Doc =   Sale GeneralDoc Money |
             SalePart GeneralDoc Money Float Parcel |
             Mortgage GeneralDoc Money |
             OriginalGrant Person Parcel Time OID

data GeneralDoc = GeneralDoc Person Person Parcel  Time Time OID
```

The code for the operations is written in *instance* declaration. Here an example:

```
instance Sales Doc where
    makeGrant p1 pa t = OriginalGrant p1 pa t unknown
    closeSale  p1 p2 pa t m = Sale (signTransfer p1 p2 pa  t) m
    establishMortgage  p1 p2 pa t m = Mortgage (signTransfer p1 p2 pa t) m
    closePartialSale p1 p2 paOld paNew t m f =
                SalePart (signTransfer p1 p2 paNew t) m f paOld
    amount (Sale d m) = m
    amount (SalePart d m f dp) = m
    amount (Mortgage d m) = m
```

4.2.2 The registry

The registry has two operations: register documents and produce the history for a right. The registry consists internally of a clock, which is strictly incremented (here simulated by an increment with every document registered) to assure that the registration events are totally ordered. The registry consists of the books, here only the actual collection of documents (in a relation indexed by document id).

The operations for clients of the registry are to register a document and to get the history to a document (the document is representing the right acquired with it). The first step in registration is stamping the document with the current time indicated on the clock in the registry and then passing it on to the books for registration. To retrieve the history one can disregard the current value of the clock and retrieve the history from the books. The same operations apply to the books, where they are executed as calls to the database (not shown here).

```
class Registries r where
    registerDoc :: Doc -> r -> r
    history :: Doc -> r -> ChainLink
data Registry = Registry Time TheBooks
instance Registries Registry where
    registerDoc doc (Registry now rs) =
            Registry (now') (registerDoc (stamp now' doc ) rs)
                    where now' = tick now
    history d (Registry now rs) = history d rs

type DocRel = Rel2 Doc
data TheBooks = TheBooks DocRel

instance Registries TheBooks where
    registerDoc doc (TheBooks  ds) =  TheBooks (store3 putOID doc ds)
    history d ds = makeChain ds (CL d [] [])
```

There are internal operations of the register, namely get the previous document to a given document (search backwards) and to find all intervening, possibly adverse transfers, to two documents. The logic for this processing is based on a filter of the collection of documents (call to *filterR2* with 2 arguments: a condition and the list of documents). The backward search looks for a document with the same name for the grantee as the name of the following grantor, regarding the same parcel (i.e. the parcel acquired previously must be the same as the parcel now to be transferred) and the document must precede the current one. In the forward direction, one searches for every document of the grantee concerning the same parcel; here the search is limited to the time between the registration of the first and the document of interest.

```
class RegistryInternals r where
    previousDocs:: r -> Doc  -> [Doc]
    interveningDocs :: r -> Doc -> Doc -> [Doc]
    oneLink :: r -> Doc -> [ChainLink]
    makeChain :: r ->ChainLink ->  ChainLink

instance RegistryInternals TheBooks where
    previousDocs (TheBooks ds) d = sort1 (<=) (filterR2 c ds)
                            where    c d' = (seller d)==(buyer d')
                                  && (parcelFrom d)==(parcelNew d')
                                  && (registered d) > (registered d')
    interveningDocs (TheBooks ds) d d2 = sort1 (<=) (filterR2 c ds)
                            where c d' = (buyer d)==(seller d')
                                  && (parcelNew d)==(parcelFrom d')
                                  && (registered d') > (registered d)
                                  && (registered d') < (registered d2)
    oneLink books doc = cls
        where cls = [ CL pd (otherSales pd)  (encumbrances pd)
                    | pd <- previousDocs books doc]
            intervDocs p  = interveningDocs books p doc
            otherSales p = filter (not.encumbrance) (intervDocs p)
            encumbrances p = filter (encumbrance) (intervDocs p)
    makeChain books cltree = exp cltree
        where exp (CL pd os es) = (CLEx pd os es  (oneLink books pd))
            exp (CLEx pd os es cls) =
                                (CLEx pd os es [exp cl | cl <- cls])
```

For a given document, several documents may be candidates for the acquisition - at least in pathological cases - and each of them must be further investigated to decide which one produces the correct chain. Therefore for a given document, all possible previous documents must be found and for each of them, the intervening documents collected and separated in adverse transfers and other transfers which reduce the ownership rights. To collect all potential documents which could be part of the chain of titles to a given document, a recursive data structure is built, which starts with a document, reconstructs all the possibly preceding documents and then expands these to the same. This is the history of a document.

The chain of titles produced from the registry must be assessed to determine if the title is valid. A chain must fulfill certain formal conditions to be valid, but then additional assessment of each instrument is necessary by a professional lawyer. This can be programmed to perform only the first type of checks.

Any branch of the tree of titles which does not lead to *original adjudication*, i.e. a transfer, granted by an authority which cannot be questioned, is not complete. The decision what an authoritative document is, depends on local customs and is - at least in the USA - often replaced by the limitation that chains are checked only for a certain length.

Second, no adverse transfers may occur in any link of the title chain.
The current ownership rights are the original rights as they follow from the authoritative title reduced for all encumbrances, sales of part of the land, mortgages erected etc.

5. Internal operation of the registry

5.1 Flow of information

This formalization just models the registry seen from the outside. Real registration offices suffer from some physical limitations in the flow of information, which must be modeled:

The contents of the document, which is necessary for entry in the register, is not immediately available. Personnel of the registry must carefully prepare the abstract, which takes time. Documents are presented for registration sometimes in rapid succession. Thus a document cannot be fully processed before the next is received, but the order of presentation for registration must be preserved. This is generally solved by stamping documents with the time of receipt and entering them very quickly in a chronological registry. Processing is then asynchronous to reception, working on the queue of documents received.

The chronological registry, typically called diary, is an important device to avoid tampering with the order of reception of documents. This was a major concern with the designers of registries in the last century as is demonstrated by overly detailed rules about keeping the diary.

To form an opinion of the current legal situation of a parcel, one must consult the history from the registry and check the documents pending for registration if they affect the parcel in question.

5.1.1Indices

A realistic registry is so large that a sequential search for all documents affecting a parcel is impossible. If the documents are registered and filed chronologically, then it is necessary to establish indices to help in the search. Indices make operations faster, but do not change the logical result from a query compared with a sequential scan. Indices are thus necessary for actual implementation, but do not contribute to the design or the logic of the problem.

6. Entities, identities and numbering methods

Much of the discussion of design of property registration systems is concerned with identifiers. There are two aspects to this discussion: the practical, directly visible, of the selection, formation and distribution of the identifiers, and the implied, theoretical one of identities. An identifier implies that two things with the same identifier are identical, at least in some aspects [Al-Taha, 1994]. Any time an object with a given identifier is referenced, one implies that the same object is referenced. This assumption is fundamental in database systems and the logic of their query languages [Reiter, 1984].

When does an object change and become a new object? This is a crucial question, which does not always have a single answer. We have a very clear understanding of the identity of a person, but also are aware that during a lifetime, this same person will change in many respects. Administrative systems define the identity of objects and which operations change the object into a new one depending on the requirements. The cadastral system further relaxes the assumption that unique names identify unique objects and requires only unique identifiers within a context: the name of a person must only be sufficient to identify him with respect to the context, i.e. the parcel affected, the set of previous and later owners, etc.

6.1 Documents

The contracts physically presented for registration are obvious entities with an identity similar to other physical objects. But documents in the legal sense, have their identity through their contents (physical copies of a single document can exist): to produce two different documents which read exactly the same is very unlikely. It is practically not possible that two sales documents have the same grantee, grantor, date of signature, parcel affected, therefore the elements of the abstract are sufficient to identify a document (one could construct a case where A sells a parcel to B in the morning, at noon B sells the same parcel back to A and in the afternoon A sells again to B - and all transferees are immediately recorded; registration fees are high enough to prevent such nonsense!). Once registered, every document gets a unique identifier, which is e.g. the book and page on which it is recorded.

6.2 Parcels

There are multiple interpretations of parcels. One might assume that the boundary of a parcel makes the parcel and any change in the boundary makes a new parcel with a new identifier. Applying this strict notion makes the logic of the registry the simplest, but conflicts with customary methods.

People tend to call a parcel the same even if a minor piece is sold or another other minor change in the boundary happened. In most practical systems, parcel identifiers remain the same despite apparent changes in the geometrical boundaries. Parcels identified in this way are changing objects and to identify a definite piece of land and a set of rights, the identifier must be given together with the date of the last boundary change (this creates a new parcel identifier, which determines uniquely a specific geometry). For most operations, the regular (undated) parcel identifiers are sufficient, but in the title search, a potential change in boundary must be considered and thus the *dated* parcel id used. (Technically a parcel identified by a *meets and bounds* description is logically sufficient, but not easy to work with).

6.3 Persons

Natural persons are not modeled as true entities in the registry, but represented by their names. For the registry, it is sufficient to establish that the grantee and the grantor of the successive transfers is the same person. Person names are used for indices on grantees and grantors. For this to work, the name must be an identifier for the person, which despite the many limitations and exceptions is a workable assumption within the context. The numerous spelling variants of a name (especially first names) and the habit to change the name of women when they marry, makes this difficult to automate and pushes for an index by parcel, sothat a chain can be established, even if the name of the owner is different

The cadastre does not require that names are globally unique. Only the identity of the grantor with the grantee of the previous transfer is checked. If the name alone is not sufficient to establish this identity, then additional evidence may be necessary (but this is relatively seldom).

In a property registry, the *unique name assumption* is possibly violated [Reiter, 1984]: there may be several persons registered as grantees or grantors with exactly the same name, but also the same person may appear under different names. With the current administrative rules, it is not possible for a registry to resolve such possible ambiguities. But there is also no need, as in any case when the identity must be established, additional documents are exchanged to establish it. But only, when necessary! It is also not possible, to find all property of a certain person within the registry; only all property which is registered with the same owner name. Laws which limit the amount of land a person can own, often encountered, cannot be enforced in the current registers, which seems not to be a major issue.

7. Conditions for a working registry

From the analysis one can deduce some conditions for a registry to work properly:
- Registration in the competent registry,
- Registration of all titles and other documents,
- Consistent use of identifiers (at least within the context of a parcel).

The first two points assure that the documents necessary are contained in the registry, the second that one can find the documents.

According to most legal systems, for registration to have full effect, it must be performed by the competent registry, which is the registry of the political subdivision of space in which the real estate is situated. In every country, a simple rule of competence of registries depending on the location is established. It follows the standard rules to determine the competent court.

The registry can only work as described if all documents relevant for the transfer of real estate are collected in a single place. This is often a problem, as many important documents are registered with the courts (or other public registries) and do not require registration with the cadastre.

To achieve completeness, the legal protection for unregistered documents can be reduced compared to registered ones. This forces registration even in a system like the USA where registration is voluntary. This does not work for documents which are already registered with other public registers. To obtain these documents, some registration systems recognize the validity of the document without registration, but request that the owner provide copies of these documents before he can transfer ownership or other rights he acquired with them.

8. Improvements of a registry

The solution for a registry of deeds as presented above is the method where the least amount of public effort is used - the registry accepts any document for registration, keeps the minimal index necessary to establish a chain of titles and leaves the construction of the chain and the assessment of the chain of titles to the client. Low public costs incur with this solution, but high private costs (i.e. high transfer costs), in total, an expensive system. In the USA, where this system is used in its purest form, only few complaints are heard and a fundamental reform of the system has not been achieved. The reason is probably the fact that from the total cost of a transfer of real estate property, the cost of property registration etc., is only a small part. In the USA the other costs, especially realtor costs, seem to be less than in other countries.

Practically, a registry of deeds is difficult to use, for one, because some documents are not included, and for one because improper documents are registered. It is customary to demand that the registrars work more careful [Dale, 1988], but without the power to refuse registering of inappropriate documents, a fundamental improvement is not possible. It is somewhat surprising to realize that *the major difference between a US registry of deeds and a continental property registration system is in the power of the registrar to reject a document presented for registration* or to demand additional documents before registering a document.

Any improvement of this registration method depends crucially on giving more competence to the registrar. Rules on how a document must be drawn up and what references it must contain are necessary and the registrar must be empowered to enforce them and to deny registration (or grant registration with less effect) to documents which do not fulfill these rules.

Such rules must be simple and formalistic to be viable. They must not demand complex interpretation of law or independent collection or evaluation of facts: the applicant must deliver documents and the registry must have formal rules to check. Every instrument registered in the registry designed so far must have a name for a transferee and transferor. This is a formal legal requirement without which a document cannot be registered. What additional checks would be desirable?

8.1 Check grantor

The registrar checks that the seller named in the current document and the buyer of the previous one are the same person The identity of the person within the local context of these two consecutive documents is established.

If the grantor is the same as the grantee, but has changed his name, or the grantee has received the title through another legal document (e.g. court order), which would not require registration, the registrar must demand that copies of these documents are produced and registered to maintain the chain of titles.

It is often demanded that addresses of the persons involved are maintained. This is not a requirement of a property registry, but other functions the registry could fulfill would require means to communicate with the owners. The difficulty is not with capturing this information initially, but to maintain it, which is only worth the effort if it is regularly used and providing benefit to the owner.

8.2 Establish backward chain by reference to previous deed

A document references the previous document, with which the transferor had acquired the real estate. If every document reliably contains this information and is checked by the registrar for correctness, then searching backwards through a chain of titles is much simplified. The procedure to check this link is very similar to the backwards search for a title. (Most registrars add to the abstracts such a reference to simplify their work, but it is not consistently done.)

8.3 Annotate forward chain with registered deed

If every document references the previous one, then the registrar can build the forward chain between documents by adding the identifier of the newly registered document next to the text of the previous (referenced) document. This is often done in traditional registries, without specific regulation, just for the benefit of the registry itself. It is a sort of additional index, maintained in the document registry. This simplifies the forward search enormously: references for all later documents exist and they must be checked only.

8.4 Check for no adverse transfers

When the registrar annotates the previous document in the register with the identifier for the newly registered document, he sees all other references to documents registered by this grantor. It is then possible to check if the grantor still holds the right to make the current transfer. The registrar must be allowed to refuse registration of documents, when the grantor has already disposed of the right. For example the registrar may detect, and then must refuse to register, a transfer of ownership if the grantor has already sold the parcel previously and is not owner anymore.

This requires that the transfer is for specific rights on specific parcels, which are identified. In practice, this is no problem, as only a few set types of rights are used and transfers are specific to assure the grantee what he receives. In principle, some legal systems allow enormous freedom in the transfer of rights to land, which, again in principle, makes it impossible for a registrar to check if the grantor has the rights he transfers.

8.5 Title registration

If the registrar reliably checks the relation of a new document with the previous ones and refuses registration of documents which are not valid, then one must ask
- if something is checked, is the result of the check recorded?
- if the check is negative, what follows?
- if the check is positive and the document registered, does this guarantee the elements checked?

If a registry starts with valid titles (e.g. adjudication from the government) and if for every transfer the relation to the previous document is checked, the identity of the persons involved established, and the reference to the parcel boundary is by reference to the previous document, it is clear that a correct chain of titles exists for each registered document.

The validity of the transfer itself may remain to be assessed. A transfer is often based on a legal contract, and this contract may not be valid. There are several legal reasons why a signed contract is not valid: fraud, error on the side of one of the contracting parties, etc. This may (depending on the applicable law) lead to invalidation of the transfer. Demanding specific forms of contracting, e.g. the assistance of a legal professional, can exclude many of the causes for an invalid contract. Excluded from registration must also be *conditional* transfer agreements, in which the transfer depends on a future uncertain event.

If the registrar does check and refuses registration of invalid documents, the title search need not be repeated and no new assessment of the validity of the title of the last owner is necessary. The registrar can carry professional liability insurance to guarantee his work in the same way a lawyer guarantees his *title opinion*. In many countries the registrar is a public official and the state guarantees his work. This would come very close to some forms of title registration. In a title registration system, the registrar does not only formally check that a chain of titles is established, but checks also if the transfer of title is effective and may guarantee the new owner his rights.

9. Conclusions

From the analysis we see that the difference in the operation of a registry of deeds and a title registration system is in the difference when and by whom the assessment of the title chain is executed. In a deeds registration system, the registrar is forced to register any document received. In a title registration system he is charged to examine every document for its relation to the chain of titles and if it effects a valid transfer. The registrar must refuse to register invalid documents.

The checks the registrar has to carry out are for one part formal as shown in the code above. For the other part, he must check the contents of the documents. Are the boundaries of the parcel in correspondence with the previous document? Is the contract valid? etc. These checks must be done quickly so that the registry is reasonably up to date. This is only possible, if the transfers are following a set pattern, and are not subject to interpretation.

9.1 A Property registration package

From the specifications here - and some extensions to include special cases not covered here - a generally usable registration system can be built. It shows how the indexes and the diary must be maintained and shows the importance of recording time. The logic of title search coded here gives the simplest case, which must remain extensible to adapt it to particular legal systems.

If such an engine is used for a deed registration system, for example in the USA, then incoming titles can be checked as a help for the registration personnel to catch errors, but deeds which do not conform must be registered, despite the doubts they raise. The title search is then performed each time a client asks for it. Such a registry could quickly give indications who is most likely the owner. In the USA, title insurance companies maintain such registries.

If the same engine is used for a title registration system, for example in Central Europe or South America, the incoming documents are checked and registration is refused if they do not conform. The registry can then answer questions about who owns a parcel authoritatively.

The result here shows that the same software can be used in both cases, as the fundamental structure and operations are the same.

9.2 Translation to conventional tools

The formalization has led to an understanding of the issues. For implementation, they must be translated to database schemata and queries. The object definitions must be split to factor out inheritance very much the same way the data statement for Document and GeneralDoc are set up here. A relational schema should then be checked for normalization. For query processing a *computationally complete* language (or a query language with transitive closure) is necessary because the computation of the title search required is recursive [Al-Taha, 1992].

Acknowledgment
I am greatly indebted to all the persons who have helped me to understand the legal aspects of property registration in Europe and the USA. I need to mention at least my teacher, Prof. H.-P. Friedrich, for the Swiss title registration system; Earl Epstein and Harlan Onsrud, who made me in long and - for them - painful discussion understand the deed registration system in the USA. The Salcedo project in Ecuador, led by Jack Rosholt and Maria-Augusta Fernandez forced me to apply what I knew then and learn more. The patient registrar from Salcedo made me finally understand how this is all carried out in practice.

Mag. Roswitha Markwart's efforts to make this presentation clear are also appreciated. Harlan Onsrud answered specific questions in the course of preparing this manuscript, and I gratefully acknowledge his contribution. I also thank an anonymous reviewer for his suggestions which - I hope - have helped me to improve the paper.

References
Al-Taha, K. 1992. Temporal Reasoning in Cadastral Systems. Ph.D. University of Maine,.
Al-Taha, K. and Barrera, R. 1994. Identities through time in ISPRS Working Group II/2, Workshop on the Requirements for Integrated Geographic Information Systems in New Orleans.
Bird, R. and Wadler, P.1989. Introduction to Functional Programming.
Dale, P F., and McLaughlin, J. D. 1988. Land Information Management. (Oxford: Oxford University Press)
Frank, Andrew U. 1994.Qualitative temporal reasoning in GIS -ordered time scales.in Spatial Data Handling in Edinburgh, Scotland, ed. Waugh, T. C., and Healey, R. G., 410-430
Frank, A. U., and Kuhn, W. 1995. Specifying Open GIS with Functional Languages in Fourth International Symposium on Large Spatial Databases - SSD 95 in Portland ME, ed. Egenhofer, M. J., and Herring, J., (Heidelberg: Springer Verlag)
Jones, M. P. 1994. Gofer - Functional Programming Environment. (Geoinformation, TU Vienna) Script
Jones, M. P.1995. Functional Programming with Overloading and Higher-Order Polymorphism.in Advanced Functional Programming, ed. Jeuring, J., and Mejer, E. (Berlin: Springer Verlag) 331-925
National Research Council. 1980. Need for a Multipurpose Cadastre. (Washington D.C.: National Academy Press)
National Research Council. 1983. Procedures and Standards for a Multipurpose Cadastre. (Washington D.C.: National Academy Press)

Reiter, R. 1984. Towards a logical reconstruction of relational database theory in On Conceptual Modelling, Perspectives from Artificial Intelligence, Databases, and Programming Languages, ed. Brodie, M.L, Mylopolous, M., and Schmidt, L. (New York: Springer Verlag), 191-233

Contact address
Andrew U. Frank
Dept. of Geoinformation E127.1
Technical University Vienna
Gusshausstrasse 27-29
A-1040 Vienna
Austria
E-mail: frank@geoinfo.tuwien.ac.at

DO-GIS - A DISTRIBUTED AND OBJECT ORIENTED GIS

B.V.S.S.D. Kiran Kumar, Prasun Sinha and P.C.P. Bhatt

Department of C.S. and E.
I.T.T.
New Delhi, India

Abstract
This paper reports "first cut" of the design of a distributed and object oriented geographic information system, **DO-GIS**, under development at I.I.T. Delhi. The paper deals with data model, architecture, query processing and data management in a GIS.

Considering that geographical data includes information about spatial reference frame, topological connections and requires maintenance of physically distributed spatial data, we argue that a distributed and object oriented GIS is the most natural choice for both data processing and management. The distribution of data (over a set of sites) has implications on the query processing, i.e. it requires query partitioning and distributed processing. **DO-GIS** architecture employs meta database and order preserving protocols to make both the distributed query processing and subsequent collation of partial results totally transparent to the user. Overall the operation is in the spirit of co-operative computing paradigm.

1. Introduction

In diverse applications, information systems are being used increasingly as a means to manage, retrieve and store large quantities of data which are tedious to handle manually. Spatial data handling is no exception. Systems for such application are known as Geographic Information Systems (GIS) with spatial data acquisition, preprocessing , management, manipulation and analysis, and output generation. Since GIS deals with spatial data which is graphical in nature each of the above components need to have a well designed Human Computer Interface (HCI).
Current computer based information systems offer very little in terms of semantic control of information processing. This is because the application semantics are difficult to map onto the descriptive information available in the input data. Object orientation promises to remedy this situation. Besides, currently GISs are rarely designed to operate with inherently distributed geographic data. Distributed GIS provide many incentives compared to monolithic systems basically taking advantage of distributedness in both management and processing of spatial data.

2. Data model

The usual paradigm for general purpose information management is the relational model. Majority of GISs implementing a relational model typically adopt a hybrid architecture where attribute data and their spatial references are stored, and managed, in independent structures. Such systems maintain the attribute data in conventional DBMS but organize and manipulate spatial data using conventional file handling techniques with some special purpose software for both the management of data and query processing.

The hybrid approach lacks rigorous management, data security and integrity control. Further it is not amenable to multiple user access and concurrency management. Also, and perhaps most importantly, it does not exploit natural distributedness of data. As Worboys [Worboys 94] argues, "If benefits are to be derived from distributed database technology, then GISs must be developed that deal with the particular complexities of distributed database management without being hindered by an approach that stores some data outside the DBMS, inaccessible to any distributed functions provided therein". In such hybrid systems query optimization is a very complicated issue as the query has to be first partitioned into spatial and aspatial components before any optimization and evaluation can be attempted [Ooi 89]. It has been shown that special characteristics of spatial data prevents development of unified extended relational systems with spatial and aspatial data under the same architecture [Healy 91].

Object oriented approach is being promoted for such areas for which the application of pure relational or hybrid technology leaves much to be desired.

2.1 Object oriented GIS

There are two broad based and opposing classes of models of geographic information [Worboys 94]. The class of *field-based* models treats such information as collections of spatial distributions, where each distribution may be formalized as a mathematical function from a spatial framework (e.g. a regular grid placed upon an idealized model of an earth surface) to an attribute domain. Patterns of topographic altitudes, rainfall and temperature fit into this view. On the other hand the class of *entity-based* models treats the information space as populated by discrete, identifiable entities (objects), each with a geo-reference.
Some of the conventional data models used in GIS data representation are *Raster Model, Spaghetti Model, Topological Model, Tesselation Models* etc. [Kasturi 89].

DO-GIS is broadly in the category of systems that describe a collection of geographic objects over a two dimensional map. Clearly each geographic object can be classified as belonging to a particular entity class such as city, lake, state, river etc. These objects are described by their associated aspatial (alphanumeric) attributes (e.g. population, name, usage etc.). On a given map geographic objects may *intersect* , may *contain* other objects, and may be *components* of other objects. Queries in a GIS are typically concerned with these spatial relationships. As an example the spatial relationship *intersect* must be used to answer the question:

"Which are the states in INDIA through which the river GANGA flows ?".

All geographic objects can be grouped into three generic spatial object classes namely, *Point* , *Line* and *Polygon*. The class hierarchy currently supported by **DO-GIS** is shown in Fig. 1. Hence it is clear that entity based geographical information models fit naturally with object orientation concepts.

2.2 Distributed GIS

Distributed databases have been shown to have several advantages over a centralized database though it means increased data management complexity. We look at why these advantages are much more relevant when we consider a GIS.

- Geographic information can be said to be inherently distributed. The cartographic organizations which deal with maintenance of geographical data are distributed all over a country. Large geographical data corresponding to unit of administration like states/districts/blocks etc. is located at the administrative headquarters. In addition various organizations collect basic data sets at their own location and exchange them with others.
- The queries in a GIS are usually based on data collected locally. The fraction of queries which access data from other locations are not large enough to justify storing all the data at each location.
- Addition of new sites as more and more regions are mapped and surveyed is easier with decentralized data management system.
- Geographic information potentially includes maps, aerial photographs, satellite images, site photographs, survey records, engineering and architectural plans, document images and tabular data [Newton 92]. When monolithic databases are maintained, all data has to be transported to all sites so that queries can be evaluated at any other site with a hierarchically more important role in organizational structure. In terms of data transmission this involves phenomenal amounts of data transfer and enormous storage requirements at all sites.
- GISs which serve emergency systems such as Fire, Storms, Floods, Tornadoes etc. need the greater reliability and speed offered by a distributed system.
- Cost of updating geographic data dominates all other costs of the GIS [Newton 92]. Periodic updates to all sites is practically infeasible. An update in a monolithic system has to be propagated to all the sites with copies; giving rise to various concurrency issues. An update in a distributed object based GIS is much easier as it is localized and is performed with lesser communication overheads.
- Sharing information among heterogeneous databases is easier with decentralized system since all data need not be transformed to a single format. Only the queries and the query results need to be transformed.

We do not address the issues realted to integration of heterogeneous GISs into a federated system but they will become very important in the near future.

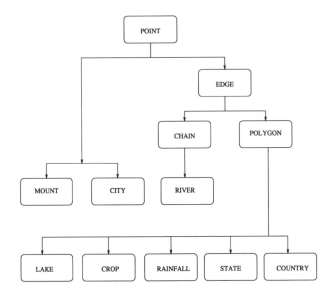

Figure 1 Class hierarchy in DO-GIS.

- GISs require interactive graphics and for real time response of queries based on local geography necessiates a decentralized GIS which can serve queries on local geography faster. Users querying remote data are usually willing to wait for a longer time.
- With the phenomenal growth of the World Wide Web (and increasing number of spatial data servers on the web such as Xerox Map Server, the Alexandria project of the Libraray of Congress etc.) it is becoming more and more evident that spatial data servers need to be distributed to effectively serve the large number of queries and also to keep pace with the huge amount of new information being added each day to the web.

Overall we can see that the disadvantages of maintaining monolithic databases at each site far overweigh the delay in serving a query based on remote data in a distributed system [Ceri 85, Newton 92].

2.3 DO-GIS system architecture

A distributed GIS needs a *global data manager* to manage the distributed database as a whole. The *global data manager* can be in a central site with all queries routed through it or can be replicated at each site [Newton 92]. **DO-GIS** uses an intermediate approach with meta knowledge servers. A Meta Knowledge Base server maps logical object names to physical sites thus providing somplified consistency, reliability, and data migration.

DO-GIS [Kiran 95] has been built over the multiprocessing environment on UNIX hosts. The message passing subsystem we have used is the ISIS [Birman 88] like environment built by Verma and Goyal [Goyal 94]. The modules of **DO-GIS** are depicted in Fig. 2. We describe each module and its basic operational role briefly.

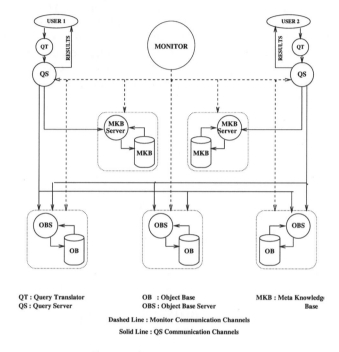

QT : Query Translator   OB : Object Base   MKB : Meta Knowledge
QS : Query Server   OBS : Object Base Server   Base

Dashed Line : Monitor Communication Channels
Solid Line : QS Communication Channels

Figure 2 Sample view of DO-GIS.

2.3.1 OB - Object Base
OB is a counterpart of a typical database and consists of the files and data structures which make up the persistent store for the objects. The data in the whole system is present in non-replicated parts at various sites. It is natural and convenient to use a non-replicated database for geographic data since it is continuosly extended and updated.

2.3.2 OBS - Object Base Server
OBS resides at each site where an **OB** is present. Its main function is to serve requests for data objects. Whenever an object is requested the **OBS** checks whether the object is in memory. If it is not it loads the object into local volatile memory. It uses various hash tables to maintain information about the objects and their location on disk in the local **OB**. Whenever an object is added to an **OB** an Object Identifier[1] is created for that object by combining information about the creator site, the class of the object and the instance count of that class at that site.

---

[1] For example, A:STATE:25 is the identifier which is assigned to an object which is 25th instance created at site A, of entity class STATE.

2.3.3MKB - Meta-Knowledge Base

MKB basically serves as a repository (or a directory) which has the information about the actual location of the objects. There can be several **MKB**s in the system and these can be active at the same time. Multiple **MKB**s are used to split the query processing load among them. The update to these **MKB**s is done by exploiting the virtual synchrony provided by the underlying platform [Goyal 94]. It is also used for targeting subqueries to relevant **OBS**s in a co-operative mode of operation. Fig. 3 captures the details of the processes and the nature of communication protocol used for all the updates. The abcast messages between Query Servers (#1, #2) and Object Base Servers (#1, #2) ensure the consistent ordering of data fetched from these object base servers.

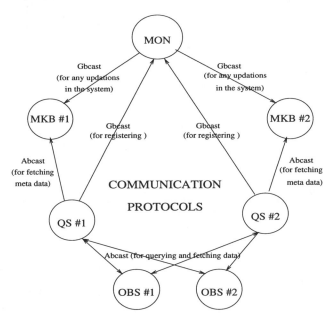

Figure 3 Communication Protocols.

2.3.4QS - Query Server

QS is at the user end and can be started up on any site irrespective of the presence of **OBS**s or **MKB**s at that site. It is used for evaluating end user queries. The evaluation of the query input by the user is done by collecting the required objects from the relevant object bases and then using the data from these objects in evaluating unit queries like equality of strings and numbers such as the name of a country, population of a state etc.

2.3.5QT - Query Translator

This is the front end to the **GEOSQL** based query. It analyzes a query and builds up the query tree. This query tree is used by the QS for evaluating the queries.

2.3.6MON - Monitor

Monitor provides a minimal level of fault tolerance and the load distribution in the system. It also serves as the system information database. Whenever a new module of the system comes up Monitor gives it the initialization data and also the information about various active modules in the system. When the Monitor detects the failure of any OBS or MKB module it sends update messages to the required modules. It also answers queries about the status of the system. In future versions of the system **DO-GIS** will provide fault tolerance by using the Monitor to detect failures of modules and starting them up automatically.

3. GEOSQL and query processing

In a GIS, the information describes entities that have a physical location and extent in some spatial region of interest. Queries in a typical GIS involve the identification of these entities based on their aspatial and spatial attributes, and spatial relationships between entities. The interface language should be powerful enough to express a query involving both spatial and aspatial components.

Structured Query Language (SQL) is by far the most popular query language but lacks constructs needed to express queries based on spatial attributes [Egenhofer 92]. We have retained the template structure of SQL and introduced the operators that reflect the GIS nature of a query. For example, consider a typical query like *"Which are the states in INDIA through which the river GANGA flows ?"*. We need to specify entity classes *state, country* and *river*, spatial attributes like the *boundary* of a state and the spatial relationships *contains* and *intersects* in the query. We have designed a query language **GEOSQL** by adding spatial constructs, operators and relations to the basic constructs in SQL viz. "SELECT", "FROM" and "WHERE". The template of a typical query in **DO-GIS GEOSQL**[2] is:

```
FROM      <ENTITY_CLASS> <VAR_NAME1> , ...
SELECT    [QUALIFIER] <VAR_NAME1>.<ATTRIBUTE>
WHERE
          <VAR_NAME1>.<ASPATIAL_ATTRIBUTE> = | < | >  <CONSTANT>
          <LOGICAL_OPER> <VAR_NAME1> <SPATIAL_OPER> <VAR_NAME2>
```

For the convenience of the end user there is heavy overloading of operators. As a result there can be queries that contain expressions such as *river* INTERSECT *state* or *crop* INTERSECT *rainfall*. One more important feature to be noted is that the end user need not be aware of the distribution of data over various sites. The query language interface makes this distribution of data transparent.

---

[2]Hereafter GEOSQL shall refer to **DO-GIS** GEOSQL.

Now expressing our example query above in **GEOSQL**:
Select the names of the states in INDIA through
which river GANGA flows

FROM STATE S ,RIVER R , COUNTRY C
SELECT S.NAME
WHERE
        C.NAME = "INDIA"
AND C CONTAINS S
AND R.NAME = "GANGA"
AND EXISTS (R INTERSECT S)

A special feature of **GEOSQL** is the CONSTRUCT-AVOID clause. Suppose one wants to explore the feasibility of constructing a facility like a tramway and wants to find the number of bridges that he may have to construct, he can express it using the CONSTRUCT clause in **GEOSQL**. The AVOID clause is used when the user desires to study an alternate path with the condition that it avoids one or more of geographic objects such as a lake, a mountain or crossing a state boundary.

A part of a GEOSQL query will clarify the use of these constructs:
A tramway is to be constructed between two cities
avoiding the lake in between

FROM REAL TRACK_LEN ,LINE L , CITY A ,CITY B,
        COUNTRY C, LAKE THELAKE
SELECT TRACK_LEN
CONSTRUCT
        L = ((A),(B))
        AVOID THELAKE
WHERE
    ......

A special class that is supported by **GEOSQL** is AMORPHOUS. A variable of this class fits in place of object of any class. This class is very useful for queries in which the classes of the objects to be selected are not known beforehand. Those classes can then be declared as AMORPHOUS and at runtime (i.e. while evaluating the query) , the actual class would be assigned to it.

3.1 Query processing

For an object oriented and distributed system the strategies used for processing a query are entirely different from those used in conventional database systems   [Egenhofer 94]. There are two ways in which the queries may be processed. The naive approach is by gathering all the required objects from relevant sites and then proceeding to solve the query.   A better approach is by sending some extra information along with the data request to the remote site so that the remote data server can do some pruning on its own and send only the selected objects. This is also the basis of co-operative computing wherein different segments of a solution are generated in a distributed manner. The co-operative framework may result in parallel evaluation of a query for which the underlying message ordering protocols can be exploited.

The major steps in query processing are described in Fig. 4. Basically the steps followed in the figure are:
1. Parse and construct the query tree.
2. Preprocess the query tree and check for its validity, i.e. ensure that scope rules are followed, variables are declared etc.
3. Extract information regarding every variable declared in the FROM clause. This information includes type declaration, all attributes referenced, subtrees which can be used to restrict the distributed search space etc.
4. Send a request to MKB for information regarding sites where objects satisfying the class declaration in the FROM clause can be found.
5. Send data requests using the appropriate protocols to sites to gather the relevant data.
6. Evaluate the query using recursive calls for the expression in WHERE clause.
7. Display the text and graphical data corresponding to the results.

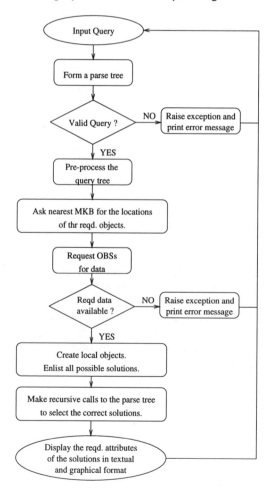

Figure 4 Query processing in *DO-GIS*.

Query processing in general involves the use of some heuristic methods for fast evaluation. In a distributed system, use of heuristics would reduce communication overheads and hence improve the performance substantially.

Various heuristics which we have identified are :-
- Object Name : Use the name of the object while requesting for the object if it is specified by a clause like O.NAME = "XYZ" where O is the object and "XYZ" is the name of the object.
- Required Attributes : Select all the attributes that are referenced in the query and fetch only those attributes of the required objects rather than fetching all the attributes of the objects from remote sites.
- Remote pruning of the data : If the query has a clause like C.POPULATION > 10000 then this information can be sent to the remote site which can then select only those objects satisfying the clause and send them.
- Overlapping computation with communication : When data required for some subtree of the query tree is being fetched computation required in other subtree can be done. Threads such as Light Weight Process library supported by SunOS operating system can be used for such a purpose.
- Parallel evaluation of query tree : Parts of tree can be sent to remote sites and the results sent back to the original site. This involves development of algorithms which collate the results from various sites to get the final result.

In **DO-GIS** we have all the above mechanisms supported by the order preserving protocols. Note that the overlapping of communication and computation and parallel evaluation of query tree are in the co-operative framework of computation.

4. An example

To explain the basic operational steps and subquery management we portray a hypothetical scenario of a GIS dealing with the geographical data of India. A logical organization of the data would be at the four metropolitan cities viz., New Delhi, Calcutta, Bombay and Madras. Each city stores maintains the data pertaining to a specified geographic region around it. Since the actual objects are not distributed over various sites, data about objects falling in two regions is arbitrarily stored at any one of the two.
Now consider the query discussed in section 3 i.e.

```
# Select the names of the states in INDIA through
# which river GANGA flows

FROM STATE S ,RIVER R , COUNTRY C
SELECT S.NAME
WHERE
        C.NAME = "INDIA"
    AND C CONTAINS S
    AND R.NAME = "GANGA"
    AND EXISTS ( R INTERSECT S )
```

First this query is parsed and processed to check the validity. The query tree constructed by the parser is shown in Fig. 5 . Now the three objects classes that are used in the query are State , River and Country. The query has two clauses which specify the names of objects Country and River. The location of INDIA and GANGA is obtained from a Meta Knowledge Base and both the objects are accessed from their locations, which in this case happens to be New Delhi. Now there is no clause in the query which provides any information about the states so a query is sent to the Meta Knowledge Base to get location of all objects of class State. So states from all four sites are accessed as shown in Fig. 6.

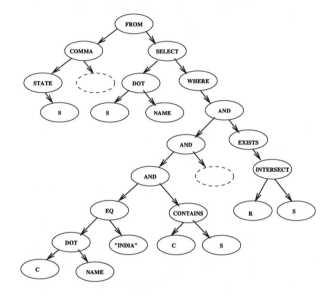

PARSE TREE FOR A SAMPLE QUERY

(The empty ellipses represent subtrees)

Figure 5 Query tree for example query.

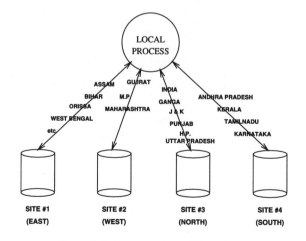

Figure 6 Data access from different sites.

Now the clauses "C CONTAINS S" and "R INTERSECT S" are evaluated and then the list of states which satisfy both the clauses is selected as the solution.

5. Conclusion

We have vindicated Worboys [Worboys 94] statement on taking advantage of natural distributedness of geographical data in designing an effective distributed GIS. The distributed character of data can be taken advantage of, if the database architecture supports distributed management and query processing. **DO-GIS** employs object oriented framework to encourage reuse of methods and captures natural data hierarchy besides its distribution. The distributed query processing is attempted in co-operative mode with adequate support from order preserving protocols.

References
[Birman 88] Birman, K.P., Joseph, T.A., Kane, K.P., Schmuck, Frank, 1988, ISIS - A distributed Programming environment User Guide and reference manual.
[Ceri 85] Stefano Ceri, Giuseppe Pelagatti, 1985, Distributed Databases - Principles and Systems, McGraw Hill.
[Egenhofer 92] Egenhofer M.J., 1992, Why not SQL ?, International Journal of Geographic Information Systems 6(2), 71-85.
[Egenhofer 94] Eliseo Clementini, Jayant Sharma, Max Egenhofer, Modeling Topological spatial Relations: Strategies for Query Processing, To appear in Computers and Graphics , Special Issue on Modeling of Spatial Data.
[Goyal 94] Arun Verma, Deepak Goyal, 1994, Design of a Shared Visual Space System for Co-operative Problem Solving. B.Tech. Thesis, DCSE, I.I.T. Delhi.
[Healy 91] Healey, R.G., 1991, Database Management Systems. In Geographical Information Systems: Principles and Applications, edited by D.J.Maguire, M.F.Goodchild and D.W Rhind , Longman, 251-267.
[Kasturi 89] Rangachar Kasturi et al., 1989, Map Data Processing in Geographic Information Systems, IEEE Computer, December 1989, 10-21.
[Kemper 94] Alfons Kemper, Guido Moerkotte, 1994, Object Oriented Database Management - Applications in Engineering and Computer Science, Prentice Hall.
[Kiran 95] Kiran Kumar, Prasun Sinha, 1995, DO-GIS: Distributed and Object Oriented Geographic Information System. B.Tech. Thesis, DCSE, I.I.T. Delhi.
[Nagy 79] George Nagy, Sharad Wagle, 1979, Geographic Data Processing. Computing Surveys, 11(2), 139-181.
[Newton 92] Newton, P.W., Zwart, P.R., Cavill, M.E., 1992, Networking Spatial Information Systems, Belhaven Press.
[Ooi 89] Beng Chin Ooi, Ron Sacks-Davis, Ken J.McDonell, 1989, Extending a DBMS for Geographic Applications. Proc. of 5th Int. Conf. on Data Engineering, Los Angeles, California, 5, 590-596.
[Worboys 94] Worboys, M.F., 1994, Object-oriented approaches to geo-referenced information. International Journal of Geographical Information Systems, 8(4), 385-399.

Contact address
Kiran Kumar, Prasun Sinha and P.C.P. Bhatt
Department of C.S. and E.
I.I.T.
New Delhi
India
E-mail: pcp@cse.iitd.ernet.in

QUALITY METRICS FOR GIS

Henri J.G.L. Aalders

Faculty of Geodetic Engineering
Delft University of Technology
Delft, the Netherlands

Abstract
In [1] and [5] a basis for a quality model is developed to be used in practical applications. Not only the quality of geometric data is considered but also the quality of non-geometric attributes. The models follow the rules that are applicable in data base information technology. In order to make all quality aspects available for practical applications this is an absolute requirement. In this presentation these requirements are discussed before presenting the metrics of quality in GIS.

1. Introduction

Geographic data is created by abstracting phenomena existing in the real world according to predefined rules that define details of what is to be represented, how it is represented and how it is described in a geographic database (see fig.1). This subjective selection and interpretation of real world objects creates an idealized view of the world called the 'nominal ground' (US: 'abstract view of the universe', French 'terrain nominal'). This interpretation is subjective for two reasons: because of its:

- object type descriptions. It is assumed that the real world is composed of a set of inter-related objects. When creating a model of the real world, the abstraction describes both these objects and their properties and as well as the relations between the objects and their properties. As such the nominal ground is a set of descriptions of properties of objects and relations that occur in the real world; they are called object(type)s, relationship(type)s (sometimes also called entities) and attributes;
- application dependency. Due to the specification only those objects are selected into the nominal ground that are of interest for the intended application(s), omitting all other objects.

Referring to the nature of all geographic objects there are different types of attributes:

- geometric; the geometric attributes refers to:
  ° position of the object, relative to some type of geo-reference;
  ° shape of the object, defined by mathematical terms e.g. interpolation types;
  ° topology, e.g. defining the neighbourhood of objects.

Geometry can be expressed both in raster or vector representation. However, in this paper the difference between these two ways of geometric object descriptions are of no importance since basically an IT object definition is used in which geometry acts in the same way as all other attributes;

- thematic, that can be either:
  ° quantitative, i.e. attribute values expressed or measured in a given unit, e.g.: width, temperature, time, etc.;
  ° qualitative, i.e. attribute values referring to one specific instance of a finite set of values, e.g.: classnames, addresses, colours, etc. A qualitative attribute can be nominal i.e. there is no order between the possible values e.g. classnames, or ordinal when there is an order between the values e.g. route numbers.

  Apart from the semantics, this type of attribute also has a value of which the range is predefined;
- descriptions in the form of text, photographs, drawings, video images, etc.

As such, entities (being the abstract representation of real world objects), attributes and relations can be defined as in table 1.

ENTITY	(id, { n attributes | n ≥ 1})
ATTRIBUTE	({ n types, n values | n ≥ 1})
RELATION	(id, {type, n entity id's | n ≥ 2, m attributes | m ≥0 })

Table 1 Definition of entities, attributes and relations.

Also quality will be attached to entities in the form of attributes. It is assumed that quality attributes describe the quality of one or more occurrences of entities and/or attributes that are all in the dataset. Each dataset, formed for the sake of quality statements, can be considered as a meta-dataset. In defining datasets one can apply set-theory as a dataset consists of a set of occurrences of entities. Because of the semantics and according to set-theory, datasets can be hierarchical subdivided into supersets, sets and subsets. An occurrence in a subset entity may be a member of an occurrence of a set-entity and an occurrence in a set-entity in its turn may be a member of an occurrence in a superset-entity. Quality follows the concept of meta-sets, i.e. (super/sub-)sets are not members of a metaset but carry the attributes of a meta-set.

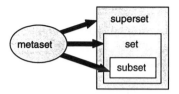

Figure 1 Meta-classes.

As such quality data is always meta-data, i.e. data about data. Quality attributes are valid for all occurrences of entities in a dataset that are related to the meta-set, i.e. quality metasets describe the quality of homogeneous entities or entity-sets.

2. Nominal ground

To define real world objects into an ideal form by position, theme and time - in order to make these objects intelligible and representable in a database -the nominal ground is defined to model the potentially infinite characteristics of objects in the real world. This process describes the abstraction of objects from the real world into an ideal concept. For modelling the quality of a dataset, it is necessary to precisely define the process that allow to derive the dataset from the real world. This process is decomposed in two steps:

- modelling, containing both the contents specification of what should be considered from the real world into the database and the abstraction of the selected objects;
- mensuration, specifying the measuring methods and the measurement requirements as well as the capturing and storing methods of data itself.

To fill the conceptual gaps between the real world and the nominal ground, and between the nominal ground and the dataset, quality indicators are to be defined for the specifications and abstraction of real world objects and for the capturing methods. The nominal ground forms the basis for this specification, against which the quantitative contents of datasets is tested. These specifications are usually defined by users or producers of geographic datasets and they depend on their intended use. Then the quality model consists of a set of quality indicators that will allow users or producers to define the relation between the specifications and the performance of geographic datasets. So, the nominal ground can be seen as the user-intended dataset in an ideal situation.

Although one can also speak of the quality of the modelling process, one should realize that the intended use is defined in the dataset contents specification. Since quality is defined as a combination of characteristics for both **expressed and intended** use, it is obvious that any quality model for an existing dataset should refer to both the quality of the data-modelling and data-capturing process(es).

3. Metaquality

The existence of the quality of each quality statement in a dataset should also be noted. In other words, to each quality indicator a meta-quality statement should be attached expressing the reliability of the quality indicator (the quality of quality). This information comprises of a:

- measure of the confidence of the quality information, indicating the level of confidence of the dataset;
- measure of the reliability of the quality information, reflecting how well the quality information represents the whole dataset concerned;
- description of the methodology used to derive the quality information, to indicate how this result was obtained;
- measure for the abstraction to account for the differences of the reality and the nominal ground. However, this measure is not relevant for factual statements concerning quality that are independent of the content of the geographic dataset.

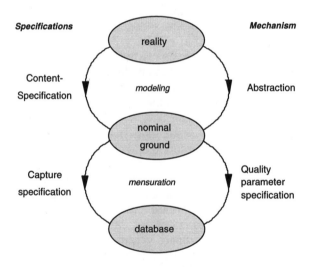

Figure 2 The role of the nominal ground.

4. Structure of a quality model

The quality indicators of data in a quality model are of four main types:

- **source** of the dataset to indicate the name and the organization responsible for the dataset, the purpose and date of the original production. Surely source information can be seen as part of the metadata of the dataset as all quality information is. Source information can also be seen as part of the "overall" dataset's lineage. However, source information is the minimum quality information that should always be available to a dataset or a data(sub)set. The attributes to the source information entity may tell the present user in general terms about the possible validity of the data. This impression of the data may be obtained from the time interval that has past since the original creation, the awareness of the reliability of the producer and who would be ultimately liable for errors in the data, as well the original purpose of the data.

- **usage**. Any previous use of a dataset by (several) other users for various applications may be a good indication of the fitness of the data for the present intended use and the reliability of the dataset in different circumstances. For that purpose, for each usage a separate statement should be given indicating the organization that has used the information, the type of usage and its fitness and any possible constraints or limitations that were imposed or found during that use. Usage of a dataset is considered to be important from the point of view of new users. If a dataset has been used for applications similar to the one envisaged by the current user than it clearly adds to the potential of the user's confidence. Also previous combinations of the actual dataset with other datasets is a firm indication of the quality of the dataset. However, usage is only useful if its information is honestly given and errors, constraints and limitations are also recorded.

- **quality parameters**, describing the measurable aspects of the

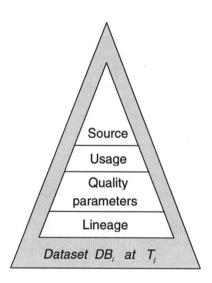

Figure 3 Pyramid of quality aspects.

performance of occurrences in a dataset for the intended use. They are described in the next paragraph;

- **lineage**, describing the process history of each occurrence of an entity that it has undergone since its original creation. For each process, a statement has to be given describing the process the entity has undergone (including details as input and methods used and possibly references to documents

containing further details), who undertook the process, when and why[1]).
Processes that have been performed to each occurrence of an entity in the dataset may refer to the type of transformations, selections, generalizations, conversions, updates, consistency checks, validations, etc. This involves the mentioning of dates of input and processes as well as the description of the processes and the person/institute responsible for each process. (For this purpose metadata-entities have to be defined and related to occurrences of entities forming the metadataset entity).

Transformations may be of the following type:

° continuous to continuous, e.g. geometric transformations, where the functions and parameters should be given;

° continuous to discrete, e.g. a re-classification schema, where thresholds (class-boundaries) and rounding-off systems should be given;

° discrete to continuous, e.g. a de-classification and a raster-to-vector conversion, where the methods used should be given;

° discrete to discrete, e.g. resampling, where the resampling rules must be specified.

Occurrences of entities may follow the quality statements of entities in meta-dataset and so obtaining their properties. This involves a rather complex logical structure within the entire datamodel for the "whole dataset".

5. Metrics for quality

A quality parameter is a specific instance of a particular quality parameter type and represents its performance characteristic. Each quality parameter type has a 'metric' which is the quantitative element of the quality parameter. Metrics may be statistical (expressed in real numbers, with or without units) or binary and in some occasions descriptions. Metrics should also be dated to indicate the currency of the result. The descriptions of the algorithms used to determine the metrics can be described as part of the metaquality methodology component.

quality feature	source	usage	lineage	quality parameters			
				accuracy	comple-teness	consis-tency	currency
dataset	text	text	text	see table 3	value	value	date
objects	- -	text	text	see table 3	- -	value	date

Table 2 Type of quality attributes for dataset and objects.

The quality attributes can be defined for both datasets as well as for single

---

[1] During its lifetime, geographic entities may have undergone many and complex processes. This includes input, transformations, compilation, generalization, etc. All contribute in some way to the present quality of the data. Therefore producers and assemblers of geographic data should provide sufficient information into the lineage for the data users in order to enable the product's capabilities judgement.

occurrences of an entity of a single attribute forming a separate dataset. According to the quality aspect triangle source, usage and lineage are described by text as indicated above. Accuracy is described below.

Metrics for completeness are defined for a dataset only and given in e.g. percentages of missing and overcomplete data (omission and commission), where the type of evaluation should be indicated as estimated, propagated (from different sources) or checked. If checked the method that lead to the result should be stated as well as whether the method is applied to a full dataset or by sampling as well as the date of checking.

Consistency includes static and dynamic consistency. Static consistency is the result of the validation of semantic constraints of data and relations, while dynamic consistency is the result of the validation of the processes that are part of the quality assurance and will not be considered in this context. The validation is a check of a constraint as described in the dataset's conceptual schemes. Its includes domain validation (on objects' thematic and geometric attributes) and relational validation. For each constraint the metric may be:
- the result of the validation process (e.g. the number of violations of a topological rule);
- the process to obtain the validation e.g. checked, assessed or calculated;
- date of checking.

Currency deals with the change of data through time: it describes the correctness of completeness and accuracy on a set of data at a certain time T. Currency probably has different meanings for the producer and the user of the dataset:
- a producer may understand currency as an update policy (which should be part of the data specification), e.g. annual, monthly, weekly, daily or continuous updates, while
- a user wants to know whether the data is still correct for the intended use at a certain time. This depends on both the update policy as on the frequency change. The frequency change can be given as a number of changes per class or aggregated in an unique value by defining weights for each class frequency change.

6. Accuracy of attributes

Accuracy describes the stochastic errors of observations on features and is an measure to indicate the probability that an assigned value is accepted. One can distinguish the:
- absolute accuracy, i.e. the accuracy against the spatial reference system;
- relative accuracy, i.e. the accuracy of positions of occurrences of entities relatively to each other.

Accuracy should be defined in terms of themes, position and time.

	metrics	
attribute type	single	multiple
quantitative valued attributes	- RMSE - σ - range - histogram - confidence level - interval	- list - error ellipse - correlation matrix - eigen value
quantitative valued attribute	- $\{ w_i, P(X=w_i) \}$ - the largest possibility m of $\{ w_i, P(X=w_i) \}$ - all values $> p$ of $\{w_1\}$ with $P(X=w_1) \geq p$ - m values $\{w_i, P(X=w_i)\}$, where $\Sigma_{i=1,m}\ P(X=w_1) \geq p$	
geometric connections	- lateral distance - curvature radius difference - point density	- σ - range - quadratic mean - bias - covariance function

Table 3.

6.1 Thematic accuracy

Thematic accuracy, describing the probability of correct or mis-classification. A distinction must be made between:

- *quantitative valued thematic attributes*, where:
  ° for single-valued attributes one may use: Root Mean Square Error (RMSE), Standard Deviation, the range (minimum, maximum) a histogram of deviations from the mean value or the confidence level or interval;
  ° for multiple valued attributes can be used: a list of accuracies for single-valued quantitative attributes, an error ellipse, a correlation matrix or function a range or the eigen value of the correlation matrix.
  These accuracy values are given in the same unit and reference system as the measured value is given;
- *qualitative valued thematic attribute* is the probability that a correct value has been assigned to this attribute, including the probability of correctly assigning alternative values.
  The possible metrics for accuracy on qualitative valued attributes can be:
  ° for both single-valued attributes and multiple valued attributes in general is a vector $\{i$ or $w_i,\ P(X = w_i)\}$, where w_i are the possible values and X is the assigned value; this corresponds to the distribution of a quantitative valued attributes.

The description of accuracy for qualitative valued attributes can reduced to:
- only the maximum probability: w_i, $P(X = w_i)$
- the largest m probabilities $\{w_i, P(X = w_i); w_j, P(X = w_j); w_k, P(X = w_k); ...\}$, where m is predefined;
- the maximum and other values where an assignment has the probability larger than the threshold p: $\{w_i; w_j; w_k\}$ with $P(X=w_n) \geq p$. This corresponds to the confidence interval for quantitative valued attributes
- the m values w_m, where $\Sigma_{i=1,m} P(X=w_m) \geq p$ and p is the predefined confidence or significance level.

6.2 Positional accuracy

Definition of geometry uses either vector or pixel geometry. In vector geometry one uses in general point-, line and area objects:
- *point objects* are defined by coordinates in a reference system (e.g. X,Y and Z). Since they are quantitative valued attributes they follow the rules mentioned accordingly;
- a *line* is given by an ordered set of points and by the connection between these points, expressed by a function to define the type of interpolation between the points. The positional accuracy of the points follow the rules of point accuracy i.e. quantitative valued attributes.
The accuracy of the connection can be described by a:
   ° distance between two lines such as the maximum lateral distance or Hausdorff distance;
   ° maximum curvature radius difference between two lines;
   ° required point density, using a regular tolerance and an interpolation type.
The other geometric element is the length of a line which is a quantitative valued attribute;
- an *area* is defined by its boundary lines, so the accuracy follows the rules for line objects, while the calculated size of the area is a quantitative valued attribute.

Raster geometry uses either grid points or pixels:
Both *pixels and grid points* are defined by a pair of grid integers and so they form a set of qualitative valued attributes following those rules.

6.3 Temporal accuracy

Temporal accuracy describes the correctness of time and updating of a data-(sub)set by:
- errors in time;
- moment of last update;
- rate of change of entities per unit of time;
- trigger value indicating the number of changes before a new version of the dataset is issued;
- temporal lapse, giving the average time period between the change in the real world in the updating of the database;
- temporal validity indicating data to be out of date, valid or not yet valid.

7. Conclusions

The implementation of the above developed quality model to describe the performance of geographic information comprises many different aspects to describe the quality of the geographic data or datasets. The basic assumption to model a database according to homogeneous data(sub)sets and considering these as new entities, makes it possible to describe the different quality aspects in terms of quality parameter types and their values that are applicable to all members of the entity occurrence. However it will make the database structure extremely complex and will extend the database size to a multiple of its original size. As yet no experience has been obtained to implement such a model completely and correctly form a theoretical and technological point of view. Implementation of such a model should learn the users whether this model has to be extended (in the sense of the completeness of the quality parameter types) and feasible.

References
CEN TC 287 WG 2, Data description: Quality, Working paper N15, August 1994.
CEN TC 287 PT05, Draft quality model for Geographic Information, Working paper D3, January 1995.
ISO, Quality. Vocabulary. Standard ISO 8402.
ISO, Guide to expression of uncertainty in measurement. Guide established by a working group supported by seven organizations, including BIPM and ISO, 1993.
Guptill, S and J. Morrison., (eds), The elements of spatial data quality, Elsevier 1995.

Contact address
Henri J.G.L. Aalders
Faculty of Geodetic Engineering
Delft University of Technology
Thijsseweg 11
2629 JA Delft
The Netherlands
Phone: +31-15-278 1567
Fax: +31-15-278 2745
E-mail: H.Aalders@Geo.TUDelft.NL

A VERIFICATION SYSTEM FOR THE ACCURACY OF

THE DIGITAL MAP[1]

Dong-Gyu Park and Hwan-Gue Cho

Department of Computer Science
Pusan National University, Keum-Jeong-Ku
Pusan, Korea

Abstract
The reliability of Geographical Information System(GIS) is mainly determined by the accuracy of spatial data stored in that system. Currently most of the spatial data are stored and processed in the vectorized format. There are several ways to obtain those vectorized data for GIS application. For example, ground surveying, aerial and terrestrial photographs, satellite photographs, graphs, charts and paper maps are typical ways of data mining. Before the era of modern GIS many commercial organizations and governments have produced geographic paper maps, which were mainly used for the construction of roads, railroads, telephone cable establishing and power line supplying. Till now lots of paper maps are kept and widely used in the general geographical works. Now it is believed to be easy and cost effective to obtain digital map data from normal paper maps by the current computer technology rather than field working. The commonly used digitization method for paper maps consists of two phases, firstly making a raster file by scanning the original paper map, secondly digitizing the raster file to make whole digital map. In this procedure it is highly required to guarantee the accuracy of digital map to the corresponding paper map. In this paper, we propose a new verification technique which compute automatically the accuracy of the digital map to the corresponding paper map. The proposed technique includes data format test, inclusion test and XOR test. We implemented our system on X Window/Motif environment with C language, and this system will be used in our national GIS construction project.

Key words
Spatial data accuracy, digital map verification, rasterization.

1. Introduction

According to works of Blakeman, Morse and Hovey, and Thapa and Burtch, data collection costs are greater than 80 percent of the total cost of GIS [Blak.1987, MorseH.1990, ThapaB.1990]. And also an issue of growing concern is whether the spatial data are reliable enough to use spatial database. Since the modern GIS requires vectorized data, it is main concern to construct spatial data easily and cost effectively from existing raster data such as paper maps [ThapaB.1990].

There are two ways in data collection. Thapa and Burtch have introduced the

---

[1] This work was supported by the development fund(1995) of Outside Plant Technology Laboratory of Korea Telecom. This system, GeoQC is a property of Outside Plant Technology Laboratory of Korea Telecom.

concept of primary and secondary methods of data collection [ThapaB.1990]. Primary methods of data collection refer to methods of data collection in which data are collected directly from the field work like ground surveying, aerial and terrestrial photographs, and satellite imaginary. Secondary methods of data collection refer to a process in which data are collected from existing documents such as maps, charts, graph, etc [ThapaB.1992]. After collecting the vectorized data, it is needed to verify the collected spatial data by comparing to the original data. If some parts of the collected spatial data are not appropriate or not reliable for using GIS, it is desirable to update or make it correct to reflect the real world situation. We have examined the verification methods for vector data made from paper map.

Currently National Geographic Institute of Korea(NGIK) has started to build national GIS. Therefore we need reliable and accurate digital maps as the base map for whole territory. For this project, NGIK decided to fully utilize all the paper maps which have been used and been proved valid in the previous application. In this paper, Section 2 will show the preliminary on digitizing works and the reasons of digitizing error. And the following Section 3 will propose our basic idea of verification, named XOR test. Section 4, we give detail algorithms for verifying some features of digital maps. Section 5, we discuss the implementation issues of our system named GeoQC and show an example of our verification. And Section 6, concludes with a discussion of future work.

2. Digitizing methods for paper maps

To obtain vector maps for GIS, it is common that a skilled human operator digitizes raster data. This digitizing method has two phases, 1) scanning paper a map to make a raster file, 2) digitizing in raster file and making a vector file. There are two types of vectorization methods, one is an automatic vectorization of raster maps, and the other is the manual digitizing with raster maps. The defects of a manual digitizing are :
- it is difficult to digitize a middle line of raster data, operator is easy to make an error when digitizing in areas where the features are dense,
- the quality of digitizing work varies, that depends on the skillfulness of each operator,
- takes much time for vectorizing because all the works are done by human operator.

And the defects of an automatic digitizing are :
- it can be only applied to raster files, which were separated in several layers, mountains, roads, building,
- it is difficult to tag out feature code of an object and attributes of an object in vector file like text, symbol, symbolized line, etc.
- it is difficult to recognize texts, symbols, symbolized lines, etc.

Semi-automatic digitizing method is one to combine these two methods and overcome the defects of these two methods. In this semi-automatic digitizing method, the operator can selectively digitize a raster line or object using vectorizing software. If a line is very simple to be applied with automated method, an operator can use automatic line tracking method. But if a line is very thin or complex or a raster object was placed with a feature code of text, he should tag out feature code of line or object. Even though semi-automatic digitizing methods are relatively good, it is also an error-prone procedure in the processing of vectorization. The most common mistakes caused by human operators are to miss objects in digitization and to assign incorrect feature codes to objects.

Errors in digitizing
Digitizing and scanning errors depend on the following factors: 1) width of the feature; 2) skill of operator; 3) complexity of the feature, 4) resolution of the digitizer, and 5) density of the features [ThapaB.1992]. And vector map can be contaminated by these types of errors: 1) illegal use of feature code, 2) illegal position of an object, 3) missing an object in raster data, 4) over digitizing an object, 5) overshooting and undershooting, 6) text misspelling or misplacement, and 7) symbol misplacement [ThapaB.1992]. As far as we know there is little techniques for measuring uncertainty properties of spatial data and GIS output.

3. A basic verification method

In this paper we consider how to measure the accuracy of digitized data which were obtained from original raster data. Even though this proposed methods may contain some disadvantages in addition to those found in the primary methods, it is desirable to find errors in secondary data collection method. In the verification procedure for raster data and vector data, manual verifications are widely used. According to this method, we compare the map of raster data with vector data, which were drawn in tracing papers, by naked eyes. Verifier examines an error of vector data by observing overlapped two maps. As manual digitizing works, manual verification method may give some errors due to the mistake of a human verifier. The problem of manual verification methods are that the results of verification are strongly determined by a skill, an eyesight, and a mental state of verifier. And also the main disadvantage is that it takes too much time to verify printed data. For these reasons, it is highly desirable to develop an automatic verification tool for verifying raster data and vector data.

The accuracy of a vector map made of a raster map
In order to measure the accuracy of the digitized data compared with its corresponding raster data, we define a similarity function $Similar()$. The

similarity function is defined as following:

$$Similar(R_1, R_2) = 0 \Leftrightarrow R_1 \equiv R_2$$

With input two raster files R_1 and R_2 this similarity function returns 0 if and only if two raster data R_1 and R_2 are exactly the same. Since the function $Similar()$ accepts only the same data format, we have to rasterize the vector file. The Boolean exclusive OR operation is the basic mechanism in our similarity measuring function.

XOR operation
XOR operation for A and B, returns 0 if two input data are equal, and returns 1 otherwise (See Table 1).

A	B	XOR
0	0	0
0	1	1
1	0	1
1	1	0

Table 1 XOR operation.

Let O_i denote original raster data, V_i denote vector data, f_r denote rasterizing function, and $\oplus$ denote XOR function. Let I be defined as following:

$$I = O_i \oplus f_r(V_i)$$

If the number of pixels in I, denoted $n(I)$, is 0, this implies that two rasterized data O_i and vector V_i are of the same shape totally. If $n(I)$ is not 0, by the property of XOR operation, we know that O_i are not same to V_i.

Rasterization vector data
Since it is required that previously defined similarity function accepts only two raster files, we must rasterize vector data. In the procedure of rasterization, it is needed to scale to fit the size of original raster data for vector data. Converting to raster images from vector data is a well-known problem in computer graphics, named as rasterization. We implemented rasterization function, f_r by applying Bresenham's algorithm.

Filtering
In scanning procedure of a paper map, raster data would be polluted by the resolution of scanner or the quality of paper, so it contains some random noises. If we simply apply XOR operation to raster data and rasterized vector data, there would be many mismatched cluster data. The cluster data means mismatched data or misdigitized line of raster data, but most of them are caused by the noise image of the raster image. To get more exact and

quantitative analysis of XORed data, we must remove these noises by applying filtering out procedure. Filtering procedure for a raster image has two phases: 1) getting clusters of each remained pixel, 2) erasing each cluster, which is smaller than the size of a given filter.

Flood fill algorithm

In order to find a cluster, it is needed to get neighbor pixel of an arbitrary pixel of a raster data file. We can define neighborhood pixel to the 4-connected or 8-connected approach. Flood fill procedure finds a neighbor pixel of an arbitrary pixel and gives coloring to examined pixel and finds recursively to the neighborhood pixel of a colored pixel. This procedure stops when no neighboring points have the specified color. If a cluster is obtained, then we compute the maximum size of cluster's width and height (the size of the minimum boundary rectangle of the cluster), then determine if small image is to be erased or not. As the size of filter grows, the number of erased noise and the size of erased noise components will grow too. But by this procedure, small objects in raster data or rasterized vector data may be erased, which is an undesirable result in this procedure. So the size of filter to be applied must be smaller than the minimum size of a symbol object in raster data or a vector data. Thus the filter size should be carefully selected and it should be constrained by the following formula.

$$\varepsilon < \min\{Object(Data_{rast})\}$$

Connected line drawing

For every digitized line in pixel based graphic system, there would be a stepwise disconnected component in rasterized vector line. If we use a small filter in this line segment then all the line segment might be treated as independent cluster. Uncarefully selected filter gives unwanted results. To avoid this result, we newly propose another line drawing algorithm to make line segments are connected in one component without stepwise dots. For this reason, we slightly modify Bresenham's line drawing algorithm. When we set a pixel, we add a connecting pixel of previous pixel. The algorithm is presented as following, where Bresenham(x_i) means a drawing point calculated by the original Bresenham's line drawing algorithm, and the input value is x_i.

[Connected Line Drawing Algorithm]
```
if( abs(x2 - x1) ≥ abs(y2 - y1) ) {
        PutPixel (xi +1, yi )
          PutPixel (xi +1,Bresenham( yi )) ;
}
else {
        PutPixel (xi , yi +1),
          PutPixel (Bresenham( xi ), yi +1);
}
```

The results are illustrated in the following figures.

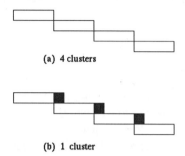

(a) 4 clusters

(b) 1 cluster

Figure 1 (a) An example generated by the original Bresenham's line drawing algorithm
(b) a line that was drawn by connected line drawing algorithm,
dotted cells denote connecting pixels.

Evaluation of accuracy value

Now we propose one measuring function for the accuracy of the digitizing procedure:

$$Accuracy(V_i) = 100 - \frac{2 * n(Cluster(I))}{\{n(Cluster(V_i)) + n(Cluster(O_i))\}}$$

If XOR file I has no cluster, accuracy value will be 100% (that means O_i and V_i are perfectly of the same feature).

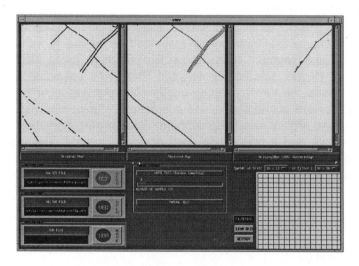

Figure 2 A snapshot of working system in XOR module in GeoQC.

Solid vector line verification

We have implemented rasterization module of a vector data using connected Bresenham's line drawing algorithm. And we apply XOR operation between original raster data O and rasterized vector data R_v, and then filtering is applied to the XORed result data, $O \oplus R_v$. An example of the system and its snapshot is shown in figure 2.

Verification method of regular thickness raster line
Though it is easy to find mismatched line segments by using simple XOR operation, a simple XORing operation gives some unexpected results. In the human digitizing procedure, it is difficult to position the center location of a raster line. And even if the raster data are digitized correctly, it is very difficult to rasterize correctly due to the truncation error of vector data in a line with small width especially.

A line with saw teeth
We propose a new line drawing algorithm which allows some artificial noises for adjusting this truncation error. A saw-teeth line drawing algorithm was derived from the original Bresenham's line drawing algorithm. This algorithm draws the outer part of a line segment to be out of joints.

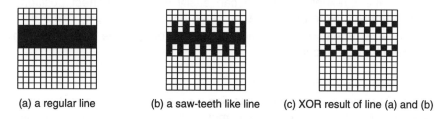

(a) a regular line   (b) a saw-teeth like line   (c) XOR result of line (a) and (b)

Figure 3 A regular line and saw-teeth like line and their XORed result.

Figure 3 shows a saw-teeth line and XORed result. The final XORed data have also a lot of mismatched dot clusters, but the width and height of all dot clusters are smaller than 2. So the mismatched dot clusters are cleared by applying an appropriate filter. Though in some worse case our saw-teeth algorithm does not guarantee to reduce the total noise of rasterized line, experiments showed that saw-teeth line gives good results in clearing noise if an appropriate filter is applied. The size of a saw-teeth also would be parameterized by a verifier according to the degree of accuracy.

4. Inclusion test

To verify objects in vector file, we can test if each object in original raster data are correctly positioned in vectorized data. In this case, we cannot verify objects in raster data which were omitted in the digitizing procedure. But we can obtain information if the feature code in vector data reflects in raster data. Now it is needed to calculate the vector file position by transforming vector line to raster coordinates. In converting vector line to raster line, it is known that hinting adjustment is required. In this case that a line in vector map has a fixed thickness, we can verify the positional accuracy of this line segment by testing ANDed number of pixels as below: put the number of existing pixel which are rasterized vector file as $n(F_r(V_i))$, the number of pixels of scanned raster file as $n(O_i)$, the number of ORed pixel denoted as $n(R_i)$

$$n(R_i) = n(F_r(V_i) \cap O_i)$$

if value $n(R_i)$ approaches the value of $n(F_r(V_i))$, two data can be considered to be highly similar. We can get corresponding testing criteria value as following:

$R_{match} = n(R_i) / n(F_r(V_i))$
/* θ **is the threshold value of matching */**
if $R_{match} < \theta$ **then**
      print("Object has an error");
else
      print("Object valid")

By tuning the threshold of θ we can easily verify the correct position of vectorized object. Note that in special case that all the value of raster file O_i may be set 1. Then we cannot apply this method, but we believe that this worst case hardly happen.

Unregular line verification method
It should be noted that unregular lines need a quite different verifying procedure. In the common paper maps, there are two types of lines, continuous lines with regular continuity and lines with discontinuity dotted lines. In maps disconnected lines are widely used in the boundary of two adjacent province. In case of a regular line, if an operator exactly digitize a center of a raster line, it is easily verified by applying XOR operation. But in case of unregular line, we cannot verify by applying XOR operation since those lines are not connected each other. The vectorized line for those disconnected lines are conceptually connected in vector maps. The characteristics of dotted line is that components with dots have arbitrary width and length and each component does not have the continuity with regular spacing. As a consequence, dotted line has a local connection between dot segments. Since sequence of dotted line has a local direction, we can estimate the position of next dot component by looking around the previously examined dots position.

Verification of unit dot segment
The characteristic of ideal dotted line is that the length of all disconnected line segments and the gap between all line segments are uniform. We denote L as the total length of dotted line, g as the total length of gap between dotted line segment, l as the length of unit dotted line segment, n as the number of unit line segments. Then it is easy to see that:

$$n = \frac{(L+g)}{l}$$

When we consider the existence marginal one starting and one ending dot segment in a dotted line, the ideal number of unit line segments are $n\pm1$. We can estimate the number of unit line segments in the following methods: follow up vectorized line and get the number of components disconnected the raster image in original map. We denote $R(l_i)$ as the image of the line l_i, and C_i as unit component image of line segment that overlapped in original map, we define the overlapped image O_i by the following equation:

$$O_i = R(l_i) \cup C_i$$

The expected number of dotted line segments can be estimated by n. In our system, GeoQC we have restricted the relaxation value to $\pm20\%$ to the ideal value. So the range of component number of O_i can be obtained as following:

$$0.8n \le n(O_i) \le 1.2n$$

The threshold value, for example 0.8 and 1.2 depends on how rigid verification is needed. For example we can verify dotted line more accurately by reapplying threshold value to 0.9 and 1.1 instead of 0.8 and 1.2.

Figure 4 Black dot is made on original raster map and several dots denote a rasterized area by a vector line for the dotted line. Note that $n(O_i) = 15$.

A verification method for symbolic line

In common paper map, symbolic line consists of continuous and regular symbols. Typical examples of symbolic lines are those for railroads and banks. The verification method for symbolic line is that: 1) positioning the center line of a railroad; 2) tracing the center line and checking the bar. The verification line follows up the raster line of the railroad in a paper map and counts the number of pixels contained in a small testing circle which is shown in Figure 5.

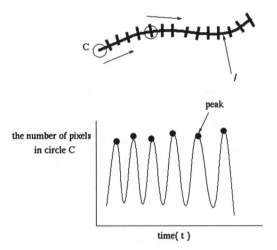

Figure 5 Railroad verifying procedure and its corresponding graph. The peak denotes the local maximum, which implies that the position was in the railroad.

The number of pixels enclosed in the circle will increase or decrease when the circle runs along the raster line l. Let us make the graph whose y axis denotes the number of pixels in the circle and x axis denotes the time for moving circle C. The distribution graph will have lots of local minimum and maximum. The number of local maximum should be equal to the number of a perpendicular bar to l. In real paper maps, the number of pixels would not be distributed uniformly. So we must normalize the number of perpendicular bars. Let I^* be ideal number of perpendicular bars, n_i be the number of local maximum, L_i be line length. And the gap between two adjacent perpendicular bars denote G. Then we can estimate the optimal number of I^* as following:

$$I^* = \frac{L_i}{G}$$

So we can verify the digitizing error by comparing number I^* with n_i. In most of actual raster data, the value I^* and n_i are not equal, because each gap between adjacent perpendicular bars is not uniform. So we have to allow small relaxation value for this situation. Therefore we suggest one relaxation variable as 0.1 in our GeoQC system.

$$|\frac{I^* - n_i}{I^*}| \leq 0.1$$

5. Implementation issues and examples of verification

We have implemented a map verification system, named GeoQC on X Window/Motif environment. Original raster data was obtained from a real paper map, and vector map is a vector formatted map digitized by a skilled operator. Let us describe an experiment. GeoQC was applied to verify a real vectorized data for the "Po-chun" province. Table 2 shows those vectorized map data file, file format and file size.

	raster file	vector file
Contour Layer	c.tif (3.7 M Byte)	p015c.dxf (2.3 M Byte)
Building Layer	p.tif (3.6 M Byte)	p015p.dxf (3.3 M Byte)
Water Layer	w.tif (3.6 M Byte)	p015w.dxf (23 K Byte)
Text Layer	t.tif (4.0 M Byte)	p015t.dxf (481 K Byte)

Table 2 An example of test data.

The file format of original map is determined in our scanning phase. There are many kinds of file formats in scanned image file such as *.gif, *.tiff, *.jpg, *.bmp, *.xbm, etc. For the internal data format of GeoQC, we newly defined "*.rast" file, which a variation of XImage structure data on X Window/Motif environment. Original map must be converted to "*.rast" file and also vector file must be converted to "*.rast". Figure 6 shows the global data flow for our GeoQC system.

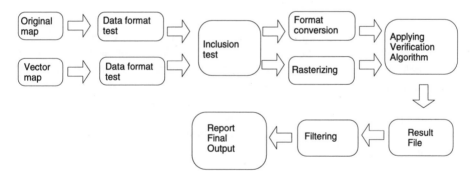

Figure 6 Flow of vector map verifying system.

We applied XOR verification and inclusion test in "Po-chun" province data, and we obtained following results.

	Total raster size	sampling area(%)	total cluster	error cluster	error ratio(%)
Contour layer	4475 x 5642	25	47943	1234	2.57
Building layer	4456 x 5618	25	37936	949	2.50
Water layer	4473 x 5641	25	8490	342	4.02

Table 3 Results of XOR verification in "Po-chun" province data.

	Sampling objects	number of errors	error ratio (%)
Contour layer	50	4	6
Building layer	50	5	10
Water layer	65	4	6

Table 4 Results of inclusion tests in "Po-chun" province data.

6. Concluding remarks

There are many methods to collect spatial data for construct GIS. It is true that data accuracy determines the reliability of GIS. So it is very important to verify the accuracy of spatial data for GIS. This paper has discussed verification methods of vector map made from paper map. We have proposed a verification algorithm of vector data for GIS and implement a vector map verification system, named GeoQC on X window/Motif environment. An inclusion test and XOR test were proposed and implemented for verifying vector data. For several types of symbolic lines and dotted lines in paper maps. we propose a novel algorithm to those lines. Experimental results showed that the proposed verification system could detect lots of digitization errors automatically.

We believe that it is highly difficult problem to verify all kinds of map symbols(e.g. grass field, beach, orchard). Since all symbols in raster data to be translated to a single point, a more reliable symbol recognition system is needed. And also it is difficult to verify text and its spell and its semantics. We hope in the future these problems could be cleared. Our GeoQC system will be used in the NGIK's GIS project as the main verification tools for vectorized maps. In this year we are trying to develop the fully automated and commercial version of GeoQC system.

Acknowledgments
The authors thank to SungYong, KIM and SongHun, BAEK in Outside Plant Technology Laboratory of Korea Telecom for their kind supports.

References
[Blak.1987] Blakeman, D. A., 1987, Some Thoughts About GIS Data Entry, GIS '87, San Francisco, ASPRS and ACMS, pp. 226-223.
[MorseH.1990] Morse, B. W., and S. T. Hovey, 1990, Data:The Foundation of a Land Information System, GIS/LIS '90 Proceedings, Anaheim, California, 7-10 November, Vol. 1, pp. 273-282.
[ThapaB.1990] Thapa, K., and R. C. Burtch, 1990. Issues of Data Collection in GIS/LIS, Technical papers, 1990 ACSM-ASPRS Annual Convention, Vol. 3, GIS/LIS, pp. 271-283.
[ThapaB.1992] Thapa, K, and J. Bossler, 1992. Accuracy of Spatial Data Used in Geographic Information Systems, Photogrammetric Engineering & Remote Sensing, Vol. 58, No. 6, 1992, pp. 841-858.

[YairY.1994]   Yair, G., and D. Yerahmiel, 1994. Automatic   Adjustment of   Line Maps,
          GIS/LSI, pp. 333-341.
[Stefanovic.1985] Stefanovic, P., 1985. Error treatment in photogrammetric digital techniques,
          ITC Journal 1985-2, pp. 93-100.
[Will.1990] William, S. W., 1990. Accuracy and small-format surveys: the influence of scale
          and object definition on photo measurements, ITC Journal 1990-1, pp. 24-28.
[Jane.1987] Jane, D., 1987. A framework for handling error in geographic data manipulation,
          ITC Journal 1987-1, pp. 73-82.

Contact address
Dong-Gyu Park and Hwan-Gue Cho
Department of Computer Science
Pusan National University
Keum-Jeong-Ku
Pusan,
KOREA, 609-735
Phone: +82-051-582-5009
Fax : +82-051-515-2208
E-mail : {dgpark, hgcho}@hyowon.cc.pusan.ac.kr

FUZZY OVERLAY ANALYSIS WITH

LINGUISTIC DEGREE TERMS

Bin Jiang[1] and Wolfgang Kainz[2]

[1]Institute of Geographic Sciences
Free University of Berlin
Berlin, Germany
[2]Department of Geoinformatics
International Institute for Aerospace Survey and Earth Sciences (ITC)
Enschede, the Netherlands

Abstract
This paper presents a fuzzy overlay model using linguistic degree terms. Special emphasis is given to the two important principles - fuzzification and combination. Three frequently used membership functions are introduced to carry out fuzzification, and four kinds of methods applied in combination are discussed. A numerical example illustrates the model. The intention is to provide a framework for fuzzy overlay operations which enhance the capacity of spatial data analysis in GIS.

Key words
Fuzzy overlay, uncertainty, spatial analysis.

1. Introduction

There are two contrasting approaches to the representation of the 'real world' in a map or a GIS database. Each has its own individual mathematical framework. Some types of real world phenomena that GISs are concerned with are fuzzy and hard to describe with precise mathematics, while others can be dealt with using precise mathematics. Much attention has been given to the issue of uncertainty (e.g., Lam 1993, Brimcombe 1993) and Lowell (1992), who pointed out that "to move beyond the mere manipulation of 'electronic paper', real-world-to-map uncertainty must be incorporated into digital spatial technologies (DSTs)". Human thinking frequently operates with certainties varying from 0 (uncertainty) to 1 (certainty). For example, in statements like "the forest coverage of some area is *high*", and "the pollution index of a certain area is *low*". Indeed, the characteristic of linguistic degree terms is never probabilistic but rather fuzzy or vague.

As GISs evolve from simple data handling systems to sophisticated spatial decision support systems (SDSSs), users want a GIS with the ability to do approximate reasoning and work like human experts. For this purpose, natural language must be introduced into spatial data analysis. In this paper, linguistic terms are incorporated into overlay operations, which is extended as a fuzzy overlay analysis model.

A set of frequently-used linguistic terms represented as linguistic degree triplets includes the following: {*least*, *intermediate*, *most*}, {*low*, *moderate*, *high*}, {*gentle*, *moderate*, *steep*}, {*small*, *medium*, *big*}. When these linguistic degree triplets are merged into overlay operations, a query is made in such terms as

"Select all areas with *low* pollution index and *high* forest coverage",

or with modifiers,

"Select all areas with *very low* pollution index and *very high* forest coverage".

If the quality of the environment is evaluated by pollution and forest coverage, the above queries can also be combined

"Select all areas regarded as *best* environment".

The above queries can also be translated into rules for expert systems, i.e.,

"If the pollution index is *very low* and forest coverage is *very high*, then the quality of the environment is *best*."

Here {notions of degree of pollution index}, {notions of degree of forest coverage} and {notions of environmental quality} are the sets of fuzzy linguistic notions regarding the observation and the conclusion spaces, respectively. Choosing one of the components of universal space of linguistic degree notions, e.g., pollution index, we have a set such as {*low*, *moderate*, *high*}, of which all the elements represent a simple linguistic notion. A composed linguistic notion (or the result of a combination) can be described formally as, e.g., "most suitable" (referring to the environmental quality). It should be noted that the following discussion is oriented toward, but not limited to linguistic degree triplets. It can be extended to more than three if needed.

There is a great body of literature on fuzzy set theory used in regional planning and spatial analysis. Leung (1982, 1985, 1987, 1988) has employed fuzzy linguistic notions to characterize spatial regions. Robinson (1988) developed a model for handling inexactness in GIS based on fuzzy logic, and Robinson (1990) demonstrated the use of fuzzy sets for representing qualitative linguistic spatial relationship such as 'near' and 'far'. Burrough (1989) demonstrated the potential of fuzzy sets for soil survey and land evaluation. Dutta (1991) addressed a generalized framework for qualitative spatial reasoning using fuzzy set theory. Altman (1994) has defined fuzzy region, and discussed the definitions of distance and directional metrics between two such regions. Lowell (1992) has treated uncertainty and uncertainty representation for natural resource phenomena.

Section 2 provides basic definitions. This is followed by a brief overview of the fuzzy overlay model. After this, two key components of the model, *fuzzification* and *combination*, are discussed. A numerical example is presented for further explanation of the model.

2. Definitions

The following definitions present key terms used in this paper.

Layer: is a set of data describing a single characteristic for each location within a restricted geographic area (e.g., soil layer, forest layer etc.).

Sublayer: (more precisely, fuzzy sublayer) is an uncertainty map derived from a layer using available membership functions for a certain linguistic term.

Fuzzification: is the operation of measuring fuzzy phenomena to a certainty degree between 0 and 1.

Combination: is the overlay operation through sublayers and thus results in different certainties assigned to a new fuzzy set. It is actually an extension of linguistic connectives 'or' and 'and' applied to fuzzy overlay.

First certainty: is the value to indicate with what certainty a unit belongs to some linguistic term, and is obtained with an appropriate membership function.

Second certainty: is the value to indicate with what certainty a unit belongs to a combination of linguistic terms, and is gained through various methods of combinations.

Membership function: Let X be the universe of discourse and x a general element of X. A fuzzy subset A of X is characterized by a membership (characteristic) function $\mu_A(x)$ which associates with each element x of X a number $\mu_A(x)$ representing the grade of membership of x in A.

α **-cuts** (Dubois and Prade 1980): Let X be a classical set of objects, called the universe, whose generic elements and a fuzzy set are denoted x and A respectively. The ordinary set of those elements whose membership value is greater than some threshold $\alpha \in (0,1)$, is defined as the α-cut A_α of A, denoted as $A_\alpha = \{x \in X, \mu_A(x) \geq \alpha\}$. In particular, a strong α-cut A_α is defined as $A_\alpha = \{x \in X, \mu_A(x) > \alpha\}$.

Extension principle (Zadeh 1975): Let X be a Cartesian product of universes, $X = X_1 \times X_2 \times ... X_N$ and $A_1, A_2, ..., A_N$ be N fuzzy sets in $X_1, ... X_N$, respectively. The Cartesian product of $A_1, ... A_N$ is defined as

$$A_1 \times A_2 \times ... \times A_N = \int_{X_1 \times X_2 \times ... \times X_N} \min(\mu_{A_1}(x_1), ..., \mu_{A_N}(x_N)) / (x_1, ..., x_N),$$

where the separator / serves to differentiate between $\mu_{A_i}(x_i)$ and x_i, and $\int_X \mu_A(x)/x$ is Zadeh's notation of a fuzzy set on X when X is not finite. A finite fuzzy set A, for instance, is expressed as $A = \mu_1/x_1 + \mu_2/x_2 + ... + \mu_n/x_n$ or more compactly as $A = \sum_{i=1}^{n} \mu_i/x_i$.

Fuzzy weighted average (Dong and Wang 1987): Let $A_1, A_2, ..., A_N$ and $W_1, W_2, ..., W_N$ be the fuzzy subsets defined on the universes $X_1, X_2, ..., X_N$, and $Z_1, Z_2, ..., Z_N$ respectively. If f is a function mapping from $X_1 \times X_2 \times ... \times X_N \times Z_1 \times Z_2 \times ... \times Z_N$ to the universe Y, then the fuzzy weighted average y is

$$y = f(x_1, x_2, ..., x_N, w_1, w_2, ..., w_N) = \frac{w_1 x_1 + w_2 x_2 + ... + w_N x_N}{w_1 + w_2 + ... + w_N}.$$

Let μ_B be the membership function of the fuzzy image B of $A_1, A_2, ..., A_N, W_1, W_2, ..., W_N$ through f. Then by the extension principle,

$$\mu_B(y) = \max_{\substack{x_i \in X_i \ w_i \in Z \\ i=1,2,...,N \\ y = f(x_1,...,x_N)}} \{\min[\mu_{A_1}(x_1), ..., \mu_{A_N}(x_N), \mu_{W_1}(w_1), ..., \mu_{W_N}(w_N)]\},$$

where μ_{A_i} and μ_{W_i} are the membership functions of fuzzy subset A_i and W_i respectively, $i = 1, 2, ..., N$.

3. Overview of the fuzzy overlay model

To discuss the fuzzy overlay model, it is best to start by thinking of a non-fuzzy overlay, as shown in figure 1, which is very similar to a fuzzy overlay in appearance. It gives a hypothetical example of a non-fuzzy overlay operation using binary set theory. Suppose there are two layers concerning forest coverage and pollution index, respectively, in the top line of the figure. The (non-fuzzy) sublayers are obtained using binary characteristic functions as second row templates of the figure. Intersection and union of these sublayers actually lead to the overlay which is shown in the bottom row templates.

In much the same way, figure 2 illustrates the basic process involved in the fuzzy overlay model. Three basic sublayers are derived from their original layer using a membership function. The contents of sub-layers consist of first certainty data calculated with membership functions against linguistic degree triplets of layers. Then an overlay operation is performed on corresponding sublayers rather than layers. This is called combination. Thus, second certainty is obtained by the combination of suitable operators.

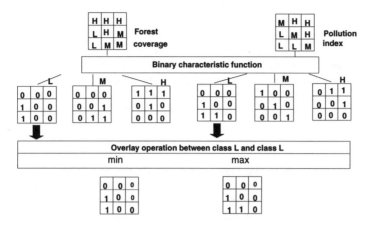

Figure 1 Hypothetical example of non-fuzzy overlay operation.

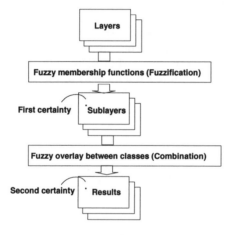

Figure 2 Basic procedures of fuzzy overlay operation.

4. Fuzzification – a solution to first certainty

According to the basic procedures of fuzzy overlay operation in the preceding section, the first step of fuzzy overlay analysis is to derive sublayers representing the set of a linguistic degree triplet such as {*low, moderate, high*}. Instead of a binary characteristic function, a membership function (or characteristic function) in fuzzy set theory describes the transition from 1 to 0 quite well.

Fuzzification, as a principle for fuzzy spatial analysis, has been discussed by Jiang and Kainz (1994), where some relevant membership functions have been defined to carry out fuzzification. With it, the existing spatial analysis may be beyond the limitation of pure numerical analysis, and extend to concept analysis which is closer to the style of human thinking. The core of fuzzification is four basic membership functions to fuzzify the individual elements into different fuzzy subsets with diverse certainties simultaneously.

From the perspective of convenience of interval analysis, Kaufmann and Gupta (1985) have given a detailed treatment of the theory of fuzzy numbers. They represent fuzzy subsets in the universe of discourse X. The fuzzy number is very useful in the subsequent combination calculation, as will be seen. For the sake of conformity, in the following discussions, some membership functions may be treated as the corresponding fuzzy numbers.

4.1 Fuzzy numbers

A linguistic degree triplet like {*low, moderate, high*}, is actually a collection of three fuzzy numbers or fuzzy subsets. Each element can be defined as a mathematical formula. But *moderate* can be regarded as the concatenation of the high and low parts. So in the following discussion, only the mathematical description of the intermediate part is presented, from which the other elements can be easily derived.

When trying to describe a linguistic degree triplet with fuzzy numbers, we identify three desired properties which determine the kind of fuzzy numbers, i.e., what element(s) belong to 0, what element(s) belong to 1, and how to transit from 0 to 1. These three points lead to three kinds of fuzzy numbers applied to fuzzification: a *triangular fuzzy number* (T.F.N.), a *trapezoidal fuzzy number* (Tr.F.N.) and a *bell-shaped fuzzy number* (B.F.N.).

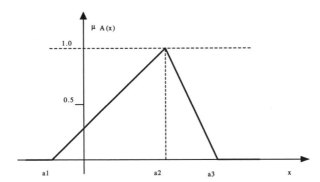

Figure 3 Triangular fuzzy number (T.F.N.).

Figure 3 illustrates a triangular fuzzy number which can be defined by a triplet (a_1, a_2, a_3). The membership function is defined as

$$\mu_A(x) = \begin{cases} 0, & x < a_1 \\ \dfrac{x - a_1}{a_2 - a_1}, & a_1 \le x \le a_2 \\ \dfrac{a_3 - x}{a_3 - a_2}, & a_2 \le x \le a_3 \\ 0, & x > a_3 \end{cases}$$

For an α-cut, we characterize the triangular fuzzy number as

$$\forall \alpha \in [0,1]: A_\alpha = [a_1^{(\alpha)}, a_3^{(\alpha)}] = [(a_2 - a_1)\alpha + a_1, -(a_3 - a_2)\alpha + a_3]$$

T.F.N. is the simplest of the three. The central concept of class is only a single value, and the transition is simply a linear connection.

In contrast to the triangular one, the trapezoidal membership function is based on the argument that the central concept of a class is often not a single value, but a range of values. That is, for $\alpha = 1$, we do not have a point but rather a flat line over an interval (a_2, a_3) as shown in figure 4. This provides another important style of function known as the trapezoidal fuzzy number (Tr. F. N.)

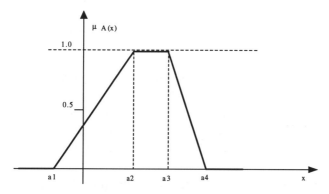

Figure 4 Trapezoidal fuzzy number (Tr.F.N.).

Thus, a Tr.F.N can be represented by a quadruplet (a_1, a_2, a_3, a_4), and the membership function is characterized as

$$\mu_A(x) = \begin{cases} 0, & x < a_1 \\ \dfrac{x - a_1}{a_2 - a_1}, & a_1 \leq x \leq a_2 \\ 1, & a_2 < x < a_3 \\ \dfrac{a_4 - x}{a_4 - a_3}, & a_3 \leq x \leq a_4 \\ 0, & x > a_4 \end{cases}$$

For an α-cut, Tr.F.N. can be defined as

$$\forall \alpha \in [0,1]: A_\alpha = [a_1^{(\alpha)}, a_4^{(\alpha)}] = [(a_2 - a_1)\alpha + a_1, -(a_4 - a_3)\alpha + a_4]$$

Neither the triangular fuzzy number nor the trapezoidal one delineate the transition zone from 0 to 1 naturally. The bell-shaped fuzzy number is more suitable to linguistic degree triplets. To this end, it should meet the following criteria, in terms of Dombi's (1990) discussion.

C1. $\mu_A(x)$ is a continuously increasing function $\mu_A(x):[0,1] \to [0,1]$
C2. $\mu_A(0)=0, \mu_A(1)=1$ (boundary condition)
C3. $\mu'_A(0)=0, \mu'_A(1)=0$ (S-shaped character)
C4. $\mu_A(x)$ is a rational function of polynomials

$$\mu_A(x) = \frac{a_0 x^n + a_1 x^{n-1} + \ldots + a_n}{A_0 x^m + A_1 x^{m-1} + \ldots + A_m} \qquad (m \neq 0)$$

C5. Find such a $\mu_A(x)$ where n+m is minimal

Theorem 1: There are no membership functions fulfilling the properties C1-C5 if $n+m \leq 3$.

Theorem 2: The minimum of n+m is 4 and the membership function is

$$\mu_A(x) = \frac{(1-\upsilon)x^2}{(1-\upsilon)x^2 + \upsilon(1-x)^2},$$

where υ is the intersection value of $y = \mu_A(x)$ and $y = x$ so υ is the characteristic value of the shape.

Theorem 3: Commonly, the membership function is written as

$$\mu_A(x) = \frac{(1-\upsilon)^{\lambda-1}(x-a)^\lambda}{(1-\upsilon)^{\lambda-1}(x-a)^\lambda + \upsilon^{\lambda-1}(b-x)^\lambda} \qquad x \in [a,b] \qquad (1)$$

$$\mu_A(x) = \frac{(1-\upsilon)^{\lambda-1}(c-x)^\lambda}{(1-\upsilon)^{\lambda-1}(c-x)^\lambda + \upsilon^{\lambda-1}(x-b)^\lambda} \qquad x \in [b,c] \qquad (2)$$

where equation (1) and (2) represent the monotonically increasing and monotonically decreasing part. Proofs can be found in Dombi (1990). Consequently, Dombi's membership function can be regarded as the optimal one due to the fact that there are two parameters λ and υ to modify the sharpness and inflection point of the membership curves, respectively (figure 5). For a specific application, the key problem is how to select appropriate values of λ and υ, to keep the result of fuzzification consistent with reality. It is commonly accepted that membership functions with an S- and bell-shape are usually used in fuzzy set theory. These shapes seem intuitively reasonable, but to date there is no theory to support such intuition.

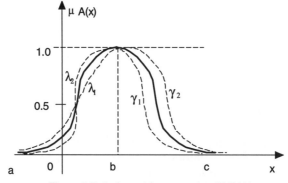

Figure 5 Bell-shaped fuzzy number (B.F.N.)

As to the shape of membership, a number of membership functions can be defined like linear, exponential, hyperbolic, hyperbolic inverse and piecewise linear functions (Sakawa 1983). The bell-shaped functions with two parameters are recognized as the more general ones.

With the fuzzification, users and systems can communicate with linguistic concepts such as *low*, *high* etc. Not limited to fuzzy overlay operations, fuzzification plays a significant role in other spatial analysis when linguistic concepts are used (Jiang and Kainz 1994).

4.2 Semantic operators (linguistic modifiers)

Semantic operators are linguistic modifiers which enable us to slightly change the qualification of certain linguistic concepts. As presented in the introduction, what a user is interested in is, for example, the area with *very low* pollution index rather than just *low*. Two main families of operators, *reinforcing* and *weakening* modifiers, can be identified according to Laforia (1992). Assuming that one element of a linguistic triplet 'A' has the membership function $\mu_A(x)$, then 'very A' and 'less A' are defined as,

$$\mu_{\uparrow A}(x) = \begin{cases} \mu_A(x), & x \geq k \\ 0, & x < k \end{cases}$$

and

$$\mu_{\downarrow A}(x) = \mu_A(x-k),$$

where "*very A*" is denoted by $\uparrow A$, and
"*less A*" is denoted by $\downarrow A$.

Obviously, there is a containment relationship between sets A, $\uparrow A$, and $\downarrow A$, this being $\uparrow A \supset A \supset \downarrow A$. According to such a relationship, it is very easy to recognize different curves indicating the reinforcing modifier and weakening modifier separately (figure 6).

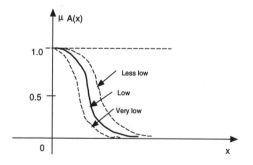

*Figure 6 Semantic operators **very** and **less**.*

In the implemented prototype system (Jiang 1996), both parameters λ and γ also act individually as a reinforcing modifier and a weakening modifier (see figure 5).

Semantic operators can be seen as the operations applied on the first certainty obtained through fuzzification. But for the whole overlay model it is the first step. Another important component involved in overlay operations is combination which is discussed in the following section.

5. Combination — a solution to second certainty

The simplest example of combination are the connectives *or* and *and* which can be treated as operators applied on the fuzzy numbers. Zadeh (1965) was the first to suggest the operations on fuzzy numbers, e.g., the unary minus operation for negation and the pointwise max and min operators or the *or* and *and* connectives. Thus instead of a query about the individual element of a linguistic degree triplet {*low*, *moderate*, *high*}, another frequently addressed query could be a combination of two elements such as

"Select all areas with *low* pollution index *and* high forest coverage"

It should be noted that areas of the query are not those that meet the condition *low* pollution index and *high* forest coverage separately, but a combination of the both conditions. The combination of *low* pollution index and *high* forest coverage is thus treated as a new fuzzy subset. Accordingly all the areas under investigation might belong to the fuzzy subset with a set of certainties which may present a distribution meeting the condition of the query.

5.1 Categorization of combination

Combination is performed between sublayers which come from layers. Due to different derivations of sublayers, combination can be categorized into two basic types. The first is between sublayers coming from the same original layer. The second is very similar to non-fuzzy overlay, i.e., the sublayers involved come from different layers.

From another perspective, combination can be regarded as the procedure of calculating the second certainty, and can be further classified in two ways as follows: (1) combination is imposed on first certainty, (2) combination is completed without using first certainty.

The concept of combination comes directly from the metaphor of natural language communication. Sometimes, instead of applying a property like *moderate* or *high* to a certain class, a user wants an intermediate answer, for example, pollution index is *'moderate or high'*. Figure 7 illustrates membership curves of *moderate* and *high* pollution indices, respectively. The cases of *'moderate or high'* and *'moderate and high'* can be represented graphically with max-min operators. In common language, if a user wants areas with *moderate* or *high* pollution index, in fact what he wants is the logical *and*, i.e., he wants to see both. The way *or* is used is for the case in which a particular value can conceivably belong either to the *moderate* or to the *high* class. Compared to the case where combination is performed among sublayers coming from different layers, this kind of combination is relatively easy to understand and to perceive, since the operation is in the same set. Here, however the emphasis is on the combination of sublayers from different layers.

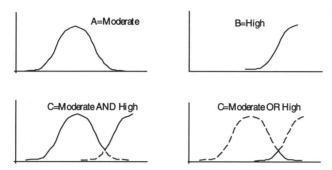

Figure 7 The concept of combination using max-min operator.

Formally, combination is very similar to logical overlay which involves finding those areas where a specified set of conditions occur (or do not occur) together (Aronoff 1989). It is originally derived from connective linguistic notions like *or* and *and*, so the operations on a fuzzy set provide basic solutions to combination. But it should be noted that combination does not include the difference between 'or' and 'and' as connectives do. In the following, a set of empirical operators to carry out combination is discussed. In addition, visualization techniques offer more possibilities for perceiving combination (Jiang et al., 1996).

5.2 Max-min operators

In the paper by Zadeh (1965), which is commonly regarded as the foundation of fuzzy set theory, definitions for union and intersection of two fuzzy sets are given as extensions of those applied to standard sets.

Union: Suppose the union of two fuzzy sets A and B with individual membership functions $\mu_A(x)$ and $\mu_B(x)$ results in a new fuzzy set C, then the membership function of C can be defined as $\mu_C(x) = \max\{\mu_A(x), \mu_B(x)\}$, or in abbreviated form $\mu_C = \mu_A \vee \mu_B$.

For a general mapping f from $X_1 \times X_2 \times ... \times X_n$ to the universe Y, the following equation applies:

$$\mu_C = \max_{i=1,2,...,m} \mu_i$$

Intersection: Suppose the intersection of two fuzzy sets A and B with individual membership functions $\mu_A(x)$ and $\mu_B(x)$ is a new fuzzy set C, then the membership function of C can be defined as $\mu_C(x) = \min\{\mu_A(x), \mu_B(x)\}$, or in abbreviated form $\mu_C = \mu_A \wedge \mu_B$.

For a general mapping f from $X_1 \times X_2 \times ... \times X_n$ to the universe Y, the following equation applies:

$$\mu_C = \min_{i=1,2,...,m} \mu_i$$

These two operators belong to the group of operators that are used to obtain second certainty from first certainty.

5.3 The use of interval analysis

Interval analysis was introduced thirty years ago with the publication of a monograph by Moore (1966). Since then this branch of applied mathematics has undergone rapid development.

The fuzzy interval method avoids the use of the first certainty to get the second one. Instead, it involves interval analysis after α-cutting the individual membership functions. Suppose there are two subsets A and B involved in combination C, and their endpoints are denoted by $\underline{A}, \overline{A}$ and $\underline{B}, \overline{B}$ respectively. The basic algorithm is presented as follows.

1. α-cut the interval of membership [0,1] into a finite number of values $\alpha_1, \alpha_2, ..., \alpha_m$. The number m determines the degree of certainty:
$$A_\alpha = [\underline{A}_\alpha, \overline{A}_\alpha]$$
$$B_\alpha = [\underline{B}_\alpha, \overline{B}_\alpha]$$
2. Multiply the corresponding intervals belonging to different sublayers in order to generate a list of new intervals $C_\alpha = A_\alpha * B_\alpha = [\underline{A}_\alpha * \underline{B}_\alpha, \overline{A}_\alpha * \overline{B}_\alpha]$
3. Multiply the values of different layers in the same location. Identify the new interval at step 2 which tightly encloses the result of the multiplication, and keep the α of the new interval as the second certainty.

The above algorithm can be easily extended to the case of combination of more than two subsets involved. A numerical example is given in section 6.

5.4 Multiple variable membership functions

In general, membership functions belong to single variable functions, i.e., when the value of a membership function is determined by only one variable. For example, a pollution index can determine the degree of pollution, and the degree of forest coverage can be calculated by a percentage of forest coverage. Combination, however, can be regarded as a multiple variable function with more than one variable involved.

Membership is defined as a function of the distance $d(x)$ between a given object and a standard (ideal), i.e., $\mu_A(x) = \dfrac{1}{1+d(x)}$.

That is to say, the nearer a given object is to the standard object, the less certainty there is of its being assigned to a fuzzy set in which the standard object is assigned full-membership. This point is supported by a day-to-day communication experience, for example, if 50 is supposed to be 'middle-aged' with a certainty of 1, then anyone whose age is near 50 belongs to the middle-aged with a higher certainty. The observation can be expanded to the case when there are more than one variables. It is what we refer to as multiple variable membership function $\mu_{AB}(x,y) = f(d(x)^{-1}, d(y)^{-1})$, where -1 indicates the inverse relation between $d(x)$ and $\mu_{AB}(x,y)$.

For the sake of simplicity, the following discussion assumes two variables x_1 and x_2, but the principle applies to cases with more variables.

According to Dombi (1990), $d(x)$ can be generalized in the following way:

$$d(x) = (\frac{1-\upsilon}{\upsilon})^{\lambda-1}(\frac{1-x}{x})^{\lambda}.$$

This can easily be extended to the case of two variables:

$$d(x_1,x_2) = (\frac{1-\upsilon}{\upsilon})^{\lambda-1}(\frac{1-x_1-x_2+x_1x_2}{x_1x_2})^{\lambda}.$$

Then, for the monotonically increasing part, we have

$$\mu(x_1,x_2) = \frac{(1-\upsilon)^{\lambda-1}(x_1-a)^{\lambda}(x_2-c)^{\lambda}}{(1-\upsilon)^{\lambda-1}(x_1-a)^{\lambda}(x_2-c)^{\lambda} + \upsilon^{\lambda-1}[(d-x_2)+(b-x_1)+(b-x_1)(d-x_2)]^{\lambda}}$$

and for the monotonically decreasing part, we have

$$\mu(x_1,x_2) = \frac{(1-\upsilon)^{\lambda-1}(b-x_1)^{\lambda}(d-x_2)^{\lambda}}{(1-\upsilon)^{\lambda-1}(b-x_1)^{\lambda}(d-x_2)^{\lambda} + \upsilon^{\lambda-1}[(x_2-c)+(x_1-a)+(x_1-a)(x_2-c)]^{\lambda}}$$

5.5 Fuzzy weighted average (FWA)

According to the experience of the overlay operation, each layer involving an overlay usually has different weights for a particular query. The case applies equally to the combination. Thus, a fuzzy weighted average could be a solution to combination, which can be denoted as follows:

$$\mu_{AB}(x,y) = \max\{\min[\mu_A(x),\mu_B(y),\mu_{w_1}(w_1),\mu_{w_2}(w_2)]\}$$

where w_1 and w_2 represent the weight for fuzzy set A and B respectively.

Based on the definition of fuzzy weighted average and Zadeh's extension principle, Dong and Wong (1987) proposed an algorithm to compute the weighted average (FWA) described in section 2 of this paper. The FWA algorithm works as follows:

1. α-cut the interval of membership [0,1] into a finite number of values $\alpha_1,\alpha_2,...,\alpha_m$. The number m determines the degree of certainty.
2. For each α_i, find the corresponding intervals for A_i in x_i and W_i in w_i, denote the end points of these intervals by $[a_i,b_i]$, and $[c_i,d_i]$, respectively, $i=1,2,...,N$. These are the supports of the α_j-cuts of $A_1,A_2,...A_N$ and $W_1,W_2,...,W_N$. If A_i or W_i is non-convex, more than one interval may result for some of the α_j. All these intervals should be treated and the result is the combined union.
3. The 2^{2N} distinctions of the 2N-ary array $(X_1,X_2,...,X_N,W_1,W_2,...,W_N)$ are constructed.
4. Compute $y_k = f(x_{k_1},x_{k_2},...,x_{k_N},w_{k_1},w_{k_2},...,w_{k_N})$, the number of the value of y is 2^{2N}. Repeat step 2, 3, and 4 for other α_j until the fuzzy number B is obtained.

As a matter of fact, the fuzzy interval method mentioned above can be regarded as a special case of FWA. In the algorithm presented above, the computation becomes very complicated and cumbersome with an increasing number of n. An improved algorithm is given by Liou and Wang (1992).

6. A numerical example

In this part, a complete example will be presented in order to explain the fuzzy overlay model, as well as some basic concepts such as first certainty and second certainty.

Let us assume two layers. Coverage percentage of forest and industrial pollution are used to evaluate the quality of the environment. It means that the quality of the environment is determined only by forest coverage and pollution index. Furthermore, suppose that the membership functions about medium forest coverage and medium pollution index are illustrated as figure 8. Now a user or an environmental specialist wants to overlay these two sublayers to get the uncertainty distribution regarding medium environmental quality.

The first step of fuzzy overlay is fuzzification, which is implemented by using available membership functions (figure 8).

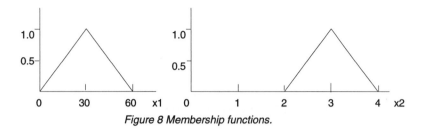

Figure 8 Membership functions.

A sublayer, medium forest coverage, is derived from the forest coverage layer with $\mu_{A_1}(x_1)$ defined as (using ratios as an example):

$$\mu_{A_1}(x_1) = \begin{cases} \dfrac{x_1}{30} & 0 \le x_1 < 30 \\ \dfrac{(60-x_1)}{30} & 30 \le x_1 \le 60 \end{cases}$$

In a similar manner, $\mu_{A_2}(x_2)$ is used to generate the sublayer intermediate pollution derived from the pollution index layer (using differences as an example):

$$\mu_{A_2}(x_2) = \begin{cases} x_2 - 2 & 2 \le x_2 < 3 \\ 4 - x_2 & 3 \le x_2 < 4 \end{cases}$$

In this step, the first certainty has been obtained by use of membership functions. In practice, Dombi's membership function can be used to determine the first certainty, which can reflect a natural language hedge.

The second step is combination. The original values are multiplied, and α-cuts are used to determine the second certainty. The step-by-step computations for eleven values of α are reproduced in table 1.

α	A1	A2	A1*A2
0	(0, 60)	(2, 4)	(0, 240)
0.1	(3, 57)	(2.1, 3.9)	(6.3, 222.3)
0.2	(6, 54)	(2.2, 3.8)	(13.2, 205.2)
0.3	(9, 51)	(2.3, 3.7)	(20.7, 188.7)
0.4	(12, 48)	(2.4, 3.6)	(28.8, 172.8)
0.5	(15, 45)	(2.5, 3.5)	(37.5, 157.5)
0.6	(18, 42)	(2.6, 3.4)	(46.8, 142.8)
0.7	(21, 39)	(2.7, 3.3)	(56.7, 128.7)
0.8	(24, 36)	(2.8, 3.2)	(67.2, 115.2)
0.9	(27, 33)	(2.9, 3.1)	(78.3, 102.3)
1.0	(30)	(3.0)	(90.0)

Table 1 Computations for eleven values of α.

The whole process presented above is shown in figure 9.

7. Conclusions

The structure of fuzzy overlay operation discussed in this paper is especially suitable for implementation by some standardized graphic user interface. This is due to their flexible user interface languages such as window, menu, icon, scrollbar etc. First of all, sublayers can be represented in individual child windows, which provide an overview about the uncertainty distribution, scrollbar can be used to simulate semantic operators and to change the magnitude of the two parameters in the bell-shaped fuzzy numbers. Operations applied to two sublayers lead to the overlay result. A prototype system under MS Windows has been completed (Jiang 1996). One of the pressing issues is the representation of certainty using color. The goal of the research is not only to measure certainty but also to represent it with the best visual effects (let the data speak for themselves).

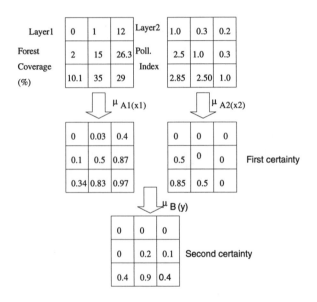

Figure 9 Illustration of the numerical example.

Even using the framework of fuzzy set theory, it is not certain that the data is error-free compared to the corresponding real world. The only thing that can be ascertained is that the data world under the framework of fuzzy set theory more closely resembles the real world. The issue of measuring such sorts of errors and error propagation is beyond the scope of this research. Compared to discrete classification, fuzzification reduces the distance between the real world and the data world. Both first and second certainty are advances in describing the real world, but these are not without problems. As can be seen, different membership functions have different effects, and the bell-shaped membership function is superior to triangular and trapezoidal ones. Second certainty can be obtained by using different operators, although the general trend of certainties is almost the same. To date, the main advantage of the fuzzy overlay modeling offers suitable operators tailored to various needs, and leave the choice to the users (different domain specialists).

Uncertainty is ubiquitous in environmental monitoring and resource management. For instance, in the tropical regions, the boundary between a lake and a swamp is extremely fuzzy, and a lake is often surrounded by a swamp. We can not be certain that a given location belongs to a lake or a swamp. We can, however, ensure with a certain percentage of certainty that the location belongs, for instance, to a lake. Thus fuzzy overlay analysis has great application prospects in environmental issues.

Acknowledgements
Thanks are due to three anonymous referees who read an earlier version of this paper and provided valuable comments.

References
Altman D. (1994), Fuzzy Set Theoretic Approach for Handling Imprecision in Spatial Analysis, Int. J. of Geographical Information Systems, Vol. 8, No. 3, pp. 371-289.

Aronoff S. (1989), Geographic Information Systems: A Management Perspective, WDL Publication: Ottawa.

Brimcombe A.J. (1993) Combining Positional and Attribute Uncertainty using Fuzzy Expectation in a GIS, Proceedings of GIS/LIS'93, Minneapolis, Minnesota, Vol. 1, pp. 72-81.

Burrough P. A. (1989) Fuzzy Mathematical Methods for Soil Science Survey and Land Evaluation, Journal of Soil Science, Vol. 40, pp. 477-492

Dombi J. (1990). Membership Function as an Evaluation, Fuzzy Sets and Systems, No.35, North-Holland, pp. 1-21.

Dong W.M. and Wong F.S. (1987) Fuzzy Weighted Averages and Implementation Principle. Fuzzy Sets and Systems, No. 21, pp. 183-199.

Dubios D. and Prade H. (1980) Fuzzy Sets and Systems: Theory and Applications, Academic Press.

Dutta S. (1991), Approximate Spatial Reasoning: Integrating Qualitative and Quantitative Constraints. Int. J. of Approximate Reasoning, 5, pp. 307-330.

Jiang B. (1996), Fuzzy Overlay Analysis and Visualization in Geographic Information Systems, PhD Thesis of University of Utrecht, The Netherlands.

Jiang B., Brown A. and Ormeling F. J. (1996), Some Perceptual Aspects of Coloring Uncertainty, Proceedings of 7th International Symposium on Spatial Data Handling, Delft.

Jiang B. and Kainz W. (1994). Fuzzification as a Basis for Fuzzy Spatial Analysis, Proceedings of Integration, Automation and Intelligence in Photogrammetry, Remote Sensing and GIS, Wuhan, pp. 294-302.

Kaufmann A. and Gupta M.M. (1985). Fuzzy Mathematical Models in Engineering and Management Science, North-Holland.

Laforia B.B.M. (1992). Fuzzy Logic and Knowledge Representation Using Linguistic Modifiers, In: Zadeh L.A. and Kacprzyk J. (eds.). Fuzzy Logic for the Management of Uncertainty, John Wiley & Sons, Inc. pp. 399-414.

Lam S. (1993). Fuzzy Sets Advance Spatial Decision Analysis, GIS World, Vol. 6, No. 12, pp. 58-59.

Leung Y. (1982), Approximate Characterization of Some Fundamental Concepts of Spatial Analysis, Geographical Analysis, No. 14, pp. 29-40.

Leung Y. (1985), A Linguistically-based Regional Classification System, In: Nijkamp P., Leitner H., and Wrigley N. (eds.) Measuring the Unmeasurable, Proceedings of NATO Advanced Research Workshop on Analysis of qualitative data, NATO series volume 20, Amsterdam, . pp. 451-485.

Leung Y. (1987), On the Imprecision of Boundaries, Geographical Analysis, No. 19, pp. 125-151.

Leung Y. (1988), Spatial Analysis and Planning Under Imprecision, Elsevier Science Publishers: Amsterdam.

Liou T.S. and Wang M.J. (1992) Fuzzy Weighted Average: An Improved Algorithm, Fuzzy Sets and Systems, No. 49, pp. 307-315.

Lowell K.E.(1992). On the Incorporation of Uncertainty to Spatial Systems, Proceedings of GIS/LIS'92, California, Vol. 2, pp. 484-493.

Lowell K. (1994), An Uncertainty-based Spatial Representation for Natural Resources Phenomena, Proceedings of 6th International Symposium on Spatial Data Handling, Edinburgh, Scotland, UK, pp. 933-944.

Moore R.E. (1966) Interval Analysis. Prentice-Hall, Inc.

Robinson V. B. (1988) Some Implications of Fuzzy Sets Theory Applied to Geographic Databases, Computer, Environment and Urban Systems, 12, pp. 89-98.

Robinson V. B. (1990) Interactive Machine Acquisition of a Fuzzy Spatial Relation, Computer & Geosciences, 16, pp. 857-872.

Sakawa M. (1983), Interactive Fuzzy Decision Making for Multiobjective Linear Programming Problems and its Application, Proceedings of the IFAC Fuzzy Information, Marseille, France, pp. 295-300.

Zadeh L.A. (1965). Fuzzy Sets. Information and Control, No. 8, pp. 338-353.

Zadeh L. A. (1975), The Concept of a Linguistic Variable and its Application to Approximate Reasoning, Part 1, 2, and 3, Information Sciences 8, pp. 199-249, 8, pp. 301-357, 9, pp. 43-80.

Contact address
Bin Jiang[1] and Wolfgang Kainz[2]
[1]Institute of Geographic Sciences
Free University of Berlin
Arno-Holz-Str. 12
12165 Berlin
Germany
Phone: +49 30 838 3892
Fax: +49 30 838 6739
E-mail: bjiang@gauss.geog.fu-berlin.de
[2]Department of Geoinformatics
International Institute for Aerospace Survey and Earth Sciences (ITC)
P.O. Box 6
7500 AA Enschede
The Netherlands
Phone: +31-53-487 4434
Fax: +31-53-487 4335
E-mail: kainz@itc.nl

STRATEGIES FOR AUTOMATED GENERALIZATION

Anne Ruas and Corinne Plazanet

Laboratoire COGIT - IGN
Saint Mandé, France

Abstract
Automating the process of map generalization goes beyond providing a GIS platform with a set of algorithms. It includes understanding the inter-relationships between objects and the informative value of each object. Generalization takes place at the local and global level. In order to have a local view of an area, we describe it by means of a *situation* that is consulted to choose a local plan (i.e. a set of operators to apply on the objects). Moreover, according to the new data base or map specifications, we utilize the concept of *constraints* to bound the problem space. These constraints are translations of the required conditions that should take into account not only the objects and but also the final state of the data base. As long as some constraints are violated, the process of generalization must carry on. A situation is described by 1) the geographical objects involved, 2) their relationships and 3) the constraint violations. In order to use a situation as a trigger, it has to be well characterized. Even when situations are well described, the chosen plan may produce undesirable results. In such a case, we need either to try another method or, if the situation is too complex, to try an alternate set of methods and to choose the one which minimizes constraint violations. In order to illustrate these principles two examples are given; one of road generalization and the other of object displacement. They are described, both in terms of data enrichment and process.

Key words
Generalization, process modeling, shape characterization, displacement process.

1. Introduction

Automating the geographical data generalization process involves the development of basic algorithms in order to transform semantic and geometric information across scales. Even if a consensus does not yet exist on the operator's definition and meaning [Rieger & Coulson 93], a large set of algorithms has been developed and implemented either on commercial platforms (e.g. Intergraph 's MGE/MG) or on research platforms (e.g. the platform *plage* developed at the IGN-France). Although some operators are still missing such as enhancement or structuration/typification ones, these platforms allow us to perfom certain levels of interactive generalization [Lecordix et al 95]. In order to reach a more fully automated environment, we need first to know when, where and how to use these algorithms [Shea & McMaster 89], then to control and manage their effects on the data. The complexity of automation is mainly due to the fact that the choice of any resolution strategy is context dependent: geographical situations are never the same. Moreover, different local strategies (i.e. a specific sequence of

319

operators applied on selected objects) may generate different solutions which have specific characteristics in terms of information content [Mackaness 94a]. From studies of the strategic aspects of generalization [Brassel & Weibel 88] [Beard 91] [Mackaness 95] we are proposing a dynamic generalization model, based on constraint violations and on the local qualification of a set of objects, represented by means of an object situation. We begin with a discussion of strategies for generalization. Then we discuss the necessity of data enrichment in order to describe precisely the characteristics of objects necessary to choose appropriate methods of generalization. This section is illustrated using line characterization and proximity relation representations. In order to illustrate strategies of generalization, we present two specific strategies: one on line generalization and another on object displacement to maintain data consistencies after line generalization. This has been implemented on an object oriented GIS.

2. Strategy for generalization

2.1 Principles

Among the generalization processes described in the literature, we   are following the one suggested by Brassel and Weibel:"*structure recognition, process recognition, process modelling, process execution and display*" [Brassel & Weibel 88, p232] complemented by the idea that " *we start with some hazy thumbnail sketch of what we want, we then source the data, apply some set of generalization operators, view the result and repeat and refine subsequence application of generalization operators in a cycle until a satisfactory solution is found*" [Mackaness 95, p1424]. The Brassel and Weibel model specifies that objects need to be correctly described in order to find the appropriate method to generalize them. It includes the geometric description of an object as well as a spatial description of a set of objects. The Mackaness model illustrates the necessity of defining a preliminary global sequence of actions and that some local solutions might not be the best, even if some spatial analysis have been performed. So, he proposes to test a set of solutions according to different criteria. In the following discussion, we propose a way of integrating these principles. It begins with the conception of a global master plan which represents the deterministic part of the process (i.e. a global schedule of the actions to perform).

2.2 Global master plan

At the lowest level, a transformation is the result of an algorithm applied to one or more objects. Our objective is to be able to choose the most pertinent set of objects and algorithms according to some set of criteria and knowledge, then to realize the transformation and to validate it by means of some sort of control algorithms. In order to guide these choices, we need a global master plan of actions in order to know the kind of transformations which have to be performed and the type of objects involved in each transformation. This global master plan must be defined according to 1- data base specifications, 2- our knowledge of the logical ordering of actions (e.g. it is pointless to simplify the geometry of objects that will be deleted) and 3- an initial understanding of the characteristics of the data base we intend to generalize (e.g. identify areas and the kind of objects which need to be modified by generalization). Table 1 is an intuitive and a non exhaustive example of a global master plan for generalizing topographic data. In this table: *"structuration"* represents an elimination of objects among a set of objects, while maintaining as well as possible their spatial distribution; *"structuring objects"* represent a set of objects among which others are organized. In urban area, main roads are maintained during the process of generalization and they can be used to partition space which gives a basis to define generalization working areas [Ruas 95]).

Task A: Simple semantic elimination of objects for each class
          (i.e. without considering the geometric properties)
Task B: Simplification of the object attributes and classification for each class
Task C: Aggregation of 'same class connected' objects
Task D: Simplification of network (such as roads and rivers):
An *example for hydrographic objects might be:*
                    1- network identification and hierarchy construction
                    2- elimination of the lowest level hydrographic arcs
                    3- structuration of maintained arcs in high density hydrographic area
                    4- geometric simplication in non dense areas
Task E: Treatment of dense areas using density criteria
                    1- Building elimination in very dense areas
                    2- Building aggregation in dense areas
Task F: Treatment of structuring objects (e.g. railway, important roads):
                    1- Simplication of most important structuring objects
                    2- Displacement between these objects
Task G: Simplification and enhancement of less important objects geometry

Table 1 Example of global master plan for generalization of 1/50000 topographic data.

This global master plan will not be enough to solve every problem. Though the earlier tasks are rather simple (i.e. elimination, semantic simplification) it then becomes more and more difficult to pre-define the kind of objects on which transformations need to be performed. One can suppose that actions which have few side effects should be first applied in order to reduce the quantity of information that need to be generalized (Tasks A to F in Table 1).
From this global master plan, we need some mechanisms in order to choose specific objects and procedures: This is what we call a *focalization process.* According to the kind of transformation to perform and a local situation description, a set of operators can be chosen. Then according to the focused objects characterisation and the chosen operators, a set of appropriate algorithms can be identified (see section 2.4). These two levels of choice are due to the difference between an operator and an algorithm: An *operator* represents a kind of transformation (e.g. aggregation, simplification), an *algorithm* is an implementation of a geometric principle to realize an operation. Thus different algorithms exist for a single operation (e.g. For line simplification we can use Douglas and Peucker or Lang or Gaussian Filtering or Jenks or Brophy or ... algorithms).

2.3 Use of constraints to guide the generalization process

2.3.1 Constraints and rules

Cartographers use knowledge in their decision making. It can be formalized as a set of condition rules (i.e. if *predicate* then *consequence*). Rules are useful either to generate knowledge (i.e. data enrichment) or to guide some local procedural choices. Thus rules either enrich the data or trigger operations, for example:
- if this line has such geometrical characteristic then it is sinuous (data enrichment),
- if this line is very sinuous then use this set of algorithms to generalize it (procedural rule),
- if this constraint has been violated then try another solution (process rule).

As observed by Kate Beard, using simple rules to manage generalization may not accomodate the flexibility required for map design. She introduced the notion of constraints which correspond to the rule predicate but does not impose the action to follow [Beard 91]. The advantage of constraints is that an action is not chosen according to a single condition but to a synthesis of conditions. Moreover, a constraint allows us to formalize fuzzy information such as " the displacement of this type of object has to be less than..". Because of geometrical and semantic specifications, many constraints are violated and *the best action is the one which minimizes constraint violations*. Potentially conflicts can exist between constraints (e.g. between shape and position maintenance). For example, in an dense urban area, building aggregation preserves position but reduces the information in terms of quantity, whereas displacements maintain the number of objects while deteriorating their position. Thus, there is a need to prioritize constraints where such conflicts of interest exist. Priorities are defined according to the specifications of the final data base or map. Changing priorities alters the relative importance of the constraints.

A constraint violation occurs when an object or a set of objects do not respect a constraint. For example an object whose area is too small violates a size constraint, two objects too close violate a proximity constraint. A constraint violation is sometimes named a conflict [Mackaness 94a]. In order to describe a set of constraint violations between objects, we introduce the notion of *situation* .

It is important to notice that constraints can be used in conjonction with rules: If an action depends on a single condition or if it is systematic (e.g. symbolization according to object nature or attribute simplification), it can be described by means of rules, otherwise constraints are used.

During the process of generalization, constraints may be used:
- to identify areas that have to be generalized, for example by evaluating the quantity and severity of constraint violations,
- to guide the choice of operators according to constraints priorities,
- to control the effect of an algorithm by detecting constraint violations on

objects after each transformation.

2.3.2 The nature of constraints

In this section we define various types of constraints and their impact on the map design process. In defining these various types of constraints we are defining the acceptable behaviour of objects and their required qualities. Kate Beard classified constraints in four classes: graphic, structural, application and procedural [Beard 91]. We concentrate our study only on constraints related to objects though the list is not exhaustive. Our objective is to understand the ontology of these constraints and how they can be used in the generalization process.

Legibility constraints:
Legibility constraints define the perceptibility threshold of objects. For map production they depend on the intended scale and our eyes perceptibility capacities. For screen visualization they depend on the pixel size. Various studies have attempted to define threshold values [Cuenin 72] [Spiess 85] [Spiess 95]. Main legibility constraints are the *minimum space between objects, the granularity of a line (i.e. the minimum detail size), the minimum length, the minimum surface, the minimum width of an area*. To detect legibility violations, it is necessary to take into account the symbology of objects.

Shape constraints:
The shape of an entity is important information that a cartographer tries to preserve. This information is often deteriorated by generalization algorithms [Plazanet 95]. We distinguish between *shape constraints of an object* (i.e. the regular shape of a railway should be preserved) with *shape constraints of a set of objects* of the same nature, such as houses alignment or hydrographic network shape. These objects constitute complex objects which are not explicitly represented in the data base but which are considered as entities by cartographers.
These constraints are generally flexible and difficult to use as the shape description remains very difficult (see section 3.1). However they should not be forgotten.

Spatial constraints:
We distinghuish between the position of each object and the position of objects one to another:
- for an object:
- the *absolute position constraint* defines the maximum authorized displacement of an object according to its nature "the displacement of a road has to be less than ...". For linear objects it can be enriched with a medium displacement value (i.e. an area).These constraints are generally flexible and can be used to control the result of a geometrical deformation.
- the *relative displacement constraint* defines an order of displacement between different type of objects (e.g. roads should be less displaced then houses)
- for a set of objects:

- the *topological constraints* define the connectivity and inclusion relations that should be preserved,
- the *proximity constraints* define the relative position constraints of a set of disjoint objects. Most of the time those are difficult to describe and to control (see section 3.2).

Semantic constraints:
These constraints depend on the data base specifications. Specifications which are not fuzzy should be represented by means of rules. Semantic constraints should be used to represent indicators such as:
- *Quantity constraints:* x% of this kind of objects should be preserved. It could be completed with contextual criteria such as: "In an urban area, x% of buildings should be preserved" thus providing a mechanism to support selection operations such as structuration.
- *Inter-classes quantity constraints:* this kind of constraint defines the priority between objects in selecting their removal in case of over density of information.
- *Functional constraints:* the objects of this type should fill a specific function: e.g. "the generalized road network should ensure the accessibility to important areas such as the railway stations or tourist areas". Functional constraints are useful for identifying objects that should be preserved during the process of generalization. Thus if an object has a specific function (e.g. a road which gives easy access to the railway station) it should be maintained.

2.3.3 Representation of constraint violations

Constraint violations allow us to indicate where an action should be performed. Some constraints are absolute (such as legibility constraints) while others have a degree of flexibility (such as quantity or shape constraints). If we want to use constraint violations as triggers, we need to represent them in the data base:
- Constraint violations related to an object can be represented by means of attributes at the object level (e.g. area too small, line too detailed) with either a flag or a quantitative value which describes the severity of the violation.
- Constraint violations related to a class of objects can be represented at the class level or by means of a specific attribute which is an indicator that should be consulted during the process (e.g. *road-quantity* with the attributes: initial, present and required quantity of objects).

A set of systematic methods (or daemons) might be used to identify the status of each object or each class, according to their constraints. So a too small house will have the attribute "I am too small" until it has been either enlarged or aggregated or deleted. On the other hand there is no simple way to represent a set of constraint violations related to a specific area. We introduce the concept of **situation** which represents a geographic area and is described by a set of attributes and structures (such as proximity relationships) related to constraint violations. For example, in order to detect proximity constraint violations within an area, proximity relationships may be described by means of structures such as a local Delaunay triangulation (see section 3.2). Thus a situation may also be linked to an object triangulation which would describe the proximity relationships within a situation. Figure 1 summarizes some propositions of constraint violations descriptions attached to a class, a specific geographical object or a situation:

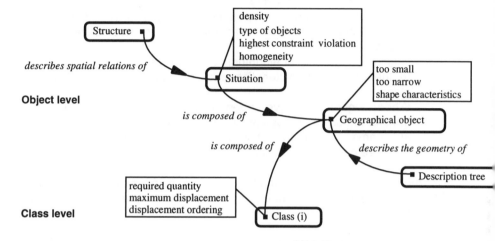

Figure 1 Representation of constraints and constraint violations.

The description of a situation is a solution that takes into account the contextual aspect. An object situation is made up of a set of geographical objects. Thus a situation is a synthesis of a set of constraint violations and has to be consulted in order to choose an appropriate plan to generalize an area.

2.4 Dynamic generalization model

According to the principles described previously, we propose a generalization model (see figure 2) guided by means of constraints, within which decisions depend on the semantic and geometric characteristics of an object or a set of objects (i.e. a situation). Additionally, the choice of the appropriate operators and algorithms requires the existence of procedural knowledge.

In summary, the model is based on:

1- The global master plan which determines a sequence of tasks to realize. It represents the deterministic part of the process. It should be defined according to the generalization specifications and some global understanding of the data. It is especially required for the first tasks of generalization (see section 2.2).
2- The selection of a situation which is made according to the kind of selected task (e.g. simplification of the hydrographic network). According to the stage of process, this focus on a specific situation is guided either by the data type or by constraint violations that need to be solved (e.g. reduction of the density of objects). During the first steps of data reduction, constraint violations can be used to find the worst conflicting situations. In solving

legibility constraints, a hierarchical space partitioning could be used to choose working areas [Ruas 95]. Such an idea opens up the opportunity for parallel processing.

3- The <u>local plan</u> which consists of finding methods for constraint violation solving, devided into two levels:

1- choice of a sequence of operators according to the situation (kind of objects, kind of constraint violations) and procedural knowledge related to operator,

2- choice of sequence of algorithms and their parameter values, according to the objects characteristics and- procedural knowledge related to algorithm.

When algorithms have been choosen, it is followed with the transformation and validation by means of quality assessment algorithms. A transformation is validated if the objects respect or minimize constraint violations.

If the transformation fails or is rejected, it is possible 1- to change the parameter value, 2- to go back to the algorithms choice, 3- to move to the local plan level.

Choice of operator or algorithm:
The experimental nature of map design means that even when a likely strategy is pursued (and armed with detailed semantic and structural knowledge plus an appropriate set of algorithms), it may still be the case that the result is not well adapted and that it is necessary to backtrack to a previous state and to try another method of resolution. Amongst possible changes would be alternate parameter values, algorithms or operators. The choice of operator, algorithm and parameter values is based on the existence of procedural knowledge (e.g. *If such a situation exists, then use this sequence of operations* or *If such an object with these characteristics has these constraint violations, then use this sequence of algorithms*). Different methods are used in order to acquire procedural knowledge (such as process tracing [Weibel et al 95]). Some of this knowledge has a high level of certainty while some of it does not. The fuzziness is increased by the difficulty of characterizing both the situation and the object geometry. In this case, a possible approach is to use alternate solutions until constraints are respected: For example, for line generalization: *solution 1*: Gaussian smoothing and Douglas and Peucker filtering, *solution 2*: smoothing with weighted average and Lang filtering, *solution 3*: local emphasizing, Gaussian smoothing and Lang filtering. Whenever methods are not clearly ordered or whenever the situation is too complex, another approach is to allow the system to try a set of methods and to choose the best one by minimizing a cost function (in terms of changes of shape, relative and absolute position, relationships between objects, etc.). As argued by Mackaness, *we need to be able to navigate through the solution space among alternate designs* [Mackaness 95, p1424].

From several tests we made on the IGN *Plage* platform, we noted that the "guided" approach might be useful for choosing operators and algorithms, while the "cost function" approach might be necessary to tune appropriate parameter values that sometimes seem difficult to predict.

In order to choose between guided and cost function approach, a *confidence indicator* related to every procedural knowledge and spatial characterization might be of great interest.

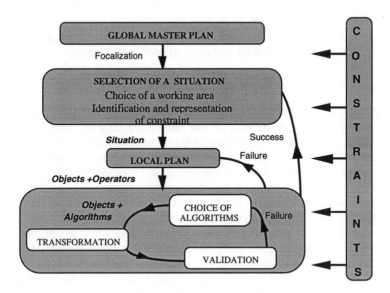

Figure 2 Generalization process model.

3. Data base enrichment for adequate choice and treatment

As the process depends on the geometric and semantic characteristics of a set of objects, they need to be either explicitly represented in the data base or extractable in the process of analysis. Much information is implicit in the data base, for example:

- the geometric description of an object is a set of consecutive coordinates. In the absence of structural and semantic information, applying an algorithm to different geometries produces inconsistencies. Most of the time the shape characteristics depend on the nature of the object. For example, a railway has a very regular curvature, a building tends to have an orthogonal shape. But even these general rules are not always specific enough. For road generalization different constraints such as relief have influenced their construction so their geometry is not sufficiently homogeneous to avoid a geometric analysis of each object [Plazanet 95].
- the spatial distribution description of a set of objects is a set of descriptions of connectivity and inclusion relationships. As the generalization process consumes space, due to the legibility constraints (see section 2.3), the proximity relations and the density of information need to be described and in some way conserved in the modeling process in order to maintain consistencies. For example proximity relations are essential to the effective management of object displacements (see section 4.2).

These points show that data enrichment is necessary to perform generalization. This information may either be computed before geometrical transformations (e.g. line characterization) or during the generalization process itself (e.g. proximity relations representation). In the following sections we give two examples of spatial characterization necessary to take an appropriate decision within the process of generalization. The section 4 integrates these characterizations for specific generalization tasks.

3.1 Line characterization for road generalization

In a road, sinuosity constitutes important information as it defines whether a road is dangerous and gives an idea of the necessary time to go from one place to another. These characteristics should be maintained during the process of generalization. Consequently there are, for sinuous roads, some important conflicts between shape maintenance and legibility constraints. In order to choose important bends to maintain and to emphasize, and those that can be removed, on-going research at the IGN-France Cogit laboratory aims at characterizing roads according to their sinuosity [Plazanet 95] [Plazanet et al 96]. As the geometry is seldom homogeneous, it is necessary to segment a line in order to accurately define sinuousity characteristics. As long as a segment is not homogeneous it has to be segmented. Each segment is qualified according to sinuosity criteria based on measurements on bends. The thiner the segmentation, the more accurate the qualification. The resulting information of a line segmentation is called a *description tree*. By analyzing the description tree while taking into account the legibility constraint violations, it is possible to select generalization operators most appropriate for a given segment of the object.

The principles of line characterization are as follows:
* global line characterization by means of a complexity measure,
* construction of a description tree:
* line segmentation according to the distribution of main inflection points and to the complexity of the line,
* characterization of each segment sinuosity according to a set of measures between inflection points (high, curvilinear length, euclidian distance),
* re-segmentation if necessary,
* at the lowest level: detection and characterization of shape singularities.

An example of a description tree is presented in the Appendix.

3.2 An example of proximity relation computation for the description of a situation

Proximity relations can be used for different tasks of generalization such as displacement (see section 4.2) or aggregation. At a higher level, they constitute important information to describe a situation (see section 2.3). There are different ways to represent them, depending on the task. We present an enriched Delaunay triangulation which is well adapted for displacement. The advantage of this structure is its capability to link an object with a set of neighbouring objects, thus it describes a network of spatial relations within a geometrical area. The same kind of data structure, named SDS, is used by [Ware et al 95] to generalize large scale data. In this work, the triangulation nodes correspond to the initial objects coordinates. Actually the choice of triangulation nodes depends on the kind of operation to proceed and their frequencies which is strongly related to the scale range. Other structures such as minimum spanning tree between objects can be used for other operations (see [Regnauld 96] for building aggregation).

The Delaunay triangulation creates a geometrical space tesselation from a set of points while minimizing the standard deviation of the triangles angles. It is a common way to find for each point its direct set of neighbours. The construction of our triangulation is based on the following rules:
1- a polygon is represented by a triangulation node, located at its centroid,
2- two polygons are neighbours if a triangulation edge connects them,
3- the proximity between two polygons is the minimum euclidian distance between their boundaries.

Figure 3 Proximity relation between two polygons.

4- a line holds a set of triangulation nodes located according to the proximity of other objects. For building displacement (see section 4.2) , a node on a line is located according to the projection of building centroids onto it. If the line is sinuous, it is first necessary to densify the number of nodes depending on the shape of the line. To do so, the local vertices of the line are identified by means of a Douglas and Peucker algorithm and added to the triangulation nodes of the line (see figure 4):

Figure 4 Proximity relations between a line and two.

5- the real distance (and its location) between either two polygons or between a polygon and a point on a line is an attribute of the triangulation edge.

We strongly believe that data enrichment is necessary to perform generalization. The above techniques to find and represent implicit information need to be extended to be able to qualify shapes and strutures such as buildings, hydrographic network, etc.

4. Examples of strategies in generalization

Having discussed the theorical issues, this section focuses on two specific implementations to illustrate the generalization model and the necessity of data enrichment during the process of generalization. These examples correspond to on going-research at the IGN France Cogit laboratory.

4.1 Mechanism to choose appropriate algorithms for line generalization

As mentioned earlier, whenever an object and an operation have been selected, we need to choose an appropriate algorithm or sequence of algorithms to generalize it. In the case of road generalization, research is being carried on on line characterization in order to make an appropriate choice of algorithms [Plazanet 95] [Plazanet et al 96]. Complementary research is on going at the GIUZ in Zürich [Reichenbacher 95] using machine learning to find an appropriate algorithm according to geometric line characteristics. The process of line generalization could be as follows:

1- Data Enrichment: Line characterization (see section 3.1)
2- Detection of granularity constraint violations according to line symbolisation (see section 2.3.2)
3- Choice of a sequence of algorithms for each segment according to the description tree analysis and constraint violations For examples:
   - A serie of sinuous bends *with legibility constraint violation* will be schematized (i.e. bend deletion)
   - A *too small* isolated bend will be emphasized
   - A *too detailed* low sinuosity segment will be slightly smoothed and filtered
4- Transformation of each segment and validation according to shape and legibility constraints
5- Treatment of segment continuity
6- Treatment of side effects (see section 4.2) and controls according to spatial constraints
7- Evaluation.

4.2 Mechanism to maintain consistencies: deformation propagation on disjoint objects

Each geometrical algorithm gives an object a new representational geometry. The consequence is that the object may use a location that might be used by other objects. Thus a geometrical transformation may create spatial constraint violations (such as overlaping or proximity violation). In order to avoid this kind of side effect, mechanisms of propagation and controls need to be developed. A mechanism of distortion propagation has been developed on *Stratège* (see section 5). In the present state, it is adapted for urban area generalization (i.e. for the displacement of disjoint polygons such as houses). This work complements other mechanisms of displacement such as [Jäger 91] in raster mode, [Nickerson 88] for lines and [Mackaness 94b] [Tallis 95] for point objects which compute radial displacements from the centroid of dense clusters. The next stage of this development will be to consider additional constraints and thereby take into account the objects' nature and their symbology.

After each line transformation, we compute distortions between the initial and the final geometry and propagate them onto the line's neighbourhood. The locations where the distortions are computed do not depend on the line segmentation but on the proximity of disjoint polygons. The problem is that a displaced polygon may overlap other objects. In order to avoid these cascade effects, we choose an area in which displacements are propagated into. If an object receives a displacement it propagates its effect into its immediate neighbourhood while decreasing the quantity of the displacement. Consequently we need to find every object's neighbouring objects. We compute a space exhaustive tesselation of the situation using Delaunay triangulation to find these relations where a triangulation node represents either a polygon or a part of a line (see section 3.2). The value of displacements are computed on the metric space from the real distance between objects and not from the distance between triangulation nodes. If objects are far one to another, the propagation stops.

The mechanism of propagation is a sequential one: Objects which have received a displacement propagate it onto the object which causes the highest proximity constraint violation (i.e. among several new locations, the one which causes the most side effect such as an overlaping). We distinguish between repulsion, attraction and radial propagations as the side effects are different. These distinctions are related to the deviation between the direction of displacement of an object and the relative position of objects:

1- when an object has been displaced towards another one, it creates a repulsion propagation,

2- when an object has been displaced on the opposite side of an object, it creates an attraction

3- when an object has been displaced along another one, it creates a radial propagation.

Only repulsions cause proximity violations. Whenever repulsions are managed the strongest attraction is treated and so on. Three methods of propagation have been developed (attraction, repulsion and radial propagations) with associated decay function. Figure 5 shows difference between repulsion and attraction:

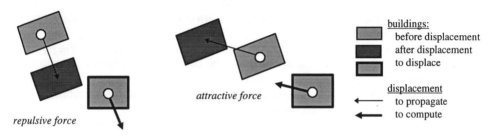

Figure 5 Repulsive and attractive forces.

So it is necessary to compute at each step the next best polygon on which a displacement is computed. Whenever an object   receives a set of displacements, they are aggregated according to their direction.
Moreover, for specific spatial configurations the logical direction of displacement creates overlapings. In this case escape directions, which attract an object towards "free areas" should be introduced.

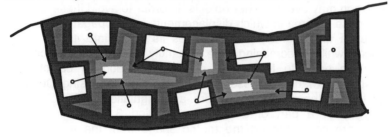

Figure 6 Computation of forces towards free areas.

This local computation of free areas can be computed by means of objects dilation (see figure 6) or by an analysis of the triangulation. It complements the propagation model by introducing directions towards which an object can be moved in a case of high space competition. It constitutes a new representation, such as the triangulation to describe more precisely the spatial relationships within the situation (see section 2.3.3). In very dense areas, an initial study of density allows us to determine if some aggregations are required before displacement propagation.

In summary, the sequence of events is as follows:
1- detection of the situation boundary, around the transformed line (see section 2.3.3),
2- situation qualification: constraint violations identification (see section 2.3.2) to know whether a propagation seems to be possible or if some aggregations are necessary,
3- data base enrichment:
   3.1 triangulation computation and classification (see section 3.2)
       - local delaunay computation
       - classification of arcs and nodes
       - computation of real distances between objects and desactivation of some arcs
   3.2 computation of displacement vectors on the transformed line
4- transformation: computation of a displacement vector for each node to displace
   - detection of the highest proximity constraint violation (see section 2.2.2): selection of the best node to displace
   - computation of forces with absorption according to distances and aggregation according to directions
   - transformation: displacement of the geographical object in relation to the triangulation node
   - control and alternate solution (see section 2.1) : introduction of escape directions if necessary

- data base update: re-compute of distances according to the displacement and change the status of the displaced node
5- global control and backtrack to another local plan in case of nconsistencies (see section 2.3)

Results of such a mechanism of propagation are shown in the Appendix 2.

5. Conclusion

Our main interest is to identify and to represent the implicit information in generalization, and to find mechanisms to generalize data according to each local situation in terms of spatial and semantical characteristics. In order to do this, the Cogit laboratory is developing a platform, named *Stratège*, in order to study the contextual and strategic aspects of generalization. Its object oriented data representation allows for the easy adaptation of the data model. The current model allows for the representation of geographical and geometrical objects, topological relationships, local proximity relations (to manage propagation), hierarchical space partitioning (to define working areas that can be used for situations) [Ruas 95], the geometry of the initial data (to control objects distortion), building clusters (to manage aggregation) [Regnauld 96] and line description trees (to represent line segmentation according to shape characteristics) [Plazanet 95]. A set of methods is related to each class of objects in order to manipulate them. This environment helps us to enrich the data structure according to every specific task of generalization. Moreover *Stratège* incorporates a rule language and facilities to create tasks which are sets of rules that can be used for specific actions. This system is useful to identify and define situations, to choose a local plan (i.e. specific objects and operators) according to a situation and to choose a set of algorithms according to objects' characteristics. This rule base system is used for displacement propagation. Consequently, the strategic principles of generalization presented in this paper are gradually being introduced, tested and refined in *Stratège*.

Appendix 1 An example of a descriptive tree of a BDCCarto® road.

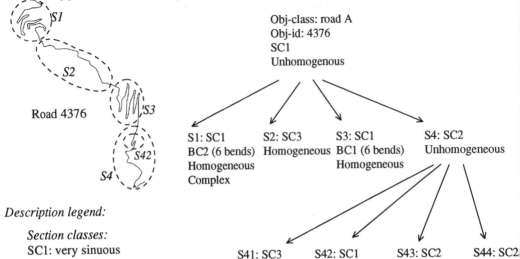

Obj-class: road A
Obj-id: 4376
SC1
Unhomogenous

S1: SC1
BC2 (6 bends)
Homogeneous
Complex

S2: SC3
Homogeneous

S3: SC1
BC1 (6 bends)
Homogeneous

S4: SC2
Unhomogeneous

S41: SC3
Homogeneous

S42: SC1
BC1
Homogeneous

S43: SC2
BC1
Homogeneous

S44: SC2
BC1
Homogen

Description legend:

Section classes:
SC1: very sinuous
SC2: barely sinuous
SC3: nearly straight

Bend shape classes:

BC1: ∧

BC2: ⟨

Appendix 2 Displacement of buildings after a line simplification from BDTopo®.

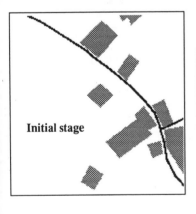

Initial stage

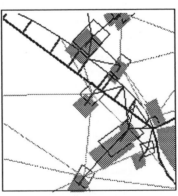

Computation and propagation
of displacement vectors
from the line distortion

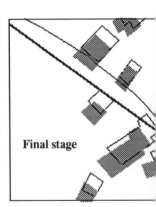

Final stage

Acknowledgement
We wish to thank William Mackaness for his helpful comments in reviewing this paper.

References
Beard MK 1991 Constraints on rule formation in BP Buttenfield and RB McMaster Map Generalization London: Longman Scientific & Technical p 121-135.

Brassel K & Weibel R 1988 A review and conceptual framework of automated map generalization International Journal of Geographical Information Systems 2(3) p 229-244.

Cuenin B 1972 Règles de lisibilité in Cartographie Générale Paris: Eyrolles Tome 1 p 164-167.

Jäger E 1991 Investigations on automated feature displacement for small scale maps in raster format Proceedings, ACI p 245-256 Bournemouth UK.

Lecordix F, Rousseau D & Rousseau T 1995 An evaluation of Map Generalizer, an Intergraph Interactive Generalization Software Package IGN & OEEPE Report.

Mackaness W A 1994 Issues in resolving visual conflict in automated map design. Proceedings, Spatial Data Handling p 325-340 Edinburgh Scotland.

Mackaness W A1994 An algorithm for conflict identification and feature displacement in automated map generalization Cartography and Geographic Information Systems 21(4) p 219-232.

Mackaness W A 1995 A constraint based approach to human computer interaction in automated cartography. Proceedings, ACI p 1423-1432 Barcelona, Spain.

Nickerson BG 1988 Automatic cartographic generalization for linear features Cartographica 25(3) p 15-66.

Plazanet C 1995 Measurements, characterization and classification for automated linear features generalization Proceedings, Auto Carto 12 p 59-68 Charlotte USA.

Plazanet C, Affholder JG & Fritsch E 1996 The importance of geometric modelling in linear feature generalization Cartography and Geographic Information Systems.

Reichenbacher T 1995 Knowledge acquisition in map generalization using interactive systems and machine learning Proceedings, ACI p 2221-2230 Barcelona, Spain.

Regnauld N 1996 Structure recognition for building generalization Proceedings Spatial Data Handling Delft, Holland.

Rieger M & Coulson M 1993 Consensus or confusion: Cartographers knowledge of generalization Cartographica 30 p 69-80.

Ruas A 1995 Multiple paraigms for automating map generalization: geometry, topology, hierarchical space partitioning and local triangulation Proceedings Auto Carto 12 p 69-78 Charlotte USA.

Shea K and McMaster R 1989 Cartographic generalization in a digital environment: When and How to generalize Proceedings, Auto-Carto 9 p 56-67 Baltimore, USA.

Spiess E 1985 Revision of 1:25000 Topographic Maps by Photogrammetric Methods. OEEPE Official Publication N°12. March 1985.

Spiess E 1995 The need for generalization in GIS environment in Müller et al GIS and Generalization London: Taylor & Francis p 31-46.

Tallis M 1995 The development of a graphical user interface for interactive feature displacement MSE in GIS dissertation Unpublished, Dept of Geography, Edinburgh, Scotland.

Ware JM, Jones CB, Bundy GL 1995 A triangulated spatial model for cartographic generalisation of areal objects Proceedings COSIT p 173-192, Semmering, Austria.

Weibel R, Keller S. and Reichenbacher T 1995 Overcoming the Knowledge Acquisition Bottleneck in Map Generalization: The Role of Interactive Systems and Computational Intelligence. Proceedings COSIT p 139-156, Semmering, Austria.

Contact address
Anne Ruas and Corinne Plazanet
Laboratoire COGIT - IGN
2, avenue Pasteur
94160 Saint Mandé
France
E-mail: ruas@cogit.ign.fr
E-mail: plazanet@cogit.ign.fr

BUILDING A MULTI-SCALE DATABASE

WITH SCALE-TRANSITION RELATIONSHIPS

Thomas Devogele, Jenny Trevisan and Laurent Raynal

IGN / COGIT
Saint-Mandé, France

Abstract
Building multiple representations is one of the key problems in GIS. To tackle this problem, we
have chosen to connect geographic data from mono-scale representations to build a multi-
scale database with scale-transition relationships. These scale-transition relationships connect
two sets of elements (classes, types or objects) representing the same phenomenon of the
real world and carry the sequence of multi-scale operations to navigate from one
representation to another. From this concept, a process has been defined to build multi-scale
databases, in three steps. The first step is dedicated to the declaration of correspondences
and conflicts between input schemata by the means of scale-transition relationships. In the
second step, conflicts are resolved and schemata are merged. Finally, the third step
corresponds to data matching, with the help of geometric, topologic and semantic information.
Scale-transition relationships between objects are created during this last step. To validate the
process, a multi-scale database has been produced from two existing mono-scale sets of road
network data. The first results of this kernel are satisfactory.

Key words
Multi-scale, multi-representation, geographic database, schema integration, data matching,
scale-transition relationship.

1. Introduction

Building multiple representations is one of the key problems in GIS [Brugger
et al. 89]. Indeed "hierarchical planning with different levels of detail for
different parts of the task, seems to be common in human way-finding" [Mark
89]. However geographic databases are very similar to maps, i.e. only one
representation at one scale is available. Nevertheless, a geographical
database does not formally include the notion of scale (the ratio between the
size of an object on the map and its real size on the ground) ; [Goodchild 91]
[Müller et al. 95] speak of **precision** (degree of detail in the reporting of a
measurement), **accuracy** (relationship between a measurement and the
reality which it purports to represent) and **resolution** (the smallest object
which can be represented). For geographic databases, it seems judicious
therefore to connect the determination of these three concepts with the notion
of scale commonly associated with a map.

So, a **multi-scale database** is a geographic database, which allows us to represent the same phenomenon of the real world at different levels of precision, accuracy and resolution. Consequently, designing a multi-scale database requires a representation mechanism of schema and data at different levels. Three methods can be used to build a multi-scale database [Govorov 95] :

- In the first method (see figure 1), a cartographic generalization[1] process is used to generate databases at smaller scales from a single database at the greatest scale. Cartographic generalization can be either automatic or interactive. For the moment, this solution seems difficult for the following reasons :
  - (i) There are no generalization tools in GIS allowing automatic generation representations less detailed than the most precise representation [Müller et al 95]. Indeed, modifications in cartographic generalization are complex, varied and intricately interwoven.
  - (ii) An entirely interactive generalization is too long and too difficult [Müller 92] and for these reasons seems inadequate for final users.
  - (iii) all the pieces of information which are wanted in the generalized representation, are not present in the most precise representation. Indeed, new themes may come on top of generalization.
- The second method consists of a mere collection of geographic databases with no link whatsoever between objects. Some rules can be used to control the choice of the database to display according to the graphic scale. Nowadays, this method is implemented in some GIS (Apic, GeoConcept, ArcView) but these rules are limited (the notion of scale is unsuitable for geographical databases). For example, it would be necessary to include other parameters such as the density of the zone to define display rules.
- Finally, in the third method (see figure 2), a multi-scale data structure is defined in order to link representations. This last solution seems the best compromise between the re-use of existing databases and their enrichment by integration. This is why we have adopted it. The aim of this integration is not to obtain a single representation but to allow interoperability between representations and to link homologous objects between the different representations of the same phenomenon of the real world. This integration is a "semantic integration" ; schemata are integrated according to the values of data. Input representations may be physically distributed databases or data sets. In the first case, we have one distributed database [Ceri and Pelagatti 87] or a federated database [Sheth and Larson 90], and in the second case we have a single database.

This article discusses the latter type of multi-scale database.

---

[1] Cartographic generalization modifies data in order to produce a simplified, or more abstracted representation, for the sake of map legibility [Lagrange and Ruas 94].

Figure 1 A single database with a cartographic generalization process.   *Figure 2 Multi-scale data structure.*

Nevertheless, the design of multi-scale data structures is difficult, due to the implicit cartographic generalization complexity [Lagrange and Ruas 94]. We have therefore opted for the concept of scale-transition relationship [Devogele and Raynal 96] defined in section 2. We will then describe the process to build a multi-scale database in section 3. Note that this process has been experimented and validated on road network databases. Section 4 offers some conclusions.

2. Designing a multi-scale database

In this section, we will first discuss some multi-scale database research given in the literature and its limitations. This will allow us to introduce the concept of the scale-transition relationship. Finally, the assets of the scale-transition relationship will be given.

2.1 Overview

Any geographic representation is an abstraction of the real world with its own concept to represent the phenomena of the real world. Designing a multi-scale database raises three kinds of problems:

- **Correspondence between abstractions**. Abstraction translate phenomena of the real world into instances of databases, by focusing only on parts of these phenomena. From research on DBMS, only the concept of views [Günther 89] [Abel et al. 94] has been suggested to overcome this problem. However, scale problems are not approached in this context.
- **Correspondence between objects from different representations**. Tree structures [Jones 91][Kidner and Jones 94][van Oosterom 95][Timpf and Frank 95] [Rigaux and Scholl 95] have been proposed to describe links between the different representations. However relationships in the real world are not always hierarchical when scale varies.
- **Defining the matching process between objects.** Data matching is a generic term to indicate methods and algorithms to search two geographic sets of data for groups of objects that represent the same part of the real world. To identify homologous objects, geometric matchings [Raynal and Stricher 94] [Edwards 94] [Gabay and Doytsher 94] have been suggested. We will show that they are insufficient in our kernel.

What is more, these three independent problems have never been coordinated.

2.2 Representations with scale-transition relationships

To design multi-scale databases, it is necessary to integrate abstractions and to connect instances. With this aim, the concept of a scale-transition relationship which uses the object-oriented data model, is proposed. The concepts used in this article are those of type, class and object.

- **Type** defines a template for all its objects. It provides a common representation for all objects of that type and a set of operations on those objects. A type can be an **atomic** type (integer, real, string, boolean and enumeration) or a **structured** type (tuple, set or list). Types are linked by **subtyping** (a type A is a subtype of another type B if an instance of A is a instance of B) and **inheritance** (a type A inherits from another type B if A shares all the behaviours of B).
- **Class** is a grouping of instances of a given type.
- **Object** is an instance of one type and is included in one class. Each object has both an invariant identifier (OID) and a value.

With these concepts, we can define what a scale-transition relationship is.

Scale-transition relationships connect two sets of elements (types, classes or objects) representing the same phenomenon of the real world and carry the sequence of multi-scale operations to go from one representation to another. Scale-transition relationships are oriented links.

Since they include multi-scale operations, scale-transition relationships are describing and storing some amount of cartographic generalization. They are more advanced structures than simple relationships between elements.

Multi-scale operations are defined in [Devogele and Raynal 95] and [Peng 96]. Two categories can be distinguished :

- Cartographic generalization operations as described in [Shea and McMaster 91] and [McMaster and Shea 92].
- Schema modification operations which are basically borrowed from the database theory, and more precisely from schema evolution and schema integration such as those defined in [Motro 87] [Scherrer et al. 93] and [Scholl and Tresch 93].

For example, in figure 3, one can distinguish :

- Scale-transition relationships between a set of buildings and a built-up area carrying the multi-scale operation : amalgamation [Shea and McMaster 91].
- Scale-transition relationships between a road section and another road section carrying the multi-scale operation : simplification [Shea and McMaster 91].

Of course, some elements may not match due to deletion in the less detailed representation.

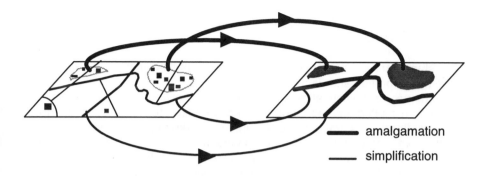

Figure 3 Representations with scale-transition relationships.

2.3 Assets of scale-transition relationships

These scale-transition relationships can describe the correspondence between schemata (type and classe definitions) and to link data from different representations. Furthermore, a multi-scale database using scale-transition relationships allows us to answer the different needs of multi-scale database users :

- From a practical point of view, this concept allows **intelligent zoom** [Timpf and Frank 95] between the different representations, and **analysis of data** from different representations. Queries can be applied simultaneously on sets of buildings and built-up areas even if these two classes are not defined in the same representation. **Navigation** between the different representations [Car and Frank 95] [Langou and Mainguenaud 94] is also feasible because scale-transition relationships authorize to go back and forth between representations.
- Information transfers between levels also become possible, giving access to the sequence of multi-scale operations between elements. This means that, at least in some cases, it is possible to **propagate an update** performed at some detailed level, to a less detailed level [Kilpeläinen 95]. Indeed, once the scale-transition relationship carries enough information (precise sequence of multi-scale operations, values of parameters,...), the update, also required at the secondary level, may be activated through the conversion from the basic level. However, it is clear that such an automation might prove to be quite difficult if complex geometric transformations or transformations involving sets of objects are needed.
- Finally for mapping agencies, the database merging favours **quality control**, and the concomitance of representations in a single structure facilitates **maintenance of consistency**. Besides, the information on quality (geometric accuracy, semantic accuracy, ...) can be easily obtained by comparison of the less detailed database, to the reference database.

In any case, building a multi-scale database allows us to derive new products and to integrate new data.

3. Process to build a multi-scale database with scale-transition relationships

Two stages are distinguished in the process to build a multi-scale database : schema integration and data matching.

*   Schema integration allows interoperability between existing databases, since a multi-database language and an integrated data description can be defined. This first stage is inspired by works of [Spaccapietra et al. 92] [Dupont 94] that use assertion-based methodology. This methodology splits the integration process into two distinct steps. The first step is dedicated to the declaration of correspondences and conflicts between input schemata by using scale-transition relationships. In the second step, conflicts are resolved and schemata are merged.
*   In the second stage called data matching, homologous objects (representing the same real world phenomenon) are connected. This process generates scale-transition relationships between objects by using semantic, topologic and geometric information.

Finally, to test the feasibility of such principles, a multi scale database kernel [Trevisan 95] has been realized from two IGN's mono-scale databases (BD CARTO® [2] and GÉOROUTE® [3]) in the area of Marne-la-Vallée (367 km of road for BD CARTO® and 991 km of road for GÉOROUTE®, the network in this area being varied and dense). This kernel has been developed in O_2 [O2 91] and GéO$_2$ [Raynal et al. 95] [David et al.93].

For the sake of easy reading the BDC acronym is adopted to designate BD CARTO® and the G acronym is adopted to designate GÉOROUTE®.

3.1 Declaration of scale-transition relationships

Declaration of scale-transition relationships is done at two levels : between classes and between types.

3.1.1 Scale-transition relationships between classes

Scale-transition relationships between classes are used to describe correspondence between **collections of instances** ; i.e. they indicate that one instance of class 1 in the schema 1 corresponds to one instance of class 2 in the schema 2. More complex correspondences are also possible.

For example, *Crossroads* classes do not correspond exactly :

*   One instance of the G *Crossroads* class corresponds to 0 or One instance of the BDC *Crossroads* class (whether it is suppressed or preserved).
*   One instance of the BDC *Crossroads* class can correspond to 1 complex

---

[2]BD CARTO® contains the basic geographical information, that is necessary for the production of 1:100,000 topographic maps, for national and regional development and network management. BD CARTO® has been created by processing existing 1:50,000 maps (scanning, vectorization, topological structuring).

[3]GEOROUTE® is a topologically structured road-planning database containing precise data for urban areas and uses BD CARTO® data for other areas. GEOROUTE® is used for transportation purposes.

crossroad in G (figure 4), which is composed of instances of *Crossroads* and *Sections* classes in G. Indeed, due to the change of scale, some instances of *Sections* and *Crossroads* classes are suppressed or are symbolised in a set of fewer objects.

BD CARTO®   GÉOROUTE®

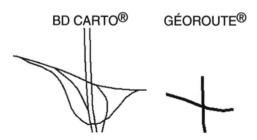

Figure 4 The same crossroads in BDC and in G.

Extended to all classes in the kernel, the scale-transition relationships are : from BDC to G (Table 1) and from G to BDC (Table 2).

BDC → G	
1 Road	1 Road
1 Section	(1-N) Sections
1 Crossroad	(0-N) Crossroads and (0-N) Sections

G → BDC	
1 Road	(0-1) Road
1 Section	(0-1) Section or 1 Crossroad
1 Crossroad	(0-1) Crossroad

Table 1 Scale-transition relationships from BDC classes to G classes.

Table 2 Scale-transition relationships from G classes to BDC classes.

Once scale-transition relationships between classes have been described, it is necessary to describe scale-transition relationships between types.

3.1.2 Scale-transition relationships between types

Scale-transition relationships between types describe differences in **structure**. These scale-transition relationships can be declared only for types whose classes correspond. Scale-transition relationships between types can be declared according to the values of data in correspondence and not according to type parameters (name, domain,...). These differences are described by multi-scale operations (see section 2.2).

For example, BDC *crossroad* type correspond to G *crossroad* type : they describe the same phenomenon of the real world. For tuples, attributes must be also link. These two tuples have three common attributes (*kind_of_crossroad*, and *toponym, geometry*). Furthermore the BDC type has one own semantic attribute: *spot_height*.

- *kind_of_crossroad* attributes of BDC and G that describe the nature of the crossroad, are enumerations (authorized values : simple crossroads, roundabout, interchange, ...). Globally, they correspond but some values have been added or suppressed. Thus, the multi-scale operation is a function of change for lists of authorized values.

-*Toponym* attributes which indicate the name of the crossroad, correspond semantically but they do not have the same structure : in BDC, toponym is a tuple (attributes: Kind, Article, Wording) while in G, toponym is a character string. These, the multi-scale operation is a function of transformation of tuple into character string.

-*Geometry* attributes correspond exactly, they represent points. There is no multi-scale operation associated with this scale-transition relationship.

-*Spot_height* attribute is only present in BDC. So, the scale-transition relationship is a deletion.

This example shows a few correspondences between types. The whole set of scale-transition relationships established in the prototype will be made available in [Trevisan 95].

3.2 Schema merging

Schema Merging is the step that unifies corresponding types and corresponding classes to build the integrated schema. In this step, the best integration techniques are chosen for each kind of scale-transition relationship.

The literature proposes several techniques to integrate schemata, depending on whether we prefer to preserve or to delete the initial schemata and on the kind of integrated schemata that we wish. In our case, we want to obtain a schema which will be able to represent all the instances of input schemata, without **loss of information** by unifying structures and will be able to **connect data** of different input representations most precisely.

To transcribe scale-transition relationships between **classes**, the integration technique is to create a relationship between types corresponding to classes. For scale-transition relationships between one class and many classes (or between many classes and many classes), the integration technique is to create one (or many) super-class(es) and the relationship is carried by the types of these super-class(es) (see figure 5). Therefore, these super-types are anchor points to scale-transition relationships between instances. Thus for *Crossroads* and *Sections* of BDC and G, super-types *Net_el_G* (Network elements of G) and *Net_el_BDC* (Network elements of BDC) have been created and a relationship *STR_Element* (scale-transition relationship between network elements) has been defined between these two super-types. All scale-transition relationships between *Sections* and *Crossroads* classes (mentioned in tables 1 and 2) are grouped in these relationships.

For scale-transition relationships between **tuples**, a technique of generalization between types has been chosen, i.e. for each pair of corresponding types (for example *Crossroad* types of BDC and G), a father type is created supporting common attributes (called *Crossroad*, in the kernel, see figure 5) and two son types are maintained for the specific attributes (called *Crossroad_G* and *Crossroad_BDC*, see figure 5).

For scale-transition relationships between enumerated types (for example *kind_of_crossroad* of BDC and G) supporting functions of change for lists of authorized values as multi-scale operation, the chosen integration technique is to create a new enumerated type. This type is defined by merging authorized values of these two types.

For scale-transition relationships between a tuple type and a primitive type (for example toponym types of BDC and G), the chosen integration technique depends on the complexity of the transformation mechanism. If the transformation process is manual, these types remain separated ; otherwise, the tuple type is preserved and the primitive type is transformed. The latter technique is applied for toponym types of BDC and G.

Once this step is finished, we have a multi-scale schema allowing us to represent input types and to represent scale-transition relationships between their instances. This multi-scale schema can be either the schema of federated databases, or the schema of a single database with several data representations. Figure 5 presents the merged schema in our kernel.

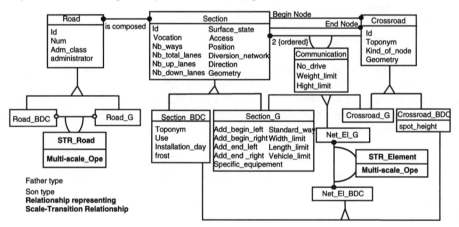

Figure 5 Multi-scale schema (OMT model [Rumbaugh et al. 91]).

This base supports two representations integrated in a single schema, but the objects themselves are not yet matched.

3.3 Data matching

Data matching is a generic term to indicate methods and algorithms to search two geographic sets of data for groups of objects that represent the same part of the real world. First, we will define the different types of data matching used, then we will describe the process realized in the kernel, before presenting some of the first results.

3.3.1 Different kinds of data matching

In the current context (a single multi-scale database), the data matching step is realized after the schema merging. Scale-transition relationships between objects are therefore constructed in a permanent manner. To match data, semantic, topologic and geometric information is used to overcome the shortcomings of simple geometric matching as presented in [Raynal and Stricher 94] [Edwards 94] and [Gabay and Doytsher 94].

- The **semantic** matching put objects in correspondence according to their semantic attributes, which are a key. Objects are matched on the value of their keys.
- The **topologic** matching uses composition or topologic relationships between the different objects to match data. If two relationships correspond, then this correspondence can be used to find homologous objects linked by this relationship.
- The **geometric** matching ; data are matched by their location with a measure of distance between objects. Other geometrical characteristics, as the direction [Gabay and Doytsher 94] have been proposed to match these data.

3.3.2 Data matching process for the kernel

These three kinds of matching have been necessary for the kernel. Furthermore, the process order is fundamental (data forming a network, many relationships exist between them). Thus three successive matching processes (see figure 6) have been defined : road matching, crossroad matching and section matching.

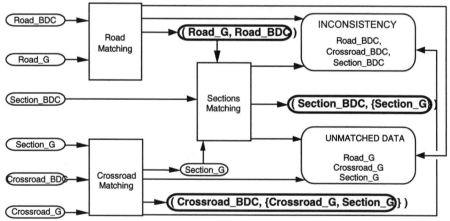

Figure 6 Data matching process for the kernel.

Road matching : This algorithm uses only semantic information (keys are attributes : number and administrator) and generates object pairs (*Road_BDC, Road_G*).

Crossroad matching : This procedure is more complex, using geometric and topologic information. Indeed, first, we automatically determine a search area[4] around each BDC crossroad (see figure 7) ; G crossroads in this area are automatically selected. Then, by using topologic relationships between crossroad and section, we form connected groups (figure 5 : 4 connected groups). We compute the connected group whose dangle sections (sections connected to the group) and bi-connected sections (sections belonging to the group) match with dangle sections of the BDC crossroad. On figure 5, we observe that group 1 matches with the BDC crossroad. This algorithm is used for all BDC crossroads and generates object pairs (*Crossroad_BDC*, {*Crossroad_G,Section_G*}.

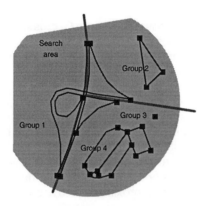

Figure 7 Example of connected group.

Section matching : Henceforth, BDC sections may be matched with G sections (previously, G sections matched with BDC crossroads must be removed). This matching is achieved in two stages. In the first stage, sections belonging to semantically matched roads are matched road by road. Then the other sections are matched. Geometric section matching is realized with the help of the Hausdorff distance [Raynal and Stricher 94] [Hangouët 95] [Hausdorff 19]. We obtain three kinds of G sections :
* Matched section : Only one distance of the BDC section from this G section is less than the matching threshold.
* Litigious section : Several distances of the BDC section are less than the matching threshold. We can not choose the homologous BDC section with distance alone.
* Unmatched section : No distance of BDC section is less than the matching threshold. Either this section has no homologue, or this section has one homologue but distance is superior to the matching threshold.

Although those last two matching algorithms are not optimal (certainly, one can integrate more semantic information), a process which has been able to match real data sets has been realized [Trevisan 95]. We can present the first results.

---

[4] A search area is the intersection between a circle and the Voronoï diagram [Yap 87] of punctual BDC crossroads.

3.3.3 First results

Results of road matching are 100% correct.
As for crossroads, a manual match was also performed to check the automatic process. For 284 crossroads, results are as follows :
- For one BDC to one G matching (191 cases), automatic matching gives good results in 84% of cases. The remaining 16% are results including the right result but also some parasitic elements.
- For one BDC to many G (49 cases) results are average. Automatic matching gives good results in 41% of cases, gives a result containing only a part of the right result in 28.5%, and gives a result including the right result in 30.5%.

No matching between BDC crossroads and G crossroads (44 cases), has been also analyzed:
- The main reason is border objects (49.4%) for which some dangle sections are missing (in BDC or in G).
- The second reason of mismatch is inconsistency (39%) ; updates are not the same and semantic or geometric quality is wrong.
- The third set of reasons are conflicts between BDC crossroads being too close (5.8%): a part of the complex crossroad can not be matched with the BDC crossroad because it is in another research area [Trevisan 95].
- The last set of causes are the translations (5.8%) between the two representations that are superior to the matching threshold.
Results of section matching are approximately 75% correct.

3.4 Generalization of this process

Our building a multi-scale database for road network data has demonstrated the feasibility of such databases from data sets. This process can be generalized whatever :
- **the data** (network, land use, topography, ...). For example, for land use data, the schema integration phase does not change much, on the other hand the data matching phase necessitates a new process [Kidner 96] (distance between surfaces, relationship of neighbourhood, ...).
- **the number of representations**. This process can be generalized to more than two input databases by using the same methods (schema integration into a single schema then, matching two by two of instances).
- **the location of representations** (data set, distributed databases). For the integration of distributed databases into a federated database, our first results will still have to be improved. Indeed, with autonomous databases, data matching will not be computed after schema merging, but during the processing of the multi-scale queries and only for the data involved. This "on the fly" process is very interesting but necessitates a very effective matching process.
- **the needs** (data analysis, quality controls, ...). According to the application chosen, the multi-scale schema will not comply with the same constraints. Therefore, the schema integration technique is different.

Building multi-scale database is an adaptable process. This process depends on the data, the needs, the number and the location of representations. This work can be also generalized for different kinds of multiple representations (temporal, thematic,...). In this cases, new operations of correspondence will have to be defined.

4. Conclusion

In this article, we have described a process to build a multi-scale database with scale-transition relationships. This is a three steps process : declaration of scale-transition relationships between schemata, schema merging and data matching. Because management of this kind of database is difficult, some problems must be first tackled (update, data manipulation language,...). This process must still be thorough, by optimizing the data matching and by working on the retrieve-and-store of multi-scale operations. We are continuing our research into the integration of geographic databases (databases from different sources or databases stemming from a generalization process).

References

D.J. Abel, P.J. Kilby and J. R. Davis (1994) The systems integration problem, in Int. J. Geographical Informations Systems Vol. 8 N° 1, pp 1-12.

B. P. Brugger, R. Barrera A. U. Frank, K. Beard and M. Ehlers (1989) Research Topic on Multiple Representations in NCGIA Initiative 3 Workshop on Multiple Representations, pages 53-67.

A. Car, A.U. Frank (1994), Modelling a Hierarchy of Space applied to Large Road Networks, in IGIS'94, LNCS n°884, pp 15-24.

S. Ceri et G. Pelagatti (1987) Distributed databases: principles & systems, McGraw-Hill.

B. David, L. Raynal, G. Schorter and V. Mansart (1993) Why objects in a geographical DBMS?, in Advances in Spatial Databases, pp 264-276.

T. Devogele, L. Raynal. (1996) Modeling a Multi-Scale Database with Scale-Transition Relationships, in Samos'96, T.Sellis D. Georgoulis Eds. pp 83-93

Y. Dupont (1994) Resolving Fragmentation Conflicts in Schema Integration in 13th Int. Conf. on The Entity Relationship Approach.

G. Edwards (1994) Characterising and maintaining polygons with fuzzy boundaries in geographic information systems, in Spatial Data Handling,Waugh and Healey Eds., Taylor&Francis, pp 223-239

Y. Gabay and Y. Doytsher (1994) Automatic adjustement of line maps, in GIS/LIS pp 233-241.

M.F. Goodchild (1991) Issue of quality and uncertainty, in Advances in Cartography, Müller Ed., Barking, Essex/ Elsevier pp 113-139.

M.O. Govorov (1995) Representation of the generalized data structures for Multi-Scale GIS, in 17th ICA/ACI Barcelonna pp 2491-2495.

O. Günther (1989) Database Support for Multiple Representations, NCGIA, in Initiative 3 Workshop on Multiple Representations, pp 50-51

C.B. Jones (1991) Database architecture for multi-scale GIS, in Auto-Carto 10 volume 6 pp 1-14.

J.F. Hangouët (1995) Computation of the Hausdorff distance between plane vector polylines, in Auto Carto 12, pp 1-10.

F. Hausdorff (1919) Dimension und äusseres in Mass. Mathematische Annalen n° 79, pp 157-179.

D. Kidner and C.B. Jones (1994) A Deductive Object-Oriented GIS for handling multiple representations, in Spatial Data Handling, Waugh and Healey Eds., Taylor&Francis, pp 882-900.

D. Kidner (1996) Geometric signatures for determing polygon equivalence during multi-scale GIS update, in Second Joint European Conference, IOS Press, pp 239-247.

J.P. Lagrange and A. Ruas (1994) Geographic information modelling: GIS and generalisation, in Spatial Data Handling, Waugh and Healey Eds., Taylor&Francis, pp 1099-1117.

B. Langou and Maiguenaud M. (1994) Graph data model operations for network facilities in a geographical information system, in Spatial Data Handling, Taylor & Francis, pp 1002-1019.

T. Kilpeläinen (1995) Requirements of a Multiple representation database for topolographical data with emphasis on incremental generalization, in 17 th ICA/ACI Barcelonna, pp 1815-1825.

D. Mark (1989) Multiple views of multiple representations, in NCGIA Initiative 3 Workshop on Multiple Representations, pp 68-71.

R. McMaster and S. Shea (1992), Generalization in Digital Cartography, Association of American Geographers.

A. Motro (1987) Superviews; Virtual Integration of Multiple Databases, in IEEE Transactions on Software Engineering 13 (7), IEEE, pp 785-798.

J.C. Müller (1991) Generalization of spatial databases, Geographical information Systems Principles and Applications, Maguire, Goodchild and Rhind Eds., Publisher Longman Scientific & Technical, pp 457-475

J.C. Müller, J.P. Lagrange, R. Weibel and F. Salgé (1995) Generalisation : state of the art and issues, in GIS and GENERALISATION, Müller, Lagrange and Weibel Eds., Taylor & Francis pp 3-17.

O2 (1991) The O2 System; in Communications of the ACM, Vol 34,n°10.

W. Peng, K. Tempfi and M. Molenaar (1996) Automated Generalization in a GIS, Geoinformatic'96, West Palm Beach pp 135-146

L. Raynal, B. David and G. Schorter (1995) Building an OOGIS prototype: experiments with GeO2, in Auto Carto 12 pp 137-146.

L. Raynal and N. Stricher (1994) Base de données multi-échelles: Association géométrique des tronçons de route de la BD Carto et de la BD Topo in EGIS /MARI'94, pp 300-307.

P. Rigaux and M. Scholl (1995) Multi-Scale Partitions: Application to Spatial and Statistical Databases in SSD '95, pp 170-183.

J. Rumbaugh, M. Blaha, W. Premerlani, F. Eddy and W.Lorensen (1991) Object-Oriented modeling and design, Prentice Hall Eds., Englewood Cliffs.

S. Scherrer,A. Geppert and K. Dittrich (1993) Schema Evolution in NO 2, Institut fur Informatik der Universitat Zurich technical report n° 93-12.

M.H. Scholl and M. Tresch. (1993) Schema Transformation without database reorganisation, in ACM SIGMOD Record, (22;1).

K.S. Shea and R.B. McMaster (1991) Cartographic Generalisation in a digital environment When and how to generalize, in Map generalization, Buttenfield and McMaster Eds., Longman Scientific & Technical, pp 103-118.

A. Sheth et J. Larson (1990) Federated database systems for managing distributed, heterogeneous, and autonomous databases,in ACM Computer Surveys 22,3 (Sept 1990), pp 183-236

S. Spaccapietra, P. Parent, Y. Dupont, (1992) Model-Independent Assertions for Integration of Heterogeneous Schemas, in Very Large DataBases Journal, 1(1), pp 81-126.

S. Timpf and U. Frank (1995) A Multi-scale DAG for cartographic objects, in Auto Carto 12, pp 157-163.

J. Trevisan (1995) Conception d'une BD Multi-échelles, ENSG Academic report, IGN Saint Mandé.

P. van Oosterom (1995) The GAP-tree, an approach to 'on-the-fly' map generalisation of an area partitioning, in GIS and GENERALISATION, Müller, Lagrange and Weibel Eds., Taylor & Francis, pp 120-132.

C.K. Yap (1987) An O(n log n) Algorithm for the Voronoï Diagram of a set of simple Curve Segments in Discete Comp. Geom. 2 journal, pages 365-393.

Contact address

Thomas Devogele, Jenny Trevisan and Laurent Raynal
IGN / COGIT
2, Avenue Pasteur
94 160 Saint-Mandé
France
Phone : +33 1 43 98 85 44
Fax : +33 1 43 98 81 71
E-mail: {devogele,raynal}@cogit.ign.fr
E-mail: trevisan@abraxa.ign.fr

VARIOSCALE TIN BASED SURFACES*

Gunnar Misund

SINTEF Applied Mathematics
Blindern, Oslo, Norway

Abstract
This paper focuses on the problem of approximating spatial objects to obtain an "optimal" resolution (or accuracy). In traditional map making, the notion of *scale* has been of vital importance, as one of the most significant components of generalization of geographic entities. However, in geographic information systems (GIS), the notion of scale becomes far more encompassing and complicated than in traditional cartography. The paper introduces an *augmented scale concept*, by considering spatial objects which are approximations with a spatially *varying tolerance*, which are referred to as *varioscale* objects. The rationale behind varioscaling is briefly discussed, and a general strategy for applying this concept on TIN surfaces is presented. A key element of the strategy is to simplify the problem to approximation with a fixed tolerance by using the "tolerance surface" to *normalize* the TIN surface. Two examples are presented, a feature based varioscaling, and a viewpoint dependent approximation.

Key words
Generalization, approximation, terrain modelling, TIN surface, variable tolerance.

1. Introduction

In many ways, the GIS field is closely related to traditional cartography, and has inherited many problem areas from the art and science of map making. From the very start of map making, a few thousand years ago (see [Bro49] for an introduction to the history of cartography), the notion of *scale* has been of vital importance. Maps are abstractions of real world phenomena, and in cartography, the abstraction process is termed *generalization*. One of the most significant components of generalization is the *scaling* of geographic entities. Typically, for a given geographic region, there exists a range of maps of different scales.

The main motivation for presenting the cartographic projection of the world in different scales is to optimize the transfer of information from the map to the user. Traditional maps have been drawn and printed on planar media, mostly paper, utilizing a wide range of sophisticated projections from the sphere-like globe. However, with the introduction of information technology in cartography, a whole new set of possibilities, challenges and problems have

---

* Partially supported by the Norwegian Research Council, project MOI.31386.

emerged. GIS offer advanced presentation, like 3D views and animation, and are capable of performing complex analysis independent of the restrictions imposed by the planar, static paper map.

In GIS, the notion of scale becomes far more encompassing and complicated than in traditional cartography. The first observation to make is that the scale concept in fact becomes meaningless, unless when referring to the actual size of presented objects on the computer display. A more suitable approach is to discuss the *resolution*, or *accuracy* of geographic objects. However, for historical reasons, we will use scale as a synonym for resolution or accuracy.

In the next section, we will outline some of the elements of an *augmented scale concept*. Section 3 focuses on how to obtain approximations of terrain models of optimal resolution. Two cases are presented in Section 4, and finally a few remarks are given in section 5.

2. Variable resolution - Augmenting the scale concept

In this section, we shall augment the traditional scale concept by introducing some elements required in efficient and flexible presentation and analysis of geographic information in a computer supported environment. We consider the scale generalization problem, i.e. how to *approximate* a given base model according to certain requirements for the accuracy of the approximant. The section will focus on the *conceptual* issues, formal details will be given in Section 3, and more elaborated examples will be presented in Section 4.

- **Traditional Multiscale Approximation.** In computer based scale generalization, we search for a sequence of approximants of a given "original", or "1:1" model of some geographic objects. The measurable "difference" between an approximant and the original should be less than a given tolerance, and each approximant corresponds to a certain scale 1 : X. In Figure 1, a multiscale sequence of a river network is depicted. Figure 1-a is the original model, and 1-b and 1-c are approximations generated by applying the Douglas-Peucker simplification algorithm [DP73] with tolerances of 15m respectively 40m.

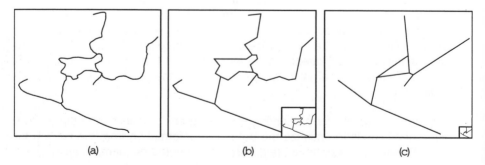

(a)   (b)   (c)

Figure 1 Multiscale network.

There exists a rich literature on how to efficiently represent sets of scale-varying cartographic entities, and this research branch in GIS is often termed *multiple representations.* See [But93] for a summary of Research Initiative 3 by NCGIA (National Center for Geographic Information and Analysis), and [MD95] for a discussion of the consistency problem of multiple representations. In scale generalization, the approximation tolerance is given as a constant value that does not vary over the model. Hence, the accuracy at all points of the map (or model), satisfies the same tolerance. In this paper, we shall focus on approximations with *variable tolerances.*

- **Visualization Dependent Approximation.** With the introduction of GIS, presentation of cartographic information is no longer restricted to the paper map. More realistic 3D models have been introduced, and the dynamic time aspect is sometimes incorporated in the underlying data models in GIS. Many users prefer 3D visualization vs. planar projections, and even with a limited 2D model, more realistic and effective views of the information are possible (see [HU94] for an overview of visualization and GIS).

  One of the characteristics of GIS is the huge amount of information required to represent a geographical model of some real world phenomena. For example, to reach a satisfactory level of detail, a terrain model might require use of a regular 10m grid. This implies that a 5km by 5km terrain needs 250.000 elevation values. To simulate the vista from a given point of view near the terrain surface, the image of the terrain model is by perspective projection mapped to a view plane defined by the viewing parameters. In the background, towards the "horizon", the relative "precision" of the terrain model is much higher than in the foreground. This may in certain contexts be regarded as a non-optimal solution. Many visualization tools have restrictions on the size of the models they are applied to. Models of excessive size may also decrease performance to a level where the response time of the system becomes unsatisfactory. To reduce the amount of data representing the terrain, we shall choose a strategy where the accuracy of the model depends on the distance to the simulated viewpoint. In other words, we want the accuracy in the foreground and the background to be *relatively* equal in the projected image of the model.

(a)   (b)

Figure 2 Visualization dependent approximation.

Figure 2 displays two views generated from a terrain model, based on a TIN (Triangular Irregular Network). The views are flat shaded to reveal the density and structure of the underlying triangulation. Figure 2-a is the original model, and 2-b is approximated with a variable tolerance to obtain a more "balanced" model which is represented with a reduced amount of data. In the projected image, the size of the triangles seems to be approximately equal. A more precise description of how to obtain optimal resolution shall be given in Section 4.2. See e.g. [GSHN94] for a discussion of variable resolution in photo-realistic visualization of natural landscapes.

- **Analysis Dependent Approximation.** Another main category of problems which will benefit from a variable scale, is a large set of spatial analysis tasks, for example water shed analysis, shortest paths methods, erosion prediction, and simulation of diffusion of toxic waste. The spatial focus of many such computations is restricted to parts of the original model, but still a "complete" model is required to carry through the analysis. Spatial analysis is often computationally intensive, and therefore we need to pay special attention to the amount of data involved, or in other words, the "size" of the problem. Therefore, it would be desirable to control the *local* accuracy of the data, in order to achieve a more satisfactory resolution of the over all model.

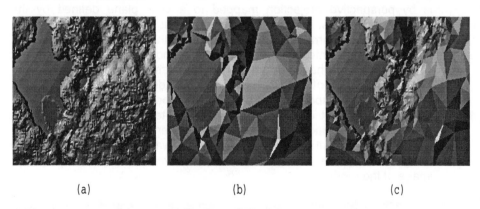

(a)   (b)   (c)

Figure 3 Analysis dependent approximation.

Three TIN terrain models are rendered in Figure 3. The "original" terrain, using all the given data, is rendered in 3-a, and the one in 3-b is a coarse approximation with a constant tolerance. The last picture, 3-c, is an approximation which emphasizes the accuracy along a network of streams, and represents the rest of the terrain at a coarse level. In the next section, we formalize the problem of obtaining location dependent accuracy by introducing the *varioscaling* concept, and then discuss some algorithmic strategies.

3. Varioscaling strategies

In this section, we shall outline some strategies for approximation of surfaces with variable tolerance.

- **Varioscale Spatial Objects.** Based on the previous discussion, we introduce some simple formalism to facilitate the further investigation of the scale problems in GIS. Roughly speaking, by a *varioscale spatial object*, we mean an approximation of a given base object, which satisfies a certain tolerance that varies over the object. Thus, the resolution of the object differs from location to location, or in other words, the scale of the object is dependent of the location. We shall give a more precise definition of varioscale *surfaces*:

- **Definition 1 (Varioscale Surface)** *Given a surface* $f : \Omega \rightarrow R$, *where* $\Omega \subset R^2$ *and a tolerance function* $\varepsilon : \Omega \rightarrow R^+$. *A surface* $g : \Omega \rightarrow R$ *is said to be a varioscale approximation of f if*

$$\mid f(x,y) - g(x,y) \mid \leq \varepsilon(x,y) \text{ for all } (x,y) \in \Omega.$$

Note that alternative methods exist for measuring the difference between surfaces. Traditional scale generalization, which approximates an object with a constant tolerance, is to be considered as a special instance of varioscale approximation. Based on this general definition, we shall develop a varioscaling strategy for terrain surfaces represented as TINs (Triangular Irregular Network).

- **Pointwise Varioscaling of TINS.** As a special case of surface approximation, we shall now consider *pointwise* varioscale approximation of surfaces defined over triangular, irregular networks (TINs). Given a set of distinct data points $V = \{v_i\}_{i=1}^{N}$, where $v_i = (x_i, y_i) \in R^2$, and an associated vector of elevation values $F = \{F_i\}_{i=1}^{N}$. Let Ω be a domain in R^2, e.g. the convex hull of V. To generate an interpolating function such that

$$f(x_i, y_i) = F_i \text{ for all } i,$$

we construct a triangulation T on V, for example a Delaunay triangulation (see [Sch87] for an an overview of triangulation methods). We define f as the piecewise linear function over Ω, where the function values are uniquely given by a linear combination of the values at the three vertices of a triangle in T. In order to obtain a pointwise varioscale approximation of f, we have given a vector of real, *positive* tolerances, $\varepsilon = \{\varepsilon_i\}_{i=1}^{N}$. The task is to find a set of $n \leq N$ new vertices $V^* \subseteq V$, such that

$$\mid F_i - f^*(x_i, y_i) \mid \leq \varepsilon_i \text{ for all } i \qquad (1)$$

where f^* is a piecewise linear function defined by a triangulation T^* on V^*.

Several schemes have been proposed and implemented to perform approximations of TINs where the tolerance function is a constant value, $\varepsilon \equiv c$. Many of these algorithms are based on iterative insertion or removal of points, where selected points are repeatedly inserted (or removed) in the current triangulation until the tolerance is satisfied (see [Flo89] for an example of an iterative approximation of Delaunay triangulations). It is indeed a trivial task to enhance existing insertion algorithms to handle variable tolerances. An additional vector of tolerances is required, and instead of checking the error in each vertex in V against a fixed value, we have to access the corresponding element of the tolerance vector.

A novel TIN varioscaling strategy is presented in [CPS95], which utilizes a so called HyperTriangulation. This is a datastructure that is iteratively build by using existing approximation methods. By applying different traversal shemes, both fixed tolerance or varioscaled approximations are derived from the HyperTriangulation. This paper focuses on methods for explisit construction of varioscale approximations. Suppose we are supplied with an existing and satisfactory implementation of an approximation scheme. By performing a simple transformation of our problem, we may utilize the existing software. We introduce the normalized function $\phi : \Omega \to R$, which is the piecewise linear function on T, interpolating the *normalized* elevation values $\Phi = \{\Phi_i\}_{i=1}^{N}$, where $\Phi_i = F_i / \varepsilon_i$ (recall that $\varepsilon_i > 0$), such that

$$\phi(x_i, y_i) = \Phi_i \text{ for all } i.$$

The problem is transformed, and we search for a subset $V* \subseteq V$, on which to construct the approximating function $\phi* : \Omega \to R$, which satisfies

$$| \Phi_i - \phi*(x_i, y_i) | \leq 1 \text{ for all } (x_i, y_i) \in V. \tag{2}$$

To obtain the approximant $f*$, we simply construct the triangulation on $V*$ with the original elevation values F, which effectively is a rescaling of $\phi*$.

To verify the procedure, we observe, since all ε_i are positive, that for all i:

$$| \Phi_i - \phi*(x_i, y_i) | \leq 1 \Leftrightarrow | F_i - \phi*(x_i, y_i)\varepsilon_i | \leq \varepsilon_i.$$

Trivially, since $F_i = \Phi_i \varepsilon_i$, we have that

$$\phi*(x_i, y_i)\varepsilon_i = f*(x_i, y_i) \text{ for all } i.$$

where $f*$ is the piecewise linear function defined over $V*$ with elevation values from F. Hence, equations 2 and 1 are equivalent. Note that even if this transformation yields a mathematically equivalent approximation problem, we are not guaranteed that a certain method will generate exactly the same approximations, respectively with a vector of tolerances, or with a constant tolerance value applied to the normalized surface. This is due to the fact that the performance of most approximation methods is highly dependent on the distribution and characteristics of the input data.

Since the normalization procedure indeed may drastically change the surface characteristics, we should expect variations, both with respect the the degree of data reduction, and to the selection of points included in the approximant.

The normalization procedure may introduce numerical instability, if the ratios between the tolerances and the elevation values are too large. However, the approximation problem is not well posed if near-zero tolerances are applied to a surface which varies over an interval of several hundred meters. Hence, we claim that in well posed problems, this potential instability is to be ignored. We summarize the normalization strategy in a simple algorithm, which finds $V^* \subseteq V$, given vectors with elevations and tolerances, and a near-zero threshold $\underline{\varepsilon}$. In addition we assume we have a TIN approximation procedure, anyApproxTIN(*V,F,c*), which returns $V^* \subseteq V$, given the vertices V, the elevation values F, and the constant tolerance c.

Algorithm 1 *Varioscale Approximation of TINs by Normalization*

```
approxTIN(V, F, ε, ε̲, anyApproxTIN())
1.       for all i
1.1.          if (εᵢ < ε̲ )
1.1.1.               εᵢ := ε̲
1.2.          φᵢ := Fᵢ/εᵢ
2.       return V* := anyApproxTIN(V, φ, 1.0)
```

---

4. Examples

In this section we shall apply the varioscaling algorithms presented in Section 3, to two cases. In the first case, we investigate how the proposed techniques can optimize a terrain model with respect to requirements of certain analysis and computations. Then we shall study how varioscaling can be utilized in *visualization* of 3D geographic data. Our overall strategy is to use existing software to the largest possible extent. In all the examples, the various approximation methods are selected from the software library SISCAT (scattered data approximation) [sis95] developed at SINTEF Applied Mathematics.

4.1 Feature based approximation

In this case, our aim is to represent a terrain surface with "optimal" resolution. The elevation model of the terrain surface is assumed to be used in erosion analysis, in order to predict water erosion along a stream network. We need as detailed elevation information as possible near the streams, but the other parts of the terrain are of less importance. We want to generate a varioscale TIN terrain model, like the one in Figure 3-a. The elevation data are obtained by digital photogrammetry, as 3D samples from a 10m by 10m grid, $F = \{F_i\}_{i=1}^{M}$, see Figure 1-a. The network information, the set $C = \{C_i\}_{i=1}^{N}$, is given as 2D data from a digitized topographic map, see Figure 1.

The first step in the varioscaling process is to construct the tolerance surface, which should have certain properties. We partition the domain Ω in two subsets, $\overline{\Omega}$, which we consider as areas of little importance in our analysis, and ½, which is the set of locations close to the stream systems. The areas should be non-overlapping, $\overline{\Omega} \cap \underline{\Omega} = \varnothing$, and should represent a complete partition, $\overline{\Omega} \cup \underline{\Omega} = \Omega$. Our strategy is to generate a set of scattered 3D data, $S = \{S_i\}_{i=1}^{\mu}$ over the domain Ω, which are to be considered as samples from the tolerance surface e. We shall use existing software for scattered data approximation to construct the function.

According to the decomposition of Ω, we search for two disjoint 3D point sets. We start with a regular grid $G = \{G_i\}_{i=1}^{\nu}$ of 2D points in Ω. To represent $\overline{\Omega}$, we select the points in G which are outside a "buffer" defined by the 2D vertices of the network C. We construct $G^* \subseteq G$ as the grid points G_i that satisfy

$$\min_{j=1}^{N} \delta(G_i, C_j) > b,$$

where b is the "size" of the buffer, and where $\delta(G_i, C_j)$ is the Euclidian planar distance between two points. We then assign the constant value $\overline{\varepsilon}$, representing the maximal tolerance, as the z value of all points in G^*. In the SISCAT library, there are methods for generating "smooth" grids from scattered 3D points. Additional constraints may be introduced, such as defining that a part of the input data should be considered as a piecewise linear curve which should be represented as a "ridge", or break-line, in the constructed grid. See [FHT, AH] for a description of the functionality of SISCAT. "Inside" the buffer, we want the tolerance to be minimal along the piecewise linear stream network, and gradually increase towards the limits of the buffer. To obtain this effect, we let the 2D network vertices C represent samples from $\underline{\Omega}$, and set the z values to equal the minimum tolerance $\underline{e}$, thus forming the new set C^*.

We apply the appropriate SISCAT method to the scattered 3D points G^* and C^*, where in addition the stream data in C^* are defined as ridge data. This results in a regular grid G^ε with user specified cell sizes. The final tolerance surface ε is constructed by bilinear interpolation of G^ε. Equipped with the tolerance function e, we proceed and construct the final TIN terrain model with variable accuracy. The SISCAT library provides us with a method for approximative Delaunay triangulation. Based on this technique, we apply the normalization scheme proposed in Algorithm 1. We summarize our first case as follows:

Case 1 *Construction of Buffer Scaled TIN*

1.   Assume we have scattered 3D terrain data, F, and a 2D network C, both defined over a domain Ω in $\mathbb{R}^2$.
2.   Create a 2D grid G on Ω.
3.   Select $G^* \subset G$ as grid points outside a buffer around C with size b.
4.   Assign a maximum tolerance $\bar{\varepsilon}$ as z values to G^*.
5.   Assign a minimum tolerance $\underline{\varepsilon}$ as z values to C, defining C^*.
6.   Define G^* and C^* as scattered 3D data, and in addition C^* as 3D ridge data.
7.   Construct a grid G^ε with a scattered data approximation method from G^* and C^*.
8.   Generate the tolerance surface ε by bilinear interpolation of G^ε.
9.   Apply Algorithm 1, approxTIN(F, ε, $\underline{\varepsilon}$, approximativeTriangulation()) to generate $F^* \subseteq F$.
10.   Construct TIN from F^*.

The process is illustrated in Figure 4. The scattered data input to generate the tolerance surface is plotted in 4-a, and the resulting e surface in 4-b. The *xy*-projection of the approximated subset F^* is displayed in 4-c along with the original stream network C. For a 3D view of the final TIN, see Figure 3.

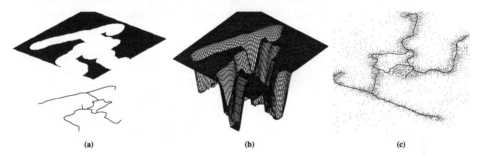

(a)   (b)   (c)

Figure 4 Construction of buffer scaled TIN.

4.2 Viewpoint approximation

We will now investigate a case where the goal is to construct an approximated TIN, which to a certain degree is optimized for efficient visualization. We use the same scattered terrain data F as in section 4.2. In addition, we have given a set of parameters which uniquely define a *view* of the terrain: An eyepoint, a look-point, and the horizontal and vertical viewing angles. We also have a maximum tolerance $\bar{\varepsilon}$, and a minimum tolerance $\underline{e}$. As in the previous case, the main challenge is to find a tolerance surface e, which should possess certain properties. The tolerance function should have a global minimum close to the eyepoint, and gradually increase relative to the distance to the eyepoint to reach the global maximum $\bar{\varepsilon}$. Outside the viewing area, the function should be identical to $\bar{\varepsilon}$. Formally, the tolerance surface is defined as follows: Let $B \subset \mathbb{R}^3$ be the 3D view volume defined by the given viewing parameters. Define $\Omega \subset \mathbb{R}^2$ as the convex hull of the *xy*-component of F. Construct a surface $f : \Omega \to \mathbb{R}^2$ from F by scattered data approximation, and consider this as the original terrain ("ground truth"). Construct a 2D grid

G, and assign a value z_i to all grid-points G_i such that

$$z_i = \begin{cases} \varepsilon_i & \text{if } G_i \in B \\ \overline{\varepsilon} & \text{otherwise} \end{cases}$$

where

$$\varepsilon_i = \begin{cases} \underline{\varepsilon} & \text{if } s \times \delta(G_i, f(G_i), e) + t < \underline{\varepsilon} \\ \overline{\varepsilon} & \text{if } s \times \delta(G_i, f(G_i), e) + t > \overline{\varepsilon} \\ s \times \delta(G_i, f(G_i), e) + t & \text{otherwise.} \end{cases}$$

The real function δ measures the Euclidian 3D distance between the eyepoint e, and a grid point G_i with a z value sampled from the original surface f. The constants s and t are used to tune the characteristics of the tolerance surface. As in the previous case, we construct the tolerance surface ε by bilinear interpolation of the grid G. See Figure 5-a for a plot of the view defined tolerance surface, including projected contours. The final TIN terrain model is constructed by applying Algorithm 1 to the original data F and the tolerance function e, and the resulting triangulation is plotted in 5-b. For a 3D view, see Figure 2.

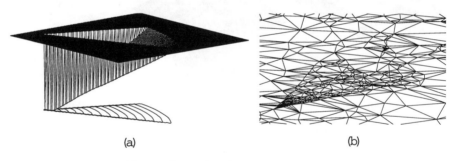

(a)   (b)

Figure 5 Generation of view dependent TIN.

We summarize the case as follows.

Case 2 *Construction of View Point Scaled TIN*

1.　Assume we have a set of scattered 3D terrain data, F, over the domain Ω, and a view volume $B \subset \mathbb{R}^3$.
2.　Create a 2D grid G on Ω.
2.　Create a 2D grid G on Ω.
3.　Assign values to the grid nodes, yielding G^*, as a function of B, the maximum tolerance $\overline{\varepsilon}$, the minimum tolerance $\underline{\varepsilon}$, and the constants s and t.
4.　Perform steps 8, 9, and 10 in Case 1.

---

5. Final remarks

In this paper, we have focused on the importance of scale in GIS. For historical reasons, scale is used as a synonym for accuracy, or resolution, of spatial objects. We have claimed that certain problems would benefit substantially from utilizing varioscale objects. Such objects are characterized by their variable resolution, i.e. that the scale is dependent on the location. We have outlined a strategy for obtaining variable accuracy in TIN based terrain models, which are of fundamental importance in GIS. In particular, we showed how to transform certain approximation problems, such that existing software for constant value approximation could be used. In two cases, varioscale approximation was demonstrated in some detail.

To fully experience the benefits of using varioscaled spatial models in GIS, varioscale approximation schemes for a broad range of types of data have to be developed. As indicated in this paper, some varioscaling problems may be transformed to a dual formulation. By finding the adequate transformation, the trivial rephrasing of the problem enables us to take advantage of a large supply of existing approximation methods. Approximation of composite objects, e.g. a terrain model with an associated stream network, is in general not trivial. A major challenge is to match the tolerances of the different objects involved. In varioscale approximation, matching of tolerances becomes even more complicated, and this problem certainly calls for further research.

Methods for generating and representing varioscale objects have been described and discussed in some papers, but the major bulk of such research is based on *refinement* techniques. In many ways, refinement is to be considered as the dual of approximation. Based on a coarse underlying model, the object is locally refined, often in real time, to meet certain resolution requirements. Advanced terrain models used in military combat simulators are often utilizing quadtree techniques to obtain variable resolution. See [DKW94] for a discussion on aspects of adaptive interpolation of terrain models on regular grids. Refinement procedures are useful in applications when only low resolution data are available. However, with the introduction of global positioning systems (GPS) and digital photogrammetry, geographical information is becoming available at high resolution and relatively low cost. Hence, efficient and flexible approximation methods are more and more in demand.

References
[AH] E. Arge and Ø. Hjelle. Software Tools for Modelling Scattered Data. In M. Dæhlen and A. Tveito, editors, Title to be decided. Submitted for publication.
[Bro49] A. Lloyd Brown. The Story of MAPS. Little, Brown and Company, 1949.
[But93] Barbara P. Buttenfield. NCGIA Research Initiative 3: Multiple Representations Closing Report. Technical report, National Center for Geographic Information and Analysis, SUYN - Buffalo, April 1993.
[CPS95] P. Cignoni, E. Puppo and R. Scopigno. Representation and Visualization of Terrain Surfaces at Variable Resolution. In R. Scateni, editor, Scientific Visualization '95 (int. Symp. Proc), pages 50 - 68, World Scientific, 1995.

[DKW94] A.R. Dixon, G.H. Kirby, and D.P.M. Wills. Towards Context Dependent Interpolation of Digital Elevation Models. Computer Graphics Forum - The International Journal of the Eurographics Association, 13(3):C-23-C-32, 1994. Conference Issue, EUROGRAPHICS '94, Oslo - Norway, ISSN 0167-7055.

[DP73] D. H. Douglas and T. K. Peucker. Algorithms for the reduction of the number of points required to represent a digitized line or its caricature. The Canadian Cartographer, 10(2):112 - 122, 1973.

[FHT] N. Fremming, Ø. Hjelle, and C. Tarrou. Surface Modelling from Scattered Geological Data. In M. Dæhlen and A. Tveito, editors, Title to be decided. Submitted for publication.

[Flo89] L. de Floriani. A Pyramidal Data Structure for Triangle-Based Surface Description. IEEE Computer Graphics & Applications, pages 67 - 70, March 1989.

[GSHN94] K.Ch. Graf, M. Suter, J. Hagger, and Nüesch. Computer Graphics and Remote Sensing - A Synthesis for Environmental Planning and Civil Engineering. Computer Graphics Forum - The International Journal of the Eurographics Association, 13(3):C-13-C-22, 1994. Conference Issue, EUROGRAPHICS '94, Oslo - Norway, ISSN 0167-7055.

[HU94] Hilary M. Hearnshaw and David J. Unwin, editors. Visualization in Geographical Information Systems. John Wiley & Sons, 1994. ISBN 0-471-94435-1.

[MD95] Gunnar Misund and Morten Dæhlen. On Consistent Modeling of Geographic Information. In Jan Terje Bjørke, editor, SCANGIS '95 - Proceedings of the 5th Research Conference on GIS, 12th 14th June 1995, Trondheim, Norway, May 1995. ISBN 82-993522-07.

[Sch87] Larry Schumaker. Triangulation methods. In Topics in Multivariate Approximation, pages 219 - 232. Academic Press, Inc., 1987. ISBN 0-12-14585-6.

[sis95] SISCAT - The SINTEF Scattered Data Library (version 2.1). Technical report, SINTEF Informatics, Oslo, 1995. Reference manual.

Contact address
Gunnar Misund
SINTEF Applied Mathematics
P.O. Box 124
Blindern, N-0314 Oslo
Norway
Phone: +47 22067961
Fax: +47 22067350
E-mail: Gunnar.Misund@si.sintef.no

METRICS AND TOPOLOGIES FOR GEOGRAPHIC SPACE

Michael F. Worboys

Department of Computer Science
Keele University
Staffs, UK

Abstract
This paper is motivated by the requirement for computer-based representations of geospatial information that parallel the structure of geographic space that humans apprehend. It is well known that such representations do not satisfy the conditions of a metric space. We consider distance and proximity relationships between entities in geographic spaces, and discuss some of the ways in which useful representations take us beyond mathematical metric spaces. We show that even though the underlying space does not satisfy the conditions for a metric space, it is still possible to induce neighbourhood topologies upon it.

Key words
Metric, proximity, locality, topology, GIS.

1. Introduction

This paper is motivated by the requirement for computer-based representations of geospatial information that parallel the structure of geographic space that humans apprehend. The emphasis of this work relates to the concepts of distance, proximity and neighbourhood. The mathematical theory of metric spaces is well-known to be inadequate as a formal foundation for distance measures in geographic spaces (see, for example, (Montello 1992)). As a simple example, metric spaces assume a symmetric distance function in that the distance from entity e_1 to entity e_2 is the same as the distance from e_2 to e_1, but geographic distance is often not symmetric (consider for example travel time).

Contextual knowledge is a key feature of human apprehension of geographic space. Different contexts provide us with quite distinct models of the surrounding space. For example, a bicyclist will have a different perception of his or her geographic neighbourhood from a driver of a wide load or an airline pilot. Even for the same person and application, distance may be perceived differently depending upon geographic location. Thus an observer in New York might perceive the distance from London to Edinburgh differently from an observer in London. Present computer systems do not generally support context-based representations. For example, almost all current GIS applications treat distance as a global relation between spatial entities, independent of user or application. This is clearly less than satisfactory when bringing in the human dimension.

This paper constructs part of the formal foundations of the distance and proximity relationships between entities in geographic space that takes account of some these issues. We examine the standard notion of a mathematical metric space, note its shortcomings for geographic space, and consider some alternative formulations. We also review some of the relevant literature on qualitative distance. Finally, we show how it is possible to define notions of locality and neighbourhood, even in spaces that deviate considerably from metric spaces, and that topologies can still be constructed.

2. Naive geography and geographic spaces

A notion that has considerable resonance in this context is that of 'naive geography', introduced by Egenhofer and Mark (1995). Naive geography is a phrase coined to encapsulate intuitive knowledge that people have about the geographic world that surrounds them. The phrase is used so as to call to mind the Naive Physics Manifesto (Hayes 1978) that sets out a similar programme for people's intuitions about their surrounding physical world. Naive geographical space, as set out by Egenhofer and Mark is two dimensional and planar, spatio-temporal, and lacking complete information but admitting multiple models. Geographic space also emphasises topology, but lacks the usual mathematical distance metric. Our naive geographic thinking often treats topological relationships as prior to measurements. We may know that the counties of Staffordshire and Shropshire share a common boundary, but be unable to reasonably estimate its length.

An important distinction is between so-called 'table-top space' and geographic space. A table-top space is viewable and potentially physically explorable from a single position. We may take a 'God's eye view' of the space, which includes the imposition on it of a global metric. Table-top spaces have been well researched in literature related to such fields as graphics, machine vision, artificial intelligence and robotics. Because table-top spaces are spatially referenced, it is natural to consider them relevant to geospatial data handling. However, geographic space differs from table-top space in several important respects, including the much decreased significance of a global view. We are immersed in geographic space, cannot see or touch its full extent from any single location, and move through its spatial and temporal dimensions to explore its domain.

With the absence of the global view, the notion of a distance relationship becomes more complex. In table-top space, the metric is globally imposed, but geographic space has no global view and therefore no such globally imposed metric. To illustrate this point, figure 1 shows a set of objects assumed to be in a table-top space. The global view sets the distance relationship between objects A and B, which then becomes fixed. On the other hand, figure 2 shows the same configuration of objects now assumed to be in geographic space. Because of the lesser importance of the global view, local context is the prime determinant of distance relationships. Entity B is large and its neighbourhood (shown as grey entities) is primarily determined by neighbouring large objects, while entity A is smaller and relates to a different neighbourhood determined by local smaller objects. We see that while A is in the locality of B, it is not the case that B is in the locality of A, demonstrating the asymmetry of the geographic distance relationship.

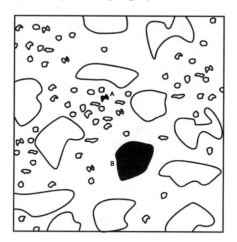

Figure 1 Distance may be context independent in table-top space.

Figure 2 Proximity is context-dependent in geographic space.

Another important distinction between table-top and geographic space is that in geographic space the distance between objects is dependent upon the time that it takes to travel from one to the other. Again, this is a direct consequence of the observer's immersion in geographic space and the lesser importance of a global view. Once more, this introduces asymmetry into the distance relationship, since in general, maybe because of traffic conditions, terrain or prevailing wind, it does not take the same amount of time to travel from A to B as from B to A. Figure 3 shows how a variable-speed road network can introduce anisotropic conditions into the space. We have given a highly simplified case, where there is a single high-speed link AB in an otherwise isotropic metric space with distance calculated according to a Pythagorean metric. For simplicity, assume that the link is so fast as to make the travel-time from *A* to *B* negligible compared with the travel-times anywhere else in the field. When travelling between two points, there is a choice whether or not to use the high-speed link. Consider points close to *X* (say, within 14 time units) in figure 3. Clearly in these cases, a traveller would do better not to use the high-speed link. However, for points near *B* it would be better for the traveller to *X* to travel to *B*, take the link (zero-time) and continue on from *A* to the destination. The isochrones are shown in the figure. The hyperbola marks the boundary between regions where it is better/worse to use the link. In the second case, to determine the travel time to X, it clearly matters in which direction the destination location is from *X* and the space is anisotropic. Anisotropic fields are common in real-world situations. Other examples that introduce anisotropic conditions are natural and artificial barriers to direct accessibility.

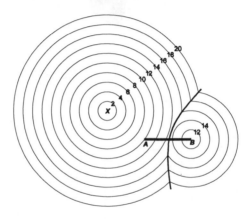

Figure 3 Isochrones surrounding the point X in geographic space.

3. Related work on geographic space and qualitative distance relationships

There is a considerable body of research on human apprehension of the large-scale space in which we live. Montello (1992) refers to this space as *environmental* space, and characterises it as the space that surrounds us and in which movement over time is usually required to gain knowledge of it. The popular term for a mental representation of an environmental space is *cognitive map*, although some writers (e.g. Tversky 1993) have pointed to shortcomings in this notion. Regarding the geometry of our mental representation of space, many authors have noted its non-adherence to the properties of a mathematical metric or at least a Euclidean space. For example, Tversky (*op. cit.*) cites psychological experiments of Sadalla *et al.* (1980) regarding the distorting effect that reference points have, leading to an asymmetric distance relationship. Montello (*op. cit.*) notes work of Gollege and Hubert (1982) on non-Euclidean metric spaces.

A characteristic of reasoning about geographic space is the incompleteness of our knowledge of it. Classical logic is formulated on the premise that knowledge is complete, therefore is inappropriate for this work. There has been some application of fuzzy logic (Zadeh 1988) in this field, but the whole ranges of non-monotonic logics are still to be explored. Qualitative spatial reasoning is discussed by Cui, Cohn and Randell (1993). Ideas applied specifically to qualitative distance are considered by Hernández, Clementini and DI Felice (1995). An approach to context-dependent proximity operators has been described by Gahegan (1995).

4. Geographic metric spaces

A point-set S is said to be a *metric space* if there exists a function, **distance,** that takes ordered pairs (s,t) of elements of S and returns a real number **distance** (s, t) that satisfies the following three conditions:

M1.   For each pair s, t in S, **distance** $(s, t) > 0$ if s and t are distinct points and **distance** $(s, t) = 0$ if, and only if, s and t are identical.

M2.   For each pair s, t in S, the distance from s to t is equal to the distance from t to s, **distance** $(s, t) = $ **distance** (t, s).

M3.   *(Triangle inequality)* For each triple s, t, u in S, the sum of the distances from s to t and from t to u is always at least as large as the distance from s to u, that is: **distance** $(s, t) + $ **distance** $(t, u) \geq $ **distance** (s, u).

The first condition M1 stipulates that the distance between points must be a positive number unless the points are the same, in which case the distance will be zero. The second condition M2 ensures that the distance between two points is independent of which way round it is measured. The third condition M3, the *triangle inequality*, states that it must always be at least as far to travel between two points via a third point rather than to travel directly.

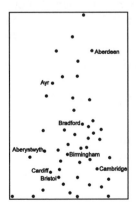

Figure 4 Forty eight British centres of population.

We have already seen that a geographic space does not admit such a metric. It is reasonable to suppose that condition M1 is satisfied by any distance function. However, context and travel-time both provide examples where condition M2 does not hold. Condition M3 does hold for travel-time metrics, but does not hold for context-related metrics. We illustrate this, and some of the earlier issues, by means of an example. Figure 4 shows 48 centres in Great Britain. Their distances, measured in miles along major roads, have been calculated (Collins, 1995), and some examples are given in table 1. These distances relate to a global view, and are here termed *objective distances*, since they take no account of users, applications or locations (except that they assume users to be travellers along major roads).

	Aberdeen	Aberystwyth	Ayr	Birmingham	Bradford	Bristol	Cambridge	Cardiff	
Aberdeen	0	445	176	416	321	492	458	490	
Aberystwyth	445	0	314	114	164	125	214	105	
Ayr	176	314	0	289	201	368	352	380	
Birmingham	416	114	289	0	110	81	100	103	
Bradford	321	164	201	110	0	189	152	204	
Bristol	492	125	368	81	189	0	155	44	
Cambridge	458	214	352	100	152	155	0	175	
Cardiff	490	105	380	103	204	44	175	0	
...									

Table 1 Part of the objective distance relationship between 48 British centres of population.

In our example, context is accounted for in the following manner. For each centre c, the mean μ_c of the distances from c to all centres is calculated. The relativised distance **reldis** from centre c to centre d is then determined by the formula:

reldis $(c,d) =$ **distance** $(c,d) / \mu_c$

Some relativised distances are shown in table 2. Note that the table is asymmetric, since **reldis** $(c,d) \neq$ **reldis** (d,c). This accords with our intuition regarding context dependent distance. For example, **reldis**(Aberdeen, Birmingham) = 1.1 and **reldis**(Birmingham, Aberdeen) = 2.6, reflecting the notion that from the perspective of Birmingham, closely surrounded by several centres, Aberdeen is relatively far away, but from the context of the relatively outlying and isolated Aberdeen, Birmingham is relatively closer.

We may also note that the **reldis** relationship does not obey the triangle inequality. For example

      reldis (Birmingham, Aberdeen) = 2.6
      reldis (Birmingham, Ayr) = 1.8
      reldis (Ayr, Aberdeen) = 0.6

and so

      reldis (Birmingham, Aberdeen) >
            reldis (Birmingham, Ayr) + **reldis** (Ayr, Aberdeen)

	Aberdeen	Aberystwyth	Ayr	Birmingham	Bradford	Bristol	Cambridge	Cardiff	⋮
Aberdeen	0	1.2	0.5	1.1	0.9	1.3	1.2	1.3	
Aberystwyth	2.1	0	1.5	0.5	0.8	0.6	1	0.5	
Ayr	0.6	1.1	0	1	0.7	1.3	1.3	1.4	
Birmingham	2.6	0.7	1.8	0	0.7	0.5	0.6	0.6	
Bradford	1.9	1	1.2	0.7	0	1.1	0.9	1.2	
Bristol	2.5	0.6	1.8	0.4	0.9	0	0.8	0.2	
Cambridge	2.3	1.1	1.8	0.5	0.8	0.8	0	0.9	
Cardiff	2.3	0.5	1.7	0.5	0.9	0.2	0.8	0	
...									

Table 2 Part of the relativised distance relationship between the 48 centres.

It will be useful in what follows to define a proximity or nearness relationship between geospatial entities. For geographic space G, define a function **nearness**, that takes ordered pairs (s,t) of elements of G and returns a real number **nearness** (s, t) that satisfies the following conditions:

1. $0 <$ **nearness** $(s, t) \le 1$
2.   **nearness** $(s, s) = 1$

The idea is that if entity y is far from entity x, then **nearness** (x, y) will have a value close to zero, while if entity y is near to entity x, then **nearness** (x, y) will have a value close to 1. Note that, as with **distance**, **nearness** is context dependent and asymmetric in general.

For our example of the 48 British centres, we may derive a nearness measure from relative distance by means of the following formula:

nearness $(x, y) =$ (**reldis** $(x, y) + 1)^{-1}$

Values of the **nearness** relationship for the some of the 48 centres are shown in table 3.

	Aberdeen	Aberystwyth	Ayr	Birmingham	Bradford	Bristol	Cambridge	Cardiff	⋮
Aberdeen	1.00	0.46	0.68	0.47	0.54	0.43	0.45	0.43	
Aberystwyth	0.33	1.00	0.41	0.65	0.57	0.63	0.50	0.67	
Ayr	0.61	0.47	1.00	0.49	0.58	0.43	0.44	0.42	
Birmingham	0.28	0.58	0.36	1.00	0.59	0.66	0.61	0.61	
Bradford	0.34	0.50	0.45	0.60	1.00	0.47	0.52	0.45	
Bristol	0.29	0.61	0.35	0.71	0.51	1.00	0.56	0.82	
Cambridge	0.30	0.48	0.36	0.66	0.56	0.56	1.00	0.53	
Cardiff	0.31	0.67	0.36	0.68	0.52	0.83	0.55	1.00	
...									

*Table 3 Part of the **nearness** relationship between the 48 centres.*

5. From metric to topology: nearness and locality

We take the point-set approach to topology and follow Henle (1979) in defining a **topological space** to be a point-set S together with a collection of subsets of S, each called a **neighbourhood** of its points, which satisfies the following properties.

T1.   Every point of S is in some neighbourhood.

T2.   The intersection of any two neighbourhoods of a point contains a neighbourhood of that point.

A major benefit of a distance function that satisfies the metric space properties is that it can be used to define a topology. Neighbourhoods in this topology are defined to be the open balls where, given a point x in the space and for real number $r>0$, an open ball $B(x,r)$ is the set of points less than the given distance r from x. The metric space properties may now be used to show that this is indeed a topology. The first property T1 above follows from the metric space property M1 (section 4). To prove the second property T2, consider two neighbourhoods $B(x_1, r_1)$ and $B(x_2, r_2)$ of point x. Suppose that **distance** $(x, x_1) = d_1$ and **distance** $(x, x_2) = d_2$. Then $r_1 - d_1$ and $r_2 - d_2$ are both positive real numbers. Let r be defined as **minimum** $(r_1 - d_1, r_2 - d_2)$. We now show that the neighbourhood $B(x,r)$ is contained in both neighbourhoods $B(x_1, r_1)$ and $B(x_2, r_2)$. In fact, we show that $B(x,r)$ is contained in neighbourhood $B(x_1, r_1)$ and appeal to symmetry. Let y belong to $B(x,r)$, then **distance** $(x, y) < r \le r_1 - d_1$, and so

$$\begin{aligned}
\textbf{distance}\,(x_1, y) &\le \textbf{distance}\,(x_1, x) + \textbf{distance}\,(x, y) & \text{(by M3)}\\
&= \textbf{distance}\,(x, x_1) + \textbf{distance}\,(x, y) & \text{(by M2)}\\
&< d_1 + (r_1 - d_1)\\
&= r_1
\end{aligned}$$

and so $B(x,r)$ is contained in neighbourhood $B(x_1, r_1)$. Thus, since T1 and T2 have been shown to hold, the open balls in a metric space serve as neighbourhoods in the induced topology.

Notice that we need all the properties of a metric space to show that a metric space has a natural induced topology based upon its open balls. Unfortunately, without the metric space properties it is not possible to define such an induced topology. Since topology is a very important aspect of geographic space, this is a great disadvantage. The remainder of this section shows how a topology may be defined in a different, but we believe, still useful manner. The key to the construction is the **nearness** function defined earlier, which we use to define a Boolean predicate $v(x, y)$, taking the value **true** when entity y is deemed to be near to entity x. Assume a fixed parameter p ($0<p<1$) and define $v(x, y)$ as follows:

$v(x, y) = $ **true** if, and only if, **nearness** $(x, y) > p$

If $v(x, y)$ = true, we will write xvy and say that *y is near to x*. Note that by property (2) of the nearness function, v is reflexive, that is for all geographic entities *a* belonging *G*, ava. However it is not the general case that v is symmetric or transitive   Table 4 shows the predicate v defined for our example with $p = 0.7$.

	Aberdeen	Aberystwyth	Ayr	Birmingham	Bradford	Bristol	Cambridge	Cardiff	:
Aberdeen	1	0	0	0	0	0	0	0	
Aberystwyth	0	1	0	0	0	0	0	0	
Ayr	0	0	1	0	0	0	0	0	
Birmingham	0	0	0	1	0	0	0	0	
Bradford	0	0	0	0	1	0	0	0	
Bristol	0	0	0	1	0	1	0	1	
Cambridge	0	0	0	0	0	0	1	0	
Cardiff	0	0	0	0	0	1	0	1	
...									

Table 4 Part of the predicate v defining a proximity relationship between the 48 centres.

By introducing the concept of *locality*, we are able to get transitivity. Let *G* be a geographic space in which a nearness relationship v is defined. Define a locality relationship λ on *G*, so that for all geographic entities *a, b* belonging *G*, $a\lambda b$ if, and only if, both of the following conditions hold:

1. avb
2.   For all *c*, if bvc then avc

If we read $a\lambda b$ as *b is local to a*, then the defining conditions above say that for *b* to be local to *a*, *b* must be near to *a* and furthermore any entity near to *b* must of necessity be near to *a*. Define the *locality* of *a* $\wedge_a$ to be the set of all entities local to *a*. Such a definition provides a means of characterising clustering of entities. We now show that localities $\wedge_a$ (*a* in *G*) have the properties of topological neighbourhoods.

It is straightforward to note that the locality relation λ is both reflexive and transitive. Reflexivity is immediate. For transitivity, suppose $a\lambda b$ and $b\lambda c$. We want to show that $a\lambda c$ obtains. Condition (1) defining $b\lambda c$ gives bvc, then condition (2) defining $a\lambda b$ gives avc.   Also, suppose cvx, then bvx by condition (2) defining $b\lambda c$, and so then avx by condition (2) defining $a\lambda b$. Thus, $a\lambda c$ holds.

To show that the localities $\wedge_a$ (*a* in *G*) are topological neighbourhoods, we must show firstly that $\wedge_a$ always contains some point of *G*, and secondly that if point *x* belongs to the intersection of two localities, then there is a locality to which *x* belongs and entirely contained in the intersection. These conditions follow from the reflexivity and transitivity of the locality relation λ. The proof is very similar to the demonstration above that a metric space induces a neighbourhood topology and is omitted.

For our example, the localities are shown in figure 5. Given the notion of a locality that has the properties of a topological neighbourhood, we have now achieved our objective of defining a topology on a geographic space having a relativised distance (or proximity) relationship. It is possible now to use this topology to define boundary, open and closed sets, connectedness and other topological properties of objects in a geographic space. Details of these constructions are the subject of current research.

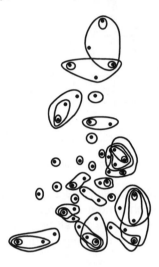

Figure 5 Locality neighbourhoods for our example set.

6. Conclusion

The research described in this paper has considered some of the formal requirements of geographic space. Notions such as context introduce complexities that move us beyond the bounds of simple Euclidean or non-Euclidean metric spaces. The main contribution of this work has been to show that even with a distance/proximity relationship that does not admit a metric space, it is still possible to define an induced topology based upon the construction of localities. This locality-induced topology can then be used, in a similar way to the more usual metric-induced topology, to explore topological properties of the space, such as boundary and connectedness. Further work is required to explore the nature of the induced locality topology. Overall, the research contributes to a wider view of human interaction with computer-based geospatial information.

References

Collins, 1995, Road Atlas Britain. HarperCollins, London.

Cui, Z., Cohn, A.G. and Randell, D.A., 1993.   Qualitative and topological relationships in spatial databases.   In Abel, D., Ooi, B.C. (eds.), Advances in Spatial Databases, Proceedings of SSD'93, Singapore, Lecture Notes in Computer Science 692. Springer-Verlag, Berlin, Germany, pp. 296-315.

Egenhofer, M.J. and Mark, D.M., 1995.   Naive geography.   In Frank, A.U. and Kuhn, W. (eds.), Spatial Information Theory, Proceedings of COSIT'95, Lecture Notes in Computer Science 988. Springer-Verlag, Berlin, Germany, pp. 1-15.

Gollege, R.G. and Hubert, L.J., 1982.   Some comments on non-Euclidean mental maps. Environment and Planning A, 14, 107-118.

Hayes, P., 1978.   The naive physics manifesto.   In D. Michie (Ed.), Expert Systems in the Microelectronic Age. Edinburgh University Press, 242-270.

Henle, M. (1979).   A Combinatorial Introduction to Topology.   Freeman, San Francisco, CA, USA.

Hernández, D., Clementini, E. and Di Felice, P., 1995.   Qualitative distances.   In Frank, A.U. and Kuhn, W. (eds.), Spatial Information Theory, Proceedings of COSIT'95, Lecture Notes in Computer Science 988. Springer-Verlag, Berlin, Germany, pp. 45-57.

Gahegan, M., 1995.   Proximity operators for qualitative spatial reasoning.   In Frank, A.U. and Kuhn, W. (eds.), Spatial Information Theory, Proceedings of COSIT'95, Lecture Notes in Computer Science 988. Springer-Verlag, Berlin, Germany, pp. 31-44.

Montello, D.R., 1992.   The geometry of environmental knowledge.   In Frank, A.U., Campari, I. and Formentini, U. (eds.), Theory and Methods of Spatio-Temporal Reasoning in Geographic Space, Lecture Notes in Computer Science 639. Springer-Verlag, Berlin, Germany, pp. 136-152.

Sadalla, E.K., Burroughs, W.J. and Staplin, L.J., 1980.   Reference points in spatial cognition. Journal of Experimental Psychology: Human Learning and Memory, 5, 516-528.

Tversky, B., 1993, Cognitive maps, cognitive collages, and spatial mental models.   In Frank, A.U. and Campari, I. (eds.), Spatial Information Theory, Proceedings of COSIT'93, Lecture Notes in Computer Science 716. Springer-Verlag, Berlin, Germany, pp. 14-24.

Zadeh, L.A., 1988. Fuzzy logic. IEEE Computer.

Contact address

Michael F Worboys
Department of Computer Science
Keele University
Staffs ST5 5BG
UK
Phone: +44782583078
Fax: +44782713082
E-mail: michael@cs.keele.ac.uk

TOPOLOGIC RELATIONS BETWEEN FUZZY AREA OBJECTS

Jeroen Dijkmeijer and Sylvia de Hoop*

Department of Surveying, Photogrammetry and Remote Sensing
Wageningen Agricultural University
Wageningen, the Netherlands

Abstract
Fuzzy set theory is used to model uncertainty and vagueness in a GIS environment. Considering topologic relations between area objects, uncertainty may arise among others because of vague geometric information or multiple- or undefined criteria used in the process of information gaining. Using this information improperly can lead to wrong conclusions about the relations between the different objects. The study in this paper focusses on two elementary topological relations among fuzzy areal terrain objects. These are the overlap and adjacency relations. Adjacency between the objects is taken as prior information, but how does that elaborate in their geometry. Functions are formulated to deal with the relations between fuzzy objects. In addition, the overlap relation is expressed in an algorithmic way. The functions yield fuzzy values.

1. Introduction

Many disciplines are exploring the use of geographic information systems (GISs), widening its range of possibilities, and at the same time, demanding new techniques. Techniques which need to be developed in the field of information technology. The ever increasing need for modelling the real world requires methods to deal with uncertain or unsharp defined entities. Traditional databases are perfectly capable of dealing with well defined-data. However, with increasing complexity of the modelled real world, traditional methods show their deficiencies. An example is the lack of facilities to work with vaguely formulated data. This becomes apparent, for example, in the field of geographic applications. The accuracy models of geodesists and terrain surveyors are useful only when they refer to geometrically well-defined objects whose characteristic points can be positioned in the terrain unambiguously (Molenaar, 1989).

This was the background of the research described in this paper. Fuzzy set theory has been used to model uncertainty and vagueness in a GIS environment. For that, an existing theory on topologic relations (Molenaar, 1995) is used in combination with fuzzy set theory (Klir and Folger, 1988;

---

*Since February 1996 employed by Q-Ray Agrimathica.

Kaufman, unknown). Because of the wide diversity of possible topologic relations and the limited extent of the research, this paper focusses on two elementary topologic relations: the overlap and adjacency relations between fuzzy area objects. The results are only derived from topologic relations, whereas the metric information of terrain objects is not considered. This in contrast to previous research, in which uncertainty and vagueness were related to metric relations and position analysis of the objects (Altman 1994, Edwards 1994). The research described in this paper is theoretical, no attempt is (yet) made to implement the concepts in the framework of a database management system or GIS, or in a rule-based knowledge system such as Prolog.

In Section 2 a short description is given of a two-level data model for area objects. The two levels can roughly be interpreted as a level of crisp thematic attributes, the object level, and a level of fuzzy geometric elements. Section 3 describes the fuzzy set approach to determine the relations between faces and objects. (Area objects are made up of one or more 2D geometric elements, called faces.) In Section 4 the functions are derived for the topologic relations between fuzzy area objects. Section 5 illustrates the functions by an example. Finally, Section 6 gives conclusions through a discussion.

2. Crisp and fuzzy information of area objects

The universe of discourse, that is part of the real world to be modelled, is represented by a finite terrain with non-movable entities. This terrain representation is recorded here according to the formal data structure for multi-valued vector maps (Molenaar, 1989; De Hoop et al.,1993). Two different semantic levels can be distinguished in the terrain representation: a geometric level describing the geometry of area objects, and a level of area objects superimposed on the geometric level (Figure 1). As Figure 1 illustrates, only area objects are distinguished in the terrain. An area object may represent a forest, grassland or, even more general, arable land. An object is geometrically described by one or more faces. The geometric data consists of metric and topologic data.

It is assumed here that area objects have crisp thematic attributes, and that the existence of an object is evident. As concerns topologic relations between area objects, however, uncertainty may arise because of vague geometric information, multiple- or undefined criteria used in the process of information gaining and errors due to inaccurate measurement. Using this information improperly can lead to wrong conclusions about the relations between the different objects. In the following sections, a number of functions are formulated which deal with the relations between fuzzy objects. These functions yield fuzzy values.

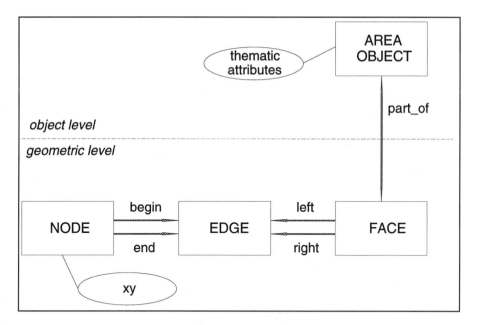

Figure 1 Formal data structure for area objects.

3. Fuzzy functions

The certainty whether a face is part of an object or not, is expressed in membership values (μs). The membership value can be proportional to a certain sample value or can be an arbitrary value based on subjective aspects. The membership value denoting the certainty of a face f_f part of an object O_a is defined as:

$$\mu_{oa}(f_f) \rightarrow [0,1] \in \Re$$

This can be redefined as

$$\text{part-of}[f_f, O_a] \equiv \mu_{oa}(f_f):O \rightarrow [0,1] \in \Re \tag{1}$$

and the complementary function as:

$$\text{Npart-of}[f_f, O_a] = 1 - \text{part-of}[f_f, O_a]. \tag{2}$$

expressing the certainty of face f_f <u>not</u> being part of object O_a.

Because many objects may be represented by the same face, the need for a function arises which displays the relation between a face and an object relative to other relations between the same face and other objects. The specific function should express the certainty that a face f_f is only part of object O_a and not of any other object.

The analysis of the sentence

   "the certainty that a face f_f belongs only to object O_a and not to any other object"

leads to fuzzy set operators, as defined in (Klir and Folger, 1988): (Please note that the linguistic formulation uses an infix notation while the fuzzy set operators use a prefix notation.)

- face f_f belongs only to object O_a <u>and</u> ...

The linguistic term `and' implies a relational `and' operation which is translated in fuzzy set terms to the minimum operator. So the operator and the first argument are now known:

   $\min\{\text{part-of}[f_f, O_a], ..\}$

The second argument can also be linguistically determined.

- (face f_f belongs) not to

implies an inversion, `not' is translated in fuzzy set terms to a `1 - operator'.

- any other object.

implies a union of all relations between a specific face and all other objects: $\max\{\text{part-of}[f_f, O_b]\}$ operator.

The complete function has now been determined. As this function describes the part-of function, relative to other part-of relations, it is called the Rpart-of function.

   $\text{Rpart-of}[f_f, O_a] = \min\{\text{part-of}[f_f, O_a], 1 - \max_{b \neq a}\{\text{part-of}[f_f, O_b]\}\}$

which can be rewritten as:

   $\text{Rpart-of}[f_f, O_a] = \min\{\text{part-of}[f_f, O_a], \min_{b \neq a}\{\text{Npart-of}[f_f, O_b]\}\}$ (3)

a and b are indices indicating all objects within the Powerset(O) (= P(O)): $a, b \in P(O)^\dagger$ $a \neq b$.

As a last step, a decision function is needed to obtain final conclusions. This function returns a boolean value. When the membership value $\mu_{Oa}(f_f)$ is greater than any other membership value $\mu_{Ob}(f_f)$ the function returns exclusively 1 (true), otherwise the function returns 0 (false):

---

† P(O) is the number of objects within the powerset, known as the cardinality of the set. P(f) represent the number of faces within the powerset, and $O_a(f)$ represents the number of faces within a certain object.

$$\text{Xpart-of}[O_a, f_f] = 1 \mid \text{part-of}[O_a, f_f] > \max_{a \neq b}\{\text{part-of}[O_b, f_f]\}$$
$$\text{Xpart-of}[O_a, f_f] = 0 \mid \text{part-of}[O_a, f_f] \leq \max_{a \neq b}\{\text{part-of}[O_b, f_f]\} \qquad (4)$$

The symbol `|' denotes `under the condition of'.

4. Topologic relations

4.1 Elementary topologic relations

The functions (1), (2), (3) and (4) only provide information on `part of' relations between faces and objects, i.e. between the geometric and object level. In the next section, topologic relations are introduced. First, topologic relations between faces are defined. Second, topologic relations are derived between objects using the earlier derived `part of' relations between the objects and faces. Ajdacency relations between faces are propagated through common edges of the faces [Molenaar '91]. A face has a left or right relation with an edge:

$le[e_i, f_f] = 1$ When face f_f has a left relation with edge e_i.
$ri[e_i, f_f] = 1$ When face f_f has a right relation with edge e_i.

Because the direction of an edge is uniquely determined, an edge can have at most one left-relation or one right-relation with a face. If two faces have respectively a left- and right relation with the same edge, the two faces are adjacent.

$$\text{adj}[f_f, f_g \mid e_i] = \min\{le[e_i, f_f], ri[e_i, f_g]\} + \min\{le[e_i, f_g], ri[e_i, f_f]\} \qquad (5)$$

Two faces f_f, f_g are adjacent when there is at least one edge e_i for which there is at least one relation $\text{adj}[f_f, f_g \mid e_i] = 1$:

$$\text{adj}[f_f, f_g] = \max_{ei}\{\text{adj}[f_f, f_g \mid e_i]\} \qquad (6)$$

So far, part-of functions have been introduced to deal with membership values of faces being part of objects. In addition, functions on topologic relations between faces have been defined. With these two functions, topologic relations between objects can be derived. Overlap relations are easily demonstrated, one face having a part-of relation with more than one object already indicates an overlap relation. Different methods to calculate the overlap relation between objects can be applied.

Indicating adjacency between fuzzy objects is more difficult. Prior information on possible adjacency, or rather disallowed overlap, between the objects is necessary. When two objects are geometrically mutual exclusive, the two objects can be at most adjacent. When a third object lies between the two objects, no adjacency relation between the former two objects is found. For example, two soil types may, after information gaining, unquestionably show an overlap relation, while we know that the two soiltypes are mutually exclusive and therefore cannot have an overlap relation, but have at most an adjacency relation.

4.2 Fuzzy overlap

For the overlap relation between two objects we can define its value as the highest value of all faces being part of one object and part of the other object:

$$\text{overlap}[O_a, O_b] = \max_{ff}\{\min\{\text{part-of}[O_a, f_f], \text{part-of}[O_b, f_f]\}\} \qquad (7)$$

Another value for an overlap relation may be found through an algorithm, which takes the size of the two objects and the membership values of the faces being part of the objects into account. To get around the problem of calculating the areas of the faces, the example given below is based on a grid in which the faces do have exactly the same area. There is no semantical difference between raster and vector geometry (Molenaar, 1994).

In figure 2, two area objects can be distinguished. A degree for the relation O_A overlapped by O_B is intuitively found by the intersection of O_A and O_B ($O_A \cap O_B$). This number is then divided with the total area of O_A. Applying this method to fuzzy grid, first the intersection between the two fuzzy objects should be calculated:

$$\Sigma \min(\mu_{Oa}(f_f), \mu_{Ob}(f_f)) = O_A \cap O_B \qquad [m^2]$$

Remark that $O_A \cap O_B = O_B \cap O_A$, so this function is commutative. To calculate overlap(O_A, O_B) the intersection $O_a \cap O_b$ is normalized with the total area A:

$$\Sigma \mu_A(f_f) = \text{Sum}(O_A) \qquad [m^2]$$

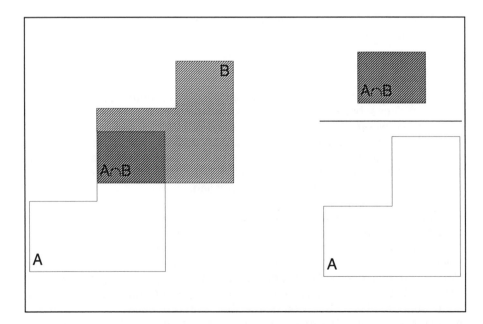

Figure 2 Degree of overlap relative to object A.

This expression does not take the area into account. To be complete, the product of the area and the membership value is calculated. The total expression for the degree of Overlap(A,B) becomes:

$$\text{Overlap}(O_A|O_B) = O_A \cap O_B / \text{Sum}(O_A) \qquad [\text{-}] \qquad (8)$$

This function is not commutative, the result is influenced by the order of arguments. An object may overlap another object for a main part, in reverse order the overlap may be only marginally. Applying this method imposes the need to identify both objects: in the example above O_A is the target object and O_B the qualifier object.

4.3 Fuzzy adjacency

Adjacency is more difficult to determine. The first question is: how is adjacency expressed? Should adjacency be expressed in fuzzy terms, or should the relation be explicitly true or false? And if there is an adjacency relation between objects, can the geometry of the border between the objects be indicated.
In the context of this reserach, the adjacency relation is here defined as a fuzzy value allowing values between zero and one:

$$\text{adjacent}[O_a, O_b] \in [0,1].$$

The same relation applies to faces. A few criteria are defined which an adjacency relation should obey.
- adjacency between objects is propagated through the faces having a part-of relation with the objects;
- when no face of an object A is adjacent to any of the faces of object B, the value of the adjacency value is zero, (no adjacency);
- the maximum value of the adjacency relation is not greater than the minimum of the greatest part-of relations of the two objects:

$$\text{adjacent}[O_a, O_b] \text{ } \epsilon \text{ } \min\{\max\{\text{part-of}[O_a, f_f]\}, \max\{\text{part-of}[O_b, f_f]\}\}$$

We assume here that adjacency between fuzzy objects is prior information. Note that the faces can only be adjacent. To determine a fuzzy adjacency between two objects, the faces should form a connection between one object to the other objects. To illustrate this, one could imagine fuzzy objects as hills. The elevation of the hills indicates the membership value of the part-of relation between a face and an object (Figure 3). Adjacency between two hills only exists when there is a path from one top of the hill to the other top of the hill without hitting the groundlevel (i.e. a relation function evaluating to zero). This may sound logical, but at first 'hitting the ground' needs to be defined.

Two different interpretations are examined in the remainder of this section. The first interpretation is focussing on the Rpart-of relation, defined in (3). Adjacency between two objects can be established, when the Rpart-of relation between a face and an object is greater than zero. In other words, as soon as the Rpart-of relation between one of the two objects and a face is zero, 'the groundlevel is hit' and no adjacency relation can be established through this face.

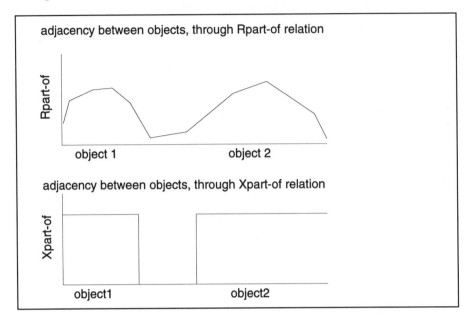

Figure 3 Adjacency between fuzzy area objects.

As mentioned before, adjacency between objects is propagated between their faces. Adjacency between two objects O_a and O_b propagated through two faces f_f and f_g is only allowed if, face f_f is part of O_a, and f_g is part of O_b, and f_g and f_f are adjacent to eachother, in formula:

$$\text{adjacent}[O_a,O_b|f_f,f_g] = \min\{\text{Rpart-of}[f_f,O_a],\text{Rpart-of}[f_g,O_b],\text{adj}[f_g,f_h]\} \quad (9)$$

As soon as for any combination of f_f and f_g the result of the function above is greater than zero, the relation is fulfilled and the adjacency relation exists. The value for the adjacency relation is then:

$$\min\{\max_{ff}\{\text{Rpart-of}[f_f,O_a]\}, \max_{fg}\{\text{Rpart-of}[f_g,O_b]\}\}\,(10)$$

In (9) the relation Rpart-of[...] can be replaced for an Xpart-of relation. There will only be an adjacency relation, when the faces have an exclusive relation with the object (no higher valued part-of relation between the same face and any other object can be found). The function yields, after substitution of Rpart-of by Xpart-of, only 1 or 0. The new formula is now directly formulated through combining (9) and (10):

$$\begin{aligned}\text{adjacent}[O_a,O_b|f_f,f_g] = \\ \min\{\text{adj}[f_g,f_h], \text{Xpart-of}[f_f,O_a], \text{Xpart-of}[f_g,O_b], \\ \max_{ff}\{\text{Rpart-of}[f_f,O_a]\}, \max_{fg}\{\text{Rpart-of}[f_f,O_b]\}\}\end{aligned} \quad (11)$$

It is not necessary to examine this function for each combination of f_f and f_g.

because after having found the first value, the value of the whole function is determined:

$$\text{adjacent}[O_a,O_b] = \text{adjacent}[O_a,O_b|f_f,f_g] \wedge f,g \in [1..P(f)], f \neq g. \qquad (12)$$

5. Example

5.1 Introduction

In the example below the two relations which have been worked out in the previous sections are illustrated. The example is kept very elementary for illustrative purposes. For the overlap relation as well as for the adjacency relation, two formulas have been derived. Every formula is illustrated here.

The example is based on a raster format, where the cells of the raster are considered as faces, the nodes lay on fixed places and this fact makes the whole map to be built of faces and edges with all the same size. The same topologic relations applicable to the vector map are applicable to the raster format as well (Figure 4).

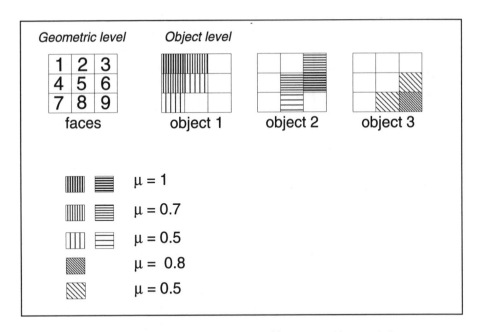

Figure 4 Example: terrain description of fuzzy area objects 1, 2, 3.

The terrain under consideration, has fixed borders, and three distinctive objects. First, the situation is considered, in which the objects are not mutually exclusive. In this case, the overlap relations are examined. In the other case, the objects are mutually exclusive (prior information), so objects can have at most an adjacency relation. The three objects in the terrain are: object 1, object 2 and object 3. In figure 4, object 1 is vertically hatched, object 2 is horizontally hatched and object 3 is diagonally hatched.

5.2 Formulating the relations

The objects are spread over 9 faces, and as can be seen in Figure 4, faces have been assigned to more than one object. First step is formulating the part-of relations for the different objects to the different faces. The part-of relations as well as the adjacent, Rpart-of and Xpart-of relations are pictured in tables:

O_a\f_f	f_1	f_2	f_3	f_4	f_5	f_6	f_7	f_8	f_9
O_1	1	0.7		0.7	0.4		0.4		
O_2			0.7		0.7	1		0.7	
O_3						0.5		0.5	0.8

Table 1 Part-of[f_f,O_a].

Adjacency between faces can also be pictured in a table:

f_f\f_g	f_1	f_2	f_3	f_4	f_5	f_6	f_7	f_8	f_9
f_1		1		1					
f_2	1		1		1				
f_3		1				1			
f_4	1				1		1		
f_5		1		1		1		1	
f_6			1		1				1
f_7				1				1	
f_8					1		1		1
f_9						1		1	

Table 2 Adj[f_f,f_g].

The information in table 1 and table 2 is directly derived from Figure 4. Using formulas (2), (3) and (4), the Npart-of[f_f,O_a], Rpart-of[f_f,O_a] and Xpart-of[O_a,f_f] respectively can be derived from table 1:

O_a\f_f	f_1	f_2	f_3	f_4	f_5	f_6	f_7	f_8	f_9
O_1	0	0.3	1	0.3	0.6	1	0.6	1	1
O_2	1	1	0.3	1	0.3	0	1	0.3	1
O_3	1	1	1	1	1	0.5	1	0.5	0.2

Table 3 Npart-of[f_f,O_a].

O_a\f_f	f_1	f_2	f_3	f_4	f_5	f_6	f_7	f_8	f_9
O_1	1	0.7		0.7	0.3		0.4		
O_2			0.7		0.3	0.3		0.3	
O_3								0.3	0.8

Table 4 Rpart-of[f_f,O_a].

O_a\f_f	f_1	f_2	f_3	f_4	f_5	f_6	f_7	f_8	f_9
O_1	1	1		1			1		
O_2			1		1	1		1	
O_3									1

Table 5 Xpart-of[f_f,O_a].

All the relations between the faces and objects being determined, the relations between the objects themselves can now be derived.

5.3 Overlap

In section 4.2 two formulas for overlap have been formulated. Using the information from table 1 the overlap relation can be calculated.

$$\text{Overlap}(O_1,O_2) = \max_{ff}\{\min\{\text{part-of}[O_1,f_f],\text{part-of}[O_2,f_f]\}\}$$

Result:

f_1	1	0	0
f_2	0.7	0	0
f_3	0	0.7	0
f_4	0.7	0	0
f_5	0.4	0.7	0.4
f_6	0	1	0
f_7	0.4	0	0
f_8	0	0.7	0
f_9	0	0	0

Total result: 0.4

Table 5a Overlap[O_1,O_2] according to formula 7.

In the formula worked out above, every possible relation is inserted, yielding 8 times the value 0 and once the value 0.4, making the total result 0.4. For the remaining relations between objects only the nonzero part-of relations are inserted. This soon finishes the overlap(O_1,O_3) relation: All combinations yield zero, as a consequence the total result gives zero.

face	$\max_{ff}\{\min\{$part-of$[O_2,f_f]$,part-of$[O_3,f_f]\}\}$		Result
f_6	1	0.5	0.5
f_8	0.7	0.5	0.5
Result			0.5

Table 5b Overlap$[O_2,O_3]$ according to formula 7.

According to formula 7, the overlap relation between O_b and O_c is the most severe overlap relation of the three possible overlap relations.

Calculating the overlap relation between two objects with Overlap(O_a,O_b) = O_AçO_B / Sum(O_a) (8) needs 6 different calculations because the order of the arguments is of interest. At first the total summations of the three objects is calculated:

Sum $O_1 = 3.2$
Sum $O_2 = 2.8$
Sum $O_3 = 1.8$

The intersection between the three objects should now be calculated:
$O_1 \cap O_2 = 0.7$
$O_1 \cap O_3 = 0$
$O_2 \cap O_3 = 0.9$

The overlap using (8) is now simply a divison by the properiate values:
Overlap$[O_1,O_2] = 0.2$ Overlap$[O_2,O_1] = 0.3$
Overlap$[O_1,O_3] = 0$ Overlap$[O_3,O_1] = 0$
Overlap$[O_2,O_3] = 0.4$ Overlap$[O_3,O_2] = 0.6$

According to formula (8) the most severe overlap relation is: O_3 covered by O_2. Followed by the combination of O_2 covered by O_3. Note that calculating the overlap for an ordinary vector map would need calculation of the area of the faces, multiplied with the value of the part-of relation with the objects.

5.4 Adjacency

The same map, pictured in Figure 4, is considered, but the objects are now taken arbitrarily mutual exclusive. The adjacency between the objects is established by using the given formulas on adjacency (9), (10) and (11). In the same example, the difference on using the Rpart-of relations vs the Xpart-of relation is demonstrated. Adjacency between objects is propagated through faces and is shown by using formula (9).The function proving the adjacency is:

adjacent$[O_a,O_b|f_f,f_g]$ = $\min\{\max_{ff}\{$Rpart-of$[f_f,O_a]$, $\max_{fg}\{$Rpart-of$[f_f,O_b]\}\}$ | $\min\{$Rpart-of$[f_f,O_a]$,Rpart-of$[f_g,O_b]$,adj$[f_g,f_h]\} > 0$

adjacent$[O_a,O_b|f_f,f_g]$ = 0 | $\min\{$Rpart-of$[f_f,O_a]$,Rpart-of$[f_g,O_b]$,adj$[f_g,f_h]\} = 0$

To define a fuzzy adjacency relation between objects every combination (cartesian product) between the faces should be evaluated theoretically. In practice however, this is not needed for two reasons.

1) Most faces are not adjacent to each other (table 2).
2) As soon as one relation yields a value greater than zero the condition for adjacency has been fulfilled. The only task left is to determine the value of the relation.

At first, the relations indicating adjacency are worked out:

adjacent[O_1,O_2]
min{adj[f_f,f_g],Rpart-of[f_f,O_1],Rpart-of[f_g,O_2]}

f_f,f_g	{adj[ff,fg],Rpart-of[ff,O1],Rpart-of[fg,O2]}			Min
f_1,f_2	1	1	0	0
f_1,f_4	1	1	0	0
f_2,f_3	1	0.7	0.7	0.7
f_2,f_5	1	0.7	0.7	0.7

Table 6a Adjacency relations between O_1 and O_2..

adjacent[O_1,O_3]
min{adj[f_f,f_g],Rpart-of[f_f,O_1],Rpart-of[f_g,O_3]}

f_f,f_g	{adj[f_f,f_g],Rpart-of[f_f,O_1],Rpart-of[f_g,O_3]}			Min
f_5,f_6	1	0.4	0	0
f_7,f_8	1	0.4	0.5	0.4

Table 6b Adjacency relations between O_1 and O_3.

adjacent[O_2,O_3]
min{adj[f_f,f_g],Rpart-of[f_f,O_2],Rpart-of[f_g,O_3]}

f_f,f_g	{adj[f_f,f_g],Rpart-of[f_f,O_2],Rpart-of[f_g,O_3]}			Min
f_5,f_6	1	0.4	0	0
f_8,f_9	1	0.4	0.8	0.4

Table 6c Adjacency relations between O_2 and O_3.

As shown from the tables 6a, 6b and 6c, all objects are adjacent to each other. Now the only problem left is determining the *strength* of the adjacency relation. This value is given by (10). In table 7 the value is deduced for adjacent[O_a,O_b]:

$$\min\{\max_{ff}\{Rpart\text{-}of[f_f,O_1], \max_{ff}\{Rpart\text{-}of[f_f,O_2]\}\}$$

f_f	{Rpart-of[f_f,O_1]}	{Rpart-of[f_f,O_2]}
f_1	1	0
f_2	0.7	0
f_3	0	0.7
f_5	0	1
$\max_{ff}$	1	1
min		1

Table 7 Membership value for adjacent(O_1,O_2).

For the combinations (O_2,O_3) and (O_1,O_3) the result in both cases equals 0.8. The fuzzy adjacency relations all yield non-zero results using formula (9) and (10). Using formula (11) gives other results as shown in the next section.

Because the analyses of adjacency in formula (11) is done in combination with the Xpart-of relations with only boolean functions as result, the conditions are unnecessary. No branched functions are needed to determine adjacency. Again the adjacency is propagated through faces. So, theoretically all combinations would need evaluation. The reasons why this is not done were mentioned earlier. The functions with the relations are given below:

$$adjacent[O_1,O_2] = adjacent[O_1,O_2|f_f,f_g] \wedge f,g \in [1..P(f)], f \neq g.$$

$$adjacent[O_1,O_2] = \min\{adj[f_f,f_g], Xpart\text{-}of[f_f,O_1], Xpart\text{-}of[f_g,O_2],$$
$$\max_{ff}\{Rpart\text{-}of[f_f,O_1], \max_{fg}\{Rpart\text{-}of[f_g,O_2]\}\}$$

f_f,f_g	adj[f_f,f_g],Xpart-of[f_g,O_2],Xpart-of[f_g,O_2], $\max_{ff}$\{Rpart-of[f_f,O_1],$\max_{fg}$\{Rpart-of[f_g,O_2]\}				
f_1,f_2	1	1	0	1	0
f_1,f_4	1	1	0	1	0
f_2,f_3	1	1	1	1	1
f_2,f_5	1	1	1	1	1
min					1

Table 8a Adjacent[O_1,O_2] calculated with formula (11).

Object O_1 is adjacent to object O_2 using formula (10). Analysing adjacency between O_2, and O_3 gives adjacent[O_2,O_3]:

$$\min\{adj[f_g,f_h],\text{Xpart-of}[f_f,O_2],\text{Xpart-of}[f_g,O_3],$$
$$\max_{ff}\{\text{Rpart-of}[f_f,O_2], \max_{fg}\{\text{Rpart-of}[f_f,O_3]\}\}$$

f_f,f_g	$adj[f_g,f_h],\text{Xpart-of}[f_f,O_2],\text{Xpart-of}[f_g,O_3],$ $\max_{ff}\{\text{Rpart-of}[f_f,O_2], \max_{fg}\{\text{Rpart-of}[f_f,O_3]\}$				
f_5,f_6	1	1	0	0.8	0
f_5,f_8	1	1	0	0.8	0
f_6,f_9	1	1	1	0.8	0.8

Table 8b Adjacent$[O_2,O_3]$ calculated with formula (11).

adjacent$[O_1,O_3]$
$$\min\{adj[f_g,f_h],\text{Xpart-of}[f_f,O_1],\text{Xpart-of}[f_g,O_3],$$
$$\max_{ff}\{\text{Rpart-of}[f_f,O_1], \max_{fg}\{\text{Rpart-of}[f_f,O_3]\}\}$$

f_f,f_g	$adj[f_g,f_h],\text{Xpart-of}[f_f,O_1],\text{Xpart-of}[f_g,O_3],$ $\max_{ff}\{\text{Rpart-of}[f_f,O_1], \max_{fg}\{\text{Rpart-of}[f_g,O_3]\}$				
f_5,f_6	1	1	0	1	0
f_5,f_8	1	1	0	1	0
f_6,f_9	1	0	1	1	0

Table 8c adjacent$[O_1,O_3]$ calculated with formula (11).

In table 8c no adjacency relation between O_1 and O_3 has been found. No adjacent relation is possible between the two objects using (11). Between the two objects, O_2 functions as a lock, no adjacency is allowed. This in contrast to the result of formula (9), in which an adjacency relation can be found.

6. Discussion

Two different topologic relations on fuzzy objects have been investigated. For each relation two functions have been introduced and illustrated. One of the overlap functions uses the fuzzy set theory, while the other one expresses the overlap in an algorithmic way relative to the total area of the objects. In the latter method the order of the arguments is of importance. This may look unlogical, but intuitively it makes sense. Two objects overlapping each other may show significant differences in areas. One object may overlap the other only marginally while in reverse order the overlap is significant. The method using the fuzzy set theory does not take the size of the objects into account. Overlapping is based on the maximum value of the part-of relations of the faces with respect to multiple objects.

For adjacency relations, two functions have been introduced. One of these is using the Rpart-of relation and the other is using the Xpart-of relation. As shown in the example the adjacent relation using Xpart-of is more strict than the Rpart-of function, because the Xpart-of function returns only one non-zero value for every face. When an object is related to a face which has a stronger relation to another object, the result is unambiguously 0. One should keep in mind that an arbitrary preference order should be defined in case more objects have the same membership value to a single face.

An overlap relation unequal to zero between two fuzzy objects is enough to conclude an adjacency between the two objects using the Rpart-of. This condition however is not enough to conclude adjacency based on the Xpart-of relation.

This paper focussed on only two topologic relations. Research remains on implementing the procedure. With that much research remains to be done on other topologic relations. The 9-intersection model of categorizing topological relations (Egenhofer and Herring, 1992) would need revision when it would be applied to fuzzy objects. Also the adjacency function needs refinement. In this paper, overlapping fuzzy objects were the main concern, in which adjacency between the objects was taken as prior information. It is impossible to deduce adjacency on objects on a certain distance of each other, and the certain knowledge of adjacency. Space between the objects (slivers) may prevent an adjacency relation and we may get lost in results as described in section 4.2.2. In this case, it is not sufficient to look at the topologic relations between mutual faces and faces and objects, but also the geometric description of the objects would need to be taken into account.

References

Altman, D., 1994 Fuzzy set theoretic approaches for handling imprecision in spatial analysis., Int. j. geographical information systems.

De Hoop, Sylvia, van Oosterom, Peter, and Martien Molenaar, 1993, Topological Querying of Multiple Map Layers. Spatial Information Theory, A theoretical basis for GIS, European Conference, COSIT'93 Italy, ed. Andrwe U. Frank and Irene Campari, pp 139-157.

Edwards, G. 1994 Aggregation and disaggregation of fuzzy polygons for spatial-temporal modelling. in Advanced Geographic Data Modelling, Molenaar, M. and Hoop S. de eds., Netherlands Geodetic Commission, New Series, Nr. 40, Delft, pp 141-154

Egenhofer Max J. and Herring, John, 1992, Categorizing Binary Topological Relationships between regions, lines and points in geographic databses. Department of Surveying Engineering, University of Maine.

Kaufman, A., unknown, Introduction to the Theory of Fuzzy Subsets, vol 1 (Academic Press).
Klir, G.J., and Folger, T.A., 1988, Fuzzy Sets, Uncertainty, and Information (Englewoood Cliffs: Prentice Hall).

Molenaar M., 1989. Towards a geoinformation theory. ITC J., (I):pp 5-11.
Molenaar M., 1991. Formal data structures and Query spaces. Konzeption und Einsatz vo umwelt-informationssystemen. Günther, O., ea. eds, Springer-Verlag, Berlin 1991, pp 340 363.

Molenaar M., 1994 A syntax for the representation of fuzzy spatial objects. in Advanced Geographic Data Modelling, Molenaar, M. and Hoop S. de eds., Netherlands Geodetic Commission, New Series, Nr. 40, Delft, pp 141-154.

Contact address
Jeroen Dijkmeijer[1] and Sylvia de Hoop[2]
[1]Department of Surveying, Photogrammetry and Remote Sensing
Wageningen Agricultural University
P.O. Box 339,
6700 AH Wageningen
The Netherlands
[2] Q-Ray Agrimathica
P.O.Box 848
3900 AV Veenendaal
The Netherlands
Phone: +31-318-543 222
E-mail: sylvia@q-ray.nl

A RUBBER SHEETING METHOD WITH POLYGON MORPHING

Mi-Gyung Cho, Ki-Joune Li and Hwan-Gue Cho

Department of Computer Science
Pusan National University
Kumjeong-Ku, Pusan
South Korea

Abstract
Building consistent database is an important task in geographical information systems. One of the difficulties in building databases comes from mismatches between digital maps, due to the diversity of database providers. In this paper, we propose a rubber sheeting method to resolve mismatches between digital maps. Our method is based on morphing algorithm designed for simple polygons. It is shown that our morphing algorithm preserves topological properties of map during transformation as well as geometric accuracy.

Key words
Digital map, GIS, rubber sheeting, morphing, Delaunay triangulation.

1. Introduction

Database in geographic information system mainly determines the quality and accuracy of its services. And it is difficult and expensive task to build a consistent database for geographic information system. One of the difficulties comes from the fact that the data sources of a geographic information system are often different. In other words, database of a geographic information system consists of data sets from diverse organizations or information suppliers. A serious problem of mismatches between databases or maps arises due to the diversity of database sources. Suppose that we want to overlap two maps on the same region but supplied by different organizations. Then, we may find mismatches between two maps.

Despite the severity of mismatch problem, very few works have been done during the last decades [11]. R. Laurini [1] has defined and rigorously investigated the mismatch problem. He has also proposed an elastic transformation or rubber sheeting method to resolve mismatches between maps. But his works are limited to the case where the mismatches occur on the boundary of maps. And P. Langlois [2] has presented an elastic transformation method by using complex numbers. Dufour has improved this method by modifying weight function [3]. These methods may be used as general elastic methods, but their accuracy is unknown and there are constraints hard to be satisfied with them. For example, the area of polygon must be unchanged even after transformation may be a constraint which cannot be satisfied by these methods. Furthermore, the types of constants differ according to the application area.

The goal of our study is therefore to develop a *rubber sheeting* method which resolves mismatches between maps, by transforming a map to another. We restrict the scope of the problem to a specific case where the method to develop will be directly applicable to a certain application. A method for general purpose rather tends to fail to satisfy any specific application areas. We are particularly interested with the case where a map is composed with a set of objects which are points, lines or polygons and the distance between objects is an important constraint to consider during rubber sheeting. Our approach is based on two methods; *morphing algorithm* and *Delaunay triangulation*. Morphing, which has been used as a computer animation technique, transforming an image to another. We propose a polygon morphing derived from morphing technique for rubber sheeting. We especially use Delaunay triangulation for polygon morphing to improve the performance of our method. This paper is organized as follows. First, we generally describe the basic idea and related works in section 2. We explain our method about polygon morphing and constrained Delaunay triangulation in section 3 and 4. Finally we conclude in section 5.

2. Rubber sheeting

Suppose that we have two different maps S and T on the same region but drawn by different organizations and that map T is more accurate than S. Then we should transform map S to T, in other words, the objects on map S onto map T. The objects on map S and T are classified into three categories;

a. $O_1 = M(S) - M(T)$
b. $O_2 = M(S) \cap M(T)$
c. $O_3 = M(T) - M(S)$, where $M(S)$ and $M(T)$ are sets of objects on
        map S and T respectively.

In order to transform map S to T, we are interested only with O_1 and O_2. First, the objects in O_2 appear in the two maps but their position or shape are different. Since we have supposed that map T is more accurate than S, each object $p \in O_2$ on map S must be identified with an object on map T and its position be corrected onto map T. Second, for each object in O_1, which is found on map S but not on T, we must find its correct position and shape on map S. Our rubber sheeting method is therefore composed of two steps as follows and this process is described by figure 1.

  step 1 : for each object $b \in O_2$, identify it with an object on map T and
             adjust its position.
  step 2 : for each object $x \in O_1$, compute its position and shape on map T.

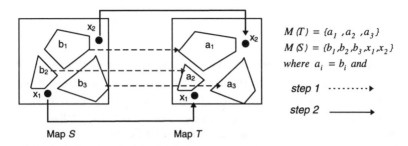

$$M\,(T) = \{a_1\,,a_2\,,a_3\}$$
$$M\,(S) = \{b_1,b_2,b_3,x_1,x_2\}$$
$$where\ a_i = b_i\ and$$

step 1 $\cdots\cdots\blacktriangleright$

step 2 $\longrightarrow$

Map S   Map T

Figure 1 Rubber sheeting.

When performing step 1, we apply polygon morphing, which will be described in section 3. The next step depends on the result of step 1, which will be taken into account to compute the position of $x \in O_1$ during executing step 2. The details of step 1 and 2 will be explained in the next sections

The main problem of rubber sheeting lies on how to transform one shape to another. This problem is closely related to morphing technique. Morphing algorithm consists of two parts; warping that two images have the same shape and then cross dissolving the warped images. While cross dissolving is simple, it is complex to determine how to warp a given image. Previous works on morphing have been concentrated on image warping method. And many related researches are still going on [4,8,9]. Our proposed method is based on field morphing [4], since all edges of polygons on map can be used as control line. In this paper, we propose a polygon morphing technique, which is derived from field morphing. Field morphing is a sort of morphing technique based on fields of influence surrounding two-dimensional control primitives. In figure 2, the position x' on destination image is derived from x on source image and two control primitives AB and $A'B'$ by *equation (1)*. The method to find the corresponding point x' of destination image from source image x is explained.

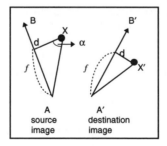

Figure 2 Method to find corresponding point X' on destination image for a given point X on source image.

(Equation 1)

$$d = AXcosa = AX \cdot Perpendicular\,(AB\,)\big/\|AB\|$$

$$f = AX \cdot AB\big/\|AB\|^2$$

$$X' = A' + f \times A'B' + d \times Perpendicular\,(A'B')/\|A'B'\|$$

where $Perpendicular\,(v)$ represents the perpendicular vector to v with the same length. The value f is the displacement along line AB, d is the perpendicular distance in point X from line AB. If f is in [0,1], then it is on line AB, otherwise it is less than 0 or greater than 1 outside of the range. By field morphing method each line has a field of influence around it, and will force points near to it to stay in the corresponding position relative to the line. The closer the points are to a line, the more closely they follow the motion of that line, regardless of the motion of other lines. Points near to a line move along with that line. In our application, points near to a line imply objects in O_1 described previous section, and the line are an element in O_2. We concentrate our efforts on finding the proper coordinates of the corresponding point on the destination map for a given point on the source map. We propose the polygon morphing technique, based on field morphing method. We will use *equation (1)* to find the corresponding point. And edges of polygons and triangles that are produced as the results of constrained Delaunay triangulation, which will be explained in section 4 is to be used as control lines.

The purpose of image morphing is to get a natural intermediate image. But the natural image may be interpreted in different way according to the type of application. Each application has its own requirements to be respected during image morphing. Although the requirements are various, we list common and important requirements of rubber sheeting. And the goal of our study is to find a rubber sheeting method which satisfies the requirements, which are as follows. First, the inside objects of polygon on the source map must be transformed into the inside objects of corresponding polygon on the destination map and vice versa. Even though we transform a convex polygon to a concave one, this requirement must be respected after transformation. The second requirement is to preserve the distance from a given point to polygon on the source map after transformation. The third requirement is that the order of the location of objects on the source map would like to be preserved on the destination map. This condition is very difficult to satisfy although the polygon to be transformed may be simple.

When a polygon on the source map and the corresponding one on the destination map, are similar, it is relatively easy to transform it. However if they are not similar, for example one polygon has a missing edge, it becomes very difficult to gain a good result. Therefore we suppose two conditions on the input maps to apply our algorithm. The first condition is that two maps must have the same number of polygons and polylines. And the second, corresponding objects must be similar with each other. To measure the similarity of two polygon A and B, we propose a similarity function $s\,(A,B)$. This function returns value between [0,1].

(Definition 1) A Similarity function

Let be $area\,(A)$ the area of polygons A. We define a similarity function $s\,(A,B)$ as follows:

$$s\,(A,B) = \frac{area\,(A \cap B\,)}{max\,\{\,area\,(A)\,,\,area\,(B\,)\,\}}$$

Though this function is different from other similarity functions in computer vision or pattern recognition, it describes the degree of similarity between two polygons in efficient way [5,6,7]. Result of rubber sheeting is sensitive to the similarity defined by this function. It has some properties of metric space as follows.

1. $s\,(A,B) \geq 0$
2. $s\,(A,B) = 1 \quad if\ A = B$
3. $s\,(A,B) + s\,(B,C) \geq s\,(A,C)\ for\ all\ A,B,\ C$

In this paper, it is supposed that the similarity should be between [0.7, 1]. If it is superior to 0.7 the corresponding polygons have to overlap a great part of area. This prevents corresponding polygons from serious deformation and from going against the requirements.

3. Polygon morphing

Before performing polygon morphing for digital map, it needs a preprocessing which corresponds to step 1 in section 2. It consists of four steps. The first step is to convert polylines on map such as rivers and streets into polygons, since polygon is more simple to handle than polyline. And next step is to find corresponding polygons between source and destination maps. The third step is for checking the similarity between each polygon on the source map and the corresponding polygon on the destination map. If two maps are quite differed, we exclude it from our process, since the quality of results could not be quarantined. At the final step, we find the number of edges that consist of each polygon and modify them to have the same number of edges between corresponding polygons. Because the edges of a polygon are used as control lines, all corresponding polygons must have the same number of edges. The first preprocessing step is quite simple to perform. We need simply to add two imaginary edges starting from the beginning point of a polyline to the beginning point of the other polyline and starting from the end point to end point so that a pair of polylines become a closed polygon. We apply the Hausdorff distance function $h\,(A_i,B_j\,)$ for the second preprocessing. We compute the Hausdorff distance between two point sets that consist of polygon A and B on map and identify the corresponding polygons with the minimum Hausdorff distance, which is explained by following equations [6,7].

$$h\ (\ A_i\ ,B_j\ ) = max\ \{\ d\ (\ A_i\ ,B_j\ ),d\ (\ B_j\ ,A_i\ )\}$$
where

$$d\ (\ A_i\ ,B_j\ ) = max\ \{\ d\ (\ x\ ,B_j\ ):x\in p\ (\ A_i\ )\}$$
$$d\ (\ x\ ,B_j\ )\ = min\ \{\ d\ (\ x\ ,y\ ):y\in p\ (\ B_j\ )\}$$
$$p\ (\ A_i\ )\qquad = \{\ p_{ij}:p_{ij}\in A_i\ ,A_i\in M\ (\ S\ )\}$$
$$p\ (\ B_j\ )\qquad = \{\ q_{js}:q_{js}\in B_j\ ,B_j\in M\ (\ T\ )\}$$
$$1\le i\ ,j\le n$$

where the notation n means the number of polygon on map and $M(S)$, $M(T)$ the source map and the destination map respectively. A_i is the i-th polygon on the source map and B_j is the j-th polygon on the destination map. The notation p_{ir} means the r-th point of the i-th polygon on the source map and q_{js} is the s-th point of the j-th polygon on the destination map. As consequence, we identify polygon A_i on the source map with polygon B_j on the destination map with the minimum Hausdorff distance. And then similarity value is computed for each pair of corresponding polygons.

At the final preprocessing, we try to adjust the number of edges of each polygon on the source map such that the number is same as that of corresponding polygon on the destination map. A pair of corresponding polygons may have different number of edges due to mismatches. In our algorithm, the number of edges of corresponding polygons must be same with each other since all edges of each polygon are to be used as control lines. Figure 3 explains this process. When we adjust the number of edges, new points may be inserted if it is necessary, and consequently an edge may be correspond to a polyline. In figure 3, two polygons have five edges after adjustment. Edges e_2, e_3, e_4, e_5 correspond to polylines e_2', e_3', e_4', e_5', respectively.

$$ratio\ (\ e_i\ ) = \|e_i\|\ /\ p\ (\ A\ ),\quad ratio\ (\ e_i'\ ) = \|e_i'\|\ /\ p\ (\ B\ )$$
$$ratio\ (\ e_i\ ) = ratio\ (\ e_i'\ )\quad for\ all\ e_i$$

where $p\,(A)\,,p\,(B)$ are respectively total perimeters of two polygons A, B

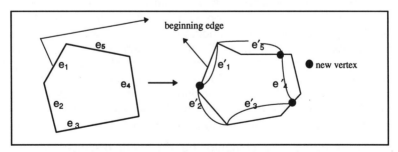

Figure 3 Adjusting the number of edges of corresponding polygons.

After the preprocessing previously explained, we find the coordinates of corresponding points by using each edge of polygon as control primitive. It is however difficult to apply *equation (1)* to find the corresponding point on the destination map, when an edge on the source map corresponds to a polyline on the destination map. *Equation (1)* in section 2.2 exclude the case where the control line is a polyline. We therefore modify *equation (1)* to consider such a case. In figure 4, it is shown to find coordinates of corresponding point by *equation (2)* when an edge corresponds to a polyline. If a point to be mapped on the source map is inside of a polygon, we can compute its corresponding position on the destination map by algorithm 1.

Figure 4 Method to find corresponding point when an edge corresponds to a polyline.

Algorithm 1 To find the corresponding point of inside point.

Input $M(S)$: source map
            $M(T)$: destination map
            x : a given point
            A_i : polygon contained x
Output   corresponding point x'

1. Find the nearest edge e_{ij} of polygon A_i from x.
2. Find edge $e'_{ij} \in A'_i$, which corresponds to $e_{ij} \in A_i$.
3. If e'_{ij} is a line segment, compute the coordinates of x' using *equation (1)*.
   else e'_{ij} is a polyline, compute the coordinates of x' using *equation (2)*.
4. If x' is not inside point of A'_i, adjust the displacement d in *equation (1)* or *equation (2)*.

(Equation 2)

if $f \leq f_1$, $X' = A' + f \times A'B' + d \times Perpendicular\ (A'B')/\|A'B'\|$

else $f_2 = f - f_1$

$X' = A' + f_2 \times B'C' + d \times Perpendicular\ (\ B'C'\ )/\|B'C'\|$

Figure 5 shows an example of rubber sheeting using algorithm 1. Two digital maps M(S) and M(T) have the same number of polygons and polylines but are deformed with each other. The gray points are the original ones on the source map, M(S). Black points on M(T) represent the corresponding points by rubber sheeting algorithm.

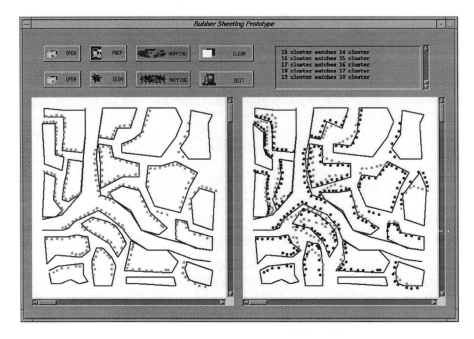

Figure 5 An example of a snapshot of program result.

The figure shows that rubber sheeting satisfies three requirements mentioned in section 2. All inside points of any polygons are transformed into the inside of corresponding polygon and vice versa. And the second and third requirements are satisfied as well except a special case where the third requirement may be broken. We will explain this special case in the final section. And when a point is outside of polygon, a sophisticated problem arises, since polygon morphing is limited to find the corresponding points inside of polygon. The reason will be explained in next section in detail. We apply constrained Delaunay triangulation to resolve the problem, which will be also explained in next section.

4. Morphing using constrained Delaunay triangulation

As mentioned in section 3, polygon morphing method allows to find the coordinates of points inside of polygon, and we cannot apply this method for points outside of polygon. The reason is that this method may give undesired result. For example, some outside points on source map are corresponding onto the inside of another polygon on destination map as shown by figure 6. This is because the corresponding points always have d displacement from control line. Even though it seems that we could solve this problem by modifying displacement d or use multiple control lines, it is limited. Instead, we propose another method for the points outside of polygon using constrained Delaunay triangulation.

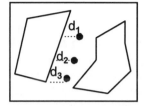 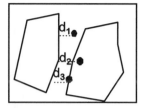

Figure 6 An example of undesired result.

We construct a constrained Delaunay triangulation for all points and line segments of polygons and polylines on map and its dual graph. The constrained Delaunay triangulation is that the circumcircle of each triangle contains no site visible from all three vertices of the triangle in its interior and the vertices of each triangle are mutually visible.

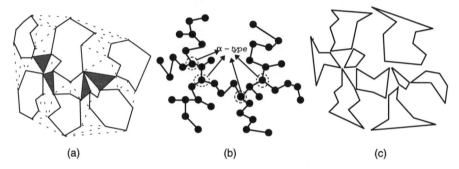

(a)   (b)   (c)

*Figure 8 (a) Constrained Delaunay triangulation: shaded triangle denote a-type tiangle
(b) Its dual graph (c) Inverse polygon.*

Vertices p_i and p_j are connected by a constrained Delaunay edge if and only if p_i is visible from p_j, and there exists a circle through p_i and p_j which has no other vertex on its boundary or in its interior visible to p_i and p_j. Because two similar maps may have the different number of points, the result of triangulation may be different. In figure 7, the result of triangulation of a map is shown, where the bold segments are polygons that exist on the map. A triangle can be shrunk it down to a vertex and connects the edge between the adjacent vertices. And the result of triangulation can be represented by a dual graph. Then we can change the problem of triangle as graph problem. There are two types of the triangles. One is a triangle whose vertices belong to different   polygons, represented as shaded area, the other is whose two vertices at least belong to the same polygon. The former is called as α-type, the latter is called as β-type.

We have supposed that two maps have the same number of polygons. All pairs of corresponding polygons may have different number of edges due to mismatches. And the number of triangles on two maps may be different. There exists an α-type triangle per three visible polygons. If two similar maps have the same number of polygons and have the same visible relation between polygons, they have the same number of α-type triangles. It can be consequently concluded that the numbers of $\alpha-type$ triangles on both maps are same even though the numbers of triangles may be different. If we eliminate all $\alpha-type$ vertices in dual graph, a dual graph is split into a number of connected subgraphs, called components. We expect that the number of components on two maps is same. If we eliminate the common edges of all triangles belonging to a component, the component will become a polygon, which we call as inverse polygon. The number of inverse polygon is same to the number of components. As consequence, the problem of outside points is reduced to a problem of inside points of inverse polygon. And we have proposed polygon morphing method for inside points in section 3. In other words, to find the corresponding point of an outside point, we first must find the inverse polygon containing a point x. And then we applies polygon morphing to the inverse polygon. In figure 9, each component of a dual graph of source map corresponds with component of the dual graph of destination map. The procedure is summarized as followings.

Algorithm 2 To find the corresponding point of outside point on destination map.

Input $M(S)$: source map
               $M(T)$: destination map
               x : outside point on source map
Output   corresponding point x' on destination map

1. Construct constrained Delaunay triangulation and its dual graph.
2. Find $\alpha-type$ vertices in dual graph.
3. Eliminate all $\alpha-type$ vertices of two dual graphs and divide into disconnected components.
4. Identify each component of the dual graph of source map with that of the dual graph of the destination map.
5. Produce inverse polygons by removing common edges for each component.
6. Find an inverse polygon or $\alpha-type$ triangle containing an input point x.
7. Apply polygon morphing with inverse polygon or $\alpha-type$ triangle containing x.

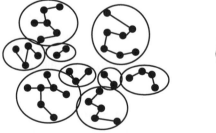 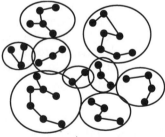

Figure 9 Nine pairs of corresponding components after eliminating all $\alpha-type$ vertices.

5. Conclusions

In this paper, we presented a new rubber sheeting algorithm for transforming the objects between two similar maps. Our algorithm is based on morphing technique and employed the original polygons and the inverse polygons as control primitives. We gave two constraints on the input digital map. A pair of maps must contain the same number of polygons and polylines, and they must be similar in a certain degree.

Our method satisfies the following three requirements. The inside points of any polygon are always corresponding into inside points of corresponding polygon on destination map. And the outside points of the source map also correspond to outside points on the destination map. It implies that our method preserves to topological properties during transformation. Second, the distance from given points to polygon of source map is preserved on destination map. And finally, the order of the position of objects on the source map is also preserved except some degenerated case. If a pair of corresponding polygon have the same number of edges, this requirement is satisfied. But if they are the different number of edges, this requirement may be violated as shown by figure 10 due to adjust edges of polygons. More work is necessary to satisfy the third requirement though the number of edges of corresponding polygons are different.

Figure 10 Counter example which does not preserve the order of the positions of objects.

References
S. Servigne and R. Launini,   "Updating Geographic Databases Using Multi-Source Information," Proc. of the third ACM International Workshop on Advances in GIS, pp. 109-119, 1995.
P. Langlois, "Une transformation élastique du plan basé sur un modéle d'interaction spatiale," Application à la géomatique," Technical Paper, MTG, University of Rouen.
P. Dufour, "Les bases de données géographiques fédèrées: continités géomètrique at topologiques," Mémoire de DEA, INSA de Lyon, June, 1995.
T. Beier, "Feature-Based Image Metamorphosis," ACM SIGGRAPH, Vol. 26, No. 2, pp. 35-42, 1994
E. Arkin, et. al, "An Efficiently Computable Metric for Comparing Polygon Shapes," IEEE trans. Pattern and Machine Intelligence, Vol. 13, No. 3, pp. 129-137, 1991.
D. Huttenlocher, et. al, "Comparing Images Using the Hausdorff Distance," IEEE trans. Pattern and Machine Intelligence, Vol. 15, No. 9, pp. 850-863, 1993.
H. Alt, et. al, "Measuring the Resemblance of Polygonal Curves," 8th annual ACM Computational Geometry, pp. 102-109, 1992.
S. Anderson, Morphing Magic, Sam Publishing, 1993.
S. Lee, et. al, "Image Metamorphosis Using Snakes and Free-Form Deformations," Proc. Of Computer Graphics, pp. 439-448, 1995.

Preparata, et. al, Computational Geometry : An Introduction, pp. 237-242, Springer-Verlag, 1985.
R. Laurini and D, Thompson, Fundamentals of Spatial Information System, Acamic Press, 1992.

Contact address
Mi-Guang Cho, Ki-Joune Li and Hwan-Gue Cho
Department of Computer Science
Pusan National University
Kumjeong-Ku
Pusan, 609-735
South Korea
E-mail: {mgcho@hyowon.cc,
E-mail: lik@spatios.cs
E-mail: hgcho@hyowon.cc}.pusan.ac.kr

A VORONOÏ-BASED PIVOT REPRESENTATION OF SPATIAL

CONCEPTS AND ITS APPLICATION TO ROUTE

DESCRIPTIONS EXPRESSED IN NATURAL LANGUAGE

G. Edwards [1,2], G. Ligozat [3], A. Gryl [1,2,3], L. Fraczak [3],
B. Moulin [2,4] and C.M. Gold [1,2]

[1] Chaire industrielle en géomatique appliquée à la foresterie
[2] Centre de recherche en géomatique, Pavillon Casault
[4] Département d'informatique, Pavillon Pouliot
Université Laval, Sainte-Foy
Québec, Canada
[3] Laboratoire d'informatique pour la mécanique et les sciences de l'ingénieur
ORSAY Cedex, France

Abstract
Different representations of space are not in general equivalent. This point is clearly illustrated in research on the generation of sketches from route descriptions given in natural language: many linguistic expressions determine only partially a spatial situation. This article explores the role played by a pivot representation based on the Voronoï diagram. We study the use of this model in the context of the translation of verbal descriptions into sketches. We show how, by combining a linguistic analysis of a route description and the Voronoï model of space, one can construct a pivot representation that integrates both the spatial and linguistic aspects of a situation. We aso examine the use of the pivot representation within the framework of the automated generation of linguistic descriptions from geographical databases. The development of a pivot representation depends on the existence of a dictionary that renders explicit the equivalence between linguistic and spatial aspects of each situation. With the help of the Voronoï model, the basis for such a dictionary has already been established for spatial prepositions in English. Extensions to other linguistic elements are under study and appear to be achievable.

Key words
Route descriptions, spatial representations, Voronoï data model, natural language representations, linguistic analysis.

1. Introduction

The problem of developing computer programmes for constructing or enriching route descriptions, such as for pedestrians in an urban environment, is complex. Furthermore, although work has been done with robot navigation in artificial environments, and with road navigation, again a simpler environment, little attempt has been made to tackle the task of providing information to pedestrians in complex, realistic urban settings.

In order for such instructions to be useful, they must be provided in a form which is familiar to the client. This requires an understanding of the cognitive procedures which people use to navigate in complex environments. Many tourist information systems today provide a bird's eye map view containing the point of origin and the destination point and indicate the route on the map. Most humans, however, when asked to provide the means to guide someone from one point to another within a city, spontaneously provide verbal instructions. Sometimes a sketch may be used to help clarify the instructions. Furthermore, when instructions are given over a phone, there is little choice but to employ verbal means. Hence, in a fully functional and useful city guide system, the computer should be able to offer linguistic route descriptions and partial sketches as well as full map views, depending on the nature of the information request and the desires of the client.

The construction of linguistic descriptions, or their enrichment based on information about the environment, is itself a complex task. A great deal of knowledge about cognitive processes is required (how do people choose which elements of the environment to include in their descriptions? what kinds of strategies for path selection are employed, when many possible paths are possible? and so on), as well as knowledge about linguistic choice (what language forms do people typically use to express spatial information in route descriptions?) and knowledge about the role of sketches as an aid to such linguistic descriptions (what elements do people choose to represent on a sketch being used in combination with a linguistic description?). The task also requires an understanding of the problems involved when manipulating spatial data on a computer, especially when an existing database is to be used to help construct the linguistic description or sketch. Finally, means must be provided to carry out qualitative "spatial reasoning" and hence to permit information to be deduced within such complex environments.

This paper surveys the state of an existing multi-disciplinary project designed to address some of these issues. The work to date has been focussed on four related areas: (1) a cognitive based study of two corpus of textual route descriptions designed to identify the key strategies and choices used by humans when giving route directions; (2) the development of procedures necessary for the tranformation of a linguistic route description to a graphical visualisation; (3) the development of methods of performing qualitative spatial and spatio-temporal reasoning about the world; and (4) the study of the use of the Voronoï diagram as a bridging structure between geographic information and linguistic representations. We shall provide a brief overview of this work, then discuss the importance of a pivot representation, and finally show how the latter can be used in the context of our project.

2. The production of a route description

The project is based on the analysis of two corpus of texts collected by A. Gryl (1995). The first corpus deals with descriptions of journeys in a semi-urban context, that of a university campus. The corpus consists route descriptions acquired from 60 subjects asked to describe three journeys which together form a closed loop (from the railway station to Building A, from Building A to the library, and from the library back to the station). The second corpus collected consists of route descriptions provided by another 30 subjects, this time for a fully urban context, that of moving between two well known departement stores, one on the left bank and one on the right bank of the Seine in central Paris. In each setting, route descriptions in both directions were obtained, in order to test for differences resulting from the order in which landmarks were traversed.

From the two corpus of route descriptions, Gryl (1995) carried out an analysis of their linguistic content and the strategies employed by the subjects in choosing routes and landmarks within the descriptions. The analysis permitted the broad outlines of a strategy for the automated generation of route descriptons to be determined, based on the most frequently used elements within the corpus of route descriptions. Hence, for example, it was noticed that landmarks were not preferentially tied to a particular direction, subjects tend to suggest the use of main traffic arteries, and they tend to limit the number of turns or decisions required by the listener to follow the trajectory. These and other results of the analysis led to a proposal for a computer implementation of the procedure in three steps:

The first, called "Where to go?", consists of the construction of a representation of the route description according to a trip strategy. This step is divided into two phases, one serving to provide a global orientation, and the other for the creation of a path between the point of departure and the point of arrival, according to a variety of criteria.

The second, "By what means?", consists of refining the representation produced by the first step according to a more precise strategy. During this step, the general descriptions are enriched via the introduction of landmarks that are usable by human beings. Moreover, the planned path is segmented into two types of elements: local descriptions and continuation paths. Local descriptions are attached to steps of the itinerary where a decision is to occur (change in orientation, continuation of the path in the presence of ambiguity, a decision to take with respect to some aspect of the environment); whereas continuation paths connect these local descriptions.

The third step, called "How to say it?", consists of translating the preceding representation into natural language. This translation uses additional information from the database, including elements such that the shape of landmarks, their names, their sizes, their colors and their intrinsic orientations.

Each of these tasks is itself an entire area of research. To carry out this program correctly implies: linking the spatial relations expressed by prepositions and other linguistic structures with corresponding cartographic configurations (Freksa 1992, Ligozat 1993, Frank 1992, Gapp 1994, Oliver and Tsujii 1994; Edwards and Moulin 1995); analyzing linguistic expressions in terms of static descriptions, movement, etc. (Herskovits 1986, Talmy 1983); identifying linguistic markers that signal special conditions related to spatial concepts (Fraczak 1994); managing the temporal aspects of language (Allen 1984, Ligozat 1991); adjusting the processing requirements according to each natural language (Dorr and Voss 1993); and improving our understanding of the cognitive abstraction process that allows one to choose certain marks in a scene in order to obtain an appropriate verbal description (Gryl 1995, Kuipers 1978). Hence a full realisation of the task is unlikely over the short term. On the other hand, this paper presents a key element in the realisation of this programme. It is our belief that with the problem of linking the spatial relations expressed in linguistic structures on the one hand and cartographic configurations on the other solved, other parts of the task become easier and the whole program may be realisable.

3. From description to sketch

As has been indicated, a verbal route description is frequently associated with a sketch of the situation in order to facilitate comprehension and help with the memorisation. In order to study the processes used in each of the two modes of expression, we have undertaken the task of developing a system allowing the automatic translation of information in the linguistic mode to a graphical format (Fraczak 1995). This work is based on the first corpus of texts collected by Gryl (1995), that dealing with pedestrian routes on a university campus.

The effort to produce a visualisation of the linguistic description led to the observation that the two modes of expression are not equivalent. Indeed, while certain kinds of information are not represented explicitly in the linguistic description, the graphical mode requires their representation. Conversely, some verbal information cannot be expressed graphically. The imbalance consists therefore of either a lack of completeness in the linguistic representation as compared to the graphical representation, or the lack of certain elements in the graphical representation as compared to the linguistic representation. Here are examples of each of these situations:

- *incomplete linguistic information (not verbalised):*

You pass in front of the cafeteria.   (to the right or to the left?)
At the church, take the pedestrian path.   (where is the path compared to
                                             the location of the church?)

- *incomplete graphic information (not figured):*

You can't miss it.
It's simple.

We have developed a translation system based on three steps: (1) a linguistic analysis; (2) the elaboration of intermediate representations (semantic and conceptual); and (3) the generation of the sketch.

First of all, the route description is divided into *sequences* that are linked via *connectors*. This is necessary to encapsulate the information to be used for the graphical visualisation. For example, in the description fragment:

it is necessary to enter the station and then to take the pedestrian walkway

there are two sequences, *it is necessary to enter the station* and *to take the pedestrian walkway*, linked by the connector *and then*. Sequences are categorized into two types: sequences which include a prescription of action and sequences which are used to indicate landmarks. Furthermore, each sequence is analyzed via the use of a grammar of route descriptions, resulting in a semantic representation. Thus, in the example shown above, one will have a prescription of action (*to enter*) followed by a second prescription of action (*to take a path*).

Following this, the semantic representation is reformulated via a route prototype of the frame type, resulting in what is called a conceptual representation. This in turn becomes the basis for the generation of the sketch - a graphical symbolic language is used to represent the various types of landmarks, progression, and changes of direction. The graphical representation constitutes nothing more, at this point, but a formalized code for visualizing the conceptual representation.

4. The necessity of pivot representations

The procedures outlined above invite a certain number of questions with regards to possible generalization within a wider context:

1) Given a fragment of a route description and its conceptual representation, can one define the compatibility of these two descriptions, or determine if one of them satisfies the constraints set by the other?
2) Given a geographical database concerning the site where the displacement is to occur, and a linguistic description, a conceptual description, or a description in the form of a sketch, can one again define the compatibility of the description with the database? If so, can one then use the database to enrich or correct the description? Or to specify certain aspects of the description?
3) Is it possible to construct part of the geographical database or modify it in a substantial way, using linguistic descriptions, sketches, or mixed forms as a data source?.

On the other hand, the systems that we have just described are characterised by two important limitations:

Certain choices are arbitrary. In the case of converting the textual description to a sketch, when the verbal description is ambiguous (is the tennis court to the left or to the right?), an arbitrary choice is required for the visualisation. Although one could argue (Riesbeck 1980) that such choices do not affect the user (i.e. when following the route, the client will perceive immediately the error and correct it), the presence of such arbitrary choices is not totally satisfactory. One would like to have a representation in which one is not forced to make such a choice or where the ambiguity of the choice is clearly marked.

The absence of true spatial semantics. The elements used for the representation are constrained only by the characteristics belonging to the linguistic level on the one hand, and to the graphical level of the other. The problem of the comparison of these elements with data representing the external world is not addressed. This is even more critical for the generation of route descriptions, where (a small but appropriate subset of) the spatial relations between elements in the database need to be extracted directly and converted into linguistic form.

Based on these observations, we suggest that a *pivot representation* is required. A pivot representation would serve to anchor the individual representations which are specific to each mode: graphical, linguistic and database. The pivot representation would consist of a structure which supports all three modes (and might be extended to support other modes as well). Most current studies seek to establish links between two of the three types of representation, but none presently permit a representation covering all three. Thus, for example, most spatial databases permit good management of spatial relations only in the presence of intersections between objects. The extension of these data structures to the representation of proximity relationships must exploit additional methods (fuzzy set theory, for example), while the shapes of objects are characterized by still another mechanism. Hence a link is possible between proximity relationships expressed in natural language and objects in the database, but a visual representation of these relationships, in the form of a sketch for example, must rely on other techniques.

5. The Voronoï model

Recent work suggests that the Voronoï model of space may provide the solution to the problem of a common representation in support of all three modes. In order to understand the reasons why, we shall provide a brief description of this model.

The Voronoï model of space (Okabe, Boots and Sugihara 1991) is based on the processing of the whole space. Each elementary map object (a point or a line segment for instance) is embedded in a *tile* (also called a Voronoï region) which is the region of space closest to the given object than any other object in the space. Voronoï regions can be determined for any arbitrary shape. They are commonly generated around points and line segments and recent work has extended them to curves and faces. The set of Voronoï regions for a set of objects in space is called the Voronoï diagram for the space and objects. It is also called a Dirichlet tesselation and the Voronoï regions are aternatively called Thiessen polygons. If the existence of a Voronoï boundary between two given objects can be established, then these objects are said to be neighbours, and we can say that they are adjacent. If the adjacency relations are represented by a line segment connecting the objects, then the set of all such line segments will form a Delaunay triangulation (for a set of points), called the dual of the Voronoï diagram. We call a *Voronoï model of space* the set of objects (half-lines and points) in two dimensional space (for the time being), their associated Voronoï regions and the dual which contains the information on adjacencies. Objects are specified as collections of half-lines and points. We use the term *half-line* to denote a side of a line or, alternatively, a line and an orientation. All lines are composed of pairs of half-lines, although it is conceivable that a single half-line might be modelled in some circumstances.

The Voronoï model differs from the more commonly used raster and vector models in many fundamental ways. Both raster and vector models are coordinate-based models or fully metric models of space (although they use different metrics). The Voronoï model is not a fully metric model of space in the sense that coordinates are not needed to determine adjacency relationships (although they are used in the computer implementation of the model). The Voronoï model of space allows definitions of neighbours which are purely based on topological information in a direct and unambiguous way. The Voronoï model of space is also fundamentally dynamic since no object may move without several other objects knowing it has done so. Object motion can be defined in terms of topological changes (changes of adjacency relations between objects) and does not need to be expressed in terms of coordinate change. The Voronoï model is hierarchical in the sense that space is nested into increasing levels of detail, each level of which is embedded in the previous level. Object and space are intimately connected. The nested and tiled nature of Voronoï space allows us to effectively remove any piece of a map and replace it with a simplified (i.e. generalized) version of the object at a lower level of detail. This is analogous to the nominalization process in natural language.

A Voronoï server has been developed by Gold (Gold 1994). It consists of a library of functions that can be used to create and to manipulate Voronoï diagrams composed of line segments and points and to make the link between these diagrams and a database. Some of the functions accessible to the user include operators for the construction of the Voronoï diagram of an arbitrary collection of line segments and points (SetFrame, AddPoint, MovePoint, AddLine, JoinPoints, DeleteObject, etc.). Other functions allow one to construct queries (NearestObject, Neighbours, Trace, Clip, BufferZone, PolygonShade, etc). Furthermore, the data structure used is fundamentally dynamic, thus allowing the modeling of events over time as well as in space. Different application projects are under development, most of which are related to cartographic applications for GIS.

Edwards and Moulin (1995) demonstrated how the Voronoï model could be used as a computer simulation of a mental model in order to represent the different spatial relationships such as one finds in prepositions in English. This article pointed out the existence of a unique correspondence between each spatial preposition and a set of configurations of the topology of adjacency links within the Voronoï model. For example, Figure 1 shows the topology configuration which corresponds to the concept "between". In fact, because the Voronoï model gives meaning to the different proximity relationships without requiring contact between objects (only contact between their Voronoï zones is required), the Voronoï model allows one to represent directly the totality of spatial relationships expressed in natural language. Furthermore, the notion of object contact, in the Voronoï model, has a preferential status compared to other relationships, because it corresponds to a particular kind of adjacency relationship between objects. Thirdly, overlapping relationships (between segments, between surfaces) are also part of the Voronoï model. Finally, it is possible to determine a gradation in the topological relationships of proximity represented by the Voronoï model, hence allowing a simple means of representing ambiguities and fuzzy elements in the categorization of different relationships. Thus the Voronoï model allows the representation of the full range of spatial relationships expressed in natural language.

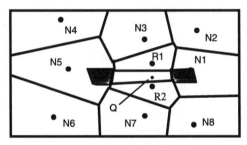

Figure 1 Situation representing the preposition "between" (Q is "between" R1 and R2 because its Voronoï zone is adjacent to the Voronoï zones of R1 and R2). Q is "Directly between" R1 and R2 if it also has both N1 and N2 as neighbours.

We have already noted that shape descriptions must exploit representational mechanisms other than those which exist within database structures. Without entering into the details (and exceeding the reach of this article), it is pertinent to note that a mechanism exists within the Voronoï model which allows one to manipulate a concept of shape. This mechanism is the geometric skeleton of an object (also known as the medial axis transform). The geometric skeleton consists of the internal boundaries of Voronoï zones of an object (it is necessary also to add the concept of radius or minimal distance between the boundary and its generator objects in order to reconstitute the skeleton according to its formal definition (Ogniewicz 1993)). Thus, the skeleton is simply another way of speaking about adjacency relationships between Voronoï regions. Using this concept, it is possible to categorize shapes, to simplify them (Ogniewicz 1993), to reconstruct them from a incomplete information (Blum and Nagel 1978) and many others operations. Furthermore, although the link between these visual properties and corresponding linguistic expressions has not been explored much, preliminary work confirms the existence of such links. Furthermore, many properties of the skeleton (categorization, generalization, reconstitution) are precisely those that one finds in the linguistic expression of shape (Landau and Jackendoff 1993).

Thus, the Voronoï model provides a spatial database structure that manages both relationships of contact (intersection, overlap, tangential contact, etc.) and relationships of proximity (close, far, between, beside, etc.). Furthermore, the model allows a direct characterization of the shapes of objects through their geometrical skeletons. The Voronoï model therefore contains a representation of objects in space which allows both an analysis of their visual and graphic elements, their linguistic representation and their manipulation within a spatial database structure. Hence this model can serve as an intermediate representation for the development of links between these three areas.

The use of the Voronoï model as a pivot representation addresses the second problem outlined in the previous section - that of providing a true spatial semantics which can be anchored in the real world. In so doing, a (partial) solution is also applied to the first problem - that of handling ambiguities in the linguistic representations. Indeed, the pivot level allows us to avoid, in many cases, making arbitrary choices in the visualisation. For example, in the case of the tennis court, the use of the Voronoï model at the pivot level allows us to represent the region associated with the expression *pass the tennis court* (i.e. the region through which one passes) rather than the tennis court itself (which can then be found to the right or to the left by the pedestrian when following the route in reality).

6. Construction of pivot representations based on the Voronoï diagram

In the case of the problem of sketch generation, we will assume that a linguistic analysis has led to a preliminary identification of the spatial context. Studies by Fraczak and Ligozat described in the third section of this article lead currently to a such result (Fraczak 1995). Thus, as we have indicated earlier, route descriptions can be decomposed into a series of sequences and connectors between these sequences. For example, consider a part of the description quoted by Fraczak (1995):

It is necessary to leave the station[1], to take the pedestrial walkway[2], to descend the walkway[3], and one arrives then at the entrance to the university[4].

This text is composed of a series of four sequences, separated by implicit or explicit connectors (in the text, sequences are indexed by a number). Furthermore, each sequence which is composed of a prescription of action uses a certain number of landmarks. For example, the first sequence corresponds to a trip between the interior and the exterior of the station (a particular type of landmark). The second sequence refers to a trip between the exterior of the station and the interior of a pedestrian walkway. Thus each sequence implies a limited spatial context.

The linguistic analysis of this description leads to the characterisation of the first sequence, as a prescription of action ("to exit") corresponding to the verb class "to progress" with respect to an object named "station". Additional processing of the linguistic aspects of the text then leads to the identification of other landmarks and qualifiers, resulting in the construction of a more complete conceptual representation. In the approach outlined earlier, the sketch itself is simply the graph formalisation of the information that is found in the conceptual representation. We propose here to enrich this formalisation by combining the linguistic and spatial aspects via a pivot representation. Two process are possible: the first presupposes the availability of a spatial database; the other, more limited, does not.

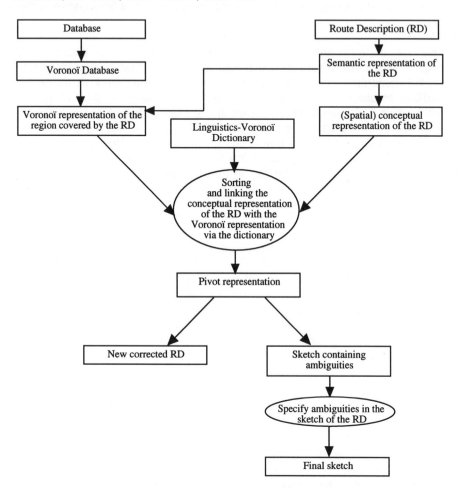

Figure 2 Flow diagram showing the processing of route descriptions and the construction of a pivot representation when a database is available.

We present first the approach which relies on the existence of a spatial database. Figure 2 presents the proposed solution. The linguistic processing schema of the route description (RD) is represented by the box at the top right. At this stage, the spatial relations and other spatial information are extracted from the linguistic description. At the top left, the database (or a part of it) is transferred into the Voronoï data structure. This step corresponds to the explicit determination of the spatial relations and other spatial information present in the database. Following this, with the help of the semantic representation, and especially its segmentation into distinct sequences, the Voronoï region directly pertinent to a particular linguistic sequence is selected.

Thus, to give a concrete example, in the case of the sequence concerned with exiting the station, Figure 3a would correspond to the original database, and Figure 3b to the Voronoï version of the database.

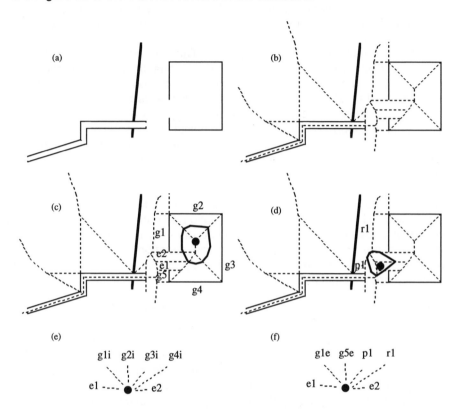

Figure 3 The process of extracting spatial relations from a geographic database. (a) the original database containing the station, section of railway and pedestrian walkway; (b) the database after conversion to its Voronoï form; (c) the situation corresponding to the location of a marker "inside" the station (the letters identify sequentially pertinent geometric objects, g for gare or station, e for entryway, p for pedestrian walkway; r for railway); (d) the situation corresponding to the location of a marker "outside" the station (and ready to move into the walkway); (e) the topology of adjacencies corresponding to the marker "inside" the station (the extension i indicates that these are interior half-lines or walls); (f) the topology corresponding to the marker "outside" the station (the extension e indicates that these are exterior half-lines or walls).

The next step consists of extracting the elements of the route description which correspond to the elements to be found in the Voronoï database and of linking these to the conceptual representation built from the relevant linguistic sequence of the route description. Figure 3a represents the relevant object (the station) and Figure 3b its Voronoï zones. Figure 3c shows the situation which corresponds to the inside of the station (state of departure for the action "exit") and Figure 3d presents the situation for the exterior of the station (state of arrival for the action "exit"). In both cases, a marker element has to be introduced in the scene so as to represent the observer in the situation. Figure 3e indicates a possible path between the two states, insuring that the indicated route is possible. It should be noted that what is important in these figures is not the exact position of the marker with respect to the Voronoï zones, but rather the topology of adjacency relations between the marker and the relevant objects (e.g. the inside walls of the station). Thus, given the situation shown in Figure 3b, it is possible to search for and find the situations shown in Figures 3c and 3d, provided the linguistic sequence has been transformed into a Voronoï topology. This can be done by means of a dictionary, giving, for each linguistic element, one or several characteristic Voronoï topologies. The basic concept of such a dictionary, at least for spatial prepositions, has already been established (Edwards and Moulin 1995). All spatial prepositions appear to be supported by this approach. Hence the pivot representation consists of the topological representation of the set of (Voronoï) adjacencies present in the scene and necessary to support or generate the linguistic descriptions, combined with the dictionary which allows linguistic structures to be built out of such topologies.

Once this step is crossed, it becomes possible to determine any errors which might be found in the existing route description, to enrich the route description with other relevant information, to compare different route descriptions, or to produce either a complete drawing of the situation and/or a sketch that presents only the information found directly to the RD (e.g. by explicitly rendering any ambiguities).

The second approach corresponds to the situation where no database is available. In this case, one may have several descriptions of the same region, but no direct way to verify their veracity. The process is similar, however, although more limited in its conclusions. The different route descriptions must be analyzed according to the linguistic steps described above. Then, in combination with the dictionary, it will be necessary to construct two pivot representations giving the spatial topology, each deduced from its own RD. Following this, it will be necessary to combine these two representations into a single representation, provided common landmarks have been identified. Here, the conceptual representation of each RD will be underdetermined to some extent, and the construction of the pivot representation is more problematical than in the first approach to the problem. Nevertheless, once a pivot representation has been obtained, the remaining steps are very similar to those described when a database is present.

Finally, the Voronoï model is also usable at each step of the process described earlier for the production of route descriptions. On the one hand, if the original database, from which the global orientation and the initial path are determined, is a Voronoï database, adjacency relations between the different elements of the region are already explicit, facilitating the initial choice. Secondly, the choice of the content of each local description can also be determined by the spatial relations present in the Voronoï representation. The result of the first two steps will be a complete path expressed via a topology (or network), segmented into a series of continuation paths and interrupted by local descriptions. With the help of the dictionary described above, the translation of the complete path into a linguistic description will be possible via an inverse, but similar, process, to that used for the analysis of the route descriptions. Thus, a pivot representation of the RD can be built and exploited for the production of the final RD. Furthermore, the cognitive segmentation of the path will be subdivided again so as to create a text having a structure that is close to that proposed by Fraczak (1995), that is, a collection of sequences and connectors, the former composed of prescriptions of action (corresponding to continuation paths) and the latter of landmark descriptions (corresponding to local descriptions). The final production of the text will therefore be based on a conceptual description of local discriptions and continuation paths which will be converted to text using AI and linguistic principles (Gryl 1995).

The approach described in the preceding pages has not yet been implemented, although many of the pieces exist in prototypical form. In particular, Gryl (1995) has developed a formal set of specifications for the different modules of code which must be developed. These specifications, however, embrace many conceptual difficulties, not all of which have been outlined in this paper (some are purely linguistic in nature). Ongoing work consists of refining these specifications by examining the conceptual issues and experimenting with small prototypes in this and related projects (e.g. Edwards and Moulin 1995).

One of the conceptual issues which has not been addressed yet is that of hierarchy of detail. It is known that the Voronoï model supports different levels of detail (Edwards 1993), but this aspect has not yet been exploited in the concept of pivot representations. The issue of hierarchy of detail is closely tied to the issue of ambiguity, because that which is ambiguous at one level of detail may be well specified at a different level of detail.

7. Conclusion

We have seen how a linguistic analysis of a route description, combined with the Voronoï model of space, can be used to construct a pivot representation that links the spatial and linguistic aspects of a scene. We have shown how this representation can serve as a basis for the comparison between different route descriptions and for the transformation of the former into graphical forms. The process depends on the existence of a dictionary that renders explicit the equivalence between the linguistic and spatial aspects with the help of the Voronoï model of space. The basis for such a dictionary has already been established for all spatial prepositions in English.

We have illustrated the use of this pivot language for the particular example of the translation of verbal route descriptions into a sketch. We have equally suggested how the pivot level could be used in the context of the automated generation of textual route descriptions from a geographical database from a cognitive perspective (a central constraint being that the descriptions obtained must be intelligable and memorisable by human clients). Furthermore, the pivot representation is useful independently of the order in which elements are introduced into the process. Thus the pivot level constitutes a general tool for applications which need to combine the graphical and linguistic aspects of space. It should be relevant in other situations of a similar nature.

Acknowledgements
Authors Edwards and Gold would like to thank the Association des industries forestières du Quebec (AIFQ) and the Canadian Natural Sciences and Engineering Research Council (NSERC) for their financial support of this work by means of the establishment of the Industrial Chair in Geomatics applied to forestry.

References
Allen, J. 1984. Towards a General Theory of Action and Time. Artificial Intelligence 23:123-154.

Blum, H., and R.N. Nagel. 1978. Shape description using weighted symmetric axis features. Pattern Recognition 10:167-180.

Dorr, B., and C. Voss. 1993. Machine Translation of Spatial Expressions: Defining the Relation between an Interlingua and a Knowledge Representation System. Proceedings of the AAAI, Washington, D.C., 374-379.

Edwards, G. 1993. The Voronoï Model and Cultural Space: Applications for the Social Sciences and Humanities. Lecture Notes in Computer Science 716:202-214.

Edwards, G., and B. Moulin. 1995. Towards the Simulation of Spatial Mental Images Using the Voronoï Model. Proceedings of the IJCAI '95 Workshop on the Representation and Processing of Spatial Expressions, IJCAI '95, Montréal, 63-74.

Fraczak, L. 1994. De la description au croquis. Mémoire de DEA, Université Paris 11, France.

Fraczak, L. 1995. Generating "mental maps" from route descriptions. Proceedings of the IJCAI '95 Workshop on the Representation and Processing of Spatial Expressions, IJCAI '95, Montréal, 75-82.

Frank, A.U. 1992. Qualitative Spatial Reasoning about Distances and Directions in Geographic Space. Journal of Visual Languages and Computing 3:343-371.

Freksa, C. 1992. Using orientation information for qualitative reasoning. Lecture Notes in Computer Science 639:162-178.

Gapp, K.-P. 1994. A computational model of the basic meanings of graded composite spatial relations in 3d space. Netherlands Geodetic Commission, Number 40, xx.

Gold, C.M. 1994. The interactive map. Netherlands GeodeticCommission 40:121-128.

Gryl, A. 1995. Ph.D. Thesis, Université Paris 11, France.

Gryl, A. and G. Ligozat. 1995. Generating route descriptions: a stratified approach. Proceedings of the IJCAI '95 Workshop on Spatial and Temporal Reasoning, IJCAI '95, Montréal, 57-64.

Herskovits, A. 1986. Language and Spatial Cognition. Cambridge: Cambridge University Press.

Kuipers, B. 1978. Modeling spatial knowledge. Cognitive Science 2:129-153.

Landau, B., and R. Jackendoff. 1993. What and Where in spatial language and spatial cognition. Behavioural and Brain Sciences 16:121-141.

Ligozat, G. 1993. Models for qualitative spatial reasoning. Proceedings of the IJCAI'93 Workshop on Spatial and Temporal Reasoning, IJCAI'93, Chambery, France, 35-45.

Ligozat, G. 1991. On Generalised Interval Calculi. Proceedings of the Ninth National Conference on Artificial Intelligence, Anaheim, California, 234-240.

Ogniewicz, R.L. 1993. Discrete Voronoï Skeletons. Hartung-Gorre: Hartung-Gorre Verlag Konstanz.

Olivier, P.L., and J. Tsujii. 1994. A Computational View of the Cognitive Semantics of Spatial Expressions. Proceedings of ACL '94.

Okabe, A., B. Boots and K. Sugihara. 1992. Spatial Tessellations - Concepts and Applications of Voronoï Diagrams, Chichester: John Wiley and Sons.

Riesbeck, C. 1980. "You Can't Miss It!": Judging Clarity of Directions. Cognitive Science 4:285-303.

Talmy, L. 1983. How Language Structures Space. Spatial Orientation: Theory, Research and Application. New York: Plenum Press, 225-282.

Contact address

G. Edwards [1,2], G. Ligozat [3], A. Gryl [1,2,3], L. Fraczak [3], B. Moulin [2,4] and C.M. Gold [1,2]
[1]Chaire industrielle en géomatique appliquée à la foresterie
[2]Centre de recherche en géomatique, Pavillon Casault
[4]Département d'informatique, Pavillon Pouliot
Université Laval
Sainte-Foy
Québec, G1K 7P4
Canada
[3]Laboratoire d'informatique pour la mécanique et les sciences de l'ingénieur
B.P. 133
91403 Orsay Cedex
France

A NON-PLANAR, LANE-BASED NAVIGABLE DATA MODEL

FOR ITS

Peter Fohl, Kevin M. Curtin, Michael F. Goodchild and Richard L. Church

National Center for Geographic Information and Analysis (NCGIA)
University of California, Santa Barbara
Santa Barbara, USA

Abstract
Digital network databases traditionally employ a fully-intersected planar data model. While this approach has been generally useful and widely accepted, it is only one of many possible paradigms. In an effort to increase the number and improve the quality of network applications, research has been undertaken to develop and employ a fully non-planar representation. Further, the use of lanes as the primary geographic element has provided a previously unattainable level of network detail. These enhancements involve identifying and utilizing new types of feature connectivity. Route finding functions will ultimately be used to provide a test of the utility of the new model, thus a discussion of the challenges posed by routing across lanes is presented. The relative efficiency of planar and non-planar representations is also discussed.

Key words
Non-planar, data modeling, ITS, lanes, networks.

1. Introduction

While the concept of the data model is used in a variety of ways by numerous disciplines, a digital geographic data model is generally defined as an information structure which allows the user to store specific phenomena as distinct representations, and enables the user to manipulate the phenomena when held in the system as data (Raper and Maguire, 1992). Given this definition, the development of such a data model requires a comprehensive knowledge of both the phenomena to be represented and the manipulations which the user intends to apply. Decisions regarding the model components limit the functionality of the database. Each unique feature representation will vary in its ability to serve the needs of any given transportation function. Based on the designers' knowledge of the pertinent transportation functions or processes that will be modeled and the variable functionality of the available data models, the appropriate representation can be chosen.

The primary purpose of this paper is to clearly outline the research and development of a non-planar, lane-based data model for Intelligent Vehicle Highway Systems (ITS). This research is an extension of the planar data modeling documented in the NCGIA Final Report (1994). This paper covers the following topics related to network data modeling:

- A review of planar data models;
- Variability in planar restrictions;
- A comparison of storage requirements for planar and non-planar features;
- The impact of data resolution on the data model design;
- The development of a non-planar, lane based data model;
- The application of network functions to the model.

2. Planar data models

The most commonly used data model for transportation networks is the fully intersected, planar data model. This is an artifact of the prevalence of several widely used, national map databases - in particular, the Topologically Integrated Geographic Encoding and Referencing (TIGER) files from the Bureau of the Census (and their precursors the DIME files), and the Digital Line Graph (DLG) series of products from the United States Geologic Survey (USGS) - and the acceptance of the fully intersected model by a number of industry leading Geographic Information System (GIS) software developers (TIGER/LineÔ Files 1992). The transportation network has been included in TIGER for a number of reasons. First, the roads are provided to assist enumerators in locating "fugitive" households. The also allows enumerators to correct and maintain the database while they use it. More importantly, the transportation features are often coincident with boundaries of the census geographic regions for which data is collected. Therefore, they must be included to allow the census geographic regions to be well defined. The need to associate census data with these polygonal areas within which individual households are located is satisfied by a fully intersected planar data model. This allows the creation of polygons based on the left and right side attributes of the line segments (including transportation segments). The widespread acceptance and use of the fully intersected model, has been reinforced by ARC/INFO and other software systems which have incorporated this paradigm into their spatial analytic tools (ESRI 1991a).

The presence of a consistent national transportation database is extremely beneficial to the transportation community. However, the limitations that fully-intersected data models place on transportation planning are considerable, and for some applications unacceptable. Of greatest concern is the proliferation of transportation line segments due to the fully intersected data model. At any intersection of linear features - whether physical or statistical - the fully intersected data model demands an intersection with a node marking the end point of each segment. This in turn requires the input and maintenance of the feature attributes for a large number of street segments, which in reality represent a single continuous feature. Often the segments have identical attributes along their lengths, or they have attributes which could be described as an uninterrupted range of values for the single continuous feature. The labor intensive process of entering and maintaining these numerous features has been responsible for introducing a substantial amount of attribute error into such databases. This is expecially true when new features or intersections are added. The need to split current features at that time demands that attributes, such as address ranges, be updated appropriately. This is not an insignificant problem. Moreover, the need to access a large number of segments for any single continuous feature adds to the selection and computation time of transportation functions over the network. The error among the feature attributes is transferred to the network functions applied over the network; these functions can only reasonably be expected to perform as accurately as the attribute data on which they rely. While the obvious positional inaccuracies of several early databases have been a continuous irritant to many users, these errors are primarily a function of the technology used for data capture rather than an artifact of the data model (Spear 1991).

These liabilities have given rise to the current interest in non-planar transportation network models. It is expected that the development of a non-planar prototype will improve on the fully intersected data model in its utility for transportation related applications. While the exploration of this type of data model has been stifled by the inertia of the previously mentioned data collection sources and analytic tools (which have a vested interest in the fully intersected planar model), this also allows for substantial freedom in the decisions made here regarding the development of the prototype.

3. Variability in planar restrictions

Non-planar networks are broadly defined as those networks which allow segments of the network to cross without a network node being located at the intersection. There is, in fact, no implicit or explicit contact between the line segments at the point of intersection. This conceptual model is common in traditional graph theory, and is, in fact, necessary when considering transport modes without well defined, physical routeways. Shipping lanes, for example, often cross in mid-ocean without any node capable of accepting attributes (beyond the approximate location) occurring at the point of intersection. The same is true for airline networks.

Therefore, a non-planar network commonly consists of links and nodes without the mandatory requirement that every intersection be associated with a node. This allows node attributes to be associated with true network intersections but limits the proliferation of unnecessary line segments and unrealistic node placement. Routing of automobiles across such a network is more accurately implemented due to the elimination of impossible turns at multiple-grade crossings (underpass/overpass). Furthermore, users often consider a fully-intersected database to be a less accurate representation of reality, with some nodes solely being artifacts of the data model.

While some unnecessary line segments are eliminated by allowing non-planar intersections, the vast majority of intersections in an urban transportation network are true planar intersections. This suggests that the traditional definition of a non-planar network allows only a relatively small improvement in database size to be achieved, while the functionality of planar systems for polygonal spatial analysis is wholly lost. This research proposes that, if the planar enforcement is going to be compromised by the data model, the benefit from this relaxed constraint should be maximized. Therefore, the non-planar data model described herein will not utilize nodes to separate line segments at any intersection. This allows each physically continuous, well defined feature to be represented by a single record in the map database. This results in a substantially more compact database with far fewer features. This, in turn, eliminates the repetition of identical attribute data associated with a single feature.

4. Data model feature comparisons

The difference in the number of features between identical networks represented through different data models is a function of the size and connectivity of the network. Thus the reduction in features through conversion from a planar to a non-planar data model will be unique to each network. However, as long as there is a single intersection where two continuous features cross (and if they cross at a point that is not an endpoint for at least one of the features) the non-planar data model will contain fewer features. Thus, if the number of features in a planar network is represented by x_p, then the number of features in a non-planar version of the same network, x_n, will be less than or equal to $x_p - 1$.

This minimum possible improvement in feature quantity is dramatically unimpressive, but the assumption of a network with only one intersection is also totally unrealistic. Commonly transportation networks are well connected, and in urban centers often take the form of a rectangular grid. Because connectivity in rectangular grid networks is consistent within the network, the reduction in features is a function of the grid or matrix size. Below are two statements of the relationship between numbers of features in planar and non-planar data models based on matrix size, and a statement of their equivalence.

4.1 Planar vs. non-planar feature numbers for an n x m matrix

Consider an nxm matrix of streets to be a set of n East-West streets and m North-South streets. The total number of non-planar features is n+m.

x_p = the number of features in a planar model
x_b = the number of features in a non-planar model

In the planar model each East-West street intersects with each m North-South street creating m-1 arcs. For all East-West streets there are therefore n(m-1) arcs. Similarly there are m(n-1) North-South arcs. Total arcs is thus:

$$n(m-1) + m(n-1) = 2mn-m-n$$

For the case where n=m:

$$x_p = 2n^2-2n$$
$$x_b = 2n$$

As n increases the number of planar arcs will increase with the square of n, non-planar arcs will grow much more slowly. An algebraic transformation allows one to determine the number of non-planar arcs, given only the number of planar arcs and the special condition that the network is a nxn matrix, for an arbitrary n:

$$x_b = 1 + \sqrt{1 + 2x_p}$$

With this equation one can find x_b given x_p if we assume that the streets are organized in a nxn matrix. This is not unrealistic for some urban areas, or for parts of many urban areas. The difference between x_b and x_p is a measure of the improvement in database storage acquired by switching to a non-planar representation.

5. External data model - lanes and spatial information

External data modeling as described by Laurini and Thompson (1992) is the process of selecting real world features for inclusion in the database. As discussed above, the continuous street features will be the smallest element included for display in the prototype data model. However, due to the variation of lane attributes within single street features it is clear that many ITS functions are dependent on the inclusion of individual lanes rather than streets as homogenous line features. For example, multiple lane streets very often have more than one direction of traffic, lanes vary with respect to turn access, lanes can restrict mode use, and lanes can be made to control speed or other driving parameters. Many secondary attributes which are crucial to ITS applications are based on these physical and mandated attributes of lanes. Among these are travel time by mode, traffic flow, and congestion. Most importantly, the lanes themselves can begin or end at any point along the continuous street feature.

Ideally, lanes would be used as display features as well as primary elements. Unfortunately, several issues of accuracy prohibit lane representation. While GPS technology purports to provide up to centimeter accuracy, these measurements are made in comparison to a benchmark on the Earth's surface. Based on known sources of error, it is generally accepted that locations on the surface of the Earth can only be absolutely known only to within 10 meters. This accuracy is not sufficient to locate a vehicle on a specific lane. Consequently, using lanes as features for display would mislead users regarding the accuracy of the data the system was receiving and dispensing. Therefore, this data model utilizes lanes as the primary element for analysis, and uses streets as the primary display feature.

The available accuracy also affects how spatial information is stored in the internal data structure. All lanes associated with a street or highway follow the same general path, and the variation from that path is not significant relative to GPS accuracy. For these reasons, there is no value in storing lanes as individual polylines. Rather, the street should be stored as a single polyline, with the spatial information for lanes being limited to start points and end points along the relevant street.

The essential fields of the lane table are given in Table 1. Each lane in the database has a unique record in this table, with a corresponding unique identifier stored in the *lane_id* field. The identifier for the street associated with each lane, that is, the polyline the lane follows, is stored in the *street_id* field. Start and end positions of each lane are stored in the *from* and *to* fields, respectively. These positions are given as linear offsets from the beginning of the street polyline.

FIELD	TYPE	DESCRIPTION
lane_id	integer	Unique lane identifier
street_id	integer	Street identifier
from	real	Start position of lane
to	real	End position of lane

Table 1 Lane table.

This method of storing spatial information for lanes is similar to the way linear attributes are modeled using dynamic segmentation in ARC/INFO. Specifically, a lane represented by reference to a particular street polyline with a start and end point is analogous to a linear event referenced to a particular route with a start and end point (ESRI 1991b). However, in ARC/INFO, the route is the basic element and the linear event is an attribute of that element. In the lane based model, the lane is the basic element, while the street polyline is used to store spatial information. For this reason, lanes must contain additional information not associated with linear events: the relationship between lanes. This relationship takes the form of lane connectivity, making the model an integrated representation of the real world, rather than a simple collection of spatial elements.

6. Feature interconnectivity

A fundamental part of any transportation network data model is the interconnectivity of the basic data elements. If a model does not store, in some fashion, how the elements are connected there can be no analysis of flow between elements. This also precludes routing across the network, a major part of ITS applications.

In the traditional planar data model, with links as the basic data element, interconnectivity is represented by nodes. Links sharing nodes are connected at those nodes (NCGIA 1994). Typically, however, there is a need for further restriction of the interconnectivity. In a model of a street network, planar enforcement can cause a node to exist where there is no actual intersection. An example of this would be a freeway overpass, where a street passes over a freeway on a bridge. Since the street and the freeway each cross over the same point on the ground, planar enforcement requires a node at that location, even thought there is no actual intersection. A second common example is the case of legal turn restrictions, where there is an actual intersection, but drivers are not legally allowed to change directions at that intersection. In both of these instances, the data model must restrict the connection between links.

One standard method used to handle this is the inclusion of a turn restriction table in the data model. This table stores, for each node, the turn restrictions for that node. This turn restriction table can also be used to store turn impedances to more realistically model the real world situation. The ARC/INFO data model uses this method for handling turn restrictions. Each node in the network may have entries in the turn table, identifying an incoming link and an outgoing link, along with information about the impedance of that connection. The maximum number of entries in the table for each node is equal to the square of the number of links sharing that node, but an entry is not required for each link pair. Missing entries are interpreted as having zero impedance, so that it is not necessary to explicitly specify zero impedance for traveling straight through an intersection. Negative impedances are used to disallow connection between two links sharing a node (ESRI 1992).

Another approach is required in the non-planar, lane based data model discussed here. Because intersections are not represented by nodes in the model, turn information must be stored differently. The intention of this model is to accurately represent traffic flow in lanes, so it is appropriate to store turn information as a lane attribute. It is not sufficient, however, to simply list turning options at the end of a lane, because it is frequently possible to turn at more than one point along a lane. In addition, the case of parallel lanes requires the model to allow turns along a linear segment of a lane. In this instance, it is possible to switch lanes at any point along the segment. These segments do not have to continue for the entire length of a lane. Legal or physical barriers may prevent lane changes for certain distances. This is commonly seen in the case of car pool lanes, or when a lane is closed due to an accident or construction (NCGIA 1994). Because of these complexities in the lane based model, the model must be able to store turn information at multiple points and along multiple segments.

The requirement of storing multiple turn possibilities for each lane makes it effectively impossible to include turn information in the same table as lanes. In a relational database this information can be stored in a separate table, similar to the turn table in ARC/INFO. The storage method differs, however, in the method of referencing the turns, because of the complexities discussed above. Two types of turns need to be stored: turn availability at a point on a lane, and linear turn availability along a segment of a lane.

Impedance points must also be introduced into the model. These points indicate places where traffic must slow or stop, even if no turn is made. In the planar data model, this is handled implicitly: even when traveling straight through an intersection a "turn" is made, which may have an impedance value. Again, in the non-planar lane based model, there is no node to use to store impedances, so we must use a different method. Similarities between a turn point and an impedance point strongly suggest that impedance points be stored with the turns, so we will consider that a third type of turn. Each of these types will be discussed in order.

6.1 Turns at points

While the lane based model is much more complex in terms of connectivity than the planar link-node model, it does have one distinct advantage: lanes only have one direction. In the planar model, movement is possible in both directions along a link, unless it is explicitly prohibited. In the lane based model, movement is only allowed in the fundamental direction of a lane. This implies an extremely straightforward method of storing turn information in a relational database. A table can be used to store a description of each turn possibility in every lane in the database. While this table is likely to be large, proper sorting should make access very efficient.

FIELD NAME	TYPE	DESCRIPTION
lane_id	integer	Turn Origin Lane
turn_id	integer	Unique Turn Identifier
position	real	Start Position on Origin Lane
to_lane	integer	Turn Destination Lane
to_position	real	End Position on Destination Lane
impedance	real	Impedance of the Turn

Table 2 Point turn table.

Consider the basic point turn table definition shown in Table 2. The six fields shown are the essential elements of a turn table in this model. The first field, *lane_id*, contains the identifier of the lane. This value references a unique entry in the lane table (Table 1). The second field, *turn_id*, is the identifier of the turn, unique within each lane. The field pair consisting of *lane_id* and *turn_id* is unique in the turn table, and can be used to reference turns globally in the database. The positional information of the turn along the lane is stored in the *position* item, giving the offset along the street containing the lane at which the turn is located.

The destination point of the turn is given in the next two items. *To_lane* indicates the destination lane, and references the *lane_id* item in the lane table. The *to_position* field specifies the location along the *to_lane* where the connection occurs. The final item in the table, *impedance*, contains the impedance value of making that turn. Additional fields can be added to this table containing time-based restrictions or impedances, or other turn attributes. If the table is sorted with *lane_id* as the primary key and *position* as the secondary key, the first turn along a lane can be found using a binary search algorithm, then turns can be processed in the order in which they actually occur in the physical network. Disallowing turns in the database can be handled quickly by "turning off" the turn by assigning a negative value to the *impedance* field.

6.2 Turns along segments

Turns along segments can represented in a fashion similar to point turns, with a table containing one entry per turn. However, differences in the turn types require a few changes in the fields. Rather than a simple turn position, the linear turns require a start position and an end position. In addition, the turn destination depends on where along the segment the turn is made, rather than being a single position on the destination. The turn also not only has an associated impedance, as with point turns, but also a travel distance required to complete the turn. A basic linear turn table definition is shown in Table 3.

FIELD	TYPE	DESCRIPTION
lane_id	intege	Lane on which the turn occurs
turn_id	intege	Unique turn identifier
start	real	Location of the beginning of the turn on the
end	real	Location of the end of the turn on the lane
to_lane	intege	Destination lane of the turn
to_offset	real	Location on the to_lane corresponding to the start_position of the turn
distance	real	Linear distance required to complete the turn
impedanc	real	Impedance of the turn

Table 3 Linear turn table.

Four of the items in this table are identical to items in the point turn table previously discussed (Table 2): *lane_id*, *turn_id*, *to_lane* and *impedance*. These items have the same meaning in both tables. The *position* and *to_position* fields in the point turn table are missing in this table, with the analogous information stored in other fields. *Start* and *end* together represent the location of the turn in the lane, with *start* giving the beginning location of turn availability and *end* giving the ending location. This indicates that the turn may be taken at any point along the lane between *start* and *end*.

In the point turn table, the destination point is given as a single location (*position*) along the *to_lane*. The situation is more complex in the linear turn table. First, the destination point is dependent on the location where the turn is actually taken, not just where the turn availability begins. Further, some distance must be traveled along the lane before the turn can be completed. The *to_offset* and *distance* fields provide the information necessary to calculate the destination point from a turn location between *start* and *end*. *To_offset* gives the location along *to_lane* that is longitudinally equivalent to *start*. In other words, *to_offset* would represent the destination point of a turn taken at start if the distance requirement of the turn was zero. Distance represents the distance requirement of the turn. With this information, the final destination point along *to_lane* can be calculated using the following formula:

Location = *to_offset* + (turn location) - *start* + *distance*.

As with the point turn table, proper sorting of the linear turn table must be maintained for efficient access. Clearly the primary key must again be *lane_id*, so that the set of turns for a lane can be quickly found. Further sorting criteria are less obvious. A common need in an application using this data model would be to find the set of turns available at a given point along the lane. If the table is sorted with start as the secondary key, all turns with a start value higher than the current position can be quickly eliminated. However, since linear turns can, and often will, extend the entire length of a lane, each turn with *start* less than or equal to the current position would need to be examined in turn.

6.3 Impedance points

The previous discussions ignore a fundamental element of transportation networks, impedance associated at a point where no turn is made. This impedance could represent a stop sign, slowdown due to merging traffic, or other legal or physical restriction. This type of impedance is handled in this data model by the use of impedance points. Because impedance points share two of the key attributes of point turns, namely a single location and an associated impedance, it is a simple matter to store them in the point turn table. The two attributes they do not share, *to_lane* and *to_position*, can be ignored or made equal to *lane_id* and *position*, respectively. Either method is sufficient to distinguish an Impedance point from a turn.

The fundamental difference between impedance points and point turns is the way that they are applied. A point turn represents an option available at that location on the lane, generally with an associated impedance, while an impedance point is automatically applied when a particular location is reached, if no turn option is taken at that point. Impedance points can also be used to represent the end points of lanes, as a point of infinite impedance. While lane endings are stored in the lane table, it may prove desirable to include this information in the turn tables as well, for some applications.

7. ITS Functions

Any research into the development of an effective digital map database for transportation purposes must accept the responsibility to provide for the integration of ITS methods and functionality into the database. It is, in fact, the inability of current data models to efficiently support this flexibility that has initiated the research at hand. According to NCGIA (1994) it is well recognized that "most of the anticipated functions of ITS rely on the existence of an accurate, dependable map database, with sufficient information on network connectivity and driving conditions to allow solution of such problems as determination of the shortest path between a given origin and destination." The map database and ITS functions will be integrated, and the ways in which to do so most effectively must be considered during the development process.

Transportation planning is, however, a broad based discipline with many specializations and countless applications. Each of these deserves a voice in determining which functions will compose a fully functional ITS, and what needs must be satisfied by the underlying digital map database. In an effort to record these diverse needs the U.S. Department of Transportation (USDOT) has produced a Program Plan for the Intelligent Vehicle Highway Systems Program. As discussed in the NCGIA Final Report, the individual systems given as potential ITS applications vary in the extent to which they will interact with the digital map database. Some will be entirely independent of the digital map database, but will rely on the communications and information transfer infrastructure that will necessarily accompany a fully functional ITS. For example, several types of collision avoidance systems and in-vehicle safety devices do not interact with the navigable map database. At most these systems will receive and transmit location information about vehicles. Trip navigation is not a concern. Other functions will be associated more closely with specific static positions on the map database but will still be relatively unconcerned with the ability to navigate across it. These include secondary information sources regarding traveler services, safety checkpoints, electronic payment stations, and various administrative functions. Those functions that do interface with the navigable map database are primarily associated with vehicle routing.

7.1 Vehicle Routing

The ITS must be able to provide an initial trip route (based on destination information provided by the user along with historical travel time data and dynamic incident information), update that route as conditions change, guide the user across the route, and inform the user of any pertinent information. Normally, shortest path algorithms developed in mathematical graph theory are used in routing applications on planar networks. Traditional planar networks correspond very well to mathematical graphs, so the use of graph theory algorithms is straightforward. A mathematical graph is an object consisting of a set of elements called vertices and a set of pairs of vertices called edges (McHugh 1990). Relating this to a planar transportation network, vertices correspond to nodes and edges correspond to links. With the non-planar model presented here, existing algorithms cannot be so directly applied.

Graph theory shortest path algorithms such as Dijkstra's (Dijkstra 1959) rely on the structure of a graph, with each edge connected to other edges only at its endpoints. This is not the case with the lane based model, where turns are possible at many points along a lane. Further, as discussed in the previous sections, lanes are frequently connected by linear lane change segments, rather than simple intersections. Because of this, routing on the lane based model requires either the development of a new shortest path algorithm or a method of fitting a mathematical graph to the lane based model so that existing algorithms can be used. The most practical approach appears to be the latter.

The chief difficulty in fitting a mathematical graph to the lane based model is dealing with continuous lane change segments. Because a turn can occur at any point along a segment, explicit representation of the segment in a graph would require an infinite number of vertices, clearly an impossibility. One solution would be to define a minimum edge length, with a continuous segment represented by a series of these edges so that a change can be made to adjacent lanes at any vertex along the series. For example, a 1 mile segment could be represented by 528 10 foot edges. To avoid artificially imposing turn restrictions, this edge length would need to be significantly less than the distance required to change lanes. For a typical highway, each lane would need to be represented, for routing purposes, by a large number of edges. The time required to complete Dijkstra's algorithm, for instance, is $O(V^2)$, where V is the number of vertices (McHugh 1990). It is easy to see that this would quickly become too large for real time routing.

An alternative to this would be to assume that a lane change is made as soon as possible if it is made at all. Using this assumption, a vertex would be placed at the first possible position along the lane, with that vertex being the only connection between the lanes. In effect, this dynamically reduces the linear turn possibility to a turn at a point. It should be noted that this assumption could be problematic in situations with large numbers of lane blockages. For instance, on a two lane road, if each lane is blocked at several points the road would be impassible without multiple lane changes. This could be avoided by allowing lane changes at points where lane conditions (i.e. congestion, construction, accidents) change drastically.

8. Conclusions

This paper represents a significant departure from the traditional planar, centerline based, transportation network. A fully non-planar representation is adopted here in order to more realistically portray human perceptions of real world transportation networks. It is clear that non-planar networks allow a substantial reduction in the number of features to be maintained in the database, and eliminates redundancies among feature attributes. The extent to which this improvement is realized is a function of the size and connectivity of the network. The inclusion of lanes as the primary geographic element allows a higher level of detail than previously achieved, and permits additional road condition information to be utilized, such as the closure of specific lanes due to accidents or construction.

At this time a prototype non-planar network has been developed and is being used to reference lane features for a test area. The test area includes a number of urban and suburban road configurations which represent local, arterial, and highway traffic conditions. Further research will include the integration of GPS spatial referencing, simulated real-time traffic conditions, and in-vehicle navigation tools to provide a dynamic routing system.

Acknowledgments
This work was completed under the CALTRANS/NCGIA Memorandum of Understanding which initiated a research project to develop the concepts, methods, and techniques of a navigable map database for ITS.

References
Dijkstra, E. W. (1959) A Note on Two Problems in Connexion with Graphs. Numerische Mathematik. Berlin. Vol. 1, No. 5, pp. 269-271.
ESRI (1991a) ARC/INFO Data Model, Concepts, & Key Terms. ARC/INFO User's Guide. Redlands, CA.
ESRI (1991b) Dynamic Segmentation. ARC/INFO User's Guide. Redlands, CA.
ESRI (1992) Network Analysis. ARC/INFO User's Guide. Redlands, CA.
Freundschuh, S. M., M. D. Gould, D. M. Mark (1989) Issues in Vehicle Navigation and Information Systems. National Center for Geographical Information and Analysis, State University of New York at Buffalo. Technical Paper 89-15.
Laurini, R., D. Thompson (1992) Fundamentals of Spatial Systems. Academic Press, San Diego, CA.
McHugh, J. A. (1990) Algorithmic Graph Theory. Prentice Hall, Englewood Cliffs, New Jersey.
NCGIA (1994) Final Report - CALTRANS Agreement 65T155. National Center for Geographic Information and Analysis, Santa Barbara, CA.
Raper, J. F., D. J. Maguire (1992) Design Models and Functionality in GIS. Computers & Geosciences. Vol. 18, No. 4, pp. 387-394.
Spear, B. D. (1991) Issues and Recommendations From the Standpoint of Transportation Analysis. Research and Special Programs Administration, U.S. Department of Transportation. TIGER Enhancement Technical Working Group, Meeting Minutes, October 27, 1991. GIS/LIS '91 - Atlanta, GA.
TIGER/Line™ Files (1992) Geographic Objects in the TIGER/Line (TM) Files. U.S. Bureau of the Census, Washington, DC.

Contact address
Peter Fohl, Kevin M. Curtin, Michael F. Goodchild and Richard L. Church
National Center for Geographic Information and Analysis (NCGIA)
University of California, Santa Barbara
3510 Phelps Hall
Santa Barbara, CA 93106-4060
USA
Phone: +1-805-893 8652
Fax:+1-805-893 8617
E-mail: fohl@ncgia.ucsb.edu
E-mail: curtin@ncgia.ucsb.edu
E-mail: good@geog.ucsb.edu
E-mail: church@geog.ucsb.edu

IMPLEMENTING A SHORTEST PATH ALGORITHM

IN A 3D GIS ENVIRONMENT

S.D. Kirkby[1], S.E.P. Pollitt[1] and P.W Eklund[2]

[1]Key Centre for Social Applications of GIS
[2]Department of Computer Science
University of Adelaide
Adelaide, Australia

Abstract
A modified version of Dijkstra's shortest path algorithm (Dijkstra, 1959) has been implemented as a series of external modules in a three dimensional (3D) Geographic Information System (GIS). When initiated the algorithm determines, using heuristically derived gradient, speed and turning values of impedance, the longest travel time for the area displayed and the shortest path between any two nominated points. The interactive 3D GIS environment allows the user to drive/fly along the traversed route conveying enhanced visual information. Consideration in the paper is given to: selection, modification and implementation of the algorithm; 3D database development and; shortest path evaluation for the study area in the city of Okayama, Japan.

1. Introduction

Efficient emergency service vehicle response to a natural disaster, is determined by the time taken to reach the affected area, the more time it takes a vehicle to reach the destination the lower the likelihood of a successful rescue operation being implemented. Given this relationship, Managers who coordinate response to a natural disaster require prior knowledge of emergency service response times in their region. Shortest path algorithms enable the emergency service response times and paths to be determined while a 3D GIS provides the spatial context for these paths via graphical display, ie. visual display of the shortest path route through a dense multi-story urban environment.

This paper considers the implementation of a shortest path algorithm in a 3D GIS system to determine emergency service response times within a portion of the city of Okayama, Japan. The topographic and economic character of Japan results in a large population being confined to small habitable land areas and unfortunately, as Japan is a country prone to earthquake activity (approximately 1500 seismic actions per year), the chances of a major natural disaster occurring in a highly populated urban area are high (CIA, 1992). Emergency service response management must therefore be prepared for such events. The city of Okayama, 300km south west of Tokyo is considered to be an appropriate urban study site.

The paper is divided into four sections:
- *Background*, discusses shortest path algorithms and 3D GIS system used;
- *Methods*, details the construction of the 3D database and implementation of the algorithm;
- *Results,* determines the shortest path route, visualises it, and ascertains algorithm efficiency and;
- *Conclusion.*

2. Background

2.1 Shortest path algorithms

Significant research has gone into developing shortest path algorithms (Dial 1969; Cherkassky et al. 1993; Ahuja et al. 1988; Bellman 1958; Dijkstra 1959; Ford and Fulkerson 1962; Gallo and Pallottino 1988; Moore 1959). This section provides a brief overview of classical shortest path algorithms with a more detailed explanation of Dijkstra's double bucket algorithm. An additional description of a method to handle multiple source nodes is also given.

2.2 Basic concepts

All the shortest path algorithms take input of a similar format, the main two being:
- a Node list;
- an Arc list.

The input is therefore a digraph, and can be represented by the standard notation $D = \{V, E\}$, where V represents the nodes and E represents the arcs. Each node has a changing potential, or *distance* value, representing the time to travel to a node.

Node labelling and scanning operations differ between the different algorithms. In this paper, each node in the network has an associated tag, S, which takes one of three values: *unreached*, *labelled*, or *scanned*. Initially, all nodes v in the network have $d(v)= \infty$, and are tagged *unlabelled*. Source nodes have "known" distances and are tagged *labelled* with $d(v)=0$.

The core of most shortest path algorithms is the *Scan* operation, which selects a labelled node as the next for processing by the algorithm (Pollitt, 1995). In this operation "children" of the passed node (all nodes that are destinations of arcs from the current node) are processed one at a time. For each node, the current shortest time to the child is compared with the sum of the time to the parent plus the time between the parent and child. If the new distance is shorter, the child's parent is updated to the current node and the child tagged *labelled*. Once all children are processed, the parent node is tagged *scanned*. The *Scan* operation will generally cause some nodes with *unreached* and *scanned* tags to become *labelled*. This indicates either a path connecting an unconnected section has been established, or a previously examined node has been visited again. The *Scan* operation continues until there are no nodes with a tag of *labelled*, whereupon the algorithm terminates. The question of labelled node selection is important as the optimisation of the selection process will be the determining factor in the efficiency of the algorithms.

2.3 Dijkstra's double bucket implementation

Dijkstra's algorithm works by selecting the *labelled* node with the current minimum distance from the source node (minimum potential) as the next node to scan. There are a number of different implementations of Dijkstra's algorithm (linear search; integral bucket structure; approximate bucket structure; double bucket structure) with the difference being the manner in which the node with the lowest potential is located (Pollitt, 1995). Essentially the bucket structure depends upon the potentials being integral. The labelled nodes are sorted into buckets, with bucket *i* containing all nodes with potential *i*. When the potential of a node changes, it is moved into the appropriate bucket. A bucket index L is maintained counting from 0. The number of buckets is determined by the constant parameter B. The double bucket structure has two levels of buckets --- *high* and *low*. If the number of low level buckets is defined to be a constant B, then the high level bucket *i* contains the *labelled* nodes with integral potential in the range *[iB, (i+1)B - 1]*. The bucket index L is kept, this time indexing the lowest non-empty high level bucket. The nodes with potential in the *range [LB, (L+1)B - 1]*, are distributed in the low level buckets. After all low level buckets are checked and the nodes scanned, L is incremented. This approach ensures that nodes are scanned at most once. This increases the efficiency of the implementation, which runs in $O(m + n(B + C/B))$, where C is the largest static travel time along a single arc.

The double bucket implementation of Dijkstra's Algorithm was chosen for its ability to scale to larger problems and for its good average case efficiency. The data set used had the following properties:

> n = 654 distinct nodes;
> m = 1636 distinct arcs;
> C = 159 seconds.

The first problem to overcome with this implementation was the assumption that distance potentials of the nodes are all integral. This was not the case with the data used. This was dealt with by casting all references to the distance potential in relation to bin positioning from *double* to *long*. This had the effect of ignoring the decimal portion of the node potential when calculating the bin in which to place the *labelled* node. Each bin therefore contained nodes with the same integral value. As the decimal portion is of much lower significance than the integral portion, this approximation is insignificant.

The other departure from the standard double-bucket algorithm addressed the use of multiple source nodes in a single data set, ie. the shortest path should be found from each node to the *nearest* source. This was implemented by cycling the algorithm through for each separate source. The result is a network of node's with each node's parent being the immediate predecessor on the shortest path back to the nearest source.

2.4 Three dimensional (3D) spatial representation

For many GIS applications, a key assumption is that all spatial data handled is referenced to a 2D Cartesian coordinate system. This convention restricts the scope of mapping the vertical dimension over terrains or within the earth as this third dimension must first be expressed as an attribute and expressed in 2D as a constant value (Raper and Kelk, 1991). In contrast, a 3D system allows the third dimension to be represented as a geometric property therefore giving the ability to represent the feature in its true geometric form. The 3D software used in this project is Vulcan[1], which appropriately stores the third dimensional Z value as a geometric coordinate along with the X and Y values.

As the aim of this study is to determine shortest path routes between locations within the study area the vector data model of space is utilised. Points represent intersections and lines the road segments. Connectivity between elemental objects in the database (ie. roads and intersections) facilitate navigation by allowing queries to be posed regarding routes through the model representative of the real world. Notably, the software does not export a line (or road segment) into an external file, rather all data concerning the road segments are stored at associated nodes. This raised the problem, where we wanted to map from the node structure contained within the GIS software to the arc structure used in our external shortest path modules.

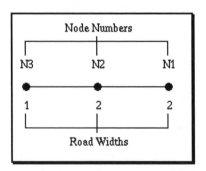

Figure 1 External module node/arc path.

The problem involved establishing the correct road widths for the arcs <N1 → N2>, <N2 → N1>, <N2 → N3>, <N3 → N2>, given the road widths at N1, N2 and N3 (Figure 1). For the case when both end nodes of an arc have the same width value (<N1 → N2>) and (<N2 → N1>) , the arc between them obviously had the specified road width. The difficult case occurs when the two end nodes had different road width values - which road width should be assigned to the road segment? To solve this problem, the arc is given the

---

[1] Vulcan is a product of Maptek Pty Ltd.

road width as specified by the *child* node. In the case of figure 1, if node N1 is the *parent* node, node N2 is the current node and node N3 is the *child* node. This ordering is taken from the corresponding order of reading the nodes from the input file. As N3 is the child of N2, its road width tag is attached to the two road segments, ie. <N2 → N3> and <N3 → N2> both have the width 1. If the arc destination node width tag were used instead of the *child's*, then <N2 → N3> would have width 1, while arc <N3 → N2> would have width 2. This is not the desired result, thus the *child's* width tag is used to calculate the widths of road segments in both directions.

Data for the road network was divided into spatial and thematic classes. The spatial being the geometric representation, ie. point, line while the thematic provides a description of the object attributes ie. a road link with vehicle speed etc. The geometric data changes much less frequently in modeling road network dynamics. Attribute data may change on very short time scales. Geometric changes will occur in certain zones after an earthquake, as evidenced by the Great Hanshin earthquake (1995), Kobe Japan. With this in mind, road networks were altered in relation to fault line locations in the study area to ascertain possible changes in shortest path route travels (Kirkby, 1995).

3. Methods

3.1 Database development

Spatial data for a portion of the city of Okayama was obtained from the local council. The data layers included; roads, footpaths, residential boundaries, urban boundaries, water bodies, drainage lines, buildings, road boundaries and spot heights. These were provided at 1:500 scale. Fault line locational information was digitised from maps provided by the Crustal Survey Division of the Geographical Survey Institute, Japan.

Data provided was in two dimensional format. To convert to 3D, a digital elevation model (DEM) with one metre contours was generated from spot heights. All surface data such as roads and buildings etc. were subsequently registered to this surface, thereby enabling automatic generation of a Z value for every point, line and polygon within the vector space. Spatial positioning error was not quantified (Kirkby, 1995). To generate three dimensional buildings from the two dimensional polygon data, each polygon was triangulated to a specified height as determined by the "story height" attribute value held within the database (Figure 2).

Figure 2 Three dimensional perspective: Okayama, Japan.

3.2 Algorithm implementation

The shortest path algorithm was implemented in external modules and integrated into the 3D GIS system. The system developed is schematically represented in Figure 3. It consists of two modules each performing a number of separate functions. They are: calculation; display. The calculation modules, represented on the left hand side of Figure 3, determine the time taken to traverse road links with the goal being to minimise the time taken to travel to any (connected) point on the road layer. The second module operation, right hand side of Figure 3, utilises the calculated data to display shortest paths between any nominated point on the area displayed on the screen.

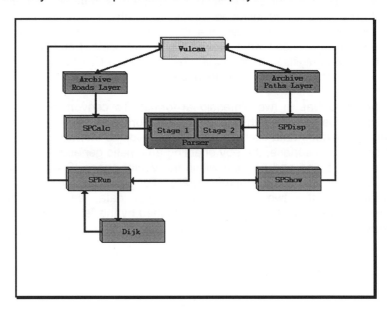

Figure 3 The integrated system.

3.3 Calculation

Three factors affecting vehicle travel time along the road network are dynamically calculated by the calculate modules, they are; travel speed, gradient and turning times (Kirkby 1995; Pollitt 1995). The speed at which a vehicle can travel along a road segment is determined by the number of lanes on the road ie., 1 lane 20 km/h; 2 lanes 50km/h; 4 lanes 80 km/h. It is calculated by dividing the arc length (straight line distance in three dimensions between the end points) by the speed value attached to the nodes for the segment. Gradient percentage is calculated using the formula:

$$gradient = 100 * \frac{\Delta z}{\sqrt{\Delta x^2 + \Delta y^2}}$$

where Δx Δy Δz represent the change in coordinates from start to end point. For uphill slopes greater than 5%, based on the percentage slope and the length of the road segment, an impedance value is added to the travel time (Kirkby 1995). There is no slope greater than 5% in the study area, but this concept is very important with further extensions to this research work eg. power to weight ratios for heavy fire fighting vehicles in mountainous terrain.

At every intersection, the change in direction of the next road segment is considered, and penalties applied. These penalties are based on:
- the number of lanes in the old road segment (the segment just traversed);
- the number of lanes in the new road segment (the segment about to be traversed);
- the turn angle between the old and the new road segments (left, right or straight).

In Japan vehicle traffic travels on the left hand side of the road, consequently right hand turns across on-coming traffic are slower than left turns. This is also influenced by the number of lanes which the vehicle wishes to cross ie., a vehicle turning right across four oncoming lanes will take longer than a vehicle turning right across one oncoming lane. The assumption here being the more lanes on a road the more the traffic flow. The turning process is represented by Figure 4.

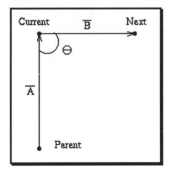

Figure 4 Turning paths.

There are three steps in the calculation:
1) The length of the vectors from parent to current (A) and current to next (B) nodes must be calculated to find the angle between them.
2) Using the cross product formula (Pollitt, 1995) (taking the x and y coordinates - gradient is already considered):

$$\|\vec{A} \times \vec{B}\| = \|\vec{A}\|\|\vec{B}\| \sin \theta$$

$$\vdots$$

$$\theta = \arcsin \frac{A_x B_y - A_y B_x}{\sqrt{A_x^2 + A_y^2}\sqrt{B_x^2 + B_y^2}}$$

If θ (the deviation of the path) is in the range -10...10 degrees then the path is essentially straight. If θ is > 10 degrees, the path turns to the left, otherwise if θ is < -10 degrees, the path turns to the right.
3) Penalty calculation (including widths of the two road segments).

The calculation for impedance is performed each time a node is scanned. If the new path is quicker than the previously recorded path to the destination node, the parent node is updated, and the new path is accepted as the new tentative best path to the node. List optimisation strategies are implemented at different stages. Following completion of the calculations the shortest path network is returned to Vulcan.

3.4 Display

The display function takes the set of calculated shortest paths (generated by the calculate modules) and returns the single shortest path to a user selected node. Data for the length of the longest path and time for the slowest path for the region displayed is automatically returned as default to the screen window.

To initiate this process the user selects the destination point from the spatial information displayed on the screen. This point is passed to the external module for processing. The node structure returned by the parser is scanned to find the node matching the user selected coordinates. A list of nodes is built up by following the parent pointer back until a source node is reached (a source node is identifiable by the parent pointing to itself). This list presents the path beginning from the destination and travelling back to the source. The order is reversed before dumping the path to the display screen.

4. Results

Shortest path route selection within the city of Okayama, visualisation of these routes and algorithm optimisation are the three output results of this project.

4.1 Route selection

The first experiment spatially identified areas with inadequate emergency service vehicle response times to the Hospital. The worst case scenario:
• Longest travel time: 13.42 minutes
To verify changes in emergency service vehicle response times to an earthquake, major roads intersecting known faults within the study area were disabled. Spatial changes to response times were obviously influenced by the fault pattern.
• Longest travel time: 26.45 minutes

These results were not validated by field work, ie. driving the selected route in a car, nor did they take into consideration traffic flow at a given time, a factor which influences response times. Nevertheless they provide an indication of potential problem areas.

The next experiment computed shortest path travel to any selected point in the study area from the hospital. Figure 5 highlights travel routes (black) to two destinations from the two exits of the hospital (black). Seamless system integration enabled this interactive module to be implemented.

Figure 5 Interactive shortest path route selection.

4.2 Traversing the route

Bishop (1994) argues GIS are devices for turning data into information with the key aspect in the transition being effective communication between the GIS, the user and other information consumers. Visualisation techniques available within interactive 3D GIS environments allow the user to not only view the data in 2D (Figure 5), but to traverse the route in 3D (Figure 6).

Figure 6 3D view of selected route.

The different perspective's between figures 5/6 are further enhanced when i) texture mapping is added to the 3D objects ie. photographs attached to buildings to give realistic perspective, and; ii) when the user interactively drives along the selected path. The conversion of this shortest path data to information to assist decision making relates to movement within the 3D environment. The interactive 3D perspective improves the users understanding of the terrain being traversed. It would be further improved if a steep slope was being traversed or if we were able to represent traffic density via an abstract form ie. "gas clouds" with changing colour density. Further research work is required to improve this data to information conversion process but visualisation techniques offer one method by which to achieve this goal.

4.3 Optimisation

The calculation of all shortest paths for the entire Okayama data set takes less than 3 seconds, disregarding the import stage back into the 3D GIS software. The project was conducted on a Silicon Graphics Indy, with 32 Mb of RAM and a 100mhz IP22 CPU. The efficiency of the implementation corresponds with the recommendations of Brown (1989).

5. Conclusion

A series of external modules has been developed to determine shortest path routes within a 3D GIS. Benefits for planning emergency service vehicle response; i) areas with poor coverage identified and; ii) paths to any designated location selected. The advantage of implementing such an algorithm in a 3D GIS environment relates to the conversion of shortest path data to graphical information enhancing the perspective of the route traversed, with further research development work on; simulating real time driving speed; quantifying and visualising traffic density; and the development of power to weight ratio functions for heavier vehicles, a more realistic modeling scenario will be implemented.

Acknowledgments
We would like to thank Masakatsu Horino and Dr. Hiroshi Murakami from the Geographical Survey Institute, Japan and Dr. Bob Johnson and Trevor Coulson from Maptek Pty Ltd. Research work was supported by the ARC Australia, the Australian Academy of Science, the Geographical Survey Institute Japan and the Science and Technology Agency of Japan.

References
Ahuja R.K., J.B. Melhorn, R.J. Tarjan (1988), Faster Algorithms for the Shortest Path Problem, Technical Report, Princeton University.
Bellman R.E. (1958), "On a routing problem", Quarterly Applied Mathematics, (16), pp 87-90.
Bishop I (1994), "The role of visual realism in communicating and understanding spatial change and process", in H.M. Hearnshaw D.J. Unwin (eds), Visualization in Geographic Information Systems, John Wiley & Sons, London, pp. 60-64.
Brown C.M.L (1989), Human-Computer Interface Design Guidelines, chapter 6, pp. 115-119, Ablex Publishing Corp.
Cherkassky B.V., A.V. Goldberg, T. Radzik (1993), Shortest Path Algorithms: Theory and Experimental Evaluation, Technical Report, Stanford University.
CIA World Fact Book (1995) (URL document)
http://www.odci.gov/cia/publications/95fact/index.html
Dial R.B. (1969), "Algorithm 360: Shortest path forest with topological ordering", Communications of the ACM, 12 (11).
Dijkstra E.W. (1959), "A note on two problems in connection with graphs" Numerical Mathematics, (1), pp. 269-271.
Ford L.R. Jnr, D.R. Fulkerson (1962), Flows in Networks, Princeton University Press.
Gallo G., S. Pallottino (1988), "Shortest path algorithms", Annals of Operations Research, (13), pp. 3-79.
Kirkby S.D. (1995), Natural Disaster Management: Using a 3D GIS for Emergency Route Planning, report to the Science and Technology Agency of Japan. 62 pp.
Moore E.F. (1959), "The shortest path through a maze" Proceedings of the International Symposium on the Theory of Switching, pp. 285-292.
Pollitt S.E.P. (1995), A 3D Spatial Information System for Emergency Routing in Okyama City, BSc (Hons) Thesis, Department of Computer Science, University of Adelaide, Australia, 81pp.
Raper J.F., B. Kelk (1992), "Three dimensional GIS", in D.J. Maguire, M.F. Goodchild and D.W. Rhind (eds), Geographical Information Systems, Principles and Application, Vol.1, pp. 299-317. Longman Scientific and Technical.

Contact address
S.D. Kirkby[1], S.E.P. Pollitt[1] and P.W Eklund[2]
[1]Key Centre for Social Applications of GIS
[2]Department of Computer Science
University of Adelaide
Adelaide, S.A., 5005
Australia
E-mail: skirkby@gisca.adelaide.edu.au
E-mail: sepollitt@gisca.adelaide.edu.au
E-mail: peter@cs.adelaide.edu.au

THE GIS WALLBOARD:

INTERACTIONS WITH SPATIAL INFORMATION ON

LARGE-SCALE DISPLAYS

John Florence, Kathleen Hornsby and Max J. Egenhofer

Department of Spatial Information Science and Engineering
and
National Centre for Geographic Information and Analysis
University of Maine
Orono, USA

Abstract
Displays of future geographic information systems (GISs) may be the size of an entire wall. This paper explores how such new technology would enable GIS users to work with information systems in completely new ways, by manipulating spatial objects or scenes with their bare hands and querying through gesture, voice, or a combination of the two. The central component of this interaction is the *WallBoard*, a GIS device whose design is based on the metaphor of an office whiteboard, with tools—both physical, such as markers and erasers, and virtual, such as lenses and measuring devices—that have similar usage capabilities to those tools associated with a regular whiteboard. The WallBoard replaces the look-and-feel of the desktop of today's personal computers. Unlike smaller-scaled devices where a user performs all interactions from more or less the same position and perspective, the WallBoard allows users to interact from three different spaces: Within *arm's length*, users may have physical contact with the objects they are manipulating; within *spitting distance* they gesture primarily; and *within sight* of the WallBoard they watch and at times interact with the WallBoard through the use of additional computing devices. Within arm's length, the WallBoard affords gesture interactions with geographic objects by selecting objects or areas, zooming in and out of a sub-area, panning, and rotating a scene.

1. Introduction

The design of today's geographic information systems (GISs) is limited by the physical size of computer displays, which are usually somewhere between 10 and 24 inches. This small display size requires objects to be shown in miniature and often restricts usage to a single person. Exceptions to this setting are recent considerations of *Virtual Worlds*, in which users get the impression of living in the same space as that which they are manipulating (Jacobson 1995). An alternative is also envisioned with the advent of new hardware technology featuring large-scale displays (Elrod *et al.* 1992), which may take up the size of an entire wall (Negroponte 1995)—three by five meters or more. Interactions with *spatial data* on such devices may provide users with new spatial experiences and enable improved collaboration among users.

This paper explores what a GIS would be like if a wall-sized version were available. Such an environment would be useful for many applications, such as for planners attempting to deal with complex planning requirements for say, an urban renewal scheme, or for electric utility managers who currently rely on huge paper maps mounted on walls. It introduces the concept of a GIS wall device, referred to as the *WallBoard*, which allows multiple users to view and interact with a large-scale, touch-sensitive display. The WallBoard is the organizing metaphor for wall-sized GIS devices, much like the desktop is for office applications running on personal computers (Smith *et al.* 1982).

Unlike smaller-scaled devices where a user performs all interactions from more or less the same position and perspective, the WallBoard supports multiple spatial perspectives and experiences. Zubin's (1989) categorization of how humans perceive objects in the physical world includes small objects that can be understood from a single perspective; objects that are too large to be manipulated with human hands and require scanning with the eyes; and larger objects that need walking through and multiple perspectives in order to be perceived. With current GIS technology, users are only exposed to the first type of experience. The large-scale display of the WallBoard, however, changes how users perceive and interact with spatial information. Where representations of geographic objects, such as buildings or forests, on desktop GISs could be manipulated by human hands and perceived at a single glance, representations of these same geographic objects on the WallBoard may be too large to be perceived from a single perspective and, therefore, require scanning back and forth. In fact, users may now have to step away from the WallBoard to see the big picture. Such experiences may be critical to evaluate correctly a model.

It is assumed that the WallBoard has full multimedia capabilities including sound, graphics, and animation. It also has all the necessary sensors to accommodate multi-modal inputs from gestures, eye-contact, and voice. This may sound fiction today, but work in progress at research labs and universities (Elrod *et al.* 1992; Cassell *et al.* 1994; Pentland 1996) indicates that such a scenario is not so far away. The principles for design of the WallBoard are based on Donald Norman's ideas of providing a sound conceptual model for design that takes advantage of natural 'mappings' between tasks the user wishes to perform and how these tasks will be undertaken on the WallBoard (Norman 1990).

This paper continues with a brief analysis of how the WallBoard contributes to collaborative spatial decision-making and presents two sample scenarios utilizing the WallBoard (Section 2). Section 3 describes the design of the WallBoard and Section 4 investigates multi-modal interactions with the WallBoard. The degree to which certain interaction modalities are either possible or not at different ranges from the WallBoard, leads to a categorization of three interaction spaces (Section 5). Section 6 studies GIS interactions—selections, zoom, pan, rotation, navigation—if the user is within an arm's length from the WallBoard. The paper concludes with a discussion of future research.

2. Collaborative spatial decision-making on the WallBoard

Collaborative spatial decision-making, such as that undertaken by a group of planners, can make full use of the capabilities offered by the GIS WallBoard. Planners regularly work together on planning projects and have collaborative needs that are currently not fully satisfied by computer technology (Densham *et al.* 1995). Planners often draw on information retrieved from a wide range of multimedia materials including maps, surveyors' reports, aerial photographs, traffic information, and other miscellaneous audio, video, and verbal information, and often more than one office needs to work off the same version of a spatial data set. Presentations by planners to various groups typically involve verbal, visual, and gestural components. They spread maps out on tables, show slides of before- and after-development scenarios, play sound recordings of noise levels at various locations, and use hand gestures pointing out "where development will take place" or "all of this area will be included" (Shiffer 1995). An intelligent, interactive, wall-mounted device in this context would improve both the collaborative spatial decision-making process and the interactions with an interested audience, as shown by the following two scenarios:

- A group of planners are working together on a proposed development project for a new shopping mall. The large size of the WallBoard allows several persons to be in front of it. As they create different scenarios, users are in close contact with the objects on the wall and manipulate them through the use of gestures and voice. To review their design they may step back to see the full effect. Collaborating at the WallBoard may create a new paradigm of work for planners as they shift from a setting of people in separate offices working on the same spatial data set, to collaborative input at the WallBoard. Other co-workers can also contribute to the process by observing from a distance and viewing what is displayed there, while still using gestures or voice to interact.
- One of the above planners gives a presentation to the local council on the proposed development. The planner conducts the entire presentation through manipulations on the WallBoard, making full use of its multimedia capabilities. He or she shows before- and after-images of the proposed development, including an animated movie of a person's view when walking along the site during construction. Buildings or other objects of interest can be added or removed as desired. The council is sitting in front of the WallBoard, in easy viewing distance. When they ask questions they may interact with the WallBoard themselves through gestures such as pointing or, if necessary, a council member may walk up to the WallBoard and, by using markers, he or she may highlight items of importance or suggest changes.

Interaction with the WallBoard should not be seen in isolation from other computing environments. Others may access the GIS operating on the WallBoard from other networked computing devices, and give input as desired.

3. The GIS WallBoard

The size of the WallBoard device could vary—up to the physical size of the wall—but for our purposes we imagine it to be similar to a large, wall-mounted office whiteboard of two by three meters. For the look-and-feel of a WallBoard, we borrow heavily from the metaphor of a whiteboard. Successful user interface designs are often based on a metaphor from a real-world example (Kuhn 1992; Mark 1992). Users expect devices to behave according to what they see and what their experience has taught them to expect from such a design. In this way a user is able to establish a mapping from the familiar to the unfamiliar. This concept builds on natural mappings that take their root in universal physical or cultural standards that are immediately understandable (Norman 1990). For the WallBoard, the familiar domain is the whiteboard, and the unfamiliar is the domain of abstract computer operations on a WallBoard. The whiteboard also provides a rationale for arranging and organizing the various tools to be used in conjunction with the WallBoard. We assume that most users know how to use a whiteboard; therefore, users will find the WallBoard simple and easy to use if it mimics the whiteboard's behavior. The layout of the WallBoard illustrates the large display area and the tool tray (Figure 1). It has been tailored for geographic collaborative applications, in a fashion similar to the LiveBoard (Elrod *et al.* 1992), a large interactive display system using a cordless pen, to facilitate interactions for generic group meetings and presentations.

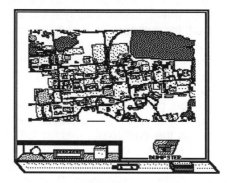

Figure 1 Layout of the WallBoard.

The principal component of the WallBoard is its large display area, where views of spatial scenes are shown as large-scale representations. Users manipulate them through various interaction modalities.

On the tool tray are actual physical objects such as markers and erasers with which the user can draw or erase objects through contact with the WallBoard. Experience from other studies with similar devices points to the success of markers or pens as collaborative tools (Moran *et al.* 1995; Elrod *et al.* 1992). The ability of anyone to approach the WallBoard and pick up a marker to highlight important areas or draw connections between related ideas, greatly enhances the collaborative potential of this device. Markers will need to provide a high degree of positional accuracy, and with a choice of color, they can be used much in the same way as people use different colored markers on a regular whiteboard to confer more meaning and provide for easier interpretation.

Virtual tools are similarly organized on a virtual tool shelf, located just above the tool tray. There may be more than one virtual tool shelf depending on how many tools are available, and users may toggle from one shelf to the next. The virtual tool shelf may include measuring devices to display lengths and areas, and lenses for locally filtering or adding information (Stone *et al.* 1994). One of the virtual tools is a sticky notepad for adhering notes, instructions, or memos to the WallBoard. The markers can be used to write on the sticky pad, and it can be attached or moved to any desired location for easy reference. Users may pick up virtual tools through gestures and move them across the WallBoard to perform an operation. These tools have been chosen in such a way that their design and use is based on real life tools or familiar icons, and so should be clear to most users.

The dumpster is the place to receive any trash—any virtual object the user wants to get rid of is simply put into it. Its functionality corresponds to the trash can on the office desktop, but its appearance was assimilated to the settings of a GIS WallBoard. This was necessary because geographic objects displayed on a WallBoard are generally large and, therefore, do not afford to be put into a trash can. The visualization of the dumpster may have to be adapted when the WallBoard is used in different cultures.

A careful study of the ergonomics associated with the WallBoard will be necessary to ensure the device is designed well and positioned at an appropriate height for use. In his article on GIS hardware design, Nick Parker (1993) reports that in reality there is no 'average' sized person, and good product design must be able to accommodate a wide range of users.

4. Interactions with the GIS WallBoard

WallBoard interactions allow users to concentrate on the tasks at hand, rather than on how to perform operations. They are supported by the WallBoard's rich interaction environment, which senses and interprets multi-modal inputs, much like humans interpret the inputs from other people. For example, if someone approaches, looks us in the eye, and says, "hello," then we know (i.e., sense through sight and sound) that the person is communicating with us and not someone else. Of course, this person may have instead shaken our hand or merely gestured hello by waving from a distance. Regardless, communicating a simple hello can be achieved in a variety of ways. Similarly, the WallBoard allows users to mix and match inputs such as those from gesture, tools, eye contact, voice, and other computing devices, into meaningful expressions.

4.1 Gesture

Gesture is a vital and expressive method for communicating, not only among people, but also between users and the WallBoard. The WallBoard accepts both touch-based gestures where the user physically touches the screen and natural, empty-handed gestures where users communicate through hand or body movements (Wexelblat 1995). Gestures are the primary method for selecting, panning, zooming, rotating, and other manipulations of geographic objects and scenes displayed on the WallBoard.

4.1.1 Touch-based gestures

Touch-based gestures for object or scene manipulations on the WallBoard, utilize a touch-sensitive screen and are based on the direct manipulation metaphor taken from haptic space (Mark 1992). Haptic space is mainly based on people's experiences in the real world through physical contact. Touch-based interactions require extending the metaphor of a whiteboard. For instance, when users rotate a 2-D scene on the WallBoard, they do this much like rotating a paper map in the physical world. This natural mapping from people's haptic space experiences with a paper map onto touch-based interactions on the WallBoard, helps to eliminate what Norman (1990) calls the *gulfs of execution and evaluation.* Touch-based gestures require the user to be close enough to the WallBoard to have direct contact.

4.1.2 Natural, empty-handed gestures

Natural, empty-handed gestures are the type of gestures that people use to complement their daily conversations. For example, when pointing in the direction of a toy box and telling a child, "When you are finished playing with your toys, put them in *there.*" Similarly with the WallBoard, users may make a gesture to choose *this* tool from the virtual tool shelf, rather than explicitly stating the tool's name or approaching the WallBoard to physically select it. Some natural, empty-handed gestures like pointing and circling, may borrow intuitively from gestures based on touch; however, empty-handed gestures may occur away from the WallBoard's touch-sensitive screen and do not require any tools such as pointers.

4.2 Tools

Physical and virtual tools simplify interactions with the WallBoard by affording explicit behavior. For instance, pen-based interactions, extended to include colored markers, afford writing and sketching directly on the WallBoard. Whether the tools are physical (markers and erasers) or virtual (lenses and sticky notes), they allow users to concentrate on the task at hand.

4.3 Eye contact

Eye contact is another valuable interaction method, because people use their vision everyday to observe interesting things in the world. It may be important, for example, to know whether a user's gestures are directed toward the WallBoard or toward another person; or identifying to which area of the WallBoard users are referring. Through tracking of head- and eye-movements, the WallBoard senses eye contact on specific areas or objects. This interaction is similar to using a pointing device, though more natural.

4.4 Voice

The most widely used mode of communication between people is voice. Users may speak freely and naturally from anywhere near the WallBoard, and their verbal interactions with the WallBoard are not constrained to a set of commands. Voice will be interpreted independently or in combination with other interaction methods. "Show all cities in Maine with a population greater than 50,000," is one example of an independent natural-language query. However, if the query were, "Show all cities in *this* state with a population greater than 50,000," the natural-language query would need to be augmented through gesture, eye contact or tool-based interaction.

4.5 Computing devices

Computing devices, such as desktop and palm-sized computers, offer alternative methods of interaction with the WallBoard. These devices extend the collaborative process to others in the workplace. For example, users too far away to gesture or to be heard may employ their personal computing devices to interact with the WallBoard, rather than moving closer to it. Such interactions that are initiated from other devices, may include connecting a palm-sized GIS to the WallBoard for direct interaction, like a sketch pad for querying, or downloading data. Additionally, some tasks in the collaborative process, like filing and other administrative duties, may be more appropriate for the desktop metaphor. While the use of such additional devices may interrupt the natural interaction style with the WallBoard, their integration enhances the collaborative process.

4.6 Multi-modal inputs

Multi-modal interactions with the WallBoard combine gesture, tools, eye contact, voice, and computing devices in one way or another. Undoubtedly, people make such combinations, and users of the GIS WallBoard will want to maintain their most natural interactions. For example, the selection of a group of objects may involve touching the desired objects directly; a combination of voice and gesture such as, "Put that there," (Bolt 1980) where a user's gesture delineates which object is to be placed where; or by a combination of voice and tool-based interactions, e.g., when the user sketches an area with a marker and says, "All parcels in *this* area." Such combinations of modalities may require the integration of interactions from multiple users, such as several users pointing at the same time, or one person pointing while the other gives a verbal instruction.

5. Interaction spaces

As one moves further away from the WallBoard, the number of interactions available decreases. For example, contact-based pen interactions would be impossible from three meters away. The observation that certain interaction modalities are either impossible or lose their effectiveness at some distance from the WallBoard, leads to a classification of interaction spaces. Feasible interactions for each of the interaction spaces are shown in Table 1.

Interaction Space	Gestures		Tools		Eye contact	Voice	Computing devices
	touch	natural	physical	virtual			
Arm's length	√	√	√	√	√	√	√
Spitting distance		√	√	√	√	√	√
Within sight							√

Table 1 Interactions in the three interaction spaces.

We refer to these spaces based on a person's bodily relation to the WallBoard (Figure 2). In order to be able to touch the WallBoard, the user has to be within an *arm's length* of the WallBoard. Farther away, the user's gestures must be precise enough to be interpreted with respect to the WallBoard (*spitting distance*). Finally, *within sight* users are essentially observers, relying on vision and hearing, and any interaction with the WallBoard from this zone is through other computing devices. Since the interaction spaces are dependent on bodily relations, the spaces do not scale up with larger meeting rooms, such that for a regular meeting room, only the range of the Within-Sight Space may increase. A small office, on the other hand, may have just Arm's-Length space.

Figure 2 Interaction spaces on the WallBoard.

5.1 Arm's length

Within an arm's length of the WallBoard, users are right up close to the WallBoard, using all of the available interaction methods including gestures, tools, eye contact, voice, and computing devices. The principal and unique interactions in Arm's-Length Space are those involving gestures and tools. Here the metaphor of a whiteboard affords the use of tool-based interactions and is extended to include interactions through touch.

In the Arm's-Length Space, objects on the Wallboard may appear much larger than the user's physical size; therefore, touch-based interactions on the WallBoard may occur differently than those on desktop GISs. For example, positioning a building on a parcel through direct manipulation on a desktop GIS may utilize a snap-dragging technique (Bier 1989), however, such an operation becomes difficult to employ on the WallBoard, where the same building may appear to be two by three meters, and the parcel on which it is to be placed is even larger. Because users are so close to the large display, it becomes difficult for them to gain perspective. In the Arm's-Length Space, users may be required to move their head back and forth, change their position within the space, or even step away from the WallBoard in order to see the big picture. Therefore, when working in the Arm's-Length Space, users may frequently shift between spaces over the course of collaboration.

Benefits of interactions in Arm's-Length Space are attributed to the touch-sensitive screen and its size. Users can interact with representations of geographic objects that are closer to their actual size. Additionally, touch-based gestures allow users to manipulate objects directly, rather than through a pointing device.

5.2 Spitting distance

The second interaction space is within *spitting distance* of the WallBoard. This space is directly behind Arm's-Length Space, where gestures can still be detected and interpreted by the WallBoard's sensors. In Spitting-Distance Space, touch-based gestures and most tool-based interactions are no longer possible because users are too far away to reach the touch-sensitive screen. The principal interactions in this space are natural, empty-handed gestures, eye contact, and voice.

In Spitting-Distance Space, objects and scenes still may appear larger than the user, however, the entire WallBoard may be perceived easily by scanning with the eyes. Whereas desktop GISs require users to look into the scene, the large size of the WallBoard, provides users with the feeling that they are actually part of the scene. This consequence provides a more intuitive and natural interaction environment for natural, empty-handed gestures. Additionally, the WallBoard offers a benefit over other environments, such as virtual reality, in that multiple users can view and interact with the same space at the same time, while observing the others' manipulations.

5.3 Within sight

Beyond spitting distance, users are *within sight* of the WallBoard. In this space, displayed information can still be viewed, but because of the physical distance from the WallBoard, the size of objects may appear smaller than a user's hand. Even if gestures were detectable from this space, they would not be precise enough to be useful; therefore, interactions from this space occur primarily through other computing devices. A benefit of interacting in the Within-Sight Space is that user's can observe other collaborators and the WallBoard simultaneously.

6. Operations within an arm's length of the WallBoard

For manipulations on the WallBoard, we focus on some fundamental GIS operations: selecting, zooming, panning, rotating, and navigating through a scene. These examples are to illustrate the nature of WallBoard manipulations in the Arm's-Length Space. Some of these operations are also possible if users are more remote from the WallBoard, but since they lack the opportunity to directly touch the display, different interaction methods may have to be employed.

6.1 Selection

Users may select spatial objects in a number of ways depending on how close they stand to the WallBoard. Within an arm's length, they may simply point to or touch any spatial object displayed on the WallBoard to select it. This interaction may require users to walk to the other end of the WallBoard, or stretch their arms above their heads. They may select a group of objects through a combination of gesture and voice, for example, "Remove *this* building … and … *this* one," while sequentially touching the objects of choice. Likewise, they may select a set of adjacent objects by drawing a circle around the area with one of the markers such that the enclosed region becomes the selected area (Figure 3a), or through a combined voice and gesture operation such as "Select all parcels *here* zoned as residential," where a user's gesture delineates the area in which parcels are to be selected.

6.2 Zoom

Zooming can be accomplished most easily when within an arm's length of the WallBoard, by using gestures. A pushing gesture towards the WallBoard, mimics pushing away a map, and results in the display zooming out. The opposite action, namely gesturing towards oneself, results in the display zooming in. These gestures are particularly effective for when the user wishes the entire display to be scaled. The zoom operation is to be an intelligent zoom (Frank and Timpf 1994), where more detailed information is brought up and a corresponding change in level of detail in object properties takes place.

To zoom on a part of the entire display, a user first draws a closed figure with a marker or makes an empty-handed gesture around a region of interest. The zoom operation is then performed by gesturing towards or away from the selected area. Although it is limited to the selected region such that the remainder of the WallBoard's display stays unchanged (Figure 3b), the boundary of the zoom area need not be sharp and a fisheye-like zoom operation with more detail in the center and continuously less toward the boundary is feasible.

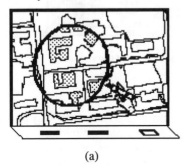   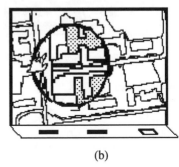

(a)   (b)

Figure 3 Zoom-in operation: (a) before zoom and (b) while zooming.

6.3 Pan

When panning, users change the field of view, while retaining its orientation, scale, and level of detail. Johnson's (1995) evaluation of user preferences for panning on a touch-controlled display indicates that users finds pushing the background easier and more intuitive to use than other methods of panning such as touching the side of the display screen or pushing the view-window over the scene. In the Arm's-Length Space, a user pans by placing his or her palm on the WallBoard and sliding it in the desired direction. The display moves as the pan takes place, giving the user the necessary feedback about the operation. Panning also applies to selected sub-areas such as the zoom area (Section 6.2), when users move the field-of-view by sliding the zoom area around on the display until it is located over the desired area. Panning is dynamic as the user interacts with the WallBoard, so the user receives immediate feedback that the operation is proceeding satisfactorily. Such selective panning acts much like using a lens or magnifying glass and passing it over an image (Stone *et al.* 1994).

6.4 Rotation

Users change the orientation of entire scenes by rotating them through direct manipulation. They place both hands on the touch-screen display and carry out a rotation motion, turning the image —clockwise or counterclockwise, back and forth (Figure 4). Continuous updates of the rotated scene provide feedback to adjust the scene to the desired orientation. Rotation can be easily combined with panning to allow for more complex, iterative displacements.

459

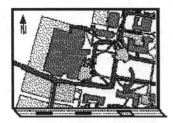 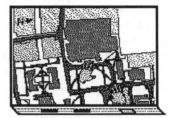

Figure 4 Rotation (a) before and (b) after rotation of the scene.

6.5 Navigation

The WallBoard is large enough to give users an impression of being part of the displayed environment. With a three-dimensional animation on the WallBoard, users may select to navigate through the space displayed in order to perceive how the environment changes as they move around. Gestures play an important role in this navigation, as users can use their hands to indicate in which direction they wish to follow (Figure 5). The display keeps pace with the action and changes dynamically as the users moves *through* the view. In this role, the WallBoard is less cumbersome than a virtual reality environment, because the user can navigate without wearing gloves, goggles, or any other special apparatus.

Figure 5 Navigation on the WallBoard.

7. Conclusions

Having a GIS available on a large-scale, wall-mounted device enables users to interact with geographic information in completely new ways. This paper explored user-interface considerations for the WallBoard, an organizing metaphor for wall-size GIS devices. Basing the design on the metaphor of a whiteboard, affords most users an immediate idea of how the WallBoard can be used. Multi-modal interactions with the WallBoard occur primarily through a combination of gestures, eye-movements, and voice. Different modalities are used depending on how close or remote a user is from the WallBoard. The availability of a certain interaction mode leads to a categorization of interaction spaces, in which users perform different types of tasks. Most challenging is the use of hand gestures to perform some of the most common GIS operations, such as selecting, panning, zooming, but also for such innovative interactions as navigating through an animated 3-dimensional scene. Such interactions are dramatically different from those used for mouse-based panning and zooming on current desktop GISs (Jackson 1990). When users stay further away from the WallBoard, they lack immediate contact with the WallBoard, but still can select objects through gestures and voice. The farther away from the WallBoard, however, the lower the accuracy of their empty-handed gestures. To compensate for distance, remote users may use hand-held pointing devices such as a laser pointer for selection. Instead of contact, the selection of individual objects is made through pointing at the same location for a certain time interval.

We have simulated interactions with a WallBoard through an animated movie on a small-scale display, using MacroMind Director on a Macintosh. Gestures were derived from interactions with a whiteboard. The study was invaluable to identifying the nature of interactions possible with a GIS operating on a wall device, and the enhanced possibilities for collaborative spatial decision-making.

The WallBoard is a spatial technology that provides a framework for studies of innovative GIS interactions, allowing for entirely new approaches to problems. The mere existence of this concept will advance our knowledge of interaction methods with GISs through comparisons with the often-so-limiting constraints of today's GIS desktop environments. At the same time, the WallBoard concept is expected to serve as a framework for specific questions related to multi-modal interactions. The design of the WallBoard is certainly only the starting point and there is a long way to go before prototyping and user testing with a comprehensive scenario. Significant theoretical advancements will be necessary to enable smooth group work with the WallBoard.

Some of the open questions include:
- What hand gestures can be identified sufficiently precise in real-time?
- When do users combine voice and gesture, and how?
- In an experiential space such as the WallBoard, is it desirable to provide simultaneously multiple views, and if so what alternatives to windows integrate well with the WallBoard?
- Which new tools beyond lenses and measuring devices will enable innovative analysis methods on the WallBoard?
- What are the requirements for a spatial database management system to support real-time manipulations on the WallBoard, including choice of data models and data structures?

Acknowledgments
This work was partially supported by the National Science Foundation for the National Center for Geographic Information and Analysis under NSF grant number SBR 88-10917. Max Egenhofer's research is further supported through NSF grant IRI-9309230, by Rome Laboratories under grant number F30602-95-1-0042, the Scientific and Environmental Division of the North Atlantic Treaty Organization, Intergraph Corporation, Environmental Systems Research Institute, Inc., and a Massive Digital Data Systems contract sponsored by the Advanced Research and Development Committee of the Community Management Staff and administered by the Office of Research and Development. We are also grateful to Doug Flewelling, Tom Bruns, and Troy Soderberg for their contributions during the early stages of this work.

References
Bier, E. (1989) Appropriate Editing Paradigms for Differently Sized Spaces, in W. Kuhn and M. Egenhofer (Eds.) Visual Interfaces to Geometry , Technical Paper 91-18, NCGIA, University of California - Santa Barbara, pp. 16-19

Bolt, R. A. (1980) 'Put-That-There': Voice and Gesture at the Graphics Interface. ACM SIGGRAPH '80, ACM Computer Graphics, 14(3): 262-270.

Cassell, J., C. Pelachaud, C. Pelachaud, N. Badler, M. Steedman, B. Achorn, T. Becket, B. Douville, S. Prevost (1994) Animated Conversation: Rule-based Generation of Facial Expression, Gesture and Spoken Intonation for Multiple Conversation Agents. SIGGRAPH '94, Orlando, FL.

Densham, P., M. Armstrong, K. Kemp, Eds. (1995) Collaborative Spatial Decision-Making: Scientific Report for the Specialist Meeting, Technical Report, National Center for Geographic Information and Analysis, Santa Barbara, CA.

Elrod, S., R. Bruce, R. Gold, D. Goldberg, F. Halasz, W. Janssen, D. Lee, K. McCall, E. Pedersen, K. Pier, J. Tang, B. Welch (1992) LiveBoard: A Large Interactive Display Supporting Group Meetings, Presentations, and Remote Collaborations. CHI '92, Monterey, CA, pp. 599-607.

Frank, A. U. and S. Timpf (1994) Multiple Representations for Cartographic Objects in a Multi-scale Tree - An Intelligent Graphical Zoom. Computers and Graphics 18(6): 823-829.

Jackson, J. (1990) Developing an Effective Human Interface for Geographic Information Systems Using Metaphors. ACSM-ASPRS Annual Convention, Denver, CO, pp. 117-125.

Jacobson, R. (1995) The Natural Representation of Mapping and Surveying Data: Applying the Virtual Worlds Paradigm. Cognitive Aspects of Human-Computer Interaction for Geographic Information Systems. T. Nyerges, D. Mark, R. Laurini and M. Egenhofer. Dordrecht, Kluwer, pp. 239-246.

Johnson, J. A. (1995) A Comparison of User Interfaces for Panning on a Touch-Controlled Display. CHI'95, Denver, CO, ACM.

Kuhn, W. (1992) Paradigms of GIS Use. Fifth International Symposium on Spatial Data Handling, Charleston, SC, IGU Commission on GIS, pp. 91-103.

Mark, D. (1992) Spatial Metaphors for Human-Computer Interaction. 5th International Symposium on Spatial Data Handling, Charleston, SC, IGU Commission on GIS, pp. 104-112.

Moran, T. P., P. Chiu, W. van Melle, and G. Kurtanback (1995) Implicit Structures for Pen-based Systems Within a Freeform Interaction Paradigm, CHI '95, Denver, CO.

Negroponte. N. (1995) Being Digital, New York, NY, Alfred A. Knopf Inc.

Norman, D. (1990) The Design of Everyday Things. New York, NY, Bantam Doubleday Dell Publishing Group, Inc.

Parker, N. (1993) Geographical Information System Hardware Design. Human Factors in Geographical Information System. D. Medyckyj-Scott and H. M. Hearnshaw, Belhaven Press, pp. 173-184.

Pentland, A. P. (1996) Smart Rooms. Scientific American 274(4): 68-76.

Shiffer, M. (1995) Geographic Interaction in the City Planning Context: Beyond the Multimedia Prototype. Cognitive Aspects of Human-Computer Interaction for Geographic Information Systems. T. Nyerges, D. Mark, R. Laurini and M. Egenhofer. Dordrecht, Kluwer: 295-310.

Smith, D. C., C. Irby, R. Kimball, B. Verplank, and E. Harslem (1982) Designing the Star User Interface. Byte 7(4):242-282.

Stone, M., K. Fishkin, and E. Bier (1994) The Moveable Filter as a User Interface Tool. CHI '94, Boston, Massachusetts, ACM, pp.306-312.

Wexelblat, A. (1995) An Approach to Natural Gesture in Virtual Environments. ACM Transactions on Computer-Human Interaction 2(3): 179-200.

Zubin, D. (1989) Natural Language Understanding and Reference Frames, in D. Mark, A. Frank, M. Egenhofer, S. Freundschuh, M. McGranaghan, R. White (Eds.) Languages of Spatial Relations: Initiative 2 Specialist Meeting Report, Technical Paper 89-2, NCGIA, University of California - Santa Barbara, pp. 13-16.

Contact address
John Florence, Kathleen Hornsby and Max J. Egenhofer
Department of Spatial Information Science and Engineering and
National Center for Geographic Information and Analysis
University of Maine
Orono, ME 04469-5711
USA
Fax: +1-207-581-2206
E-mail: {jflorenc, kathleen, max}@spatial.maine.edu

VISUALIZATION OF UNCERTAINTY IN METEOROLOGICAL

FORECAST MODELS

Elizabeth Dirks Fauerbach, Robert M. Edsall, David Barnes and
Alan M. MacEachren

Department of Geography
Penn State University
University Park, USA

Abstract
This paper describes a prototype visualization environment designed to facilitate an understanding of, and comparison among, meteorological forecast models. Emphasis is on tools that allow an analyst to compare predictions of three models, both to one another and to the actual weather that they predict. The system allows for surface pressure patterns (an area feature) and storm centers (a point feature) to be examined as they correspond or diverge over time. Our discussion of the prototype includes attention to the role of animation for exploratory analysis of geographically and temporally referenced information, interactive controls for dynamic maps, and methods for representing uncertainty (reliability) of georeferenced forecasts.

1. Introduction

A plethora of georeferenced data exist in the earth and atmospheric sciences. These data are derived from both field measurement and models. Typically, they include a temporal as well as a spatial component. A primary goal of data collection or generation in this context is an understanding of change in spatial patterns over time. Visualization (in the form of static maps) has been used throughout the history of science as a primary method for constructing knowledge about change in environmental attributes. Today, dynamic visualization tools offer the potential for new insight into existing questions related to environmental change and for the identification of new questions.

In the context of scientific visualization more generally, georeferenced visualization has been characterized as emphasizing the private interactive exploration of spatial information with a goal of revealing unknowns (MacEachren, 1994). From this perspective, the project presented here might be described as a prototypical visualization application. Its focus is on the exploration of model-derived atmospheric data covering relatively short temporal duration (i.e., meteorological time scales) and continental spatial extent. Specifically, the prototype presented is designed to facilitate an understanding of the spatio-temporal reliability of weather forecast models. This prototype allows an analyst to compare results of three models B to one another and to the actual weather that they attempt to forecast. Description of our prototype will be preceded with a brief review of relevant developments in georeferenced visualization.

2. Background

Scientific visualization is a research area that integrates developments in theory, methods, and tools for representation with specific areas of scientific application. Thus, two distinct literatures are usually relevant to any visualization research, one related to representation theory, methods, and tools and the other related to the scientific questions that the methods and tools are designed to explore.

2.1 Representational issues

Three interrelated representational issues are relevant to the project described here: the application of animation to depiction of temporal data, the role of interactive controls for data exploration by experts, and methods for symbolizing data uncertainty or reliability.

- *Animation*
Much of the recent literature on both scientific and cartographic visualization promotes animation as a valuable tool for both the "visual thinking" and "visual communication" ends of the research sequence (DiBiase, 1990; Peterson, 1995). As noted by several authors, animation has the advantage of representing the dynamic elements of a process, as well as more closely approximating the real world process itself (DiBiase, et al., 1991; Dorling and Openshaw, 1992; Kraak and MacEachren, 1994). Another contribution of animation is its ability to show the correlations and relationships of many phenomena (Karl, 1992; DiBiase, et al., 1994). This ability is grounded in application of one or more "dynamic" variables. Three variables were initially identified by DiBiase, et al. (1992), duration, rate of change, and order. MacEachren (1995) subsequently added three additional dynamic variables to this set, display date ("date" in animation time, typically specified as the frame number), frequency (the logical dual of duration), and synchronization (applicable only when two or more features are being represented).

Atmospheric phenomena in particular can be represented effectively using the dynamic variables. The complexity and fluidity of meteorological and climatological processes along with their continuity across space and time lend themselves to the increased flexibility and power of animation (Koussoulakou, 1994). In the context of climate models, DiBiase, et al. (1992) demonstrated that use of the dynamic variable order can be a particularly powerful tool in understanding model uncertainty as it relates to agriculture. They generated an animation in which precipitation predictions of five climate models were depicted along with an indication of prediction uncertainty (as assessed by dispersion around the mean prediction). When the animation was "played" in order of increasing uncertainty (rather than in temporal order), the earth system scientists for whom the animation was created noticed (for the first time) that model uncertainty was greatest in the spring planting season B precisely the time at which certain predictions are most important.

- *Interactivity*

Like animation, the ability to interactively manipulate map parameters (whether static or animated) and to interactively query the database behind the map offers the potential for fundamental changes in how maps are used in scientific thinking (MacEachren and Ganter, 1990). A considerable body of research directed to interactive data exploration has been conducted in the field of exploratory data analysis (EDA) (see, for example, Cleveland and McGill, 1988; Buja, et al., 1991). Several EDA methods have been successfully extended to exploratory *spatial* data analysis. For example, Monmonier (1989) has extended the concept of scatterplot "brushing" into the spatial domain with his "geographic brushing" and MacDougall (1992) has combined the techniques of "brushing" and "linking" to develop a dynamic variant on cluster analysis. A related EDA technique, "focusing" (a method in which subsets of data are highlighted) has been implemented for exploring data depicting environmental change and metadata representing the certainty of spatial and temporal interpolation of these data (MacEachren, et al., 1993).

Some effort has also been directed to combining animation and interactive control in the exploration of spatial (and spatio-temporal) data. For non-temporal spatial data, the addition of interactive controls to an animated "flyby" facilitates the use of that animation as a tool for exploring the character of terrain (Moellering, 1980) as well as the relationships such as those between terrain and vegetation (Kraak, 1994). Dorling (1992) has applied similar methods (that he terms animating space) to exploring maps of enumerated social and economic data. Egbert and Slocum (1992) have implemented related interactive controls in a tool designed to allow exploration of choropleth maps. In their case, animation was used not to shift the viewer's focus from one place to another, but from one category of attribute to another, through a sequencing of information by data category. For georeferenced temporal data, Monmonier (1990) added the concept of temporal brushing to his earlier geographic brushing. With hypermedia environments (e.g., MacroMedia *Director* or animation tools for the World Wide Web, such as Sparkle), at least one aspect of temporal brushing has become common place B the ability to change the pace of an animation and to move forward and backward, frame-by-frame or at any speed the user desires.

One of the more innovative combinations of interactivity and animation is that associated with Buttenfield and Weber's (1994) concept of "proactive graphics for exploratory visualization." They define proactive graphics as those in which "..., users initiate queries and steer data presentation in a manner consistent with the associative power of the human intellect" (Buttenfield and Weber, 1994, p. 8). To illustrate their concept of proactive graphics, they developed a prototype that allows an analyst to interactively control an animated representation of tree growth over time. The prototype actually includes many individual animations (of maps and graphs at different scales of analysis) that play out simultaneously (in the background) with the user being able to change scale, location, or representation form of the visible display (from map to graph, to diagram) as time progresses.

- *Uncertainty*

With the proliferation of digital data, metadata (or data about data) has become increasingly important. Uncertainty is a key facet of that metadata. To represent uncertainty it must be defined, both generally and specifically. A step toward such a definition is a categorization of components. These include data quality and variability in relation to spatial, attribute, and temporal aspects of the data (MacEachren, 1992). As part of the U. S. Spatial Data Transfer Standard (SDTS) data quality is further subdivided into positional accuracy, attribute accuracy, logical consistency, completeness, and lineage (Fegeas, et al., 1992). Variability can arise from spatial, attribute, and/or temporal aggregation in which a single characterization is used to represent a group of phenomena that may not be homogeneous (MacEachren, 1992). Variability introduced through aggregation is, of course, directly related to the resolution of that aggregation (Buttenfield, 1993). With the general components of uncertainty identified, the next step in the production of a representation of uncertainty is to determine which aspects are relevant to a specific instance. Once this decision is made, estimates of uncertainty must be matched to appropriate visualization tools and methods.

Visual representation of uncertainty is receiving increasing attention (e.g., Beard and Buttenfield, 1991; MacEachren, 1992; Fisher, 1994; Buttenfield and Beard, 1994; van der Wel, 1994), but remains an area of georeferenced visualization with many unanswered questions. Two major issues can be identified. One is that of symbolization. MacEachren (1992) and others (e.g., McGranaghan, 1993) have examined Bertin's graphic variables for use in representing uncertainty and have added new variables, notably *saturation* (i.e., purity) of color and *clarity*. The latter can be further broken down into crispness, resolution, and transparency (MacEachren, 1995). The second major issue in representing uncertainty is that of display types or user interfaces. These include single bivariate maps, map pairs or multiple maps, sequential presentation, and interactive displays (MacEachren, 1992). Each of these has their own advantages and disadvantages. Some are obviously not suited for animated maps, others are impossible to implement with static maps.

Little attention has been directed to representation of uncertainty over a time series (see MacEachren, et al., 1993 and Mitasova, 1993 for initial attempts). With an animated map, showing phenomena that change over time and uncertainty that also varies with time the choice of display type is limited. Map pairs are often impractical, due to the limited screen size of computers. Sequential presentation and toggling between maps of data and maps of uncertainty would probably be confusing for time series data. Bivariate and multivariate maps seem most suitable for this type of representation. These methods have the advantage of having everything on one map thus allowing the largest possible display given screen size. In addition, the map can be shown as a smooth animation in which viewers can observe the change in uncertainty at the same time they are studying the data. Multivariate maps can, however, easily become complex and confusing (McGranaghan, 1993) and, as Buttenfield (1993) has found, can require viewer training. One goal of our prototype is to determine whether animated bivariate data/metadata maps are practical.

2.2 Atmospheric science issues

The development of interactive and animated computer displays of atmospheric data provides a unique opportunity to explore large data fields with significant spatial and temporal variation in a holistic and qualitative way. Efforts have been underway for several years to exploit the power and memory of supercomputers and software to enable users to more readily manage, analyze, and visualize the large data sets in the earth sciences (Schiavone and Papathomas, 1990). In earth system science, map animation has been combined with statistical analysis to develop evidence that supports the equilibrium hypothesis explaining long term vegetation change as a response to continuous climate forcing (Prentice et al., 1991). The massive data collection and dissemination rates in present-day meteorology require visualization techniques that provide a synthetic and systematic way of making sense of hundreds of data values at thousands of spatial and temporal data points. Simulations of prototypical severe storms (Wilhelmson, et al. 1990; Hibbard and Santek, 1989), diagnostic re-creations of significant meteorological events (Hibbard, et al. 1989), or more general representations of mesoscale, synoptic, and global forecasts (Grotjahn and Chervin, 1984; Max, 1993) all have benefitted from creative visualization and animation techniques. Advanced packages such as McIDAS (Hibbard and Santek, 1989), 4DCAP (Neeman and Alpert, 1990), and Savvy3D (Leidner, 1995) have been used to perform a variety of relevant tasks for meteorologists including 3D representations of multiple variables, interactive contour mapping, and flybys (through simulated storms and storm systems) which allow visualization from multiple perspectives.

Most of the atmospheric visualization techniques developed have been for display of large sets of data from either observations or models. Little consideration has been given to the representation of uncertainty associated with these data sets. Such information is especially critical for a forecast, where the quality of the predictions is as important if not more so than the predictions themselves. Treinish (1992) employs simple elegant visualization techniques to qualitatively correlate winds, temperature, and Antarctic ozone depletion, though his method stops short of a quantitative mapping of correlation statistics (e.g., residuals). Mapping information about uncertainty of a forecast, for example, by quantitatively describing the differences among simultaneous forecasts from three distinct models, could lead forecasters to uncover important but hidden information about the relative performance of each of the models over time. As noted above, such an approach has already been demonstrated to be successful in the context of climate models (DiBiase, et al., 1992). In a short-term forecast, information about differences in model predictions could affect public preparedness for a storm or heat wave. For a longer-range forecast, this metadata could profoundly affect economic and political policies on climate issues such as global warming, agriculture, and disaster relief. The coupling of the most advanced visualization techniques with formulation and manipulation of meteorological and climatological metadata is a crucial step toward development of improved forecasts and analyses of atmospheric phenomena.

3. The prototype

As noted above, animation of environmental data has been used to examine change in data reliability over time (MacEachren, et al., 1993 and Mitasova, 1993). Animation of model results have been used to help understand the interaction between vegetation and climate change (Prentice, et al., 1991) and to assess model reliability by comparing several models over time (DiBiase, et al., 1992). Interactive manipulation of a dynamic display has been applied to the study of change in forest structure over time (Buttenfield and Weber, 1994). Our project builds upon all of these efforts to develop a prototype that includes integration of interactive control with animation, exploration of output from spatio-temporal models, and the representation of model reliability.   The resulting system enables users to compare individual meteorological models with each other, to explore the relationship between an average of the models and the amount of divergence from that average, and to contrast the models with the actual weather pattern of the same time period. The models explored in this prototype are short-term meteorological forecast models from the National Meteorological Center (NMC).

3.1 The models

Three models, the **eta**, **NGM**, and **spectral** are released simultaneously as the NMC's short-term (two-day) forecast product. Each is a result of a highly complicated computer algorithm which assimilates a fine grid array of observations and, each through a slightly different technique, produces forecasts for North America.   The three models differ in resolution, physical parameterizations, and numerics, and use varying lengths of developmental data (Vislocky and Fritsch, 1995). As a result, the models perform differently in different situations; for example, Junker, et al. (1989) found that the NGM (nested grid) model is less reliable than other existing models when predicting winter precipitation when the polar jet stream is in a certain orientation. In light of the large body of literature discussing the variable performance of each model (e.g. Vislocky and Fritsch, 1995; Caplan and White, 1989; Petersen, 1992), we expected that the models would be significantly divergent in their forecasts, especially toward the end of the 48 hour period.

Each of the three models is released by the NMC as a series of static maps, each representing one "time slice" of the 48 hour period. There are eight maps for each model produced every six hours; each map is a forecast of conditions at a time which is an integral multiple of six hours from the time of release. Each forecast is stored as a grid of points, equally spaced on a base map of North America, at a spatial resolution of approximately one point per 50 km, with 30 points in the latitudinal dimension and 40 points in the longitudinal. Any finer spatial resolution using data exclusively from these models would necessarily involve interpolation. An array of meteorological variables is stored at each point, including surface and upper air winds, temperature, humidity, and pressure. As a result, each time-step of each model consists of 1200 grid points, each with several forecasted attributes.

3.2 The case

We have compiled the 24 forecast data sets (eight data sets from each of the three models) from one release time, 7:00 am on February 2, 1995, for our project. During the 48-hour period following the forecast, an important storm developed over the Tennessee valley and rapidly intensified as it moved over the warm waters off the U.S. east coast. The storm brought blizzard-like conditions to the major cities of the northeast, including Philadelphia, New York, Boston, and Buffalo. Such a storm is called a "Nor'easter" by meteorologists because of its northeasterly track as it deepens off the Atlantic coast.

The parameter we deemed most useful for representing this event was surface pressure. *Isobars* (lines of equal pressure) reveal a great deal about weather conditions at a particular point; in general, surface winds circulate clockwise around an area of high pressure, counterclockwise around a low, and parallel to the surface isobars. Closely packed isobars are indicative of high winds, widely spaced isobars are indicative of light and variable winds. The location of the "low" or "high" can also have a bearing on the moisture content of the atmosphere; for example, a low positioned off the Carolina coast will be likely to bring moist easterly winds to the middle Atlantic and Chesapeake regions. If, as in the case of a classic "Nor'easter", a wintertime cold high is positioned over Labrador, the clockwise winds associated with it will bring cold air to the northeast U.S., and any moisture transported into this cold air (from the Atlantic low, for example) will fall as snow.

A focus on isobars and the pressure patterns they represent is also of practical significance for us as cartographers, since a time series of isarithmic depictions allows us to *track* the movement of the systems in a fluid way in our animation. Other parameters such as temperature and humidity are not as indicative of air mass (e.g. storm) *movement*, and thus would be less informative in an animation of the development of "Nor'easter".

3.3 Procedures

As mentioned above, the data from the models is formatted in an array of 40 columns and 30 rows. Each array represents a grid of pressure values at a particular time, the total area of which corresponds to the area of the continental United States and some of southern Canada. We received data for each model in six-hour increments, as well as the *actual* pressure values for 7 a.m., February 2 (time zero) and for every subsequent 12 hours within the forty-eight hour time period.

In order to create animated isoline maps from the data files, temporal interpolation between time snapshots generated by the models was required. This interpolation was necessary to create a sufficient number of frames to yield a smooth animation. These interpolations could be done easily in Microsoft *Excel,* but first the file formats needed to be manipulated slightly in order for *Excel* to read the files as a spreadsheet with 40 columns and 30 rows. Once translated into *Excel*, the data were linearly interpolated across time. Three intermediate data sets were generated between each modeled time-step, creating data files for every 1.5 hours. In addition, since we were also interested in the predictions of the three models together, a grid of mean values (of the three models) was calculated for each 1.5-hour time period.

Once all of the files for each model had been interpolated, and the average values of the three models had been calculated, individual files were brought into *DeltaGraph*. In *DeltaGraph,* isoline maps were created from each of the data grids, for all of the models and for the actual data. These maps consisted of solid black isolines with a contour interval of 2 millibars. There were several considerations that led us to choose isolines for representing the pressure surfaces. First, isolines are commonly used for this purpose in meteorology and therefore experts are familiar with them, thus our computer display maps look similar to the static maps meteorologists are used to viewing. Second, because of limitations in the software used (MacroMedia *Director*), especially in the use of color palettes and the way *Director* maps colors to the palette when files are imported from other applications, it was impractical to use filled contour surfaces for a large number of classes. Third, it is difficult for people to distinguish very many gradations in color. And, fourth, we needed a method of representation that would not be confused with our representation of uncertainty. Here, we followed DiBiase, et al. (1994) in using different symbol dimensions (lines versus area shading) to depict the two attributes of our bivariate maps (pressure forecast for each grid cell and uncertainty of those forecasts).

Once all of the files had been converted into isoline maps, each map was saved as a PICT file, so that it could be brought into *Director.* First, a movie of each of the three models was created, yielding three different movies of pressure predictions. To make the movies, the isoline PICT for each 1.5-hour time period was imported into *Director* as a cast member; the first cast member was a map of the actual 7 a.m. data. The cast members could then be played in sequence, with each member occupying one frame of the animation. This yielded a relatively smooth animation, illustrating the predicted progression of the isobars over the forty-eight hour period. Next the average maps (represented with isolines in cobalt blue) were also animated, creating a movie of the combined predictions of the three models.

The final step in this particular animation was to introduce some representation of the spatial distribution of uncertainty. We decided to define uncertainty in this case as the standard deviation among the three models at each grid point for each time slice. Thus, we have a predefined location and time, with an estimate of attribute accuracy made for each time and place. Where the models are in close agreement, there is little uncertainty in forecast pressure; where the models diverge greatly, the uncertainty of forecast pressure is large. We represented attribute uncertainty across the grid for each time slice using shaded isolines of standard deviation, displayed underneath the blue isobars. Because of software and perceptual limitations discussed earlier, the number of categories we could use to depict uncertainty was limited, but most analysts are likely to want only a few categories depicted. Use of two dissimilar methods of representation, area shading (shaded isoline maps) and unshaded isolines, allowed us to overlay the two while retaining visual separability of the representations.

The shaded isoline maps of standard deviation were created in a similar fashion to the isoline maps of average pressure described above. PICT files from *DeltaGraph* were imported into *Director* as cast members in frame positions that coincided with the synchronous isoline frames. The result is a movie that simultaneously displayed the average of the model predictions and the uncertainty of those predictions as a dynamic bivariate map.

To provide further information about the path of the Nor'easter, the position of the lowest surface pressure at the center of the storm was located and denoted by point symbols of three colors to represent the three models. In order to compare the three predictions of the path and timing of the storm, an animation was created that shows three different predicted storm "tracks" generated by using *Director* to leave a linear trail behind as the point symbol representing the storm center shifts over time. The tracks of these storms can be played individually, or together, facilitating comparison among the three predictions.

3.4 Results

The completed animation package allows the user to view several variables simultaneously or separately, which can reveal otherwise hidden characteristics of the data. For example, when the uncertainty information is displayed with the average isobars, it is clear that the models disagree widely in the extreme northwestern corner of the map, off the coast of Victoria Island in British Columbia. Upon investigation, this disagreement was found to be a result of inherent differences in each model's methods of resolving the intense deep low pressure systems which form regularly in the Gulf of Alaska. Also supported by the animation is the supposition that maritime climates are not handled as well by the models as continental climates, as evidenced by the significant uncertainty of the Nor'easter track as it reaches the warm waters of the Atlantic.

The animations of the models' storm tracks, displayed simultaneously, reveal that a key component of uncertainty surrounding the storm is temporal discrepancy among the models. For instance, as the storm moves across the Appalachians, the uncertainty of the prediction is attributable not to differences in the predicted location of the storm center but rather to differences in the rate at which the storm will move along the predicted path. Here, the display shows the lack of synchronization in the model predictions. The spectral model carries the storm toward the Atlantic much more rapidly than do the other two models, but the storm centers predicted by the NGM and eta models eventually catch up to that of the spectral.

A feature of the system that provides further insight into the data is the interactive control over the animation provided by *Director*. The ability to change the pace of the movie or to step through the animation one frame at a time reveals details about the data that would be difficult if not impossible to detect otherwise. This was clearly illustrated in the early stages of the system's development, when initial animations of the fluid isolines appeared jumpy, indicating data processing errors. Stepping through the animation allowed us to easily pinpoint the aberrant frame, which would have been difficult to locate in a table or in a series of static maps.

4. Discussion

Since we made the decision to represent the model output as an average of the three short-term forecast models, meteorologists Vislocky and Fritsch (1995) have investigated the forecasting performance of an average of two of these three models and compared this average to the performance of each model individually. They have found that such an average (termed *consensus*) in fact outperforms each model alone. They call for a "strategy for statistically combining available forecast products rather than relying upon the single most superior product" (p. 1157). Though we had been unaware of their findings when making our initial data processing and representation choices, we are of course pleased that our project by its nature employs such a strategy.

Although emphasis in developing our prototype has been on particular spatial and temporal scales and on a particular kind of model derived data, the animation principles, interactive interface tools, and reliability representation methods implemented in the prototype are relevant to a broader range of environmental change applications. Since the interactive controls on our animations are similar to those provided by many hypermedia tools (including some available for use with the World Wide Web), results of building and assessing our prototype are relevant to a range of visualization applications well beyond those build within the *Director* environment.

As with most prototype visualization tools, the one described here is probably more valuable as a device for identifying unanswered questions than it is as an application tool that will facilitate scientific thinking. At this point, the prototype (due to the Macromedia *Director* multimedia development environment it was designed within) is "hardwired" to a particular data set. To be put into practice as a tool for scientific research, the basic operations described above would need to be implemented in a dataBdriven environment in which analysts can easily apply the tool to any data set of interest (including non-meteorological data). In addition, more flexible interactive controls are needed that will allow analysts to change symbolization used to depict each model, their combination, and the variation among them. We anticipate making these and other improvements as the project evolves.

References

Beard, M. K., B. P. Buttenfield, 1991. NCGIA Research Initiative 7: Visualization of Spatial Data Quality. NCGIA.

Buja, A., McDonald, J. A., Michalak, J. and W. Stuetzle, 1991. Interactive data visualization using focusing and linking. Proceedings, Visualization '91, IEEE Conference on Visualization San Diego, CA, pp. 156-163.

Buttenfield, B. P.,1993. Representing Data quality. Cartographica, special issues on Mapping Data Quality, 30(2&3): 1-7.

Buttenfield, B. P. and M. K. Beard, 1994. Graphical and geographical components of data quality. Visualization in Geographic Information Systems. London, John Wiley & Sons. D. Unwin and H. Hearnshaw, (ed.) 150-157.

Buttenfield, B. P. and C. R. Weber, 1994. Proactive graphics for exploratory visualization of biogeographical data. Cartographic Perspectives 19: 8-18.

Caplan, P. M. and G. H. White, 1989. Performance of the National Meteorological Center's Medium Range Models. Weather and Forecasting 4: 391-400.

Cleveland, W. S. and M. E. McGill, 1988. Dynamic Graphics for Statistics, Belmont, CA: Wadsworth & Brooks/Cole.

DiBiase, D., 1990. Visualization in the Earth Sciences. Earth and Mineral Sciences (Bulletin of the College of Earth and Mineral Sciences, the Pennsylvania State University), 59(2): 13-18.

DiBiase, D., J. Krygier, C. Reeves, A. M. MacEachren and A. Brenner, 1991. Animated Cartographic Visualization in Earth System Science. Proceedings, 15th International Cartographic Conference, Bournemouth, England, 23 Sept. - 1 Oct., 1991, ICA. 223-232.

DiBiase, D., MacEachren, A. M., Krygier, J. B. and C. Reeves, 1992. Animation and the role of map design in scientific visualization. Cartography and Geographic Information Systems 19(4): 201-214.

DiBiase, D., C. Reeves, J. Krygier, A. M. MacEachren, M. von Weiss, J. Sloan and M. Detweiller, 1994. Multivariate display of geographic data: Applications in earth system science. Visualization in Modern Cartography. Oxford, UK, Pergamon. A. M. MacEachren and D. R. F. Taylor ed. 287-312.

Dorling, D., 1992. Stretching Space and Splicing Time: From Cartographic Animation to Interactive Visualization. Cartography and Geographic Information Systems 19(4): 215-227, 267-270.

Dorling, D. and S. Openshaw, 1992. Using Computer Animation to Visualize Space-Time Patterns. Environment and Planning B: Planning and Design 19: 639-650.

Egbert, S. L. and T. A. Slocum, 1992. EXPLOREMAP: An exploration system for choropleth maps. Annals of the Association of American Geographers 82(2): 275-288.

Fegeas, R. G., J. L. Cascio and R. A. Lazar, 1992. An overview of FIPS 173, The spatial data transfer standard. Cartography and Geographic Information Systems 19(5): 278-293.

Fisher, P., 1994. Hearing the Reliability in Classified Remotely Sensed Images. Cartography and Geographic Information Systems 21(1): 31-36.

Grotjahn, R. and R.M. Chervin, 1984. Animated graphics in meteorological research and presentations. Bulletin of the American Meteorological Society 65(11):1201-1208.

Hibbard, W. and D. Santek, 1989. Visualizing large data sets in the Earth Sciences. IEEE Computer 22:53-57.

Hibbard, W., L. Uccellini, D. Santek and K. Brill, 1989. Application of the 4D McIDAS to a model diagnostic study of the Presidents' Day cyclone. Bulletin of the American Meteorological Society 70(11): 1394-1403.

Junker, N. W., J. E. Hoke and R. H. Grumm, 1989. Performance of NMC's Regional Models. Weather and Forecasting 4: 368-390.

Karl, D., 1992. Cartographic Animation: Potential Research Issues. Cartographic Perspectives 13(Fall): 3-9.

Koussoulakou, A., 1994. Spatial-temporal analysis of urban air pollution. in Visualization in Modern Cartography MacEachren, A.M. and D. R. F. Taylor (ed). London: Pergamon, pp. 243-269.

Kraak, M.-J., 1994. Interactive modeling environment for three-dimensions maps: Functionality and interface issues. Visualization in Modern Cartography. A. M. MacEachren and D. R. F. Taylor (ed). London: Pergamon. 269-287.

Kraak, M.-J., A.M. MacEachren, 1994. Visualization of spatial data's temporal component. SDH94 (Sixth International Symposium on Spatial Data Handling), Edinburgh, Scotland, 5-9 September, 1994, IGU, pp. 391-409

Leidner, M., 1995. Pennsylvania State University, personal communication.

MacEachren, A. M., 1992. Visualizing Uncertain Information. Cartographic Perspectives 13(Fall): 10-19.

MacEachren, A. M., 1994. Visualization in modern cartography: setting the agenda. in Visualization in Modern Cartography MacEachren, Alan and D. R. F. Taylor (ed). London: Pergamon, pp. 1-13.

MacEachren, A. M., 1995. How Maps Work. New York: Guilford Press.

MacEachren, A. M. and J. H. Ganter, 1990. A pattern identification approach to cartographic visualization. Cartographica 27(2): 64-81.

MacEachren, A. M., Howard, D., von Wyss, M., Askov, D. and T. Taormino, 1993. Visualizing the health of Chesapeake Bay: An uncertain endeavor. Proceedings, GIS/LIS '93, Minneapolis, MN, 2-4 Nov., 1993, pp. 449-458.

MacDougall, E. B., 1992. Exploratory analysis, dynamic statistical visualization, and geographic information systems. Cartography and Geographic Information Systems 19(4): 237-246.

McGranaghan, M., 1993. A Cartographic View of Spatial Data Quality. Cartographica 30 (2&3): 8-19

Max, N., R. Crawfis and D. Williams, 1993. Visualization for climate modeling. IEEE Computer Graphics and Applications 13(4): 34-40.

Mitasova, H., 1993. Multidimensional interpolation, analysis and visualization for environmental modeling. Proceedings, GIS/LIS '93 Minneapolis, MN, 2-4 Nov., 1993, pp. 550-556.

Moellering, H., 1980. The real-time animation of three dimensional maps. American Cartographer 7: 67-75.

Monmonier, M., 1989. Geographic brushing: Enhancing exploratory analysis of the scatterplot matrix. Geographical Analysis 21(1): 81-84.

Monmonier, M., 1990. Strategies for the visualization of geographic time-series data. Cartographica 27(1): 30-45.

Neeman, B.U. and P. Alpert, 1990. Visualizing atmospheric fields on a personal computer: application to potential vorticity analysis. Bulletin of the American Meteorological Society 71(2): 154-160.

Petersen, R., 1992. Comparisons of CFM and NGM lifted index calculations. Weather and Forecasting 7: 536-541.

Peterson, M. P., 1995. Interactive and Animated Cartography. Englewood Cliffs, NJ: Prentice Hall.

Prentice, I. C., Bartlein, P. J. and T. I. Webb, 1991. Vegetation and climate change in eastern North America since the last glacial maximum. Ecology 72(6): 2038-2056.

Schiavone, J.A. and T. V. Papathomas, 1990. Visualizing meteorological data. Bulletin of the American Meteorological Society 71(7):1012-1020.

Treinish, L.A., 1992. Correlative data analysis in the Earth Sciences. IEEE Computer Graphics and Applications May: 10-12.

van der Wel, Franz, R. M. Hootsmans and F. J. Ormeling, 1994. Visualization of data quality. in Visualization in Modern Cartography MacEachren, Alan and D. R. F. Taylor (ed). London: Pergamon, pp. 313-333.

Vislocky, R. and M. Fritsch, 1995. Improved model output statistics forecasts through model consensus, Bulletin of the American Meteorological Society 76(7): 1157-1163.

Wilhelmson, R., B.F. Jewett, et al., 1990. A study of the evolution of a numerically modeled severe storm. The International Journal of Supercomputer Applications 4(2):20-36.

Contact address
Elizabeth Dirks Fauerbach[1], Robert M. Edsall[2], David Barnes[3] and Alan M. MacEachren[4]
Department of Geography
Penn State University
302 Walker
University Park, PA 16802
USA
[1]E-mail: efd106@psu.edu
[2]E-mail: edsall@essc.psu.edu
[3]Email: dbarnes@essc.psu.edu
[4]E-mail: alan@essc.psu.edu

Dynamic graphics to accompany this paper can be found on the WWW at:
http://www.gis.psu.edu/MacEachren/MacEachrenHTML/febm.html

SOME PERCEPTUAL ASPECTS OF COLOURING

UNCERTAINTY

B. Jiang[1], A. Brown[2] and F.J.Ormeling[3]

[1]Institute of Geographic Sciences
Free University of Berlin
Berlin, Germany
[2]Department of Geoinformatics
International Institute for Aerospace Survey and Earth Sciences (ITC)
Enschede, the Netherlands
[3]Cartography Section, Faculty of Geographical Sciences
University of Utrecht
Utrecht, the Netherlands

Abstract
This paper is intended to provide more insights into intuitive perceptual aspects of colouring uncertainty, based on the perception experiments conducted. In total, five different kinds of colour scales and four bivariate colour schemes were investigated for the representation of uncertainty and data combined with uncertainty respectively. It is concluded that instead of the saturation scale attempted in some previous investigations, a lightness scale starting with white is relatively easily perceived as one of the best solutions. It is shown that a random dot scale, as an alternative to colour scales, provides an intuitively good understanding of uncertainty. It is also shown that bivariate schemes using a random dot scale on both axes, or a random dot scale on one axis and a tint scale on the other, have advantages over spectrally encoded schemes for visualizing data combined with uncertainty, as well as for overlaying of attribute information of different phenomena.

Key words
Uncertainty, visualization, colour variables, perception.

1. Introduction

Colour is one of the most important visual variables for communicating map information. In the past few years, an increasing body of literature has focused on colour for the visualization of uncertainty (e.g., Brown and Van Elzakker 1993, MacEachren 1992, Howard and MacEachren, 1995). Many assumptions have been made regarding colouring uncertainty. However, little attention has been paid to the perceptual aspects, e.g., how users interpret uncertainty encoded with colour.

Schweizer and Goodchild (1992) examined whether or not connotative implications of grey are effective in displaying data uncertainty on a computer screen. Their study was conducted on the basis of the HSV colour system. Figure 1 shows that in this system the visual variables lightness and saturation are not used as independent variables (as they are, for example, in the Munsell system). This could be one of the reasons for the conclusion that there is not an intuitive association between data quality and value (lightness) as displayed by the HSV colour model, when value is used to represent data quality and saturation is used to represent data quantity.

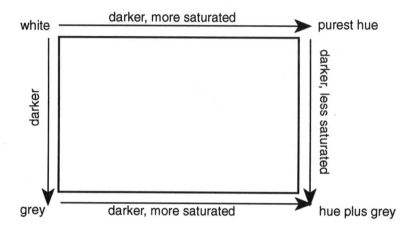

Figure 1 Lightness and saturation change of a bivariate colour scheme.

The basic principle used by Schweizer and Goodchild (1992) is that of bivariate representation. However, Olson (1981) concludes that bivariate choropleth maps are not intuitive schemes, and Tufte (1983) called them 'puzzle graphics'. It is conceivable then that Schweizer's subjects did not understand, in the first place, the basic concept of a bivariate map legend.

Brown and Van Elzakker (1993) considered that the interpretability of bivariate schemes could be improved if the colour variables were used strictly independently. Their investigation was designed for categorical area information, with the variable hue used to represent the category and the variable saturation (with lightness being kept constant) used to represent the uncertainty of the information. The hypothesis that this scheme would be intuitively perceptible was not, however, tested by experiment.

The goal of the research described by this paper was to investigate the perception of colour used to represent uncertainty. Two aspects of uncertainty representation were investigated: uncertainty in the transition zones between adjacent classes in numerical, classified data, and the representation of data combined with uncertainty. For the former, various colour scales are the possible representation methods; while the bivariate colour schemes can be used to represent the combined uncertainty.

2. Colour variables

Three colour variables have been discussed for uncertainty visualization (Brown and Van Elzakker 1993). In the following, one more variable, random dots, is included for the sake of this study, although strictly speaking it is not a colour variable.

Hue is what we mean when we refer to colours by names such as red, green, blue etc. Physically, it is the visual stimulus provided by a dominant wavelength or combination of wavelengths. Hue is referred to as colour in Bertin's system of visual variables (1983). Hue is usually used to represent qualitative information in cartography.

Lightness is another colour sensation which is also mentioned as one of the seven visual variables. Lightness is defined as the lightness or darkness of achromatic and chromatic colours. *Value* is a common synonym for lightness, used for example in the Munsell system and by Bertin. The visual variable lightness has ordered and selective perception properties, so it may well be used to represent ordered or relative quantitative information.

There are two kinds of lightness scales. One is lightness(+), starting from white and ending with the purest colour. Another is lightness(-), starting from black, and ending with the purest colour. These scales can be thought of as tint and tone scales respectively.

Saturation is the third colour sensation: it is the term given to describe the relative purity of a colour. Although it is not included as a separate visual variable in Bertin's system, more and more arguments are being put forward to treat it as such (Brown and Van Elzakker 1993; MacEachren 1994). A possible reason why it was not included in the original seven is the fact that it is rather difficult to use it in practice as an independent variable to represent quantitative information, since changes in saturation are often accompanied by changes in value.

There is a need to make a distinction between a true saturation scale and the lightness(+) or the lightness(-) scale. A true saturation scale has constant lightness from grey at one end to the purest colour at the other. In a lightness(+) or lightness(-) scale, in addition to the change in lightness, saturation increases along the scale from the white or black end respectively to the pure hue end.

Random dots are similar in form to texture in Bertin's system of visual variables. By a variation in the density of random dots differences in given data values can be conveyed. By changing their density, random dots are as useful as lightness and saturation in representing quantitative or ordered information. A higher density of dots gives the sensation of a darker and purer colour, and a lower density provides the opposite sensation. On a white background, random dots therefore appear as very coarse tints, and they can be used in much the same way.

Grey. In the achromatic (grey) scale, the variables hue and saturation do not exist, only the lightness variable.

3. Principle of colouring uncertainty transition

In many chorochromatic maps, different colours are used to represent different numerical or ordered data classes, referring to areas (polygons). Examples of the former include average slope or average pH. Examples of the latter include terms like 'clayey sand', 'sandy clay', 'moderately eroded' and 'severely eroded.

In the case of strictly numerical classes, Boolean logic is often applied to decide the class of each area. In practice, however, the data collection methods are very rarely precise or accurate enough to allow the strict application of Boolean logic. The usual situation is that, for data values lying near a class boundary, we cannot be sure as to which class the data actually should belong. Data values falling exactly on the boundary have a 50% chance of belonging to the class on either side. There exists therefore a certainty transition zone astride each class boundary, in which this chance increases to 100%. The width of the zone depends on the quality of the original data. Usually, classes are chosen in such a way that these transition zones do not overlap: i.e. the classification is certain near the centre of each class.

In the second case, that of ordered data classes using verbal descriptions, the problem is even greater. Now we have to deal not only with possible inaccuracies in the data, but also with the 'fuzzy' linguistic concepts. When exactly does a 'clayey sand' become a 'sandy clay'?

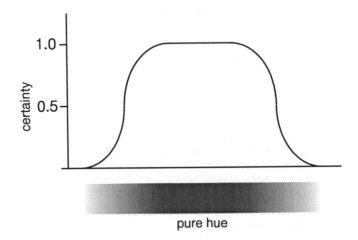

Figure 2 Colouring uncertainty transition (original in colour).

The certainty transition zones are often represented gaphically by membership function curves, though Hootsmans (1996) suggests that straight lines might be more suitable in many cases. Figure 2 shows an example of symmetrical hypothetical curves on either side of a class. Certainty of membership of any particular data value in the class decreases from 1 in a range around the centre to 0.5 at the class boundaries to 0 at some distance outside the class. This can be represented visually by, for example, a tint scale in which the colour changes from pure hue at the centre of the class to white for zero certainty, i.e. absolute certainty of non-membership. If this technique is used, it entails producing a separate map for each data class. For clarity, a different hue is used for each class. Note that due to the limitations of monochrome reproduction, the tint scale is represented in Fig. 2 by a grey scale. The necessity to produce separate maps is one of the main disadvantages of the method, especially in the case where (expensive) hard copies are required.

Instead of a tint scale, i.e. a lightness(+) scale, it may be possible to use other colour scales, e.g. a lightness(-) scale, a saturation scale, a grey scale, or a random dot scale. One of the objectives of this research was to find which of these scales is intuitively perceived to be the best for the purpose.

4. Principle of colouring combined uncertainty

In the visualization of uncertainty information, one of the key issues is how to merge data quality and attribute data values together for informed exploration. MacEachren (1992) addressed three alternative solutions to the problem of representing the map data and data quality: map pairs, sequential presentation, and bivariate maps. In a later article, Howard and MacEachren (1995) suggest in addition the use of a type of bivariate display in which symbol overlays are used to represent data quality.

A second issue concerns linguistic uncertainty, e.g., the actual meaning of the words *high, medium* and *low*. We may want to overlay two sublayers based on these linguistic notions e.g., a combination of *high(severe)* erosion with *low* crop suitability. The representation methods mentioned above (map pairs, sequential representation and bivariate maps) may also be used to represent this combination. Some reports are available of tests carried out on two of these methods, i.e. map pairs and bivariate maps.

4.1 Map pairs and multi-window

In their design of a graphic user interface, Goodchild et al. (1994) once used multiple juxtaposed windows to display different class memberships after image classification. According to the principle of spatial visual mixture, human beings do have the ability to mix two maps in their minds and generate a mental structure about the combination of the two maps. The recognition of coherent spatial patterns does, however, require a high degree of   spatial correlation (Burrough et al, 1996). Goodchild et al. did not investigate the perceptual efficiency of multiple displays, but it seems possible that they provide only a general rather than a detailed visual combination. Howard and MacEachren (1995) suggest that displaying map pairs can be the first step, after which the analyst can select among various options for overlay and merging of maps.

4.2 Bivariate colour schemes

Bivariate colour schemes are used to visualize bivariate data in a detailed combination, and they can be encoded in different ways. It seems probable that the particular colour encoding system used has considerable influence on the interpretability of the bivariate colour scheme.

Olson (1981) investigated several spectrally encoded schemes as used by the US Census Bureau on paper maps. Her conclusions are that the test subjects did not in general find the maps too complex, but they did find single maps on each subject easier to read than the bivariate maps. She also concludes that a clear and prominent legend is necessary for the interpretation of bivariate maps, and that test subjects could not spontaneously order the colours into the legend arrangement used. This appears to indicate that such spectrally encoded schemes, which in effect use only the variable hue in a consistent way, are not intuitive. According to Tufte (1983), such schemes (which he called 'puzzle graphics') must be interpreted through a verbal rather than a visual process.

A second type of bivariate scheme is based on a combination of a tint scale of a hue along one axis and a grey scale along the second axis as used by Schweizer and Goodchild. We have already seen that this is not an intuitive scheme, when used to represent both data quantity and data quality. For comparison purposes, a scheme of this was type was tested in the experiments described by this paper.

Another type of bivariate scheme is based on lightness (+), i.e., tint scales of two different hues along mutually perpendicular axes. The corner opposite to white is then the subtractive mixture of the two hues. For example, it is blue if the hues used are cyan and magenta, and it is black if the hues used are red and green. Brewer (1994) refers to this type of scheme as a sequential / sequential scheme. This type of scheme was also tested in our experiments.

A modification of this idea is to use random dot (tint) scales of the two hues instead of a percentage tint scale. In this case, the maximum percentage coverage of the random dots is about 50%, since the dots of the two hues should not overlap. The corner opposite to white is perceived not as a uniform grey but as a mixture of, for example, red and green dots. Clearly, this kind of display was not possible in the days of analogue cartography. This scheme was among those tested in this study, since it was felt that it might be easier for inexperienced users to interpret than the percentage tint version. The reasonable extension of the randomly encoded bivariate scheme is to combine a random dot scale with a lightness(+) colour scale. Furthermore, extension to trivariate cases appears to be possible without perception problems. A different technique, not pursued further in this research, is to depict the relative proportion of two or more phenomena by intermingled stripes or a regular pattern of square cells. When white stripes or cells are included, the absolute values of the mapped phenomena can also be shown by the width of coloured stripes or the number of coloured cells. This type of display can, however, create visually disturbing patterns.

5. Perception tests

5.1 Subjects

Cartographic students (n = 30), from 16 different countries (4 Asian, 7 African, 2 European, 3 American) participated in the perception tests on uncertainty visualization. They all had graduate training in cartography. The basic assumption is that if those subjects with a solid background in cartographic visualization could not correctly interpret certain visualization schemes, it would be at least as difficult for most other people, which would mean the visualization schemes would be unacceptable. The reverse, however, does not necessarily hold true.

5.2 Research hypotheses and questionnaire

Five hypotheses concerning the visualization of uncertainty were considered for the perception tests.

Hypothesis 1: On a computer monitor, both lightness and saturation scales can be used to visualize linguistic uncertainty (e.g. the linguistic degree triplet, *low, medium, high*), but the lightness(+) scale is better for this purpose than the saturation scale.

Hypothesis 2: If quantitative area information is visualized according to the principle of "The more, the darker" or its opposite "the more, the lighter", then saturation can be used to visualize uncertainty according to the principle of "the more certain, the purer".

Hypothesis 3: It should be possible to visualize the uncertainty of linguistic degree triplet area information by a technique in which each class is represented by randomly distributed dots of a single hue, and the uncertainty is visualized by an overaly of random dots of another hue.

Hypothesis 4: Fog is a significant metaphor for the visualization of uncertainty. A fog effect can be produced by blurring boundaries on a map of linguistic degree triplet area information combined with the use of the principle "the closer to white, the less certain". i.e., use of the lightness(+) scale.

Hypothesis 5: With suitable uncertainty visualization schemes, it is easy to set up a mental structure for linguistic notions. Overlay results can be perceived intuitively, if suitable bivariate colour schemes are used.

Based on these hypotheses, a questionnaire was constructed for the purpose of the experiment. There was a logical structure behind the questionnaire. Questions 1 to 10 examined whether or not colour and random dot scales can be used in colouring uncertainty transitions. So the question is, what are the most suitable scales? Questions 1 to 5 referred to colour and grey scales presented as continuous variables in a single rectangle. In questions 6 to 10, these continuous scales were divided into a large number of classes (10) and used on sample choropleth maps, which contained no actual boundary lines.

Question 11 examined the fog effect on an intuitive visualization. In this case, the lightness(+) scale was used again, but with deliberately blurred boundaries on the map. The last four questions dealt with the intuitive ease, or otherwise, of interpreting bivariate maps, with a view to their use in presenting quantitative data simultaneously with information on data quality. The choropleth maps used for this test were different than those of the previous part, but again no boundary lines were shown.

5.3 Procedure

All the perception tests were carried out using a computer monitor for the display. Since all the displays were constructed using a white background, the appearance was not very different from printed hard copies. Once a preliminary questionnaire was drafted and the test maps produced, a pilot test was conducted using 6 students. On the basis of the pilot test, the time allocation for each question was obtained, and some changes were made in the wording of questions for the sake of clarity. For instance, to avoid possible confusion with double negative type terminology, i.e. 'low uncertainty', all relevant questions were worded in terms of certainty in the final version of the questionnaire. For each of the six main sessions, 5 subjects received the test, which they completed using a computer with a 14 inch colour monitor. First, they were informed about the purpose of this experiment which was to evaluate the visualization schemes, not their map skills. Sufficient time was made available for them to understand each question. Every effort was made to avoid bias due to the learning effect. In particular, different groups of subjects were presented with the displays in a different sequence, and the hues employed for some of the tests were also changed between the scales and the maps.

6. Results and discussion

Figure 3 shows a monochrome version of the colour scales used in the first part of the questionnaire. In this version, the hue is represented by the fine screen grey, while the grey scale (i.e. tints of black) is represented by the coarse screen. On the monitor, hue scales and grey scales appear equally smooth, and on colour printer output, the same screen ruling is used for the scales of both the hue and the black. Table 1 below shows which end of each scale was selected by the respondents as the high certainty end.

A general conclusion from this test is that the scales which intuitively are best perceived are the lightness(+), random dot and grey scales. The variable value is dominant in these scales, reinforced by the variable saturation in the case of lightness(+) and random dots. In the lightness(-) scale, the variable value conflicts with the variable saturation, but it is the value (in terms of blackness) which dominates. In this particular scale the hue used was yellow, so there was a large lightness contrast between the two ends of the scale. In the saturation scale, there was little value difference between the two ends, but the hue used in this case was red. This hue may have given a 'warning' impression, a conclusion which was reinforced by the results of the tests using maps.

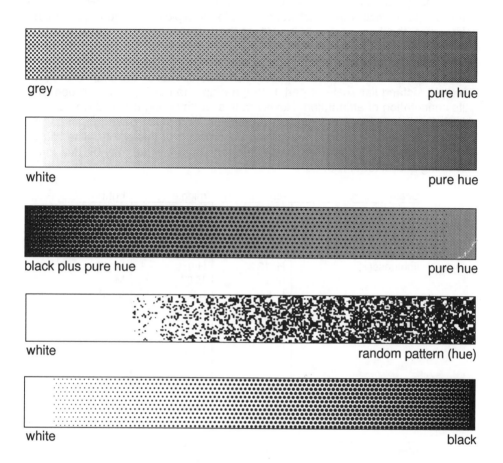

Figure 3 Different colour scales for uncertainty visualization.

Scales	Left end	Right end
saturation	70	30
lightness(+)	7	93
lightness(-)	70	30
random dots	7	93
grey	7	93

Table 1 Percentage selection of high certainty end.

In the part of the questionnaire dealing with sample maps, in each case three map areas were selected from near the two ends and the middle of the scale used. The areas were indicated on the maps by the letters A, B and C, at random. The respondents were asked to indicate for each area the degree of certainty: high (H), medium (M), and low (L).

Generally speaking, the map situation produced less unanimous responses than the questions relating to the simple scales. This is to be expected, since the maps exhibit the patchwork of colours typical of choropleth maps. However, the best scales of part 1 of the questionnaire remained the best scales here too (Table 2), with the lightness(+) scale as the best scale of all. The table has been arranged in the sequence left-middle-right of the scales of Fig. 3. Note that the saturation scale performed better in the map than in the simple rectangular scale of part 1. In the map, the hue used was green, with its connotation of affirmation. However, it is worth noting that red was used in both the lightness(+) and random dot scales, and these gave in general the expected results, despite the warning connotation of this colour. The slightly lower score for the dark (red) end of the random dot scale may be due to the warning character of red.

Scale	left	middle	right
saturation	H 27 M 27 L 46	H 10 M 63 L 27	H 63 M 10 L 27
lightness(+)	H 13 M 3 L 84	H 0 M 97 L 3	H 87 M 0 L 13
lightness(-)	H 72 M 4 L 24	H 4 M 73 L 23	H 24 M 23 L 53
random dots	H 20 M 0 L 80	H 6 M 87 L 7	H 74 M 13 L 13
grey	H 10 M 20 L 70	H 7 M 77 L 16	H 83 M 3 L 14

Table 2 Percentage selection for colour scales used on sample maps
(underlined letters are the expected choices).

One question related to the use of the fogginess metaphor. The lightness (+) scale map with sharp edges was compared to a 'foggy map' i.e., the same map with blurred edges. The respondents were asked to compare these maps by answering a multiple choice question. Few responded with the expected answer, that the map with blurred boundaries is based on less certain data. The reason could be that the respondents (trained cartographers) associated blurring with low quality mapping rather than low quality data.

The last four questions related to the interpretation of bivariate map schemes. The purpose was to find out the extent to which respondents were able to interpret these schemes correctly, without being able to refer to a legend. Four different schemes were used, as described (in words) in Figure 4.

In order to reduce the number of variables under consideration, and to concentrate on the perception aspects, the questions used with these maps made no reference to quality. A typical question was as follows:
"High soil erosion is visualized by the varying density of red dots and low ground water salt content is visualized by the green scale. Four areas, A, B, C and D are indicated on the map. Please describe briefly below your interpretation of each area in terms of soil erosion and ground water content." The areas chosen were near the corners of the bivariate schemes.

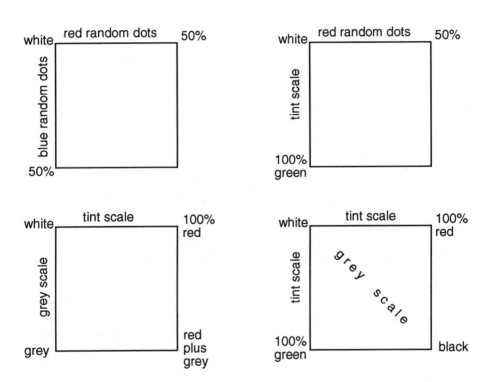

Figure 4 The four bivariate schemes used in the perception tests.

Some of the questions used the combination of surface water pollution with air purity. Also, various combinations of "high" and "low" were used. This part of the questionnaire gave less clear results than the earlier parts. The respondents did not always understand the questions, and the answers were sometimes ambiguous. Also, with so many variables being tested, and a high proportion of unusable answers, the sample was on the small side.

Tables 3 and 4 show typical responses. In each case, the underlined figures indicate the expected responses. Note that the only difference between these maps is that in the first case, blue random dots are used to indicate high ground water salt content, while in the second case blue random dots are used to indicate low ground water salt content. The performance of the first map was clearly better, supporting the view that the concept of 'darker equals lower' should be avoided.

In addition to this, a few other conclusions could be reached with some confidence. The bivariate schemes using tints of red and green and tints of red plus grey respectively gave the poorest results. In the absence of a legend, most respondents appeared to have difficulty in conceiving which colours would lie along the diagonal of these schemes. The schemes using random dots, either twice or in combination with a tint scale, performed better (see for example Table 3).

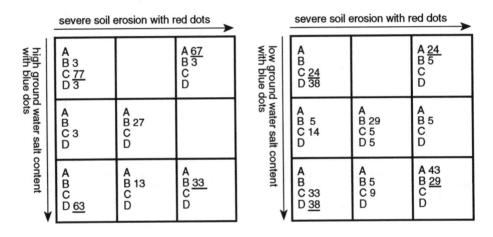

Tables 3 and 4.

In all the schemes, the actual choice of hue probably had some influence, particularly the use of red, as discussed above. Clearly, more research is required before definite conclusions can be drawn.

7. Conclusions

Successful cartographic visualization depends on correct perception of the map displays. Much of the perception research conducted to date has been based on paper maps, although in more recent research there has been more emphasis on the perception of screen displays. The question arises as to how many of the conclusions based on experiments with one display medium can be carried over to another. In the case of this research, a monitor was used as display medium, but the maps were designed on a white backgound to give a reasonable similarity to maps on paper. The research presented displays to respondents with no explanatory legends, in order to test intuitive perception.

It is rather difficult to combine certainty information with quantitative map data referring to areas, since both are quantitative, and the cartographer would prefer to use the same visual variable, lightness, for both. One approach, which particularly suits the flexibility of screen displays, is to present a separate map for each data class, and for each area to indicate the level of certainty with which that area belongs to that class. The first part of the research was an attempt to find which colour order schemes are intuitively best perceived, for this purpose. The most successful are lightness(+) scales (i.e. tint scales), grey scales and scales with an increasing density of a random dot pattern. The dark ends of these scales are almost always taken to mean the most certain data. A saturation scale, with little or no change in lightness, was less successful, which tends to contradict earlier work of one of the authors (Brown and Van Elzakker, 1993). In the second part of the research, the same colour order schemes, divided into classes, were applied to choropleth maps. The results confirmed in general the tests on the continuous scales. Comparing the results, however, it was noted that the hue selected had a large influence on the intuitive perception of the saturation scale. The use of red, with its warning connotation, may be taken to mean 'the purer the red, the greater the uncertainty', while when green is used, the understanding is 'the purer the green, the greater the certainty'.

The problem with this display principle is that it requires a large number of maps. It may be a suitable technique to use in the data analysis phase, but it is less suitable for a general overview. For this situation, experiments were conducted on four bivariate colour schemes, used on choropleth type maps. Bivariate schemes using tints of two colours, that is spectrally encoded schemes, were less successful than those using two random dot scales, or a random dot scale superimposed on a tint scale.

The bivariate schemes were not tested explicitly on the concept of uncertainty. A reasonable conclusion which can be drawn from all the tests is, however, that a good technique to represent data plus uncertainty for quantitative area maps (e.g. choropleth maps) is to use a tint scale to represent the quantitative data classes, overlain with a random dot pattern to indicate certainty. Green would be a suitable colour to use for the random dots, with its connotation of agreement. In this type of display, the random dots should not occupy more than 50% of the area, for greatest certainty, in order not to obscure the underlying tint. This type of representation requires that the tint be removed under the random dots, in order to retain the hue of the dots.

The visual variables used in this investigation apply to both paper maps and screen displays. However, other sensory variables such as animation (e.g. blinking) and sound apply only to 'soft copy' output. If the user does not require paper copies, then these other sensory variables could be used to provide information on aspects of spatial data such as data quality or certainty.

References

Bertin J. (1983), Semiology of Graphics, Graphics Press.

Brewer C. A. (1994) Color Use Guidelines for Mapping and Visualization, In: MacEachren A. M. and Taylor D. R. F. (eds.) Visualization in Modern Cartography, Pergamon Press, pp. 123 - 147.

Brown A. and van Elzakker C. P. J. M. (1993), The use of colour in the cartographic representation of information quality generated by a GIS, Proceedings of ICC'93, pp. 707-720.

Burrough P. A., van Gaans, P. F. M. and Hootsmans, R. M. (1996), Continuous classification in soil survey: spatial correlation, confusion and boundaries, *Geoderma* (in press).

Goodchild M. F., Chih-Chang L. and Leung Y. (1994), Visualizing Fuzzy Maps, In: Hearnshaw H. M., and Unwin D. J. (eds.) Visualization in Geographical Information Systems, Chichester: John Wiley & Sons.

Hootsmans R. M. (1996), Fuzzy Sets and Series Analysis for Visual Decision Support in Spatial Data Exploration, Ph.D. Thesis, University of Utrecht, The Netherlands.

Howard D. and MacEachren A. M. (1995), Constructing and Evaluating an Interactive Interface for Visualizing Reliability, Proceedings of 17th International Cartographic Conference of the ICA, Barcelona, pp. 320-329

MacEachren A. M. (1992), Visualizing Uncertain Information, Cartographic Perspectives, No. 13, pp.10-19.

MacEachren A. M. (1994), Time as a Cartographic Variable, In: Hearnshaw H. M., and Unwin D. J. (eds.) Visualization in Geographical Information Systems, Chichester: John Wiley & Sons.

Olson J. M. (1981), Spectrally Encoded Two-variable Maps, Annuals of the Association of American Geographers, Vol. 71, No. 2, pp.259-276.

Schweizer D. M. and Goodchild M. F. (1992), Data Quality and Choropleth Maps: An Experiment with the Use of colour, Proceedings of GIS/LIS'92, pp. 686-699.

Tufte E. R. (1983), The Visual Display of Quantitative Information, Cheshire: Graphics Press.

Contact address

B. Jiang[1], A. Brown[2] and F.J.Ormeling[3]
[1]Institute of Geographic Sciences
Free University of Berlin
Arno-Holz-Str. 12
12165 Berlin
Germany
Phone: +49 30 838 3892
Fax: +49 30 838 6739
E-mail: bjiang@gauss.geog.fu-berlin.de
[2]Department of Geoinformatics
International Institute for Aerospace Survey and Earth Sciences (ITC)
P. O. Box 6
7500 AA Enschede
The Netherlands
Phone: +31 53 4874 501
Fax: +31 53 4874 335
E-mail: brown@itc.nl
Enschede, the Netherlands
[3]Cartography Section, Faculty of Geographical Sciences
University of Utrecht
P. O. Box 80115
3508 TC Utrecht
The Netherlands
Phone: +31 30 2531 373
Fax: +31 30 2540 604
E-mail: F.Ormeling@frw.ruu.nl

GENERALIZED OPERATORS FOR SCULPTING DIGITAL

TOPOGRAPHIC SURFACES

Caroline Westort

Department of Geography
University of Zürich
Zürich, Switzerland

Abstract

Geographic Information Systems (GIS's) have evolved with the ability to handle the large datasets characteristic of landscape, but have heretofore focused on the display and analysis of elevation data, rather than on active tools for manipulating the geometry of digital topographic surfaces. Manipulation strategies in many GIS's allow only either 'local' or 'global' changes to a digital terrain model: A local change is a change that happens to a single vertex; a global change is one that affects the entire topographic dataset, e.g., scale changes, which exaggerate the Z value [Ervin and Westort '95]. What is generally missing is the specification of a coherent set of tools for the *regional* scope of action. This paper surveys the range of non-digital and digital surface manipulation methods in order to specify a generalized toolset which meets the need for regional geometric control and which is conceptually suitable to implement. The toolset consists of four parts: (1) Generalized operators: carving/swathing, drilling/pulling, smoothing/roughening, cutting/joining, (2) A parameterized topographic primitive library: e.g., swales, berms, hills, etc., (3) Dynamic operators: e.g., narrow/widen, straighten/twist, raise/lower, and (4) Generic functions borrowed from Computer Aided Design (CAD): transformations (rotate, translate, re-scale), selection, position placement, input/output. The implementation plan is to abstract the four toolbox components into classes and to develop a prototype iteration of a landscape sculpting process using elements from each essential class.

Key words

Digital sculpting, DTM manipulation, landscape architecture, terrain generalization, surface deformation.

1. Introduction

Computers are used extensively in many fields to assist the designing process. Their implications for the design of topography are equally dramatic. The computer can act as manager and imaging device [Fischer '85], but also as a means to combine different and previously uncombined manipulation methods for a digital surface. Domains such as civil engineering, landscape architecture and planning, geology, geomorphology, and the military, among others, routinely use cut and fill calculations, visibility analyses, and sectional profiles to analyze data. As the number of terrain data bases grows, the need and demand to provide easier, faster, and more flexible methods to modify original static elevation data has increased. And as more participants become involved from different domain areas, each set of users must be given the opportunity to modify the terrain to suit their own unique and specific scenarios.

Heretofore, the regional scope of action has been neglected in arenas traditionally concerned with digital surface manipulation. Geographic Information Systems (GIS's) have evolved with the ability to handle the large datasets characteristic of landscape, but heretofore have focused on the display and analysis of elevation data rather than on active tools for manipulating topographic geometry. Manipulation strategies in many GIS's allow only either 'global' or 'local' changes to a digital terrain model. Generally missing is specification of a coherent set of tools for the *regional* scope of action [Ervin and Westort '95].

While some operations exist to limit the scope of activity to a mask of pixels or a specific polygon, these techniques are not especially useful for achieving direct geometric control over a DTM region. *Geometric control* is comprised of the following seven criteria for optimally useful surface sculpting tools, revised from [Ervin and Westort '95]:

1) form any surface geometry/shape
2) operate at a 'regional scope of action'
3) dynamism /uniformity
4) provide visual feedback to a user
5) be quantitatively accurate
6) handle easily
7) have a quick response time (real time?)

Most approaches addressing the notion of regional geometric control in the GIS arena have been application specific, and also algorithmic and technical in emphasis. This paper surveys, as broadly as possible, non-digital and digital surface manipulation moves in order to first present the range of surface manipulation methods. The functional behavior of these tools, rather than their algorithmic or technical particulars, is highlighted, and from this their geometry-determining attributes isolated. Rather than aspire to be comprehensive in scope, the operations covered attempt to present main functional types in order to generalize them. To generalize, in the context of this paper, refers not to simplification of geometric features relative to cartographic scale changes. Rather it refers to an abstraction, and later an encapsulation, process for isolating geometry-determining operator functionality independent of model scale.

2. Non-digital methods

Non-digital methods of topographic manipulation include traditional sculpting, particularly the deformation of malleable materials like clay, contour line manipulation, maquette or model-making, and the behavior of agricultural machines and tools.

2.1 Traditional sculpting

Traditional sculpting is complex and often involves an assortment of tasks, materials, and intentions. The term sculpture is derived from the Latin *sculptura*, from *sculpere*, to carve or cut out of stone [Rich '47]. Common use of the term sculpture has a broader meaning, however, and embraces an abundance of materials and methods. The following is an inclusive list of *traditional sculpting activities*:
(1) carving, or chipping away at a stone, or other surface material [Rich '47], [Economist '95] (includes classical technique of pointing [Rich '47], [Duncan and Law '89])
(2) combinations of objects [Rich '47],[Kurmann '95a], [Duncan and Law '89]
(3) casting/moulding [Economist '95], [Rich '47]
(4) folding [Kurmann '95b]

The term *modeling* in the context of traditional sculpture refers to sculpting activity 1 and/or 2, conducted with a malleable material, like clay. Modeling in the context of landscape architecture or architecture, by contrast, refers to *model-making*, which typically involves sculpting activity number 2. In this latter case, maquettes are made at various levels of detail or scale to explore different design alternatives. Despite this distinction in terms, all of the above sculpting activities usually assume it is an object being created -- to be viewed 'in the round', i.e., from all angles. Characteristic to sculpting or modeling topography, however, is the model's continuous and relatively planar form (i.e., 2.5-D instead of full 3-D [Weibel and Heller '91]). The sculptural genre, 'relief' sculpture, whether *intaglio* or *basso-*, *mezzo-*, or *alto-relievo* [Rich '47], emphasizes the form and intent of sculpting planes, or continuous surfaces, versus objects. Activities 1 and 2 in particular intuitively lend themselves to landform relief manipulation if one assumes the metaphor of modeling clay. Functionality for these two activities are the ones further generalized for this application.

In other domain areas, control over surface geometry is accomplished by the *wielding* of a *tool* by a user, where wielding encompasses grasping a *handle* and *manipulation* of it to execute a *task* over a specifiable *area*[Krishnan, '94]. The field of robotics defines a sculpting task as an *operation* on a *target* by a tool, or *operator*, which results in a visible geometric change to a surface. This project borrows from both of the above definitions to define a tool as an operator. An operator takes a *geometric element* (i.e., a shape), and via a *function*, imposes a geometric change to another geometric element that functions as the target (eg. a terrain model). Scope of action and uniformity of a particular operator are important attributes of geometric control:
• *Scope of action* is the area of influence of a particular operator. In terms of DTMs, example scopes of action include *local* scope, which incurs changes to the value of only one point (or vertex) in a model, a *global* scope, which operates on all vertices in a particular data set, and a *regional* scope. Regional scope consists of a *move*, which is an incremental geometric operation performed between two points, or a *pass* which is a collection of moves. See [Ervin and Westort '95] for further discussion of this distinction. Regional scope means the area of influence

is user-specifiable in shape and size. A regional scope may also consist merely of a line, for example.

- *Uniformity* refers to the number of geometry-altering operations able to be performed simultaneously by one operator. If a particular operator smoothes while it carves, for example, it has a uniformity of minus two (-2) because two operations are being performed at the same time. If only carving were being done, on the other hand, uniformity would be *static* and have the value one (1). A uniformity generates a static change, while uniformity less than one dynamic. Geometric control is greatly enhanced by a user's ability to define the degree of uniformity a given operator will assume. Static or dynamic changes may behave in either a *relative* or *absolute* manner. Algebraic relationships such as >, <, >=, <=, maximum, minimum, average, etc. are examples of operators able to take on variable uniformity attributes because they are usable singly or in combination. The following are additional example operations, which if used in conjunction with, say, algebraic relationships would involve uniformity less than one. They are presented in pairs because members exhibit relative opposite behavior:

   twisting/straightening
   narrowing/widening
   lowering/raising

2.2 Contour lines

Contour lines are a 2-dimensional graphic technique used to represent 3-D landform on paper. They are lines which represent the same elevation along their entire length and commonly have regular vertical separations. They are the most popular and widely used method of representing landform on paper because they allow both the display of other information on the same map, and the designer to approximate the elevations of other points with relative accuracy [Landphair, et al '88]. Topographic manipulation via contour lines is typically a manual process which:

1) sets some preliminary elevations and lines of slope, including fixed floor elevations, main drainage pattern, and critical spot elevations, and then
2) draws connections between zones of major importance staying within functional, environmental, and legal constraints [Landphair, et al '88].

Regional changes to the landform via contour lines require a re-working of the elevations in the area of interest, followed by a reconfiguration of all major connections and lines relative to the change. Most regional changes require a subsequent global re-working of the line representations, making geometric modifications via this method labor-intensive, and therefore rather inflexible.

2.3 Model-making

Landscape design in most domains typically use layered models constructed out of cardboard or wood as site topographic models. Individual layers are cut out along contour lines superimposed on a piece of cardboard of a thickness that is to scale. The layers are then piled up and glued together. Geometric changes to these models involve first revision of the contour map, and then re-cutting and re-compilation of the individual layers accordingly. Because this method works via contour line abstraction, it too is a relatively inflexible means for regional geometric modifications.

2.4 Agricultural equipment & gardening tools

The following is a list of common earth-moving equipment or tools whose functionality is specific to movement of topography or soil [Foley, et al '87]:

- Bulldozer - a shovel, or blade, is moved along a path with the earth piling up before it. Typically used for digging, also used for levelling terrain and filling.
- Shovel - of various shapes and sizes; used for digging and filling soil
- Rake - used for levelling and roughening up soil surface texture
- Plow - of various shapes and sizes; used essentially to turn over the topsoil of earth; both for smoothing and roughening surface soil
- Dump-truck - used for transporting soil, fill
- Trowel - used for digging and filling, but essentially for smoothing
- Garden hose - used for smoothing and removing soil from a surface
- Wheel-barrow - used for transporting soil

2.5 Discussion

In transposing non-digital techniques to the digital context, one wants to focus on those tools offering optimal flexibility in terms of geometric control. Of the non-digital methods, traditional sculpting techniques and the behavior of agricultural equipment and gardening tools offer familiar and flexible operations for achieving desired surface geometry. Contour lines and layered models derived from a contour line map, though familiar to many people, are intuitive only to experienced experts, and are relatively inflexible at modifying regions of a landform surface. The behavior of agricultural equipment and tools can be said to exhibit the same operations on a surface as traditional sculpting activities 1, and carving or chipping away at a surface and combining objects, with the difference being that the equipment behaves on the real earth at a 1:1 scale. In considering only the moves made to a malleable clay surface and those made by agricultural machines and tools, it can be said they are identical and consist of the following generalized sculpting operators which expand on, and make more explicit, traditional sculpting activities 1 and 2:

- carving/swathing -- cutting or filling, respectively. Direction of operation is relatively parallel to the target surface plane
- drilling, piling/pulling -- cutting or filling, respectively. Direction of operation is relatively perpendicular to the target surface plane. (Includes the classical sculpting technique pointing which bores holes into a   stone down to specified depths and then the excess stone chipped away.)
- smoothing/roughening -- surface texture relaxed or exaggerated respectively, either with respect to a pattern, or randomly.
- cutting/joining -- detaching or adding portions of a model, respectively.

The above operations are presented in pairs to describe the relative duality of each member function. As discussed in the next section, in the digital domain many of the methods described merely attempt to mimic the above non-digital operations and simply automate them. Some digital methods, however, augment non-digital ones. By combining the recognizable and familiar components of non-digital methods together with, for example, expanded notions of uniformity and dynamic capabilities of 'digital-only' characteristics, 'functionally-richer' operators are conceivable and able to be generalized and implemented. In this way sculpting methods as we know them in the 'real-world' may be expanded and augmented in the digital arena.

3. Digital methods

Digital 3-D model manipulation strategies for topography specifically have evolved chiefly in the CAD and GIS digital domains [Ervin and Westort '95]; the former focusing on the parameterization of objects, the latter on combinations of spatial attribute information with digital topographic models in several representations (e.g., TINs, grids, contour lines).

3.1 Geographic Information Systems (GIS)

In the context of GIS, DTM sculpting is usually referred to with the term manipulation, which is included as one of the following principal tasks of a comprehensive digital terrain modeling system [Weibel and Heller '90]:
• DTM generation
• DTM manipulation
• DTM interpretation
• DTM visualization
• DTM application

Typical DTM manipulation tasks referred to in [Weibel and Heller '91]'s description include: editing, filtering, joining and merging, and converting. While all of these tasks impact model geometry, the following operators are commercially available for regional geometric control regardless of TIN or grid/lattice data structure:
filtering: Uniformity of one. For a rectangular selection set, apply a function uniformly over it. Example filtering functions applied to elevation data include:
• increase or decrease values by a set amount, i.e., exaggerate or depress all values.
• smooth or randomize all values
• screen all points for points above, below or between particular values
masking: Uniformity of one. For a given DTM, block out areas which the user desires to be left unchanged. Typically, a filter is then run over the model which can change the geometry of the surrounding areas. The geometry of the masked area remains unchanged.
contour lines: Variable uniformity. Typically manipulating DTM geometry via contour lines is an automation of manual techniques on paper (see Contour Lines above under Non-Digital Methods). Set critical points, slopes, etc. and then automate the computation of lines. Adjustments require revision of

relevant point elevations, lines, and perhaps slopes, and then a re-computation of all contour lines.

3.2 Computer Aided Design (CAD)

Landform design using 2-D CAD systems relies heavily upon the representation of 3-D form as contour lines; an abstraction well suited to representation, but as already discussed, cumbersome for design. Manipulating contour lines effectively with CAD systems remains a daunting challenge, requiring spline curves and geometric constraints that have nowhere yet been satisfactorily packaged for -- much less mastered by -- landscape designers [Ervin and Westort '95].

An obvious advantage of digital/virtual methods is the ability to directly manipulate a 3-dimensional representation. The following list, adapted from [Mealing '94] summarizes some of the 3-dimensional topographic modeling tools available in modern CAD software packages:

Data structures
a. B-Rep-- Boundary Representations -- The surface of an object is 'polygonized' and a description stored as a list of vertices, lines joining the vertices, and list of faces. (e.g. Triangulated Irregular Networks (TINs))
b. Primitives-- Library of simple, generic, 3-dimensional models (cube, sphere, cylinder, cone, torus, wedge, plane and others). These primitives can be scaled, translated and rotated within the application, often both interactively (such as with a mouse), and by numerical input.
c. CSG -- Constructive Solid Geometry -- An object is represented as a combination of simple primitives such as cube, sphere and cylinder. These basic solids are used as building blocks for more complex objects by means of a system which uses Boolean combinations (union, intersection, and difference) to describe the logical operations of adding two objects, subtracting one from another, or defining the overlap between two objects.
d. Voxels -- Volume modeling. Spatial occupancy enumeration divides 3-dimensional space into cubic units called voxels, or 3-dimensional pixels.
e. Swept Forms -- a 2-dimensional (XY) section 'swept' along a third (Z) dimension.
f. Extrusion -- a template is swept in a direction orthogonal to the plane in which it lies.
g. Surface of Revolution -- a 2-dimensional template, closed or open, rotated about an axis.
h. Skin -- the ability to construct a 'skeleton' of a form and then wrap a surface skin around it to create an object.
i. Patches -- same as skin, except using boundaries as the skeleton.

Geometric modeling with primitives (b above) such as cones and cylinders in the 1970's was considered promising, but by itself is currently deemed too limited for most landform sculpting. Yet implementation of the parameterized primitive approach is the data structure most widely found in the fields of architecture [Kurmann '95a, Mitchell '91], and with other approaches to digital sculpting [O'Rourke '88, Ursyn '93, Fisher '85]. Methods of geomorphologic

[Pike '88] and landscape architectural categorization of landform types [Landphair, et al '88] are also consistent with this approach. One disadvantages, however, is that topography is a continuous field, and libraries of topographic primitives, or forms, are distinct pieces and therefore difficult to synthesize seamlessly into a surface without simply placement on top of the original model. Another drawback to this approach is that the geometry of a topographic region must be known in advance for it to be parameterized and say, put into a library of forms. Such predictability of shape is often not possible.

Nevertheless, the primary strategy in CAD echoes traditional sculpting activity number 2 through composition of parameterized object primitives. Applying this method to topography, [Landphair, et al '88] presents the following forms, derived from contour line signatures. It is a potential initial set of 3-D topographic primitives to parameterize and have available in a library for integration into a continuous surface:

glen	ravine	hogs back ridge
river bottom	knoll	knob
butte	camel back ridge	bay and promontory
meadow	swale	fan
saddle	berm	mound/hill

[Ervin and Westort '95] is one attempt at implementing berm, swale, box, and mound forms with this paramaterized 'primitive' approach. In addition to 3-D parametric objects present in the library, it would also be logical for there to be 2-D forms as well, which could serve as geometric elements of an operator. Because CAD's approach exists in an 'object-realm', instead of a planar, continuous surface realm, the following transformations are typically provided by CAD programs to act on object primitives:

CAD transformations
revolve
translate
re-scale

Additional functionality characteristic of CAD programs, but which are less object-bound and more generic in their utility are the following sets of functions:

Selection   -- select points, lines, global areas, parameterized objects.

Input/output   -- read in, read out, open, close, save, scripting
Location   -- go to center, mid-point, end-point, intersection of a portion
                              of a model

3.3 Physics based

Physics based modeling determines behavior and form of objects and surfaces by physical properties. The motivation for physical research is the desire to visualize results [Foley, et al '87]. Most physics based modeling is constraint-based, and methods which deal with continuous surfaces seem to consist of those which use physical forces to elastically deform surface geometry [Terzopoulos and Hong '94]. Significant is that this method allows one to see the deformation respond 'live' to the operation being performed, in [Terzopoulos and Hong '94]'s case the drilling/pulling sculpting operation. This is also so for [Li '93] which utilizes a conservation of mass algorithm to mimic real-world behavior of soil as it is modified. Soil's movement is seen to assume an angle of repose in front of a bull-dozer blade as it is also being plowed. The disadvantage to these techniques with respect to topographic sculpting is again the absence of flexibly attained geometric control over the resultant surface. In the former technique, neighborhoods of points are adjusted which respond in intuitive ways to a user's pushes and pulls, yet the ability to control the different kinds of simultaneously sculpting operators, carving coordinated with elastic pulling, for example, limits flexible control over geometry. This is also the case with [Li '93], where realism of soil movement as it is plowed receives priority over user control over geometry resulting from moves. Significant, however, is that both the [Terzopoulos and Hong '94] and [Li '93] approaches result in a uniformity greater than one. Surface geometry changes appear to occur dynamically in response to a user's intuitive expectations. The goal, then, is to synthesize the notion of increased uniformity, or dynamism, together with other functionality offered by digital strategies to control regional surface geometry. The following section reviews research efforts directly preceding the strategy proposed in this paper to accomplish this.

3.4 Previous own research

Research efforts in regional digital topographic sculpting have been initiated by [Baer '96] and [Ervin and Westort '95]. The former emerges from a cartographic generalization context, the latter from landscape architecture. [Baer '96] has implemented digital equivalents of the following real-world hand-tools to edit rasterized and shaded digital surfaces: chisel, iron, and a particle-gun. All of the tools utilize a gridded matrix window at either a punctual, regional, or global scope of action. The values contained in the cells of this window specify in plan the geometric footprint of elevation values desired. This window matrix passes over a planimetric view of the original surface following the speed and direction of a mouse-defined path, leaving the specified matrix values behind. In this way, the carving and swathing operations of traditional sculpting activity 1 are achieved interactively. Variants on the chisel tool may also achieve drilling and pulling effects, while smoothing is accomplished via the iron tool which lowers the elevation values of the original surface relative to the matrix window's values. The particle gun fills only some of the cells in the window grid according to a particle distribution density, leaving the others empty, and giving the original surface a speckled, or roughened, geometry. [Baer '96] also provides mathematical

formulas with which to define pure geometric forms, like a cone, hemisphere, etc., along with opportunities to modify these formulas. The prototype landform design system of [Ervin and Westort '95] also borrows the simple raster array data structure common to GIS systems. The CAD modeling strategies skin, sweep, and Boolean combinations of Constructive Solid Geometry modeling are adapted and called a 'virtual bulldozer', to consist of three main parts:
1. An "extrude" tool, made of a "path" and a "blade"
2. A library of parameterized primitives
3. Boolean and algebraic combinations between primitives and extrusions.

At the same regional scope of action the simple metaphor of a two-dimensional profile, or 'blade', is swept along a three-dimensional trajectory, or 'path', leaving behind a modelled digital topographic surface. Conceptually similar to [Baer '96]'s, this 'path' and 'blade' system takes a simple rectangular blade and moving it along a path constrained to straight lines and large radius curves. A 'blade' is defined as a user-definable number of points that define a 2-dimensional profile. A 'path' is defined as a user-definable number of 3D points, i.e., a 3D polyline. The result of moving the blade along each segment of the path, from one point to the next , represents a move, while the collection of moves along the path represents a pass' A real bulldozer always cuts relative to the existing surface, as its treads are confined to travelling over the surface. For this virtual bulldozer, the points of a path may have no z-coordinate specified, in which case the tool will cut relative to the existing terrain, taking its base Z value from the terrain data. If a Z-value is specified in the path, the blade cuts along the absolute path given. A single path may have both specified and unspecified Z-values, allowing a mixed path. By making the Z-coordinate optional along the path, the virtual bulldozer has the ability to either accommodate or ignore existing topographic conditions of a site; to both mimic and go beyond the abilities of a real bulldozer. A library of topographic primitives, such as mound, berm, swale, torus, box, etc., each with their geometric properties parameterized is the second component of [Ervin and Westort '95]'s system. Primitives and extrusions can be combined into a single landform, using algebraic operators such as +, -, max, min, average, etc. These are the combinational operators which have the potential to make the individual sculpting operators take on additional uniformity.

Both [Baer '96]'s and [Ervin and Westort '95]'s approaches, like most other digital methods mimic real-world tools and can in fact achieve any desired geometry on a 2.5 D surface. The means for doing so may not meet the above stated criteria for geometric control, however. [Baer '96]'s and [Ervin and Westort '95]'s approaches are consistent with each other in that they both result in a uniformity of one. [Baer '96]'s approach, as a fully realized system, is mouse-driven, and therefore more interactive. The steadiness and velocity of the user's hand are the greatest determinants of resultant geometry, however. Repeated passes over a model with one tool, i.e., doubling back on the mouse's path, for example, offers only an additive relationship with the elevation values already there from a previous pass. The static uniformity of having only an additive combination available could be augmented by a greater range of combinatorial operators introduced in [Ervin and Westort

'95]'s strategy, especially when more than one operation is employed at one time. In this way greater model-fidelity could also be achieved if the dependence on hand-mouse motion were augmented. [Ervin and Westort '95]'s system is a batch system by contrast, with no links to a mouse.

3.5 Discussion

It is discernible from the above digital methods which ones go beyond merely automating non-digital surface manipulation techniques. The following apply to a regional scope of action and augment non-digital methods:
- Parameterized topographic primitives
- Quantitative accuracy -- computing power (more for analysis of slope, and cut & fill)
- Combination operators (more)
- Generic CAD functionality
  - transformation
  - selection
  - input/output
  - position location

A synthesis of the essential functionality which combines the digital and non-digital realms is attempted in the next section.

4. Generalized operators: a toolset

The notion of geometric control in the digital medium, therefore, has to do with providing the user with means to achieve any geometric form in recognizable and flexible ways. Borrowing geometry-determining functionality both from non-digital and digital realms results in a toolset consisting of the following four component sets:
I.   Sculpting Operators
carving/swathing
drilling/pulling
smoothing/roughening
cutting/joining
berm
II.   Library of Parameterized Topographic Geometric Forms
swale
retaining wall
retention basin
III.   Dynamic operators
twist/straighten
narrow/widen
raise/lower
Algebraic combinatorial operators
Absolute, Relative behavior

IV.   Generic CAD Functionality

Transformations	(revolve, translate, re-scale)
Selection	(points, lines, global areas, parameterized objects)
Input/output	(Read in, read out, open, close, save, scripting)
Location	(go to center, mid-point, end-point, intersection of a portion of a model)

The sculpting operators under I, and dynamic operators under II are toolbox components borrowed from non-digital methods offering a the familiar and recognizable functionality. III and IV offer functionality found primarily in the digital realm, and when used in combination with other toolbox components at a user-specifiable scope of action and uniformity offer new and useful sculpting possibilities.

5. Implementation notes

The above generalized categorization of operation types suggests an object-oriented implementation approach and class organization. The following root classes are apparent:

Root class:	Derived Classes:
CElevModel	- Ctarget
COperator	- CExtrude
	- CDrill-Pull
	- CSmooth-Roughen
	- CSolid-Operation
CGeometry	- CPoint-Based
	- CParameterized
CFunction	- CFilter
	- CAlgebraic
	- COperatorSpecific

The above classes are to be implemented in a *manipulation environment* that consists of three top-level tasks: Selection, Execution, Output. The subtasks are organized according to the following decision outline.

1. Selection
      A. Which Target?
            1) Which model?
            2) Which geometric element?
      B. Which Scope of Action?
            a) Which Point/Vertice?
            b) Which Region?
                  1. Along which Line?
                  2. Which Polygonal area?
                      a. Local
                      b. Global (entire dataset)

C. Which Operator?
    1) Which geometric elements (fr. library)?
    2) Which geometric parameters (specified through dialog box or via command line)?
    3) Which function?
        a) algebraic?
        b) filter
        c) operator specific

    4) What relationship between target, operator and function?
        a) static
        b) dynamic
        c) absolute
        d) relative

2. Execution of Target/Operator/Function relationship

3. Output
    A. Temporary
        1) Display
        2) File
    B. Permanent File

Acknowledgements
The author wishes to acknowledge Stephen Ervin, Martin Heller, faculty and students of the Landscape Architecture Department, Interkantonales Technikum-Rapperswil, and Werner Thie for fruitful discussions, and especially Robert Weibel and Geoff Dutton for reading and commenting on the draft manuscript. This research is supported by the U.S. Army European Research Office, Contract No. N68171-94-9153.

References
Baer, Hansrhdi; 1996: Interaktive Berbeitung von Gelaendeoberflaechen; Konzepte, Methoden, Versuche; Ph.D Dissertation, Geographisches Institut, Univerist@t Zhrich.
Duncan, J.P., Law, K.K, 1989: Computer-Aided Sculpture; Cambridge University Press.
Economist Magazine; April 30th 1995: "Science & Technology: Artisans of the tiny" p. 117, Winterthur.
Ervin, Stephen M.; Westort, Caroline Y.; 1995: "Procedural Terrain; A Virtual Bulldozer"; Proceedings - CAAD Futures '95; Singapore, September.
Ervin, Stephen; 1994: "Digital Terrain Models", Landscape Architecture Magazine, January.
Fisher, Robert N., Masters, Raymond J.; 1985: "Computer-Aided Sculpture: Visual and Technical Considerations", LEONARDO, Vol. 18, No.3, pp. 133-143.
Foley, J.; van Dam, A.; et. al.; 1987: Computer Graphics Principles and Practice; Addison Wesley Publishing, U.S.A..
Klenin, N.I., Popov, I.F., Sakun, V.A.; 1985: Agricultural Machines; Theory of Operation, Computation of Controlling Parameters and the Conditions of Operation, (transl. from Russian), Amerind Publishing Co. Pvt. Ltd., New Delhi.
Krishnan, Radha; 1994: "Planning Wielding Strategies for Dynamic Tasks"; Proceedings of the SPIE - The International Society for Optical Engineering; Vol. 2244, pp 151-62, for Conference: Knowledge-Based Artificial Intelligence Systems in Aerospace and Industry 5-6 April,Orlando, FL, USA.
Kurmann, David; 1995: "Sculptor - A Tool for Intuitive Architectural Design"; Proceedings CAAD-Futures'95, September, Singapore.
Kurmann, David; 1995: Conversation with Caroline Westort, ETH, Dept. Architecture, Computer Aided Design Laboratory --Honggerberg, Zhrich, Switzerland.

Landphair, Harlow C.; Klatt, Fred Jr.; 1988: Landscape Architecture Construction; Chapt. 2, Grading and Earthwork; Elsevier Science Publishing Co., Inc.,. 24-46.

Li, Xin; Moshell, Michael; 1993: "Realtime Graphical Simulation of Soil Manipulation"; Institute for Simulation and Training, University of Central Florida, 14.April.

Mealing, Stuart, 1994. Three Dimensional Modeling and Rendering on the Macintosh, Oxford, England.

Mitchell, William J.; 1991: Digital Design Media, Van Nostrand Reinhold, NY.

O'Rourke, Michael; 1988: "Sculpting with Computer Graphics: An Approach to the Design and Fabrication of Abstract Sculpture"; LEONARDO, Vol. 21, No. 4, pp. 343-350.

Pike, Richard J.; 1988: "The Geometric Signature: Quantifying Landslide-Terrain Types from Digital Elevation Models"; Mathematical Geology, Vol. 20, No. 5, pp. 491-511.

Rich, Jack C.; 1947: The Material and Methods of Sculpture; Oxford Univ. Press -- NY.

Terzopoulos, D.; Hong, Q.; 1994: "Dynamic NURBS with Geometric Constraints for Interactive Sculpting"; ACM Transactions on Graphics, Vol. 13, No. 2, Pp. 103-136, April.

Ursyn, Anna; 1993: "Planks, Programs and Art: Computer Graphics as a Sculptural Tool", LEONARDO, Vol 26, No. 1, pp. 29-32.

Weibel, R. and Heller, M.; 1991: "Digital Terrain Modeling", in Maguire, Goodchild & Rhind (eds.) Geographical Information Systems: Principles & Applications. London: Longman Scientific.

Weibel, R., and Heller, M.; 1990: A Framework for Digital Terrain Modeling. Proceedings of the Fourth International Symposium on Spatial Data Handling. Zurich: Vol. 1, pp. 219-229.

Contact address

Caroline Westort
Department of Geography
University of Zürich
Winterthurerstrasse 190
8057 Zürich
Switzerland
E-mail: westort@geo.unizh.ch

ENCODING AND HANDLING GEOSPATIAL DATA WITH

HIERARCHICAL TRIANGULAR MESHES

Geoffrey Dutton

Department of Geography
University of Zürich
Zürich, Switzerland

Abstract
Planetary geocoding using polyhedral tessellations are a concise and elegant way to organize both local and global geospatial data that respects and documents locational accuracy. After a brief review of several such spatial referencing systems, topological, computational, and geometric properties of one of them are examined. The particular model described in the remainder of the paper – the *octahedral quaternary triangular mesh* (O-QTM) – is being developed to handle and visualize vector-format geodata in a hierarchical triangulated domain. The second section analyzes the geometric regularity of the model, showing that its facets are relatively similar, having vertices spaced uniformly in latitude and longitude, and areas that vary by less than 42 % from their mean sizes. Section 3 describes some fundamental operations on this structure, including mapping from geographic coordinates into O-QTM addresses and back again, filtering map detail through the triangular hierarchy and associating locations that are close together, but in different branches of the tree structure. The final section outlines and illustrates a recent application of O-QTM to map generalization, using its multi-resolution properties to enable multiple cartographic representations to be built from a single hierarchical geospatial database.

1. Hierarchical polyhedral modeling of planetary locations

Mapmakers and others have attempted to model the earth as a polyhedron for many years, going back to at least the time of the German artist Albrecht Dürer (1471-1528), whose drawings of polyhedral globes appear to be the first instance of thinking about mapping the planet in this way. In the late 19th and early 20th centuries, a number of cartographers, such as Cahill (Fisher and Miller, 1944; also used by Lugo and Clarke, 1995 in their variant of QTM), reinvented this idea, projecting the land masses of the Earth to various polyhedra, then unfolding their facets into flat, interrupted maps. The best known of these is R. B. Fuller's Dymaxion projection, dating from the early 1940's (Unknown, 1943). Originally based on a cubeoctohedron, the Dymaxion Map was then recast as an icosahedron, oriented to the earth in a way that minimized the division of land areas between its 20 facets. Fuller devised a projection method -- only recently well-enough understood to implement digital algorithms for it (Gray, 1994; also see Snyder, 1992) -- that has remarkably little distortion.

Pre-computer polyhedral projections all were based on platonic or other simple shapes, and were intended as amusements or devices that let a paper globe be unfolded to lie flat. They did not attempt to deal with more complicated cases involving subdivision of polyhedral facets into smaller ones that would more closely fit the figure of the Earth and have more inherent accuracy. Fuller's geodesic domes are examples of how a physical polyhedron can be subdivided to give it greater physical strength per unit weight and span larger areas than any simple polyhedron is capable. Fuller apparently never treated the Dymaxion map in the same manner, probably because he viewed it as a physical structure that was nearly optimal for display of global thematic data, and bringing it to the next level of complexity (either 60 or 80 facets, depending on the method of subdivision) would make the map unwieldy to manufacture and manipulate, and achieve no particular benefit. Now, it's a different world.

The digital revolution in mapping and associated geoprocessing techniques have freed map-makers from such practical constraints and physical limitations, and over the past 30 years a number of approaches have blossomed for using polyhedra to index and display spatial data on a world-wide basis for a variety of purposes. Several such models are summarized in order to indicate how they are alike and in what respects their form and properties differ.

1.1 The Octahedral Quaternary Triangular Mesh (O-QTM) framework

Inspired by Fuller's maps and domes, the author developed a global digital elevation model in 1983 that divided a planet into facets defined by a concentric octahedron and cube. (Dutton, 1984). In 1988 this model was revisited, revised and recast as a tool for spatially indexing planimetric data in a geographic information system (Dutton, 1989). The cube was discarded, but the octahedron remained as a geometric basis that roots a forest of eight quadtrees containing roughly equal triangular quadrants (facets) that approximate a sphere quite closely after only a few subdivisions. Figure 1a depicts the basis of the O-QTM model, an octahedron embedded in a spheroid. Figure 1b illustrates the quadrant numbering scheme, using a map projection that renders every facet at each level as an isosceles right triangle. Use of this projection simplifies the computation of addresses, as figure 3 shows.

An early implemention of QTM was done by Goodchild and Yang (1991) in an NCGIA hier-archical spatial data structure (HSDS) project. It used a different facet addressing scheme from the above, centering on encoding and manipulating data in the multilevel triangular raster defined by the QTM grid. Twelve- and 15-neighbor chain encoding methods were used to identify connected components and enable intersection and dilation in a triangular raster. Visualization software was also created to let users at workstations orient a globe and zoom in to regions at various levels of detail. A similar, independent project was undertaken at NASA Goddard Space Center around the same time.

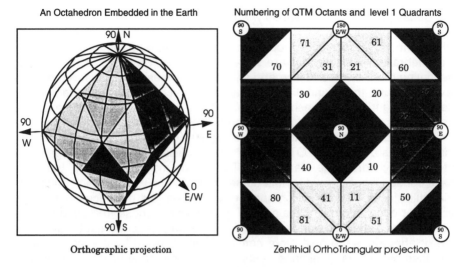

An Octahedron Embedded in the Earth   Numbering of QTM Octants and level 1 Quadrants

Orthographic projection   Zenithial OrthoTriangular projection

Figure 1a and Figure 1b.

The sphere quadtree (SQT) of Fekete (1990) recursively decomposes facets of an icosahedron into four triangular "trixels" to spatially index geodata (satellite imagery) and manipulate it directly on the sphere. The addressing scheme for SQT is similar to that for QTM, but relations between its 20 individual trees are more complex than those of QTM's 8 facets. Algorithms for finding neighbors, connected component labelling and other basic operations were developed, but further SQT development seems not to have taken place. Besides these closely-related approaches, Lugo and Clarke (1995) have applied QTM itself to indexing and compressing digital terrain models, and Barrett (1995) -- also of NASA Goddard -- has proposed using QTM to index astronomical catalogues (looking outward from Earth rather than inward). Otoo and Zhu (1993) developed a "semi-quadcode" variant of QTM, which was claimed to be optimized for efficient spatial access and dynamic display of data on a sphere.

Many discussions of this topic tend to hang up on the choice of an initial polyhedron, and then on what subdivision method should be used. The model explored in this paper is rooted in an octahedron, for reasons that are explained next. This is not to discount any other polyhedral approaches, which also make sense given the types and constraints of the applications their models are intended to support. But the principal one with which we are concerned – managing vector GIS data for multi-resolution analysis and display – centers on identifying and negotiating conflicts among vertices and edges of digital descriptions of point, line and area features. Because the focus of this work is in the vector domain we do not maintain and manipulate (hierarchical or other) raster images of map data. As section 4 shows, we substitute QTM addresses for coordinates of map features, then analyze these codes at lower levels of resolution to simplify features and negotiate conflicts among their vertices to retrieve consistent, simplified feature descriptions at smaller scales.

1.2 Consequences of different polyhedral bases

All polyhedral triangulations other than regular tetrahedra, octahedra and icosahedra have facets that are non-equilateral or vary in size and shape, becoming more diverse in size and shape as the mesh densifies. As this is a law of nature that cannot be defeated, why choose an octahedron for developing a triangular mesh on a sphere? Any of the platonic solids might do as well, or even less regular polyhedra, such as a cubeoctahedron or a rhombic dodecahedron (divided into triangles). Why does this choice matter, inasmuch as descendent QTM facets will vary in size and shape, regardless of what polyhedron one starts with? The choice may be arbitrary, but does have consequences. If we restrict ourselves just to polyhedra which have only triangular faces, each of which has the same area (but not necessarily equal angles or edge lengths), there are still quite a few to chose from, especially if truncated and stellated ones are included. The simplest case is a tetrahedron; the octahedron is the next more complex one, followed by the icosahedron, and then by various archimedean solids. Given that each facet of the original polyhedron roots a quadtree of triangles that grow smaller and more heterogeneous as they subdivide (see table 3), two consequences in particular need to be considered:

- The larger the spherical area subtended by the initial polyhedral facets, the more will the areas of sub-facets vary at any given detail level or map scale;
- The greater the number of initial polyhedral facets, the more often will map features cross facet boundaries (involve traversing two or more data trees).

These phenomena, while inevitable, are both undesirable, because each tends to complicate certain kinds of computations. The first one penalizes simpler shapes such as tetrahedra and octahedra by making the variance of error of facet areas greater than necessary. The second one complicates data management of shapes such as icosahedra by partitioning them into larger number of rooted trees, each with three edge neighbors and at least four vertex neighboring trees in the forest. These other trees must be examined when a feature or a search path crosses an edge or vertex a neighbor shares with the tree of the facet currently being traversed.

These two constraints offset one another, so that one can trade areal equality for structural simplicity. The balance to be struck could depend on the application for which QTM data is to be used. For example, an environmental sampling regime may assume equal probability of inclusion of random points in a sampling grid, which means that the areas of all facets should be as nearly equal as possible (which generally requires map projections). An early example of this approach to tessellation was the work of Wickmen et al (1974). A better-known example is documented in White et al (1992), describing the EMAP hierarchical data model developed for the US Environmental Protection Agency. Statistical considerations (minimizing distance variance between sample points) pointed toward using a basis polyhedron with a large number of facets, and the one selected (a *truncated icosahedron*, familiar as a *soccerball*, and also as the *Fullerene* carbon molecule) has 32 of them, 20 hexagons and 12 pentagons. Figure 2 shows this shape, its soccerball variant, and how it is used in the EMAP program. Orienting the polyhed-ron in this particular way allowed the 48 conterminous US states to fit in one hexagonal facet; but inevitably, this strategy dissected other nations and territories in inconvenient ways.

Truncated Icosahedron	Soccerball	EMAP Model

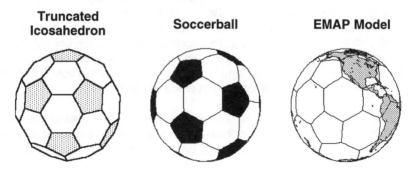

Illustration courtesy of J. Kimmerling, Oregon State University

Figure 2.

Structural simplicity is greatest in a tedrahedron, which has 4 facets, 6 edges and 4 vertices. Each facet has 3 edge neighbors and 3 vertex neighbors (both the same), a total of 3 unique neighbors. The EMAP soccerball, on the other extreme, has a more complicated set of neighborhood relations; the hexagonal facets have 6 edge and no additional vertex neighbors, and the pentagonal facets have 5 neighbors each. The average number of neighbors is thus   (20*6 + 12*5)/32, or 5.625. An icosahedron has exactly 9 total neighbors per facet, and an octahedron has 6 neighbors.

Another way to quantify this is to compare the total number of edge neighbors, a measure of structural complexity that indexes the number of expected transitions between facets. This number ranges from 6 for a tetrahedron to 12 for an octahedron to 30 for an icosahedron to 90 for a truncated icosahedron. When one considers how relatively often it may be necessary to relate data in adjacent trees, the larger polyhedra start to look less attractive, and an octahedron may be a wise compromise, with only 12 edge adjacencies. Table 1 summarizes the parameters discussed above; *N/F* means neighbors per facet; *VN* is edge neighbors; *EN* is edge neighbors; *TN* is total neighbors.

Basis Shape	V	E	F	VN/F	EN/F	TN/F
Tetrahedon	4	6	4	2	3	3
Ocathedron	6	12	8	3	3	6
Icosahedron	12	30	20	5	3	9
"Soccerball"	60	90	32	5.625	5.625	5.625

V = Vertices; E = Edges; F = Faces.

Table 1.

509

1.3 Basic properties of the octahedral framework

Perhaps the most compelling reason for using an octahedron as a basis for a QTM is not its topological properties, but the fact that it can be readily aligned with the conventional geographic grid of longitude and latitude. When this is done, its vertices occupy cardinal points and its edges assume cardinal directions, following the equator, the prime meridian, and the 90th, 180th and 270th meridians, making it simple to determine which facet a point on the planet occupies. Each facet is a right spherical triangle. Except for the one at the South Pole, all vertices are located in ocean areas, minimizing node adjacency problems for most land-based geospatial data. Table 2 defines the octahedral facets when vertices are at cardinal points.

O-QTM numbers octants from 1 to 8, proceeding clockwise in the northern hemisphere from the prime meridian, then continuing in the same direction in the southern hemisphere. One simple function that computes octant numbers is:

OCT = (1 + LON div 90) - 4 * (LAT - 90) div 90

This will yield an incorrect result at the South Pole but nowhere else. Negative longitudes must be complemented prior to computing octants. Given the number of an octant, it is easy to identify neighboring ones, which meet along octant edges except for the North neighbors of octants 1-4 and the South neighbors of octants 5-8, which meet only at the poles:

```
EAST_NEIGHBOR(OCT)    = 1 + (OCT + 8) mod 4 + (4 * (OCT div 5))
WEST_NEIGHBOR(OCT)    = 1 + (OCT + 6) mod 4 + (4 * (OCT div 5))
NORTH_NEIGHBOR(OCT)   = 1 + (OCT + 9 - 2 * (OCT div 5)) mod 4
SOUTH_NEIGHBOR(OCT)   = 9 - (OCT + 9 - 2 * (OCT div 5)) mod 4
                        - (2 * OCT mod 2)
```

The last two functions can easily be modified to eliminate vertex neighbors, if there is no need to handle such transitions. Overall, the octahedron has some useful properties and no significant disadvantages, other than generating subdivisions that have greater areal variation than do figures having a larger number of facets. An analysis and summary of QTM areal variation and its consequences is presented in section 2.

Octant	MinLon	MaxLon	MinLat	Maxlat
1	0	< 90	> 0	90
2	90	<180	> 0	90
3	180	<270	> 0	90
4	270	<360	> 0	90
5	0	< 90	-90	0
6	90	<180	-90	0
7	180	<270	-90	0
8	270	<360	-90	0

Table 2.

2. Areal inequalities in the O-QTM tessellation

We embed *QTM*'s initial octahedron in a sphere by placing its six vertices on the surface of a unit sphere (which may readily be scaled to Earth radius), such that the distance of the six octa vertices from the center of the octahedron (sphere) is unity. All other points (along octa edges and within octa faces) lie closer to the center than do the vertices, with the centroids of the eight facets lying at the smallest radius enclosed by the octahedron. As the QTM structure develops, it blossoms into a multifaceted polyhedron, having 32, 128, 512, 2048 ... faces. This section discusses the size and shapes of these planar facets, all vertices of which touch a circumscribed sphere, but whose edges are geodesic lines (chords through the sphere).

Unlike geodesic domes and related polyhedral world models such as described in section 1, O-QTM does not follow great circles on the sphere in decomposing facets. When a facet is subdivided, the latitudes and longitudes of pairs of its vertices are averaged to yield edge midpoint locations. Except along edges of the original octahedron (lat or (lon mod 90) = 0), these midpoints do not coincide with locations on great circles that connect existing vertices. Bisecting a facet edge with a normal vector, then extending the normal to the surface will also bisect the latitudes and longitudes defining the endpoints of that edge *only* along great circles, not along small ones (parallels). In addition, halving latitudes and longitudes creates a tessellation with certain asymmetries; east-west edges -- being parallels -- are straight, but the other two sets of edges are not straight and only roughly parallel. As a result, most of the triangles in the QTM network have different shapes and larger areas than would spherical triangles defined from the same set of vertices. Still, the chord length of any given edge segment will be identical to that of a great circle passing through the same endpoints, and can be computed using a standard formula (Snyder, 1987):

$$\sin(arc/2) = \{\sin^2[(lat1\text{-}lat2)/2]+\cos(lat2)\cos(lat1)\sin^2[(lon2\text{-}lon1)/2]\}^{1/2}$$

The chord length (in radians) is twice the value of this expression. Using this result, the areas of facets can be tabulated via the formula:

Area = $(s(s\text{-}a)(s\text{-}b)(s\text{-}c))^{1/2}$, where a, b, and c are the chord lengths, and s = (a+b+c)/2 (half the perimeter).

Plane facet size statistics based on these formulae for the first five QTM levels are present-ed in Table 3. Note how the total area of the figure rapidly approaches that of a sphere, and that the ratio of maximum to minimum facet area stabilizes at 1.83... Other indicators of uniformity are the decline of normalized deviations with level, and the decline of kurtosis values at higher levels from that of a normal distribution to values characterizing a uniform one (Bulmer, 1967, p. 64).

Area property	Level 1	Level 2	Level 3	Level 4	Level 5
Facets*	32	128	512	2048	8192
Min Area	0.28974	0.07554	0.01917	0.00481	0.00120
Max Area	0.43301	0.13188	0.03466	0.00877	0.00220
Max/Min Area	1.49448	1.74583	1.80803	1.82328	1.83333
Mean Area	0.32555	0.09346	0.02424	0.00612	0.00153
Median Area	0.28974	0.09342	0.02387	0.00611	0.00152
Std Dev	0.07164	0.01612	0.00354	0.00083	0.00020
Kurtosis	1.74493	2.91530	2.99556	2.44369	2.15422
Total Area	10.41775	11.96281	12.41066	12.52712	12.55654
Sphere Area	12.56637	12.56637	12.56637	12.56637	12.56637
% of Sphere	82.90183	95.19705	98.76091	99.68765	99.92174

*Statistics for columns are based on n = Facets/8 (single octant).

Table 3.

3. Computational properties of QTM identifiers

An O-QTM location code consists of an octant number (from 1 to 8) followed by up to 30 quaternary digits (from 0 to 3) which name a leaf node in a triangular quadtree rooted in the given octant. Central facets have an ID of 0, and corner facets take the ID of the vertex de-fining them; when a vertex appears, its number is assigned as six minus the sum of the IDs of the endpoints of the bisected edge, as figure 3 shows. Each QTM digit doubles linear pre-cision, identifying a specific facet having slightly more than one-fourth the area of its parent. For example, the 18-level QTM ID for the building housing the Geography Department at the University of Z_rich is 1133013130312301002; this encodes the geographic location 47_ 23' 48" N, 8_ 33' 4" E within about 60 meters(roughly the length of the building, and close to the limit of precision obtained from measuring on a 1:25000 scale topographic map), occupying a QTM facet about 1,000 meters square. Finer resolution, when available, is expressed by adding low-order quaternary digits to the ID; 30 digits will encode locations to an accuracy of about 2 cm., and can be expressed as one 64-bit word (Dutton, 1996). Note that these identifiers have a different sequence of quaternary digits in Goodchild and Yang's O-QTM system, and are different still in Fekete's I-QTM system (the former would identify the same triangular facet, giving it a different name; the latter would identify a facet having entirely different vertices and which would be 60% smaller in area for a given number of digits).

One basic operation of any geocoding scheme is determining proximity of points. To date no complete "native" solution for determining distance between two QTM IDs has emerged, but heuristics have been developed for closely-related computations. For example, one often needs to know if two QTM facets share an edge or vertex at some level of detail. There are three distinct cases:

1. Both facets are interior cells (both IDs terminate in 0)
2. One facet is an interior cell (its ID terminates in 0)
3. Neither facet is an interior cell (neither ID terminates in 0)

The first case rules out edge adjacency, as 0 cells only have non-zero neighbors (but not conversely). The second case is also trivial, because any 0 cell's neighbors will also be its siblings, and their IDs will be identical except for the last digit. The third case is more interesting; while we cannot easily affirm that two such facets are neighbors, it is easy to identify many cases where they are not. First, the QTM numbering scheme guarantees that all facets that share a common vertex must have the same terminal digit, either 1, 2 or 3 (six facets will share a vertex except at the six vertices of the initial octahedron, where four do). Second, two facets will be adjacent only if their QTM IDs differ in exactly one digit, and as the prior sentence asserts, this cannot be the terminal digit as long as case 2 is not true (this property is also one exhibited by Fekete's SQT model). Unfortunately, the converse is not true; many IDs that differ in only one digit are not adjacent. Regardless, if more than one digit of two IDs differs, one can be sure the two facets are not neighbors. A geometric (rather than lexical) approach to solving this problem is described in section 3.2.

3.1 Conversion from geographic coordinates to O-QTM IDs

Computing a quaternary address for a geographic location involves recursively identifying child facets of the octant containing the location, and giving each one visited an appropriate name from the set {0,1,2,3}. Goodchild and Yang (1992) project octants to equilateral triangles, and apply algorithms that require exponentiation and irrational arithmetic. The approach used here projects to right isosceles triangles, and requires only linear arithmetic, which could be performed on properly-scaled integers. The Zenithial OrthoTriangular (ZOT) map projection (Dutton, 1991) is used to project the 8 octants to a square; the North pole occupies the center, and the South pole maps to the four corners (see fig. 1b). Each octant is a right isosceles triangle, as is each QTM quadrant at every level of detail. ZOT space uses *Manhattan Metric*, so that distances are always the sum of x- and y-displacements. Figure 3 illustrates this method across two levels of detail.

New nodes in the triangular mesh appear at edge midpoints and are assigned numbers by subtracting the sum of the nodes of the bisected edge from 6, as figure 3b shows.

3.2 Conversion from O-QTM IDs to geographic coordinates

It is possible to directly convert a QTM ID back to latitude and longitude at any level of detail it contains, within the accuracy that level of detail denotes. The location computed is an arbitrary point within a triangle, normally chosen as its centroid. One method for doing this uses the same algorithm described in figure 3, computing triangle vertices in ZOT space, but skipping the point-in-triangle test, as a QTM ID specifies what triangle to describe next.

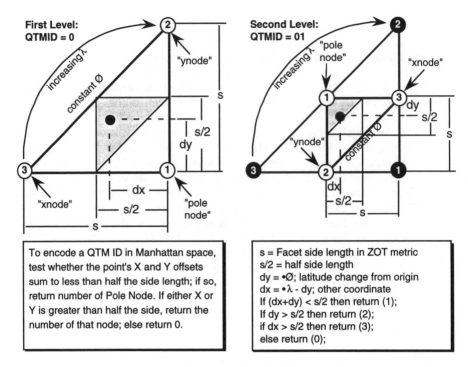

Figure 3a and Figure 3b.

The ZOT centroid of the leaf facet is then de-projected into latitude and longitude. A more direct and faster method to compute centroids of facets at the desired level of detail, uses a 3-axis coordinate system. The axes connect octant vertices to their opposite edge midpoints (with a local origin at 30_N/S, 45_E/W) to define locations in each of the spherical right triangles, as figure 4 shows. Starting at the root, each digit is mapped to movement along one of these axes. However, because the central cell of each group of four siblings is always coded as 0, encountering a 0 digit denotes no movement, only a reversal of future direction along all axes. Movement along axes are accumulated in three registers, then projected to the North-South ("1") axis and scaled to yield a latitude, which is used in computing longitude. The following function definitions summarize this approach:

VOID QTMtoAxes(QTMID qid, INT qlen, INT level, BOOL vertex, REAL axes[3])

The function returns no value, placing its result in *axes*; it will compute a meaningful set of axial distances from any valid set of input parameters. The role of the *vertex* parameter is discussed below. The *axes* are converted to a latitude and a longitude by the functions:

REAL AxesToLat(REAL axes[3])
REAL AxesToLon(REAL axes[3], lat)

In *QTMtoAxes*, a distance parameter is initialized at 1/12 the mean circumference of the Earth, and is divided in half at each level. The same parameter is used in moving along each of the three axes. However, were we to model an ellipsoid rather than a sphere, we could generalize this method by implementing a *vector* of three distance steps (one for each axis, scaled according to an ellipsoid model). The *axes* vector that *QTMtoAxes* returns is then used to compute first a latitude, then a longitude for the QTM ID being evaluated, at the specified level of detail.

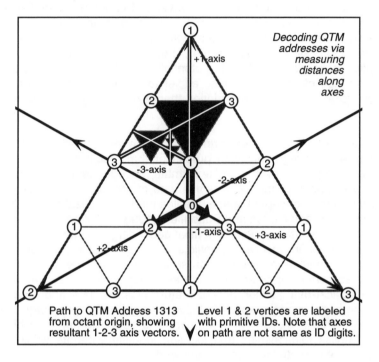

Figure 4.

We now return to the problem of determining if two QTM IDs represent neighboring facets or not. We wish to evaluate vertex as well as edge neighbors, because applications we pursue (for example, cartographic generalization) must identify nearby points that may not lie in the same quadrant or whose QTM IDs do not share a parent (even a com-mon ancestor). The above func-tions can do this when the *vertex* parameter of *QTMtoAxes()* is set to TRUE. Then, (unless the ID ends in zero) one more statement is executed. This causes the loca-tion of the vertex named by the last ID digit to be returned rather than the centroid of the facet (zero-coded (central) facets are not associated with any vertex other than their centroids, and thus have no vertex neighbors). While the axis vectors may differ considerably, they will evaluate to the same latitude and longitude for each of the six facets that share this vertex. Effects of round-off errors can be cancelled by comparing returned pairs of latitudes and longitudes to one another within a tolerance related to QTM level of detail. Locations in abutting octants require no special handling if longitudes are offset and axes initialized appropriately in each octant.

4. Application to cartographic generalization

The information content of QTM IDs is greater than the coordinates they represent, because the model denotes scale and accuracy as well as position. By encoding each coordinate of a cartographic database into QTM at levels of resolution appropriate to the source data (and these can change from one layer, one feature or even one vertex to another), a *multi-resolution representation* results. We consider this to be a potentially more efficient alternative to storing *multiple representations* of a set of features (i.e., a separate database for each scale), because QTM provides a unified description of model data that is potentially capable of rendering that data across a range of scales. It is also easier to maintain and edit cartographic features stored in a unified description than a set of multiple representations would be.

Current research involves exploring the above and other algorithms in a feature-based carto-graphic data processor that receives source data from GISs via cartographic exchange file formats, encodes features or layers into QTM representations, and extracts this data at source and smaller scales to test how well this concept of space handles map generalization. As described in Dutton (1996) and Dutton and Buttenfield (1993), QTM encoding provides a number of properties beyond hierarchical geocoding that can help identify and potentially assist in resolving conflicts for map space because of scale change. This requires additional data structures to achieve, but perhaps not as many as current approaches, such as Jones et al (1992), which uses a non-hierarchical triangulation across feature classes. This and other object-based methods model explicit relationships of map geometry (e.g., by topological simplices). Our approach models space rather than objects; it exploits implicit relationships that location identifiers denote within a multi-resolution global spatial reference framework.

Figure 5 illustrates an early result of using QTM to generalize vector map features. This set of polygons (the Swiss canton of Schaffhausen) had been previously filtered via the Douglas algorithm to 306 segments, of mean length 0.56 km. Analysis of this data indicated that all its details could be encoded in 18 QTM digits, so that it is represented at roughly 1:100 000.

The QTM-filtered representations decrease detail to 300 points (0.58 km) at QTM level 16, to 290 points (0.59 km) at level 15, to 258 points (0.65 km) at level 14, and to 200 points (0.79 km) at level 13. Typically, the filtering process results in little point reduction near the limit of resolution, with more changes appearing 2 or 3 levels down, then tapering off again after 5 or 6 detail levels.

QTM Generalization Results at Four Scales
Schaffhausen Canton, Switzerland

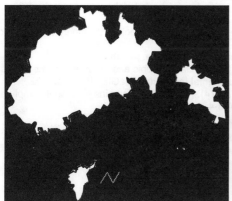

The approximate scales are based on the
assumption that the smallest line detail
that can be read off a map is 0.2mm
which is about 0.5 point on a printed page

Lines in this figure are drawn at 0.5 point

A map of Switzerland
that would fit on this
page would be drawn
at a scale of 1:2 M
(QTM level 14)
1:2M

1:4M

Figure 5.

The useful range of detail depends on various aspects of both the digitized data and its application, but would not normally exceed a ratio of about 1:250. Also, more sophisticated approaches to retaining, displacing and eliminating polyline vertices in the QTM domain are possible than the experiment illustrated above used, and will be explored during the course of our project.

References
Barrett, P. (1995). Application of the linear quadtree to astronomical databases. Astronomical Data Analysis Software and Systems IV, ASP Conf. Series, v. 77.

Bulmer, M.G. (1967). Principles of statistics, 2nd ed. Cambridge MA: MIT Press.

Dutton, G. *Geodesic modelling of planetary relief*. Cartographica. Monograph 32-33, Vol. 21 no. 2&3, 1984; pp. 188-207.

Dutton, G. Modelling locational uncertainty via hierarchical tessellation. Accuracy of Spatial Databases (M. Goodchild & S. Gupta, eds). Taylor & Francis, 1989. pp. 125-140.

Dutton, G. Zenithial Orthotriangular Projection. Proc. Auto-Carto 10. Falls Church, VA: ACSM/ASPRS, 1991, pp. 77-95.

Dutton, G. & B.P. Buttenfield. (1993). Scale change via hierarchical coarsening: Cartographic properties of Quaternary Triangular Meshes. Proc. 16th Int. Cartographic Conference. Cologne, Germany, pp. 847-862.

Dutton, G. (1996) Improving locational specificity of map data - a multi-resolution, metadata-driven approach and notation. Int. J. GIS, 10:3, pp. 253-268.

Fekete, G. (1990). Rendering and managing spherical data with sphere quadtree. Proc. Visualization '90. New York: ACM.

Fisher, I. and O.M. Miller (1944). World Maps and Globes. New York: Essential Press.

Fuller, R.B. (1982). Synergetics: Explorations in the geometry of thinking. New York: Macmillan, 2 vols.

Goodchild, M. and Yang Shiren (1992). A hierarchical data structure for global geographic information systems. CVGIP, 54:1, pp. 31-44.

Gray, R.W. (1994). Fuller's Dymaxion Map. CaGIS 21:4, pp. 243-246.

Jones, C.B., J.M. Ware and G.L. Bundy (1992). Multiscale spatial modelling with triangulated surfaces. Proc. Spatial Data Handling 5, vol. 2, pp. 612-621.

Lugo, J.A. and Clarke, K.C. (1995). Implementation of triangulated quadtree sequencing for a global relief data structure. Proc. Auto Carto 12. ACSM/ASPRS, pp. 147-156.

Otoo, E.J. and Zhu, H. (1993). Indexing on spherical Surfaces using semi-quadcodes. Advances in Spatial Databases. Proc. 3rd Int. Symp. SSD'93, Singapore. pp.510-529.

Snyder, J.P. (1992). An equal-area map projection for polyhedral globes. Cartographica 29:1, pp. 10-21.

Snyder, J.P. (1987). Map Projections - A Working Manual. US Geological Survey Prof. Paper 1395, Wahington: USGPO.

Unknown (1943). R. Buckminster Fuller's Dymaxion World. Life, March 1.

White, D., J. Kimmerling, and W.S. Overton (1992). Cartographic and geometric components of a global sampling design for environmental monitoring. CaGIS, 19:1, pp. 5-22.

Wickman, F.P., Elvers, E. and Edvarson, K. (1974). A system of domains for global sampling problems. Geografisker Annaler, 56A 3-4, pp. 201-211.

Contact address
Geoffrey Dutton
Department of Geography
University of Zürich
Winterthurerstrasse 190
CH-8057 Zürich
Switzerland
E-mail: dutton@geo.unizh.ch

LOSSY COMPRESSION OF ELEVATION DATA

Wm Randolph Franklin[1] and Amir Said[2]

[1]Electrical, Computer, and Systems Engineering Department
Rensselaer Polytechnic Institute
New York, USA
[2]DENIS - Faculty of Electrical Engineering
University of Campinas (UNICAMP)
Campinas, Brazil

Abstract
Generic lossy image compression algorithms, such as Said & Pearlman's, perform excellently on gridded terrain elevation data. We used 24 random 1201 x 1201 USGS DEMs as test date. Lossless compression required 2.12 bits per point on average, with a range from 0.52 to 4.09. With lossy compression at a rate of 0.1 bits per point, the RMS elevation error averaged only 3 meters, and ranged from 0.25 to 15 meters. Even compressingdown to 0.005 bits per point, or 902 bytes per cell, gave an indication of the surface's characteristics. This performance compares favorably to compressing with Triangulated Irregular Networks. If it is desired to partition the cell into smaller blocks before compressing, then even 32x32 blocks work well for lossless compression. However, for lossy compression, specifying the same bit rate for each block is inadequate, since different blocks may have different elevation ranges. Finally, preliminary tests suggest that the visibility indices of the points are robust with respect to even a quite aggressive lossy compression.

Key words
Lossy compression, Triangulated Irregular Network, terrain compression, elevation compression, visibility index.

1. Introduction

Earlier, in Franklin (1995), we studied lossless compression of DEM (Digital Elevation Model) terrain data. We showed that image processing algorithms work surprisingly well. On the average elevation data compressed losslessly to 2 bpp (bits per point),although this ranged from 0.5 to 4 bpp in our test cases. This compares favorably to storing the data as binary integers, in 16bpp, or even using gzip, at 4 bpp on the average.

The best algorithm was by Said & Pearlman (1995), Said &Pearlman (1993b), Said & Pearlman (1993a), Said (1994). There are several variants; here we use progcode. Progdecd, is a separate decoder program, which facilitates validating the results. Progcode can now handle arbitrary sized rectangular files with 1 or 2 bytes per point. It is a progressive resolution method, which first calculates an approximation to the file, and then successive refinements, so that the final result is exact. Progdecd can report what the mean squared error would be if the file was compressed lossily at various bitrates. We use this in this paper.

The main question with storing elevation data is how to compress it. The two main competitors are the Triangulated Irregular Network (TIN) and the regular array or grid. Accepting the TIN means accepting the principle of lossy compression since an exact TIN would require about a triangle for every point. Since each triangle requires many bytes, this would increase storage by perhaps an order of magnitude over simply storing all the points.

Since we've accepted that compressing elevation data may be lossy, the next question is whether the TIN is the right way To go. This paper answers that question, for many applications, "No". The TIN does have other advantages, such as that ridge and streamlines can be explicitly encoded. Nevertheless, it would not be as compact as compressing the DEM unless sophisticated, theoretical, data structures such as succinct graphs were used, (Jacobson 1989). This is a possible future research area. Alternatively, a curved surface could be fitted to each triangle in the TIN to increase the accuracy. Still, the TIN would seem better than a compressed array only if some criterion other the elevation accuracy were desired.

The general relevance of these results is that compression research in image processing in computer science is important in terrain data representation. This is something that was not totally obvious a priori, since the data are so different. This is important since so many resources are being thrown at the compression problem in image processing. We might expect even better results in the future.

However, this does follow the trend in computer science, in hardware at least, of general purpose solutions often being better than special purpose ones, because of the greater amount of resources devoted to the general problem. This is why most of the machines developed in all the following special purpose categories have failed: Lisp machines, floating point processors, database engines, special graphics engines, and parallel machines.

2. Review

Weibel (1992) filters gridded DEMs in various ways. He uses global filtering doing smoothing as in image processing by convolving with a 3 x 3 or 5 x 4 filter. He compares this with a selective filtering to eliminate points that do not add anything to our characterization of the surface. He tests a 220 x 390 elevation grid to see whether generalization changes essentials of the terrain, such as hill shading and RMS error. Shea & McMaster(1989) also discuss generalization.

Chang & Tsai (1991) found that lowering DEM resolution hurt the accuracy of the calculated slope and aspect of the terrain. Carter(1992) shows that the 1 meter resolution particularly affects the aspect, causing a bias towards the four cardinal directions, and suggests smoothing the data. Lee, Snyder & Fisher (1993) analyze the effect of elevation errors on feature extraction. Fisher(1993) considers the effect on visibility.

One operation often performed on terrain data is visibility determination, (De Floriani & Magillo 1994). Puppo, Davis, deMenthon & Teng (1994) use a parallel machine to convert a DEM to a TIN. They scale the elevation to 8 bits and perform experiments on grids of up to 512 x 512, reporting results for a 128 x 128 grid. For example, for a 30 meter accuracy 497 of the 16,384 points are selected. This TIN is then used to calculate line-of-sight-communication in De Floriani, Magillo & Puppo(1994).

Drainage pattern determination is another frequent DEM operation, as described by McCormack, Gahegan, Roberts, Hogy & Hoyle (1993). Skidmore (1990) extracts properties of a location, such as being on a ridge line, from a DEM. Franklin & Ray (1994)and Ray (1994) do visibility calculations on large amounts of data.

The use of a linear quadtree with 2-D run-length encoding and Morton sequences is discussed in Mark & Lauzon (1984). The storage can be about 7 bits per leaf. Waugh (1986) critically evaluates when quadtrees are useful, while Chen & Tobler (1986) find that quadtrees always require more space for a given accuracy than a ruled surface. Dutton (1990) presents a region quadtree based on triangles, not squares. This quaternary triangular mesh defines coordinates on a quasi-spherical world better than a planar, Cartesian, system does. Leifer & Mark (1987) use orthogonal polynomials of order up to 6 and quadtrees for a lossy compression of three 256 x 256 DEMs. Their work anticipates ideas used in wavelets and in the best current image processing methods.

3. Compression algorithm overview

Here are some of the ideas in the Said & Pearlman compression algorithm. If x_1 and x^2 are two adjacent heights, we can transform them to $l = \lfloor (x_1 + x_2)/2) \rfloor$ and $h = x_1 - x_2$, which take the same space to store. If x_1 and x_2 are similar, then h will be small, and could be compressed into fewer bits.

The algorithm applies the above transformation to each pair of heights down each column of the array, and stores all the h's before the l's in each transformed column. Now, the bottom half of the transformed array will tend to have smaller numbers. Then each row of the transformed array is similarly transformed, leaving differences of differences of heights in the lower right quadrant. In the upper left quadrant, each number is the average of a 2 x 2 block of the original image. Next, this whole process is repeated on this half-resolution image, on the resulting quarter-resolution image, and so on. Finally, the resulting coefficients are compressed with, say, Huffman coding, which uses fewer bits to represent numbers that occur more frequently.

This method is better than well-known and widely used JPEG algorithm, which is not state-of-the-art any longer. Some states are in Franklin (1995).

4. Lossy compression experiments

Our test data are 24 level-1, 3 arc-second DEM files, or cells. We picked the first cell starting with each letter of the alphabet. They are listed in Table 1 , and shown in Figure 1 on the next page. We use only each cell's elevation data, which is a1201 x 1201 array of points, taking 2,884,802 bytes with the elevations stored as 2-byte integers.

The work was greatly facilitated by our environment, consisting of Unix workstations, the Apogee C++ compiler, and software such as Rokicki & Berry (1994), Bradley (1994), and Poskanzer (1993).

Cell Name	Stdev of Elevations	Gzip Size, Bytes	Progcode Size, Bytes	Progcode rate, Bits per Point
Aberdeen E	36.	222,245	167,629	0.93
Baker E	377.	1,395,574	626,961	3.48
Caliente E	335.	1,264,671	534,495	2.96
Dalhart E	87.	431,241	262,995	1.46
Eagle Pass E	88.	494,974	273,267	1.52
Fairmont E	34.	367,819	240,675	1.33
Gadsden E	74.	872,802	440,551	2.44
Hailey E	516.	1,566,137	610,836	3.39
Idaho Falls E	145.	455,373	270,683	1.50
Jacksonville W	7.7	120,413	93,909	0.52
Kalispell E	343.	1,421,867	682,352	3.78
La Crosse E	49.	1,028,266	612,162	3.40
Macon E	22.	508,171	302,247	1.68
Nashville E	37.	801,073	414,536	2.30
O'Neill E	74.	593,771	324,296	1.80
Paducah E	27.	581,892	318,113	1.76
Quebec E	140.	853,427	374,115	2.07
Racine E	17.0	134,653	94,068	0.52
Sacramento E	695.	1,467,837	737,945	4.09
Tallahassee E	26.	427,257	266,154	1.48
Ukiah E	514.	1,107,751	562,100	3.12
Valdosta E	9.3	219,191	166,315	0.92
Waco E	25.	492,679	291,135	1.61
Yakima E	305.	1,259,666	500,679	2.78
Average	166.	753,698	382,009	2.12

Table 1 The 24 lossless compression test cases.

First, let's see how lossless compression handles these cells. When compressed with the common text-compression program gzip using the default setting, the average resulting file size is 754KB, while with progcode, the average size is 382KB. The test cases vary considerably in compressibility, but gzip and progcode tend to track each other in this respect. Both of them track the standard deviation of the elevations from the mean elevation for that cell. The numbers are somewhat different in Table 1 than in Franklin (1995) since there we were using a 1024 x 1024 subset of each cell, while here we are using the whole 1201 x 1201 cell.

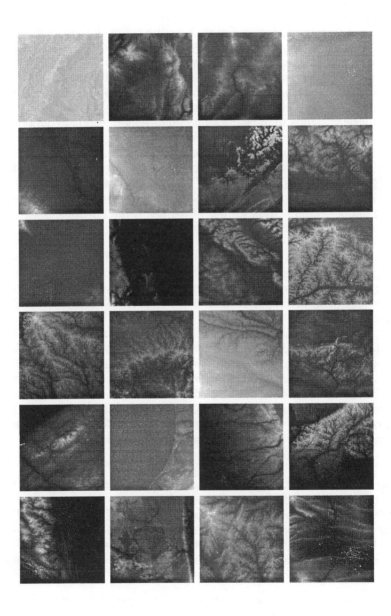

Figure 1 The 24 sample USGS DEMs.

How good is lossy compression? Our measure of goodness was the RMS error in elevations in the lossily compressed cell, compared to the original, as reported by progdecd. Image processors tend to report compression rates as bits per pixel (or point), or bpp. A cell compressed to 1 bpp would take 1201 x 1201/8 = 180300 bytes.

We tested each cell at bit rates of 0.005, 0.010, 0.015, and then from 0.02 up in steps of 0.02 bpp to the lossless compression rate, which is shown in Table 1. The results are plotted in Figure 2 on the following page for the first 12 cells; the second 12 are similar. In these plots, one line shows the RMS error for one cell compressed at the different bit rates. The letters in a column near the left of each plot show the standard deviation of the elevations of the cell whose name starts with the letter. For example, h is at 516 since that is the standard deviation of the elevations of *Hailey E*. We see that the lossy compressibility tends to track the standard deviation, although not perfectly.

These plots show that lossy compression is surprisingly good. One of the worst of the 24 cells is Hailey E, whose elevations have an standard deviation of 516. The low compressed rate of 0.1 bpp gives an RMS error of 12.2, or 2.3% of the elevation standard deviation. The extremely low compression rate of 0.01 bpp, or 1803 bytes for the whole file, gives an RMS error of 10% of the elevation standard deviation.

Study the plot to see the average behavior of the test cells. Compressing to 0.1 bpp gives an RMS error of 3 meters, and 0.01 bpp gives an RMS error of about 10 meters. These are not bad considering the accuracy standards of the original data, which are described in Carter (1989) and Walsh, Lightfoot & Butler (1987).

What do the lossily compressed cells look like? Figure 3 shows the original Hailey E cell. Figures 4 to 7 show the cell compressed to 0.1, 0.03, 0.01, and 0.005 bpp, respectively. Admittedly, much information is lost in printing these images. Since at most about 100 different levels of intensity are visible, the elevation range, which is 954 to 3600 meters, means that any error under 26 is probably invisible. To transmit as much information to the reader, all the greyscale images in the paper have been histogram-equalized to enhance their contrast and make the details more visible.

Even though the RMS error is small, perhaps the errors might be badly distributed. Figure 8 shows the absolute differences of the elevations between the 0.03 bpp compression and the original image. The errors are distributed fairly evenly. Ten percent of the pixels are in error by less than 14 meters, 50% by less than 37, 90% by less than 82, and the worst error is 224 meters. This compares favorably to the range of elevations, which is 2646 meters.

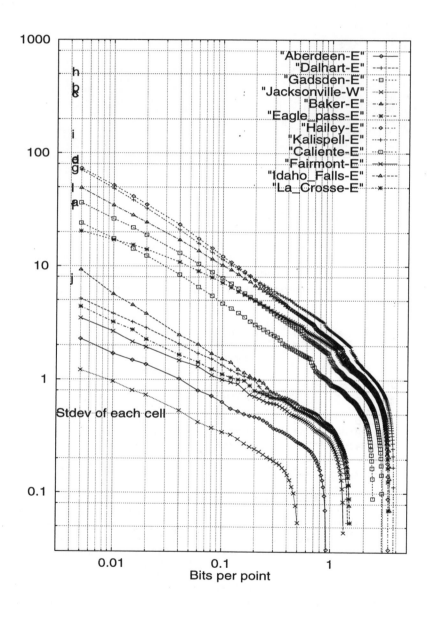

Figure 2 Lossy compression of first 12 DEMs.

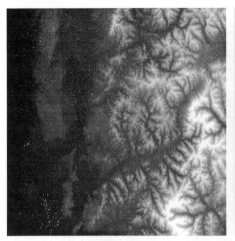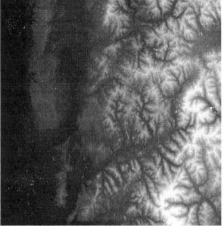

Figure 3 Hailey E cell and Figure 4 Compressed to 0.1 bpp.

5. Effect of partitioning before compression

Suppose that we partition the cell into blocks before compression so that during interactive use, we need to uncompress only the current block. Two complications might occur: the total size of the compressed blocks might be larger, and the compressed blocks might have considerably varying sizes.

Block Size	Number of Blocks	Total Size of Compressed Blocks	Ratio of Tota Size to Size of Original Compressed Blocks	Standard Error of Compressed Block Sizes	(Max/Mean) of Compressed Block Sizes
1024x1024	1	1082800	1.0	0	1
512x512	4	1022153	0.94	0.352	1.37
256x256	16	1000048	0.92	0.376	1.42
128x128	64	1007686	0.93	0.387	1.46
64x64	256	1022742	0.94	0.424	1.5
32x32	1024	1080857	1.00	0.432	1.51
16x16	4096	1274001	1.18	0.411	1.49

Table 2 Effect of partioning before compression with gzip.

Block Size	Number of Blocks	Total Size of Compressed Blocks	Ratio of Tota Size to Size of Original Compressed Blocks	Standard Error of Compressed Block Sizes	(Max/Mean) of Compressed Block Sizes
1024x1024	1	413438	1.00	0	1
512x512	4	414033	1.00	0.371	1.38
256x256	16	418353	1.01	0.379	1.46
128x128	64	427790	1.04	0.383	1.52
64x64	256	450031	1.09	0.399	1.53
32x32	1024	511382	1.24	0.388	1.53

Table 3 Effect of Partitioning Before Compression With Sp_Compress

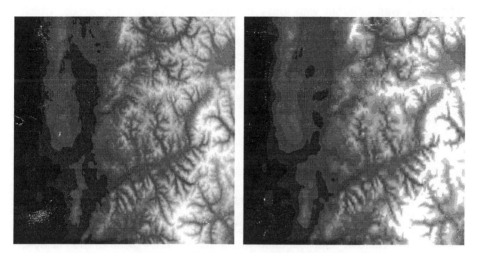

Figure 5 Compressed to 0.03 bpp and Figure 6 Compressed to 0.01 bpp.

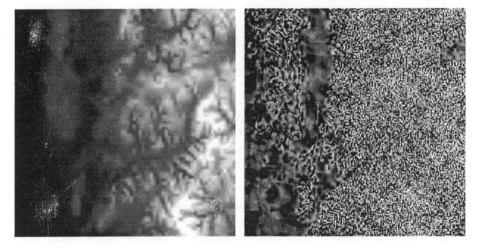

Figure 7 Hailey E cell, compressed to 0.005 bpp and Figure 8 Errors of the 0.03 bpp compression.

To test this on lossless compression, we took a 1024x1024 block of the Hailey-E cell, and subdivided successively into blocks of sizes 512 x 512 down to either 32 x 32 or 16 x 16. We then compressed these blocks with both gzip and sp_compress. We looked at the total size of the compressed blocks, the ratio of that to the size of the compressed original single 1024 x 1024 block, the standard error (standard deviation divided by mean) of the compressed sizes, and the ratio of the largest compressed block to the average of all the compressed blocks. The results are shown in Tables 2 and 3. The short answer is that partitioning before lossless compression is ok. Viswanathan & Nagy (1992) also observes this for scanned images of documents.

The story was different with lossy compression. We used progcode, which takes as input the desired compression rate, in bits per point. Since all the compressed blocks of a given resolution will be the same size, we used the root mean square error (RMSE) of the points as a quality measure.

Block Size	Number of Blocks	RMSE of RMSE of Compressed Blocks	Median of RMSE of Compressed Blocks
1024x1024	1	26.5	26.5
512x512	4	61.2	(5+84)/2=45
256x256	16	66.7	(10+75)/2=43
128x128	64	76.7	(24+61)/2=43
64x64	256	106	(45+47)/2=46
32x32	1024	254	(96+97)/2=97

Table 4 Effect of artitioning before lossy compression.

We used progdecd to measure the RMSE of each block, then combined them, as shown in Table 4. Our test case was a 1024 x 1024 piece of the Hailey-E cell, compressed at 0.3 bpp. Partitioning before lossy compression considerably increases the RMSE.

This is caused by using the same compression rate for each block. 0.3 bpp may be fine for a flat region, while quite inadequate for a mountainous block. A few mountainous blocks with a high RMSE raise the RMSE for the whole cell. The median of the RMSEs of each block does not change so much, for block sizes between 64 and 512. However there is an odd effect for block sizes between 128 and 512. Since there are an even number of blocks in any particular partition, the median RMSE is the average of the two RMSEs of the two middle blocks in sorted order. For those three partitions, the block just before the middle has a much smaller RMSE than the block just after. We show this in the table by showing the two numbers that are averaged to get the median.

This suggests that, while 0.3 bpp is more than adequate for exactly half of the blocks in those partitions, it is quite insufficient for the other half of the blocks. The obvious solution is to compress each block to have the same RMSE, not the same size, or alternatively, to choose an RMSE proportional to the standard deviation of each block's elevation. This is an area of future research.

6. Effect of lossy compression on visibility analysis

Presumably we want the elevation data in order to perform some operation, such as calculating visibility or drainage patterns. Therefore, how much lossiness can we tolerate before these essential properties of the data are damaged? We tested the visibility index for Hailey E compressed to 0.03 bpp using los, an approximate visibility index program of Franklin & Ray (1994). It fires 16 rays out from the observer and counts how many points on these rays are visible. At larger distances from the observer, it also skips points along each ray. The effect is to measure the visibility of a sample of the targets within the range.

Figure 9 shows the visibility indices of the points in the original Hailey-E cell. For each of the 1201 x 1201 points in turn, we assumed an observer at that point, 10 meters above the surface. The observer was looking for targets within a range of 50 points that were 10 meters above their local surface. The fraction of possible targets visible from that observer is called the visibility index of that point. A brighter point has a higher visibility index. Let each visibility index range from 0% (for an observer who can't see any potential target) to 100% (for an observer who can see them all). In this case, the median visibility index is 50%, with the 25- and 75-percentiles at 38% and 62%.

Figure 10 shows the visibility index of each point in the Hailey-E cell when lossily compressed to 0.03 bpp, or only 5409 bytes for the whole cell. The agreement is quite good. The chief differences are a loss of some fine detail and errors in the flatter regions. Some of this might be an artifact of the approximations in our visibility program.

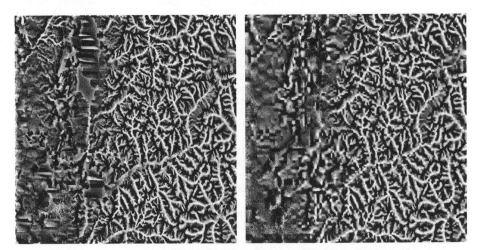

Figure 9 Visibility indices of points in the Hailey E cell and Figure 10 Visibility indices for the 0.03 bpp compressed Hailey E cell.

Since detail is lost in the printed images, how does the visibility index compare point-by-point? We took the point-by-point difference of the above two data sets. The median difference in the visibility indices was 5.7. The 25-percentile was at 2.3, while the 75-percentile was at 11. Therefore even a compression by a factor of 100 from the lossless rate does not seriously hurt the data for the purposes of visibility index computation.

A less aggressive compression is even better. We tried Hailey-E again, compressed only to 0.3 bpp, and differenced the visibility indices. The 25-percentile of the differences is 0.75, the median is 1.5, and the 75-percentile at 3.8.

We are now studying whether this good performance generalizes to other data.

7. Conclusion

The Said & Pearlman lossy image compression algorithm progcode is an excellent method for compressing gridded elevation data, or DEMs. For any desired accuracy, this method compresses much better than a TIN is likely to do, and is also simpler. If the data is to be partitioned into smaller blocks before compression, then each block should not be compressed the same, but the compression rate for each block should be adjusted to that block's elevation variance. Lossy compression also seems to preserve visibility indices. For the Hailey-E cell, even a lossy compression to 0.03 bits per point, or a factor of 100 better than the lossless compression rate, which itself was one-fifth of the size of the original binary file, did not change the visibility indices of the points to a large extent.

References

Bradley, J. (1994), `xv 3.10a _ interactive image display for the X window system', ftp://ftp.cis.upenn.edu/pub/xv/.

Carter, J. R. (1989), Relative errors identified in USGS gridded DEMs, in `Autocarto', Vol. 9, pp. 255-265.

Carter, J. R. (1992), `The effect of data precision on the calculation of slope and aspect using gridded DEMs', Cartographica 29(1), 22-34.

Chang, K.-T. & Tsai, B.-W. (1991), `The effect of DEM resolution on slope and aspect mapping', Cartography and Geographic Information Systems 18(1), 69-77.

Chen, Z.-T. & Tobler, W. (1986), Quadtree representations of digital terrain, in `Proceedings, Auto-Carto London', Vol. 1.

De Floriani, L. & Magillo, P. (1994), `Visibility algorithms on DTMs', Int. J. Geographic Information Systems 8(1), 13-41.

De Floriani, L., Magillo, P. & Puppo, E. (1994), `Line of sight communication on terrain models', Int. J. Geographic Information Systems 8(4), 329-342.

Dutton, G. (1990), Locational properties of quaternary triangular meshes, in K. Brassel & H. Kishimoto, eds, `4th International Symposium on Spatial Data Handling', Vol. 2, Zürich, pp. 901-910.

Fisher, P. F. (1993), `Algorithm and implementation uncertainty in viewshed analysis', International Journal of Geographical Information Systems 7, 331-347.

Franklin, W. R. (1995), Compressing elevation data, in `Fourth International Symposium on Large Spatial Databases - SSD '95', Portland, Maine, USA.

Franklin, W. R. & Ray, C. (1994), Higher isn't necessarily better: Visibility algorithms and experiments, in T. C. Waugh & R. G. Healey, eds, `Advances in GIS Research: Sixth International Symposium on Spatial Data Handling', Taylor & Francis, Edinburgh, pp. 751-770.

Jacobson, G. (1989), Foundations of Computer Science, Vol. 30, chapter Space-efficient static trees and graphs.

Lee, J., Snyder, P. K. & Fisher, P. F. (1993), `Modeling the effect of data errors on feature extraction from digital elevation models', Photogrammetric Engineering and Remote Sensing 58, 1461-1467.

Leifer, L. A. & Mark, D. M. (1987), Recursive approximation of topographic data using quadtrees and orthogonal polynomials, in N. R. Chrisman, ed., `Autocarto 8: Proceedings Eighth International Symposium on Computer-Assisted Cartography', ASPRS and ACSM, Baltimore, pp. 650-659.

Mark, D. M. & Lauzon, J. P. (1984), Linear quadtrees for geographic information systems, in `Proceedings of the International Symposium on Spatial Data Handling', Vol. 2, Zürich, pp. 412-430.

McCormack, J. E., Gahegan, M. N., Roberts, S. A., Hogy, J. & Hoyle, B. S. (1993), `Feature-based derivation of drainage networks', Int. J. Geographic Information Systems 7(3), 263-279.

Poskanzer, J. (1993), `netpbm _ a toolkit for conversion of images',
    ftp://wuarchive.wustl.edu/graphics/graphics/packages/NetPBM/. (based on the
    Portable BitMap package).

Puppo, E., Davis, L., de Menthon, D. & Teng, Y. A. (1994), `Parallel terrain triangulation', Int.
    J. Geographic Information Systems 8(2), 105-128.

Ray, C. K. (1994), Representing Visibility for Siting Problems, PhD thesis, Electrical,
    Computer, and Systems Engineering Dept., Rensselaer Polytechnic Institute.

Rokicki, T. & Berry, K. (1994), `dvips - postScript driver for TeX',
    tp://labrea.stanford.edu/pub/dvips558.tar.Z.

Said, A. (1994), An image multiresolution representation for lossless and lossy compression.
    (submitted).

Said, A. & Pearlman, W. A. (1993a), A new fast and efficient image codec based on set
    partitioning in hierarchical trees. (submitted), presented in part at the IEEE
    Symposium on Circuits and Systems, Chicago, May 1993.

Said, A. & Pearlman, W. A. (1993b), Reversible image compression via multiresolution
    representation and predictive coding, in `Proceedings SPIE', Vol. 2094: Visual
    Commun. and Image Processing, pp. 664-674. email: amir@densis.fee.unicamp.br,
    pearlman@ecse.rpi.edu.

Said, A. & Pearlman, W. A. (1995), `(new image coding and decoding programs)', ftp://ftp.
    ipl.rpi.edu/pub/EW_Code/.

Shea, K. S. & McMaster, R. B. (1989), Cartographic generalization in a digital environment:
    When and how to generalize, in `Autocarto', Vol. 9, pp. 56-67.

Skidmore, A. K. (1990), `Terrain position as mapped from a gridded digital elevation model',
    Int. J. Geographic Information Systems 4(1), 33-49.

Viswanathan, M. & Nagy, G. (1992), Characteristics of digital images of scanned articles, in D.
    D. et. al., ed., `Proc. SPIE/IS&T Machine Vision Applications in Character Recognition
    and Industrial Inspection', Vol. 1661, SPIE, San Jose, pp. 6-17.

Walsh, S. J., Lightfoot, D. R. & Butler, D. R. (1987), `Recognition and assessment of error in
    geographic information systems', Photogrammetry Engineering and Remote Sensing
    53, 1423-1430.

Waugh, T. (1986), A response to recent papers and articles on the use of quadtrees for
    geographic information systems, in `Proceedings of the Second International
    Symposium on Geographic Information Systems', Seattle, Wash. USA, pp. 33-37.

Weibel, R. (1992), `Models and experiments for adaptive computer-assisted terrain
    generalization', Cartography and Geographic Information Systems 19(3), 133-153.

Contact address
Wm Randolph Franklin[1] and Amir Said[2]
[1]Electrical, Computer, and Systems Engineering Dept., 6026 JEC
Rensselaer Polytechnic Institute
Troy
New York 12180-3590
USA
Phone: +1 (518) 276-6077
Fax: +1 (518) 276-6261
E-mail: wrf@ecse.rpi.edu, http://www.ecse.rpi.edu/homepages/wrf/
[2]DENSIS - Faculty of Electrical Engineering
University of Campinas (UNICAMP)
Campinas, SP 13081
Brazil
E-mail: amir@densis.fee.unicamp.br

A TYPOLOGY OF CONSTRAINTS TO LINE SIMPLIFICATION

Robert Weibel

Department of Geography
University of Zürich
Zürich, Switzerland

Abstract

Line simplification is a popular generalization operator which has traditionally been a major focus of research in automated generalization of spatial data. Many algorithms for simplifying linear cartographic features or mosaics of polygons (polygonal subdivisions) have been developed over the past three decades. Surprisingly, however, the overwhelming majority of these algorithms is based on very restricted criteria, neglecting significant constraints which are well established in cartographic practice. As a consequence, these techniques produce results which can only be used for minimal simplification (i.e., for small scale reduction factors), and fail badly when more significant scale changes are required. Only a handful of algorithms has been published which either attempt to clean up problems created by conventional algorithms in a post-processing operation, or enforce at least some topologic constraints. In order to solve more complex problems such as the simplification of polygonal subdivisions, which is a frequently occurring requirement in GIS operations, a more comprehensive approach needs to be developed. This paper thus attempts to identify further constraints in relation to established cartographic principles, in order to form a basis for the development of extended line simplification algorithms. Key elements for the implementation of such methods are discussed, including supporting data structures and algorithmic components. It is observed that many of these basic building blocks are already available in the literature. This paper should offer a framework for their integration and the development of the missing components.

Key words

Generalization, line simplification, polygonal subdivisions, constraints, cartographic principles.

1. Introduction: Is line simplification really that simple?

During three decades of past research on automating the generalization of spatial data, most work has concentrated on line generalization. This focus is readily justified by the fact that the majority of map features are either directly represented as lines (e.g., road centerlines, streams), or form polygons which are bounded by lines (e.g., administrative regions, soil polygons, forest stands). Of the various generalization operators which may be deemed relevant for the generalization of linear features — selection, simplification, smoothing, merging, refinement, exaggeration, and displacement (McMaster and Shea 1992) — by far the most part of past research has concentrated on line *simplification*. Again, there are good reasons for that. Simplification reduces the amount of line detail and thus visibly contributes significantly to the 'generalization effect'. If line simplification is implemented as a point reduction algorithm (which is the usual case), it automatically reduces data volume. Simplification algorithms are also highly useful for eliminating high frequency detail on lines digitized by continuous point sampling (stream mode) or scan-digitization. Finally, line simplification is a less complex operator compared to others such as merging and displacement, which also explains some of its attraction as a research topic.

As a result, a seemingly countless number of line simplification algorithms has been developed over the past three decades, and alternative solutions are still being published (e.g., de Berg et al. 1995). Commonly, simplification algorithms start with a polyline C made up of two endnodes and an arbitrary set of vertices V. C is then turned into a simplified polyline C' by reducing the number of vertices V to V', while keeping the endnodes fixed. V' is thus a subset of V, and no further vertex locations are introduced, or vertices displaced. The classical criteria which guide the point reduction are the following: 1) minimize displacement and distortion (e.g., no vertex of C' should be further away from C than a maximum error e); 2) minimize V'; and 3) minimize the computational complexity.

Substantial research has concentrated on assessing which of the algorithms developed for simplification would best satisfy the above criteria (White 1985, McMaster 1987). It is surprising, however, to find that these objectives have never been extended over all these years, with very few exceptions. Important constraints of line generalization and simplification have thus been consistently neglected, leading to a plethora of algorithms which produce clearly inferior results if applied to real generalization situations. It is easy to show, for instance, that widely used simplification algorithms may generate self-intersecting lines since they do not include mechanisms to avoid such inconsistencies (Muller 1990). Similarly, these algorithms cannot prevent overlaps of neighboring lines from being created.

In recent years, some researchers have addressed the problem and developed specific solutions to overcome particular weaknesses. Muller (1990) has proposed a post-processing procedure to clean up self-intersections generated by commonly used line simplification algorithms lacking topologic control. A small group of authors has proposed new algorithms which avoid self-intersections from the outset. Wang and Muller (1993) have done so for complex coastlines with estuaries and mounds which are bound to lead to problems when processed by traditional methods. Li and Openshaw (1992) have developed an algorithm based on a mixed raster and vector approach which — by virtue of its underlying raster structure — implicitly (but not explicitly) avoids self-overlaps. The approach to polygon mosaic (subdivision) simplification published by de Berg et al. (1995) is certainly the most advanced and comprehensive one to date. Not only does it avoid self-intersections and overlaps of neighboring lines, it is also the first approach to consider objects from other feature classes: The algorithm can ensure that all points of a point set P stay within the polygons they originally fall into. For instance, cities stay within the same country polygons rather than changing sides as a result of line simplification.

While these more recent developments are encouraging, we argue that there are even more criteria which would need to be taken into account in order to develop more meaningful line simplification procedures. This paper thus sets out to explore further constraints to line simplification as well as ways to implement them. The following questions are dealt with in the subsequent sections:

- What constraints to line simplification can be derived from cartographic principles?
- How can these constraints be managed?
- What are possible elements to implement such constraints?

In our discussion, we will focus primarily on the problem of simplifying subdivisions of the plane formed by polygon mosaics (*subdivision simplification*). Polygonal subdivisions form a very frequent data type in many GIS applications: administrative or political boundaries (countries, voting districts, census tracts, etc.), soil, landuse, and many other categorical maps are made up of polygon mosaics. It is clear, however, that many of the criteria relevant for polygonal subdivisions apply similarly to other simplification problems.

2. Constraints to line simplification

This section attempts to identify the various constraints that apply to line generalization, and more particularly to polygonal subdivision simplification. After providing an initial classification into overriding types of constraints, three sets of detailed constraints are discussed — *within a line, within a feature class,* and *between feature classes* — and remedial actions described which are normally taken in cartographic practice.

2.1 Constraint types

Four types of constraints may be distinguished. Three of these relate to the basic aspects of data modeling and can be formally defined, while the fourth is dictated by aesthetic and perceptual criteria and thus more complex to operationalize.
- *Metric constraints:* These constraints are mainly influenced by aspects of perceptibility such as minimal separability, minimal size, or minimal width. Very importantly, metric constraints also require that symbolization effects (i.e., increased relative line width at smaller scales) are taken into account (cf. Fig. 1).

Figure 1 Self-coalescence due to symbolization. Line without symbolization (left); line symbolized at 1:25,000 (middle); line at 1:50,000 (right).

- *Topologic constraints:* These relate to maintenance of topologic consistency, including avoidance of self-intersections, mutual overlaps, containment of point features, etc.
- *Semantic constraints:* This constraint type relates to semantic modeling, and includes criteria such as the preservation of class memberships, or the domain of existence in the spatial context (e.g., rivers on slopes instead of in valleys are out of context).
- *Gestalt constraints:* Beyond the consideration of metric, topologic, and semantic constraints, a good generalization should also take into account shape and Gestalt-related perceptual criteria such as maintenance preservation of original line character or of the distribution and arrangement of map features. While such principles are routinely used in conventional cartography, they are often hard to translate into operational terms and can be expected to be considerably more demanding than other constraints in terms of database preparation and structural analysis. Note that Gestalt principles can only be met if the other constraint types are satisfied; this is a prerequisite.

The above grouping into constraint types is driven by the general requirements of digital and graphical representation. A further subdivision of constraints is possible, which distinguishes three sets of constraints according to the *constraint extent* (i.e., the range of features which are involved): 1) constraints which only affect an individual line *(within line)*, 2) those which involve other lines within the same layer or feature class *(within feature class)*, and 3) constraints which affect more than one layer or feature class *(between feature classes)*. (Note that since we are dealing with polygonal subdivisions a layer is equivalent to a feature class). These detailed constraints will be discussed below and are summarized in Table 1.

2.2 Constraints within a line

(1) *Avoid imperceptible crenulations* (metric): Small, imperceptible crenulations on a line caused by scale reduction or symbolization (increased line width) should be eliminated (or alternatively exaggerated). This is the main reason triggering generalization.
(2) *Avoid self-coalescence* (metric): Coalescence (McMaster and Shea 1992) among segments of a line must be avoided. Coalescence on a line may occur due to symbolization (Fig. 1) or as a result of simplification (Muller 1990). In both cases, the situation is detected if the minimal separability distance between consecutive bends is not met. This is countered in manual cartography by either eliminating or exaggerating (widening) the bend that causes the problem.
(3) *Minimize shape distortion* (metric): Distortion of the original line should be minimized. That is, line length and angularity should be preserved as far as possible, and vector and areal displacement minimized (McMaster 1987, Cromley and Campbell 1991). Obviously, while this constraint is generally valid, it must be relaxed in favor of constraints 5 and 2 if greater scale reduction takes place (Plazanet 1995).
(4) *Avoid self-intersection* (topologic): A polyline should remain a polyline after simplification. That is, no intersections of the line with itself should

occur, and no loops be formed. Like all topologic problems listed here, self-intersection is an artifact of automated line simplification and would never occur in manual cartography.

(5) *Preserve original line character* (Gestalt): The intrinsic character expressing the generating process of a feature represented by a cartographic line should be preserved. For instance, since administrative boundaries are usually based on surveyed lines, their angular shape should be preserved in the line generalization process, whereas soil boundaries commonly exhibit smoother shapes which should equally be maintained.

2.3 Constraints within a feature class

(6) *Minimum polygon size* (metric constraint): No polygon should fall below the perceptibility limits after scale reduction. In manual cartography, this problem is answered by two alternative solutions: Either the polygon in question is eliminated, or — if the region it represents is of particular importance — it is exaggerated by increasing its size above the minimal area (Fig 2).

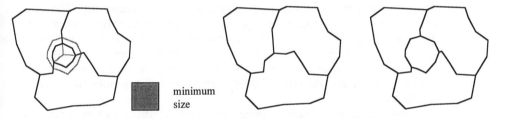

Figure 2 Minimum polygon size. If a polygon falls below the minimum size (left), it can either be eliminated (middle), or enlarged (right). The second case only applies to polygons marked as semantically significant.

(7) *Minimum polygon width* (metric): No part of a polygon should be so narrow as to fall below the minimum line separability distance (coalescence). In manual cartography, the situation is cleaned up by widening the polygon or parts of it, or — if the polygon is small and insignificant — by eliminating it entirely.

(8) *Preserve size ratios between polygons* (metric): The size ratios between the polygons of a feature class should be maintained as far as possible. For instance, in a landuse map (making up a polygonal subdivision equivalent to a feature class), the total area of each category (forest, built-up areas, etc.) should change as little as possible in relation to each other. This is particularly important if some categories consist only of small polygons, while others include mainly large polygons.

(9) *Avoid intersections between lines* (topologic): No lines or polygon boundaries within a feature class should intersect as a result of simplification. This (frequent) problem only occurs in automated line simplification as a consequence of erroneous algorithms.

(10)*Maintain polygon topology* (topologic): Polygon topology of the original map should not be destroyed during simplification, including the preservation of chain-node connectivity, polygon adjacency, and enclosure of islands. This constraint is related to the avoidance of intersections between lines of the same feature class.

(11) *Preserve categorical membership* (semantic): No polygon should change its category due to simplification. As an indirect consequence of topologic problems (intersecting lines, inconsistent polygon topology), polygons may be assigned wrong attributes and thus change category.

(12) *Maintain overall visual balance* (Gestalt): Lines of a particular feature class are successfully generalized if the overall pattern, distribution and density of map features is maintained, and if the degree of generalization is homogeneous across the map. While a substantial part of this requirement is satisfied if the above constraints (particularly 8) are met, it also includes aspects which cannot be subsumed in the other constraint types and are hard to formalize.

2.4 Constraints between feature classes

The constraints outlined in this section are all aiming eventually at optimizing the interplay of map elements (Imhof 1982) which is the ultimate objective of map design.

(13) *Preserve size ratios between feature classes* (metric constraint): The size ratios between polygons across feature classes should be maintained. This criterion is similar to the preservation of size ratios in a single class, only that it extends over several classes. Note that this constraint also depends on the specifications of the target map or database. If some feature classes are dropped entirely, this constraint should be relaxed accordingly.

(14) *Preserve proximity relations* (metric): Proximity relations between objects of different feature classes should be maintained. For instance, if a road follows a lake as a mainly parallel line, a cartographer will attempt to maintain this aspect in the generalized map (Fig. 3a). Likewise, the relative position of a point in a polygon should be preserved (Fig 3b). It is not entirely sufficient to simply ensure that the point remains in the polygon, as is the case in the method developed by de Berg et al. (1995).

Figure 3 Preserving proximity relations.
a) Preserving distance relations between neighboring lines: the road must pick up the simplified shape of the lake.
b) Preserving distance relations between a polygon boundary and a point: the point needs to be displaced to remain adjacent to the boundary.

(15)*Preserve feature containment* (topologic): If a polygon contains point features or other polygons (islands), this relationship must be preserved. For point features, two cases must be distinguished: depending on their nature and significance, points remain either fixed (e.g., survey points) or they may be displaced to a certain extent to accommodate modifications of the polygon geometry (buildings, cities at small scales). The second case is more frequent in cartography, since only measured points at large scales are considered to be fixed points, while at smaller scales, the priority of point features is low and they may be displaced away from line features. Point displacement can also be used to preserve proximity relations (see above; Fig 3b).

(16)*Shared line primitives* (topologic): If two polygon features or a polygon and a line feature belonging to different feature classes share a common boundary, this relationship should be maintained during line simplification. For instance, if a forest parcel extends to a road along one of its bounding chains, then, if the river (i.e., the higher priority feature) is simplified, the forest parcel (the lower priority feature) should pick up that modification (Fig. 4). Note that this problem is similar, but not equivalent to the preservation of proximity relations (14).

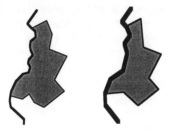

Figure 4 Shared line primitives. The river and the forest share a common boundary. Thus, if the rivers is simplified, the forest boundary must change accordingly.

(17)*Domain of existence in the spatial context* (semantic): No feature should be placed out of context as a result of simplification. For instance, a river which flows on a slope or hilltop because of a generalization of the digital terrain model is out of context. As a second example, contours on small scale visualizations loose their meaning. Frequently, semantic inconsistencies are indirectly caused by other problems.

(18)*Interplay of elements* (Gestalt): The degree of generalization should be equal, proximity and topologic relations should be maintained, and the priorities should be clear among the map feature classes. More than anything else, this criterion requires as a prerequisite the satisfaction of all other constraints, and is also hard to formalize. Nonetheless, it is clear that an aesthetically pleasing and readily legible map image is the premier goal.

Constraint Extent	Constraint Type			
	Metric	Topologic	Semantic	Gestalt
Within line	1, 2, 3	4	N/A	5
Within F.C.	6, 7, 8	9, 10	11	12
Between F.C.	13, 14	15, 16	17	18

Table 1 A summary of constraint types vs. constraint extent.

2.5 Constraints satisfied by existing algorithms

Some of the above constraints are already met by existing line simplification algorithms. The *avoidance of imperceptible crenulations* (1) is usually taken care of reliably by all line simplification algorithms, by virtue of the simplification process. The *minimization of shape distortion* (3) is satisfied to a varying degree by present algorithms. For instance, while tolerance band algorithms ensure that no point on the simplified line lies further away from the original line than the tolerance *e*, other criteria such line length or areal displacement may be equally important (McMaster 1987). The approach by de Berg et al. (1995) goes beyond that, and *avoids of self-intersections* (4), *avoids intersections between lines* (9) of a feature class (and as a result of that also *maintains proper polygon topology*, 10), and *preserves feature containment* (15). All other constraints, most notably the metric ones, are not met by any of the existing algorithms.

3. Managing the constraints

Most of the above metric and Gestalt constraints would clearly imply a departure from the commonly used definition of line simplification set forth by McMaster and Shea (1992) which states that vertices on a line are only eliminated, not displaced. However, the quality of results of such extended line simplification methods would be improved considerably, regardless of conformity with existing terminology. Since line simplification is still central to this approach, we will continue to call the operator *extended line simplification*, rather than line generalization. As a matter of fact, we suspect that the definition of simplification by McMaster and Shea has been influenced by the available algorithms for this purpose, neglecting the fact that simplification of shapes as it is practised in manual cartography can also be achieved by other means.

The purpose of this section is to discuss how the numerous constraints listed above can be managed. With all constraints, it is basically up to the user to 1) define their limits and parameters (e.g., a displacement tolerance *e*, the minimum polygon size, etc.), and 2) to specify to what extent they are enforced. Depending on the given feature classes, scales, and map purposes, there are sometimes good reasons to relax or even ignore a particular constraint. All constraints are not equally important in general and particularly not in the context of specific applications. Finally, there is a tradeoff between some of the constraints, which necessitates that a scheme for managing the constraint sets is developed, including priority orderings and interactive control.

Constraints may be prioritized in two ways: 1) according to their effect on the final result (*cartographic relevance*), and 2) with respect to the ease with which they can be formalized and implemented (*ease of implementation*). The latter issue will be dealt with in the next section, while the discussion in this section focuses on cartographic relevance.

The cartographic priorities with respect to constraint sets are quite obvious. Since the ultimate goal of map design is to achieve harmony between the various elements of a map, constraints between feature classes take precedence over those within individual feature classes and within lines. If a constraint between different feature classes is active, then all affected lines of the individual classes must satisfy that constraint.

The generic priority ordering of constraint types is from Gestalt constraints (highest priority) to semantic, topologic, and geometric (lowest priority). Map design aspires to produce visually pleasing graphics, thus making Gestalt criteria a top priority. As was explained above, however, Gestalt constraints are very hard to formalize. Initially, the certainly simplifying assumption is thus reasonable to consider their fulfillment to be the result of meeting all other constraint types (thus initially ignoring the presence of Gestalt constraints). Likewise, the semantic constraints listed above are indirectly a consequence of observing topological and metric constraints. Thus, the only constraint types which need to be explicitly considered are — at least *initially* — topological and geometric constraints, among which the former clearly take precedence over the latter ones. Topologic consistency is indispensable for accurate maps and databases; any inconsistencies are obvious errors of existing algorithms. Metric constraints are harder to handle than topologic ones since they require the specification of thresholds. It is thus not surprising that the approach by de Berg et al. (1995) only considers topologic constraints, while neglecting metric problems such as line coalescence or imperceptibility. Metric constraints are mainly dictated by the objective of increasing the legibility of cartographic products, and often induce displacement. It is therefore necessary to also specify displacement hierarchies among feature classes. Generic displacement priorities can be found in the literature (Imhof 1982, Lichtner 1979). For instance, transportation features are usually displaced away from hydrographic features, while buildings are displaced from roads.

To manage the various constraints — their existence, priority, and parameters (for metric constraints) — it appears useful to develop a *constraint editor* which provides an interactive environment in which the user can specify, evaluate, and modify the constraints (Weibel 1991). Default constraint sets could be compiled automatically on the basis of the properties of the data involved (point, line, polygon; single vs. multiple feature classes; feature class relations; etc.), as well as the controls of generalization (scales, map purpose). Further default criteria and priorities can be contributed by previously acquired procedural cartographic knowledge (Weibel et al. 1995).

4. Elements for implementation

Work on implementing some of the more important constraints discussed in section 3 has recently been started. Implementation is based on two observations. Firstly, some constraints are more cartographically relevant than others, and some are easier to formalize and implement. *Initially*, it thus makes sense to concentrate on cartographically relevant constraints which can be formalized accurately. Secondly, the literature reveals that many of the basic building blocks necessary for implementation are already in existence (although in isolation from each other, but few attempts have been made to date to identify these isolated elements and integrate them into a comprehensive strategy.

Among those types of constraints which are cartographically relevant and tractable — topologic and metric constraints (cf. section 3) — topologic constraints generally pose less problems in terms of formalization and implementation. Metric constraints relating to size criteria are also relatively easy to specify, while distance-related constraints (coalescence, proximity relations) are harder to formalize. Constraints within individual lines or feature classes are more convenient to handle than constraints between different feature classes which require extended data modeling facilities (e.g., a mechanism for sharing geometric primitives) as well as more effort in database preparation (e.g., construction of shared primitives, specification of displacement hierarchies). Given this, we propose the following phasing for the implementation of constraints:
- First phase: constraints 1, 3, 4, 6, 8, 9, 10, 13, 15
- Second phase: constraints 2, 7, 16
- Third phase: constraint 14

Gestalt constraints (5, 12, 18) and semantic constraints (11, 17) are not explicitly considered in this initial phasing due to the reasons given above. Some elements for implementation are given in the remainder of this section.

Auxiliary data structures: The polyline data structure commonly used in GIS to represent lines and polygons is not very powerful in terms of supporting constraints which need a more synoptic view, in particular constraints involving topologic and proximity relations. Auxiliary data structures may provide additional support. A very rich data structure for this purpose has been proposed and implemented by Jones et al. (1995). The so-called Simplicial Data Structure (SDS) is based on a constrained Delaunay triangulation of the vertices of all objects of all feature classes of a given dataset. Edges of polygonal or linear features act as linear constraints to the Delaunay triangulation. Given the web of triangle edges which connects feature vertices in a closest point arrangement, complex topological and proximity constraints can be readily implemented. Since a planar triangulation is used, overlaps of triangle edges generated as a result of generalization alert of topological inconsistencies. Additionally, the triangulation provides a frame in which to detect and resolve complex metric problems such as coalescence and proximity relations. A similar triangulation-based data structure is used by Ruas (1995). Her approach, however, uses feature centroids instead of the complete set of feature vertices to build the triangulation.

If less complex constraints are to be satisfied and the data volume is large, use of a complete triangulation may be too expensive. De Berg et al. (1995) have shown that the majority of topologic constraints can be solved efficiently and effectively based solely on the original polyline representation. Auxiliary data structures are mainly beneficial if complex metric (distance-related) constraints are involved. A simple supporting data structure is the raster model, which may be helpful to detect and resolve local distance-related problems such as line coalescence or displacement of a moderate degree (Jäger 1991). Li and Openshaw (1992) have used a raster structure in which the cell size was equal to the minimal visual separability in relation to display scale in order to mimic the resolution of human vision and develop line generalization algorithms based on this paradigm. Finally, uniform grids are also a popular data structure to support spatial search.

Topologic constraints (4, 9, 10, 15, 16): Much of the discussion relating to topologic constraints has already been led in the preceding paragraph. The approach presented by de Berg et al. (1995) deals with topologic constraints within lines and within feature classes (4, 9, 10) as well as feature containment between feature classes (15) successfully and efficiently. The method proposed by Muller (1990) has the advantage that it can also handle metric problems to some extent (coalescence), and that it can be applied as a post-processing operation on previously simplified data (useful for cleaning of existing databases). Finally, a triangulation-based approach such as those by Jones et al. (1995) and Ruas (1995) hold the potential of providing the most comprehensive solution to deal with topologic constraints in combination with distance-related metric constraints.

Shared primitives between feature classes (16) is a constraint which has not yet received much attention in the generalization literature, with notable exceptions (Ruas and Lagrange 1995, Jones et al. 1995). It mainly requires extensions to the basic topological data model across feature classes. It must be possible to distinguish between geometric primitives (points, lines, polygons) and the features they make up (e.g., forest parcels, roads), and to share primitives between features of different classes via a referencing mechanism.

Perceptibility (1): The elimination of imperceptible line crenulations is ensured by all existing line simplification algorithms and presents no further problem.

Shape distortion (3): An approach to this sub-problem based on a dynamic programming paradigm which yields noninferior (optimal) solutions to line simplification with respect to arbitrary shape distortion criteria (e.g., maximizing line length, minimizing vector displacement, etc.) has been presented by Cromley and Campbell (1991). Although this method effectively minimizes shape distortion in comparison to conventional heuristic line simplification algorithms, it is also computationally expensive, a factor which may be prohibitive if large datasets are involved. As a pragmatic approach, heuristic algorithms which have been shown to create acceptable distortions in empirical studies (e.g., McMaster 1987) may be used. Some heuristic algorithms also implicitly include optimization characteristics, since they retain points based on local distortion measures. An example of this is the line simplification technique by Visvalingam and Whyatt (1993), which tends to minimize areal displacement. Finally, local shape distortion can be better controlled if lines are subdivided into ssections of homogeneous sinuosity prior to generalization (Plazanet 1995).

Size-related constraints (6, 8, 13): Size-related constraints are rather straightforward to deal with. Since it only affects small polygons, constraint 6 (minimum polygon size) can simply make use of the polygon centroid (cf. Fig. 2). If the polygon is not marked as significant, it is eliminated by extending the incident chains of the adjacent polygons to the polygon centroid. In the case of islands, the polyline is simply removed. Conversely, if the polygon is marked as significant, its size may be exaggerated by radial projection of the nodes away from the centroid beyond the minimum size. The minimum size criterion is applied *before* the actual line simplification. It helps reducing potential topological problems and enhances simplification because small, spurious polygons need no longer be considered.

Preservation of size ratios (constraints 8, 13) follows line simplification. If the changes of size ratios between feature classes due to simplification exceed a specified threshold, small polygons of feature classes which excessively lost total area are exaggerated beyond the minimum polygon size according to the procedure for constraint 6 (Fig. 2).

Coalescence (2,7): Resolution of coalescence problems almost invariably involves displacement of some sort. The approach by Muller (1990) works directly on the polyline representation, and applies a displacement propagation model involving a distance decay function similar to those developed by Lichtner (1979) or Nickerson (1988) in regions of coalescence. A report of other research on displacement and distance decay functions can be found in Mackaness (1994). As mentioned above, it may be beneficial to use auxiliary data structures to detect and resolve such conflicts. Li and Openshaw (1992) have presented a hybrid algorithm which makes use of a raster at a resolution equal to the minimal visual separability distance. This algorithm implicitly cleans up coalescence problems, but also has a smoothing effect. It is thus suited to treat features of undulating nature, but probably less so to process angular features (e.g., administrative boundaries). Clearly, triangulated data structures (Jones et al. 1995, Ruas 1995) could be beneficially exploited to detect and remove line coalescence, but depending on line complexity and data volume, they may create excessive overhead with respect to the gain that may be achieved compared to the above more traditional methods.

Proximity relations (14): With respect to proximity relations, two cases must be distinguished: point-to-line and line-to-line relations. Both cases involve displacement, but displacing points is easier because shape preservation is not an issue. Point-to-line relations can be handled relatively easily by radial displacement (Fig. 3b). Methods for handling line-to-line proximity relations are traditionally based on displacement propagation models which apply distance decay functions to determine the direction and length of displacement vectors emanating from the vertices of the lines (Nickerson 1988). It can be assumed that a triangulation-based auxiliary data structure would provide a better structure to establish fields of displacement vectors in conjunction with displacement functions.

5. Conclusions

Although line simplification is a popular generalization operator, present algorithms are not living up to their potential since they neglect important constraints imposed by cartographic principles. For lack of better alternatives, users of GIS and cartographic systems today are condemned to using such algorithms to perform scale changing operations on cartographic line and polygon data. If scale reduction requires more than minor coordinate weeding, however, detrimental results may be generated.

We have discussed fundamentals required for the development of more comprehensive approaches to line simplification, in order to extend the usability of this potentially powerful operator, with particular emphasis on the simplification of polygonal subdivisions. The basis is formed by an analysis of the various constraints which can be specified from cartographic principles, as well as a scheme for managing the priorities of constraint sets. Elements of data structures and algorithms to implement these constraints in extended line simplification methods have also been discussed. It is our firm belief that many of these elements are at least partially already in existence, although developed in isolation from one another. Thus, this paper should provide a framework for the integration of individual building blocks. It is obvious, however, that integration and harmonization of these isolated elements (which are often based on different paradigms) will be non-trivial, and that only a full implementation can show the benefits of each element in relation to others.

Acknowledgments
Many thanks are due to Geoff Dutton, Martin Heller and Matthias Bader for fruitful discussions, as well as Caroline Westort for reading the draft manuscript. I am also grateful for the useful comments provided by one of the reviewers. This research is part of project «Methods for Knowledge Acquisition and Evaluation in Cartographic Generalization» supported by the Swiss National Science Foundation under contract 2100-043502.95.

References

Cromley, R.G., and Campbell, G.M. (1991): Noninferior Bandwidth Line Simplification: Algorithm and Structural Analysis. Geographical Analysis, 23(1), 25-38.

de Berg, M., van Kreveld, M., and Schirra, St. (1995): A New Approach to Subdivison Simplification. Technical Papers, ACSM/ASPRS Annual Convention, 4 (Auto-Carto 12): 79-88.

Imhof, E. (1982): Cartographic Relief Presentation. Berlin, etc.: Walter de Gruyter.

Jäger, E. (1991): Investigations on Automated Feature Displacement for Small Scale Maps in Raster Format. Proceedings 15th ICA Conference, Bournemouth, 1: 245-256.

Jones, C.B., Bundy, G.Ll., and Ware, J.M. (1995): Map Generalization with a Triangulated Data Structure. Cartography and Geographic Information Systems, 22(4): 317-331

Li, Z., and Openshaw, S. (1992): Algorithms for Automated Line Generalization Based on a Natural Principle of Objective Generalization. International Journal of Geographical Information Systems, 6(5): 373-389.

Lichtner, W. (1979): Computer-Assisted Processes of Cartographic Generalization in Topographic Maps. Geo-Processing, 1: 183-199.

Mackaness, W.A. (1994): An Algorithm for Conflict Identification and Feature Displacement in Automated Map Generalization. Cartography and Geographic Information Systems, 21(4): 219-232.

McMaster, R.B. (1987): The Geometric Properties of Numerical Generalization. Geographical Analysis, 19(4): 330-346.

McMaster, R.B., and Shea, K.S. (1992): Generalization in Digital Cartography. (Resource Publications in Geography). Washington, D.C.: Association of American Geographers.

Muller, J.-C. (1990): The Removal of Spatial Conflicts in Line Generalization. Cartography and Geographic Information Systems, 17(2): 141-149.

Nickerson, B.G. (1988): Automated Cartographic Generalization for Linear Features. Cartographica, 25(3), 15-66.

Plazanet, C. (1995): Measurement, Characterization and Classification for Automated Line Feature Generalization. Technical Papers, ACSM/ASPRS Annual Convention, 4 (Auto-Carto 12): 59-68.

Ruas, A. (1995): Multiple Paradigms for Automating Map Generalization: Geometry, Topology, Hierachical Partitioning and Local Triangulation. Technical Papers, ACSM/ASPRS Annual Convention, 4 (Auto-Carto 12): 69-78.

Ruas, A. and Lagrange, J.-P. (1995): Data and Knowledge Modelling for Generalization. In: Muller, J.-C., Lagrange, J.-P., and Weibel, R. (eds.): GIS and Generalization — Methodology and Practice. London: Taylor & Francis, 73-90.

Visvalingam, M., and Whyatt, J.D. (1993): Line Generalisation by Repeated Elimination of Points. Cartographic Journal, 30(1): 46-51.

Wang, Z., and Muller, J.-C. (1993): Complex Coastline Generalization. Cartography and Geographic Information Systems, 20(2): 96-106.

Weibel, R. (1991): Amplified Intelligence and Rule-Based Systems. In: Buttenfield, B.P., and McMaster, R.B. (eds.): Map Generalization — Making Rules for Knowledge Representation. London: Longman, 172-186.

Weibel, R., Keller, St., and Reichenbacher, T. (1995): Overcoming the Knowledge Acquisition Bottleneck in Map Generalization: The Role of Interactive Systems and Computational Intelligence. Lecture Notes in Computer Science (Proceedings COSIT 95), 988: 139-156, Berlin: Springer-Verlag.

White, E.R. (1985): Assessment of Line Generalization Algorithms Using Characteristic Points. The American Cartographer, 12(1): 17-28.

Contact address

Robert Weibel
Department of Geography
University of Zurich
Winterthurerstrasse 190
8057 Zurich
Switzerland
Fax: +41-1-362 52 27
E-mail: weibel@geo.unizh.ch

A SPATIAL MODEL FOR DETECTING (AND RESOLVING)

CONFLICT CAUSED BY SCALE REDUCTION

J. Mark Ware and Christopher B. Jones

Department of Computer Studies
University of Glamorgan
Pontypridd, UK

Abstract
Reduction in the scale at which cartographic data are plotted is usually accompanied by spatial conflicts that must be resolved as part of the process of map generalisation. In this paper we describe a set of search procedures that are used to detect the situation of proximal conflict, whereby objects become too close to be clearly resolved visually. The procedures exploit the Simplicial Data Structure (SDS) which represents the map data by a constrained Delaunay triangulation, the constituent edges and triangles of which are used to define higher level map objects. The neighbourhood relationships of the triangulation facilitate efficient detection of the nearest neighbouring objects to any given object. These search procedures can be combined with a set of SDS-based operators for resolving spatial conflict in map generalisation.

Key words
Automated generalisation, conflict detection, proximity relations, constrained Delaunay triangulation.

1. Introduction

When digital map data are displayed at a smaller scale than that of the original representation, graphical conflicts frequently arise. The problem is a familiar one in cartography and reflects the fact that, assuming map symbols for points and lines remain the same size, then they occupy a proportionally greater area of the map than at the original scale. Hence there is a competition for the diminishing available map space (Mark, 1990). On the displayed map, this manifests itself by the fact that nearby objects may become too close to be clearly distinguishable, or indeed they may overlap each other. In general the map becomes excessively cluttered. The process of map generalisation is specifically concerned with modifying the map content to suit the space available, and the purpose of the map, and includes procedures such as object elimination, reduction (or simplification) of detail in line and area symbols, exaggeration, collapse, amalgamation of nearby objects, typification of graphic form, and displacement of graphically conflicting objects (Shea and McMaster, 1989). At a recent workshop on map generalisation (Barcelona 1995), three aspects of generalisation were identified as priority areas for research. These were: assessment of the quality of generalisation alternatives; knowledge formalisation; and detection and resolution of conflicts. This paper addresses the last of these areas, specifically that of the detection of conflict, which is a necessary prerequisite to its resolution, by means of one or more of the generalisation operators.

We are concerned with conflict that is due to objects being in too close proximity to each other. Any pair of disjoint objects lying within a defined tolerance distance Dtol of each other will be regarded as in conflict (Figure 1). Note that the value of Dtol will depend both upon the purpose of the derived map, and the class or type of objects being considered. An example of the latter of these distinctions might be that Dtol would differ when considering the separation distance between two buildings and, say, the distance between a building and a road.

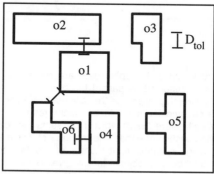

Figure 1 Subsequent to scale reduction certain objects may come into conflict with each other. Here, o1 is in conflict with objects o2 and o6, while o6 is also in conflict with o4.

The specific problem addressed in this paper is: given a set of objects O and a target object o (o ∈ O), find all other objects in O which lie within a distance Dtol of o. The problem of finding nearest neighbouring objects is a familiar one in computational geometry, and its solution can be facilitated by the various spatial indexing methods used in geographical information systems. In this paper we present a proximal search procedure, for solution of the nearest neighbouring polygonal object problem, that is based on constrained Delaunay triangulation of the vertices and edges of the geometric data. Because of the very high degree of explicit connectivity between neighbouring vertices in the triangulation, high performance is obtained in proximal search, though at the cost of storage overheads for the triangulation. The storage overheads are however justified in the light of the considerable benefits that the triangulation data structure brings in solving other aspects of map generalisation. This includes implementation of operators such as those of amalgamation and displacement, which may be used in resolving conflict once it has been detected (Ware et al. 1995; Bundy et al. 1995).

In the remainder of the paper we describe in Section 2 the advantageous properties of triangulations, before describing in Section 3 the Simplicial Data Structure, which is used for implementing the conflict detection procedures, that are described in Section 4. Experimental results are described and evaluated in Section 5, which is followed by concluding remarks in Section 6.

2. Spatial characteristics of triangulations

The spatial characteristics of triangle-based data structures and their usefulness to a variety of applications involving spatial data have been recognised for some time. Frank and Kuhn (1986) describe the potential use of triangulated cell complexes for the explicit maintenance of topological relations between a set of two dimensional point locations. These benefits are reiterated by Egenhofer and Jackson (1989) who define a set of maintenance operations for insertion of nodes, lines and polygons within a 2D simplicial complex. Evidence of the benefits of triangulations in determining proximal relations is borne out by their extensive use in representing terrain elevation data in the form of triangular irregular networks (TINs) (Peucker et al.). In particular, the Delaunay triangulation and its dual, the Voronoi diagram, have been well documented in relation to the determination of neighbourhood relations between point locations (Shamos and Hoey 1975; Preparata and Shamos 1988; Aurenhammer 1991).

2.1 Proximity relations between points, lines and polygons

Given a set of points S, the Voronoi diagram delimits space into polygonal regions, each associated with a single point p (p ∈ S) and lying closer to p than to any other point in S.
The Delaunay triangulation of S can be found by joining each point p to those points lying in Voronoi polygons adjacent to the Voronoi polygon associated with p. For a particular point p, one of the edges thus formed will connect to p's nearest neighbour (Preparata and Shamos 1988).

The basic Voronoi diagram has been extended to include regions surrounding points, lines and polygons in what are referred to as generalised Voronoi diagrams (Lee and Drysdale 1981). Here, the boundary segments, consisting of straight edges and parabolae, between the proximal regions serve to identify the set of objects nearest to a single object, since these boundaries are shared with the proximal regions of the nearest objects. Gold (1994) has highlighted the potential of the generalised Voronoi diagram to maintain neighbourhood relations in geographic databases and has made use of their properties for generating buffer zones and in a novel technique for digitising 2D polygonal maps. Other applications include the creation of medial axis transformations to generate the skeletons of polygonal objects (Kirkpatrick 1979) and path planning in robotics (Aurenhammer 1991).

2.2 The Constrained Delaunay triangulation

Unlike the Delaunay triangulation of a set of points, the edges of a constrained Delaunay triangulation, in which constrained edges correspond to object boundary edges, do not always connect nearest neighbouring objects. Such a nearest neighbour graph can be constructed as the dual of the generalised Voronoi diagram. Unlike the dual of a simple Voronoi diagram, its nodes represent map elements that may be either vertices or the edges that connect vertices of map objects, and hence they do not correspond directly to single points in 2D space. However, for a constrained Delaunay triangulation in which straight line constraining segments are similar in length to the separation distances between objects, the triangulation is such that the vertices of neighbouring objects are usually connected by edges of the triangulation. This observation follows from the fact that if vertices on the boundary of one object are near to those on the boundary of another object, and there is no intervening constraining edge, they are bound to be connected due to the equiangularity properties of the constrained Delaunay triangulation.

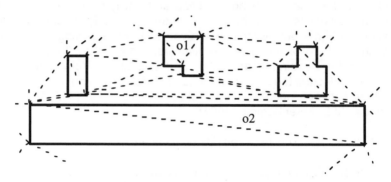

Figure 2 Part of a constrained Delaunay triangulation of a set of objects. Constraining edges (i.e. object boundaries) are shown as solid lines, other edges are shown as dashed lines. The figure shows that there is no direct edge connectivity between object o1 and its nearest neighbour, o2.

Exceptions to the direct edge connectivity between nearby objects can occur where part of a long constraining edge approaches close to the vertices of the boundary of another object (Figure 2). It has been shown that, in these exceptional cases, the nearest neighbour to an object can be found efficiently by a breadth-first triangulation-based search procedure that progresses from the externally connected triangles of that object (Jones et al. 1995).

3. The Simplicial Data Structure

The Simplicial Data Structure, or SDS, is used to represent the collection of planar polygonal objects O and regions of free space F (lying between and, or, within objects with holes) which make up a map M. The primary component of the SDS is a constrained Delaunay triangulation T in which every boundary vertex of each object oi in O corresponds to a triangle vertex and every boundary edge of each object oi in O serves as a constraining edge. The triangulation T forms the basis of a hierarchical description of M in which: each object oi and free space region fi is defined by an unique object identifier and references to the triangles of T which lie within its boundary; each triangle ti of T is described by an unique triangle identifier, references to each of its three constituent edges, plus a reference to the object or free space region within which it lies; each edge ei is described by an unique edge identifier and references its start and end vertices; and each vertex vi stores an unique vertex identifier plus x and y co-ordinate values. Additionally, each edge description contains a flag indicating if it is a real edge (i.e. an edge forming part of an object boundary) or a virtual edge (i.e. an edge lying either within an object or within a free space region), plus supplementary topological information in the form of references to the two triangles (or, the single triangle if the edge lies on the boundary of M) to which the edge belongs. An example map and its SDS representation are shown in Figure 3.

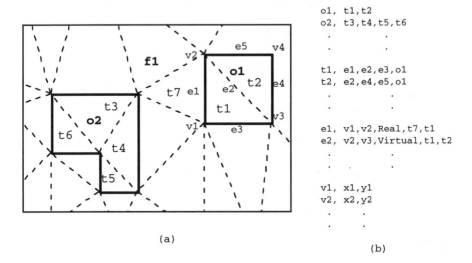

```
o1,  t1,t2
o2,  t3,t4,t5,t6
  .              .
  .              .

t1,  e1,e2,e3,o1
t2,  e2,e4,e5,o1
  .              .
  .              .

e1,  v1,v2,Real,t7,t1
e2,  v2,v3,Virtual,t1,t2
  .              .
  .              .

v1,  x1,y1
v2,  x2,y2
  .      .
  .      .
```

(a)

(b)

Figure 3 A map (a) and its SDS representation (b). The portion of the map shown contains two objects (o1 and o2) and part of a free space region f1. Real edges are shown as solid lines, virtual edges as dashed lines.

3.1 Terminology

If a triangle ti lies within an object oi then ti is said to belong both to oi and O, and is referred to as an object triangle. Similarly, a triangle lying within a free space region fi is said to belong both to fi and F, and is referred to as a free space triangle. The notation of^{ti} is used to denote the object or free space region to which ti belongs.

Two objects, oi and oj, that share a common vertex or edge are said to be contiguous. This definition is refined by reference to the extent of contiguity, i.e. whether a vertex (vertex-contiguous) or an edge (edge-contiguous) is shared. If oi and oj are separated by free space triangles, then they are said to be proximal. The proximal relationship is refined into proximal-n relationships, where n is the minimum number of triangles in free space separating the two objects. If oi and oj are minimally separated by a single free space triangle then they are described as being first order proximal (i.e. proximal-1). Note that the contiguous relationship is equivalent to the proximal-0 relationship. The contiguous and proximal terminology is extended to encompass relationships between any SDS component type. For example, two triangles that share a common edge are said to be edge-contiguous, while an object and a triangle that share a common vertex are said to be vertex-contiguous.

3.2 Halo Triangles

To support the development of procedures to identify conflict situations, we introduce the concept of the halo triangles Th^{oi} of an object oi. These are triangles that share a vertex with oi but which do not belong to oi. A procedure, FindHaloTriangles, which finds Th^{oi} is shown in Figure 4.

```
FindHaloTriangles(IN : oi; OUT : Thoi)
Initialise Thoi to empty
Initialise Processed_Vertex_List to empty
For each triangle t ∈ oi
    For each vertex v ∈ t
        If (v ∉ Processed_Vertex_List)
            Temp_Triangle_List := TriangleVertexConnect(v, t)
            For each triangle t1 ∈ Temp_Triangle_List
                If (t1 ∉ oi) and (t1 ∉ Thoi)
                    Add t1 to Thoi
                Endif
            Endfor
            Add v to Processed_Vertex_List
        Endif
    Endfor
Endfor
End FindHaloTriangles
```

Figure 4 The procedure FindHaloTriangles which returns the halo triangles (Th^{oi}) associated with object oi.

The procedure is primarily concerned with finding triangles which are vertex contiguous with oi (i.e. triangles that share a vertex with oi). Any such triangle which does not belong to oi is added to Th^{oi}, provided it has not already been added. FindHaloTriangles makes use of the procedure TriangleVertexConnect which, given a vertex v and an initial triangle t to which v belongs, returns a list of the identifiers associated with all triangles to which v belongs. It achieves this by initially identifying an edge e of t to which v belongs. This edge has two adjacent triangle pointers associated with it, one which points to t, the other which points to the triangle t1 adjacent to t and sharing edge e, and hence sharing vertex v. The triangle t2 adjacent to t1 and sharing v (and not t) can be found in a similar way. The process of finding adjacent triangles continues to rotate about v until the original triangle t is found.

The halo triangles of oi are of interest in that they provide a convenient way of identifying its contiguous and proximal-1 neighbouring objects. These neighbours are found by examining each of oi's halo triangles in turn. If a particular halo triangle th belongs to an object oj then oi and oj are contiguous. If, however, th belongs to F, then each of the objects contiguous to its vertices needs to be found. It follows that each of these objects (with the exception of oi itself) is at least proximal-1 to oi. The contiguous neighbours of oi are referred to as its Real Neighbour Set (*RNoi*), while oi's proximal-1 neighbours are referred to as its Virtual Neighbour Set (*VNoi*).

RNoi and *VNoi* are collectively referred to as oi's Connected Neighbour Set (*CNoi*).

3.3 The Tolerance Set

Given an object oi in O, all other objects in O which lie within a tolerance distance Dtol of oi are referred to as oi's Tolerance Set (*Toi*). Note that when the distance between two objects is mentioned in the text, it is, unless otherwise stated, the minimum Euclidian distance that is being referred to.

3.4 The Conflict Set

The collection of objects in O which lie within a distance Dtol of oi and are at least proximal-1 (i.e. not contiguous) to oi is referred to as the Conflict Set of oi (*Coi*). We choose to omit oi's contiguous neighbours from its conflict set since the distance (in viewing co-ordinates) between oi and such neighbours has remained constant (i.e. zero) between scales, and therefore no conflict due to a reduction in minimum separating distance will have occurred.

4.   Detecting spatial Conflict

We now describe a method for identifying the conflict set of a particular object. The method makes use of the SDS. Its efficiency is reliant on the explicit proximity properties inherent to the constrained Delaunay triangulation (upon which the SDS is based).

4.1 Finding the Conflict Set

The Conflict Set of oi is found in three stages. Firstly, the Real Neighbour Set of oi is found, using the technique described in section 3.2. Next, we find the Tolerance Set of oi. *Toi* is found   by application of the procedure FindToleranceSet (Figure 5) and works on the basis of finding objects within tolerance of individual boundary edges of oi. The procedure begins by initialising *Toi* to empty. The boundary edges of oi are then found, each of which is processed in turn. For a particular boundary edge e this process begins by placing e on a previously empty queue (Edge_Queue) which will contain triangle edges awaiting processing. Edges are removed from this queue in turn and processed. For a particular edge e1 this process involves a series of actions upon each of its adjacent triangles. If we consider a triangle t adjacent to e1 then the initial action entails tests to ascertain if t belongs to oi or if t has already been tested. If either of these tests prove to be true then no further processing of t is required. Alternatively, if both tests return a negative result, then more work upon t is required. Firstly, tests are carried out to check if o^t belongs to O and if o^t does not already belong to *Toi*. If each of these tests prove to be true then o^t is added to *Toi*. Secondly, each edge e2 of t is examined in turn. If e2 is not already in Edge_Queue (i.e. is not already awaiting processing) and has not already been processed and lies within a distance Dtol of oi, then e2 is added to Edge_Queue and will be subsequently processed. The number of edge to edge distance calculations is minimised by firstly checking if edge e2 is contiguous with oi. If this is the case then it follows that e2 is at a distance zero from oi. Note that Edge_Queue will always only contain edges which are within a distance Dtol of the boundary edge e currently being examined. Finally, *Coi* is formed by making a comparison of *Toi* with *RNoi*. All objects belonging to *Toi* but not belonging to *RNoi* make up *Coi*.

FindToleranceSet makes use of three other procedures, GetObjectEdges, EdgesContiguous and EdgeDistance. GetObjectEdges returns the list of real edges relating to a given object o (i.e. the boundary of o). These edges are found by examining the triangles belonging to o. EdgesContiguous returns a Boolean value indicating if two given edges are contiguous. EdgeDistance returns the minimum distance between two edges.

```
FindToleranceSet(IN : oi, Dtol; OUT : Roi)
Initialise Toi to empty
Object_Edges := GetObjectEdges(oi)
For each edge e in Object_Edges
   Initialise Edge_Queue to empty
   Initialise Edge_Tested_List to empty
   Initialise Triangle_Tested_List to empty
   Add e to Edge_Queue
   Do while (Edge_Queue not empty)
      Dequeue edge e1 from Edge_Queue
      Add e1 to Edge_Tested_List
      For each triangle t adjacent to e1
         If (t ∉ o1) and (t ∉ Triangle_Tested_List)
            If (of$^t$ ∈ O) and (of$^t$ ∉ Toi)
               Add of$^t$ to Toi
            Endif
            For each edge e2 ∈ t
               If (e2 ∉ Edge_Queue) and (e2 ∉ Edge_Tested_List)
                  If (EdgesContiguous(e2,e))
                     Add e2 to Edge_Queue
                  Else
                     If (EdgeDistance(e2,e)<=Dtol)
                        Add e2 to Edge_Queue
                     Endif
                  Endif
               Endif
            Endfor
            Add t to Triangle_Tested_List
         Endif
      Endfor
   Enddo
Endfor
End FindToleranceSet
```

Figure 5 The procedure FindToleranceSet which returns the list of the objects (Toi) which lie within a distance Dtol of oi.

4.2 Classification of Neighbours and Conflict

In exceptional cases (as in Figure 2) there may exist one or more objects (including, perhaps, oi's nearest neighbour) which do not belong to *CNoi* but which lie nearer to oi than one or more of the objects in *CNoi*. These objects are termed oi's Unconnected Neighbour Set (*UNoi*). The collection of objects belonging to either *CNoi* or *UNoi* is called oi's Neighbourhood Set (*Noi*). The various neighbour sets associated with oi are illustrated in Figure 6.

By means of a slight alteration to the procedure FindToleranceSet, so that minimum separating distances between oi and objects in *Toi* are recorded, it is possible to form the Immediate Conflict Set (*ICoi*), made up only of the subset of objects in *Coi* which also belong to *Noi*. Also, all objects in *CNoi* which lie within a distance Dtol of oi form the Connected Conflict Set of oi (*CCoi*). The reason for classifying conflict in this way relates to the techniques that are adopted when resolving the conflict. For example, the object merge procedures detailed in (Ware et al. 1995; Jones et al. 1995), which make use of the triangles intervening between objects to assist in object amalgamation, require that objects being merged be at most proximal-1 (i.e. oi can only be merged with objects contained in *CCoi*). Similar procedures are envisaged for application between oi and those members of *ICoi* which do not belong to *CCoi*. However, it would not appear sensible to adopt such an approach to conflict resolution between oi and those members of *Coi* which do not belong to *ICoi*. Any such operation would in itself be likely to cause further conflict (due to a further violation of Dtol or the introduction of overlapping objects) because of a high possibility of the existence of a third object lying between, or partly between, the two objects being merged.

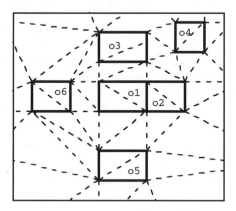

Figure 6 The neighbour sets associated with an object, o1. RNo1 = {o2}, VNo1 = {o3,o5,o6},
CNo1 = {o2,o3,o5,o6}, UNo1 = {o4} and No1 = {o2,o3,o4,o5,o6}.

5. Experimental results

The procedures described in previous sections have been implemented in C on a Sparcstation 10. The implementations have been tested and validated using large scale (1:1250) topographic map data from the UK Ordnance Survey. Dtol is set to 0.64mm for all calculations, this being a typical value used by the OS when producing 1:10000 maps. Three data sets are used, each representing a 500m x 500m area, and consisting of objects representing mainly buildings and roads. Performance results are shown in Figure 7.

Data Set	No. of Objects	No. of Points	No. of Real Edges	No. of Triangles	Average Size of Toi per Object	Average Size of Coi per Object	Average No. of Edge Distance Calculations per Object	Average CPU Time used per Object (s)
1	627	4959	6833	40409	2.89	1.82	217.09	0.00078
2	657	7810	10303	72869	3.37	2.27	628.55	0.00455
3	146	2237	2677	21902	2.21	1.25	652.51	0.00730

Figure 7 Results from experiments to find conflict situations resulting from a reduction in map scale. Note that edge to edge calculations refer to comparisons between both real and virtual edges in the triangulation, and that the average number of real edges per object are approximately 10.9, 15.7 and 18.3 for data sets 1, 2 and 3 respectively.

These results appear satisfactory, especially when considering that much of the information derived during this process (such as the halo triangles of each object) can be directly used in the solution to many of the conflict situations found (Ware et al. 1995). The authors have found difficulty in comparing these results with other methods, due mainly to the rather arbitrary nature of the Dtol value. However, it has been shown (Jones et al. 1995) that similar SDS proximity procedures, which relate to locating the nearest object edge to an arbitrary point, do compare favourably with existing spatial indexing data structures. The average number of edges examined to find the nearest edge in experiments with topographic data was in the range 7.4 to 17.9. Hoel and Samet (1991) report on nearest line segment searches in a PMR quadtree. Their experimental results show the average number of line segments examined to find the nearest segment to a point average between 29.67 and 37.22.

5.1 Theoretical evaluation

For the purposes of conflict resolution, we have shown that our main concern is, for a given object oi, to find *ICoi*. In an ideal triangulation, for the purposes of proximal search, the lengths of real edges would be similar to the separation distances between neighbouring objects. This would ensure that in a constrained Delaunay triangulation the nearest vertex to an edge was just as likely to belong to a neighbouring object as to the boundary of the object itself. This would result in a high probability of pairs of neighbouring objects being connected directly by edges of the triangulation. In other words, *ICoi* becomes identical to *CCoi*. This condition did in fact hold with the data used in our experiments. In such circumstances, a simple analysis of performance can be carried out in relation to the number of edge to edge distance calculations carried out on average per object.

Let the average number of edges per object in O be n. It can be shown that the average number of halo triangles per object oi will be 4n+1. Making use of the property of Delaunay triangulations that the number of edges emanating from a vertex does not, on average, exceed 6 (Preparata and Shamos 1988), we expect that, on average, each vertex of oi will have 6 contiguous triangles (each also contiguous with oi). At least one of these triangles will belong to oi, and is therefore not a halo triangle. Also, n-1 points of oi (say, points 2 to n) can be thought of has having at least one contiguous triangle which is also contiguous with the immediately previous point. Therefore, the total number of halo triangles associated with oi will, in the worst case, average 4(n-1) + 5, which equals 4n+1. Each halo triangle of oi will always have at least 2 edges that are vertex-contiguous with oi. Its third edge will have an equal likelihood of being either contiguous with oi or proximal-1 with oi. The only edge to edge distance calculation to be carried out will be in the case of the proximal-1 edge. On average, each halo triangle will therefore require 0.5 distance calculations to ascertain if any of its edges are within tolerance of oi. Therefore, for an object oi, the average total number of distance calculations required to find *ICoi* is, at most, (4n+1)/2. Experiments to find *CCoi* for each of the objects in data sets 1, 2 and 3 produce average numbers of edge to edge distance calculations per object of 12.45, 16.49 and 17.98 for the three data sets respectively. The results fall well within worst case averages derived using the (4n+1)/2 formula, which are 22.3, 31.9 and 32.1 respectively.

6. Concluding remarks

The paper has presented a spatial data structure and associated search procedures that detect spatial conflict in a map. Practical experiments and a theoretical analysis confirm that the approach, which is based on the use of a constrained Delaunay triangulation, is very favourable for proximal search in general, particularly when the typical length of triangle edges is similar to the typical separation distance between objects. The experiments are limited in that Dtol is fixed for all calculations. In practice Dtol should change in accordance with the type of objects being processed. Future versions will incorporate this facility. It is also envisaged that this work will be harnessed with the conflict resolution procedures referred to earlier to provide a means of detecting and then resolving conflict automatically.

The data model described incurs a storage overhead in the form of the triangulation T and the triangle references stored by each object description. Note, however, that in applications that employ large spatial databases it may not be necessary to store the explicit triangulation and triangle references permanently. Jones et al (1994) have presented procedures for local construction of triangulations from a spatial database containing point, line and areal features. Thus local triangulations (i.e. local SDSs) can be constructed for user-specified windows for purposes of user interaction and for purposes of potentially complex transformations, such as map generalisation.

Acknowledgement
The authors are grateful to the UK Ordnance Survey for supplying digital map data for research purposes.

References
Aurenhammer, F. 1991, "Voronoi Diagrams - A Survey of a Fundamental Geometric Data Structure", ACM Computing Surveys, Volume 23, Number 3, pp345-405.
Barcelona 1995, Workshop on Progress in Automated Map Generalisation, Barcelona, Spain, September, 1995.
Bundy, G.Ll., Jones, C.B. and Furse, E. 1995, "A Topological Structure for the Generalisation of Large Scale Topographic Data", Innovations in GIS 2, Taylor and Francis, pp19-31.
Egenhofer, M. and Jackson, J. 1989, "A Topological Data Model for Spatial Databases", Symposium on the Design and Implementation of Large Spatial Databases, Santa Barbara, USA, pp271-286.
Frank, A. and Kuhn, W. 1986, "Cell Graphs: A Provable Correct Method for the Storage of Geometry", Second International Symposium on Spatial Data Handling, Seattle, USA, pp411-436.
Gold, C.M. 1994, "Three approaches to automated topology and how computational geometry helps", Sixth International Symposium on Spatial Data Handling, Edinburgh, UK, pp145-158.
Hoel, E.G. and Samet, H. 1991, "Efficient processing of spatial queries in line segment databases", Advances in Spatial Databases: Proceedings of the Second Symposium on the Design and Implementation of Large Spatial Databases, Lecture Notes in Computer Science 525, Springer, pp237-256.
Jones, C.B., Bundy, G.Ll. and Ware, J.M. 1995, "Map Generalisation with a Triangulated Data Structure", Cartography and GIS 22(4), pp 317-331.
Jones, C.B., Ware, J.M. and Bundy, G.Ll. 1995, "Proximity Relations with Triangulated Spatial Models", Computer Studies Research Report CS-95-4, University of Glamorgan, UK.
Jones, C.B., Kidner, D.B. and Ware, J.M. 1994, "The Implicit TIN and Multiscale Spatial databases", The Computer Journal, Volume 37, Number , pp43-57.
Kirkpatrick, D.G. 1979, "Efficient computation of continuous skeletons", Twentieth Annual IEEE Symposium on Foundations of Computer Science
Lee, D.T. and Drysdale, R.L. 1981, "Generalisation of Voronoi diagrams in the plane", SIAM Journal on Computing, Volume 10, Number 1, pp73-87.
Mark, D.M. 1990, "Competition for map space as a paradigm for automated map design", Proceedings GIS/LIS '90, Anaheim, California, pp97-106.
Peucker, T.K., Fowler, R.J., Little, J.J. and Mark, D.M. 1978, "The Triangulated Irregular Network", ASP/ACSM Digital Terrain Models (DTM) Symposium, pp516-540.
Preparata, F.P. and Shamos, M.I. 1988, "Computational Geometry", Texts and Monographs in Computer Science, Springer-Verlag.
Shamos, M.I. and Hoey, D. 1975, "Closest-point problems", Sixteenth Annual IEEE Symposium on Foundations of Computer Science, pp151-162.
Shea, K.S. and McMaster, R.B. 1989, "Cartographic generalisation in a digital environment: when and how to generalise", Proceedings Auto-Carto 9, Baltimore, ACSM/ASPRS, pp56-67.
Ware, J.M., Jones, C.B. and Bundy, G.Ll. 1995, "A Triangulated Spatial Model for Cartographic Generalisation of Areal Objects", Lecture Notes in Computer Science, edited by A.U. Frank and W. Kuhn, Springer, pp173-192.

Contact address
J. Mark Ware and Christopher B. Jones
Department of Computer Studies
University of Glamorgan
Pontypridd
Mid Glamorgan CF37 1DL
UK
E-mail: jmware@glam.ac.uk

HOW GENERALIZATION INTERACTS WITH THE

TOPOLOGICAL AND METRIC STRUCTURE OF MAPS

Giuliana Dettori and Enrico Puppo

Istituto per la Matematica Applicata, C. N. R.
Genova, Italy

Abstract
In this paper we discuss how to define in a mathematical framework the set of operations that cartographers call generalization. First we make a clear distinction between topological and metric aspects of maps, and describe suitable mathematical tools for treating both cathegories. Then we analyse the basic operations of generalization, pointing out whether each one affects the metric or topological structure of a map. Finally we discuss how generalization operations can fit into the described mathematical framework. This is intended as a step towards a global approach to automatic generalization.

Key words
Map generalization, topology, shape, abstract cell complex, homotopy.

1. Introduction

It is well known that the structure of maps can be suitably represented by means of planar graphs, possibly with dangling edges, multiple edges, self-loops and isolated vertices. Semantic information is associated with points, edges and faces [3, 6, 7]. The problem of representing and automatically manipulating these graphs is rather complex and is of extreme importance in Geographic Information Systems. The production of maps at different scales based on an original, extremely accurate and detailed map, is a major problem, both when several maps at different scales are stored in a multiresolution system, and when a representation at a required scale is computed when necessary.

A suitable approach to map processing should carefully distinguish between topological aspects, concerning the map structure and mutual relations of its entities, and metric ones, concerning local shapes. These two components are influenced in different ways by scale changes and can be treated more suitably in different mathematical environments.

The process of representing maps at smaller scales is known as generalization. It is correctly considered a non-algorithmic task since it involves both exact rules and as yet unquantifiable considerations. The difficulty lies in the fact that moving to smaller scales introduces more and more symbolisation - from maps where precise metric information is available to maps where symbolic aspects are prevalent - while at the same time there is a need to preserve the topological consistency of different maps representing the same area.

Müller et al. [8] correctly note that two conceptual levels of map generalization should be distinguished, i.e. cartographic and model-oriented generalization. Cartographic generalization consists of a series of transformations at graphic representation level that are traditionally performed by cartographers. Model-oriented generalization is required in contexts which are not strictly limited to graphical representation, as in the case of GIS. Here a map is first of all a model that includes numerical data, relations between them and semantic information, and great importance is given to the consistences of internal data through increasing levels of abstraction. The tools currently available for automated generalization have mostly been developed to mimic the cartographer's behaviour in manual generalization. Stress has largely been put on algorithmic aspects, without ever addressing the whole problem systematically in the digital domain; hence cartographic and model-oriented aspects have often been confused. Our aim is to contribute to clarifying objectives and characteristics of generalization within a formal framework where all its aspects can be treated.

In a previous paper [9], we addressed the issue of building a formal model for multiresolution spatial maps by separating metric and topological aspects and modelling them with suitable mathematical tools. Based on those results, we discuss here the operations of generalization by clearly separating topological, metric and semantic aspects, and trying to understand what aspects of a map are involved in each operation, and how. Although we do not claim that our analysis is complete, nor that it encapsulates all possible cases, we believe it to be an important step towards automatic generalization.

2. Separating topological and metric aspects in a map

2.1 Accuracy and resolution

Two maps at different scales can differ both in accuracy and resolution. Reduced *accuracy* means that the spatial extent and location of an object in a map is a less precise approximation of the extent and location of the entity represented by that object. Variations of accuracy change the shape of a map, since measurements become less reliable, but should not change its topology. On the other hand, a decrease in *resolution* means that some entities are no longer represented, or are represented in a simpler form. This is the case, for instance, when erasing the shape corresponding to a town, or marking it only with a point. Variations of resolution, besides changing the shape, obviously change the topology of a map as well.

When passing to a smaller scale, the resolution and/or accuracy of a map also need altering for drawing reasons, since a map too dense in information would be difficult to trace and read. Another reason for reducing the amount of information displayed by the map is to highlight some of it: while high-accuracy maps carefully describe fairly restricted areas, low-accuracy maps aim to point out only some aspects of a more extensive area, and this can best be done by eliminating some features while overloading others with symbolic value. The relation between the degrees of resolution and accuracy definitely cannot be determined solely by the scale or dimensions of a map, but rather by its semantics. Let us think, for instance, of a map representing the highways of a country: only highways are shown, and these are represented with considerable accuracy by means of thick lines over-proportioned to the scale of the map; most other roads and features are eliminated, even though they are actually of comparable dimension. A map of the same country at the same scale could, on the other hand, aim to give general information on the country, and hence show many more details, possibly with lower accuracy. The difference and mutual independence of these two features is one of the reasons for separating the processing of metric and topological aspects. Another reason is that shape and topology can best be treated by means of different mathematical tools.

2.2 A formal tool to control accuracy

The problem of approximating a line with a simpler one has been tackled by many authors who adopt more or less heuristic approaches, generally under the headings of line simplification or polygonal approximation [6]. Basically, the problem has been tackled in two different ways, either by controlling the error introduced by the approximation, or by minimizing some feature of the approximating line, such as its length or its number of joints. However, though these are usually based on mathematical procedures, neither is developed within a rigorous mathematical framework, which is our aim in this paper.

Homotopy represents a rigorous mathematical tool for controlling the accuracy of shapes. Intuitively, this is based on the observation that a geometric object o is represented by an entity e at precision $\varepsilon \geq 0$ if and only if o is contained in a region obtained by "fattening" e by ε, i.e., in $r_{e, \varepsilon} = \{x \mid d(x, e) \leq \varepsilon\}$, where d denotes the Hausdorff distance in $\Re^2$.

Formally, we express this idea by specifying a continuous transformation that maps the object o into the entity e representing it, without leaving the fattened region $r_{e, \varepsilon}$. The formal definition is given on lines, and extended to regions through boundary lines (see [9] for details).

A line l can be defined in parametric form by means of a continuous function $l: I \to \Re^2$, where the domain I is the unit interval [0,1] on the real line and the codomain is a subset of the real plane $\Re^2$ containing all points of the line. Let f and g be two lines. A *homotopy* between f and g is a continuous function $H: I \times I \to \Re^2$ such that $H(x,0) = f(x)$ and $H(x,1) = g(x)$. Such a homotopy H defines a continuous deformation from f to g. For $\varepsilon \geq 0$, a homotopy H is called an ε-homotopy if
$$\forall x \in I, \, \forall y \in I, \, d(H(x, y), H(x,0)) \leq \varepsilon \, [10].$$

The ε-homotopy represents a much stronger condition than simply assuming *g* inside the region fattening *f*, since it requires any portion of *g* to remain inside the fattened region of the corresponding portion of *f*. This scheme can be implemented in different ways, by using any function H that satisfies the above conditions.

2.3 A formal tool to control resolution

Relations between pairs of plane spatial entities have been widely studied and classified in the literature (for a survey, see [2]). A map can be regarded as a disjoint covering of a portion of the plane $\Re^2$, with a collection of atomic entities. This fact suggests that, if necessary, lines and regions should be subdivided so that no pair of open regions intersect, and no line intersects the relative interior of any other line [3, 5, 7, 9]. In this way, the possible relations between pairs of entities in a map are highly simplified. This subdivision eases processing operations and does not change the semantics of any entity in a map, while emphasising semantic differences. Let us take for example a road that crosses a river by means of a bridge: extra boundary points and lines delimiting the intersection area between the two strips will be introduced, thus dividing the road into "road to the left of the river", "road on the bridge", "road to the right of the river"; not only is this semantically correct, it also provides the chance to associate more precise information to map entities.

In order to capture completely the topological structure of a map while making an abstraction of its shape, we describe it as an abstract cell complex [9, 10], where points are the entities of order 0, lines of order 1, and regions of order 2. This approach allows us to process a map as a topological space, thus exploiting the theoretical results of topology, and defining some criteria of structural consistency between two maps of the same area at different scales.

Formally, an abstract cell complex is a triple $\Gamma = (C, <, ord)$, where *C* is the set of cells (in our case points, lines and regions), < is a strict partial ordering on *C* called *bounding relations*, and *ord*: $C \to \{0,1,2\}$, called the *order function*, is such that for every pair of cells in *C*, γ_1 and γ_2, $\gamma_1 < \gamma_2 \Rightarrow ord\,(\gamma_1) < ord\,(\gamma_2)$. Based on this definition, all topological relations in a map can be translated into relations between cells of an abstract cell complex: $\gamma_1 < \gamma_2 \Leftrightarrow \gamma_1$ *inside, bounds*, or *coveredBy* γ_2.

Given two maps Γ and Γ' of the same area at different resolutions, we model the *topological generalization* of Γ into Γ' as a surjective mapping $F : \Gamma \to \Gamma$'. For every object γ describing an entity in Γ, its image $F(\gamma)$ describes the same entity in Γ'. Such mapping F is *valid*, that is it describes an acceptable generalization, if for any two cells γ_1 and γ_2 whose relation in Γ is not *disjoint*, the relation between $F(\gamma_1)$ and $F(\gamma_2)$ is also not *disjoint*. This means that *F* is valid if map features are contracted rather than expanded. In topological terms, this is expressed by the condition that F be a continuous mapping in the finite topology, i. e. in the topological space underlying abstract cell complexes (see [4, 9] for details).

Possible generalizations of an entity γ of Γ are: *preservation*, when the entity preserves its original order (i.e. a point is mapped to a point, a line to a line, a region to a region); *contraction,* when its order is decreased (i.e. a line is mapped to a point, a region to a point or to a line); *immersion*, when it is mapped, together with some other entity, to an object of higher order (i. e. a point is mapped to a region or a line, a line to a region). Since the composition of two continuous functions is itself a continuous function, all acceptable generalizations can be obtained as a sequence of atomic simplifications. This reduces the problem, since only basic operations need to be analysed.

Changes are monotonical, passing from relations expressing more definite separations, like *disjoint*, to others expressing an increasing degree of closeness, down to *equal* which expresses the contraction of several elements into one, thus ensuring the production of simpler maps.

3. Generalization operations

One problem that makes analysis of generalization operations difficult is the fact that it has been treated in non-homogeneous way by different authors, as concerns both the number and selection of the analysed operations, and the terminology used. However, map generalization is generally considered to be based on six operations: simplification, selection, symbolisation, exaggeration, displacement, aggregation [8]. In this paper, we overload these words, using them both for cartographic operations and for the mathematical mappings that can express them.

Simplification eliminates a number of points describing a line. In a formal computerized framework, it is necessary to distinguish topological points (which mark distinctive features characterising the shape of entities) from points that are introduced to draw smooth lines mathematically, since a free-shaped line is defined by the vertices of a polygonal or spline. Simplification acts on points introduced for drawing reasons. This operation is the most common one, and the easiest to understand, since it is obvious that at smaller scale many representation points are no longer necessary. Thus, it is not surprising that this aspect is the most often addressed by unsophisticated systems, which largely reduce generalization to simplification. Hence, this operation changes only the shape of a map, not its topology.

The effect of *selection* is the elimination of some entity or group of entities considered irrelevant at a given scale, either because they are too small or semantically irrelevant in the context of a considered map. Hence, this operation affects the topology of a map and must also be driven by the semantic information associated with the entities, besides metric computations.

When applying *symbolisation*, a semantically relevant feature of small dimension is represented by a suitable symbol rather than by a sketch of its actual shape, in order to emphasise its character and location. When reduced to a symbol, an entity with a spatial extent can structurally be assimilated to a point or a line; hence the topology is affected if the inverse image of that entity in the original map was of a different order. The shape is also affected, since symbols, by their very nature, do not allow accurate metric evaluations.

Exaggeration also seeks to preserve entities that would be too small for the context of a map, but are of semantic importance. The entities generalized through exaggeration lose part of their geometric meaning and acquire some symbolic significance, while respecting the structural relations with the other entities in the map. For instance, a road can be exaggerated by means of a thick strip, in order to mark its importance rather than its real dimensions, but it should still touch the same towns, and pass through the same regions, otherwise both the topological relations and the semantics of the map would be affected. Hence, this operation does not influence the topology of a map, but only its shape, and must obviously be driven by semantic information.

Another tool for emphasising elements that would disappear at a small scale is *displacement*. This moves entities to slightly different positions in order to avoid overlapping or excessive closeness that would make elements unreadable. A classic example is that of a road parallel to a river or a railway line. At larger scales, it would not be possible in principle to distinguish the two elements, since they should be represented by two superimposed lines of irrelevant width; in actual maps, a displacement is made by inserting a strip separating the two elements, so that both of them are perceptible. Sometimes this operation might appear to be an exaggeration of a region, if a strip of land actually exists between the two considered entities; however, rather than seeking to preserve map topology through a scale change, displacement is actually aimed at preserving visual perceptibility, as we argue in the following section. However, displacement can not be seen as an exaggeration when a strip does not exist in reality and it is created only to ensure visual separability. In this case, the map topology is affected even in a non-monotonic way. Another classic example of displacement is that of two adjacent cities, which can be drawn by two clearly separated symbols though they actually share a common boundary. As in the previous example, this displacement changes the topology in a non-monotonic way. Moreover, the metric structure of maps is always changed, because displacement obviously involves metric modifications.

The last operation of generalization is *aggregation*, which also proves to be the most difficult one both from the logical and mathematical points of view; in fact, though usually applied in practice, it is often avoided in papers on generalization. Aggregation consists in grouping together entities that, because of their dimension, would individually disappear, but when taken as a whole are still significant. Regions to be aggregated can be either adjacent or not. The first case is easier to tackle, since it is equivalent to a selection operation which eliminates the lines that separate regions, leaving only their external shell. The case of non-adjacent entities is more complex, and quite common in practical applications. Typical examples are groups of islands along a jagged coastline or small lakes in countries like Canada or Finland. At first sight, it may seem that this operation could be obtained by eliminating some entities and exaggerating others. However, this seldom happens in hand-made maps, and is only done when one of the considered features is of much greater importance than the surrounding ones, thus deserving explicit representation. In most real cases, the new entity introduced cannot logically be matched with any of the previous ones, but rather has the meaning of a set of them. This means that not only are the topology and shape affected, but also the semantics of the map, and this explains why the operation is conceptually difficult.

4. Do generalization operations fit into our theoretical framework?

The analysis of generalization operations carried out in the previous section underlines the fact   that selection, symbolisation, and aggregation are essentially topological; simplification and exaggeration are exclusively metric; and that displacement is primarily metric, but can involve topological changes in some cases. A different classification is possible if we want to stress the distinction between conceptual, or model-oriented, aspects and cartographic ones: simplification, selection, aggregation and symbolisation are conceptual operations, while displacement and exaggeration are cartographic ones. In Table 1, generalization operations are subdivided according to these two possible classifications.

Topological operations can usually be considered prior to metric ones, since metric changes need to preserve topological constraints, while topological changes are by definition shape-independent. Moreover, as suggested in [8], conceptual operations can be considered prior to cartographic operations, since it is possible to maintain a conceptually generalized model where metric aspects are only partially evaluated, and to complete generalization by specifying shapes fully only for the purpose of drawing maps.

	model-oriented	cartographic
only metric aspects	simplification	exaggeration, displacement
topological aspects	selection, aggregation, symbolisation	displacement

Table 1 Classification of generalization operations according to two different perspectives; topo-logical changes indirectly imply metric changes as well, but not vice versa. Displacement appears in both rows on the one hand it may give rise solely to metric changes, and on the other affects both metric aspects and topology.

In the framework described in Section 2, we made a distinction between operations that decrease resolution, affecting the topology, and operations that reduces accuracy, affecting the geometry. A reduction in accuracy is controlled through ε-homotopy, while a decrease in resolution is further broken down into two classes of operations: contractions and immersions. Continuity of mappings and ε-homotopy are the sole, formally specified means needed in our framework to control the consistency of generalization.

Here, we seek to match our approach with the six reviewed operations, while taking into account all the above considerations. Since our framework was originally conceived for conceptual generalization, matching is neater for conceptual operations, while the framework needs relaxation rules to adapt to some cartographic operations. The difficulty in fitting some examples to our framework helps highlight the lack of consistency of some commonly performed operations.

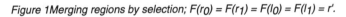

Figure 1Merging regions by selection; $F(r_0) = F(r_1) = F(l_0) = F(l_1) = r'$.

Selection directly matches immersion: when a feature disappears from a map at lower resolution, it is ideally immersed into some larger entity surrounding it, as in the example in Fig. 1.

Similarly, *symbolisation* (Fig. 2) matches with contraction, though a finer classification is made in our framework. Indeed, symbolisation is intended as a general term for a transformation that reduces the conceptual detail in describing an entity. We differentiate between the different ways in which detail is reduced by considering the order of objects representing an entity at different resolutions. The most common case of symbolisation is that of a region (e.g. a municipality) which is substituted by a symbol associated to a single location in space. In conceptual generalization, this is equivalent to mapping the region into a point (region-to-point contraction). The extent of the symbol used, which remains unevaluated, is relevant only in the subsequent drawing phase. Let us now consider the example of a river whose width is relevant at a relatively high resolution, but becomes irrelevant at a lower resolution: in this case, an area representing the river at higher resolution is mapped into a line (region-to-line contraction). We have made a different kind of symbolisation by abstracting some aspects of the shape of the river (the width), while preserving others (the waterway), perhaps with lower accuracy.

Finally, let's consider a bridge located at the point where a road crosses a river: when the width of the river is relevant, the length of the bridge is also relevant (though the width of the road might not be), hence the bridge is represented by a line; when the length of the bridge becomes irrelevant, it will simply be symbolised with a mark at the intersection between the lines representing the river and the road. Hence, a line at higher resolution is mapped into a point (line-to-point contraction).

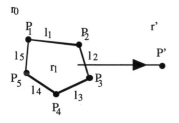

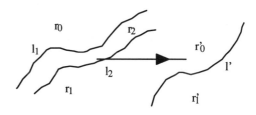

$F(e)=P'$, where $e= r_1, P_1, P_2, P_3,$ $P_4, P_5, l_1, l_2, l_3, l_4, l_5; F(r_0)=r'.$

$F(r_0)=r_0'; F(r_1)=r_1'; F(r_2)=F(l_1)=F(l_2)=l'.$

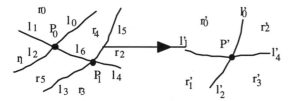

$F(r_0) = r_0'; F(r_1) = r_1'; F(r_2)=r_2'; F(r_3)=r_3'; \quad F(l_1) = l_1'; F(l_4) = l_4';$
$F(l_0) = F(l_5) = F(r_4) = l_0'; \ F(l_2) = F(l_3) = F(r_5) = l_2'; \ F(P_0) = F(P_1) = F(l_6) = P';$

Figure 2 Symbolisation by means of a continuous function.

$$F(r_0) = F(r_1) = F(l) = r'$$

Figure 3 Aggregating adjacent regions by immersion of boundary line.

Unless it is handled properly, *aggregation* may cause problems with topological mapping. Indeed, while aggregation of adjacent entities (like crop fields) is an immediate consequence of immersion of line boundaries (see Fig.3), a function mapping a group of disjoint entities into a single entity (see Fig.4a) would map a non-compact entity into a compact one, with possible undesired side effects on the semantic level. We can overcome the semantic and structural inconsistency by forcing a subdivision of the region containing the aggregated entities into two parts (see Fig. 4b), according to the semantics: one part is the surrounding region, while the other contributes to form a "shell" containing all entities to be aggregated. For instance, let us suppose that the three shapes in Fig. 4 are navigable lakes that need to be aggregated into one lake. If the shell region r_4 is not mapped into r' as well, the whole entity r' would be labelled as "lake" and "navigable", and this could lead a query system to incorrect answers. In the mathematical model, the extra boundaries between r_0 and r_4 must be defined from the very beginning and the proper semantics associated to them, but these are not traced at cartographic level in the map at higher resolution. At lower resolution, all entities in the shell are merged through line immersions that eliminate all lines internal to the shell; this directly corresponds to an abstraction of the difference between lakes and separating land.

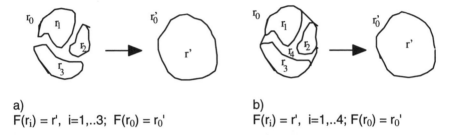

a)   b)
$F(r_i) = r', \ i=1,..3; \ F(r_0) = r_0'$ $F(r_i) = r', \ i=1,..4; F(r_0) = r_0'$

Figure 4 Aggregating regions by introducing conceptual extra lines to preserve semantic soundness.

Simplification is directly controlled through ε-homotopy. Once a threshold value for the accuracy error is given, any line satisfying the corresponding ε-homotopy is an acceptable simplification. In cartographic practice, it is also important that a lower accuracy representation also be simpler in some respects. Whether a simpler line is obtained or not, depends on the algorithm employed for its computation. The aim of ε -homotopy is to ensure that the accuracy of the new map is above an assigned threshold. If it is necessary to breach this rule to improve map readability, this should be done by applying an exaggeration or displacement to the result of the simplification.

Exaggeration is conceptually more complicated than simplification, since it corresponds to an expansion of an entity rather than to an abstraction: hence it somehow goes against the general trend of information reduction in the map. The conflict here is between the semantic importance of a feature and its insignificant size. However, this operation can still be covered in our framework: it is not in contrast with continuous mapping, since it is not a topological operation; it is compatible with homotopy in that it applies this approach in reverse, to enlarge rather than contract a line. It requires the relaxation of the metric constraint at a given scale so that objects whose semantic importance prevails over their small size can be enlarged even beyond the current threshold.

Displacement presents greater difficulty since it can change the topological relations between map elements in a non-monotonic way, which is inconsistent with the functioning of continuous mapping. However, displacement is essentially a cartographic operation and this issue can be solved at drawing level. In fact, it is not necessary to store data variations describing a displaced entity, but only to record the necessary drawing displacement values in the associated semantic information. For instance, in the case of the two adjacent towns mentioned in the previous section, we could introduce a point called "the centre of the town" as a topological entity (Fig. 5); this need not be drawn as long as the scale of the map allows representation of the shapes of the two towns, and would become visible as the scale increases. A similar role could be played by a conceptual entity called "centre of the road" in the case of two parallel roads. Of course processing at cartographic level will require a suitable drawing procedure, since the displacement to be applied must be evaluated according to the scale and measure of the map, and to the positions of the adjacent entities in the map.

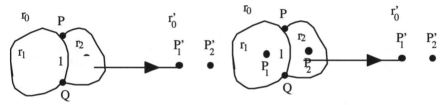

$$F(r_1) = P_1'; \; F(r_2) = P_2'; \qquad\qquad F(r_1) = F(r_2) = r_0;$$
$$F(P) = ?; \; F(Q) = ?; \; F(l)=? \qquad\qquad F(P_1) = F(P_1'); \; F(P_2) = F(P_2')$$

Figure 5 Displacement: introducing a topological centre for regions r_1 and r_2 allows displacement to be expressed by means of a continuous function

5. Concluding remarks

We have shown that basic generalization operations can be interpreted in the context of the mathematical framework we proposed in [9], where generalization is controlled by means of ε-homotopy and continuous mapping between cell complexes. Topological operations are actually handled all together in our general framework. The mechanism that causes selection, symbolisation, or aggregation is strictly dependent on the scale limits of existence of each entity in each order. A given generic entity can be represented alternatively as a complex structure of objects (submap), as a region, a line, a point, or, finally, may be ignored. Whether and how the entity is represented depends on the resolution of the map. Performing a topological generalization corresponds to deciding, in a consistent way, whether and how each entity will be represented at a certain resolution. Given a map at the highest possible resolution (reference model), it is possible to implicitly define a whole set of generalizations by giving atomic objects existing in that map ranges of persistence in each order, i.e. the range of resolutions in which that object will appear as a region, as a line, or as a point. If this is specified consistently, it will encompass all immersions and contractions necessary for topological generalization. The rules for specifying the presence of an object in each order must be of both geometric and semantic nature. While a purely geometric approach would be manageable automatically, it is less evident how the relation between resolution and representation of objects could be automatically assigned on a semantic basis.

Acknowledgement
This work was partially supported by the Italian National Research Council's Coordinated Project "Models and Systems for Handling Environmental and Territorial Data".

References
[1] Bertolotto, M., De Floriani, L., Puppo, E.: Multiresolution topological maps. In Advanced Geographic Data Modelling - Spatial Data Modelling and Query Languages for 2D and 3D Applications, M. Molenaar, S. De Hoop (eds.), Netherland Geodetic Commission, *Publications on Geodesy - New Series,* N.40, pp.179-190
[2] Clementini, E., Di Felice, P.: A comparison of methods for representing topological relationships. Information Sciences, 3, 1995, 149-178.
[3] De Floriani, L., Marzano, P. Puppo, E.: Spatial queries and data models. In Spatial Information Theory - A theoretical basis for GIS, A.U. Frank, I. Campari (Eds.), Lecture Notes in Computer Science **716**, Springer-Verlag, 1993, pp.113-138
[4] Dettori, G., Puppo, E.: Simplification of combinatorial maps through continuous mapping. Technical report, 3-95, Istituto per la Matematica Applicata, C.N.R., Genova, Italy
[5] Herring, J.: TIGRIS: topologically integrated GIS. Proceedings Autocarto 8, ASPRS/ACSM, Baltimore, MD, pp.282-291, March 1987
[6] McMaster, R., Automated line generalization, Cartographica, 24, 1987, pp.74-111.
[7] Molenaar, M.: Single valued vector maps - a concept in GIS. Geo-Informations-systeme, **2** (1), 1989
[8] Müller, J.C., Weibel, R., Lagrange, J.P., Salgé, F.: Generalization: state of the art and issues. In Gis and Generalization: Methodology and Practice, J.CMüller, J.P.Lagrane and R.Weibel eds., Taylor and Francis 1995, pp.3-17
[9] Puppo, E., Dettori, G.: Towards a formal model for multiresolution spatial maps. Advances in Spatial Databases (Proc. SSD95), M.J. Egenhofer, J.R.Herring eds. Lecture Notes in Computer Science, N.951,1995, pp.152-169
[10]Rourke, C.P., Sanderson, B.J.: Introduction to Piecewise-linear Topology. Springer-Verlag, 1972

Contact address
Giuliana Dettori and Enrico Puppo
Istituto per la Matematica Applicata, C. N. R.
Via De Marini, 6
16149 Genova
Italy
Phone: +39-10-6475 1
Fax: +39-10-6475 660
E-mail:{dettori,puppo}@ima.ge.cnr.it

THE EXTENSIONAL UNCERTAINTY OF SPATIAL OBJECTS

Martien Molenaar

Department for Geo-Information Processing and Remote Sensing
Wageningen Agricultural University
Wageningen, the Netherlands

Abstract
Spatial objects have a geometric and a thematic description. Both can be uncertain in the sense that there is no full confidence that they are correct. Traditionally the geometric uncertainty has ben expressed by variances of the coordinates and measures like epsilon-bands expressing the possible discrepancies between the real object boundary and its digitised representation. The thematic uncertainty can be expressed by variance of the attribute values of the object or by fuzzy measures. The thematic uncertainty of objects will often have a spatial effect in the sense that it is difficult to decide where the object boundary is. This effect will be called the extensional uncertainty of the object. The geometric uncertainty mentioned before indicates then how accurate the shape of a boundary and the position of boundary points can be measured, once a decision has been made where the boundary is. The concept of the extensional uncertainty of spatial objects will be explained in detail, some examples related to remote sensing and photo interpretation will be given to illustrate it.

Key words
Fuzzy objects, extensional uncertainty, topology, data model, syntax.

1. Introduction

The determination of the spatial extend of geo-objects is generally approached through the boundaries, or more precisely through the position of the boundary points. The analysis of the geometric uncertainty of the objects is therefore often based on accuracy models for the coordinates of these points. The epsilon band method is well known in this context. Yet the solutions for handling this problem are not found satisfactory though because the geometric uncertainty of geo-objects is not only a matter of coordinate accuracy, i.e. it is not only a problem of geometry, but it is also a problem of object definition and thematic vagueness, see also the discussions on this topic in e.g. [Chrisman 1991] [Burrough&Frank 1995] and see the contribution of e.g. Coulclelis, Fisher, Lagacherie and Poultier to [Burrough&Frank 1996]. This latter aspect can not be handled by a geometric approach alone. This becomes apparent when mapping is not done in a vector structured geometry as for landsurveying and photogrammetry. Remote sensing works in raster structure, there the uncertainty is considered to be a thematic aspect where likelihood functions express the degree of a pixel belonging to a thematic class. Image segments can then be formed of adjacent pixels falling under the same class. If these segments represent spatial objects then the uncertainty of the geometry of

571

these objects is due to the fact that the value of the likelihood function varies per pixel.

The syntactic approach for handling spatial object information as presented in [Molenaar 1994 and 1995] makes it possible to distinguish three types of statements with respect to the existence of spatial objects:
- an *existential statement* asserting that there are spatial and thematic conditions that imply that an object O exists,
- an *extensional statement* identifying the geometric elements describing the spatial extend of the object,
- g*eometric statement* identifying the actual shape, size and position of the object in a metric sense.

These three types of statements are intimately related, the extensional and geometric statement imply the existensional statement and if an object does not exist it can not have an spatial extend and a geometry. The existential statement often relates to the uncertainty of thematic information though, that is not explicit in the other two statements. The geometric statement also implies the extensional statement, often the actual geometry of the object is derived from the extensional description. The object detection through image interpretation is in fact an example of the formulation of extensional statements. These three types of statements can all have a degree of uncertainty and although these statements are related they give us different perspectives that may help us to understand the different aspects of uncertainty in relation to the description of spatial objects. In the following chapters we will concentrate on the uncertainty related to the extensional and geometric statements and their mutual effects. When object uncertainty is analyzed in terms of extensional statements we will call that the *extensional approach*, when it is done through geometric statements we will call that the *geometric approach*. No explicit attention will be paid to the thematic description of spatial objects, but the effect of thematic uncertainty on the geometric description of objects will be discussed at several occasions.

The syntactic approach of [Molenaar 1994 and 1995] shows that the vector and the raster geometry have a similar expressive power. That implies that the handling of spatial uncertainty should in principle also be the same for both geometric structures, so that it must be possible to combine or even unify the boundary oriented and the pixel oriented approaches of the previous paragraphs. This assumption will be elaborated in this paper. Through the syntactic approach it will be possible to identify where uncertainty plays a role in the spatial description of objects and what its semantic implications are.

2. Relationships between the geometric elements of a planar graph and terrain objects

For GIS applications the geometry of spatial objects should be described by means of discrete geometric elements. These elements are then entities that can be represented in a database. The following chapters describe how the elements of a planar graph can be used to represent the geometry for 2-D terrain descriptions (see [Molenaar 1994 and 1995]).

2.1 Edges and faces in a planar graph

The numbers of nodes and edges of a planar graph and the faces that they define in a planimetric situation comply with Eulers number. These geometric elements can be interpreted as cell-complexes, so that the algebraic topology of these complexes can be applied. We will use the terminology of planar graphs here though and explain a notation expressing their relationships.

Each edge will always have one face at its lefthand side and on at its righthand side. These relationships will be expressed by the following functions:
- *Edge e_i has face f_a at its left-hand side $\rightarrow Le[e_i, f_a] = 1$, otherwise = 0.*
- *Edge e_j has face f_a at its right-hand side $\rightarrow Ri[e_j, f_a] = 1$, otherwise = 0.*

With these functions we can define $B[e_i, f_a] = Le[e_i, f_a] + Ri[e_i, f_a]$

If the value of this function = 0 then e_i is not related to f_a, if the value = 2 then the edge has the face at both sides and is thus inside the face, if the value of this function = 1 then the edge has the face at only one side so that it must be part of the boundary. The boundary of a face f_a is then

boundary $\quad \partial f_a = \{ N_{\partial fa}, E_{\partial fa} \}$
$$N_{\partial fa} = \{ n_j \mid n_j \in e_i, e_i \in E_{\partial fa} \}$$
$$E_{\partial fa} = \{ e_i \mid B[e_i, f_a] = 1 \}$$

The geometry of a face is fully determined by its boundary. The adjacency of two faces will be expressed by the function *ADJACENT[f_i , f_j]* this function has the value = 1 if the boundaries of the two faces have at least one common edge, it has the value = 0 otherwise.

Position information
Information about the location of geometric elements is contained in the coordinates that can be assigned to the nodes. For a 2-dimensional case an x and y coordinate can be assigned to each node n_i, that is X_i and Y_i. Through the topologic relationships defined earlier these coordinates implicitly determine the position of the edges and faces and also the shape of the faces.

Cell rasters
Cell rasters are a special case of this topologic structure in the sense that the cells are rectangular faces, so that each face has a boundary consisting of four edges and four nodes. Four cells meet at each node, except at the nodes at the outer boundary of the raster. There are no nodes and no edges inside the cells. This interpretation of cell rasters has been explained in more detail in [Molenaar 1994 and 1995]. The fact that rasters can be handled by the same syntactic approach as has been formulated here for the vector structure implies that both geometric structures can in principle handle similar situations. This means that the following elaborations will be valid for both structures.

2.2 Area objects in a vector map with a planar graph structure

So on the one hand there are the sets of geometric elements defining the geometric structure of the map, on the other hand there are the sets of terrain objects to be represented. Relationships between these data sets can be defined to link geometric elements to the objects. That will be done for area objects in the next sections because our discussions will concentrate on the representation of fuzzy area objects.

The link between geometric elements and area objects

If a face f_f is part of an area object O_a this will be represented by

$$Part2[f_f, O_a] = 1$$

The notation *Part2* means that this is a relationship between 2-dimensional entities. If the vector map represents overlapping area objects then one face could be part of several objects, but also each object will contain one or more faces. Therefore this is a many-to-many relationship. The relationship between an edges, a face and an area object in figure 1 can be found through:

$$Le[e_i, O_a \setminus f_f] = MIN(Le[e_i, f_f], Part2[f_f, O_a])$$

the edge has the object at its lefthand side if both arguments of the minimum operator have the value one, that is why the minimum operator is applied. Then we evaluate

$$Le[e_i, O_a] = MAX_{ff}(Le[e_i, O_a \setminus f_f])$$

If there is a face for which the first function has the value = 1 then the edge e_i has object O_a at its left-hand side then $Le[e_i, O_a] = 1$ otherwise = 0, therefore the maximum over f_f is evaluated for the first function. Simmilarly we have
$Ri[e_i, O_a \setminus f_f] = MIN(Ri[e_j, f_f], Part2[f_f, O_a])$ and
$Ri[e_i, O_a] = MAX_{ff}(Ri[e_i, O_a \setminus f_f])$

If edge e_j has object O_a at its right-hand side then $Ri[e_j, O_a] = 1$ otherwise = 0. For the relationship between an edge and an area object we can use the function

$$B[e_i, O_a] = Le[e_i, O_a] + Ri[e_i, O_a]$$

the interpretation of the values of this function is similar to what we had for the relationship between an edge and a face. The boundary of the area object is then

$$\text{boundary} \qquad \partial O_a = \{ N_{\partial O_a}, E_{\partial O_a})$$
$$N_{\partial O_a} = \{ n_j \mid n_j \in e_i, e_i \in E_{\partial O_a} \}$$
$$E_{\partial O_a} = \{ e_i \mid B[e_i, O_a] = 1 \}$$

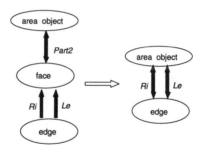

Figure 1 Transition from edge-face-object relationships to edge-object relationships.

3. Uncertainty aspects of the spatial description of objects

The topic of discussion of this section is how the semantics of fuzzy spatial objects can be handled in the syntax of section 2. To establish the spatial extend of an object two types of decision should be made:
- first we should decide which geometric elements will be linked to the object, the spatial extend of the object is determined through the identification of the faces of the object with their edges and nodes,
- secondly the position of the object is determined through the coordinates that will be assigned to the nodes.

Area objects represented as fuzzy fields
If the objects O are fuzzy in the sense that their spatial extend is uncertain then this can be expressed by the fact that the function Part2[f, O] does not only take the value 0 or 1, but that it can take any value $0 \leq$ Part2[f, O] ≤ 1 as in figure 2. The faces should be defined so that the function is homogeneous per face. The fuzzy set of faces related to an object O_a with an uncertain spatial extend is

$$Face(O_a) = \{ f_j | \ Part2[f_j, O_a] > 0 \}$$

In this notation the geometric description of the objects is organised per object. Alternatively we can define a notation where the geometric description of the area objects is organised per face. If uncertainty must be taken care of then we can define the set of area objects that have a fuzzy relationship with a face f_j

$$AO(f_j) = \{ O_a | Part2[f_j, O_a] > 0 \}$$

With this notation it is possible to interpret a fuzzy area object as a fuzzy field in the sense that per face f_j the function Part2[f_j, O_a] is evaluated for each object of the map, in case of a raster the function is evaluated for each cell of the raster, see figure 2.

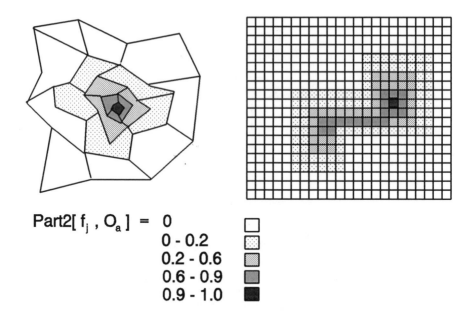

$$\text{Part2[} f_j \text{ , } O_a \text{] } = \quad \begin{array}{l} 0 \\ 0 - 0.2 \\ 0.2 - 0.6 \\ 0.6 - 0.9 \\ 0.9 - 1.0 \end{array}$$

Figure 2 Objects represented as fuzzy fields: a) in a vector map and b) in a raster map.

When the fuzzy variant of Part2[f, O] is used in stead of the crisp version in the evaluation of the functions Le[e, O] and Ri[e, O] according to section 2.2, then we obtain also fuzzy variants of these latter functions.

The conditional boundaries of fuzzy objects
The *shape and position* information is mainly contained in the coordinates of the nodes of the object boundaries. Because of the uncertainty of the functions Le[ ] and Ri[ ] the boundary of objects can not be determined in the simple way of section 2.3.1, for fuzzy objects boundaries can only be established for specified certainty levels. Let the fuzzy set of faces related to object O_a with certainty level c be

$$Face(O_a|c) = \{ f_j | \ Part2[f_j, O_a] \geq c \}$$

With this set we can define the conditional function
$$Part2[f_j, O_a | c] \qquad = 1 \Rightarrow f_j \in Face(O_a|c) $$
$$\qquad\qquad\qquad\qquad = 0 \Rightarrow f_j \notin Face(O_a|c)$$

and for the relationships between edges, faces and objects

$$Le[e_i, O_a|f_j, c] = MIN(Le[e_i, f_j], Part2[f_j, O_a|c])$$
$$Ri[e_i, O_a|f_j, c] = MIN(Ri[e_i, f_j], Part2[f_j, O_a|c])$$

These conditional functions are not fuzzy, they take the values $= 0$ or $= 1$. Through these functions it can be verified whether an edge is conditionally related to the object by evaluation of the functions

$Le[e_i, O_a | c] = MAX_{fj} [Le[e_i, O_a | f_j, c]]$
$Ri[e_i, O_a | c] = MAX_{fj} [Ri[e_i, O_a | f_j, c]]$

These are also crisp functions that take the value = 1 if there exists a relationship and = 0 otherwise. With the conditional function

$B[e_i, O_a | c] = Le[e_i, O_a | c] + Ri[e_i, O_a | c]$

the boundary of O_a at certainty level c can be found

$\partial_C O_a = \{ e_i | B[e_i, O_a | c] = 1 \}$

The precision of the position and shape of these boundaries is then a function of the accuracy of the coordinates of the nodes. This accuracy depends mainly on the measuring procedure and on the idealisation accuracy of the face boundary, i.e. the accuracy of the identification of the face boundary. In rasters the node positions have been defined by the cell geometry, so that position and shape accuracy of the boundaries at specified certainty levels follows directly from the accuracy of the raster datum.

4. Examples of the extensional approach for the determination of the uncertainty of objects

An important step in the extensional approach is the evaluation of the function Part2[f,O] to establish the link between faces and objects. Several examples will be given how values for this function could be obtained.

Remote sensing image classification
In remote sensing applications the spectral classification of the image results in a class label and a likelihood value per pixel. In [Shi 1994] this information is combined with information about the uncertainty of a measured object boundary to evaluate for each position the joint uncertainty that it belongs to the area covered by the object **and** that it belongs to a certain cover class.

We will follow a different approach here and use the uncertainty of the cover classes to evaluate the extensional uncertainty of objects. If these classes represent cover types then mutually adjacent pixels belonging to the same class can be merged to form image segments, which can be considered to represent spatial objects with a characteristic ground cover. The pixels of an image can be considered as the cells of a raster, let p_{ij} be such a pixel. The value of the likelihood function for a cover class could then be used to express the uncertainty of each pixel that it belongs to the cover class, let this function for pixel p_{ij} and class C_k be $L[p_{ij}, C_k]$. The value of this function will be between 0 and 1. Each pixel p_{ij} will be assigned to the class C_k for which the likelihood function has a maximum value:

let $D[p_{ij}, C_k]$ be a decision function with
$\quad D[p_{ij}, C_k] = 1 \quad if\ L[p_{ij}, C_k] = MAX_{Ci}(L[p_{ij}, C_i])$
$\quad D[p_{ij}, C_k] = 0 \quad otherwise$

An object O_a with covertype C_k will then be represented by an image segment that fulfils the following two conditions

- *for all pixels $p_{ij} \in O_a$ is $D[p_{ij}, C_k] = 1$*
- *if $p_{kl} \in O_a$ <u>and</u> ADJACENT$[p_{kl}, p_{ij}] = 1$ <u>and</u> $D[p_{ij}, C_k] = 1$*
  then $p_{ij} \in O_a$

This definition of D[p,C] guarantees that the objects will not overlap, but if we use instead L[p,C] > t_l where t_l is some lower threshold, then the objects might overlap, i.e. some pixels could be assigned to mote then one object.

The adjacency relationships between pixels has either the value = 0 or = 1, so that the uncertainty of the relationship between a pixel and the object is due to the function L[p,C]. Therefore we have

if $p_{ij} \in O_a$ then Part2$[p_{ij}, O_a] = L[p_{ij}, C_k]$

Image classification will hardly ever result in a value L[p,C] = 1. For very high values of the likelihood function the assignment of the pixels to the object is as good as certain, therefore this rule could be modified in the sense that an upper threshold t_u is introduced so that

if $p_{ij} \in O_a$ and L$[p_{ij}, C_k] > t_u$ then Part2$[p_{ij}, O_a] = 1$

A display of these certainty values for a particular object will have an appearance like the raster in figure 2. We see that remote sensing image interpretation gives implicitly an extensional approach for the evaluation of uncertainty of spatial objects.

The boundary of O_a at certainty level c is then $\partial_c O_a = \{ e_i | B[e_i, O_a | c] = 1 \}$. The position of this boundary is fully determined by the geometry of the raster, uncertainty with respect to position is only due to uncertainty in the datum definition of the raster.

Photo interpretation
Photo interpretation results generally in the delineation of the boundaries of spatial objects. In many applications like vegetation or landuse mapping these objects are fuzzy, so that repeated interpretations lead to varying object boundaries. We will assume that the objects form a complete segmentation of the image, i.e. they define a partition. Each face of the segmented image will then be assigned to exactly one object in each interpretation. For each face f_i it is decided for which object O_a the function Part2$[f_i, O_a] = 1$.

Middelkoop and Edwards analyzed the uncertainty of such interpretations through an overlay of the results of several repetitions. [Middelkoop 1990] first rasterized the interpretation results, whereas [Edwards 1994] worked with an overlay of polygons. We will compare these two methods here.

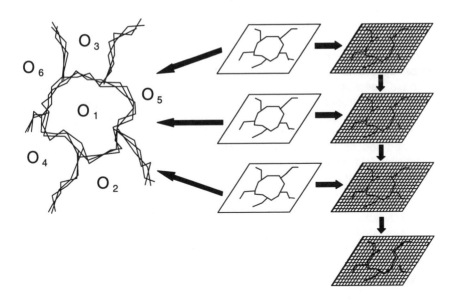

Figure 3 The overlay in vector and raster format of the results of several photo interpretations.

Middelkoop had an image interpreted by thirteen persons [Middelkoop 1990]. Fifteen classes had been defined in advance and each interpreter was asked to identify regions in the image and to assign class labels to these regions. The output of each interpreter was converted to high resolution raster. These rasters were overlaid as in the righthand side of figure 3; for each cell the frequency of each of the fifteen classes was counted, and relative frequencies were computed from these. The relative frequencies finally served as a certainty measure for a function $L[f_{ij}, C_k]$ relating a class C_k to cell f_{ij}; these measures can then be used as before as the evaluation of $Part2[f_{ij}, O_a]$ relating a cell f_{ij} to an object O_a with cover class C_k. We see now in this method that although the interpreters drew boundaries between regions of areas under the different classes, these operation can be seen as assigning areas to classified regions. The further analysis of the uncertainty of the boundaries between regions follows then from the values of the $Part2[ ]$ function for the faces in the transition zone between regions. Middelkoop developed an extensional approach for the evaluation of the uncertainty of spatial objects, he did this through an overlay of the rasterized interpretations.

Edwards overlaid the polygons that resulted from the interpretation of an image by thirteen persons as in the lefthand side of figure 3, see [Edwards 1994]. In this way he was able to identify a set of boundary segments for each pair of adjacent regions. Then he estimated the location of an average boundary segment for each pair of regions and he defined dispersion measure for the locations of the interpreted boundaries with respect to these average boundaries. In his approach he considers the location of the boundaries as fuzzy or uncertain, the fuzziness of the objects is then a consequence of the fuzzy boundaries. He developed an approach that is geometric rather than extensional, he did this through a polygonal representation of his objects.

An extensional approach for the evaluation of uncertainty objects in a vector representation

We have seen two examples of how rasters could be used to formulate an extensional approach for the evaluation of uncertainty of spatial objects. We also saw one example of a geometric approach by means of polygons overlays. Because of the stated syntactic similarity of vector structured and raster structured descriptions of spatial objects in [Molenaar 1994 and 1995] it should be possible to formulate an extensional approach also for vector descriptions. This will be explained now.

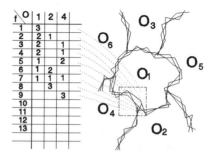

Figure 4 Frequency counts or face-object relationships.

The method has been illustrated in figure 4. This figure shows that faces are created in the overlay of the boundaries identified in three interpretations. For each of these faces we can trace back how it is related to the objects in each interpretation. This has been done for some of the faces in the rectangle in the figure, the frequency of their occurrence in each object has been counted, similar to what Middelkoop did for the raster cells. Again relative frequencies can be computed and these can be used as values for the function $Part2[f_i, O_a]$ relating a face i to object a. Three interpretations have been overlaid so that for each object O_a there are faces for which this function has a value = 1 and faces with a value = 0 and faces with intermediate values (in this example 2/3 and 1/3). If we consider O_1 then boundaries $\partial_c O_a$ can be found for the certainty levels c = 1/3, 2/3 or 1. Due to the process for the evaluation of the function $Part2[f,O]$ these boundaries are concentric.

5. Conclusion

The extensional approach concentrates mainly on the problem of deciding per object which faces, with their edges and nodes, describe its spatial extend. This implies the decision where to choose the boundary of the object. The identification of the object boundary is uncertain due to the fact that the spatial extend of the object is uncertain in the sense that the values of the function Part2[f,O], relating the faces to the object, is not just 0 or 1, but it varies between 0 and 1. If a choice has been be made about the lower threshold t_l for the uncertainty level of the faces that are still to be considered part of the object, then the conditional boundary $\partial_c\,O$ with $c = t_l$ can be used as the object boundary. The extensional uncertainty of this boundary can be expressed through the fuzzy functions Le[e,O] and Ri[e,O] which should be evaluated for the edges of the boundary.

In photo interpretation the decision about the spatial extend of the objects is often separated from the measurement of the geometry of the object boundaries, these are often digitised after the interpretation has been done. In landsurveying boundary identification and measurement are generally combined in one process, but that still leaves that fact that these are two operations. The uncertainty about the geometric position of a boundary can often be ecpressed by means of epsilon-bands, see [Chrisman 1991] and [Shi 1994].

The possibility that an edge e is part of the boundary of O is then the possibility that e is adjacent O on one side and adjacent to "not O" on the other side. Let L = Le[e,O] and R = Ri[e,O] and let L > R. The possibility that e is adjacent to O is then the possibility that e has O on the lefthand side or at the righthand side, that is MAX(L,R) = L (in this example). The possibility that the edge has "not O" at the lefthand side is 1-L, for the other side this possibility is 1-R. For the possibility that "not O" is right or left is then MAX(1-L, 1-R) = 1-R. If the edge is indeed part of the boundary then it should have O at the lefthand side and at the same time "not O" at the righthand side, this possibility is then MIN(L, 1-R). The value of this function should be evaluated for all edges of the boundary. The result of this evaluation combined with the value of t_l could then be used to define a certainty measure that this boundary is a correct choice, that is then in fact a measure to evaluate whether the spatial extend of the object has been identified correctly. When a boundary with a satisfactory certainty level has been accepted then the geometric accuracy of the boundary can be evaluated, e.g. by means of an epsilon band. For crisp objects we will find that L = 1 and R = 0 so that MIN(L, 1-R) = 1, there we can be certain that the identification of the boundary is correct. There is no uncertainty about the spatial extend of the object, only the geometric description might be inaccurate or not precise. The geometric uncertainty is then dominant. This situation generally occurs in cadastral and topographic surveys where man made objects are mapped.

If the spatial extend of objects is very uncertain, i.e. if the object identification is very fuzzy, then a clear identification if the boundary will be difficult. For such objects several zones might be identified:
- the area inside $\partial_{c=1}\,O$ that is the area that certainly belongs to the object,

- the area outside $\partial_{c=0}$ O that is the area that certainly does not belong to the object,
- the area between these two boundaries, the transitional zone where the certainty varies between 0 and 1.

If the transitional zone is significantly larger than the size of the epsilon bands that represent the geometric accuracy of the digitised boundaries then the extensional uncertainty dominates the geometric uncertainty. This is the case is many survey disciplines mapping natural objects, it also occurs often in many types of small scale mapping. An intermediate situation occurs when the transitional zone has about the same size as the epsilon bands. Then it is difficult to distinguish between the extensional and the geometric uncertainty, although they are semantically quite different.

Several uncertainty aspects of the description of spatial objects have been discussed. The two main categories in this paper were the extensional and the geometric uncertainty. The extensional uncertainty is generally a spatial effect of fuzziness of the thematic description of the objects, it is therefore strongly related to the thematic class of the objects. The geometric uncertainty is due to a combination of object geometry and measuring procedure, where the object geometry is in most cases related to its thematic class. A quality model for spatial objects should therefore take the object characteristics as expressed in their thematic description into account. A quality model should also refer to the measuring or observational process for data collection, taking into account that in many cases the uncertainty of the thematic and geometric data are strongly interrelated.

References

[Burrough Frank1995] Burrough, P.A. and A.U. Frank:Concepts and paradigms in spatial information: are current geographical information systems truly generic?, Int. Journal GIS, 1995, vol.9, no.2, pp101-116.

[Burrough&Frank1996] Burrough, P.A and A.U. Frank, eds: Geographic objects with indeterminate boundaries. Taylor and Francis, London, 1996.

[Edwards1994] Edwards,G.: Aggregation and disaggregation of fuzzy polygons for spatial-temporal modelling. In: Advanced Geographic Data Modelling, Molenaar, M. and S. de Hoop eds., Netherlands Geodetic Commission, New Series, Nr. 40, Delft, 1994, pp141-154.

[Chrisman1991] Chrisman, N.R.: The error component in spatial data, in: Geographical Information Systems, Maguire, D.J. ea (eds), Longman, Harlow U.K., 1991.

[Middelkoop1990] Middelkoop, H.: Uncertainty in a GIS: a test for quantifying interpretation output. ITC-Journal, 1990-3, pp225-233.

[Molenaar1994] Molenaar, M.: A syntax for the representation of fuzzy spatial objects. In: Advanced Geographic Data Modelling, Molenaar, M. and S. de Hoop eds., Netherlands Geodetic Commission, New Series, Nr. 40, Delft, 1994, pp155-169.

[Molenaar1995] Molenaar, M.: A Syntactic Approach for Handling the Semantics of Fuzzy Spatial Objects. In: Geographic Objects With Indeterminate Boundaries, P.A. Burroug and A.U. Frank (eds), Taylor and Francis, London (1995) pp207-224.

[Shi1994] Shi Wenzhong: Modelling positional and thematic incertainties in integration of remote sensing and GIS. ITC Publication Nr 22, ITC, Enschede, 1994.

Contact address
Martien Molenaar
Department for Geo-Information Processing and Remote Sensing
Wageningen Agricultural University
P.O box 339,
7500AH Wageningen
The Netherlands
Phone: +0317-482910
Fax: +0317-484643
E-mail:Martien.Molenaar@wetensch.lmk.wau.nl

MODELING ELEVATION UNCERTAINTY IN GEOGRAPHICAL

ANALYSES

Charles R. Ehlschlaeger[1] and A. Shortridge[2]

[1]Department of Geography
University of Cincinnati
Cincinnati, USA
[2]National Center for Geographic Information and Analysis
University of California
Santa Barbara, USA

Abstract
This work uses a stochastic simulation approach to explore the impact of uncertainty in digital elevation data within the context of the least-cost path algorithm. A method for deriving accuracy parameters from higher accuracy data, taking into account the impact of spatial autocorrelation, is explicated. It is demonstrated that the spatial complexity of map generalization, when propagated through non-local GIS operations, can be too problematic to model without employing Monte Carlo stochastic simulation methods explicitly for each application of the operation.

1. Introduction

Digital elevation models (DEMs) enjoy a range of application in the geosciences. Two widely available data sets from the United States Geological Survey (USGS) are of particular interest for this study. Complete coverage of the contiguous United States is available as 1-degree DEM files with elevation values at 3-arc second intervals. While 30 meter spot elevation data (7.5' DEMs) offer higher accuracy standards and greater sampling density, they do not provide complete national coverage and they are somewhat more difficult and costly to obtain [USGS, 1992]. This research develops a method to model the difference between collocated 3 arc-second DEM data and 7.5' DEM data. Assuming the 7.5' data is adequate to address a particular spatial analytical problem, the researcher may be able to employ the model in similar, nearby areas for which 7.5' data is not available.

The standard data error model we employ treats the elevation data surface Z as being equal to the actual surface plus a disturbance term, ε:

$$Z = Actual + \varepsilon \qquad (1)$$

Because we are dealing with continuous spatial phenomena, the terms in (1) are not scalars, but fields. By modeling the parameters of the field ε, one can gain insight into the impact of uncertainty on spatial applications.

Spatial analysis is subject to uncertainty, due in part to fact that the datasets upon which they operate are the outcome of a process of discretization and generalization. In terrain data, uncertainty about the actual value at any particular spot on the continuous landscape is a factor of the distance that spot is from the nearest data points, the variation of the terrain between the data points, and the accuracy of the elevation measures in the dataset [Isaaks & Srivastava, 1989]. As the dataset's depiction of the terrain surface becomes more generalized, that is, as the sampling interval increases, uncertainty about elevations in the larger gaps between data values increases. The importance of calculating what effect this generalization will have on any particular application is widely recognized [Zhang & Montgomery, 1994].

Here, we present an approach to accomplish this. First, the error model referred to previously is developed to allow for the creation and analysis of these surfaces. Second, error parameters are derived for an error probability distribution function (p.d.f), and surfaces are created matching this distribution. Finally, each surface is used as input for a particular GIS function, in this example, the least-cost path algorithm. By studying the distribution of the results across a wide range of sample error surfaces, conclusions can be drawn about how the uncertainty in the spatial data affects the result of the analysis for this particular function, upon this particular dataset. This approach is known as unconditional stochastic simulation in geostatistic literature [Openshaw, 1979].

Alternatives exist which attempt to assess the impact of uncertainty in spatial analysis. [Lanter & Veregin, 1992; and Veregin, 1994] have suggested that analytic models of error propagation can be calculated with coefficients that can be calibrated with the data set to make this assessment for a variety of spatial analysis. This work suggests that, for some sets of applications, the simulation approach may be preferred, and an example of this methodology using the last-cost path routine is employed to examine the spatial impact of uncertainty on spatial applications.

2. Uncertainty in elevation data

We model the DEM not as a collection of $n \times m$ different spot elevation samples, but instead as one single sample of the entire actual elevation surface. The approach adopted by this research defines a DEM, or any surface of continuous attributes with z values contaminated by error, as a random outcome with a particular probability density function (p.d.f.) [Isaaks & Srivastava, 1989]. The p.d.f. represents a mean surface and a random field function defining the potential deviations from the mean surface. In this way, the DEM is characterized as one of an infinite number of equally potential realizations of the actual elevation surface.

Uncertainty in elevation data has been categorized as containing both horizontal (positional accuracy) and vertical components (attribute accuracy). However, positional and attribute accuracy generally cannot be separated. Measurement error may be due to an incorrect elevation value at the correct location, or a correct elevation for an incorrect location, or some combination of these. To measure this compound error, more information is required about the nature of the surface. Typically, a set of data of sufficient quality for the specific application is used to check a collocated subset of the data of interest. For examples, see [Shearer, 1990; Bolstad & Stowe, 1994; and Adkins & Merry, 1994]. By examining the differences, parameters for the p.d.f. can be estimated. For the sake of clarity, the term error is used throughout this work when referring to the model. The actual quantity being measured, however, is the difference between two sets of data modeling the same spatial phenomenon, one of which we assume to be close enough to the true surface to be useful for the spatial application of interest.

As an example, suppose we are interested in test area A, for which we have coarsely sampled elevation dataset 1. The total extent of dataset 1 includes area B, which is also covered by finer resolution dataset 2. If we assume that the measurement error, measured by the difference between datasets 1 and 2, are similar between areas A and B, we can develop estimates for the parameters of the error model – the mean and standard deviation -- for dataset 1.

These parameters represent the mean and standard deviation of the uncertainty in the data set of interest, as captured by the data set of higher accuracy. Using these statistics, a p.d.f. model is developed. This model creates random fields with a gaussian distribution matching the mean and standard deviation parameters derived from the difference map, with the spatial autocorrelative characteristics of the spatially dependent uncertainty.

$$E(\mathbf{u}) = \mu(\mathbf{u}) + \mu(\mu(T)) + s^2(\mu(T))\varepsilon_1 + (\mu(s^2(T)) + s^2(s^2(T))\varepsilon_2)Z(\mathbf{u}) \tag{2}$$

where $E(\mathbf{u})$ is a realization of the higher quality (7.5') elevation data using the more generalized 1-degree DEM $\mu(\mathbf{u})$, T is a group of sets of spatially uncorrelated sample points, ε_1 is a random variable with mean 0.0 and variance 1.0 perturbing the realization's mean uncertainty, ε_2 is a random variable with mean 0.0 and variance 1.0 perturbing the realization's standard deviation, and $Z(\mathbf{u})$ is a random field perturbing all points u within the realization. The following expressions define the remaining terms in (2):

$$\mu(\mu(T_j)) = \sum_j \frac{\mu(T_j)}{j}, j \subset T \tag{3}$$

$$s^2(\mu(T_j)) = \sum_j \frac{(\mu(T_j) - \mu(\mu(T_j)))^2}{j-1} \tag{4}$$

$$\mu(s^2(T_j)) = \sum_j \frac{s^2(T_j)}{j} \tag{5}$$

$$s^2(s^2(T_j)) = \sum_j \frac{(s^2(T_j) - \mu(s^2(T_j)))^2}{j-1} \tag{6}$$

and

$$\tag{7}$$

$$Z(\mathbf{u}) = \frac{\sum_v w_{u,v}\varepsilon_v}{\sqrt{\sum_v w_{u,v}^2}}, w_{u,v} = \begin{cases} 1 & : & d_{u,v} \leq F \\ \left(1 - \dfrac{d_{u,v} - F}{D - F}\right)^E & : & F < d_{u,v} < D, u \subset \mathbf{u}, v \subset \mathbf{v} \\ 0 & : & d_{u,v} \geq D \end{cases}$$

where equation (3) is the average mean for all sets; (4) is the variance of the mean values for all sets; (5) is the average variance for all sets; (6) is the variance of the variances for all sets; and (7) specifies the random field with spatial dependence parameters where $Z(\mathbf{u})$ is a point on the random field Z with a theoretical mean of 0.0 and theoretical variance of 1.0, $\mathbf{u}$ is the set of points that comprise the resulting random field, $\mathbf{v}$ is the set of points affecting $Z(\mathbf{u})$, $w_{u,v}$ is the spatial autocorrelative effect between points u and v, ε_v is a random variable with a mean of 0.0 and variance of 1.0, $d_{u,v}$ is the distance between u and v, D is the minimum distance of spatial independence, E is the distance decay exponent, and F is a parameter that adds flexibility to the probability distribution function model fitting process. Matching the spatial autocorrelation using $Z(\mathbf{u})$ is important in order to capture the terrain texture observed in the higher resolution dataset [Goodchild, 1986; Theobald, 1989], and was implemented as the GRASS command r.random.surface [Ehlschlaeger & Goodchild, 1994]. The result of performing this analysis is a set of parameters defining the p.d.f. for modeling the uncertainty of the elevation surface.

We use this p.d.f. to generate a stochastic simulation employing a series of Monte Carlo drawings of realizations. By adding each random field realization to the data set in the area of interest, one creates a collection of alternative, equally probable models of a spatially distributed variable, in this case, uncertainty or error [Deutsch & Journel, 1992]. Instead of one sample (the original data set) to analyze, the researcher has dozens or hundreds of samples with which to study error propagation. This approach allows one to transmit the effect of generalized data from the initial spatial data sets through the stages of analysis to a large collection of potential results. The distribution of these results can be analyzed to formally assess the impact of uncertainty in the base data on the spatial application [Openshaw, 1979]. A number of studies [Fisher, 1991; Lee et. al., 1992] discuss error simulation methodology using spatially autocorrelated surfaces.

There have been efforts to develop more generalizeable error models by building an analytical expression with a series of coefficients that model error propagation for a particular GIS analysis [Veregin, 1994]. Examples of spatial analyses for which error propagation would be desirable include viewshed generation, crop growth, and least-cost path estimation. A distinction can be made between two types of spatial analysis. In type 1 analyses, spatial interactions play no role (GIS operations are local to individual cells), while spatial interaction characterizes type 2 analyses (GIS operations are on a neighborhood or larger basis [NCGIA, 1990]. Research into deriving error propagation equations has focused on type 1, since these can be solved analytically [Heuvelink et al., 1989]. Research into solving for type 2 operations has been undertaken [Lanter & Veregin, 1992], but a concern with the expression approach for type 2 analyses is that translating the numerical result of the mathematical expression into a concrete statement of the likely impact of the data generalization may not capture the complex spatial nature of the effect spatial data generalization will have on the application [Theobald, 1989].

3. Least-cost path

The GRASS GIS employs a three step process to determine the least-cost path. First, a cost surface must be generated. Each cell of the cost surface represents the cost of traversing the shortest distance across that cell. Second, an accumulated cost map is created using the r.cost command [Awaida, 1991], the cost surface, and the coordinates of the start location. This map represents the accumulated cost to reach a cell from the start location. Finally, the command r.drain generates the least-cost path by traversing the accumulated cost map to the finish location [Khawaja, 1991]. This work uses elevation data to develop the cost surface, but in practice many data layers representing many routing objectives may be considered [Huber & Church, 1985].

Research in corridor location problems has recognized the limitations of returning only a single option. Potentially attractive alternatives exist, and approaches for identifying identifying these have been developed. One is to represent the estimated cost of traversing a path visually. [Church et al., 1992] implements a graph-based technique to determine the costs of alternative routes to the least-cost route. This is easily accomplished in the GRASS GIS by adding the accumulated cost map from the start location to the accumulated cost map from the finish location. This approach allows one to visually assess the degree to which the character of the terrain data restricts the least-cost path.

Figure 1 Accumulated Cost Surface, Goleta Test Area.

This illustration is a map for which each cell is an estimate of how much it would cost to traverse the landscape from the start location to the finish location assuming the path went through that cell. The information in this map gives a good "sense" of the likely routes traversed, but cannot provide an estimate of the likelihood that a particular cell is part of the least-cost path. This approach does not take direct account of terrain data uncertainty, which is the focus of this research.

The following experiment illustrates the stochastic simulation methodology discussed earlier, and provides an example of a spatial data application for which the impact of spatial data generalization on the input data set produces variation in the application that is too case-specific and complex to derive from "shortcut" methods. Note: All of the source data, shell scripts, and source code from this study are available to be downloaded at:

http://geo.swf.uc.edu/cgi-bin/wrap/chuck/SDH96/

4. Experiment

This study's digital elevation models cover a large study area around Santa Barbara, California. The 1-degree DEM, USGS quad los_angeles-w, extends from 119 to 120 degrees west of Greenwich, and from 34 to 35 degrees north. The terrain is characterized by mountainous topography extending inland (to the north and east) from the Pacific Ocean, punctuated by river valleys and narrow coastal plains. A sample of six USGS 7.5' DEMs provide "base" data throughout the area covered by the 1-degree DEM.

A rectangular test area several kilometers on each side was defined within this DEM. It extends from the coastal plain at the town of Goleta in the south across the Santa Ynez mountains to the Santa Ynez river valley in the north. This test area contains portions of four 7.5' DEMs, none of which were used to determine the p.d.f. statistics.

The 1-degree DEM was imported into GRASS 4.1, the GIS which performed the bulk of the spatial analysis for this study. The data was stored as a raster and reprojected to universal transverse mercator with a square cell resolution of approximately 94 meters on a side. The raster was resampled to 30 meters using GRASS' thin-plate spline algorithm [Mitasova & Mitas, 1993]. Differences between resampled 1-degree DEM and 7.5' DEM cells were calculated for the areas covered by the six 7.5' DEMs outside of the test area.

The correlogram for this data was calculated to determine the range of spatial dependence using r.1Dcorrelogram [Ehlschlaeger, 1996]. We found that spatial dependence persisted to 3,210 meters. To determine the p.d.f. for the stochastic simulation, 100 sets of spot locations were selected from the areas covered by the six 7.5' DEMs outside of the test area. Each set of samples contained between 75 and 90 points, all of which were separated by more than 3,210 meters. For each set, p.d.f. statistics -- mean and standard deviation -- were derived. The mean average difference value (from (3)) was - 8.60 meters, indicating that, on average, the 1-degree DEM underestimated the 7.5' DEM by 8.6 meters in the sample locations. Equation (4) was 10.36 meters, indicating the spread of the central tendency measure of the distribution around -8.60. Equation (5), the average standard deviation, was 132.55 meters, and the standard deviation of this measure − equation (6) -- was 11.34 meters.

After testing 135 sets of random field parameters for $Z(\mathbf{u})$, $D = 4600$, $E = .07$, and $F = 200$ were found to best represent los_angeles-w using the six 7.5' DEMs outside the study area. The p.d.f. was implemented with the difference parameters indicated above:

$$E(\mathbf{u}) = \mu(\mathbf{u}) - 8.6 + 10.36\varepsilon_1 + (132.55 + 11.34\varepsilon_2)Z(\mathbf{u}) \qquad (8)$$

A total of 247 equally probable potential elevation surfaces of the test area were created using this p.d.f.. A least-cost path routine was performed on each of these realizations. Two endpoints (at the same coordinate-referenced location in each dataset) were chosen, and the algorithm was tasked with finding the path that minimized a cost function incorporating distance, elevation changes, and absolute elevation value. Therefore, the selected route will be as direct as possible while attempting to avoid traversing steeper slopes and higher elevations. The endpoints were selected such that the intervening terrain would include these high cost attributes.

5. Results

The resulting costs associated with each of the 247 potential elevation surfaces generated trails form an approximately bell-shaped distribution with a mean of about 70,000 units, ranging from a minimum of less than 50,000 units to a maximum of over 90,000 units. Using the actual 7.5' DEM data in the study area, the trail would cost 61,263 units. The path on the original, non-perturbed 1-degree DEM surface cost 58,175 units. The lower estimate is a result of the lack of texture in the coarser, 1-degree DEM data, since sub-grid resolution land surface variation is not picked up by the path algorithm for this data. The stochastic method described above has "added texture" by perturbing the dataset to match the differences observed in the 7.5' data for other regions on the same 1-degree DEM.

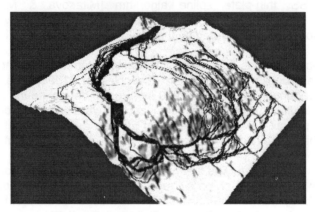

Figure 2 Path distribution across the test area.

While the histogram of path cost is approximately gaussian, the spatial distribution of the 247 least-cost paths across the test area is not. In fact, there are at least two distinct corridors through which many of the paths concentrate, but dozens of other routes were approximated by more than one perturbed surface. This complex pattern is dependent upon the nature of the spatial autocorrelation in the data, and the degree to which the fine-scale resolution of the higher quality data affects the ruggedness of the surface. By studying this distribution, the researcher interested in error propagation gains a greater appreciation for the complexity of the problem, and the application-oriented user is provided with a set of alternative routing corridors rather than a single (and probably incorrect) path.

An important point must be made about the model parameters resulting from the observed differences between the datasets. The extremely high mean standard deviation parameter (132.55 m) may be due in some large part to registration error. Aligning two data sets of the same area that are originally in different projections and different datums, each subject to unknown positional error, is not a trivial problem. We chose to concentrate our efforts on developing the model for this test case rather than focusing on data import issues. In areas of rugged relief, as in our study area, positional errors in registration of several 10's of meters can result in very large vertical differences, even between two otherwise identical datasets. This potential registration error does not invalidate the approach. However, it is factored into the overall variation between the datasets.

6. Discussion

The experiment developed here demonstrates a method for stochastically exploring the spatial complexity of the outcome of a GIS operation on uncertain data. The spatial distribution of the paths in this example does not fall neatly into a pattern that could be simply modeled using epsilon bands around the routes chosen on the original, non-perturbed data sources, or by employing previously defined models of error propagation on the data source and GIS function. Were we to change the test area, source data, or the study region, the uncertainty model parameters (mean and standard deviation of uncertainty, degree of spatial dependence of uncertainty) would create new and complex patterns, affecting the output of the GIS routing algorithm in unpredictable ways. Only by repeating the analysis with the new data could we feel certain that the effects of uncertainty were being properly modeled.

Several extensions to the basic design will improve the basic model. The difference mean and standard deviation parameters, key for the random field generator, were kept constant across all elevation values for this work. Uncertainty was treated as if it did not vary as a function of elevation, slope, or aspect. However, there may be elevation or elevation derivative-based trends in either the mean or standard deviation models that could be modeled. For example, empirical studies have generally shown a positive relationship between DEM measurement error and slope. If such a trend can be measured between the 7.5' DEM data and the 1-degree DEM data, mean and/or standard deviation values for the model could be modified and improved based on the slope of each location in the study. This work also treated all 7.5' DEMs equally, even though several different methods were used to generate the 7.5' DEMs -- each with different data generalizations and measurement error magnitudes. A more precise model would take into account the type of 7.5' DEM the application was designed for, or only use the highest quality 7.5' DEMs for the higher accuracy maps.

Modeling elevation data uncertainty is a difficult task. The characteristics of map generalization manifest themselves on the least-cost path algorithm in ways too complex to be captured without explicitly modeling the interactions between many sample realizations of the data and the GIS algorithm. The approach presented in this work allows the user to accomplish precisely this. We chose the least-cost path algorithm for this study because it represents a common GIS operation on extended neighborhoods of cells. Other GIS commands within this class of non-local GIS operations include viewshed analysis, watershed delineation, and buffering algorithms. While local GIS operations (such as overlay and reclassification functions) need not require unconditional stochastic simulation to understand uncertainty, complex GIS algorithms and applications need methods such as these to explicitly model the impacts of spatial data uncertainty.

Acknowledgments
We thank Dr. Michael Goodchild and Dr. Joel Michaelsen for their comments and advice on this research, but note that any errors and omissions in this paper are entirely the fault of the authors.

References

Adkins, K.F., C.J. Merry. (1994). Accuracy Assessment of Elevation Data Sets Using the Global Positioning System. Photogrammetric Engineering & Remote Sensing, 60(2), 195-202.

Awaida, A. (1991). r.cost MAN page. GRASS4.1 Reference Manual. U.S. Army Corps of Engineers Construction Engineering Research Laboratory.

Bolstad, P.V., T. Stowe. (1994). An Evaluation of DEM Accuracy: Elevation, Slope, and Aspect. Photogrammetric Engineering & Remote Sensing, 60(11), 1327-1332.

Church, R. L., S. R. Loban, K. Lombard. (1992). An Interface for Exploring Spatial Alternatives for a Corridor Location Problem. Computers and Geosciences, 18(8), 1095-1105.

Deutsch, C., A. Journel. (1992). Geostatistical Software Library and User's Guide. Oxford University Press.

Ehlschlaeger, C.R. (1996). r.1Dcorrelogram MAN page. GRASS4.2 Reference Manual. U.S. Army Corps of Engineers Construction Engineering Research Laboratory. (http://geo.swf.uc.edu/cgi-bin/wrap/chuck/SDH96/).

Ehlschlaeger, C.R., M.F. Goodchild. (1994). Uncertainty in Spatial Data; Defining, Visualizing, and Managing Data Errors. Proceedings of GIS/LIS 1994, Phoenix, AZ, 246-253.

Fisher, P. F. (1991). Modelling Soil Map-unit Inclusions by Monte Carlo Simulation. International Journal of Geographical Information Systems, 5(2), 193-208.

Goodchild, M.F. (1986). *Spatial Autocorrelation.* CATMOG 47. Norwich, UK: Geo Books.

Heuvelink, G. B. M., P. A. Burrough, A. Stein. (1989). Propagation of Errors in Spatial Modelling with GIS. International Journal of Geographic Information Systems, 3(4), 303-322.

Huber, D. L., R. L. Church. (1985). Transmission Corridor Location Modeling. Journal of Transportation Engineering, 111(2), 114-130.

Isaaks, E. H., R. M. Srivastava. (1989). An Introduction to Applied Geostatistics. New York: Oxford.

Khawaja, K. (1991). r.drain MAN page. GRASS4.1 Reference Manual. U.S. Army Corps of Engineers Construction Engineering Research Laboratory.

Lanter, D.P., H. Veregin. (1992). A Research Paradigm for Propagating Error in Layer-Based GIS. Photogrammetric Engineering & Remote Sensing, 58(6), 825-833.

Lee, J., P.K. Snyder, P. F. Fisher. (1992). Modeling the Effect of Data Errors on Feature Extraction from Digital Elevation Models. Photogrammetric Engineering & Remote Sensing, 58(10), 1461-1467.

Mitasova, H., Mitas, L. (1993). Interpolation by Regularized Spline with Tension 1, Theory and Implementation. Mathematical Geology, 25(6), 641-655.

NCGIA. (1990) NCGIA Core Curriculum. Ed. M. F. Goodchild & K. K. Kemp. Santa Barbara: NCGIA.

Openshaw, S. (1979). A Methodology for Using Models for Planning Purposes. Environment and Planning, 11, 879-896.

Shearer, J.W. (1990). The Accuracy of Digital Terrain Models. Chapter 24, Terrain Modelling in Surveying and Civil Engineering. G. Petrie, T.J.J. Kennie, Eds. Whittles Publishing, London. 315-336.

Theobald, D. M. (1989). Accuracy and Bias Issues in Surface Representation. Chapter 9, Accuracy of Spatial Databases. M.F. Goodchild, S. Gopal, Eds. Taylor & Francis, London. 99-106.

U.S. Geological Survey (National Mapping Division). (1992). National Mapping Program Technical Instructions: Standards for Digital Elevation Models. U.S. Dept. of the Interior.

Veregin, H. (1994). Integration of Simulation Modeling and Error Propagaton for the Buffer Operation in GIS. Photogrammetric Engineering and Remote Sensing, 60(4), 427-435.

Zhang, W., D. R. Montgomery. (1994). Digital Elevation Model Grid Size, Landscape Representation, and Hydrologic Simulations. Water Resources Research, 30(4) 1019-1028.

Contact address
Charles R. Ehlschlaeger[1] and Ashton Shortridge[2]
[1]Assistant Research Professor
 Department of Geography
 University of Cincinnati
 Cincinnati OH 45221-0131
 USA
 E-mail: chuck@geo.swf.uc.edu
[2]National Center for Geographic Information and Analysis
 University of California,
 Santa Barbara CA 93106-4060
 USA
 E-mail: ashton@ncgia.ucsb.edu

AN INVESTIGATION INTO THE USE OF MEDIAN INDICATOR

KRIGING TO ASSIST IN POST ACCIDENT RADIATION

ASSESSMENT

Jackie Carter and Stuart A. Roberts

School of Computer Studies
University of Leeds
Leeds, UK

Abstract
This paper reviews the geostatistical method called median indicator kriging, and assesses its suitability for analysing post-accident deposition data in the context of the nuclear industry. Following an accidental release of radioactivity to the environment decisions will be made based on the information available to those organisations responsible for assessment of the radiological impact of the accident. The median indicator method is applied to various datasets, generated from a gaussian plume atmospheric dispersal model over an area 40 km by 40 km, in order to produce results in a form suitable for use by decision makers. The results show that the method can be customised to include intervention levels, which are used in determining countermeasures to be implemented following a release of radiation to the environment. Incorporating intervention levels into the method is shown to reduce the uncertainty of the estimates produced. An analysis of the results is conducted such that at each location on the grid the probability of a predicted value exceeding a given threshold is compared to the known value at that location, thus enabling the possible misclassification of values to be determined. It is shown that with as few as 50 data values the method gives useful results when intervention levels are used. These preliminary results suggest that the method could be used as a suitable tool for an emergency response system.

1. Introduction

1.1 Median indicator kriging

The full theory on indicator kriging (IK) can be found elsewhere (see Deutsch and Journel 1992; Journel 1983; Isaaks and Srivastava 1989). The overview presented below is based on that given in Deutsch and Journel. An understanding of the geostatistical interpolation method known as kriging is also required for a full understanding of the method (see same references).

An indicator variable I is defined as follows:

$$I(u;z) = \begin{cases} 1, & \text{if } Z(u) \leq z \\ 0, & \text{otherwise} \end{cases}$$

for some cut-off value, z.

Kriging of the indicator variable results in an estimate that is the best least squares estimate of the conditional expectation of the indicator random variable for cut-off z. It can be shown that the conditional expectation of the indicator variable is equal to the conditional cumulative distribution function of the random variable $Z(\mathbf{u})$; that is:

$$E(I(\mathbf{u};z)|(n)) = F(\mathbf{u};z|(n))$$

where $F(\mathbf{u};z|n)$ represents the conditional cumulative distribution function (the cumulative distribution function conditional upon a specific set of data, (n)).

Thus the result of applying the kriging algorithm to the indicator data is a least squares estimate of the ccdf. It is important to note that indicator kriging does not aim to supply an estimate of either the unsampled value, $z(\mathbf{u})$, or the indicator value, $i(\mathbf{u})$, but rather it aims to provide a model of uncertainty about $z(\mathbf{u})$. Since the approach of indicator kriging does not use the (estimated) parameters of the ccdf but estimates directly the values of the ccdf for various cut-offs, the method is said to be non-parametric.

The measure of spatial continuity for each set of indicator values at each cut-off level is captured in the geostatistical tool known as the variogram, or more precisely, the indicator variogram. The cut-off values are usually chosen with the result that the corresponding indicator variograms are different from each other (different data ranges are expected to display different spatial structures). Median indicator kriging is a simple and fast procedure which uses the same variogram for all cut-off values. This can be done when the sample indicator variograms appear proportional to one another (they exhibit the same shape), or when the threshold of interest is close to the median cut-off value, and when the relative nugget effect over various cut-off values is not too dissimilar (Deutsch and Journel, 1992). In this case, the sample indicator variogram at the median cut-off, $z = M$, such that $F(M) = 0.5$, is used. At the median cut-off the indicator values are evenly distributed (same number of zeros as ones), with no outliers.

The main issues involved in the use of the method of indicator kriging are thus which cut-off levels to use and whether to produce a measure of spatial variability for each cut-off used, or to use one measure that will adequately describe the continuity at each cut-off level. The trade-off is between accuracy and both computational and human effort. Median indicator kriging is used in this work as the effort required to model several variograms might not be available in a real situation.

From the ccdf produced for each cut-off the following values can be calculated: the mean value of the cumulative distribution function (called the 'e-type' estimate) for each grid location; the probability of exceeding a threshold value which may or may not be one of the cut-off values used in the production of the ccdf.

2. Post accident radiation assessment

Following an accidental release of radioactivity to the environment decisions will be made based on the information available to those organisations responsible for assessment of the radiological impact of the accident. One approach to dealing with the consequences of a radiological accident is to use a physical mathematical model to predict the dispersion, transport, deposition and resuspension of the radioactive material into the environment (see for example: Mikkelsen and Desiato 1993; Shershakov et al 1993; Renier et al 1993; Clarke et al 1979; Sohier et al 1993; French and Smith 1993). This approach, used in many systems, varies in terms of the sophistication of the models used and whether monitoring data is integrated into the prediction process. The job of atmospheric dispersion models is primarily to provide information to assist with the assessment of the severity of the situation in the early phase of an accident when data collected are few, but they have a significant role to play in determining the locations where data should be collected (McColl, 1995).

A complementary approach is to use monitoring data, as it becomes available, to provide as complete and accurate a view of the situation as possible. This approach is likely to be of greater use in the longer term when the cloud of radioactive material has passed over, and deposition readings and dose rate measurements are collected. These readings take the form of a sampling of a continuous field at discrete points in space and time. The spatial reference is available as a grid reference and temporal information is usually the date on which the measurement was taken. Normally the sampling is irregular in space and is likely to be concentrated in the region where readings are expected to be highest. Radiological monitoring and information systems are now established in many countries (RIMNET, 1993; Weiss and Leeb, 1993; Chino et al 1993) and tools for the analysis and display of the data collected are required. Recent work has considered the use of spatial statistical techniques for the analysis of post-Chernobyl data (Kanevsky, 1993) and the use of such techniques in emergency response information systems merits further investigation.

The main aim of the experiments presented in this paper is to investigate the use of the geostatistical technique known as median indicator kriging for the assessment of post accident radiation data. Geostatistical techniques have been investigated as a means of producing estimates of contamination levels at unknown locations from a set of known sample values (Carter and McLaren, 1996). The ordinary kriging method does not present information in the most appropriate form for the purposes of decision making based on intervention levels. In the context of radiation assessment it is important to be able to place a level of uncertainty on any estimate that is to be used directly in the decision making process in order to justify the decision that is made on the basis of such estimates. If the uncertainty is deemed too large it might be necessary to collect more data before the decision can be made with confidence. Geostatistical methods such as indicator kriging prioritise the modelling of uncertainty of the underlying distribution, and are therefore worthy of investigation. The output from this method is also in a form more suitable for conveying information to decision makers, and the method can be tailored to deal with intervention levels of interest for radiation assessment.

Indicator kriging is a non-parametric technique in the sense that the method does not assume a model for the underlying distribution. It copes well with data which may be from a highly skewed distribution (Fytas et al, 1990). The method deals with classes of data defined by the user, and interpolation is performed for each of these classes. For the particular application being studied, the decision making is based on agreed intervention levels so these form a basis for selecting the classes. The method can be tailored for radiation assessment purposes by using intervention levels as cut-off values, thus defining the indicator variable. This step can be repeated for several intervention levels. The method thus produces as output a probability that an intervention level will be exceeded based on the information given as input, and this for each intervention level used in the analysis.

The method might be more appropriate for the analysis of datasets that span several orders of magnitude (typical of radiological datasets) due to the fact that that it takes into account the changing spatial structure at different cut-off levels. In addition, since it produces an output that quantifies the uncertainty associated with the estimation procedure it is more suitable for the needs of assessment for radiation protection.

Isoprobability maps can be produced from the method for a given threshold of interest. Alternatively maps representing the mean-type estimate according to the value derived from the resulting conditional cumulative distribution function can be produced.

A dataset representing deposition readings of Caesium-137 (^{137}Cs) was generated from a Gaussian plume atmospheric dispersion model for the experiments. Further details describing the dataset can be found in the following section.

The experiments presented in this paper were designed with the following aims:

- To produce maps, based on randomly selected sample sets of varying size taken from the full dataset of (i) probability contours of exceeding a threshold value representing a possible intervention level, and (ii) estimates of the distribution of the ground deposition.
- To ascertain the effect of the following on the estimates of the uncertainty: the number of samples used; the number of cut-offs used; which cut-off values were used.

3. Methodology

3.1 Datasets used

A problem with the development of analysis tools for radiation assessment following a nuclear release is that they are unlikely to be tested under conditions representing a real situation. Emergency exercises can help in overcoming this. The gaussian model chosen for the work in this paper is the one employed for emergency exercises. We cannot be sure that this model gives realistic data with respect to variability, so the results in terms of reliability (see Tables 2 and 3) must be viewed as preliminary.

A reference dataset of 1681 regularly gridded samples, at 1 km intervals, of ^{137}Cs deposition data was generated from a Gaussian plume atmospheric dispersal model (Clarke, 1979), where the release was 2.70E14 Bq of ^{137}Cs, at ground level with weather conditions Pasquill category F.

Initially an area of 80 km (downwind of the release) by 80 km (cross-wind) was considered. The full spatial distribution of the data clearly showed the expected teardrop shape of the deposition. From this dataset, an area was taken which represents a region starting at 40 km from the source and extending up to 40 km downwind and 20 km in either direction in the off-wind direction. This area might represent one that would be considered in the first few days following an accidental release. Different courses of action might be recommended immediately following a release of radiation, but these fall into three main categories: sheltering, evacuation and the administration of stable iodine tablets. Longer term countermeasures include relocation and /or decontamination (Morrey 1994a and 1994b). Intervention levels of dose, related to these courses of action, are used by organisations responsible for the dissemination of information and advice in the UK (Morrey et al 1996; NRPB 1986).

The spatial distribution of the data for the region considered in this paper can be seen in Figure 1. The values displayed on the contours correspond to the nine quantile values of the dataset.

Single datasets of 50, 100, 150 and 200 values were randomly selected from the dataset of 1681 values.

3.2 Indicator Variograms

Variograms corresponding to the nine quantile values of the dataset of 1681 values were produced. These were standardised by dividing by the indicator data variances in each case, in order that they could be compared across cut-off levels. Using comparisons by eye, the median cut-off indicator variogram was found to be not noticeably different in structure from variograms for all but the two extreme cut-offs. This justifies the decision to work with median indicator kriging since the extreme cut-off values are beyond the extreme intervention levels.

For each of the datasets an indicator variogram was produced for the median cut-off value. This indicator variogram was modelled using a spherical structure and the parameters for the fitted model were used in the kriging equations (Deutsch and Journel, 1992; Isaaks and Srivastava, 1989). The parameters for the four different models are given in Table 1.

3.3 Analysis

In order to assess whether median indicator kriging is providing accurate estimates based on our knowledge of the 'true' underlying distribution, we have adopted a method of measuring the number of false negatives and false positives. The details of the way in which these values are calculated are given in the results section below. We have further refined this idea by introducing 'strong' and 'weak' false estimates. These concepts make use of a limiting value on the predicted probability of a value exceeding some threshold. In this work, the limiting value is chosen to be 50% because for the particular case study decision making could go ahead based on estimates that were accurate to within 50%. This limiting value could, of course, be changed for other applications.

Regularly gridded data (1681) 40 by 40 km

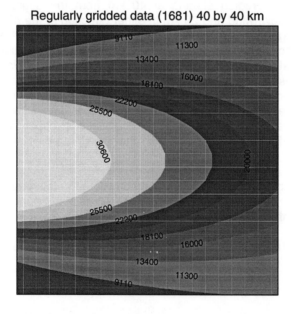

Figure 1 Dataset of 1681 deposition values for a region 40 km by 40 km.

number of data	range	sill	nugget
50	22.0	1.27	0.60
100	21.5	1.04	0.36
150	23.0	1.56	0.10
200	20.7	1.65	0.00

Table 1 Parameters used from the model fits to the indicator variograms.

4. The experiments

Four sets of experiments were performed. In all cases the model of spatial continuity that was used was the one described by the parameters reported in Table 1 according to the dataset used.

In the first experiment, for each of the four datasets median indicator kriging was performed using all nine cut-offs, corresponding to the nine quantile values of the conditioning dataset. Results were displayed as probability maps of a threshold being exceeded and maps of e-type values (mean of the resulting ccdf at each location). The threshold corresponded to a value which was not one of the cut-off values used in the production of the ccdf.

In the second experiment, for each of the four datasets the number of cut-offs used for median IK was reduced to 5. These five cut-off values corresponded to the 0.1, 0.3, 0.5, 0.7 and 0.9 quantile values of the conditioning dataset used. Output was again produced in the form of probability maps for exceeding a threshold value not corresponding to one of the cut-off values used, and maps of e-type estimates.

In the third experiment, for each of the four datasets only three cut-off values were used. These cut-off values corresponded to the 0.2, 0.5, and 0.8 quantile values of the dataset used. Output produced was the same as for the first two experiments.

The fourth experiment was designed to test the effect of including (hypothetical) intervention levels (ILs) in the procedure. In this case, three of the cut-off values used represented intervention levels; the lowest represents a value below which no intervention would be recommended, the middle a value for which intervention might be recommended, and the third a value above which intervention would almost certainly be recommended. For each dataset, the ILs were compared to the quantile values for that dataset, and two more cut-off values were chosen to lie above and below the three ILs. The same output was produced as for the other experiments using the middle threshold value of 21500 Bqm^{-2}.

5. Results

The results reported here focus on the probability values produced as these present the main interest in terms of the initial aims of the work.

The results of the conditional probabilities produced for each experiment can be analysed by comparing these values with the actual values at each grid location of the reference dataset. The misclassification of results will be most serious in the case when a value at a location is predicted to have a high probability of exceeding a threshold, when it is known not to, and when the reverse is true. The predicted results at any grid location that coincides with the reference dataset can be one of four cases: it is correctly predicted to be less than the threshold; it is correctly predicted to be greater than the threshold; it is misclassified as a false negative; it is misclassified as a false positive.   These categories can be further subdivided and the following classification is used to analyse the results.

A. TRUE NEGATIVES: Locations which are correctly estimated as being less than the threshold (predicted probability of 0);
B. TRUE POSITIVES: Locations which are correctly estimated as being higher than or equal to the threshold (predicted probability of 1);
C. DEFINITE FALSE NEGATIVES: Locations which are higher than the threshold value but which have a predicted probability of 0 of exceeding it;
D. STRONG FALSE NEGATIVES: Locations which are higher than the threshold which have a predicted probability in the interval of (0.0, 0.5) of exceeding it;
E. WEAK FALSE NEGATIVES: Locations which are higher than the threshold with a predicted probability in the interval of [0.5, 1) of exceeding it.
F. WEAK FALSE POSITIVES: Locations which are lower than the threshold value but which have a predicted probability in the interval of (0, 0.5) of exceeding it;
G. STRONG FALSE POSITIVES: Locations which are lower than the threshold but which have a probability in the interval of [0.5, 1) of exceeding it ;
H. DEFINITE FALSE POSITIVES: Locations which are lower than the threshold but which have a predicted probability of 1 of exceeding it.

The results of the analysis according to these classes are presented in tables 2 and 3 below. Figures given represent percentages of the region for which each of the classes were calculated. The reference dataset has 33% of the grid values above the threshold of 21500 Bqm^{-2} and 67% below this value. In tables 2 and 3 columns B, C, D and E thus sum to give the 33% of values which are greater than the threshold, and columns A, F, G and H sum to give the remaining 67%.

The method is shown to produce no definite false negatives or positives as would be anticipated. Also, as would be expected, increasing the number of conditioning data reduces the uncertainty of the resulting estimates. As more data become available, the proportion of the region for which data values are known to be less than the threshold value increases. For this case (class A), there is no difference in the results between using three or five cut-off values. The use of nine cut-off values however improves the certainty of locations having values less than the threshold, and using the required threshold as a cut-off value in the IK method further improves the accuracy of the estimation.

Using only 50 data values the best result that can be obtained for correctly predicting the values which are less than the threshold is when the intervention level is used. Using 200 values and the intervention level, on the other hand, improves the prediction, such that only 1% of the grid is misclassified as strong false positives and only 2% is misclassified as strong false negatives.

Similar results, in terms of increasing the number of data values, can be seen for the case of accurately predicting with increased certainty values that lie above the threshold. In the cases reported here, using five cut-off values offers an improvement on using only three, but there is little to be gained from using nine instead of five. Using an intervention level as a cut-off value again improves the results, but this improvement appears to be only marginal.

In all cases, using only 50 data values leads to poor results for accurately predicting the locations where values lie above the threshold. These preliminary results indicate that more than 100 data are required to predict the values which definitely lie above the threshold accurately.

Strong false negatives, which are probably cause for the greatest concern, are highest in cases when only 50 conditioning data were used. Increasing the number of cut-off values used reduces the percentage of false negatives produced, and incorporating intervention levels into the method reduces this even further. Note that even when 200 data are used, using only three cut-off values results in a relatively high percentage of the grid being predicted as strong false negatives. Increasing the number of cut-off values used eliminates this effect. The best results in terms of the lowest percentage of strong false negatives occurring was in all cases when ILs were employed. The percentage of weak false negatives produced tends to fall as the number of conditioning data increases. This result corresponds to the level of certainty of values greater than the threshold increasing, as noted above.

Strong false positives are few in all cases, the worst case occurring when 150 data are used with only three cut-off values. The percentage of weak false positives are high in all cases where only 50 data values are used. Using five cut-offs offers little improvement on using three, but using nine cut-offs does reduce the percentage of false positives, and has the added effect of increasing the certainty of a value being greater than the threshold.

Number of cut-offs	Number of data values	A True Negatives	B True Positives	C Definite False Negatives	D Strong False Negatives	E Weak False Negatives	F Weak False Positives	G Strong False Positives	H Definite False Positives
3	50	2	1	0	10	22	65	0	0
3	100	22	2	0	2	29	44	1	0
3	150	28	6	0	3	24	34	5	0
3	200	37	13	0	7	13	30	0	0
5	50	2	3	0	7	23	65	0	0
5	100	22	6	0	4	23	45	0	0
5	150	28	12	0	5	16	39	0	0
5	200	38	20	0	2	11	29	0	0
9	50	10	1	0	6	26	57	0	0
9	100	30	6	0	2	25	37	0	0
9	150	44	12	0	3	18	22	1	0
9	200	48	19	0	3	11	19	0	0

Table 2 Results for experiments 1, 2 and 3 for exceeding a threshold of 21500 Bqm^{-2}.

Number of cut-offs	Number of data values	A True Negatives	B True Positives	C Definite False Negatives	D Strong False Negatives	E Weak False Negatives	F Weak False Positives	G Strong False Positives	H Definite False Positives
5	50	14	3	0	4	26	53	0	0
5	100	33	8	0	1	24	33	1	0
5	150	49	17	0	3	13	17	1	0
5	200	58	22	0	2	9	8	1	0

Table 3 Results for experiment 4 (using ILs) for exceeding a threshold of 21500 Bqm^{-2}.

6. Discussion

The results of these experiments indicate very clearly that in order to use the method of median indicator kriging, cut-off levels of interest to the problem should be incorporated into the procedure. In all cases reported, utilising intervention levels in the production of the ccdf reduced the uncertainty of the estimates produced. This would be expected, but in terms of customising the method for use in radiation protection this is confirmed by the results reported here. An important issue in presenting the results of an analysis to decision makers is to be able to communicate the level of confidence that can be placed on any estimate produced. By analysing the misclassification of the resulting estimates in these experiments, we have shown that median indicator kriging produces useful estimates based on the conditioning data used. Even in the worst case scenario, the prediction of strong false negatives are very few. It is also likely that if a region were incorrectly classified in this way that decisions taken would be influenced not only by the predicted values for this region, but by values (measured and predicted) at neighbouring regions and additional information known to the decision maker.

The results of these experiments indicate very clearly the following in the use of median indicator kriging as an estimation tool for assisting with post accident radiation assessment:

- Utilising intervention levels reduces the area of uncertainty;
- Using all quantile values is the next best option;
- The extent of the region of uncertainty is a function not only of sample size but also of the number and type of the cut-off values used in the method;
- If small datasets are used, the best results are obtained by utilising ILs;
- The method provides a very rapid way of providing probability estimates associated with a threshold of interest being exceeded.

Although the results are limited to a small number of datasets, we have shown that the use of ILs in the production of the ccdf is very important. The threshold value reported here lies in all cases between the 0.6 and 0.7 quantile values. As median indicator is used there is thus no cause for concern that the resulting estimate would be poor based on a model of continuity derived at the median value of the dataset. It may not be appropriate to use median IK if the threshold values are not close to the median value, in which case variograms should be produced which are close to these values. In this paper we consider output for only one threshold, but it is likely that more than one will be required in a real situation. What is clear is that the choice of cut-off levels has more influence on reducing the area of uncertainty than increasing the number of data points. For radiation protection work this is important. By customising the method to include intervention levels we can consider its use when there are few data values available.

Some discussion from these results is required for a consideration of the chosen classes. As the distinction between 'strong' and 'weak' is based on a 50% probability value, it is not possible to see from the results presented in tables 2 and 3 what the proportion of probabilities attached to estimates within these classes is like. In fact, there is no reason why this classification needed to be chosen, and it would be interesting to see how the results were affected if the definitions of strong and weak were changed, or further subdivided. In particular, it might be interesting to investigate what the 50% probability value should be changed to in order to give no strong false negatives, and to ascertain the effect that this has on the corresponding false positives.

However, on the basis of these results consider what would happen if a decision were to be based on requiring a probability of 50% or greater that a value were expected to exceed the threshold used (conditional to the data available); thus, if a location has a 50% or higher probability that the threshold value is exceeded action will be taken. The results show that at worst, 5% of the grid would be misclassified as having contamination levels greater than the threshold, and 10% would be misclassified as having contamination levels less than the threshold value. The remainder of the region would, according to this criteria, be correctly classified. Increasing the number of data used and incorporating intervention levels into the method reduces these figures to 1% and 2% respectively.

7. Conclusion

This paper has served to review the median indicator kriging method for spatial data analysis in the context of an application arising in the nuclear industry. A characteristic of the application is the need for estimates for field values based on relatively few irregularly spaced readings in 2-D space. The application is also characterised by the need for the quantification of uncertainty of estimates relative to pre-defined cut-off values (intervention levels).

The results of these experiments have raised some interesting issues. The output from median indicator kriging is in a form that could usefully be employed by decision makers, seeking to confirm their opinions of the situation following a nuclear release, based on expert knowledge and information already available. These preliminary results indicate that this tool could be of use in an emergency response system.

Geostatistical techniques provide methods which are well suited to the analysis of the types of data that are available following a nuclear release. We have shown here that the technique of median indicator kriging merits further investigation. What is important for an analysis tool to be used during the post-release phase of a nuclear emergency is that there are sufficient data values to render the tool usable, and that the analysis is timely. Median indicator kriging used with intervention levels on small datasets has proved to be useful for producing output that could help with the decision making task. The method gives good results when compared to the known underlying distribution of data generated from a model. If cut-off levels are used which correspond to intervention levels of interest to the problem misclassification can be minimised. Using nine cut-off values corresponding to the nine quantile values of the data distribution is the next best option.

Further experiments are planned with a view to obtaining more general conclusions; these include generating more datasets to cover the many different spatial configurations. Also in preparation are plans to repeat similar experiments using datasets of real data values that describe more realistically the pattern of deposition following a radiological release.

Finally, the results of these experiments are encouraging to all those working with data which requires a way of quantifying the uncertainty associated with the method used.

Acknowledgements
The authors would like to acknowledge the support of the NRPB in the funding of the CASE studentship to J Carter, and for supplying the code from which the dataset used was generated. We would also like to thank Helen Wright for her help in developing the interactive variogram modelling tool.

References
Carter, J. And McLaren, F.E., 1996, An Investigation into the Applicability of Geostatistical Techniques for Estimation of Contamination Following an Accidental Release of Radioactivity. (In preparation).

Chino, M., Ishikawa, H. And Yamazawa, H., 1993, SPEEDI and WSPEEDI: Japanese Emergency Response System to Predict Radiological Impacts in Local and Worldwide Areas Due to a Nuclear Accident, Radiation Protection Dosimetry : Decision Making Support for Off-site Emergency Management (G.N. Kelly and G. Fraser eds.) Proceedings of a Workshop held in Schloss Elmau, Bavaria, October 1992. Vol. 50, Nos. 2-4, pp.145-152 Nuclear Technology Publishing, UK

Clarke, R.H., 1979, NRPB R91 The First Report of a Working Group on Atmospheric Dispersion. A Model for Short and Medium Range Dispersion of Radionuclide Release to the Atmosphere. HMSO (London).

Deutsch, C.V. and Journel, A.G., 1992, GSLIB Geostatistical Software Library and User's Guide. Oxford University Press.

French, S. and Smith, J., 1993, Using Monitoring Data to Update Atmospheric Dispersion Models with an Application to the RIMPUFF Model, Radiation Protection Dosimetry : Decision Making Support for Off-site Emergency Management (G.N. Kelly and G. Fraser eds.) Proceedings of a Workshop held in Schloss Elmau, Bavaria, October 1992. Vol. 50, Nos. 2-4, pp.317-320 Nuclear Technology Publishing, UK

Fytas, K., Chaouai, N.-E. and Lavigne, M, 1990, Gold Deposits Estimation Using Indicator Kriging, CIM Bulletin, February.

Isaaks Edward. H. and Srivastava R. Mohan, 1989. An Introduction to Applied Geostatistics. Oxford University Press.

Journel, A.G. 1983, Non-parametric Estimation of Spatial Distributions, Mathematical Geology, Vol. 15, No 3, pp 445-468.

Kanevsky, M., Arutyunyan, R., Bolshov, L., Linge, I., Savel'eva, E. And Haas, T., 1993, Spatial Data Analysis of Chernobyl Fallout. 1. Preliminary Results. Pre-print NSI-23-93 Moscow, Nuclear Safety Institute, September..

McColl, N. 1995, Radiological Emergency Monitoring and Information Systems, in NRPB Coursenotes on Planning for Radiation Emergencies, University of Surrey, UK.

Mikkelsen, T. and Desiato, F., 1993, Atmospheric Dispersion Models and Pre-processing of Meteorological Data for Real-time Application, Radiation Protection Dosimetry : Decision Making Support for Off-site Emergency Management (G.N. Kelly and G. Fraser eds.) Proceedings of a Workshop held in Schloss Elmau, Bavaria, October 1992. Vol. 50, Nos. 2-4, pp.205-218 Nuclear Technology Publishing, UK

Morrey, M., 1994a, Interventions During the Early Phase, Fourth Training Course on Off-site Emergency Response to Nuclear Accidents, June 27-July 01, SCK.CEN, MOL (Belgium).

Morrey, M., 1994b, Consequences of Accidental Releases of Radioactive Materials, Fourth Training Course on Off-site Emergency Response to Nuclear Accidents, June 27-July 01, SCK.CEN, MOL (Belgium).

Morrey, M., et al, 1996, Decision Aiding Systems for the Management of Post-Accidental Situations (To be published).

NRPB, 1986, National Radiological Protection Board, Derived Emergency Levels for the Introduction of Countermeasures in the Early to Intermediate Phases of Emergencies Involving the Release of Radioactive Materials to Atmosphere. NRPB-DL10. Chilton, UK.

RIMNET, 1993, Nuclear Accidents Overseas. The National Response Plan and Radioactive Incident Monitoring Network (RIMNET) Phase 2. Department of the Environment (HMSO).

Shershakov, V.M., Borodin, R.V. and Kosykh, V.S., 1993, Radioecological Analysis Support System (RECASS), Radiation Protection Dosimetry : Decision Making Support for Off-site Emergency Management (G.N. Kelly and G. Fraser eds.) Proceedings of a Workshop held in Schloss Elmau, Bavaria, October 1992. Vol. 50, Nos. 2-4, pp.181-184 Nuclear Technology Publishing, UK

Sohier, A., Van Camp, M., Ruan, D. and Govaerts, P., 1993, Methods for Radiological Assessment in the Near-Field During the Early Phase of an Accidental Release of Radioactive Material, Using an Incomplete Database, Radiation Protection Dosimetry : Decision Making Support for Off-site Emergency Management (G.N. Kelly and G. Fraser eds.) Proceedings of a Workshop held in Schloss Elmau, Bavaria, October 1992. Vol. 50, Nos. 2-4, pp.321-326 Nuclear Technology Publishing, UK

Weiss, W. And Leeb, H., 1993, IMIS - the German Integrated Radioactivity Information and Decision Support System, Radiation Protection Dosimetry : Decision Making Support for Off-site Emergency Management (G.N. Kelly and G. Fraser eds.) Proceedings of a Workshop held in Schloss Elmau, Bavaria, October 1992. Vol. 50, Nos. 2-4, pp.163-170. Nuclear Technology Publishing, UK

Contact address
Jackie Carter and Stuart A. Roberts
School of Computer Studies
University of Leeds
LS2 9JT
UK
E-mail: {jackie, sar}@scs.leeds.ac.uk

NEW DEVELOPMENTS IN REGIONALISATION

FOR THE GIS ERA[*]

Mike Coombes[1], David Atkins[2] , Colin Wymer[3] and Stan Openshaw[4]

[1]NE.RRL, CURDS
[2]Geography Department
[3]Planning Department
Newcastle University
Newcastle upon Tyne, UK
[4]Geography Department
Leeds University
Leeds, UK

Abstract
Regionalisation methodology has so far been limited by needing to choose a single 'least worst' run of a single method on a single dataset. The new spatial data handling potential of GIS - together with hugely increased computing power and the attractions of a more 'fuzzy' approach - call for new developments. A practical challenge to the development of new methods has emerged in Britain where there is need for locality boundaries for census researchers with a wide range of interests: the analysis of a single dataset could not provide adequate boundary definitions for this diverse set of researchers. The paper begins by outlining recent developments which have led to a new 'best practice' regionalisation algorithm which can be customised for application to different datasets. The key innovation is then the creation of synthetic data, which allows the information present in any number of sets of boundaries to be brought together. The paper goes on to detail the initial application of this form of data integration to define localities in Britain - and also notes that a neural computing method has been devised to provide an alternative method of building on this innovation.

Key words
Locality boundaries, regionalisation algorithms, data integration, synthetic data, neural computing.

1. Background

The point of departure for this research was the belief that social science researchers in Britain, and census users in particular, are hampered by the absence of 'off-the-shelf' definitions of localities and city regions - and, indeed, other features of the geography of Britain as basic as the distinction between urban and rural areas. Many census researchers are comparing data for different areas without the confidence that the areas are genuinely comparable as geographical entities.

These difficulties affect both qualitative and quantitative social science research. Examples in qualitative studies were presented in numerous locality studies, for which the definition of the areas for the research was recognised to both reflect a particular understanding of what "locality" meant, and thereby also to shape the findings (see for example, Cooke 1989). For quantitative analyses, the importance of study area boundaries has been formalised as

[*] This research was sponsored by ESRC (project H507255129).

the "modifiable areal unit problem" (Openshaw 1984). In short, any set of areas will influence the results of an analysis - especially if the analysis leads to outputs such as ranking which emphasise relatively minor statistical differences between areas.

The primary areas for census data output in Britain - as in most other countries - are local authorities. These areas are currently under revision, but it is already clear that the new set of boundaries will be no more satisfactory than the old for census researchers. The boundaries of some towns and cities continue to embrace quite large tracts of rural land (eg. Leeds and Carlisle), while others do not even include the whole of their own built-up area (eg. Leicester and Norwich). Map 1 shows the 1991 District boundaries, in the London region, readily illustrating their inconsistent relationship to built-up areas (shaded in the map).

To summarise: the need for appropriate definitions of localities and city regions is, if anything, increased by recent local government area changes, while the continuing evolution of Britain's urban system requires that a newly up-to-date set of boundaries is required for contemporary researchers. An additional challenge is that the interests of census researchers are as diverse as their understanding of the term "locality" which is the core concept here. This project aims to avoid the reductionism of previous approaches which have tended to focus almost exclusively on the labour market as the definitive form by which localities are defined. The two spatial data handling challenges are to create a set of meaningful and valid 'general purpose' locality definitions for a diverse set of census researchers and, in order to do so, to devise an innovative regionalisation method which is much less narrowly based than previous approaches.

2. Methods

The preliminary task was to review recent developments in regionalisation methods. The literature is relatively slight, partly because much of the effort in technical regional science has shifted to the extension of Geographic Information System (GIS) techniques - but as yet GIS developments have scarcely reached the regionalisation field. Three main 'traditions' were identified:

1. the longest standing approach is rooted in manual methods and involves a multi-step procedure which typically starts by identifying central cities and moves out to assign other areas to these foci - the official Metropolitan Area definitions in the USA are the classic example;

2. more statistical approaches derive from a numerical taxonomy approach - there are a number of alternative 'black box' methods ranging from cluster analysis to regionalisation-specific algorithms (e.g. graph theoretical techniques), but they typically revolve around a single procedure seeking to maximise a statistical criterion which represents the objectives set for the definitions (e.g. "maximise internal coherence, subject to a minimum size"); and

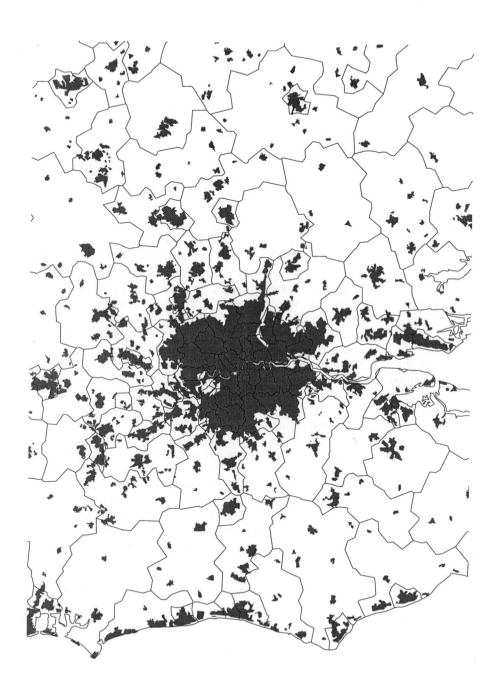

Map 1 1991 Local Authority Districts.

3. a hybrid of the above two alternatives, adopting a multi-step approach which is based on a traditional understanding of cities as foci for hinterlands, but which uses more statistical methods and criteria with successive stages of the analysis in order to ensure that final boundaries all meet strictly pre-defined objectives, and can be 'optimised' in relation to these objectives.

The findings of the literature review were that there had been few recent major new statistical approaches which were of relevance here. With regard to the first tradition, there was on-going research into Metropolitan Area definitions in the USA (Dahmann & Fitzsimmons, 1995), but none of the canvassed alternatives advanced much beyond the familiar multi-step approach, involving no statistical optimisation. Table 1 summarises the key dimensions of the current debate in the USA on the definition of (non-)metropolitan areas. It can be seen that the only point of agreement among the four contributors to Dahmann & Fitzsimmons (1995) is that the current approach is unsatisfactory in not covering the whole country. With the exception of Adams - whose remarkable contribution argues for a move back to more simplistic methods - there is a recognition that a more sophisticated form of analysis is needed in order to obtain more appropriate definitions while also achieving a more scientific, consistent, definitional procedure. Frey & Speare explicitly cite the methodological developments which followed on from the definition of Functional Regions in Britain (Coombes et al, 1986). Even so, there is little evidence in Table 1 that researchers in the USA generally have grasped the extent and implication of those developments, because there is still a widespread adherence to large building block areas (viz. Counties), and to unreliable indicators (eg. density measures), and to procedures without any element of statistical optimisation.

At the same time, there has been little advance in purely statistical regionalisation methods (ie. the second type of approach identified above). It is in labour market area definition that some of this tradition's emphasis on optimisation has been taken forward. A recent review (Eurostat 1992) concluded that the European Regionalisation Algorithm (ERA) provided the most flexible and reliable form of local labour market area definition. Table 2 sets out the key requirements for these definitions: ERA was found to be the one method to fully meet these requirements. This hybrid method (ie. type 3 in the list above) is derived from the Functional Region definitions, was modified to identify the Travel-to-Work Areas shown in Map 2 - the only official statistical areas in Britain defined by scientific methods (Coombes et al, 1986) - and has subsequently been further developed and enhanced.

Yet even the ERA software, as the 'best practice' algorithm, cannot resolve some inherent limitations of definitions based on a single run on a single dataset. In the first place, of course, any single dataset will provide an inappropriate basis for definitions which will be used for many different purposes. Analyses of commuting flows may offer the most widely accepted guide to the general pattern of linkages which make up a locality, but commuting patterns alone can only be a poor proxy for some important aspects of a rounded approach to locality definitions (eg. allegiance to the

primary features of the method	current (1989) approach	alternatives in Dahmann & Fitzsimmons (1995)*			
		Adams	Berry	Frey & Speare	Morrill
'building blocks' [approx. equivalents in Britain]	Counties [Districts]	Counties	Zip Codes [Postcode Sectors]	Counties, with urban areas separate	Counties (Zip Codes to 'fine tune')
over-riding factors	New England analysed differently	State borders not to be crossed	consistency to replace ad hoc rulings	need more sophisticated procedure	evolution of current approach
criteria for urban centre	mainly population size and density	population density	housing unit density	centres not pre-defined	housing unit density
criteria for 'hinterland' assignment	commuting to central city above a pre-set level	density rather than commuting	part of the same media region; commuting	commuting clusters (not centres and hinterlands)	commuting to urban centre above a pre-set level
whether classifies whole country?	no	yes	yes	yes	yes
nature of other 'tiers' of areas	areas grouped in more urbanised regions	areas to be sub-divided by density	areas to be sub-divided in some regions	differentiated hierarchy	differentiated hierarchy
* the entries in this table are greatly simplified to focus on key differences					

Table 1 Alternative methods of US (non-)metropolitan area definition.

Principle	practice
OBJECTIVES	
1. Purpose	statistically-defined areas appropriate for policy
2. Relevance	each area an identifiable labour market
CONSTRAINTS	
3. Partition	every building block in 1 and only 1 area
4. Contiguity	each area a single contiguous territory
CRITERIA in descending priority	
5. Autonomy	self-containment of flows maximised
6. Homogeneity	areas' size range minimised (eg. within fixed limits)
7. Coherence	boundaries to be reasonably recognisable
8. Conformity	alignment with administrative boundaries preferable
SUMMARY	
9. Flexibility	method must perform well in very different regions

Table 2 Principles for local labour market area definitions.

historic identities represented by old county borders).

The second problem is that any regionalisation is only one run out of the many which were possible. No matter how good the regionalisation algorithm, the very best it can aim for is optimality in terms of a limited selection of criteria (eg. Table 2). These criteria will include a number of basic parameters - such as average, minimum and/or maximum size - and choosing alternative values on these parameters would lead to different results. With most algorithms (including ERA) there are also technical parameters which provide the mechanism by which the expertise and experience of the analyst contributes to produce more optimal results. In short, regionalisation procedure is far from deterministic. A key reason for this built-in 'flexibility' is the fact that any analysis of over 10,000 'building block' areas is likely to produce suboptimal (or even positively paradoxical) results in some areas. A set of slightly differing runs may well, between them, include an appropriate boundary for each part of the country, but each individual run will include some apparent anomalies. The traditional response has been to manually adjust the less appropriate boundaries of the single 'least worst' run while, over time, algorithm developments gradually restrict these rather unscientific inputs by reducing the proportion of areas whose results are paradoxical. Yet the advances in GIS and the move to fuzzy processing should help to liberate regionalisation from these limitations. The challenge is to utilise the new levels of processing power to move beyond the single 'least worst' run on a single dataset. As a response, the innovation here has been to split the whole regionalisation procedure into two phases:

• compile numerous analyses from numerous datasets, and then
• collate the results from these analyses within a single synthesis analysis.

Given that ERA provides an appropriate algorithm for the first phase of this strategy, it is the second phase which still calls for new technical developments.

The solution devised here centres on creating "synthetic data" which provides the basis for phase two of the method by using as input the initial, phase one, analyses. Each of these analyses produces a classification of all parts of the country (viz. the 10,529 wards (sectors in Scotland) in the 1991 Census). Such a classification identifies which of these 'building block' areas are grouped together as a single 'region' in this set of local labour market areas (or migration regions or whatever the classification represents). Thus the key information in each classification can be reduced to binary data by taking each pair of areas and identifying whether they are ("1") or are not ("0") classified into the same region. In this way, a classification list - which assigned a region number to each of the 10,529 areas - is re-expressed as a binary matrix of 10,529*10,529 cells (although the matrix is in fact symmetrical, so only half of it is needed). For example, if area B was in the same region as area C but in a different region to area D then the cell BC would take the value 1 while cell BD would be 0 (nb. cell CD would also take the value 0).

The crucial benefit from re-expressing each separate classification in this binary form is that these matrices can then be cumulated to produce the synthetic data needed. For example, if three 'input' analyses were collated in this way, the value in each cell of the synthetic data matrix would vary from 0 (for any pair of areas which were not in the same region according to any of those analyses) up to 3 (i.e. for any pair of areas which all three analyses had put in the same region). In GIS terms, it is analogous to layering the sets of boundaries on top of each other and counting the number of layers in which there is no boundary between each pair of areas. It can be seen that this approach provides an assessment of the 'strength of evidence' that two

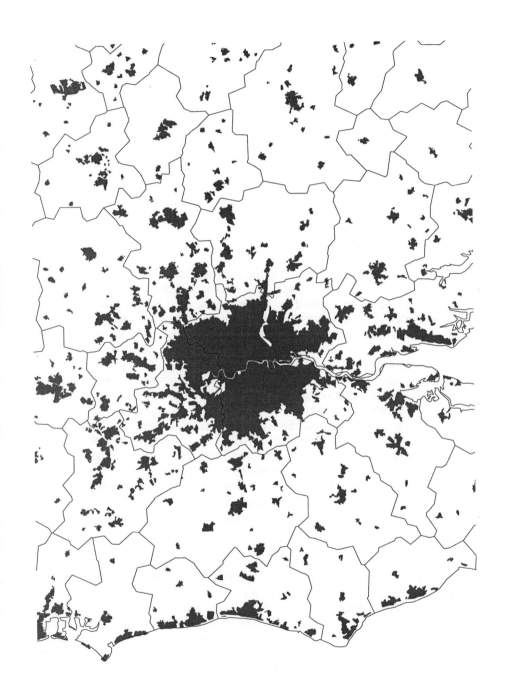

Map 2 Travel-to-Work Areas.

areas should be grouped together. The final synthetic dataset is, then, an ideal basis for the second phase of the definitional procedure - and it can be analysed with a version of ERA which has optimised for this purpose. Other forms of analysis of the synthetic dataset have also been evaluated.

The methodological innovation of creating synthetic data removes the technical limitations which arise from relying upon a single analysis of a single dataset. In particular, something of a 'fuzzy' approach can be adopted, by initially carrying out more than one form of analysis of the same dataset and accepting that each of these may provide a valid insight into the patterns lying within that dataset. Each of these analyses can then provide a separate input classification to the synthetic dataset, which will automatically identify those findings which these analyses have in common and also those which on they differ.

3. Application

For the substantive challenge, the huge benefit of the synthetic data method is the ability to draw upon analyses of different datasets. Virtually all previous regionalisations have centred on the analysis of a single dataset of flows between areas (most usually commuting flows, but sometimes migration). The synthetic data, however, can draw upon the evidence of many different sets of flows. As a result, localities can be defined not only from several analyses of the 1991 Census commuting and migration datasets, but also taking into account an analysis of a confidential dataset on the 'journey-to-bank' patterns of account holders with a major High Street bank. There are further advances here which the synthetic data draws upon:
- the commuting and migration datasets can also be disaggregated into distinct groupings so that, for example, the pattern of migration by older people can be considered separately and so is neither 'drowned out' nor 'dominating' the rather different patterns of other age groupings; and
- the synthetic dataset has also been enriched by taking as a further form of input a range of existing sets of boundaries - such as local authority areas - because these are also indicative of which areas might be better kept together and which kept separate.

In short, the methodological advance provided by creating the synthetic data, within the two phase regionalisation method, removed the limitations which are inherent in all 'single analysis of a single dataset' regionalisations. These limitations were partly technical, because so constrained an approach was almost certain to be sub-optimal. More importantly, however, the limitations were in terms of the inputs - the types of flow, and other forms of relevant evidence such as existing boundary sets - which could be drawn upon for the analysis. The new approach now shifts the emphasis to seeking and selecting the information which will be relevant as input to the definition of localities.

The first practical task was to collect together information on many different facets of localities. In the event, it was possible to bring a large volume of relevant information together - although it is not possible to claim this data necessarily constitutes a comprehensive coverage of the locality concept. Table 3 includes brief descriptions of the 39 sets of boundaries which have been compiled as phase one of the project. Several general points can summarise this huge undertaking:

- the disaggregation of the census commuting - see Maps 3 and 4 - and migration datasets provides a major bulwark against the analysis being dominated by, say, information on middle class men of middle age;
- the commuting analyses, for example, stem from two different types of ERA run to ensure that the results are less sensitive to any specific form of analysis;
- the banking zones were 'weighted' by being entered into the eventual synthetic dataset four times - this ensured that this rare evidence on people's 'journey-to-services' was not overwhelmed by the evidence on commuting or migration patterns (while also demonstrating another methodological option which this approach can support).

A critical problem facing the compilation of boundaries from different sources and eras is that their original definitions are in terms of incompatible building block areas. Table 3 shows that the project worked with four principal sets of building blocks - although this should not disguise the dependence here upon earlier work by the project team and colleagues (for example, Owen et al (1986) who compiled the majority of the boundary sets which were accessed in terms of 1981 SAS areas). The efforts of Atkins et al (1993) were also essential through their provision of the link between 1981 SAS areas and 1991 SMS areas. The linkage actually takes place at a lower level of area (the ED/OA) for which the nearest equivalent in the other period's set of areas is found.

4. Results

The result of this data compilation activity in phase 1 of the project has been the synthetic data (generated along the principles outlined in the preceding section of this paper). The ERA software can then be readily customised to focus on the inherent geographical structure in the synthetic data (as shown in Map 5 by the localised clustering of the larger 'flows'). The first step becomes grouping together all pairs of the c.10,000 'building block' areas which are grouped together in at least 60% of the input sets of areas.

The major part of the ERA software is an iterative adjustment process which ensures that all the localities which are eventually defined meet the specified criteria. In general, it is assumed that the objective is to maximise the number of separable areas capable of meeting these criteria - the aim is not to maximise the areas' values on these criteria, because this would lead to grouping areas which could be kept separate. The standard ERA approach is to set absolute minimum values on, say, two key concerns for that analysis. To be acceptable, an area must meet both minimum requirements and at least one of the higher 'target values' which have been set on each of the two key measures (or to be close enough to both 'target values' to pass a test for

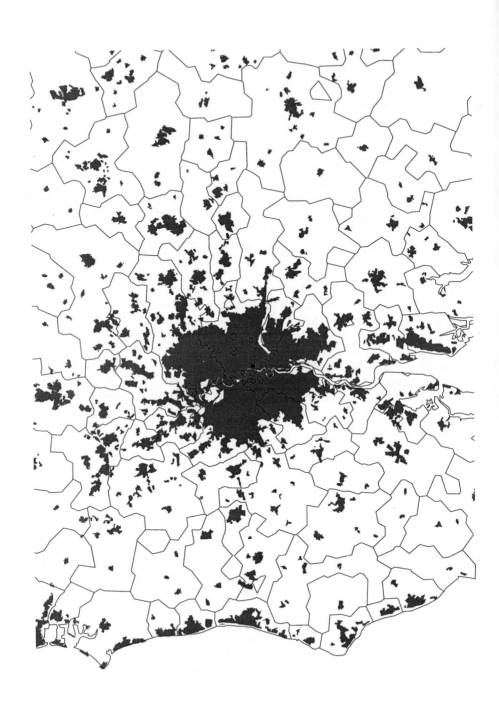

Map 3 Part-time worker local labour markets.

Map 4 Male worker local labour markets.

the 'trading off' of the value on one issue against that on the other). The standard ERA criteria are size of population and self-containment (i.e. the relative lack of linkage across the proposed locality's outer boundary).

Inputs' 'building block' areas	Official boundaries	boundaries derived from prior regionalisations
Postcode Sectors	Postcode Districts Postcode Areas	banking zones <nb. these were given more 'weight' than other inputs>
Census (1981) 'small areas'	Parliamentary Constituencies Enterprise Council areas Job Centre Areas Local Education Authorities Districts/Boroughs (of 1971) Counties (of 1951)	Functional Regions Functional Zones Local Labour Market Areas Metropolitan Regions Travel-to-Work Areas <nb. these are also the current Official boundaries for Regional Policy>
Census (1991) 'small areas'	Districts (of 1991) Counties/Scottish Regions (of 1991)	migration areas of: all people women men all of pensionable age adults 45-p.a. adults 30-44 adults 16-29 children 1-15 wholly migrating households
Census (1991) wards		local labour markets of: all workers women men part-time workers full-time workers car commuters public transport users other mode users commuting clusters of: all workers professionals/managers semi-professionals non-manuals skilled manuals no/low skilled manuals

Table 3 Inputs to the Synthetic Data for the initial application.

In this analysis, the population minimum was set at 7,000 (equivalent to that used by Coombes et al 1986), with the target value at 50,000 (cf. Coombes et al. 1982). The self-containment is measured in terms of the synthetic data: the indicator is assessing how for the proposed boundary is separating areas which were grouped together by many of the input sets of areas. Localities so defined (as illustrated in Map 6) do appear to conform to 'common sense' expectations that most such areas should generally include a main town - with perhaps some nearby satellite towns - together with its rural hinterland.

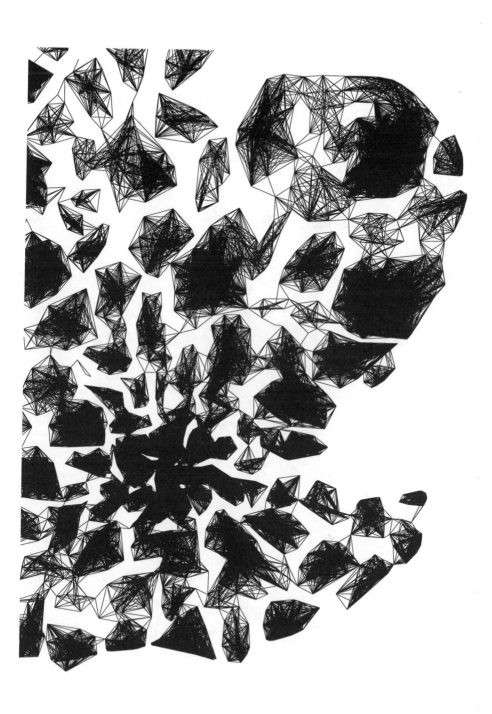

Map 5 Principal synthetic flows.

Map 6 Localities.

It is not coincidental that the number of Localities defined here (307) is quite similar to the 281 Local Labour Market Areas of Coombes et al (1982), the 322 Travel-to-Work Areas (TTWAs) which were defined with the 1981 Census data, and the number of local authorities which look likely to emerge from the current revisions. From a statistical viewpoint, it is around this level of breakdown at which the advantage of greater detail, from further sub-division of areas, begins to be outweighed by the disadvantage of reduced reliability and increased instability of data for areas with smaller populations (especially for sources whose samples are substantially less than 100%). From a more geographical viewpoint, further disaggregation tends to produce less comparable areas because smaller areas will inevitably be more often exclusively suburbs or rural areas or some other component of the wider urban-centred mixed areas which would be more meaningfully kept together for comparison with other integrated areas.

Of course, there is a nearly infinite number of different ways in which 10,529 building block areas can be combined, into approximately 300 groupings. Thus the numerical similarity between the Localities and other sets of areas is potentially misleading: a comparison of Map 6 with Maps 1 and 2 demonstrates that these boundaries are distinct from those earlier areas - even though they had formed part of the input to the synthetic data. It is not appropriate here to dwell on a description of the Localities' boundaries, so a few key points only will be made:
- unlike Functional Regions and TTWAs, the Localities provide a breakdown of Greater London (and one which, unlike the District boundaries, is consistent with the treatment of the other conurbations);
- unlike Districts and TTWAs, the Localities all tend to include at least one identifiable urban centre; and
- unlike all other sets of boundaries, the Localities are defined by reference to the latest Census evidence of interaction patterns - and also to a range of other evidence including the unique journey-to-service information of the banking zones.

5. Alternatives

A parallel line of analysis in the project took a rather different approach to the definition of area groupings. The 39 input classifications shown in Table 3 were translated into a matrix with 10,529 rows (ie. the 1991 Census areas) and 17,115 columns (ie. one for each 'region' in each of the classifications). Each cell was filled with "0" except, for each of the input classifications, a "1" was placed in the one column for the 'region' to which that row's 1991 area belonged. This immensely sparse matrix was then analysed by a Kohenen type of self-organising map (Openshaw 1994). This unsupervised neural net classifier was run on a Sunsparc 10/41 unix workstation. After 8 days' processing time it converged on a solution in which there are 214 area groupings (illustrated in Map 7). This form of analysis has shown that a radically different approach to regionalising the (unaggregated) synthetic data can produce a set of areas which are in fact quite similar to the Localities in many parts of the country.

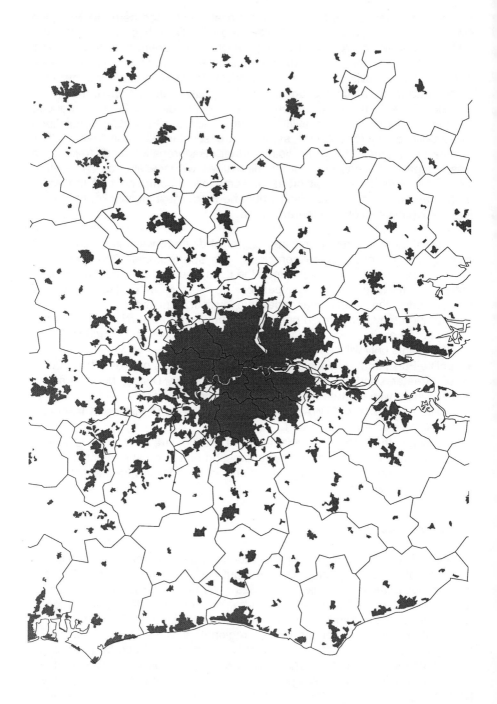

Map 7 Neuroclassification.

Further research along these lines can build on this software's advantages of objectivity and flexibility, while considering ways of ensuring that the resulting area groupings meet pre-specified objectives for the analysis. At the same time, there is every reason to believe that the aggregated synthetic dataset - used here for the ERA analysis - has tremendous potential for further forms of spatial analysis and for visualising spatial association.

At this stage, it seems clear that two main lines of research have been opened up by this research. The more obvious focus is on the new Locality definitions themselves: for example, documenting their relationship to other sets of areas (e.g. the emerging new local authority areas), and exemplifying their use for a diversity of social science enquiries. The other line of research to pursue flows out of the innovation of the synthetic data. Examples of research questions arising are:
- which of the input sets of boundaries was most influential in shaping the Locality boundaries which emerged from analysing this data;
- how would the synthetic data methodology cope with adding in non-contiguous classifications (e.g. area profiling systems such as those created in Openshaw, 1994);
- how sensitive are the results obtained from the synthetic data to this particular mix of input regionalisations (eg. what would be the effect of including some very much larger groupings such as Standard Regions); and
- are there other ways in which the Synthetic data can be analysed, given its distinctive features - such as the fact that it is a symmetrical matrix?

More fundamentally, perhaps, the potential of neural computing methods for spatial data handling has only been hinted at by the application outlined here.

References
D J Atkins, M E Charlton, D F L Dorling, C Wymer, 1993 "Connecting the 1981 and 1991 Censuses" NE.RRL Research Report 93/9 CURDS, University of Newcastle upon Tyne.

P H Cooke (ed.), 1989 Localities: the changing face of urban Britain Unwin Hyman, London.

M G Coombes, J S Dixon, J B Goddard, S Openshaw, P J Taylor, 1982 "Functional Regions for the population census of Britain" in D T Herbert & R J Johnston (eds.) Geography and the urban environment (vol.5) Wiley, London.

M G Coombes, A E Green, S Openshaw, 1986 "An efficient algorithm to generate official statistical reporting areas: the case of the 1984 Travel-to-Work Areas revision in Britain" Journal of the Operational Research Society 37, p.943-953.

D C Dahmann, J D Fitzsimmons (eds.), 1995 "Metropolitan and nonmetropolitan areas: new approaches to geographical definition" Population Division Working Paper 12 US Bureau of the Census, Washington DC.

Eurostat, 1992 Study on employment zones (E/LOC/20) Eurostat, Luxembourg.

S Openshaw, 1984 The modifiable areal unit problem GeoBooks, Norwich.

S Openshaw, 1994 "Neuroclassification of spatial data" in BC Hewitson RG Crane (eds.) Neural nets: applications in geography Kluwer, Boston (Mass).

D W Owen, A E Green, M G Coombes, 1986 "Using the social and economic data on the BBC Domesday interactive videodisc" Transactions of the IBG NS11, 305-314.

Contact address

Mike Coombes
NE.RRL (NorthEast Regional Research Laboratory)
CURDS (Centre for Urban & Regional Development Studies)
Newcastle University
Claremont Bridge
Newcastle upon Tyne
NE1 7RU UK
Phone: +44-191-2228014
Fax: +44-191-2329259
E-mail: Mike.Coombes@ncl.ac.uk

A QUALITATIVE MODEL OF GEOGRAPHIC SPACE

Thomas Bittner

Department of Geoinformation
Technical University of Vienna
Vienna, Austria

Abstract
In this paper a qualitative model of geographic space is proposed. The model consists of two types of primitives, covers and regions, and relations between them. A cover is a set of regions that form a partition of space. Spatial entities are represented by regions. The primitives correspond either to the field view or to the entity-oriented view of geographic space. The primitives and relations between them are formalized in terms of topology. The relation of a region with respect to a cover is represented formally in terms of topological relations between the region and elements of the cover. This set of relations can be considered as a qualitative representation of a region within a cover. This representation qualitatively specifies the location of a spatial entity, abstracted by a region, within its environment, abstracted by a cover.

Key words
Qualitative spatial reasoning.

1. Introduction

In GIS natural, physical, biological, or social phenomena are considered regarding their relation to their spatial environment. Those phenomena for example are temperature, soil type, soil wetness, and slope of terrain. Laurini & Thompson characterize the treatment of such phenomena in GIS in the following way (Laurini and Thompson 1994): Natural, physical, biological phenomena are often continuously variable and continuously distributed over space. Social phenomena may be considered as spatially continuous at a particular scale. These phenomena are usually treated by the field view of space. Continuous distribution of phenomena over space is the principal element of this view of space.

> "...areal units are delimited somehow so that boundaries can be drawn for polygonal entities, producing land use zones, and the like. The attribute information is carried by the area elements, and the boundaries reflect varying assumptions of change between adjacent units. A sharp break or discontinuity is generally assumed, without any possibility of providing a mathematical statement for the differences, and varying assumptions are made about variation within the units" (Laurini and Thompson 1994) pg. 89.

The need for a sharp break or discontinuity at the boundaries is due to the representation of the boundaries by means of geometric objects, that are formalized analytically in terms of Euclidean geometry. Analytical Euclidean geometry is an abstract formal system that is based on measurement along coordinate axes. Abstract spaces are no real spaces. They only exist in our minds. Sometimes there exists a correspondence between abstract and real space, for example between Euclidean space and table-top spaces (Nunes 1991). There does not exist a correspondence between the natural phenomena in geographic space (Mark and Frank 1995) and abstract geometric primitives in Euclidean space (Couclelis 1992). This becomes obvious in the contradiction between the assumption of continuity of distribution over space and the 'need' for discontinuities or sharp changes at the boundaries that have to be describable in terms of geometric primitives.

For the formalization of natural phenomena in geographic space we need theories of spaces that do not constrain their models to sharp bounded geometric objects. In this approach topological spaces are used for the formalization. Topological spaces allow us to represent boundaries without forcing them into geometric primitives by declaring metric positions.

Similar problems occur if we apply the entity-oriented view to natural phenomena in geographic space. In this view space is conceptualized as a set of entities that correspond to phenomena described by a pattern of attributes. The spatial extend of an entity is considered as one of its attributes. If we represent for example an area of pollution in terms of geometric primitives using analytical formalization of Euclidean geometry we are required to draw sharp boundaries where in nature are continuous changes and/or the precise metric position of the boundary is unknown. A similar problem is to decide where a valley ends and a mountain begins (Couclelis 1992). In nature there are neither entities nor fields. The distinction between different views of space is due to different conceptualizations of phenomena in geographic space made by men (Couclelis 1992). How geographic space is conceptualized depends for example on the geographic scale of the space under consideration (Zubin 1989), the purpose of the conceptualization (Couclelis 1992), and the nature of phenomena (examples above). GIS can be seen as the implementation of a formal theory about geographic space (Mark and Frank 1995). In this theory one should take the different conceptualizations of geographic space into account. Since the phenomena in the geographic world are interrelated independently of their conceptualization by men, models based on their different conceptualizations should be interrelated too. Therefore the representations of both views should be related to each other and formalized in a unified theory. In this paper I propose a representation scheme that incorporates both views of geographic space. The formalisation is based on topology and results in a qualitative representation (Freksa 1991). Since this representation does not rely on coordinates and geometric primitives, there is no need to measure and to represent unknown metric boundary positions.

I concentrate on problems where the application of the field view results in a regional subdivision of space. Thus the application of the field view results in a set of regions that are either disjoint or meet each other, which is called a *cover*. A *cover* corresponds to the concrete spatial structure of a particular environment, extracted by application of the field view on particular aspects of this environment. The spatial extend of entities is abstracted by non-overlapping regions. Relations between a *cover* and a region are defined qualitatively in terms of topology. This relations topologically specify the location of a region with respect to a structured set of regions - the *cover* .

This approach differs from other formalizations of geographic space in the following points:

*   This approach is based on topology. Thus it is qualitative and no (coordinate) measurement is involved.
*   The cover structure that corresponds to a field view of a particular environment is used as a reference system for the representation of knowledge about spatial entities within this environment. Thus spatial entities are represented formally within a concrete space rather than within an abstract space.
*   Qualitative spatial reasoning usually takes abstract relational structure of the spatial domain into consideration (i.e. topological relations, conceptual neighborhood) (Egenhofer 1991; Frank 1991; Freksa 1991; and Freksa 1992; Hernandez 1994) rather than the concrete structure of a particular environment, for example, in terms of a *cover* structure (Bittner 1995).

The paper is organized as follows: In the second section the conceptual model is introduced. In the third section basic point set topological notions and binary topological relations between regions are explained. Covers, regions and relations between them are formalized in the fourth section. A summary and examples are given in the last sections.

2. The conceptual model

The conceptual model is intended to cover some aspects of geographic space and its conceptualization that are relevant for GIS. In this model geographic space is modeled as a set of two types of primitives. Instances of the primitives of the model are created by the conceptualization of phenomena in geographic space. Corresponding to the two views of geographic space the two basic types of primitives in this model are (Figure 4):

*   Entity-regions
    The spatial extend of a geographic entity is represented by one region or more regions that may have holes but do not overlap. The representation of geographic entities by regions is due to the assumption that all objects in real space have an extension in three dimensions which projections on Earth's surface result in regions.
*   Cover
    The field view of a particular attribute varying continuously over space is represented by a set of regions that either meet each other or are disjoint and form a partition. This set of regions is called a *cover*. Within each region the attribute values belong to the same equivalence class in the attribute domain. The regional partition of space corresponds to the equivalence classes in the attribute domain.

Instances of these primitives are objects in that sense that they are characterized by the prototypical object qualities of discrete identity, relative permanence of structure and attribute, and the potential of being manipulated. The important point is that not the soil-structure of a particular environment is the object, but the cover that represents a particular view of the soil structure is the object.

In this model geographic space is considered as being relative (Nunes 1991). Thus it consists of primitives and relations between them.

The geographic space is modeled by :

- a set of covers, each cover corresponding to the field view of one attribute varying continuously with respect to Earth' surface;
- a set of entity-regions, corresponding to spatial entities;
- relations between covers;
- relations between a cover and entity-regions;
- and relations between entity-regions.

Geographic phenomena that can be represented by a cover are for example soil classifications, political boundaries, ownership... . Buildings, parcels, pollution areas, lakes, ... can be represented as entity-regions. Relations between covers correspond to results of overlay operations in current coordinate based GIS. Relations between a cover and an entity-region are for example containment of an entity in a region of the cover ('Vienna lies in Austria' where Austria belongs to the cover 'Countries in Europe'). These relations specify location. Relations between entity-regions are for example the binary topological relationships (e.g. disjoint, meet,...). In Euclidean geometry based GIS covers are represented by layers of polygons, where regions are represented by polygons which consist of boundary-segments that are shared by neighboring polygons.

3. Point set topology and binary topological relationships

The primitives of the model and the relations between them are defined in terms of point set topology. Their formal definition is based on the point set topological notions of interior and boundary (Spanier 1966). The definitions are based on the notion of open sets. The interior of a point set Y, denoted by $Y°$, is the union of all open sets that are contained in Y. The *closure* of Y, denoted by $\bar{Y}$, is defined by the intersection of all closed sets that contain Y. The *boundary* of Y, denoted by δY, is the intersection of the closure of Y and the closure of the complement of Y. A *separation* of Y, $Y \subseteq X$, is a pair A, B of subsets of X satisfying the condition $A \neq \emptyset$ and $B \neq \emptyset$; $A \cup B = Y$; and $\bar{A} \cap B = \emptyset$ and $A \cap \bar{B} = \emptyset$. Y is said to be disconnected, otherwise Y is said to be connected. A *region* is a non-empty connected set in $\mathbf{R}^2$ (Egenhofer and Herring 1990).

Topological Relationships between regions in $\mathbf{R}^2$ were formalized by Egenhofer and Herring (Egenhofer and Herring 1990). The topological invariance of the intersection of the boundary and the interior of two regions are used to define topological relationships between them. Egenhofer and Herring show that from the 16 combinatorically possible intersection pattern only eight can exist between two regions with codimension 0. These are geometrically interpreted and identified with the topological distinguishable relationships between two regions without holes in $\mathbf{R}^2$. The eight topological relations are : disjoint, meet, equal, inside, covered by, contains, covers and overlap. In this approach a similar technique is applied to define topological relationships between a cover and an entity-region.

4. Relations between covers and regions

4.1 The cover-structure

4.1.1 The definition of cover-structures

Let *cover* be a set of regions in $\mathbf{R}^2$ forming a finite regional partition. $\mathbf{R}^2$ is covered by the union of all cover-regions and the complement of this union. A region A is a non-empty connected set of points. It consists of two disjoint subsets: the interior, denoted by A° and the boundary, denoted by δA, such that A=A°∪δA and A°∩δA=∅. All regions that belong to a cover are either disjoint or meet each other, where disjoint means that they do not have points in common and meet means that they share common boundary points.

The boundary δA of each region consists of one or more than one *boundary segments*, denoted δA#. Each boundary segment is a connected subset of the boundary of a region. A boundary segment δA$_B$ is the non-empty intersection of the boundaries of two adjacent regions A and B. There are boundary segments which belong to the outer boundary of the *cover*, denoted δA$_\varnothing$. They are shared with the complement of the *cover*. δA# denotes an arbitrary boundary segment δA#⊆ δA (either δA$_B$ or δA$_\varnothing$). If there exist more than one separated boundary segments between two neighboring regions or a region and the complement of the *cover*, then they are uniquely numbered.

A *cover-structure* is the representation of a *cover*. It is the set of all pairs of boundary segments and the corresponding interior of a region that belong to the cover. Thus for each boundary segment δA$_B$=δB$_A$ two pairs (A°,δA$_B$) and (B°,δB$_A$) exist within the *cover-structure*.

Definition 1 (*cover-structure*) :

Let *cover* be a finite regional partition, A∈ *cover*, and (A°,δA#) a pair, called interior-boundarysegment-pair , where

- A° denotes the interior of the region A,
- A# denotes a boundary segment of the region A.

A *cover-structure* is the set of all pairs (A°,δA#) definable for regions of a cover:

cover-structure =$_{Def}$ {(A°,δA#) I A ∈ *cover* & δA#⊆δA }.

4.1.2 Cover-structures vs. cell complexes

Cover-structures and cell-complexes (Alexandroff 1961; Frank and Kuhn 1986) are both formalizations of regional partitions of space. Cell-complexes have a simple and regular structure. Due to the regular structure of cell-complexes in general no one to one correspondence between the cells and units of the conceptual model can be established. Thus meaningful units have to be approximated by sets of cells. The elements of the cover-structures directly correspond to the smallest meaningful units of the conceptual model : boundary segments and interior of regions that constitute a cover. Equivalent to the point set topological definition the *cover-structure* could be defined in terms cell-complexes based on algebraic topology.

4.2 Relations Between a Cover and a Region

Entity regions in this formalization are single regions without holes. In Definition 1 pairs $(A°, \delta A_\#)$ are used as primitives of the *cover-structure*. The elements $A°$ and $\delta A_\#$ refer to subsets of a region A. A belongs to the *cover*. The region C represents a spatial entity which consists of the union of the two disjoint subsets: interior and boundary, $C = C° \cup \delta C$. The topological relations between a *cover* and an entity-region are defined in terms of topological relations between sets denoted by pairs $(A°, \delta A_\#)$ of the *cover-structure* and the entity-region C. The topological relation between a pair $(A°, \delta A_\#) \in$ *cover-structure* and an entity-region C is defined in terms of the content of the intersection of the following subsets:

- emptiness / non-emptiness of the intersection of the interior of A and the interior of C : $A° \cap C°$,
- emptiness / non-emptiness of the intersection of the boundary segment $\delta A_\#$ of A and the interior of C : $\delta A_\# \cap C°$,
- emptiness / non-emptiness of the intersection of the interior of A and the boundary of C : $A° \cap \delta C$,
- emptiness / non-emptiness of the intersection of the boundary segment $\delta A_\#$ of A and the boundary of C : $\delta A_\# \cap \delta C$.

Since we do not consider the whole boundary δA of A, we cannot handle C as a whole. If we intersect $(A°, \delta A_\#)$ and C then a number of separated subsets of C can be created (Figure 1). We define a *connected intersection* $\underline{C}$ of C as a connected subset of C which is created by the intersection of C with the interiors and the shared boundary segment of two neighboring *cover*-regions. The set of all connected intersections $\underline{C}$ of C that can be created by the intersection of C with neighboring *cover*-regions or with a *cover*-region and the adjacent complement of the *cover* is called *connected intersection set* of C, denoted by C *. Each connected intersection $\underline{C}_i \in C^*$ consists of the union of two disjoint subsets $\underline{C}_i°$ and $\delta \underline{C}_i$, where $\underline{C}_i° \subseteq C°$, and $\delta \underline{C}_i \subseteq \delta C$.

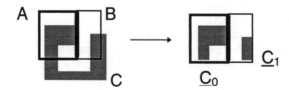

$\underline{C}_0, \underline{C}_1 \in C^*$

Figure 1 Connected intersections $\underline{C}_0$ and $\underline{C}_1$ of a region C that are created by the intersection of C with the interior of two neighboring regions A and B and their common boundary segment $\delta A_B = \delta B_A..$

We define the topological relations between a cover and an entity-region as the set of topological relations between the elements of the connected intersection set of the entity-region C, $\underline{C}_i \in C^*$, and the elements $(A°, \delta A_\#)$ of the *cover-structure*. Before we can define this set of topological relations formally we need to define the topological relations between a single connected intersection and a single interior-boundarysegment-pair.

4.3 The Definition of CR-Relations

The topological relations between a pair $(A°, \delta A_\#) \in$ *cover-structure* and a connected intersection $C_i \in C^*$ of an entity-region C are defined in terms of the content of the intersection of the four sets $A°$, $\delta A_\#$, $C_i°$, and δC_i. There are 16 combinatorically possible intersection pattern (Table 1). These pattern are used for the definition of the topological relationships:

number	$\delta A_\# \cap \delta C_i$	$\delta A_\# \cap C_i°$	$A° \cap \delta C_i$	$A° \cap C_i°$	classification
∅	∅	∅	∅	∅	CR-empty
1	¬∅	∅	∅	∅	R3
2	∅	¬∅	∅	∅	L0
3	¬∅	¬∅	∅	∅	L1
4	∅	∅	¬∅	∅	L2
5	¬∅	∅	¬∅	∅	L3
6	∅	¬∅	¬∅	∅	L4
7	¬∅	¬∅	¬∅	∅	L5
8	∅	∅	∅	¬∅	L6
9	¬∅	∅	∅	¬∅	E0
10	∅	¬∅	∅	¬∅	E1
11	¬∅	¬∅	∅	¬∅	E2
12	∅	∅	¬∅	¬∅	R0
13	¬∅	∅	¬∅	¬∅	R1
14	∅	¬∅	¬∅	¬∅	E3
15	¬∅	¬∅	¬∅	¬∅	R2

Table 1 The intersection table.

The intersection pattern can be divided into four classes. The pattern in class one, {L0...L6}, do not exist for non-holed entity-regions in 2D (Bittner 1996)(Lemma1,Lemma2). The second class, {E0...E3}, contains the pattern that imply that the boundary segment $\delta A_\#$ under consideration is a proper subset of the connected intersection C_i. The pattern E1 and E3 imply that the boundary segment $\delta A_\#$ is a proper subset of the interior of the connected intersection C_i. The third class of pattern {R0,R1,R2,R3} imply a non-empty intersection between $(A°, \delta A_\#) \in$ *cover-structure* and $C_i \in C^*$. The boundary segment $\delta A_\#$ is not contained within the connected intersection C_i. Class four contains the pattern that corresponds to the empty intersection between $(A°, \delta A_\#)$ and C_i and is called CR-empty. In Figure 2 possible geometrical realizations of the relations {R0..R3,E0..E3} are shown.

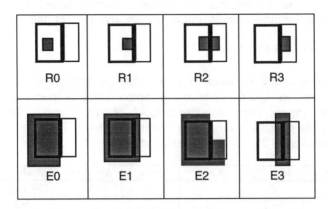

Figure 2 Geometric Interpretation of the CR- relations. The relations hold for the intersection of the dark entity-region C and the interior of the two cover-regions that meet each other (A denotes the left and # denotes the right region) and their common boundary ($\delta A_\#$) .

Now we can express the topological relations between a pair $(A°,\delta A_\#) \in$ *cover-structure* and a connected intersection $\underline{C}_i \in C^*$ in terms of CR- relations:

Definition 2 ((C)over element - entity (R)egion - relations) :

The topological relationships, between a pair $(A°,\delta A_\#)$, that denotes the interior and one boundary segment of a cover-region A, and a connected intersection $\underline{C}_i \in C^*$ of an entity- region C, that correspond to the realizable pattern of the content of the intersection of the sets $A°$, $\delta A_\#$, $\underline{C}_i°$, and $\delta\underline{C}_i$ (Table 1), are called *CR-relations*:

CR-relations $=_{Def}$ { R0, R1, R2, R3, E0, E1, E2, E3, CR-empty }.

A CR-relation is a binary relation that holds between a pair $(A°,\delta A_\#)$ and a connected intersection $\underline{C}_i \in C^*$:

$<(A°,\delta A_\#), \underline{C}_i > \in$ CR-relations $=_{Def} <(A°,\delta A_\#), \underline{C}_i > \in R$ & R$\in$ CR- relations

The set of all CR-relations that can be defined for an entity-region with respect to a *cover* represents the topological relationship between the *cover* and the entity-region. This set of relations specifies the location of the entity-region C within the *cover-structure*. Each of these sets of CR-relations corresponds to the equivalence class of all regions that have topologically equivalent relations to the same elements of the *cover-structure*. They have equivalent location within the *cover*. Consequently we have a n-to-1 mapping between entity-regions and CR-relation sets (location) with respect to a *cover-structure*.

4.4 Representation of the location of regions with respect to cover-structures

Now we apply the notion figure and ground (Talmy 1983) from Linguistics to entity regions and cover-structures. Entity regions are identified with the figures and a cover-structure is considered as a representation of the ground. Let us now define a structure called figure-ground-position:

Definition 3 (figure-ground-position) :
    A figure-ground-position is a triple ($(A°,\delta A_\#)$, $\underline{C}_i$, R) where
- $(A°,\delta A_\#)$ denotes the interior and a boundary segment of a region which belong to a cover,
- $\underline{C}_i$ denotes a connected intersection $\underline{C}_i \in C^*$ of an entity region C,
- R denotes the CR-relation which holds between the region-parts denoted by $(A°,\delta A_\#)$ and $\underline{C}_i$:
        $<(A°,\delta A_\#), \underline{C}_i > \in$ R , R∈ CR- relations;

and sets of figure-ground-positions, called figure-ground-location:

Definition 4 (figure-ground-location) :
    The set LOC $_{cover}$(C) of figure-ground-positions ($(A°,\delta A_\#)$, $\underline{C}_i$, R) between a cover of regions (*cover*) and a entity-region C , that has the following properties :
        LOC $_{cover}$ (C) = { ($(A°,\delta A_\#)$, $\underline{C}_i$, R) | $(A°,\delta A_\#) \in$ *cover-structure* & $\underline{C}_i \in C^*$ & R ≠ CR-empty }
    is called figure-ground-location of the region C within the *cover*.

If the identity of the *cover* is clear the figure-ground-location of the entity-region C with respect to the *cover* is denoted by LOC(C).

4.5 Binary topological relations and CR- relations

From a single CR-relation between a pair $(A°,\delta A_\#)$ and a connected intersection $\underline{C}_i \in C^*$ of a region C, $<(A°,\delta A_\#)$, $\underline{C}_i > \in$ R, R∈{R0...R3,E0...E3, CR-empty}, in general no unique conclusions about the topological relations between the regions A and C can be drawn. The reason therefore is that only a segment of the boundary of the region A namely $\delta A_\#$ is considered. We have to split the region C in connected intersection sets. This has the advantage that the results we get are more specific with respect to the boundary segment explicitly considered. As a consequence the results we get are less specific with respect to the whole regions A and C. In the special case $\delta A_\# = \delta A$ we can handle C as a whole and we get the intersection table of the 4-intersection and thus the topological relations between two (non-holed) regions. This happens for example if the region A is a hole in the region B. C is assumed to be non-holed.

CR- intersection	4-intersection
$<(A°,\delta A_B)$, C> ∈ R0	<C,A> ∈ inside
$<(A°,\delta A_B)$, C> ∈ R1	<C,A> ∈ covered by
$<(A°,\delta A_B)$, C> ∈ R2	<A,C> ∈ overlap
$<(A°,\delta A_B)$, C> ∈ R3	<A,C> ∈ meet
$<(A°,\delta A_B)$, C> ∈ CR-empty	<A,C> ∈ disjoint
$<(A°,\delta A_B)$, C> ∈ E0	<A,C> ∈ equal
$<(A°,\delta A_B)$, C> ∈ E1	<C,A> ∈ contains
$<(A°,\delta A_B)$, C> ∈ E2	<C,A> ∈ covers
$<(A°,\delta A_B)$, C> ∉ E3 (C non-holed)	does not exist

Table 2 Correspondence of CR-intersection and 4-intersection for the case $\delta A_\# = \delta A$.

In the general case $\delta A_\# \subset \delta A$ we do not know whether or not C intersects the rest of the boundary $\delta A \backslash \delta A_\#$ because it is not represented in the pattern of a single CR-intersection. The three cases in Figure 3 are not distinguishable in terms of a single CR-relation.

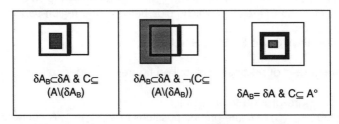

$\delta A_B \subset \delta A$ & $C \subseteq (A \backslash (\delta A_B))$ $\delta A_B \subset \delta A$ & $\neg(C \subseteq (A \backslash (\delta A_B)))$ $\delta A_B = \delta A$ & $C \subseteq A^\circ$

Figure 3 Three configurations of an entity-region C (■) and a cover-region A (□) that are not distinguishable in terms of the CR-relation : $<(A^\circ, \delta A_B), C> \in R0$.

In order to get unique results, we have to consider all boundary segments of the regions under consideration. If we want to calculate the topological relations between A and C, we have to consider the subset of figure-ground-positions of LOC(C), which elements refer to the interior of A. The topological relations between the cover-region A and the entity-region C can uniquely be derived from this set of figure-ground-positions   (Bittner 1996)(Lemma 3).

4.6 Example

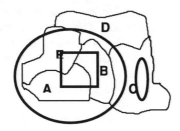

Figure 4 A cover of regions { A, B, C, D, E } and entity-regions □ , ○ , and ()

cover   = { A, B, C, D, E }
cover-structure = { $(A^\circ, \delta A_\varnothing)$, $(A^\circ, \delta A_B)$, $(A^\circ, \delta A_E)$, $(B^\circ, \delta B_\varnothing)$, $(B^\circ, \delta B_C)$, $(B^\circ, \delta B_D)$, $(B^\circ, \delta B_E)$,
                  $(B^\circ, \delta B_A)$, $(C^\circ, \delta C_\varnothing)$, $(C^\circ, \delta C_B)$, $(C^\circ, \delta C_D)$, $(D^\circ, \delta D_\varnothing)$, $(D^\circ, \delta D_B)$, $(D^\circ, \delta D_C)$,
                  $(D^\circ, \delta D_E)$, $(E^\circ, \delta E_\varnothing)$, $(E^\circ, \delta E_A)$, $(E^\circ, \delta E_B)$, $(E^\circ, \delta E_D)$ }

□ = F, LOC(F) = { $((A^\circ, \delta A_E), F, R2)$, $((A^\circ, \delta A_B), F, R2)$, $((A^\circ, \delta A_\varnothing), F, R0)$, $((B^\circ, \delta B_\varnothing), F, R0)$,
                  $((B^\circ, \delta B_C), F, R0)$, $((B^\circ, \delta B_D), F, R0)$, $((B^\circ, \delta B_E), F, R2)$, $((B^\circ, \delta B_A), F, R2)$,
                  $((E^\circ, \delta E_D), F, R0)$, $((E^\circ, \delta E_\varnothing), F, R0)$, $(E^\circ, \delta E_A), F, R2)$, $((E^\circ, \delta E_B), F, R2)$ }

() = G, LOC(G) = { $((C^\circ, \delta C_B), G, R0)$, $((C^\circ, \delta C_D), G, R0)$, $((C^\circ, \delta C_\varnothing), G, R0)$ }

○ = H, LOC(H) = { $((A^\circ, \delta A_\varnothing), H, E1)$, $((A^\circ, \delta A_B), H, E1)$, $((A^\circ, \delta A_E), H, E1)$, $((B^\circ, \delta B_\varnothing), H, E1)$,
                  $((B^\circ, \delta B_C), H, E1)$, $((B^\circ, \delta B_D), H, E1)$, $((B^\circ, \delta B_E), H, E1)$, $((B^\circ, \delta B_A), H, E1)$,
                  $((C^\circ, \delta C_\varnothing), H, R2)$, $((C^\circ, \delta C_D), H, R2)$, $((C^\circ, \delta C_B), H, E3)$, $((D^\circ, \delta D_\varnothing), H, R0)$,
                  $((D^\circ, \delta D_E), H, R2)$, $((D^\circ, \delta D_B), H, E3)$, $((D^\circ, \delta D_C), H, R2)$, $((E^\circ, \delta E_\varnothing), H, R2)$,
                  $((E^\circ, \delta E_A), H, E3)$, $((E^\circ, \delta E_B), H, E3)$, $((E^\circ, \delta E_D), H, R2)$ }

The *cover* in Figure 4 consists of the cover-regions A, B, C, D, E. This *cover* for example could correspond to the soil-structure of a particular environment. The *cover* is formally represented by the *cover-structure*. Each boundary segment is a representation of a boundary between two neighboring areas of different soil-types. Each interior refers to an area which soil-type is assumed to be homogenous with respect to a particular classification. One can see that there is redundant information in the set of figure-ground-positions. This is due to symmetric properties of some CR-relations.

F, G, and H correspond to phenomena conceptualized as entities. Let G be a lake, H a forest and F an area of pollution. H could also be element of a *cover* 'vegetation-type'.

Each figure-ground-position within a figure-ground-location LOC() specifies the location of the entity represented with respect to one interior-boundarysegment-pair. From LOC(G) it can easily be derived that the lake G is totally contained in the region corresponding to the area of soil type C. The forest-area H has parts that are outside the *cover* (e.g. $((A°,\delta A_\varnothing),H,E1) \in$ LOC(H)). For this parts we do not know the soil type. There are positions within the figure-ground-location of H that have the CR-relation E1: for example $((B°,\delta B_A),H, E1)$. This indicates that the boundary between soil type B and A runs somewhere inside the forest H. In $((B°,\delta B_A),H, E1)$ is further encoded that the boundary of the forest H does not cross the soil-area B. The forest H contains the soil-type-areas A and B: A *cover* region is contained in an entity-region if all figure-ground-positions of the figure-ground-location of the entity-region that refer to the *cover* region, contain a CR-relation that indicates that the whole boundary segment of the *cover* region is contained in the entity-region.

Both entity-regions G and H intersect the *cover* region C. Thus both figure-ground-locations contain figure-ground-positions that refer to the interior of the *cover* region C: $((C°,\delta C_B),H, E3)$ and $((C°,\delta C_B),G,R0)$. We cannot further specify the relations between G and H since no further information is available. It may be that the lake lies within the forest or outside the forest, or the edge of the forest runs along one bank of the lake, ...

From their figure-ground-locations we can derive that F and G must be disjoint and that F and H do intersect. The fact that F and G are disjoint implies that the pollution has not reached the lake yet. Whether or not it can reach it may depend on the soil-types B and C and the boundary between them.

5. Summary and ongoing work

The point of this paper is that a figure-ground-location formally and qualitatively characterizes the location of a geographic entity within its concrete environment. The emphasis on a concrete environment is due to the concern of geography with real spaces that exist on earth and are meaningful for human beings rather than on abstract spaces existing only in human minds. Since 'The Environment' as a whole can neither be conceptualized nor be processed on a computer, the field view of 'The Environment' under consideration is applied in order to extract relevant aspects and their spatial structure. In this particular approach we concentrate on problems where the application of the field view results in a regional partition of space. Since we further concentrate on natural phenomena, we only consider topological structural aspects. This is due to the indeterminate character of the boundaries established by the application of the field view of natural phenomena of space. Topology does not force us to specify unknown metric positions. We model the resulting structure by a set of regions that are either disjoint or meet each other and call it *cover*. We use point set topological notions of interior and boundary of a region to formalize *covers*. The formalization of a *cover* results in a *cover-structure*. A *cover-structure* consists of a set of interior-boundarysegment-pairs. A boundary segment is a connected subset of the boundary of a region that belongs to a *cover*. Each boundary segment is shared either with a neighboring *cover* region or with the complement of the whole *cover*.

A geographic entity is represented by a set of non-overlapping regions. For simplification in this formalization only single, none-holed regions are considered.

Location is the relation between a *cover* and an entity-region. It is formalized as a set of relations of parts of the region and parts of elements of the *cover* denoted by elements of the *cover-structure*. This set of relations is called figure-ground-location. The elements of a figure-ground-location are called figure-ground-positions. A figure-ground-position is a triple connecting an element of the *cover-structure* $(A°\delta A_\#)$, an element of the connected intersection set of the entity-region C, $\underline{C}_i \in C^*$ and the CR-relation that holds between the sets that are referred by $(A°\delta A_\#)$ and $\underline{C}_i$. The set of CR-relations covers the whole domain of topologically (in terms of the content of the intersection of subsets) distinguishable configurations that can occur between the interior of a region, a boundary segment of this region and a non-separated subset of another region.

The cover is finite. Thus the *cover-structure* is finite. The set of CR-relations is finite too. We use point set topology based on infinite abstract point sets only for the formal definition of the primitives and the relations between them. The resulting formal structures used for representation are finite sets and each element is directly related to an element of the conceptual model. On this level we only process sets of tuples representing *cover-structures* or representing relations between regions and *cover-structures*.

Thus we have a tuple-based representation of geographic space that takes the concrete topological structure of a particular environment under consideration into account. The qualitativeness of the representation is due to the definition of the CR-relations as topologically equivalent configurations.

Advantages of figure-ground-locations with respect to their application in GIS are:

- The representation consists of sets of tuples which can be easily stored in tables of relational database systems. Reasoning can be formalized in terms of operations on these sets of tuples and thus be transformed to relational algebra (Codd 1970) and implemented for example in SQL.
- Figure-ground-locations can be easily derived from geometric representations or from a small set of samples.
- Spatial and non-spatial data can be directly related to each other without reference to implicit geometric data structures.

A limit of figure-ground-locations is that they only preserve a subset of the topological relations between the entity-regions depending on the structure of the cover (example : relations between regions G and H cannot be specified).

Ongoing work will further formalize the conceptual model. Therefore topological relations between different *cover-structures* will be formalized. The composition of these relations and CR-relations will be used to transform figure-ground-locations between different covers. Furthermore operations on figure-ground-positions will be used to derive topological relations between the corresponding regions. This will be compared and combined with techniques based on relation composition.
If the structure of the cover follows a regular pattern (e.g. the pattern proposed in (Frank 1991) for direction reasoning) then topological reasoning and reasoning with cardinal direction becomes integrated in the same framework.

Acknowledgments
I gratefully thank Andrew Frank for valuable comments and review of earlier versions of this paper.

References
Alexandroff, P. (1961), New York, NY, Dover Publications.
Bittner, T. (1995), Levels of Spatial Inference, Department of Geoinformation, TU-Vienna
Bittner, T. (1996), A Formal Theory about Covers and Regions, Department of Geoinformation, TU-Vienna.
Codd, E. F. (1970), "A Relational Model of Data for Large Shared Data Banks" 13(6): 377 - 387.
Couclelis, H. (1992), Theories and Methods of Spatio-Temporal Reasoning in Geographic Space, Springer.
Egenhofer, M. (1991), Advances in Spatial Databases, SSD, Springer.
Egenhofer, M. and J. R. Herring (1990), 4th International Symposium on Spatial Data Handling, Zurich, Switzerland, IGU.
Frank, A. U. (1991), Auto-Carto10, ACSM & ASPRS.
Frank, A. U. and W. Kuhn (1986), Second International Symposium on Spatial Data Handling, Seattle, WA, IGU.
Freksa, C. (1991), Qualitative Spatial reasoning. D. Mark and A. U. Frank, Kluwer Academic Press.
Freksa, Chr. (1992), Using Orientation Information for Qualitative Spatial Reasoning. A. U. Frank, I. Campari and U. Formentini, Springer.
Hernandez, D. (1994) , Springer.
Laurini, R. and D. Thompson (1994), Academic Press.
Mark, D. and A. Frank (1995), "Experiential and Formal Models of Geographic Space." .
Nunes, J. (1991), Geographic Space as a Set of Concrete Geographical Entities. D. Mark and A. U. Frank, Kluwer Academic Press.
Spanier, E. (1966), New York, McGraw-Hill Book Company.
Talmy, L. (1983), How Language Structures Space. H. Pick and L. Acredolo. New York, NY, Plenum Press.
Zubin, D. (1989), Languages of Spatial Relations: Initiative Two Specialist Meeting Report, NCGIA Technical Paper.

Contact address
Thomas Bittner
Department of Geoinformation
Technical University of Vienna
Gusshausstrasse 27-29
A - 1040 Wien
Austria
E-mail: bittner@geoinfo.tuwien.ac.at

TOWARDS A MULTIPLE PLACEMENT LAND SUITABILITY

EVALUATION FRAMEWORK

Howard T. K. Lee and Zarine Kemp

Computing Laboratory, University of Kent
Canterbury, UK

Abstract
Land evaluation techniques help decision-makers find sites for locating entities such as public utilities, residential, agricultural, commercial and industrial areas. The suitability of a given site depends upon the relationship of that site with other features or entities in the environment. Location decision making is a vast area where different approaches solve different problems. Land suitability evaluation techniques arguably encompass the most generic approach to location analysis but do not generally consider the placement of more than one new entity at any one time. On the other hand, location-allocation methods are especially suited for multiple placements but require specific mathematical formulations where different circumstances necessitate different constructions. This article highlights the implications of time on multiple placements and the effects of dependencies between newly placed entities. We present a unified framework to clarify these issues and provide a generic approach for problem solving.

Key words
Land suitability evaluation, location-allocation, multiple criteria decision making.

1. Introduction

Location problems are hugely varied [Birkin et al 1996]. There are many factors which can affect the placement of a new entity and even more ways to resolve it. The notion of suitability is context dependent and we enforce no restriction upon the influencing factors; all that is assumed is the existence of some formalisation and an applicable assimilation function which provides a quantitative measure of suitability.

This article is not a survey of all existing location decision making methods nor about identifying the exact determinants of suitability. What this article addresses are the issues involved in location analysis from a problem-oriented perspective. We present a conceptual *representation* of multiple placements and a generic method of *resolution* for location problems. This approach facilitates navigation through the extensive research in location decision making and provides a comparison of existing methods. Let us consider an example of siting two new schools, whose aim is to serve the maximal number of children from Ashford and Canterbury.

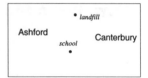

Figure 1a - Placement of a school . Figure 1b - Placement of two schools.
Figure 1c - Placement of school & landfill.

If we considered each school at a time irrespective of the other, then the first school would be ideally placed in between Ashford and Canterbury (figure 1a). Unfortunately, we are left with the dilemma of effectively siting the second school. Clearly, the solution lies in figure 1b, where both schools are taken into account *simultaneously*. It is evident that the placement of one entity without due consideration of the as yet unplaced entities may result in a suboptimal placement scheme.

Let us now examine another example involving the placement of a school and a landfill site where the school still abides by the criterion described earlier and the landfill must (i) serve both Ashford and Canterbury and (ii) be located a reasonable distance from any public zones such as the school. Unlike the previous example where the siting of each school was solely derived from satisfying the demands of each area, now it is necessary to also consider the relationship between the two new facilities. Whilst human judgment may be applied to simple cases like this (figure 1c), it is not possible with large and complex problems.

The examples above illustrate two principal determinants that can potentially affect the placements of a set of new entity sites:
- *External relationships* - Existing environment factors which influence the placements
- *Internal relationships* - Dependencies between the new entities

It has been assumed that all the sites are located simultaneously but there may be *delays* between the placements. It is important to understand the implications. For instance, with the example involving two schools; where do we place the first school if the second school is not scheduled to be built for another 2 years? Is it acceptable for children to travel to Ashford whilst awaiting the construction of a school in Canterbury? If the first school is placed exactly between the two counties, where will the second school go? We are back to the original problem! Delayed multiple placements is an important aspect of any geographically-oriented planning process since there are likely to be expenditure constraints during any one given period and it is frequently desirable to account for change in the environment.

2. Framework of multiple placements

Given only one entity to place, generating a *land suitability map* [Hopkins 1977] best depicts the problem. Unfortunately, when we start to consider two or more placements, it may be necessary to tentatively fix the positions of certain entities in order to evaluate the suitability of another entity. This process is repeated until all the entities are suitably placed, which is difficult to apply using land suitability evaluation methods. Location-allocation techniques [Hillsman 1984; Chhajed 1993; Ghosh and Harche 1993; Densham and Armstrong 1994] are much more suited for multiple placements. Despite their differences, they are essentially a *search* process. A land suitability map reflects an exhaustive search and is only suitable if the search domain is not too large. Linear and non-linear programming methods are generally able to find optimal or near optimal solutions without an exhaustive search but make certain assumptions about the search domain. Nevertheless, an efficient search process is necessary for very large problems.

A few definitions:are given here to clarify the subsequent discussion:
* *Scenario* - A possible placement of an entity.
* *Entity* - We assume everything in the environment are entities. They may be physical like a house, or conceptual like political boundaries. An entity is characterised by attributes.
* *Attribute* - A factor influencing the suitability of a given scenario. Profitability, aesthetics, safety, location and slope are a few possible attributes. Attributes are often derived from other attributes and often be structured hierarchically. The value of an attribute is determined by the placement of an entity in relation to all other features in the vicinity. Some influential features affecting profitability are the surrounding population and competitors. With aesthetic value, one could measure the number of obstructing features in the scenery and so on.
* *Suitability* - This is obtained by a function assimilating all the attributes concerned. The assimilation process is defined by a suitability function.

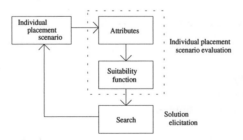

Figure 2 Framework for individual placements.

Figure 2 illustrates the framework for placing one entity. Each candidate site is a scenario and the implications of that placement is reflected by a set of attribute values. Function v assimilates the attributes and outputs a suitability measure.

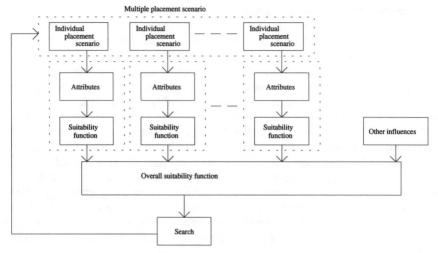

Figure 3 Framework for multiple placements.

The multiple placement framework (figure 3) extends the individual placement framework (figure 2) to allow for the simultaneous consideration of several entities. If the new entities do not affect each other whatsoever (i.e. they do not influence the attributes of one another) then we can place them without concern by iteratively applying the individual placements framework. However, when they do affect each other's attributes then it is necessary to find an overall suitability measure that takes into account all the individual entities. In addition, there may also be other influences such as an "overall budget" or an "overall demand" to satisfy.

2.1 The suitability function

Even without a search algorithm, a suitability evaluation function can be a very useful tool for evaluating different scenarios. The suitability functions are defined as follows:

$$\text{suitability measure} = v_e(a_1, a_2, \ldots) \rightarrow [0,1] \tag{1}$$

$$\text{overall suitability measure} = v_{overall}(v_1, v_2, \ldots, v_N, other) \rightarrow [0,1] \tag{2}$$

where,

e is an entity
a is an attribute
N is the number of entities

The number of entities N may of course be a parameter for search. For instance, one could ask what is the most profitable number of new outlets to open given a certain level of consumer demand.

Although there are many ways to define the suitability functions, two schools of thought dominate the existing literature:
- Multiple criteria decision making (MCDM) methods.
- Rule-based methods.

2.1.1 Multiple criteria decision making (MCDM) methods

MCDM techniques [Hwang and Masud 1979; Vincke 1992; Keeney and Raiffa 1993; Yoon and Hwang 1995] embrace both multiple attribute decision making (MADM) and multiple objective decision making (MODM) methods. There are no standardised definitions for an attribute and objective in the literature, suffice to say it is sometimes difficult to distinguish between the two terms. However, there are distinctive methodological differences in MADM and MODM techniques.

Multiple Attribute Decision Making (MADM)

MADM methods come in a wide variety of forms. The simple additive function is the most commonly used:

$$v \Rightarrow \sum_{i=1}^{M} w_i u_i (x_i) \qquad (3)$$

where,

$$\sum_{i=1}^{M} w_i = 1$$

w_i is the weighting associated with attribute i
$u_i \rightarrow [0,1]$ is the utility function associated with attribute i
x_i is the attribute

The aim is to maximise equation (3). Weights reflect the relative importance of a given attribute. The utility function enables different types of attribute to be compared by converting all attribute values to some common frame of reference. The determination of weights and utility functions are an important part of the decision making process.

Unfortunately, the additive model requires *preferential independence* between the attributes. This means if we preferred a set of attributes p to another set q, then even if we changed an attribute that is equal in both p and q to a different value (both still the same), our preferences must remain the same. This restriction may be circumvented by either redefining the attributes or employing different assimilation methods such as the multiplicative method. Both additive and multiplicative approaches assume the attributes are allowed to compensate one another, which may not necessarily be valid. Dominance and sequential elimination on the other hand are non-compensatory but they are difficult to quantify. As a compromise, constraints may be placed upon the attributes in order to restrict the compensatory effects.

MADM methods are well-suited for declaring the suitability function of individual placements though they require a fair amount of deliberation amongst the attributes. MADM may be applicable to the multiple placements suitability function if the entities are identical or very similar but becomes difficult to apply with different types of entities.

Multiple Objective Decision Making (MODM)
MODM techniques are generally based upon comparisons with "ideal" solutions, which may be arbitrarily defined or identified by exhaustive search. The suitability function assimilates the degree to which each of its objectives is satisfied.

A simple technique is:

$$v \Rightarrow \sum_{i=1}^{M} w_i \left(x^* - x_i \right) \tag{4}$$

where,

$$\sum_{i=1}^{M} w_i = 1$$

w_i is the weighting associated with objective i
x^* is the "ideal" objective value
x_i is the value for objective i

Although very similar to the simple additive method in MADM, the MODM version does not require preferential independence. The aim is to minimise equation (4). MODM may be employed for both individual [Pereira and Duckstein 1993] and multiple suitability functions.

2.1.2 Rule-based methods
Rule-based methods characterise the suitability function as a set of rules. Rules may be applied either in a normative or descriptive manner. The former renders the problem into an algorithmic exercise involving a set of operations. Descriptive rule application on the other hand only states what is desired or not desired. When descriptive and normative applications are combined, they form a prescriptive application.

Normative rule application
Map overlays and transformations [Burrough 1986, Tomlin 1991] are the best examples of normative rule application. They take the general form:

$$v \Rightarrow f\left(f\left(\ldots f\left(a_1, a_2, \ldots\right)\right)\right) \tag{5}$$

where,

f is an arbitrary transformation operator
a is an attribute on a map

Computationally, normative rule application is by far the most efficient since it incorporates both the suitability aspect and the search process as a single function, though this is not necessarily the most effective method because it imposes an exhaustive search. However, we focus only on descriptive or prescriptive methods because they capture expert domain knowledge more explicitly than normative methods.

Prescriptive rule application
Although this is perhaps the most intuitive approach towards the specification of a function, it is up to the designer to ensure there are enough rules to capture all the possible circumstances. Fuzzy theory has played a major role in developing rule-based systems [Kosko 1992, 1994] and has been applied to land suitability evaluation.

$$v \Rightarrow \sum_{fuzzy} \{Rule\} \qquad (6)$$

where,
$Rule =$ IF ... THEN ...

Rule-based approaches can capture very complex relationships and are therefore represent a powerful yet easy to use methodology. However for most practical problems, the number of rules needed to ensure completeness and accuracy is quite large. This is very much in contrast to MCDM methods which are efficient at capturing the general relationships but not the specific cases. Naturally, a combined MCDM and rule-based approach could be very beneficial [Hopkins 1977] which is partially supported by the proposed framework since each suitability function may be defined using any of the methods described, but whether we can have a single function integrating both MCDM and rule-based techniques seems to be an issue for further research.

2.2 The search algorithm

In general, a good search algorithm is capable of finding a solution or solutions without an exhaustive search. Unfortunately, without more information of the search domain and without an exhaustive search, the solution cannot be guaranteed to be optimal. We briefly describe several search techniques:

- *Gradient ascent* examines its current suitability value and alters one of its parameters (e.g. location, number of new entities) slightly and re-evaluates the new scenario. It continues the search path until no improvement can be found.
- *Random search* samples a number of random parameters, evaluates them and selects the best solutions. This works without knowledge of the search domain but the accuracy of the solution is dependent upon the number of samples. Random search is simple to implement but inefficient.
- *Genetic algorithms* are analogous to Darwinian natural selection and perform well on virtually any complex search domain with little or no domain knowledge.
- *Simulated annealing* is analogous to the annealing process used to strengthen the construction of certain materials. In the beginning of the search where the "temperature is high", the probability of accepting an inferior search path is relatively high but as the temperature is reduced, only the best search paths are retained. This is obviously much better than

a simple gradient ascent, potentially faster and more accurate than a genetic algorithm but requires some domain knowledge.

- *Tabu search* is by far the most effective approach since it never backtracks. If a search path is deemed inferior to another search path, the inferior path will never be examined again. Obviously, this approach requires in-depth knowledge of the search domain.

The problem with gradient ascent is that it only works with linear search domains. A nonlinear search domain is one in which it is sometimes necessary to investigate a poorer scenario from the existing one in order to find an even better scenario. Clearly, one would expect most search domains in our context to be nonlinear. The rest of the algorithms are capable of dealing with nonlinearity but progressively require more and more search domain knowledge.

The following are also important design factors to consider:
- During each iteration cycle, the values of strongly dependent attributes should be carefully manipulated.
- There are likely to be several optimal or near optimal solutions. The algorithm should keep a record of them since those with particularly contrasting scenarios identify different alternatives to the decision maker.

3. Extending the framework with time

Even without the consideration of delays between multiple placements, time can play an important role in individual placements. Naturally, there may be scenarios that fare better in the long term than in the short term and vice-versa; e.g. a decision maker could choose a supermarket site that has excellent short term gains but poorer long term gains relative to another site that has good long term gains but low short term profits. This is due to the fact that some attributes are dependent upon time (variable costs) whilst others are independent of time (e.g. fixed costs).

Let us consider a hypothetical situation involving the placement of two new supermarket branches *a* and *b*. A supermarket is placed according to its overall revenue and is affected by a simple model based on population and distance. Figure 4a illustrates the case when branch *a* was placed first then followed by branch *b*, i.e. *a* did not take into account *b*. Figure 4b illustrates the case when both branches are taken into account and placed simultaneously.

Figure 4b is obviously the most profitable scheme where both branches are placed in the two cities. However, if only one branch is to be placed then the most profitable location is for argument's sake between the two cities, but this scenario will be suboptimal when the second branch is built. These examples illustrate that a relationship exists between individual placements and multiple placements. That relationship is of course time. If we wanted to place two entities but there is a very long delay between them then one approaches the situation described by figure 4a since the later branch can be ignored but as the delay decreases until the placements are simultaneous, we have a situation analogous to figure 4b.

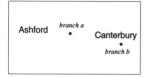

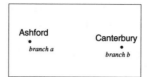

Figure 4a Branch a placed   Figure 4b Simultaneous
before branch b.   placement of two branches.

Two different temporal influences can be distinguished here. The first one depends on how far in the future we wish to plan whereas the second, described by the figures 4a and 4b is associated with the delay between placements. The next subsection describes a temporal framework to encapsulate these two influences.

3.1 Relevance period

All entities and their respective suitability functions are evaluated within a time interval called the *relevance period* which is defined by a *conception* time and *planning horizon*. *Duration* is the difference between conception time and the planning horizon. The relevance period is considered an attribute. Since attributes may derive values from other attributes, some attributes may derive their values from time (e.g. variable costs) whilst others remain independent of time (e.g. fixed costs).

$$relevance\_period \Rightarrow [conception, planning\_horizon]$$
$$duration = planning\_horizon - conception$$

In order to support a meaningful notion of time, the following is enforced upon the framework:
• Every entity must have one and only one relevance period.

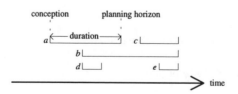

Figure 5 Relevance periods of several entities.

The planning horizon is merely a mechanism for varying the duration so that any time dependent attributes (e.g. variable costs, revenue) may derive their values accordingly in contrast to time independent attributes like fixed costs. The planning horizon may of course be utilised as a search parameter. This accounts for decision processes addressing gains in the short and long term.

Much more interesting are the implications of conception time which concerns the issue of existence as depicted in figures 4a and 4b. Let us consider two special cases: (i) if the conception times of a set of entities are all the same then we have the basic multiple placement problem where the entities are sited concurrently; and (ii) if some entities which are to be placed in the future have a very long delay to the entity being placed in the present moment, we ignore the entities in the future and concentrate on the individual placement at hand. The point of contention arises when delays are in a zone of partial influence, where the delay is not long enough to be treated as the simultaneous multiple placement problem but not short enough to be ignored. Clearly, this is only an issue if there is some sort of a dependence between the entities, i.e. one or both of them have attributes derived from the other. If two entities do not affect each other whatsoever, then it does not matter whether there is a delay between their placements. To take into account different conception times, the attributes of equation (1) is amended as follows:

$$\text{suitability measure} = v_e\left(exists(e,d)a,...\right) \tag{7}$$

where,

    $exists \rightarrow [0,1]$
    a is derived from an attribute of entity d

If entity d is not significant to the evaluation of entity e then all attributes that are derived from d are ignored because $exists = 0$. If entity d is significant because there is little or no delay with the placement of e then $exists = 1$ and the evaluation of e takes d into full account. The function exists may of course adopt any value between 0 and 1, thereby indicating partial influence.

    IF $delay(e,d) \leq 0$
    THEN $exists(e,d) = 1$ (8)
    ELSE $exists(e,d) = D\bigl(delay(e,d)\bigr)$

    IF $conception(d) = conception(e)$
    THEN $exists(e,d) = exists(d,e) = 1$ (9)
    ELSE $exists(e,d) \neq exists(d,e)$

where,

    $delay(e,d) = conception(e) - conception(d)$
    $D \rightarrow [0,1]$
    d, e are any arbitrary entities

Whereas the multiple placements framework extension permitted the number of entities to be a search parameter, the temporal extension allows for two more search parameters. We just observed that conception time and planning horizon affect attributes differently, each of which may be a parameter for search.

With this framework in mind, it is possible to apply a few search optimisations. For instance, any two mutually independent entities may be sited concurrently regardless of their conception times. If an entity e is independent of d (either because there are no inherent dependencies or the delay is very long) but d is dependent on e, then the search process should be divided into two distinct stages. The first stage locates entity e but not d. Once e is placed, the second stage will include entity d.

This completes our description of the key issues such as *location, number of new entities, planning horizon* and *conception time* within a multiple placement land suitability evaluation framework.

4. Conclusion

Location problems are hugely varied and it is quite apparent there does not exist a unified framework which can be used to facilitate planning and decision making processes. This paper is a step towards that goal. We examined the issues involved multiple placements and proposed a generic framework which covers a search domain involving location, number of new entities, planning horizon and conception time. It also reveals the temporal relationship between individual and multiple placements. However, there is still much scope for further research:

- A more rigorous treatment of the ideas is needed to establish correctness and completeness in relation to existing location decision problems, as well as to facilitate future research.
- We have not prescribed any techniques for constructing the existence function (8). Quantifying the temporal relationship between individual and multiple placements is a difficult task and is very much context dependent.
- The extension of the suitability function to take into account delays by (7) has been simplified. Since attributes (section 2) may take on a hierarchical structure, (8) will need to be applied to some specific part of the attribute hierarchy.
- In practice, an efficient search process is very important. Further research could establish some sort of a dependency graph formalisation in order to determine the number of search stages and the parameters associated with each stage.

There are few restrictions imposed upon the framework and this permits a number of extensions. For instance, one could introduce several "overall" suitability functions to reflect the contention between several competitors for a limited supply of resources. Generally, we believe this is a powerful framework that can facilitate better decision making in a spatiotemporal context.

Acknowledgments
The first author, Howard Lee, gratefully acknowledges the financial support from the E. B. Spratt bursary granted by the Computing Laboratory, University of Kent at Canterbury, UK.

References
Birkin M., Clarke G, Clarke M, and Watson A. (1996) Intelligent GIS: Location decisions and strategic planning. GeoInformation International.
Burrough P. A. (1986) Principles of geographical information systems for land resources assessment. Oxford, UK: Clarendon Press.
Chhajed D., Francis R. L. and Lowe T. J. (1993) Contributions of operations research to location analysis. Location Science. vol. 1, no. 4, pp. 263-287.
Densham P. J. and Armstrong M. P. (1994) A heterogeneous processing approach to spatial decision support systems. Proceedings of the 6th International Symposium on Spatial Data Handling. vol. 1, pp. 29-45.
Ghosh A. and Harche F. (1993) Location-allocation models in the private sector: Progress, problems and prospects. Location Science. vol. 1, no. 1, pp. 81-106.
Hillsman E. L. (1984) The p-median structure as a unified linear model for location-allocation analysis, Environment and Planning A. vol. 16, pp 305-318.
Hopkins L. D. (1977) Methods for generating land suitability maps: A comparative evaluation. Journal of the American Institute of Planners. vol.43, no. 4, pp. 386-400.
Hwang C.L. and Masud A. S. M. (1979) Multiple objective decision making - methods and applications: A state of the art survey. Springer-Verlag.
Keeney R. L. and Raiffa H. (1993) Decisions with multiple objectives: Preferences and value tradeoffs. Cambridge University Press.
Kosko B. (1992) Neural networks and fuzzy systems: A dynamical systems approach to machine intelligence. Englewood Cliffs, NJ: Prentice Hall.
Kosko B. (1994) Fuzzy systems as universal approximators. IEEE Transactions on Computers. vol. 43, no. 11, pp. 1329-1333.
Pereira J. M. C. and Duckstein L. (1993) A multiple criteria decision-making approach to GIS-based land suitability evaluation. International Journal of Geographical Information Systems. vol.7, no. 5, pp. 407-424.
Tomlin C. D. (1991) Cartographic modeling in Maguire D. J., Goodchild M. F., Rhind D. W. Geographical information systems: Principles and applications. vol. 1, pp. 361-374. Harrow: Longman Scientific & Technical.
Vincke P. (1992) Multicriteria Decision Aid. Chichester: Wiley.
Yoon K.P. and Hwang C.-L. (1995) Multiple attribute decision making: An introduction. Sage University Paper series on Quantitative Applications in the Social Sciences, 07-104, Thousand Oaks, CA: Sage.

Contact address
Howard T. K. Lee and Zarine Kemp
Computing Laboratory
University of Kent at Canterbury
Canterbury, Kent CT2 7NF
UK
E-mail: T.K.H.Lee@ukc.ac.uk
E-mail: Z.Kemp@ukc.ac.uk

SPATIO-TEMPORAL INTERPOLATION BY INTEGRATING

OBSERVATIONAL DATA AND A BEHAVIORAL MODEL

Ryosuke Shibasaki[1] and Shaobo Huang[2]

[1]Institute of Industrial Science
University of Tokyo
Tokyo, Japan
[2]Center for Environmental Remote Sensing
Chiba University
Chiba, Japan

Abstract
Spatio-temporal interpolation to generate voxel-field data in space-time domain from observational data is indispensable to many spatio-temporal analyses and visualization of dynamic spatial objects. However only very primitive interpolation methods such as nearest neighbor interpolation based on Voronoi diagram are proposed for nominal or "class variable" data such as land use data. In interpolating nominal data with these primitive methods, we cannot make use of knowledge on spatial or temporal patterns or behaviors of the object. The authors propose a spatio-temporal interpolation scheme for generating a voxel-field of nominal data under the framework of optimization of likelihood. The likelihood is computed from the fitness to both observational data and expected patterns/behaviors described by a behavioral model or rules specific to the object. Any model which provides likelihood or probability to a given spatio-temporal pattern can be used in this framework. For the optimization of likelihood, a genetic-algorithm (GA) was combined with the Hill-climbing (HC) method to increase the efficiency and reliabitliy of optimization. Through some experiments, it is demonstrated that GA/HC based interpolation method can generate voxel-fields which fit both the observational data and to the knowledge of it's behavior and that the reliability of interpolation can be evaluated quantitatively in terms of the maximal likelihood.

1. Background and objectives

Temporal or dynamic analysis of spatial data are needed in various fields such as environmental systems analysis. A fundamental problem faced by users is the difficulty in generating spatio-temporal data fields (3D or 4D voxel field) through interpolation of observational data. This comes from the fact that observational data from multi-sources often have sparse or biased distributions, and different forms (point, edge, polygon and solid in a spatio-temporal space), resolutions and accuracy/reliability.

In several fields, to improve reliability of spatio-temporal interpolation/ extrapolation in generating quality data, models and/or equations describing a mechanism and structure underlying a spatial or behavioral pattern are integrated with observational data.

Integration methods for data and models have been mainly developed for continuous variables such as temperature and precipitation in meteorological and oceanographic studies. These methods are known as 4DDA (Four Dimensional Data Assimilation).

For nominal or class variables such as land use types, only relatively primitive interpolation methods have been available, such as nearest neighbor interpolation. We propose a method of integrating models and nominal observational data from multiple sources under the framework of optimization of likelihood of spatio-temporal events. For optimizing the likelihood, genetic algorithm (GA) is combined with classical Hill climbing method. Experimental results demonstrate that GA with HC can be successfully applied to the integration.

2. Genetic algorithms (ga) and nominal variable interpolation

2.1 Introduction of Genetic Algorithm (GA)

Genetic algorithms have been developed by John Holland and his colleagues as an approach to optimization problems. The search algorithms are based on the mechanisms of natural selection and evolution of natural genetics. The approach combines survival of the fittest among string structures with a structured but randomized gene exchange to form a search algorithms with some of innovative flair of human search (D.E.Goldberg, 1989).

Genetic algorithms are computationally simple and powerful in their search without restrictive assumptions about search spaces. In a simple genetic algorithm, five basic aspects should be considered; the representation or coding of the problem, the initialization of the population, definition of the evaluation function, the definition of genetic operators, and the determination of parameters.

2.2 Optimization Scheme for Nominal Variable Interpolation

Most natural properties in magnitude vary along a continuous scale. Spatial continuity and temporal continuity are intuitive assumptions which provides rationale for interpolating observational data (M.A.Olover, 1990). However, knowledge and rules governing spatio-temporal patterns and behavior of geographic objects (e.g. environmental systems) are now being rapidly accumulated and represented by many simulation models. They can provide more robust and quantitavie basis for interpolating observational data, though many of the models still may not be sufficiently accurate and reliable. On the other hand, it can be said that reliability of results estimated from model simulation can be improved by combining reliable observational data. Integration of observational data and models (GCM etc.) are conducted in meteorology as daily routine. There have not been no such attempt to extend the idea of integration to more generic geographic objects.

It is reasonable to assume that spatio-temporal events or "voxel-field" of nominal variables which are estimated should maximize likelihood under given observational data and behavioral models, if we suppose that observation is a probabilistic event and behavioral models are structured and probabilistic a priori knowledge on behavior of the object phenomenon. Observational data and behavioral models/rules can be integrated in the process of maximizing the likelihood of spatio-temporal events.

Since searching for the most likely spatio-temporal voxel-field of nomimal data is a typical combinatorial optimization problem, we introduce the genetic algorithm as a optimization scheme.

3. Application of ga for integrating behavioral models and observational data to class variable interpolation

3.1 3D representation of an individual (coding)

In the following discussion, three dimensional array is defined to represent an individual (Figure 1) in space-time domain. The horizontal plane represents 2D space and vertical dimension represents temporal dimension.

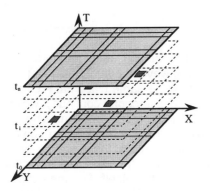

Figure 1   Representation of individual.

3.2 Initialization of population

An initial population for a genetic algorithm is usually chosen at random; one random trial is made to produce each individual. All members of initial population are chosen automatically by same procedure so that the expected value of each member of initial population is same. In addition we use cubes of 1*1*1, 2*2*2 and 3*3*3 pixels as the initial unit for the initialization of population to increase the efficiency of procedure.

3.3 Definition and computation of individual's fitness

3.3.1 Spatio-temporal behavioral models/rules of class variable data
Any type of behavioral models can be used for the GA-based interpolation if they can determine the probability of every possible behavior/transition of nominal or class variables. For nominal variable data, possible changes of a class at one pixel is basically defined by the probability of the changes from one class to another. A simple example is a Markov chain, where transitional

probability is determined only by the previous class. In addition, the probability can also be affected by the combination of classes in the neighborhood. In this study, we used a model where transitional probability is determined by the combination of classes in the neighborhood. Test data have five classes.

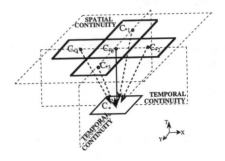

Figure 2 3D spatial-temporal relation of pixel based class-variable data.

In our experimental model, spatial and temporal relations affect the transitional probability in three ways (Figure 2). The first is "spatial continuity", based on the assumption that the same class data tends to continue in the spatial dimension. The second is temporal continuity, which is an extension of the spatial continuity to the temporal domain. The third is expansion-contraction relations based on the assumption that some data classes have higher possibility to expand their area at the next time-slice, while others tend to contract. The temporal change in the pixel with non-contractible class type will be determined by the pixels class itself. And the temporal landuse change in the pixel with contractible type will be determined by the class of the pixel and classes of its expansible neighbors.

3.3.2 Definition and computation of fitness of an individual

Fitness of an individual is defined by the combination of behavioral fitness and observational fitness. Behavioral fitness is defined as the combined probability of change events of nominal variables under the condition that these changes follow a given probabilistic behavioral model or rule.

Observational fitness can be defined as the combined probability that the observational nominal values occur under probabilistic functions of observational errors or uncertainties. Observational probability can be determined by accuracy, resolution and frequency of observation. Overall fitness can be computed by multiplying behavioral fitness and observational fitness. Thus, behavioral/structural models and observational data can be integrated by optimizing the overall fitness.

1) Behavioral fitness:

As shown in Figure 2, let $P_V(C_{P,T}, C_{P, T+1})$ as the probability of changes from landuse class C_{*0} to landuse class C_* considering both the temporal continuity and the expansion-contraction effect of its neighbours ($P_{TC}(C_{*0}, C_*)$), and $P_{SC}(C_{*0} \sim C_{*4})$ as the probability of spatial continuity. If we assume that $P_{TC}(C_{*0}, C_*)$ and $P_{SC}(C_{*0} \sim C_{*4})$ are independent, we can compute behavioral fitness of each individual according to following formula:

$$\text{FITNESS}_{\text{(behavioral fitness)}} = \prod_{P=1}^{N_P} \{ \prod_{T=1}^{N_t} P_V(C_{P,T}, C_{P,T+1}) \}$$

$$= \prod_{P=1}^{N_P} \{ \prod_{T=1}^{N_t} P_{TC}(C_{*0}, C_*) P_{SC}(C_{*0}, C_{*1}, C_{*2}, C_{*3}, C_{*4}) \}$$

where N_P : is the pixel number,

N_T : is the temporal slice number,

$C_{P,T}$: is the landuse class of the cell on the

Pth pixel at the T time slice;

For the class change probability with spatial continuity, $P_{SC}(C_{*0} \sim C_{*4})$, we set values according to following five neighboring pixel's status along the spatial dimension, which form a set of behavioral rules: 1) whether or not classes in all neighbouring pixels are equal; 2) whether or not classes in 4 neighbouring pixels are equal; 3) whether or not classes in 3 neighbouring pixels are equal; 4) whether or not classes in 2 neighbouring pixels are equal; 5) whether all classes in 5 pixels are unequal.

To calculate the probability of class changes under the temporal continuity/expansion-contradiction effect, $P_{TC}(C_{*0}, C_*)$, three possible changing patterns of landuse classes in spatial-temporal distribution are picked up and listed in the Figure 3. The probabilities of those cases can be determined by integrating probability of class changes in Markov chain, $P_M(C_{*0}, C_*)$, and expansion probability of the class-types into neighboring pixels (Table 1).

2) Observational fitness:
Observational fitness can be computed with the following formula.

$$\text{Observational Fitness} = \prod_{n=1}^{N_O} P_{Obs}(C_{P,T}, C_{P,T,Obs})$$

$P_{Obs}(C_{P,T}, C_{P,T,Obs})$: Probability that observational
value Cp,T,Obs is given when actual
value is C,P,T

Observational probability can be determined mainly by the accuracy of observation. Observational location and time/frequency can be represented by locating observational pixel in the two dimensional string. Spatio-temporal resolutions can be represented by setting an aggregation formula over the range of observation.

3) Total fitness:
Total fitness is computed from behavioral fitness and observational fitness by the following formula.

Total fitness = Behavioral fitnesss * Observational fitness,

or ln(Total fitness) = ln(Behavioral fitness)+
ln (Observational fitness)

Value of Invasion	C_*	C_{*i1} $P_M(C_{*0},C_{*i1})$	C_{*i2} $P_M(C_{*0},C_{*i2})$	$C_{*i1}=C_{*i2}$ $P_M(C_{*0},C_{*i1})$	Others $P_M(C_{*0},C_*)$
C_{*i1}	C_{*i2}				
Yes $(\alpha_{C_{*i1}})$	Yes $(\alpha_{C_{*i2}})$	0	0	Invasion	0
Yes $(\alpha_{C_{*i1}})$	No $(1-\alpha_{C_{*i2}})$	Invasion	0	Invasion	0
No $(1-\alpha_{C_{*i1}})$	Yes $(\alpha_{C_{*i2}})$	0	Invasion	Invasion	0
No $(1-\alpha_{C_{*i1}})$	No $(1-\alpha_{C_{*i2}})$	Markov Chain	Markov Chain	Markov Chain	Markov Chain

Notice: 1> Supposed C_{*i2} and C_{*i2} are expansible (i1, i2 = 1 ~ 4);

2> $\alpha_{C_{*i1}}$ and $\alpha_{C_{*i2}}$ are defined as expansion speed of C_{*i2} and C_{*i2}

Table 1 Behaviors of class changing in Case 3.

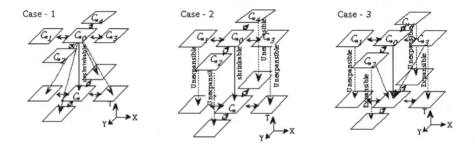

Figure 3 Possible changing patterns of landuse classes in temporal and spatial distributions.

3.4 Definition of operators

3.4.1 Reproduction

Reproduction is a process in which individual strings are copied according to their objective function values or the fitness values. Copying strings according to their fitness values means that strings with a higher value have a higher probability of contributing one or more offspring in the next generation (Figure 4). This operator may be viewed as an artificial version of natural selection, a Darwinian survival of the fittest among string creatures.

There are several proposals for selecting survival individuals. The most basic schemes are the roulette wheel scheme, the deterministic sampling and the elitist scheme. In order to efficiently find the best solution in the search space, we adopted the selection scheme based on the combination of the deterministic sampling and the elitist scheme. The selected survival possibility in next generation of each individual is calculated as in the deterministic sampling. And the best individual is kept into the next generation as in the elite scheme.

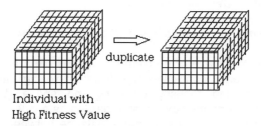

Figure 4 Reproduction of individual.

3.4.2Crossover

The crossover operator first randomly mates newly reproduced individuals in the mating pool. Then it randomly locates a window with random size for a pair of individuals. Finally, the contents of individuals within the window are swapped to create new individuals (Figure 5).

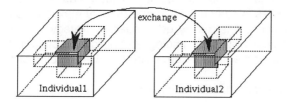

Figure 5 Crossover of individuals.

3.4.3Mutation

Mutation operator plays a secondary role in the simple GA by occasionally altering the value in an individual position (Figure 6).

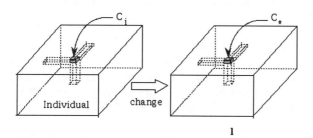

Figure 6 Mutation of one individual.

4. Improvement of the search in gas

4.1 Hill-Climbing method to improve the efficiency of genetic algorithm

Searching a complex space of problem resolutions often involves a tradeoff between two apparently conflicting objectives: exploiting the best solutions currently available and robustly exploring the space (Lashon Booker, GA&SA). Generic algorithms have been regarded as a class general purpose search strategies that strike a reasonable balance between exploration and exploitation. The power of these algorithms is derived from a very simple heuristic assumption: that the best solutions will be found in regions of the search space containing relatively high proportions of good solutions. The problem is that, if the complex space of problem resolutions become larger and larger, the population size and the generation size have to be increased at same time. The efficiency of GA is one of the obstacles to real-woird applications of the GA.

Hill climbing is a good example of a search strategy that exploits the best among known possibilities for finding an improved solution. Although Hill-Climbing strategies is easy to trap in one of local maxima more far away from the optimal solution, it is a very good search strategy that exploits the best among known possibilities for finding an improved solution. In our research we investigate the potential for combining the Hill-Climbing strategy with GA.

4.2 Maintenance of population diversity

Despite the demonstrated advantages of GAs and high performance of most implementations, it still fails to live up to the high expectations engendered by the theory. The problem is that any implementation uses a finite population or set of sample points. Estimates based on finite samples inevitably have a sample error associated with them. Repeated iterations of algorithm compound the sample error and lead to search trajectories much different from those theoretically predicted. The most serious phenomenon is the premature convergence.

The premature convergence is caused by early emergence of an individual that is better than the others in the population, although far from optimal. Copies of this structure may quickly dominate the population. Search continues then but is concentrated in the vicinity of this structure and may miss much better solutions elsewhere in the search space.

To avoid the premature convergence, one has to avoid the loss of population diversity. Although, reducing the reproduction number cannot always eliminate the premature convergence, it can be used as a simple way to reduce the rapid convergence. In our research, we therefore limited the duplicated number of individuals less than two; if individualÕs expected duplicated number is larger than two, we set it equal to two.

5. Experiments

The test program of GA/HC for spatio-temporal interpolation of pixel based landuse data was coded with C language and was run on SPARC/station2. Small spatio-temporal datasets are used in the experiment to check the behavior of the GA/HC based interepolation under different conditions. The test data size of individual had been defined with 20 pixels * 6 time-slices for two dimensional case, while 11 lines * 11 columns * 6 time slices for three dimensional case. The first and last time-slice in the individual are supposed to be sampl(observational) data and all middle time-slices should be estimated by the interpolation. In these experiments, we set the generation size of GA/HC to 2000, which was large enough to get stable results. The probability of crossover operation was defined as 0.7, while the probability of mutation operation was relative small in natural population, so that we used 0.01 as the probability of mutation. Two dimensional and three dimensional experiment results of GA/HC, in which the individual has largest fitness value are presented in Figure 7. In the model used for this experiment, smooth transition/expansion has the largest fitness or likelihood.

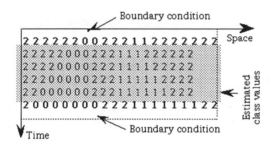

Figure 7a Result of GA/HC (2D case).

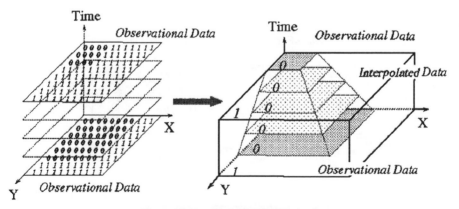

Figure 7b Result of GA/HC (3D case).

Several experiment results are compared in Figure 8. From this figure, we can find that the larger the population size is, the faster the stable result can be obtained, and the closer the stable result tend to best solution. Comparison of GA/HC with GA for spatio-temporal interpolation of landuse class variable data is presented in Figure 9. GA/HC has much higher efficiency than GA for spatio-temporal interpolation because much higher fitness value can be obtained in younger generations.

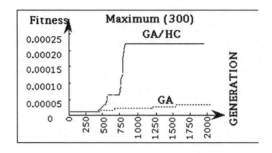

Figure 8 Comparison of GA with GA/HC.

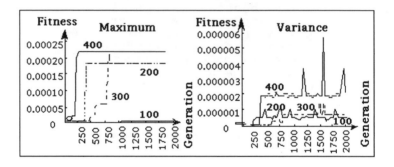

Figure 9 Effect of pupolation & generation size.

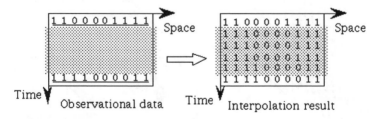

Figure 10 Interpolation result (overlapping case).

In Figure 10, observed location of class '0' in the first time slice and the last time slice are overlapping spatially. The interpolation result naturally merges class '0' together, forming a band of class '0'. On the other hand in Figure 11, while class '0' is not overlapping, it is demonstrated that the most likely interpolation do not connect class '0' together, and that a case forming a band of class '0' apparently has lower likelihood, though it appears reasonable.

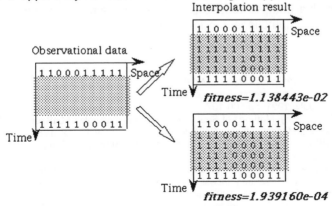

Figure 11 Interpolation results in a non-overlapping case.

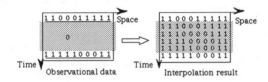

Figure 12 Interpolation result with additional observational data.

In Figure 12, another observational data is given at the middle. In this case, the most likely spatio-temporal pattern of class changes has a band of class '0'. These variations in the resulting patterns suggest that the interpolation method integrating   observational data and behavioral models/rules can estimate the most likely voxel field under different conditions.

6. Conclusion and future prospects

In this study, a spatio-temporal interpolation scheme is proposed for raster nominal data which can integrate observational data with behavioral models under the framework of the maximum likelihood of spatio-temporal events. Genetic-Algorithm/Hill-Climbing can be successfully applied to the combinatorial optimization of nominal voxel-field data. Our conclusions are summarized as follows:

1) GA/HC can be very rigorous because it can generate the most likely spatio-temporal distribution of class variables under observational data and a behavioral model;
2) Hill-Climbing method can be effective method to greatly improve the efficiency of GA.

Although our proposed GA/HC appeared effective for spatio-temporal interpolation, it is only an initial attempt to apply GA in the field of spatio-temporal interpolation. In future reseach, we intend to apply GA/HC for larger nomimal variable data, to reduce the speed of premature convergence and achieve higher computational efficiency of GA/HC.

References

[1] Bramlette, M.F. (1991): Initialization, Mutation and Selection Methods in Genetic Algorithms for Function Optimization, Proc. of the 4th. Conf. on GA, R.K.Belew and L.B.Booker (Editors), July, 1991, pp.100-107.

[2] Burrough, P.A. (1986): Methods of Spatial Interpolation, Principle of GIS for Land Resource Assessment, Monographic on Soil and Research Survey, No.12, 1989, pp.147-166.

[3] Davis, L. (1987) : Genetic Algorithms and Simulated Annealing, Pitman Publishing, 128 Long Acre, London WC2E 9AN, 1987.

[4] Eshelman, L.J. and J.D.Schaffer (1991): Preventing Premature Convergence in Genetic Algorithms by Preventing Incest, Proc. of the 4th. Conf. on GA, R.K.Belew and L.B.Booker (Editors), July, 1991, pp.115-122.

[5] Gold, C.M. (1989): Surface interpolation, spatial adjacency and GIS, Three Dimensional Applications in Geographic Information System, Edited by J.Raper, Taylor & Francis Ltd., 1989, pp.22-35.

[6] Shibasaki,R., T.Ito and Y.Honda (1993): Integration of Remote Sensing and Ground Observation Data for Developing Global GIS, Proc. of SEIKEN Symposium, Vol.12, Aug. 1993, pp.263-277.

[7] Shibasaki,R., M.Takaki and Y.Honda (1994): Spatio-temporal Interpolation Using Multi-Source Data for Global Dataset Development, COMMISSION III, IGWG III/IV; 1994.

[8] Huang, S.B. and R. Shibasaki (1995):Development of Generic Algorithm / Hill-climbing Method for Spatio-temporal Interpolartion, Sixth symposium on functional graphics and geographic infomation systems ,IIS, Univ. of Tokyo, 1995.

Contact address

Ryosuke Shibasaki[1] and Shaobo Huang[2]
[1]Institute of Industrial Science,
University of Tokyo
7-22-1 Roppongi
Minato-ku
Tokyo 106
Japan
E-mail: shiba@shunji.iis.u-tokyo.ac.jp
[2]Center for Environmental Remote Sensing
Chiba University
1-33 Yayoi-cho
Inage-ku
Chiba 208
Japan
E-mail: shaobo@rsirc.cr.chiba-u.ac.jp

MODELLING TROPOSPHERIC OZONE DISTRIBUTION

CONSIDERING THE SPATIO-TEMPORAL DEPENDENCIES

WITHIN COMPLEX TERRAIN

Wolfgang Loibl

Austrian Research Centre
Seibersdorf, Austria

Abstract

The article describes a model to derive hourly maps of tropospheric ozone distribution out of ozone monitoring network data within mountainous areas. This method is based on the fact that there is a strong dependence of tropospheric (ground level) ozone concentrations on relative altitude above the valleys that is changing daily and hourly. The core of the model is an elevation/daytime-dependence function that is automatically parameterised by day-specific monitoring data. Using this function and a digital elevation model ozone concentration surfaces can be modeled for every specific hour throughout the summer-half-year. Remaining local deviations of the monitored data are interpolated and added as deviation surface to the ozone concentration surface. Hourly available model results include the seasonal and day-specific influences as well as the regional influences caused by local emission - or weather situations. Validity tests show that the spatio-temporal dynamics of ground level ozone concentrations were calculated with high accuracy and spatial resolution. The model is applied since 1992 to create hourly maps of day specific ozone distribution maps for Austria. Since summer 1995 the model is integrated into the "Austrian Ozone Monitoring Network"-visualisation interface and maps are published by the Federal Environment Agency Austria as WWW-pages for every day.

Key words

Ozone, interpolation model, spatio-temporal dependence, elevation dependence, Austria.

1. Introduction

1.1 Comments on air quality monitoring and ozone mapping

Increased tropospheric ozone concentration is injuring human health and environment. Goal of the project was to develop a model that allows the automatic production of hourly maps of the spatial ozone exposure.

Air quality monitoring results in point data that reflect the actual air quality at the monitoring stations. To get information about the spatial distribution of air pollution the concentrations between monitoring stations have to be interpolated. Mapping of tropospheric ozone needs a special treatment. Ozone is a secondary air pollution gas. Its concentration is increasing and decreasing very fast due to ozone precursor concentration and intensity of solar radiation as described in chapter 1.2. In mountain regions ozone concentration patterns are very much dependent due to vicinity of ozone precursor emission sources and the vertical mixing of ozone and its precursors . This causes high ozone concentration variation within small distances (and few hours). So monitoring sites in mountain regions reflect only the very local situation of ozone and precursor situation.

Thus usually utilised (horizontal) distance depending interpolation models like "inverse distance weighting" or the often discussed Kriging or Co-Kriging models cannot be used to produce valid ozone distribution maps especially for mountain regions. The main reasons are:
- horizontal distance, indicating spatial influence-decrease, is surpassed by the vertical distance (above valley ground) that indicates ozone concentration increase,
- mountain ridges are interfering the spatial influence due to a barrier-effect,
- at least variogramm analysis within Kriging applications must be supervised because of spatio-temporal dependencies on ozone distribution that are changing constantly (See also Sen, 1996). Even if Kriging would lead to valid maps, they could not be generated without personal control.

The model presented in this article generates ozone concentration surfaces considering these spatio-temporal dependencies and allows unsupervised parameterisation for automatic calculation of valid ozone distribution maps during the summer half-year.

1.2 Spatio-temporal ozone dynamics in mountain regions

Tropospheric ozone is photochemically produced during hours of high solar radiation in regions with high concentrations of ozone precursor emissions (e.g. NO_x and hydrocarbons). These emissions are caused by several anthropogenic activities (industrial processes, car traffic, solvent use, etc.) which are in mountain regions usually concentrated in the valleys. Production of ozone is reinforced in valleys during noon hours at hot summer days. Vertical flux of air leads to transportation of ozone up to mountain regions. During night hours mountain-valley wind systems usually change direction and temperature inversion occurs. This enhances consumption of ozone by anthropogenic pollutants (i.e. NO) in the valleys. At higher elevations (above the valleys) ozone concentrations remain high due to lack of pollutants (*Broder and Gygax 1985*).

These reasons lead to characteristic ozone concentration variations. The diurnal variation in valleys are high - the average ozone concentration curve during a day is bell-shaped with maximum concentrations during noon hours and minimum concentrations in morning and night hours. At higher elevations diurnal ozone variations are low with constantly high levels throughout the day and night hours (*Puxbaum et al. 1991*). Ozone concentration in relation to elevation increase shows different ozone concentration gradients depending on daytime. During night and early morning hours there is a strong gradient with low ozone concentrations in the valleys and high concentrations in the mountain regions. During noon and afternoon hours there are high ozone concentrations either in valley or mountain regions. Other local factors also influence levels and temporal dynamics of ozone concentrations but in mountain regions they are also indicated by the vicinity to valley bottoms and by daytime. Thus the relative elevation above valley bottomand the daytime remain as the dominant factors reflecting the influence on spatial ozone distribution in mountain-valley systems (*Loibl et al. 1991, 1993, 1994*).

2. The ozone distribution model for mapping spatial ozone distribution out of ozone monitoring data

The idea of the model is to make use of the elevation dependence of ozone. The approach needs a digital elevation model of the study area and a (sufficiently) dense monitoring network.

The model is <u>not</u> an interpolation model based on a function of horizontal distance, that estimates concentration depending on the values of the neighbouring monitoring sites. Our model is based on a function that includes vertical distance (to the valley bottom) and daytime. It estimates the spatial concentration pattern dependent on the elevation pattern similar to a regression model. The function generates hourly standard ozone/elevation curves showing the different ozone concentration gradients depending on elevation at different daytimes (e.g. the increase of ozone concentration with increasing elevation during night hours). Hourly ozone concentration measurements are used to include seasonal and day-specific influences to the expected ozone concentration (for any specific hour) by fitting the hourly standard ozone/elevation curves. The day-specific curves and a digital elevation model of the study area are used to generate day-specific ozone concentration surfaces per hour. The remaining deviations of the observed data from the calculated ones reflect the local influences per hour. Hourly deviation surfaces are calculated by interpolation of the local deviations. The integration of the deviation surfaces into the ozone concentration surface leads to actual hourly day-specific ozone distribution maps of the study area considering the local influences. (*Loibl et al. 1991, 1992, 1993*). The model is performed by several steps as described in the sub-chapters 2.1 to 2.6.

2.1 Calculation of the relative elevation of the study area

The relative elevation has important influence to the ground level ozone concentrations in mountain-valley systems. It is defined as the vertical distance from the valley bottom to the monitoring sites and further to every point within the study area. The relative elevation is calculated as the difference between the absolute elevation (above sea level) of the monitoring station and the elevation of the bottom of the nearest major valley. To calculate a relative elevation surface out of a digital elevation model (DEM) the valley bottom elevation was assumed as the lowest absolute elevation within a distance of 5 km around every DEM - grid point. The relative DEM is generated through cell-by cell-calculation of the difference between the elevation of the original DEM-grid-value and the lowest within a search radius of 5 km. DEMs mentioned within the article are always DEMs of relative altitude. (The used DEM has a spatial resolution of $1 \times 1 km^2$.)

2.2 Calculation of a standard ozone/elevation/daytime function

The diurnal variation of ozone concentration depending on daytime and elevation is described by an analytical function. It was approximated by analysing scatterplots of ozone concentrations against relative elevation of the monitoring sites. Therefore ozone concentrations from about 100 ozone monitoring stations of the Austrian monitoring network were collected through several episodes of high ozone concentrations between 1991 and 1993 and averaged for every hour. The Austrian monitoring network covers an area of 84.000 km^2 with elevations between 200 and 3800 m and has about 125 monitoring sites in 1996. The highest monitoring site has a relative elevation of 1650m but the function can be used for any meaningful relative elevation. At higher relative elevations only little change in ozone/ daytime/ elevation dynamics will be reflected. (*Loibl et al. 1991, 1992, 1993*).
Equation (1) shows the basic function:

$$O_{std}(re,t) = A(t)^* B(re,t) + C(re,t) \qquad\qquad (1)$$

The sub-functions *A, B* and *C* are:

$$A(t) = a_1 + a_2 . e^{-(t-a_3)^2 . a_4} \qquad\qquad (2)$$

$$B(re, t) = ln\ (b_5.re + b_1 + b_2 . e^{-(t-b_3)^2 . b_4}) \qquad\qquad (3)$$

$$C(re, t) = \frac{c_1}{1-(t-c_2)} . \frac{re}{c_3} \qquad\qquad (4)$$

with O_{std} = standard ozone concentration
  re = relative elevation
  t = daytime
  a_n, b_n, c_n = parameters approximated with the 1991 to 1993 ozone monitoring data

Function (2) describes daytime and intensity of the morning ozone-depletion in the valleys. Function (3) describes intensity of ozone concentration increase due to elevation and the daytime of the ozone maximum. Function (4) corrects the mountain ozone concentration to be constant instead of increasing during afternoon hours. The approximation was started with one function for every hour. Later all 24 functions were integrated into the general function (1) using the daytime as constant value responsible for the hourly change of the ozone gradient depending on elevation. The complete function (1) generates a 2-dimensional trend surface of ozone concentration reflecting the spatio-temporal dependencies of ozone on elevation and daytime (fig. 1). "Slicing" the surface parallel to the time-(x)-axis one gets the diurnal variation for the elevation level where the surface is "cutted". "Slicing" the surface parallel to the elevation-(y)-axis one gets ozone/elevation curves for the day time at which the surface is "cutted". These hourly curves are the base information to generate day-specific hourly ozone concentration distribution maps.

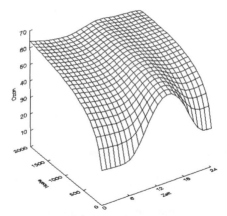

Figure 1 Trend surface of the standard ozone/day-time/elevation function.

2.3 Calculation of day-specific ozone/elevation curves

The hourly standard ozone/elevation curves describe the (idealised) standard dependence of ozone concentrations on relative altitude of a typical summer day. Measured concentrations usually deviate from those because of seasonal, day-specific and local influences. In order to apply the model for different synoptic meteorological patterns or ozone precursor concentration variation, the hourly standard ozone/elevation curves must be modified for every day. This is performed by adding a deviation curve to the standard ozone/elevation curve. Such a deviation curve is calculated using the deviations of the measurements of the specific hour from the elevation-specific standard concentration and averaging the deviations for 3 elevation ranges. The elevation ranges are 0 to 100 m, 200 to 600m and 1100 to 1650m (1650 m is the highest relative elevation of a monitoring station). These 3 average deviations are transferred to 3 defined "fitting elevation levels" (0m, 400m and 1650m) to fix the start point and 2 intermediate points of the hourly "deviation curve". To complete the curve the deviations for all relative elevations (points along the curve) are linear interpolated using the 2 average deviations of the neighbouring fitting elevation levels (0 and 400m or 400 and 1650m). For relative elevations above 1650m the 1650m-deviation is kept constant. The addition of the deviation curve to the standard ozone/elevation curve results in a day-specific ozone/elevation curve. (Fig. 2 shows the standard and the day-specific ozone/elevation curve for 8:00 and 14:00h of 5th July 1993).

2.4 Calculation of day-specific spatial ozone distribution

The expected ozone distribution is calculated cell-by-cell recognising the elevation of the DEM - grid cell and copying the elevation-specific ozone concentration out of the day-specific ozone/elevation curve to the output grid. The result is an ozone concentration surface with the spatial resolution of the elevation model. It reflects the day-specific ozone distribution of the study area but does not include locally caused influences (e.g. local high precursor emissions or thundershowers).

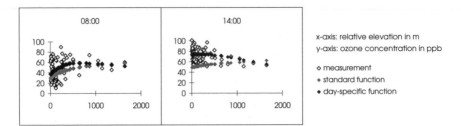

Figure 2 Ozone concentration and ozone/elevation curver (July 5, 1993).

2.5 Interpolation of regional deviations considering elevation dependencies

The remaining deviations of the monitoring data from the day-specific ozone concentrations indicate the local influences to the overall ozone concentration. To include these influences into the ozone concentration surface, they have to be interpolated. It is assumed that the influences causing the deviations decrease with growing distance to a given monitoring site.

Until now the elevation of the monitoring data was relevant, now the location of the measured ozone data is considered. The grid-cells of the day-specific ozone concentration surface that correspond to the locations of the monitoring sites are used to calculate the deviation surface. But this deviation surface cannot be calculated by traditional spatial interpolation procedures because there is again a deviation dependence on altitude. The lower the relative elevation is, the higher the deviations are due to higher local influences at valley monitoring stations caused by various anthropogenic activities (*Loibl et al.* 1993). Most monitoring sites are located in the valleys near emission sources. If the high positive and negative deviations at valley stations (see fig. 3) are spatially interpolated "over" mountain regions, the high deviations would be extended to regions where these valley-specific influences have no effect. Therefore the (usually) high deviations from valley monitoring sites have to be interpolated as low deviations over mountains. On the other hand deviations from mountain monitoring sites must be interpolated as they are, because those deviations reflect regional influences without very local variation. They are causing the same deviations in valley - and mountain regions. The deviation interpolation is performed by following steps:

(A) Standardising the deviations to elevation independent deviations
This standardisation is performed using an elevation-dependent weighting function considering the relative elevation of the monitoring sites. To derive the function, the absolute values of the deviations are plotted against the relative elevation with deviations along the y-axis and relative elevation of the monitoring sites along the x-axis. The weighting function was approximated by generating a curve that follows the shape of the upper hull of the scattered average deviations in x-axis-direction as shown in fig. 4. The function generates elevation dependent weighting factors that are used to generate elevation independent deviations *(Loibl et al., 1993).*

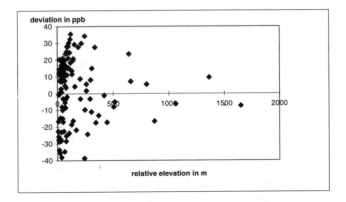

Figure 3 Ozone concentration deviation from day-specific curve of July 5, 1993, 8:00h against elevation.

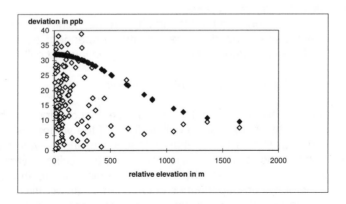

Figure 4 Absolute deviations on July 5, 1993, 8:00h against elevation weighting function curve.

(B) Spatial interpolation of the elevation independent deviations

Spatial interpolation is performed by inverse distance-weighting (IDW), assuming that the influences of data points decline with increasing distance from the point being estimated *(Shepard, 1968)*. The IDW-algorithm does not lead to such smooth surfaces as e.g. spline interpolation, but it always produces interpolation surface values not higher or lower as the surrounding monitoring sites in the vicinity and it produces identical surface values as the observed monitoring data at the grid cells where the monitoring sites are located .

(C) Elevation dependent re-extending of the elevation independent deviation surface

To include the elevation effects to the deviation surface the values have to be re-extended cell-by-cell, using the inverse weighting factors of the relative elevation of the corresponding DEM-grid cell.

(D)Spatial interpolation of the original deviations

Deviations from mountain monitoring sites must always be interpolated as they are, but step (C) will lead to higher deviations in valleys near mountain monitoring sites. Their (lower) deviations are extended by the inverse weighting factor due to low elevations. Thus an interpolation of the original deviations has to be performed which is done by IDW too.

(E) Final deviation surface

Now one cell value of the two deviation surfaces has to be selected to create a final deviation surface. Original deviation surface grid-cells should be selected if influenced by mountain monitoring stations while elevation dependent deviation surface grid-cells should be selected if influenced by valley monitoring sites. The selection is performed cell-by-cell selecting the smaller value of both surfaces. It is assumed, that the valid deviation is always that one which is closer to the expected day-specific ozone concentration:

- If there are valley-monitoring sites in the vicinity, the elevation dependent deviation surface values are lower in higher elevated grid cells than the original deviation surface values.
- If there are only mountain-monitoring sites in the vicinity, original deviation surface values are lower than the elevation dependent deviation surface values that are extended in the neighbouring valley grid cells.

2.6 Generation of actual ozone distribution maps

The day-specific ozone concentration surface is combined with the final deviation surface by simple cell-by-cell addition of the values inside the grids. The resulting map shows the spatial distribution of the elevation-dependent ozone concentrations for the specific day and the specific daytime including the regional deviations caused by local influences. Due to the curve-fitting step the ozone/elevation-dependence curves can be automatically adapted to any day-specific elevation-dependence for every random hour and day, at least during a period from early spring to late autumn.

Fig. 5 presents as example two ozone distribution maps of Austria showing the distribution patterns at 8:00 and 14:00h for the 5th July 1993. The 8:00h-map shows the very high dependence of the ozone concentration on the elevation in morning hours, the elevation pattern reflects the ozone concentration pattern: along the valleys and in the plains there is low ozone concentration, while in the mountain areas ozone concentration is remarkable higher. The 14:00h-map shows typical noon hour ozone distribution. The highest ozone concentrations are in the eastern part of Austria with the highest settlement density, industrial- and traffic density.

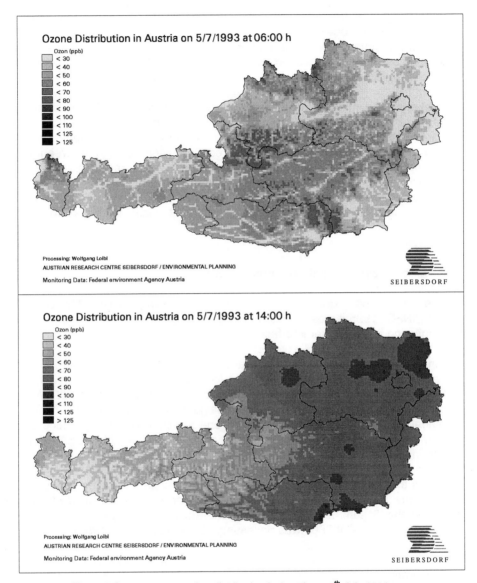

Figure 5 Ozone concentration distribution in Austria on 5th July 1993.

3. Comments on the validity of the results

The validity of the function and the curve fitting process can be easily examined by scatterplots showing the deviations of the measurements from the model results (along y-axis) plotted against the relative altitude (along x-axis). The deviations will usually scatter symmetrically along the x-axis from low to high elevated areas as shown in fig 3. *(Loibl et al., 1993).*

The validity of the maps is depending on their spatial representativity. At the monitoring sites the calculated ozone concentration will certainly be identical to the measured ones because of the inclusion of the deviations. To test the spatial representativity of the results for regions without monitoring stations, maps have been calculated with and without the exclusion of several monitoring sites.

Spatial representativity is proved if both maps show the same or nearly the same results. The comparison of such maps shows just little deviations - 0 to a few ppbs during hours of high ozone concentration (usually between 60 to 100 ppb). During morning hours (with concentrations usually between 10 to 40 ppb), when local ozone concentration reacts more sensitively to anthropogenic influences, the deviations are a little higher but usually below 10 ppb *(Loibl et al., 1993)*. (This test is only significant if there are no monitoring stations excluded reflecting great local influence to ozone formation.)

4. Model development and applications

The model was developed 1990 to 1994 using ESRI's GIS-Software ARC/INFO. 1995 it was rewritten in C and integrated into the Austrian ozone monitoring network user interface.

Since 1995 it is used by the Federal Environment Agency Austria (FEEA) to generate daily ozone distribution maps for several hours during summer days. Since April 1996 these daily maps are published constantly on a WWW-page of the FEEA (homepage: http://www.ubavie.gv.at/). Since summer 1996 the model is applied within a statistical ozone forecast model to produce short-term ozone forecast maps. They will be also published by the FEEA - WWW-site in the near future. 1995 and 1996 the model was used to create AOT40-maps for 1993 and 1994 as Austrian contribution to the UN-ECE Critical Loads and Levels Program to estimate the spatial pattern of the yearly ozone exposure (AOT40 = accumulated concentration over a threshold of 40 ppb during the 6 summer months). To generate AOT40-maps for 2 years the calculation of about 8800 was necessary *(Loibl, 1995, Loibl 1996)*.

References

Broder B., Gygax H.A., 1985, The influence of locally induced wind systems and formation on the effectiveness of nocturnal dry deposition of ozone, Atmos. Environment 19, 1985, pp.1627-1637.

Loibl W., Orthofer R. et al. ,1991, Ozon in Österreich Teil 2: Verteilungsmodelle und Flächen-interpolationsverfahren zur synoptischen Darstellung in Kartenform. OEFZS-A-2177, Seibersdorf.

Loibl W., Züger J., Kopcsa A., 1992, Flächenhafte Ozonverteilung in Österreich für ausgewählte Ozonepisoden 1991. Plausibilitätsanalyse der Ozonmeßdaten. OEFZS A-2436, Seibersdorf.

Loibl W., Züger J., Kopcsa A. , 1993, Darstellung des Ozonverlaufs während der Ozonepisoden 1992 und Analyse der Stationen je Überwachungsgebiet auf redundante Information. OEFZS A-2783, Seibersdorf.

Loibl W., Winiwarter W., Kopcsa A., Züger J., Baumann R., 1994, Estimating the spatial distribution of ozone concentrations in complex terrain using a function of elevation and day time and Kriging techniques; in: J. Atmos. Environment; Vol 28, No.16 pp 2557-2566.

Loibl W., Züger J., 1994, Überprüfung der Parameter des Verfahrens zur Generierung von Ozonbelastungskarten anhand ausgewählter Tage der Ozonepisoden 1993. OEFZS A-3208, Seibersdorf .

Loibl W., 1995, Modelling the Spatial Distribution of Critical Levels of Ozone Considering the Influence of Complex Terrain. UN-ECE Technical meeting on Mapping of Critical Loads /Levels, Helsinki. 1995. Republished 1995 as Seibersdorf Report OEFZS-A--3434. Seibersdorf.

Loibl W., 1996, Spatial modelling of accumulated ozone exposure in alpine regions considering daytime and elevation-dependence - Comparison of different ozone exposure patterns. in: Proceedings of the Workshop on Critical Levels for Ozone in Europe: Testing and Finalising the Concepts". UN-ECE Technical Meeting. University of Kuopio, Finland 1996.

Puxbaum H., Gabler K., Smidt S., Glattes F., 1991, A one year record of ozone profiles in an alpine valley, J. Atmos. Environment 25A-9, 1991, pp 1759 - 1666.

Shepard D., 1968, A two-dimensional interpolation function for irregular spaced data. Proceedings 23rd ACM National Conference pp.517-524.

Sen Z., 1996, Discussion of "Geostatistical Analysis and Vizualisation of Hourly Ozone Data" presented by Casado L. et al., 1994 . J. Atmos. Environment; Vol 30, No.2, pp 353-354.

Contact address
Wolfgang Loibl
Austrian Research Centre
A-2444 Seibersdorf
Austria
Phone: +43-2254-780-3875
Fax: +43-2254-780-3888
E-mail: loibl@zdfzs.acs.ac.at

TEMPORAL EXTENSIONS FOR AN OBJECT-ORIENTED

GEO-DATA-MODEL

Andreas Voigtmann, Ludger Becker and Klaus H. Hinrichs

FB15, Informatik
Westfälische Wilhelms-Universität
Münster, Germany

Abstract
The various needs of GIS applications require an extensible data model capable of handling both time dependent spatial and non-spatial data. OOGDM is an object-oriented data model developed to serve as a general base for GIS applications. Two- and three-dimensional raster- and vector-based data is supported by the data model. A SQL-like query language and a data definition language have been developed for OOGDM. In this paper, we describe recent extensions to OOGDM to incorporate time into the data model and the query language.

Key words
Database, temporal data model, temporal query language.

1. Introduction

Temporal databases have been an active field of research in the past decade [KI93]. Although most research dealt with temporal extensions to the relational data model, the work on temporal object-oriented databases becomes more and more important. The terminology for temporal databases has been defined in [Je93].

For spatial databases and geographic information systems (GIS), the incorporation of time is an important task. Spatial data is often highly time dependent (e.g. temperatures, vegetation coverage, or economic information related to states). Current systems lack the ability to process these types of temporal information. The incorporation of the time domain into spatial databases is currently investigated from both the database and the GIS community. An overview can be found in [ASS93].

The Object-Oriented Geo-Data Model OOGDM [VBH95] is an extensible object-oriented data model for geo-applications. Based on the modelling concepts of object-oriented data models, a hierarchy of abstract classes has been designed as a base for the development of GIS applications. OOGDM supports both 2D and 3D GIS-applications as well as raster- and vector-based data. Concrete GIS-applications specify user-defined classes based on the OOGDM hierarchy. T/OOGDM is an extension of the OOGDM data model incorporating time. The basic class hierarchy of OOGDM has been extended by time-dependent attributes. For user-defined classes, temporal support can be added as required by the application.

Temporal query languages for temporal data models have been developed for relational and object-oriented models. Usually, these languages are extensions of traditional languages, e.g., [Sn93] extends Quel, [NA93] and [Sa93] extend SQL, and [OM93], [RS93], [Sn95] extend SQL-like query languages for object-oriented data models. Recently, efforts have been made to design a standardized temporal query language TSQL2 [TSQL2] for temporal relational databases. This language is based on the SQL-92 standard. For object-oriented databases, there is still no consensus about a SQL-like query language. The *Object Database Standard* [ODMG93] proposes the SQL-like query language OQL for object-oriented databases. However, this language has not yet become a standard. The SQL-like query language OOGQL for OOGDM is based on OQL. The query language T/OOGQL extends OOGQL by temporal extensions comparable to TSQL2.

The paper is organized as follows: In section 2 we briefly review OOGDM. Further details and examples can be found in [VBH95] and [VBH96]. Section 3 describes the temporal extensions to OOGDM. We first give a short overview of the principles underlying temporal data modelling. Afterwards, we describe the extensions for T/OOGDM with respect to the data model and the query language. In section 4, we present an example showing the temporal features of T/OOGDM and its query language. Section 5 concludes this paper.

2. Overview of OOGDM

2.1 Basic structure

The core of OOGDM is a hierarchy of classes. The top-level and therefore most abstract class of our data model is *spatial_object*. It represents a set of attributes and operations common to all instances which are stored in the GIS-database. The class *spatial_object* is specialized into two different subclasses – *feature* and *geo_object* (see figure 1). Geo-objects are used to describe complex objects of the real world. They are composed of *features*, other *geo_objects*, and atomic values. A feature represents a single geo-abstraction. It is either elementary or composed of several features (class *feature_set*).

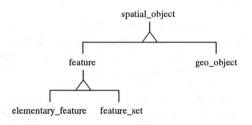

Figure 1 Top-levels of the inheritance tree.

2.2 Elementary features

An elementary feature contains a description of the geometry related to this abstraction. All concrete information which is required in a specific application area must be added by providing a type parameter for the thematic information or by deriving subclasses from the predefined classes. Based on the geometry of the objects, methods for basic operations can be provided which are useful in all GIS application areas. We distinguish the elementary features by their spatial dimensionality.

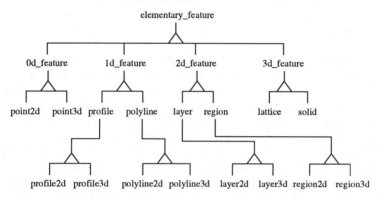

Figure 2 Child-classes of elementary features.

Except for the zero-dimensional elementary features which are described by the class 0d_feature we distinguish between vector-based and raster-based representations. One-dimensional elementary features are either raster-based *profile* objects or vector-based *polyline* objects. Two-dimensional elementary features are specialized into *layer* objects which are used for storing raster-based data or into *region* objects which are used to store vector-based data. A three-dimensional elementary feature is either a raster-based *lattice* object or a vector-based solid object. We assume that all vector-based representations describe a contiguous part p of the data space, i.e., the information associated with these features is known for all points in p, whereas the raster-based representation obviously describes a discrete part of the data space.

To support both 2-D and 3-D GIS-applications we have to distinguish whether an elementary feature is used in a two- or in a three-dimensional data space. Hence, we introduce corresponding subclasses of *0d_feature*, *profile*, *polyline*, *layer*, and *region*. Figure 2 introduces the names of these classes and presents the complete hierarchy of classes for elementary features.

2.3 Geometric and topological operations

The basic classes of OOGDM introduced above represent spatial data. To support processing of this data, a set of geometric and topological operations is required. Geometric operations include a *distance* operation, *length*, *area*, or *volume* operations for one-, two-, or three-dimensional spatial objects, a center operation returning the *center* of an object, and a *bounding_box* operation returning the minimal bounding box of an object.

Topological predicates have been discussed intensively, e.g. [EF91], [ET92], [CFO93]. For our purposes, we use the following set of five basic topological predicates: *disjoint*, *touch*, *in*, *cross*, *overlap* proposed by [CFO93]. The predicates can easily be adapted to support 3-dimensional data. Based on these predicates corresponding topological functions *touch*, *cross*, and *overlap* can be defined which construct for each pair of objects satisfying the topological relationship the resulting intersection object.

2.4 Feature sets

Elementary (geo-) features are not sufficient to describe the objects used by a GIS. There is often the need to model compound features, e.g., a river consisting of many segments or a network of rivers. The class *feature_set* is used to represent these types of features. Similar to the class *elementary_feature*, we perform a subdivision of *feature_set* according to the spatial dimensionality, the kind of the resulting object, and the kind of the elementary features being composed. Since objects consisting of components of different types are described by geo-objects, feature sets are homogeneous. A feature set combining objects of 2-D space belongs to 2-D space, and a feature set combining objects of 3-D space belongs to 3-D space.

In figure 3 we show the definitions for the class *feature_set* and its subclasses. It is obvious that 0-D elementary features can only be used to define raster-based feature set objects. A set of 0-D features or a set of profiles can describe a profile in 2-D or 3-D space, a layer in 2-D or 3-D space, or a lattice. The corresponding classes are given in parts (1) and (3) of figure 3. A set of polylines is a network in 2-D or 3-D space (part (2)). The classes of part (4) describe sets of regions. A set of layers can define a layer in 2-D or 3-D space or a lattice (part (5)). A set of solids is described by the class *solid_set* of part (6). A set of lattices can define another lattice. Such feature sets are denoted by the class *lattice_3d_set* in part (7).

All classes of feature sets which combine raster-based objects of class x basically inherit from *feature_set<x>* and from the class of elementary feature constructed by the set. The first class provides the methods for processing sets and the second class describes the methods and the geometry for the resulting feature. Due to multiple inheritance the feature sets are subclasses of an elementary feature class. Hence, we may also use feature sets to define new feature sets. For example, if we use a set of *point2d* to construct a *layer2d_0d_set*, we can use this set as an element of a *layer2d_2d_set* object.

2.5 Geo-objects

Geo-objects are complex entities of the real world. They are defined as a tuple containing various information based on elementary features, feature sets and additional (non-spatial) data. Examples for such complex objects are cities, countries, rivers, roads, and even maps.

class feature_set <T -> feature> : feature { **relationship Set <T>** members; };
(1) Sets of 0-D features:
      class profile2d_0d_set <T -> point2d> : feature_set <T>, profile2d;
      class profile3d_0d_set <T -> point3d> : feature_set <T>, profile3d;
      class layer2d_0d_set <T -> point2d> : feature_set <T>, layer2d;
      class layer3d_0d_set <T -> point3d> : feature_set <T>, layer3d;
      class lattice_0d_set <T -> point3d> : feature_set <T>, lattice;
(2) Sets of polylines
      class network2d <T -> polyline2d> : feature_set <T>;
      class network3d <T -> polyline3d> : feature_set <T>;
(3) Sets of profiles
      class profile2d_1d_set <T -> profile2d> : feature_set <T>, profile2d;
      class profile3d_1d_set <T -> profile3d> : feature_set <T>, profile3d;
      class layer2d_1d_set <T -> profile2d> : feature_set <T>, layer2d;
      class layer3d_1d_set <T -> profile3d> : feature_set <T>, layer3d;
      class lattice_1d_set <T -> profile3d> : feature_set <T>, lattice;
(4) Sets of regions
      class region2d_set <T -> region2d> : feature_set <T>;
      class region3d_set <T -> region3d> : feature_set <T>;
(5) Sets of layers
      class layer2d_2d_set <T -> layer2d> : feature_set <T>, layer2d;
      class layer3d_2d_set <T -> layer3d> : feature_set <T>, layer3d;
      class lattice_2d_set <T -> layer3d> : feature_set <T>, lattice;
(6) Sets of solids
      class solid_set <T -> solid> : feature_set <T>;
(7) Sets of lattices
      class lattice_3d_set <T -> lattice> : feature_set <T>, lattice;

Figure 3 Definition of feature_set and its subclasses (methods omitted).

The class *geo_object* has a *geometry* attribute of class *elementary_feature*. This relationship can be redefined in classes inherited from *geo_object* to adapt the geometry for the purposes of the new class. Due to inheritance, each object of a class derived from *geo_object* has an attribute *geometry*. Hence, geometric operations can be applied to any object, and no specification of the attribute describing the geometry of an object is required. This is an important issue for the query language OOGQL [VBH96].

3. Temporal extensions for OOGDM

3.1 Time in database and geo-information systems

The integration of time into database systems has been an active field of research for several years. Relational and object-oriented temporal data models have been developed ([ASS93], [ÖS95], [Sn92], [Sn95]), and corresponding query languages have been proposed. Recently, an attempt to define a standardized temporal query language for temporal relational databases which is based on the SQL-92 standard has been made [TSQL2]. Temporal databases handle time with respect to *valid-time*, i.e., the time when the event occurs in the real world, and *transaction-time*, i.e., the time when the event was stored in the database. In contrast to valid- and transaction-time *user-defined time* is handled like other attribute types.

Valid- and transaction-time are orthogonal and have different semantics. A data model supporting valid-time is called *valid-time model*, and a model supporting transaction time is called *transaction-time model*. Data models supporting both valid- and transaction-time are called *bitemporal models*. Valid-time models can be bounded or unbounded. Events in the future are supported in an unbounded valid-time model. On the other hand, transaction-time is always bounded by the time the database was created and the current time.

In an object-oriented data model, timestamps can be attached either to objects or to attributes. A timestamp may be a *chronon* (i.e., a non-decomposable time-interval of a fixed, minimal duration), an *interval* (i.e., the period between two instants of time which is represented by a set of contiguous chronons), or a *temporal element* (i.e., a set of intervals). [Sn92], [Sn95], and [TCG93] give a comprehensive survey of temporal aspects of databases.

Integrating time into geo-data models has become more and more important in the world of GIS. Geo-data depends on its spatial properties and on the time it has been recorded. Besides the spatial "where" and the thematic "what" queries a geo-database must support a "when"-operation. A geo-database should be able to answer queries like *"Select the borders of all countries of Europe in 1900"* or *"Show the development of urbanization in the area of 'Münster' from 1980 to 1990"* (see also [La92], [Pe95]).

Temporal support for small (geo-)objects can be realized by the methods described above for non-spatial database systems. However, in general geo-objects have a considerable size. If a large object changes over time, a simple temporal model like the *snapshot model* [La92] replicates the unchanged and the changed data. Since such an approach is unsuitable for large data sets, the *grid model* and *the vector model* ([La92], [Pe95]) store only those parts of an object which change at a given point in time. More sophisticated spatio-temporal data models like ESTDM [Pe95] even provide a compression of data over time. As mentioned in [Wo94], it is important to apply the timestamps to small units of geographic objects.

Our extensions to OOGDM to incorporate time into the data model and the query language consider the aspects mentioned above. In this paper we do not discuss implementation specific aspects but concentrate on the description of temporal extensions to OOGDM at the user-level. Implementation specific details are subject of a forthcoming paper.

3.2 Extending the data model

In T/OOGDM we consider valid-time and transaction-time only. However, each user can add his specific classes for user-defined time. We first consider valid-time.

3.2.1 Valid time

In T/OOGDM the timestamps are applied to simple types, i.e., *byte, short, long, float, double, boolean,* and *string*, or to relationships referencing other objects, i.e., T/OOGDM provides a kind of attribute-timestamping concept. As described in [VBH95] objects are considered as containers for several attributes. Objects cannot be attributes of another object, they may only be referenced via relationships. Hence, there are no timestamps for whole objects. This applies to objects of OOGDM- and user-defined classes, and special collection classes *like Set, List, Bag,* and *Array*. However, relationships can be timestamped allowing a user to assign different objects of a class to another object in the course of time.

The object definition language of T/OOGDM has been extended by a new keyword *timestamped* to identify timestamped attributes or relationships. Consider the following example:

```
class myClass{
        timestamped attribute short v;
        timestamped relationship Set<String> w;
        timestamped relationship Bag<class_with_timestamped_attribute> x;
        timestamped relationship class_without_timestamped_attributes y;
        timestamped relationship class_with_timestamped_attribute z};
```

The first attribute of *myClass* is a timestamped short integer attribute. Consider attribute w: the set of strings is timestamped but the individual element strings are not, i.e., there are relationships to different sets of strings over time. Obviously, a collection can contain objects having timestamped attributes (attribute x). Similarly, a user can timestamp a relationship to an object with or without timestamped attributes (attributes y and z).

Often data is recorded at a specific point in time and is only valid at this point in time, i.e., before and after recording it is not (exactly) known. An example are temperatures at a specific location. This type of data requires a timestamp describing exactly the time of recording. We model this by a *chronon-timestamp*. Other types of data are valid during specific time intervals. For example, the vegetation coverage of a given area may contain data of type "subarea A is covered by forest since April, 1st 1990" or "subarea B was covered by grass from June, 20th 1985 to March, 1st 1995". This should be modelled by an *interval-timestamp*. Interval timestamps can be unbounded (first example) or bounded (second example). If the timestamp is unbounded, the interval timestamp will be closed when the attribute value is modified the next time.

To satisfy the requirements of various GIS-applications T/OOGDM supports chronon-timestamps and interval-timestamps. In the object definition language of T/OOGDM the keywords *chronon_timestamped* and *interval_timestamped* specify whether chronon-timestamps or interval-timestamps shall be used. The keyword *timestamped* used above is a synonym for *interval_timestamped*. Interval-timestamps are the default since a chronon-timestamp can be represented by an interval containing a single point on the time-line. Of course, timestamping of attributes or relationships should only be used when it is required by the application. For example the name of a city should not be timestamped since usually it does not change over time, however, the city's population should be timestamped.

3.2.2 Transaction time

As mentioned above, transaction-time has different semantics as compared to valid-time. Transaction-time is controlled by the database itself since it describes the time data has been stored, deleted, or modified in the database. Hence, a user cannot control transaction-time. In contrast to valid-time, timestamps for transaction-time can only be intervals. Suppose, an object is inserted into the database at transaction-time a, and is modified or deleted later at transaction-time b. This object has the interval-timestamps [a, b] for the transaction-time. The timestamp for transaction-time of a newly inserted object is not closed (b equals *now*).

For transaction-time, we chose to use object-timestamps instead of attribute-timestamps. Object-timestamping leads to higher storage cost, but for transaction-time a clear distinction should be made between objects which are active in the database and objects representing previous states of the database. Object-timestamping for transaction-time allows an easy rollback to a previous state of the database. Moreover, since (non-temporal) object-oriented databases allow versioning of objects (e.g., ObjectStore [LLOW91]), we can use this mechanism to support transaction-time easily.

To reduce the storage cost the user may decide to create the database without transaction-time. In addition, the storage cost can be controlled by the length of the transactions. E.g., if all updates of a layer describing a vegetation coverage are subsumed in a single transaction, the corresponding object is replicated only once.

3.2.3 T/OOGDM

In this section we discuss the necessary modifications to the predefined OOGDM classes.

Elementary features

We first consider class *elementary_feature* and its subclasses. The geometry attribute and the attribute storing the associated thematic data are not timestamped by default. To timestamp attributes of elementary feature classes each usage of an elementary feature class can be qualified by the *keywords timestamped, interval_timestamped,* or *chronon_timestamped*, e.g.:

```
class myClass{
        relationship timestamped region2d<...> a;
        timestamped relationship timestamped layer2d<...> b; };
```

The first relationship is not timestamped and references a *region2d* object having timestamped attributes as discussed below. The second relationship is timestamped and references a *layer2d* object having timestamped attributes.

Now we consider how the attributes of an elementary feature class should be *timestamped* if the class is qualified by timestamped-keywords. The geometry is the key property of any feature object. The data attribute stores associated thematic data, and the database must be able to record the change of data in the course of time. If the elementary feature class is qualified as timestamped, both the geometry and the data attribute are timestamped. For raster based elementary features each element of the array associated with the data attribute is timestamped, providing a grid model like approach. Since elementary features are generic classes, the data attribute can be a user-defined class containing timestamped attributes or relationships. Hence, there may be inconsistencies between the timestamps of attributes of the user-defined class and the timestamp on the elementary feature's data attribute.

Feature sets
For feature sets, there is one relationship *members* referring to a set of elementary features. This relationship is *timestamped* if the class is qualified by timestamped-keywords.

Geo-objects
The attributes and relationships contained in geo-objects are application dependent. All these attributes and relationships as well as the geometry relationship predefined by class *geo_object* may be timestamped.

Comparison to other spatio-temporal data models
T/OOGDM addresses the main requirements for a spatio-temporal data model as discussed in section 3.1. Since timestamps are applied to elementary features, we can record changes for components of geographic objects [Wo94]. By default vector-based features use a snapshot approach. However, since OOGDM supports alternative representations for elementary features [BVH96], other representations can be chosen (e.g., to realize the vector model [La92]). For raster-based elementary features, an approach similar to the grid model [La92] is used to provide data reduction for large data sets which change over time. By using an alternative representation, we may even integrate a data model providing data compression [Pe95] for raster data.

3.3 Extending the query language

To support the new temporal features of T/OOGDM, we extend OOGDM's query language. This extended query language called T/OOGQL is a full spatio-temporal SQL-like query language. The basic features are similar to ODMG's query language OQL [ODMG93], and the basic syntax for temporal extensions is similar to TSQL2 [TSQL2].

3.3.1 The select-statement
A query in T/OOGQL is defined by
> **select** (**snapshot**) (**distinct**) <select - part>
> **valid** <valid-time projection>
> **from** <from - part>
> **where** <where-part>

The select-, from-, and where-parts of the query are similar to the non-temporal spatial query-language OOGQL as described in [VBH95] and [VBH96]. The main changes are the valid-clause, the new keyword snapshot in the select-part, and temporal predicates in the where-part. The temporal predicates are discussed below. The keyword snapshot has the meaning as defined in TSQL2: If the keyword is not present in a query, the query result contains the associated timestamps for valid- and transaction-time. If the keyword snapshot is present these timestamps are removed from the query result - the output is a snapshot of the result.

The valid-clause is used for valid-time projection, as discussed in [TSQL2]. The result is computed as described by the query without valid-clause, and afterwards the timestamps of the result are projected on the range of time defined by the valid-clause.

OOGQL already introduced a transparent handling of spatial attributes ([VBH95], [VBH96]), i.e., we can apply geometric or topological operations to objects without referencing the geometry attribute. This methodology is also applied to the timestamps. This is an advantage as compared to the handling of timestamps in TSQL2: For example the comparison of two valid-time timestamps for tuples d and e in a TSQL2 query requires access to the timestamps:

> where ... <non-temporal comparison> ... and valid(e) contains valid(d)

In T/OOGQL the valid-time timestamps of two objects, attributes, or relationships d and e can be compared by applying the valid-time predicate to the objects:

> where ... <non-temporal comparison> ... and e vt_contains d

Furthermore, temporal operations can be applied to complete objects, even though timestamps are defined for attributes only. If a temporal operation is applied to an object, an implicit object-timestamp is generated for the object which is derived from all attribute-timestamps.

3.3.2 Temporal predicates and operations

T/OOGQL is a spatio-temporal query language. Hence, there may be a naming conflict for some temporal predicates and operations and geometric or topological predicates and operations (e.g. the predicate contains). To distinguish geometric and topological operations and temporal operations the following naming schema has been selected:

- Temporal predicates accessing a valid-time timestamp are named starting with 'vt_'. Since valid-time is the default temporal dimension supported by T/OOGDM, these predicates and operations are available with a preceding 't_', too. For example the temporal predicate contains is used as t_contains or vt_contains for valid-time.
- Temporal predicates accessing a transaction-time timestamp are named starting with 'tt_'. For example the temporal predicate contains is used as tt_contains for transaction-time.

The temporal predicates available in T/OOGQL are shown in figure 4. Further predicates, like *follows, during, starts, finishes*, etc. [RP92] can be expressed by the predefined predicates or can be realized via the extension mechanism provided for T/OOGDM.

Besides the temporal predicates, the following constructors and operations are available:

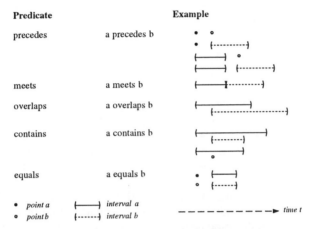

Figure 4 Temporal predicates for T/OOGQL.

- First() and Last() return the oldest and latest timestamp associated with an attribute, object, or relationship.
- FirstValue() and LastValue() return the oldest and latest value associated with an attribute, object, or relationship.
- Begin() and End() return the start time and the end time of an interval-timestamp.
- Period(s,e) and Period(d) return a period having start-time s and end-time e and a period of duration d.
- day(), month(), and year() are constructors for interval-timestamps describing a day, month, and year.
- date(), time(), and datetime() are constructors for chronon-timestamps of a specified date, time, and date, and time.
- years(n), months(n), days(n), hours(n), mins(n), and secs(n) return a time-span having a duration of n years, months, days, hours, minutes and seconds.
- now, current_time, and current_date are timestamps denoting the current time and date.

The method vt() (resp. tt()) returns the valid-time- (resp. transaction-time) timestamp of the associated timestamped attribute, relationship, or object. vt() and tt() can be used to compare timestamps using the standard comparison-predicates =, <, ..., or for the retrieval of valid- or transaction-time. For each timestamped attribute and timestamped relationship of a class, methods vt() and tt() exist. Furthermore, class *spatial_object* defines these methods for each class of T/OOGDM's class hierarchy. Hence, an implicit object-timestamp is available for each class.

3.3.3 Constructors and modification statements in T/OOGQL

In order to support time, constructors and OOGQLs' modification statements must be modified. We first describe the syntax for constructors in OOGDM and then consider the modification statements *apply, delete,* and *insert* (cf. [VBH96]).

In T/OOGQL constructors support the definition of valid-time for each temporal attribute, i.e., to construct an object of class city we may use the following object constructor:

> city (geometry: polygon(...*polygon description*...) valid ...*time specification*...,
>   population: 150000 valid ...*time specification*...,
>   name: "Münster", ...)

Since the data model of T/OOGDM is attribute-timestamped, it is appropriate to specify the valid-time for each temporal attribute instead of defining a timestamp for the whole object. If there is no valid-timestamp given for a temporal attribute, chronon-timestamped attributes receive the timestamp *current_time* and interval-timestamped attributes start at *current_time*.

The apply-statement remains unchanged from OOGQL. As described in [VBH96] attributes may only be changed by modification methods in OOGQL. Obviously, the update-/modification-methods for temporal attributes must have a parameter specifying the valid-time for the new attribute value. Hence, an additional valid-clause in the apply-statement is not necessary.

The delete-statement of T/OOGQL is augmented by an optional valid-clause, i.e., the syntax for the delete statement is given by

> delete ... from ... where ... valid ...

If the valid-clause is not present, the specified objects are deleted starting *at current_time*. Otherwise, the specified objects are deleted in the valid-time-range given by the valid clause.

Similar to the delete-statement, the insert-statement has been augmented by a valid-clause:

> insert ... from ... where ... valid ...

For the insert-statement, the (optional) valid-clause can be considered as a valid-time projection similar to the valid-clause of the select-statement. If non-snapshot objects are inserted (i.e., the insert-statement contains a temporal constructor or a temporal subquery supplying timestamped objects) and the valid-clause is absent, the objects are inserted and the timestamps remain unchanged. Otherwise, a valid-time projection using the range of time specified by the valid-clause is performed. E.g., let A be a temporal attribute having timestamp t and let t' be the time specified by the valid-clause. Then for the insertion of A we obtain timestamp $t'' = t \cap t'$. If snapshot objects (i.e., objects retrieved by a snapshot-subquery) are inserted all temporal attributes receive timestamps as specified in the valid-clause. If the valid-clause is missing, default valid-time timestamps are associated to temporal attributes: chronon-timestamped attributes receive the timestamp *current_time*, interval-timestamped attributes the interval starting at *current_time*.

```
class vegetation_map : timestamped layer2d<String> {
        Short coverage (String vegetation_type, region2d query_region,
                              time query_time) const;
        Set<Short> coverage (String vegetation_type, region2d query_region,
                                   time_interval query_time) const; };
class temperature : chronon_timestamped point2d<Short> { };
class elevation : point2d<Short> { } ;
class elevation_map : timestamped layer2d_0d_set <elevation> {
        elevation min_height (time t) const;
        elevation max_height (time t) const;
        Double avg_height (time t) const;
        elevation min_height (time_interval t) const;
        elevation max_height (time_interval t) const;
        Double avg_height (time_interval t) const; };
class river_pollution : timestamped profile2d<Short> {
        Short min_pollution(time t) const;
        Short max_pollution(time t) const;
        Short min_pollution(time_interval t) const;
        Short max_pollution(time_interval t) const; };
class river : geo_object {
        relationship polyline2d geometry;
        attribute String name;
        relationship Set<river_pollution> pollution; };
class railroad : geo_object {
        relationship polyline2d geometry;
        timestamped attribute Short no_of_tracks;
        timestamped attribute Short max_speed; };
class city : geo_object {
        relationship timestamped region2d geometry;
        attribute String name;
        chronon_timestamped attribute Long population;
        relationship Set<temperature> temp;
        relationship state belongs_to inverse state::cities;
        relationship vegetation_map vegetation; };
class state : geo_object {
        relationship timestamped region2d geometry;
        attribute String name;
        chronon_timestamped attribute Long population;
        relationship Set<city> cities inverse city::belongs_to;
        relationship city capital;
        relationship Set<vegetation_map> land_use;
        relationship Set<elevation_map> topography;
        relationship Set<river> rivers; };
```

Figure 5 Class definitions for the example application.

4. An application example

In this section, we present an example application of T/OOGDM and its query language T/OOGQL. The example is derived from [VBH95] and extended by the timestamped attributes. Based on this example, we discuss some T/OOGQL queries showing the temporal facilities of the query language and the integration of spatial and temporal processing of data.

4.1 Class definitions

We consider the classes for the geo-objects city, state, river, and railroad. Figure 5 shows the class definitions of the mentioned geo-objects as well as some auxiliary subclasses. Note that the database for the described information could also be defined using a different set of classes. For a detailed discussion of OOGDM's data definition language we refer to [VBH96].

4.2 Query examples

T/OOGQL can be used similar to SQL and OQL for retrieving data without temporal constraints. The following query is used to retrieve the population counts of all cities:

select snapshot struct (n : c.name, p : c.population) **from** city c

The query above is a simplified version of the following query, which explicitly retrieves the latest population count of the cities stored in the database:

select snapshot struct (n : c.name, p : LastValue(c.population)) **from** city c

To retrieve the population counts of all cities in 1980, we have to select the city objects containing a population attribute having an associated (valid-time) value during 1980:

select snapshot struct (n : c.name, p : c.population) **from** city c
where Year(1980) t_contains c.population

It is also possible to access the transaction-time of attributes. To retrieve the population counts of all cities in 1980 stored at that time in the database we use the following query:

select snapshot struct (n : c.name, p : c.population) **from** city c
where (Year(1980) vt_contains c.population) **and** (Year(1980) tt_overlaps c
    or Year(1980) tt_contains c)

The queries below show how to retrieve data of a specified valid-time and how to combine the temporal and spatial capabilities of T/OOGQL.
Retrieve the average temperature within a area of 10km around the center of "Münster" measured on July, 30th, 1995:

select snapshot avg(t.get_data()) **from** temperature t, city c
where c.name = "Muenster" **and** distance(t, center(c)) <= 10
    and Day(30, "July", 1995) t_contains t

Which railroads in Germany have been renewed in 1980 or later and now allow a maximum speed of at least 180km/h.

```
select snapshot r from railroad r, state s
where s.name = "Germany" and (r cross s or r in s) and r.max_speed >= 180
    and Begin(Year(1980)) vt_precedes r.max_speed
```

Which states have elevation maps that were updated after July, 15 1990?
```
select s.name from state s where exists em in s.topography :
    Date(15, "July", 1990) vt_precedes em
```

List all cities where the coverage with "Wood" increased in the city area by at least 5 percent during the last 5 years?
```
select c.name from city c
where c.vegetation.coverage("Wood", c.geometry, now)
    >= min(c.vegetation.coverage("Wood", c.geometry, Period(now-years(5),
    now))) + 5
```

Which river of the database had the worst pollution during 1970 and 1985?
```
query_time := Period(Begin(Year(1970)), End(Year(1985)));
select r.name from river r
where exists rp in r.pollution: rp.max_pollution(query_time) >= max (select
    max_pollution(query_time) from river_pollution)
```

To retrieve the valid-time of a selected object, we can use the vt() method of T/OOGQL, for example to query the time when the last population count occurred in France:
```
select s.population.vt() from state s where s.name = "France"
```


5. Conclusions

We have presented temporal extensions to OOGDM, our object-oriented data model for geo-information systems. We support both valid-time and transaction-time. While the former records the changes in the real world, the latter records the changes of the database and enables the user to compare current query results with query results of the past. The extensible SQL-like query language of OOGDM has been modified to incorporate the temporal domain. The extensions follow the proposals for a temporal SQL for relational databases [TSQL2]. Similar to the spatial dimension, the temporal dimension integrates seamless into the query language.

OOGDM is currently being implemented in a first prototype version [BVH96] on top of the object-oriented database system ObjectStore [LLOW91]. The first prototype is developed without temporal support. However, the next step is an extension of the implementation by valid-time. Transaction-time will be supported in future, since we feel that valid-time is more important for the users of our prototype.

In this paper, we did not investigate implementation specific issues of the temporal extensions to OOGDM. This will be subject of a forthcoming paper. Current and future research has to concentrate on specific problems in the field of temporal databases, e.g., temporal indices, the integration of spatial and temporal indices, and temporal joins.

References

[ASS93] K.K. Al-Taha, R.T. Snodgrass, M.D. Soo: Bibliography on Spatiotemporal Databases, ACM SIGMOD Record, 22(1), March 1993, 59 -67.

[BVH96] L. Becker, A. Voigtmann, K. Hinrichs: Developing Applications with the Object-Oriented GIS-Kernel GOODAC, Bericht Nr. 4/96-I, Institut für Informatik, Westf. Wilhelms-Universität Münster, Germany, 1996.

[CFO93] E. Clementini, P. DiFelice, P. van Oosterom: A Small Set of Formal Topological Relationships Suitable for End-User Interaction, SSD'93, LNCS 692, 277 - 295.

[EF91] M.J. Egenhofer, R.D. Franzosa: Point-set topological spatial relations, Int. Journal of Geographic Information Systems, 5 (2), 1991, 161 - 174.

[ET92] M.J. Egenhofer, K.K. Al-Thaha: Reasoning about Gradual Changes of Topological Relationships, Int. Conf. GIS, 1992, Springer, LNCS 639, 196 - 219.

[Je93] C.S. Jensen et al.: A Consensus Glossary of Temporal Database Concepts, Technical Report 93-2035, Inst. for Electronic Systems, Dept. of Computer Science, Aalborg-University, Denmark, Nov. 1993.

[Kl93] N. Kline: An Update of the Temporal Database Bibliography, ACM SIGMOD Record, 22(4), 1993, 66 - 80.

[La92] G. Langran: Time in Geographic Information Systems, Taylor & Francis, 1992

[LLOW91] C. Lamb, G. Landis, J. Orenstein, D. Weinreb: The ObjectStore Database System. Communications of the ACM, 34 (10), 1991, 50 - 63.

[NA93] S.B. Navathe, R. Ahmed: Temporal Extensions to the Relational Model and SQL, In: A.U. Tansel, et al. (eds.): Temporal Databases - Theory, Design, and Implementation, Benjamin-Cummings, 1993, 92 - 109.

[ODMG93] R.G.G. Cattell (ed.): The Object Database Standard: ODMG-93, Morgan-Kaufman Publishers, 1994.

[OM93] L.M. Oliveira, C.B. Medeiros: Managing Time in Object-Oriented Databases, Technical Report DCC-14/93, Dept. of Computer Science, Universidade Estadual de Campinas, Campinas SP, Brasil.

[ÖS95] G. Özsoyoglu, R. T. Snodgrass: Temporal and Real-Time Databases: A Survey, IEEE Knowledge and Data Engineering, 7(4), 1995, 513 - 532.

[Pe95] D.J. Peuquet, N. Duan: An event-based spatiotemporal data model (ESTDM) for temporal analysis of geographical data, Intl. Journal on Geographical Information Systems, 9(1), 1995, 7-24.

[RP92] J.F. Roddick, J.D. Patrick: Temporal Semantics in Information Systems - A Survey, Information Systems, 17(3), 1992, 249 - 267.

[RS93] E. Rose, A. Segev: TOOSQL - A Temporal Object-Oriented Query Language, In R.A. Elmasri, V. Kouramajian, B. Thalheim (eds.): Proc. 12th Int. Conf. on ER-Approach, Arlington, TX, 1993, LNCS 823, 122 - 136.

[Sa93] N.L. Sarda: HSQL: A Historical Query Language, In: A.U. Tansel,et al. (eds.): Temporal Databases - Theory, Design, and Implementation, Benjamin-Cummings, 1993, 110 - 140.

[Sn92] R. Snodgrass: Temporal Databases, In A.U. Frank, I. Campari, U. Formentini (eds.), Proceedings of the International Conference on GIS: From Space to Territory, LNCS. 629, 22-65.

[Sn93] R. Snodgrass: An Overview of TQuel, In: A.U. Tansel, et al. (eds.): Temporal Databases - Theory, Design, and Implementation, Benjamin-Cummings, 1993, 141 - 182.

[Sn95] R. Snodgrass: Temporal Object-Oriented Databses: A Critical Comparison, In W. Kim (ed.): Modern Database Systems, Addison-Wesley/ACM Press, 1995, 386 - 408.

[TCG93] A.U. Tansel, J. Clifford, S. Gadia, S. Jajodia, A. Segev, R. Snodgrass (eds.): Temporal Databases - Theory, Design, and Implementation, Benjamin-Cummings, 1993.

[TSQL2] R.T. Snodgrass (ed.): The TSQL2 Temporal Query Language, Kluwer Academic Publishers, 1995.

[VBH95] A. Voigtmann, L. Becker, K. Hinrichs: An Object-Oriented Data Model and a Query Language for Geographic Information Systems, Bericht Nr. 15/95.-I, Institut für Informatik, Westf. Wilhelms-Universität Münster, Germany, 1995.

[VBH96] A. Voigtmann, L. Becker, K. Hinrichs: A Query Language for Geo-Applications, Bericht Nr. 5/96-I, Institut für Informatik, Westf. Wilhelms-Universität Münster, Germany, 1996.

[Wo94] M.F. Worboys: Unifying the Spatial and Temporal Components of Geographical Information, SDH'94, Edinburgh, UK, 1994, 505-517.

Contact address
Andreas Voigtmann, Ludger Becker and Klaus H. Hinrichs
FB15, Informatik
Westfälische Wilhelms-Universität
Einsteinstr. 62
D-48149 Münster
Germany
Phone & fax: (++49) 251 / 83 - 3755
E-mail: {avoigt,beckelu,khh}@math.uni-muenster.de

VIRTUAL REALMS: AN EFFICIENT IMPLEMENTATION STRATEGY FOR FINITE RESOLUTION SPATIAL DATA TYPES[1]

Volker Muller[1], Norman W. Paton[2], Alvaro A.A. Fernandes[3], Andrew Dinn[1] and M. Howard Williams[1]

[1]Department of Computing and Electrical Engineering, Heriot-Watt University, Edinburgh, UK
[2]Department of Computer Science, University of Manchester, Manchester, UK
[3]Department of Mathematical Sciences, Goldsmiths College, London, UK

Abstract
A realm is a planar graph over a finite resolution grid that has been proposed as a means of overcoming problems of numerical robustness and topological correctness in spatial database systems. While the realm structure and the spatial algebra that is associated with it provide a range of desirable facilities for modelling spatial information in database systems, widespread exploitation will only be practical if efficient implementation strategies are identified. This paper shows how data types can be supported efficiently over *virtual realms*, where the finite resolution grid is not stored explicitly, but is generated only partially and as needed. This approach avoids the considerable storage space overheads associated with the original proposal for the implementation of realms, and provides overall runtime performance that often improves upon that of less space efficient implementation strategies.

Key words
Spatial databases, spatial data types, finite resolution, object management.

1. Introduction

The notion of *realm* is presented in (Guting, 1993, Guting 1995a), along with an associated collection of spatial data types and operations, known as the ROSE algebra. The ROSE algebra supports a wide range of operations over data types representing *points, lines* and *regions*, and has been designed in such a way that all operations are closed (i.e. yield results that can be operated on by the algebra), and can be efficiently implemented, as described in (Guting, 1995b). However, the overall efficiency of a spatial database based upon the ROSE algebra depends not only upon the efficiency of the operations of the algebra, but also on the way in which the underlying realm has been constructed.

A comprehensive description is given in (Guting, 1995b) of how the ROSE algebra operations can be implemented, and data structures and algorithms for implementing realms are described in (Guting, 1993). In that proposal, the realm is stored explicitly, and data in the structures that represent the realm is replicated in data structures that support the operations of the ROSE algebra.

---

[1] This work was done when the first author was visiting, Heriot-Watt University from the University of Karlsruhe, supported by an Erasmus scheme coordinated by the EU HC & M Network ACT-NET.

697

The approach presented in this paper can be used directly with the implementations of the ROSE algebra operations in (Guting, 1995b); the focus is thus on improving upon the implementation of the underlying realm presented in (Guting, 1993). The principal characteristic of the approach presented in this paper is that, unlike that in (Guting, 1993), the realm is not explicitly stored (i.e. it is *virtual* - fragments of realm are only generated temporarily, and as needed by operations that update the database). As the implementation in (Guting, 1993) is expensive in terms of space used, considerable overall savings in storage space for ROSE algebra objects are obtained. The question that suggests itself, given the reduced space overhead of virtual realms, is what impact this has on the runtime performance of systems built on top of realms. It is shown in this paper that, although the two approaches have different performance profiles, the virtual realm supports runtime performance that is at least comparable with that offered by stored realms.

This paper is structured as follows. Section 2 describes realms and their relationship to spatial data types in more detail; section 3 summarises the implementation used by (Guting, 1993) and presents the virtual realm approach; section 4 compares the performance of the two approaches; and section 5 presents some conclusions.

2. Context: realms and the ROSE algebra

A realm, an example of which is depicted in figure 1, is a set of points and non-intersecting line segments over a finite discrete grid. All the spatial data values of an application have realm points and line segments as their components, so the realm contains the complete underlying geometry of an application. Figure 2 shows two *regions* objects *(A,B)*, a *lines* object *(C)* and a *points* object *(D)* which can be constructed from the points and line segments in the realm depicted in figure 1.

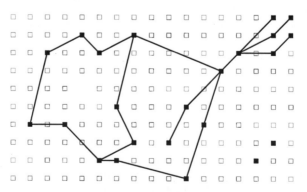

Figure 1 Example of a realm.

To represent a realm in the computer, integer representations and error-free integer arithmetic are used. This is made practical by exploiting algorithms that use finite-resolution computational geometry to characterise the inevitable distortion of the geometry when contiguous space is mapped onto discrete space (Greene, 1986). The essence of this approach is that the ambiguities associated with inexact representation of spatial concepts are resolved when information is inserted into the database, which allows all future processing to assume that information is represented precisely with respect to the finite resolution grid, thereby simplifying processing and avoiding error propagation.

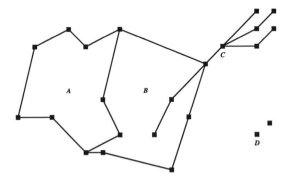

Figure 2 Spatial values which can be built from the realm.

The insertion-time resolution of ambiguities involves the following steps:
- All points from the application are mapped to points in the realm; if two points are close together, it is resolved at insertion time whether or not they are to be represented by the same realm point.
- All line segments from the application are mapped onto a sequence of one or more intersection-free line segments in the realm. In the process, lines may be split and new realm points introduced to represent intersections involving the inserted line segments. Thus all intersections between points and line segments and between groups of line segments are identified and resolved at insertion time. The size of the error introduced when resolving intersections is bounded by the *envelope* of the line segment, which is, informally, the set of grid points adjacent to the line segment. As an example, figure 3 shows two segments, *s1* and *s2*, one of which has just been inserted into the realm. As the realm records intersection-free line segments at finite resolution, it is necessary to identify a realm point to act as the intersection of the two segments, and to redraw the component segments so that they pass through this point. An example of such a redrawing is given in figure 4, where the segments *s1* and *s2* from figure 3 are redrawn so that they intersect at realm point *p1*. A formal description of the redrawing process is given by (Greene, 1986).

A realm forms part of a layered architecture which supports the data types and operations of the ROSE algebra. The ROSE algebra supports the types *points, lines* and *regions*, which in turn are associated with a comprehensive, if conventional, collection of spatial operations defined in (Guting, 1995a).

Algorithms for the implementation of the ROSE algebra operations are detailed in (Guting, 1995b), along with the data structures used to represent the associated types.

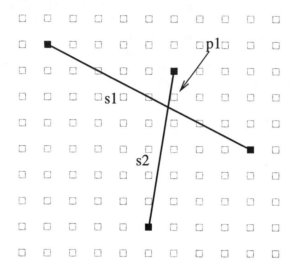

Figure 3 Two segments in a realm before changes are computed.

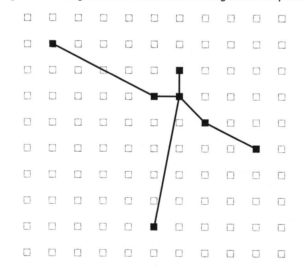

Figure 4 Two segments in a realm after changes are computed.

2.1 Stored realms

This subsection gives an overview of the implementation strategy proposed for the ROSE algebra and the underlying realm in (Guting, 1993), (Guting, 1995b).

Figure 5 shows the relationship between the data types of the ROSE algebra and the underlying realm for a *lines* spatial object. The *lines* object consisting of the segments *S1, S2, S3* and *S4*. Note that in the original data, *S1* and *S3* could have been joined as a single segment; if so, the partitioning would have taken place when the intersection with *S2* was resolved on insertion into the realm is represented by an ordered sequence of half segments, as this supports efficient implementation of spatial operations using plane sweep algorithms, etc (Guting, 1995b). The half segments are named in the figure with the suffix *L* representing the leftmost end of the segment and *R* representing the rightmost. This representation is scanned from left to right by algorithms to identify, for example, where two *lines* objects intersect. To support efficient updating of the ordered sequence of half segments, the halfsegments are themselves indexed using an AVL tree (a balanced tree searching structure) to support rapid search and update. The representation of the spatial data types described so far is independent of the use of a stored or a virtual realm.

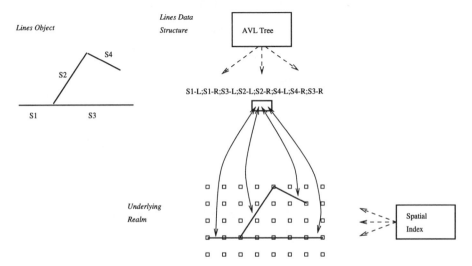

Figure 5 A lines data object and its stored representation.

In the representation used in (Guting, 1993), the realm is stored explicitly, and every realm object (*point* or *segment*) both references and is referenced by every spatial object (*points*, *lines* or *regions*) of which it is part. However, the geometry (i.e. coordinates describing the absolute locations of points and line segments) is stored redundantly in the realm and in the ordered collection of half segments, so that it is not necessary to access the realm during the execution of ROSE algebra operations. As a result of this redundant storage of the geometry, the realm only needs to be accessed when operations take place that change ROSE algebra objects. It is not made clear in (Guting, 1993) exactly what data structures are used to implement the realm, but information in the realm does have to be indexed to support efficient update of the realm.

2.2 Virtual realms

The implementation proposed here uses the same data structures and algorithms as (Guting, 1995b) to implement ROSE algebra operations, but the realm is not stored explicitly. Instead, all that is held in the database is the ordered list of half-segments described above, and fragments of realm are constructed only temporarily, to identify and resolve intersections, when ROSE algebra objects are updated. Essentially, whenever a ROSE algebra object R is updated, all objects that intersect the minimum bounding rectangle of R are retrieved, and the section of realm relating to these objects is constructed temporarily to resolve interactions involving these objects and R . When such interactions have been resolved, the changes made to R and nearby ROSE algebra objects are stored, and the recently constructed fragment of realm is discarded. The consequences for implementors of the choice of virtual or stored realms are considered more fully in section 3.

3. Implementations: stored and virtual realms

As the role of the realm is to provide a context for the resolution of ambiguities when mapping continuous space onto discrete space, and the underlying realm is not used directly when implementing ROSE algebra operations, the principal operations that are supported on the realm involve the insertion of points and segments. This section outlines how operations for inserting segments are implemented in both stored and virtual realms, as insertion of a segment is similar to, if somewhat more complex than, insertion of a point. The relative performance of the approaches is considered in section 4.

3.1 Segment insertion in the stored realm

The algorithm presented in figure 6 stored for inserting a segment into a realm is essentially that given in (Guting, 1993); differences are in presentation rather than substance.

InsertSegmentS takes two input parameters, a segment *s*, and the realm *R* into which *s* is to be inserted. The outputs are an updated realm *R'* and a set of ROSE algebra objects *RO* that have to be updated to reflect the changes to the realm. This propagation of updates is required to ensure that the information stored in the spatial data types is consistent with that in the realm, using the structures in figure 5.

The algorithm performs the following steps:

[Step 1:]

Initialise the set of ROSE objects that have to be revised to the empty set.

[Step 2:]

Check to see if the segment s is already in the realm; if it is, then the realm need not be changed. If the segment is not already in the realm, it has to be inserted; when doing this, it is necessary to identify which points and segments in the realm are in close proximity with the new segment.

Lines (16-20) identify the consequences of having a point in the envelope of the new segment. The first consequence is that the segment must be split into two, as it is considered to go through any point within its envelope. The fact that the segment is to be split is marked by inserting a hook (Greene, 1986) (a short directed line segment) from the new segment to the grid point. The second consequence is that any further segments in the realm that intersect the hook must also be redrawn.

```
(1)   algorithm InsertSegmentS
(2)   inputs:        s: Segment
(3)                  R: Realm
(4)   outputs:       R': Realm
(5)                  RO: set of RoseObjects
(6)
(7)   Step 1: Initialisation
(8)          RO := ∅
(9)
(10)  Step 2: Find nearby realm objects
(11)         if s already in R then
(12)                R' := R
(13)                return
(14)         else
(15)                SR := ∅
(16)                for each point p in the envelope of s do
(17)                       create a hook h from s to p
(18)                       for each segment v intersecting h do
(19)                              create a hook from the intersection of h and v to p
(20)                              SR := SR ∪ {v}
(21)                       for each segment t intersecting s do
(22)                              create a hook h from the intersection
(23)                              of s and t to the nearest grid point
(24)                              SR := SR ∪ {t}
(25)                              for each segment v intersecting h do
(26)                                     create a hook from the intersection of h and v to p
(27)                                     SR := SR ∪ {v}
(28)         end-if
(29)
(30)  Step 3: Redraw hooked lines
(31)         RS := ∅
(32)         for each segment t in {s} ∪ SR do
(33)                RS := RS ∪ segments created when redrawing
(34)                       t using the procedure of [4]
(35)
(36)  Step 4: Update realm
(37)         R' := R minus all segments in {s} ∪ SR
(38)         R' := R' plus all segments in RS
(39)         RO := spatial objects of redrawn segments
```

Figure 6 Segment insertion algorithm for the stored realm.

Lines (21-27) identify the consequences of having a segment in the realm intersect the new segment. Where this is the case, both the intersecting segments have to be redrawn through the realm point nearest to the intersection; this need for redrawing is again marked using hooks. Any further segments intersecting the hook are identified in lines (25-27) and are themselves marked for redrawing.

[Step 3:]

All segments that need to be redrawn have been identified by Step 2. These segments are then redrawn using the algorithm of (Greene, 1986), which is not stated here to save space, and because it is applied similarly with both the stored and the virtual realm.

[Step 4:]

The changes identified are applied to the realm, and the ROSE algebra objects that have to be updated are recorded in *RO*.

3.2 Segment insertion in the virtual realm

The algorithm presented in figure virtual indicates the steps that are involved in constructing a portion of a virtual realm in response to the insertion of a segment. In this algorithm, as with the algorithm in figure stored, it is assumed that the end points of the segment have already been inserted.

InsertSegmentV takes one input parameter, the segment s that is to be inserted. The output is the set of ROSE algebra objects *RO* that have to be updated to reflect the changes to the realm.

The algorithm performs the following steps:

[Step 1:]

Initialise the set of ROSE algebra objects *RO* that are affected by the insertion, and the virtual realm *VR*.

[Step 2:]

The spatial values that have components that are in close proximity to the inserted segment have to be identified. It is assumed that all *points, lines* and regions objects are stored using a spatial index (in the implementation, an R-tree is used), and that fast access is thus supported to all spatial objects that overlap a given minimum bounding rectangle (MBR).

Lines (12-23) identify the *lines* and *regions* values that have components (segments) that either need to be redrawn as a result of the insertion of *s*, or that force *s* to be redrawn. If it transpires that any existing spatial object contains a segment identical to *s* (line 14), then there is no need to proceed further with the construction of the virtual realm. The interactions that are important are: intersection of *s* with existing segments (line 17) and the discovery of endpoints of existing segments in the envelope of *s* (lines 19 and 21). Any such interactions are   recorded in the virtual realm *VR*.

Lines (24-29) identify the *points* values that have components that are within the envelope of *s*, and store them in *VR*.

[Step 3:]

The nature of the interactions between the objects in the virtual realm and s have to be worked out, and such redrawing as is necessary performed. For each point in *VR*, this involves creating a hook to s and identifying all segments in *VR* that intersect the hook (lines 32-35). For each segment in *VR*, this involves creating a hook from the intersection with s to the nearest grid point, and also identifying all segments in *VR* that intersect the hook (lines 36-40).

```
(1)    algorithm InsertSegmentV
(2)    inputs:        s: Segment
(3)    outputs:       RO: set of RoseObjects
(4)
(5)    Step 1: Initialisation
(6)            RO := ∅
(7)            VR := ∅
(8)
(9)    Step 2: Identify nearby objects for insertion into realm
(10)           SR := ∅
(11)           SLR := lines and regions objects with MBR that overlaps s
(12)           for each spatial value sv in SLR do
(13)                   for each segment t in sv do
(14)                           if t = s then
(15)                                   delete VR
(16)                                   return
(17)                           elseif intersect(t,s) then
(18)                                   VR := insertRsegment(t,sv,VR)
(19)                           elseif t.p1 in envelope of s then
(20)                                   VR := insertRpoint(t.p1,VR)
(21)                           elseif t.p2 in envelope of s then
(22)                                   VR := insertRpoint(t.p2,VR)
(23)                           end-if
(24)           SP := points spatial values with MBR that overlaps s
(25)           for each ps in SP do
(26)                   for each point pr in ps do
(27)                           if pr in envelope of s then
(28)                                   VR := insertRpoint(pr,VR)
(29)                           end-if
(30)
(31)   Step 3: Perform redrawing of objects in virtual realm
(32)           for each point rp in VR do
(33)                   create a hook h from s to rp
(34)                   for each segment t in VR intersecting h do
(35)                           create a hook h from the intersection of h and t to rp
(36)           for each segment rs in VR do
(37)                   create a hook h from the intersection of
(38)                   s and rs to the nearest grid point rp
(39)                   for each segment t in VR intersecting h
(40)                           create a hook h from the intersection of h and t to rp
(41)           RS := ∅
(42)           for each segment t in VR do
(43)                   RS := RS ∪ segments created when redrawing
(44)                           t using the procedure of [4]
(45)           RO := spatial objects of redrawn segments in RS
(46)
(47)   Step 4: Tidy up
(48)           delete VR
```

Figure 7 Segment insertion algorithm for the virtual realm.

Lines (42-45) then perform redrawing of the portion of the realm affected by the insertion of *s* using the algorithm of (Greene, 1996), and record the ROSE algebra objects affected by the changes in *RO*.

[Step 4:]
As the realm is only created partially and temporarily, VR is then deleted.

4. Performance: comparing stored and virtual realms

This section presents an informal comparison of the performance of stored and virtual realms for the insertion of a segment, as this operation is both common and representative of realm based activity. Significant to a broader consideration of the performance of a spatial database based upon the ROSE algebra, however, is the fact that most ROSE algebra operations never access structures at the realm level, and thus will perform in an identical manner for both stored and virtual realms.

The precise storage saving that derives from exploiting virtual realms is not straightforward to state, depending as it does upon the data structures used to implement both approaches and the patterns of data stored. We note, however, that the overall space saving is likely to be significant, as space is saved not only in the realm itself, but also in linking the realm to the structures used to support the spatial data types. We estimate that the storage space occupied by spatial information described using the types of the ROSE algebra will be reduced by at least 30 when virtual realms are used in preference to stored realms.

In considering the relative performance of stored and virtual realms, it is assumed the realm resides on secondary storage, as will certainly be the case in any realistic applications. Both approaches require the use of a spatial index - in the case of the stored realm, this is assumed to be an index on the realm, in the case of the virtual realm, this is assumed to be an index on the spatial types built using the realm.

Figure 8 summarises the tasks performed when inserting a segment using the stored and the virtual realms. The *Difference* column indicates whether the virtual realm increases (+) or decreases (-) the cost of performing input-output (I/O) or processing (CPU) tasks.

The results presented in the table can be summarised with respect to the effect on I/O costs and CPU time:

[I/O:]
Considerably more I/O activity is required with the stored realm than with the virtual realm. This is to be expected, in that realm information must be read from and written to disk when it is stored, whereas this is not required when the realm is virtual. The only entry which shows the virtual realm imposing an increased I/O load is when spatial values in close proximity to the inserted segment are read from disk. This shortcoming of the virtual realm is partly compensated for later, as many of these spatial values must be read in when propagating updates from the realm to associated points, lines and regions values.

Separate Realm	Virtual Realm	Difference
Retrieve required nodes of spatial index (many entries)	Retrieve required nodes of spatial index (fewer entries)	-I/O
Retrieve segments and points from MBR of inserted segment(possibly many)	Retrieve points, lines and regions spatial values from MBR of inserted segment (possibly many, possibly large)	+I/O
-	Scan all points, lines and regions spatial values, found above, to construct the virtual realm	+CPU
Compute changes in the realm	Compute changes in the virtual realm	-
Delete changed segments from disk	-	-I/O
Write new segments to disk	-	-I/O
Compute changes to index(possibly several deletions) (possibly many insertions)	Compute changes to index (max. 1 deletion) (max. 1 insertion)	-CPU
Write changed index to disk	Write changed index onto disk	-I/O
Retrieve the spatial values of which the changed segments are elements (possibly many)	-	-I/O
Replace changed segments in the spatial values of which they are elements	Replace changed segments in the spatial values of which they are elements	-
Write changed spatial values to disk	Write changed spatial values to disk	-
-	Delete virtual realm	+CPU

Figure 8 Performance comparison of stored and virtual realm for insertion of a segment.

[CPU:]
The virtual realm is more expensive than the stored realm in terms of CPU time. This is also to be expected, as additional processing is required to build and destroy the virtual realm, which is a main memory data structure.

Overall, it can be argued that the virtual realm is likely to provide faster processing than the stored realm, as the performance improvement resulting from the reduction in disk activity is likely to be more than enough to compensate for the increased CPU activity associated with the building of the virtual realm.

5. Conclusions

This work is being carried out in the context of a project which is seeking to extend the kernel of the deductive object-oriented database system ROCK & ROLL (Barja, 1995) with facilities to support efficient handling of spatial concepts (Abdelmoty, 1994, Paton, 1996). Such research requires efficient support for a coherent collection of spatial concepts that can be implemented efficiently. The ROSE algebra seemed to us to provide a range of facilities that made it highly suitable for incorporation into the kernel of a database system. However, the space overhead associated with explicit storage of the realms was a source of considerable concern. The notion of a virtual realm was introduced to overcome the space problems associated with stored realms, while retaining the advantages of the realm in terms of numerical robustness and topological correctness. As demonstrated in section 4, this space saving has not been acquired at the cost of slower runtime performance, as reduced overall I/O activity means that systems based on the virtual realm will often perform more quickly than those based upon stored realms.

At time of writing, the virtual realm has been implemented in C++ for use with the secondary storage management facilities of the EXODUS extensible database system (Carey, 1990). Work is underway on the integration of the virtual realms and the associated ROSE algebra operations with the ROCK & ROLL deductive object-oriented database.

Acknowledgements
We are pleased to acknowledge the support of the EPSRC initiative on Architectures for Integrated Knowledge Manipulation Systems (grant GR/J99360) for funding this work, and of Neil Smith and Glen Hart of Ordnance Survey as the Industrial Partners.

System Availability
The ROCK & ROLL system is freely available over the Internet. It is anticipated that the first public release of the spatially extended ROCK & ROLL system will be in March 1997. For details, see WWW page http://www.cee.hw.ac.uk/Databases/dood.html.

References
A.I. Abdelmoty, N.W. Paton, M.H. Williams, A.A.A. Fernandes, M.L. Barja, and A. Dinn. Geographic Data Handling in a Deductive Object-Oriented Database, In D. Karagiannis, editor, Proc. 5th Int. Conf. on Databases and Expert Systems Applications (DEXA), pages 445-454. Springer-Verlag, 1994.
M.L. Barja, A.A.A. Fernandes, N.W. Paton, M.H. Williams, A. Dinn, and A.I. Abdelmoty. Design and Implementation of ROCK & ROLL: A Deductive Object-Oriented Database System, Information Systems, 20:185-211, 1995.
M. Carey, D. DeWitt, G. Graefe, D. Haight, J. Richardson, D. Schuh, E. Shekita, and S. Vandenberg. The EXODUS Extensible DBMS Project: An Overview. In S. Zdonik and D. Maier, editors, Readings in Object-Oriented Databases, 1990. Morgan Kaufman Publishers, Inc.
D. Greene and F. Yao. Finite-Resolution Computational Geometry. In IEEE Symp. on Foundations of Computer Science, pages 143-152, 1986.
R.H. Guting and M. Schneider. Realms: A Foundation for Spatial Data Types in Database Systems. In D. Abel and B.C. Ooi, editors, Proc. 3rd Int. Conf. on Large Spatial Databases (SSD), Lecture Notes in Computer Science, pages 14-35. Springer Verlag, 1993.
R.H. Guting and M. Schneider. Realm-Based Spatial Data Types: The ROSE Algebra. VLDB J., 4(2):243-286, 1995 (a).
R.H. Guting, T. de Ridder, and M. Schneider. Implementation of the ROSE Algebra: Efficient Algorithms for Realm-Based Spatial Data Types. In Proc. 4th Int. Symposium on Large Spatial Databases (SSD). Springer-Verlag, 1995 (b).
N.W. Paton, A.I. Abdelmoty, and M.H. Williams. Programming Spatial Databases: A Deductive Object-Oriented Approach. In D. Parker, editor, Innovations 3 (Proc. GIS Research UK (GISRUK)). pages 75-84, Taylor and Francis, 1996.

Contact address
Volker Muller[1], Norman W. Paton[2], Alvaro A.A. Fernandes[3], Andrew Dinn[1]
and M. Howard Williams[1]
[1]Department of Computing and Electrical Engineering,
 Heriot-Watt University
 Riccarton
 Edinburgh
 UK
 E-mail: [andrew,howard]@cee.hw.ac.uk
[2]Department of Computer Science
 University of Manchester,
 Oxford Road
 Manchester
 UK
 E-mail: norm@cs.man.ac.uk
[3]Department of Mathematical Sciences
 Goldsmiths College
 New Cross
 London SE14 6NW
 UK
 E-mail: a.fernandes@gold.ac.uk

MANAGING SPATIAL OBJECTS WITH THE VMO-TREE

Weiping Yang and Christopher Gold

Chaire industrielle en géomatique Centre de recherche en géomatique
Université Laval
Québec, Canada

Abstract

An object-oriented approach is described to model a set of 2D vector data, depicting points, lines, and polygons. These primitive spatial objects are embedded in Euclidean space and are topologically related with the dynamic Voronoi diagram. A *spatial object condensation* technique is used to partition the Voronoi diagram into smaller subsets. Each subset corresponds to a bounded subspace called a *condensed object* which can contain primitive and smaller condensed objects. Condensed objects are managed and manipulated by the *VMO-tree* (Voronoi based Map Object Tree), a novel data model describing relationships between condensed objects. The VMO-tree can be constructed by either top-down, bottom-up, or combined approaches. Spatial objects in the VMO-tree are presented in different levels of detail. Topological structures for every condensed object are integrated by the structure of the tree. Spatial searches can be performed vertically along edges of the tree or horizontally inside a topological structure of each node object. It is not required that all nodes of the tree are loaded in memory at one time. Any node object is an independent map and can be worked with separately. Each map sheet corresponds to a disk page and has a contiguous memory occupation.

1. Introduction

Contemporary mapping applications require that spatial data be handled with flexibility. This need can be imposed on all aspects of processing data, such as: viewing, storing, manipulating, analyzing, and communicating. Ideally, any interesting data can be organized as an operable unit or object, the change of focus back and forth amongst objects should be smooth and automatic, and the system should allow both outlined and detailed views. From the data structure point of view, these requirements imply that the spatial data be topologically structured in a hierarchical fashion, with varying levels of detail. Each node accommodates a region of arbitrary polygonal shape which again contains isolated points, lineal features, and smaller regions of different complexity, and the nodes at each level correspond to disk pages.

The paper describes an object-oriented dynamic data model satisfying the above needs. It first summarizes previous contributions to spatial data handling and sets the perspective view of this research. An alternative approach based on the kinematic Voronoi diagram [Gold 1990; Roos 1991; Gold et al. 1995] is then proposed. Disadvantages of the primitive Voronoi diagram applied to large data sets are pointed out. They are overcome by introducing the *spatial object condensation* [Yang and Gold 1995] concepts. The major contribution of this paper, the VMO-tree class, is described as a dual structure managing condensed objects.

2. A review of spatial data structures

Realizing that it is the quantity and geometric complexity of spatial data that make the task of management and operation intricate, research over the past two decades has aimed to moderate the problem by decomposing a data set into smaller ones. This results in numerous geometric data structures which are broadly categorized into three families: trees (i.e. K-d tree [Bentley 1975], Quadtrees [Samet 84]); buckets (i.e. Grid Files [Nievergelt et al. 1984, EXCELL [Tamminen 1983]); and linear orderings (i.e. Morton code [Morton 1966], Spiral order [Mark and Goodchild 1986]). Tree structures recursively decompose space. Depending on decomposition schemes used and types of geometric primitives concerned, trees can have different shapes and the contents of intermediate nodes vary as to whether geometric elements participate in the partition. Leaves of a tree can contain a single object or a bucket of multiple objects. Bucket structures are not hierarchical: space is flatly divided into rectangles of different size. Each rectangle has a location code as its address. The criteria guiding the divisions vary depending on the distribution of spatial objects and on the dynamics of a structure being used. The storage for indices to buckets is generally smaller than that for trees because each bucket contains more objects. Linear orderings transform a k-dimensional problem into a one-dimensional one. They originate from scanning and storing pixels of images into linear lists and are extended to index points and buckets. Other ordering schemes are designed to preserve spatial contiguity.

Classical geometric data structures are constructed for 0-cell objects. Variants of the original data structures have been developed to accommodate complex objects. For example, Matsuyama et al. [1988] modify the K-d tree to index axes-parallel rectangular blocks, where three types of geometric primitives are recorded. The PM Quadtree [Rosenberg 1985], is used to partition a polygonal map. This type of adapted data structure is still based on the *space decomposition*. Another type of tree structure is based on *object decomposition*, where the dividing criteria are oriented to objects themselves. This kind of data structure is suitable for representing polylines and polygons in different levels of detail. Nodes of trees record the result of a line generalization process, with lower ones containing partial objects of finer resolution. Examples of object based trees are the Strip tree [Ballard 1981], the Arc tree [Günther and Wong 1989], and the BLG-trees [van Oosterom 1990]. Examples of data structures that support generic geometric objects include the R-tree [Guttman 1984], the R^+-tree [Faloutsos et al. 1987], the Field-tree [Frank and Barrera 1989], the Cell-tree [Günther 1988], and the BANG files [Freeston 1987].

While geometric data structures provide a facility for data storage and access, they cannot form the sole data structure for a spatial information system. In addition to managing and accessing geometric data, a spatial information system must be able to answer topological queries about topological relationships among spatial entities. Examples of the topological relationships include adjacency, containment, neighbourhood, and connectivity. Although it can be argued that geometric data structures preserve certain topological relationships in their subdivision, they are not designed to efficiently answer all topological queries. Quick response to topological queries is the purpose of using topological data structures in designing a spatial information system. Well known topological structures, as are widely used in current vector based commercial GISs, are based on the *planer graph* topological data model. Topological relationships of a given data set, representing points, lines, and polygons, can be calculated and assembled after all the data is present.

We are left with the situation that, on the one hand, research on geometric data structures is directed towards the hierarchical, representing objects in different levels of detail, and partitioning space or objects into disk pages; and on the other hand, widely used topological data structures are still based on a flat, indivisible planer graph model. In a hybrid system, where both geometric and topological data models are employed, complex links between the two must be built. It would be difficult to maintain the integrity of topological relationships if update occurs in geometric data structures.

Recent research on spatial data models tackles the problem by developing generic spatial data models with a hierarchical topology. Worboys [1993] proposes a canonical representation of areal objects which explicitly captures connectedness and region inclusion. Without providing an implementation structure, the paper formalizes the definition of classes of areal objects and operations upon these objects. Bertolotto and De Floriani [1994] report a HPEG (Hierarchical Plane Euclidean Graph) for a multiresolution representation of a region. The HPEG recursively decomposes regions into smaller ones. Region boundaries at one level are simplified by creating ε-homotopys with line generation algorithms, where ε specifies the radius of a band convoluted from a chain of edges. Each smaller region in HPEG is a PEG whose features are refined with smaller horizontal error. In order to support navigation in the hierarchy of PEGs, the boundary information of a PEG must be recorded and the links to the direct refinements of a PEG must be maintained. The HPEG provides a way of browsing a map at different levels of resolution. The data structure implementing the HPEG is based on the DCEL [Preparata and Shamos 1985], an encoding of an edge-oriented planar subdivision. It is not clear how a PEG be stored in a separate disk page and operated upon as an independent object in a dynamic environment.

Starting in Section 3, we propose an alternative implementation of a hierarchical topological data model. This data model is aimed at producing a set of meaningful map objects, such that each node of the hierarchy has both geometric definition for its embedded components and a complete topology for the components. In addition, each node corresponds to a disk page and can be worked with in both linked and unlinked fashions. We believe that the ability to represent maps as manageable objects in both spatial and memory subspaces is a key feature in object-orientation.

3. The dynamic Voronoi diagram

Given a set of n ($n > 1$) objects S = {points, line segments} in the R^2 Euclidean space, the dynamic Voronoi diagram, $V(S)$, is constructed incrementally [Gold 1990; Roos 1991; Gold et al. 1995] which partitions the plane into *Voronoi regions*, v, such that for two distinct objects s_i, $s_j \in S$, the dominance of s_i over s_j is defined as the subset of the plane being at least as close to s_i as to s_j :

$$D(s_i, s_j) = \{x \in R^2 \mid \delta(x, s_i) \le \delta(x, s_j)\},$$

for δ denoting the Euclidean distance function. $D(s_i, s_j)$ is a closed half plane bounded by the bisector of s_i and s_j, denoted by $B(s_i, s_j)$:

$$B(s_i, s_j) = \partial D(s_i, s_j) = \{x \in R^2 \mid \delta(x, s_i) = \delta(x, s_j)\}.$$

The Voronoi region $v(s_i)$ is the portion of the plane lying in all of the dominances of s_i over the remaining objects in S:

$$v(s_i) = \bigcap_{s_i \in S - \{s_j\}} D(s_i, s_j).$$

If two bisectors $B(p, q)$ and $B(p, r)$ (p, q, and $r \in S$) intersects, the intersection is called a *Voronoi point* and the bisector delimited by two consecutive Voronoi points is called a *Voronoi edge*. Two objects are *neighbours* if they share a common Voronoi edge (possibly extended to infinity). The Voronoi point is at an equal distance to at least three objects and is therefore the circumcentre of a circumcircle defined by these objects.

If S is a point set, $v(p)$ for $p \in S$ is a (possibly unbounded) convex region intersected by all half-planes containing p and delimited by bisectors between p and other members in S. This nice property does not hold for a general Voronoi region if S contains points and lines. The bisector between a point and a line is no longer a straight line but a locus of points bisecting them. While the definition of the dominance $D(p, q)$ still holds, the bound $B(p, q)$ may become a parabola. Figure 1 shows the Voronoi diagram (grey) of points and lines (dark).

3.1 The Delaunay triangulation

The Voronoi edge separating objects p and q indicates that p, q are neighbours. A straight line linking p, q can represent this relationship. For e Voronoi edges we can have the same number of straight lines. Each Voronoi point corresponds to one triangle formed by three objects on the circumcircle[1]. The union of these triangles is the dual structure of the Voronoi diagram, the Delaunay triangulation $D(S)$. Each straight line linking two objects is called a *Delaunay edge*. For S = {points, line segments}, the corresponding Delaunay edges between a line and a point can be graphically implemented by linking the point and the middle of the line segment. Figure 2 illustrates a Delaunay triangulation (dotted) for some lines and points (solid).

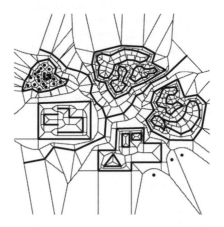

Figure 1 A Voronoi diagram of points and lines.

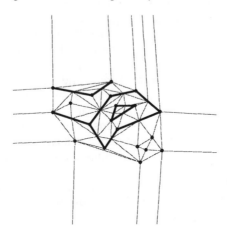

Figure 2 A Delaunay triangulation of points and lines.

Unlike its counterpart for a point set, the Delaunay triangulation for a set of points and lines is not necessarily *equiangular* [Sibson 1977]. It simply becomes a graph representing adjacency relationships between objects. Nevertheless, the circumcircle tangent to three vertex objects is object-free, i.e. none of the given objects in S is contained in its interior.

---

[1] By saying so we assume that S is in general position, i.e. there are no cocircular cases and not all objects in S are collinear. These degenerate cases can be handled by some additional rules.

3.2 The memory space related problems

The *memory space* is referred to as the computer memory allocated to store objects and their topological structures. In our case, the topological structure is the Delaunay triangulation. Two problems can be observed in applying the dynamic Voronoi diagram as a fundamental data model for a GIS, within the framework of incremental construction. Firstly, compared with other geometric and topological data models used in most current GISs, the Voronoi diagram consumes more memory space. In addition to storing objects and their co-ordinates, the structure must maintain the triangle network. For a complicated polygonal map composed of hundreds of thousands of short lines, the size of the triangulation could be extremely large. Secondly, since the Voronoi diagram is generated incrementally, the input of objects does not necessarily follow a particular order. This may result in an instance of the data structure with poor spatial indexing. That is, objects and triangles that are close in space may be stored far apart in memory. This makes the ordering of the space according to some pattern impossible. It follows that no matter how big a map is, all the object and topological data must be loaded in memory in order to ensure the presence of the neighbourhood around any area of interest.

Large memory use with unnecessarily detailed objects and their relationships lowers the overall performance of the system and hinders operators from concentrating on more important properties of a data set. Managing spatial data in a non-splittable fashion fails to meet the requirements for modern data structures which must be dynamic, hierarchical, with varying levels of detail, and divisible into disk pages [van Oosterom 1993]. If a part of the Voronoi diagram of detailed objects can be taken away from the organizing structure and represented by some compound objects, the number of objects present in memory would be smaller. Again, if the separated compound object preserves its original topological partitions of the space, it can be loaded into memory at any time as an independent workspace. The whole space can then be managed seamlessly as if there is no partition. Furthermore, if scattered indices to objects and their local topological relationships in a memory subspace can be aggregated, they can be more efficiently managed and retrieved. These expectations can be achieved by introducing the concept of condensed spatial objects.

4. The spatial condensation technique

The spatial object condensation technique [Yang and Gold 1995] has two aspects. It firstly partitions the Voronoi diagram into subsets such that the topological structures for subsets of objects in the interior of partitioned subspaces are independent of each other. Topological relationships between partitioned subspaces are preserved in the mutual partitioning boundaries. Secondly, it separates partitioned subsets from the original memory space into different memory spaces which can be stored on separate disk pages.

During this process a transformation is applied to spatial objects and Delaunay triangles such that if they are close in space, they will be stored close by in the memory. This eliminates large gaps between indices of spatially contiguous triangles and objects in the original memory space.

The general procedure of condensing a spatial object starts with a predefined partitioning boundary around a cluster of primitive spatial objects, some of which form closed polygons. The partitioning boundaries are not necessarily required to be simple loops. They can be complex loops with other lines incident to them. It is assumed that there exists a heuristic approach that can identify desired clusters with closed boundaries to be partitioned. Besides line objects, map frames as well as constrained Delaunay edges also constitute boundaries for partitioning thematic polygons, map sheets, and clusters of points.

The difficulty of partitioning a subset of topological structures lies in its computer implementation. Most topological data structures explicitly encode certain adjacent or incident relationships of one object to others. For example, the DCEL based data structures contain boundary and co-boundary information. Each edge structure encodes polygons on both sides. This is also true for the Delaunay triangulation. In our implementation, indices of adjacent triangles are encoded in each triangle. Especially, two adjacent triangles, sharing a common edge collinear with the partitioning boundary, contain indices referring to triangles in another subspace. Two subspaces are therefore topologically linked through the two *bordering triangles*. In order to partition a topological structure into two independent subsets, the topological linkage has to be cut off. Since a bordering triangle is in the interior of a subspace, it is not stable in a dynamic environment. *Critical triangles* inside a partitioning boundary around both subspaces are developed to prevent triangles in the interior of a partitioned subspace from having indices referring to triangles in the exterior of the subspace. Only critical triangles have indices representing triangles in other subsets. Critical triangles will never be changed and are stable. Their indices can be modified so that they represent critical triangles belonging to the same subspace as the boundary. Figure 3 illustrates critical triangles inside a line object boundary in subspace S_i. The arrows depict indices referring triangles in the exterior of S_i. Figure 4 shows the result of modifying these indices to refer to triangles in S_i.

The next step of the condensation technique is to identify objects and their topological structure in one subspace. Once identified, they can be written into a newly allocated memory space and later removed from the original one. Unfortunately, due to an incremental construction, indices for one subset of the topological structure may not be contiguous. This makes sequential access to the subset impossible. A spatial traversal has to be employed. In our implementation, we use the flood fill algorithm to traverse the topological space in a partial order. For each triangle and object visited, instead of copying its original index to the new memory space, a new consecutive index number is assigned to it. This eliminates large gaps in memory space between two spatially adjacent objects.

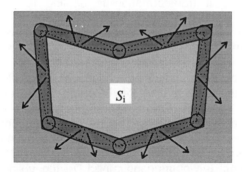

Figure 3 Indices in critical triangles referring to triangles in the exterior of S_i.

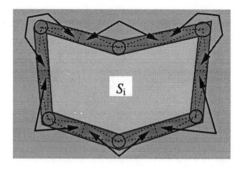

Figure 4 Modifying indices to refer to triangles on the boundary of S_i.

As an example, five clusters of complex objects are partitioned (Figure 5) from the map in Figure 1. Each partitioned object is a map of the same type as its ancestor. It can be worked with independently. Even smaller objects can be created from it. As is illustrated in Figure 6, three more objects are created from the object labelled "C". It can be shown that map modification and spatial queries in the interior of a partitioned subspace X_i are *self contained*. That is, no subsets of the topological structure in the exterior of X_i have to be involved in these operations.

5. The VMO-tree

The spatial object condensation creates complex polygonal objects. Each object resides in a separate memory space and can be saved on a storage medium. The portion of its memory space in the original map will then be released. In this section, we describe the relationship between the new object and the one from which it is created. A new class named *VMO-tree* (Voronoi based Map Object Tree) is introduced. The VMO-tree is a graph representation of the relationships of map objects. The prime roles of the VMO-tree are managing complex objects and serving as a platform for transactions upon these objects.

5.1 The data model for the VMO-tree class

The objects created by condensation naturally form a hierarchical relationship where the object from which a partition is made is the *parent* and the newly partitioned object is the *child*. A parent can have a number of children and each child itself can have children. When a complex polygon is condensed, the partitioning boundary is duplicated in both parent and child objects. At the parent level, the subspace $Y \subset R^2$ enclosed by the boundary is a simple polygon with all detail suppressed. Topologically, the subspace Y is equivalent to a closed disc while the subspace $X \subset R^2$, enclosing Y, is open. If we attach a label to the internal boundary of the parent, the enclosed subspace represents the condensed object. At the child level, the detail is expanded and the external side of the boundary has the same label as that of its parent.

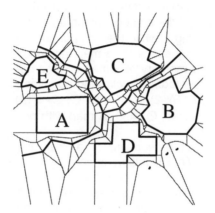

Figure 5 A partition of the map in Figure 1.

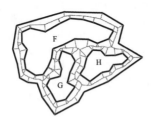

Figure 6 A partition of object C in Figure 5.

In what follows, four aspects of the data model for the VMO-tree class will be identified. They are namely: components of the class; data structures representing components and their relationships; operations on components; and constraints over operations.

5.1.1 The Components of the VMO-Tree Class

Definition 1: The *outline image* is the partitioning boundary left on the subspace X from which the space partitioning occurs. The *object outline* is the partitioning boundary duplicated on the partitioned subspace Y whose extent is confined in the enclosure of Y.

Definition 2: The class *map* is a set $M = (P, L, R, A)$, where P is a class of points, L a class of lines including outline images of condensed objects, R a class of regions with closed boundaries, and A a class of condensed objects. Within the four subclasses, the point class is the most primitive one, the other three classes are aggregated from relatively simpler ones. For example, the compositions of R and A are $R = (P, L, R_s)$ with R_s denoting smaller regions covered by R, and $A = (P, L, R, A_s)$ with A_s denoting smaller condensed objects in A, respectively. Since A and M have the same composition of subclasses, they belong to the same map class. Therefore, a condensed object is also called a map in a recursive way. Both R and A represent areal objects. They differ in that the components in an instance $r \in R$ are present in a map instance containing r, while the components in an instance $a \in A$ are dropped from the memory space for a map instance. Only the outline image of a is preserved in the memory of a map instance. A map may not have all classes of objects presented. The following combinations are valid cases: 1) $M = (P, L, R, \varnothing) = (P, L, R)$, 2) $M = (P, \varnothing, \varnothing, \varnothing) = (P)$, 3) $M = (P, L, \varnothing, \varnothing) = (P, L)$, and 4) $M = (\varnothing, \varnothing, \varnothing, \varnothing) = \varnothing$. A map with no components is called a *null map* or *null object*.

Definition 3: An *object embedding* of a map object is a layout of its composing objects in R^2 Euclidean space. For a single map, the object embedding disallows general intersection between objects, following the concept of the *single valued vector map* [Molenaar 1989]. This ensures the planarity of the map. The object embedding can be dynamic. That is, the components of the map object can be changed. However, any change must be confined to the extent of the object outline.

Definition 4: A *topological embedding* of a map object is a realization of topological relationships over the object embedding of the map. The topological relationships are specified in the topological data structures. Note that the topological embedding of a map object does not apply to the interior of condensed objects in the map because the object embeddings of these objects are suppressed. However, the outline image of a condensed object is topologically embedded in the map object. The topological embedding is also dynamic, due to the dynamic property of the object embedding.

5.1.2 The VMO-Tree Structure

Definition 5: A VMO-tree is a graph $G = (N, E)$ with node set N and edge set E. The node set N is of the map class M and each node has a label. For any pair $<x,y> \in N$, an edge $e \in E$ exists between x and y if the object embedding of x contains the outline image of y and y has both object and topological embeddings, and both of them are confined in its object outline duplication. The node x is called a parent, and the node y is a child.

The construction of a VMO-tree can be either top-down, bottom-up, or a combination of both. There are two scenarios when the top-down strategy is used. In either case, the root node is first created. One scenario of the top-down approach condenses a null object for each smaller area such that the whole study region is completely partitioned by the outline images of these null objects. These null objects are added into the VMO-tree at level 1. Each null object can then be worked on independently and even smaller objects are added in the tree at lower levels (Figure 7).

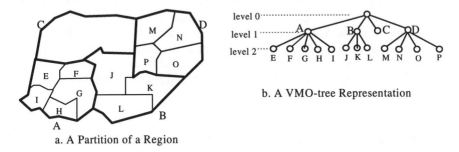

a. A Partition of a Region

b. A VMO-tree Representation

Figure 7 The top-down partition and its VMO-tree.

Another scenario of the top-down construction works with the root map and adds components in it. Some heuristic approach can be taken to aid the decision on when and where a cluster of components can be condensed. Whenever a new object is dynamically condensed, the node representing the object is inserted into the tree.

The bottom-up strategy works in a reverse way where the smallest distinct objects are first composed as maps separately. There are also two ways of constructing the tree. The first one creates the root and inserts into it all the object nodes from the first step. These objects will be at level 1. This results in a flat tree where all children have the root as the parent. Then a generation process takes place which aggregates smaller objects into bigger ones. When a bigger object n aggregated from m smaller objects at level i is formed, the object n becomes the parent of the m objects. Note that the aggregation happens at level i-1. The new node representing object n will therefore be at level i and the m children are dropped to level i +1. The m pointers of the original parent at level i - 1 will be released and the tree becomes narrower by m - 1 branches at level i. The generation process can be performed heuristically.

The second method of bottom-up tree construction generalizes more complex objects without knowledge of the root. When an aggregate object is created, it serves temporarily as a root of a smaller tree. A forest can exist during the whole process. A bigger tree is composed by combining a number of smaller trees.

From a cartographer's point of view, the top-down approach works from smaller scale, larger sized objects towards larger scale, smaller objects. The deeper down in the VMO-tree, the more details an object can expose. On the contrary, the bottom-up approach works from larger scale, smaller sized objects towards smaller scale, larger objects. The higher a node is in the VMO-tree, the more abstract it becomes.

The combined construction repeatedly uses partition and aggregation processes. The description is straightforward and is omitted from this paper. Inserting an existing condensed object into the VMO-tree and aggregating smaller condensed objects into a bigger one are compound operations each of which takes several steps. The major steps for insertion duplicate the outline image of the new object in the proper map object represented by the parent node and attach the new child object to the tree. For aggregating, the outline of the new object can have the same extent as that defined by the union of outlines of the smaller objects, or a bigger one, depending on the heuristic decision.

5.1.3 Constraints upon the VMO-Tree Construction

The VMO-tree is a topological structure mapped from partitioning of another topological space. In order to make VMO-tree well defined, the constraints guiding partitions must be defined. Let $A = \{a_1, a_2, ..., a_n\}$ be a finite set of condensed objects contained in a map object. A is the parent and a_i (i = 1, 2, ...n) are children. These constraints state:

Constraint 1: The number of condensed objects contained in one map object corresponds to the number of children under the node representing the map object.

Constraint 2: For a_i, $a_j \in A$, (i ≠ j), $a_i^{\circ} \cap a_j^{\circ} = \varnothing$. That is, interior spaces of any two distinct condensed child objects must be disjoint at the same level. Overlapped or partially overlapped subspaces enclosed by outline images are disallowed. The non-empty intersection of distinct member of A may only happens on the outline images.

Constraint 3: $\bigcup_{a \in A} \text{outline}(a) \subset \text{outline}(A)$. That is, the union of outline images of all children must be confined in the outline of the parent object.

5.2 Operations over the VMO-Tree

There are five categories of operations over the VMO-tree: building operations; set-theoretic operations; spatial queries; the operations linking object attributes; and operations for network communications. The first three categories are spatial operations. More formal and detail discussions about some of these operations can be found in Worboys [1992]. We are currently developing a prototype of the VMO-tree classes and the format for their operations. Detailed descriptions are in preparation. For now, the functionality of each operation group is list below.

The building operations insert and delete nodes from the tree. They can be activated either interactively from the graphics interface or from behind the scene where only the tree and the node concerned are given. In either case, the outline image of the inserted node must be embedded in the space represented by the tree, or the embedded outline image must be removed from the space when deleting. The building operations also contain basic editing tools which allow modification over components of a node object.

The set-theoretic operations work on either single or multiple themes. These include calculating unions, intersections, and complements, with respect to given themes and objects. New instances of the VMO-tree can be generated by a set operation. An example of this calculates the intersection of two objects, each from a VMO-tree instance. A third instance of the VMO-tree can result if the calculation results in a non-empty subset of objects.

Spatial queries can be greatly facilitated by the VMO-tree, which provides a fast search along edges of the tree, and the topological embedding within each spatial object, which enables navigation and searches over components of a map object. This ability to search objects vertically and horizontally resembles that of the B^+ tree.

How to store and manage attribute data of spatial objects is still an open question. The majority of GISs implementing a relational model typically adopt a hybrid architecture. Due to the special characteristics of spatial data, future options have two major directions: extending relational systems with specialized functionality for geo-information; or moving to radically new approaches, such as object-oriented systems [Worboys 1994]. The option of adding specialized functions to relational databases not only can speed up the development, but also conforms to the reality that most attribute data are maintained by relational databases.

The VMO-tree manages map objects, some of which may reside in or be accessed from remote sites. The object networking functions have to take care of the data security, integrity and concurrency controls, and dynamic communications. A client-server architecture can be designed to allow node objects of the VMO-tree to be linked to client applications.

6. The interface for the VMO-tree operations

One of the objectives of the spatial data handling system based on the VMO-tree is that any operations on spatial objects should be made easy. This can be achieved by designing a well defined visual interface. For example, the VMO-tree can be represented by a visual component, following the visual programming paradigm. The VMO-tree component responds to some events triggered by users, and the methods built in the VMO-tree class can be activated by events. Figures 8, 9 and 10 illustrate a prototype interface from which some typical operations can be facilitated.

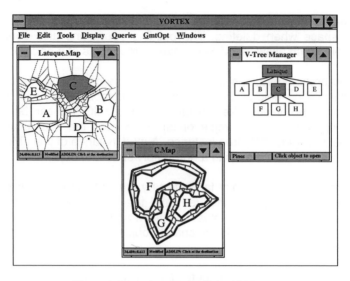

Figure 8 Accessing objects in the VMO-tree.

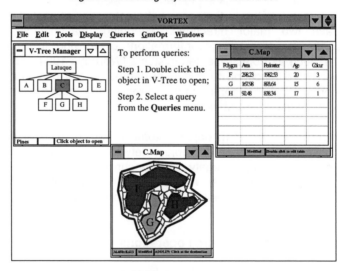

Figure 9 Viewing attributes of the VMO-tree objects.

7. Conclusions and future work

In this paper, an object-oriented spatial data model for managing maps is proposed. The objects are condensed from the primitive Voronoi diagram, each having a separate memory subspace corresponding to disk pages. Condensed objects are managed by an acyclic VMO-tree which is one-one correspondent to the map partitions. The VMO-tree can be built by either top-down subdivision, or bottom-up aggregation of map objects, or a combination of the two. The construction constraints ensures that the correspondent tree is well defined.

The current specifications of the data model are preliminary. The authors are developing detailed descriptions of relationships of spatial components, the operations, the boundary conditions for dynamic modifications, and the visual interface of the data model. The attribute data storage and networking issues will be further studied in the near future.

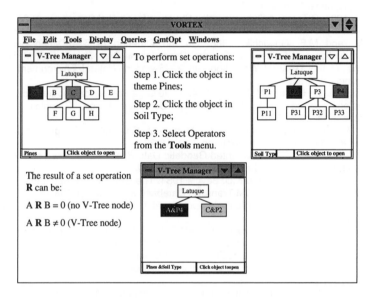

Figure 10 Set-theoretic operations through the VMO-tree.

Acknowledgements
The authors acknowledge the support by the Natural Sciences and Engineering Research Council of Canada (NSERC) and the Association de l'Industrie Forestière du Québec (AIFQ).

References
Ballard, D.H. 1981. Strip Trees: A Hierarchical Representation for Curves. Communications of the ACM, 24(5), 310-21.

Bentley, J.L. 1975. Multidimensional Binary Search Trees Used for Associative Searching. Communications of the ACM, 18(9), 509-17.

Bertolotto, M. and L. De Floriani 1994. Multiresolution Topological Maps. in Molenaar, M. and S. de Hoop (eds.), Advanced Geographic Data Modelling: Spatial Data Modelling and Query Languages for 2D and 3D Applications. Publications on Geodesy, New Series No. 40, Delft, Netherlands, 179-90.

Faloutsos, C., T. Sillis, and N. Roussopoulos 1987. Analysis of Object Oriented Spatial Access Methods. ACM SIGMOD, 16(3), 426-39.

Frank, A. and R. Barrera 1989. The Field-tree: A Data Structure for Geographic Information System. in Symp. on the Design and Implementation of Large Spatial Databases, Santa Barbara, California, Berlin, Springer-Verlag, 29-44.

Freeston, M.W. 1987. The BANG File: A New Kind of Grid File. in Proc. ACM SIGMOD Conf., San Francisco, California, 260-69.

Gold, C.M. 1990. Spatial Data Structures: the Extension from One to Two Dimensions, in: L.F. Pau (ed.), Mapping and Spatial Modelling for Navigation, NATO ASI Series, F, No. 65, Springer, Berlin, 11-39.

Gold, C.M., P.R. Remmele, and T. Roos 1995. Voronoi Diagrams of Line Segments Made Easy. In Proc. 7th Canadian Conference on Computational Geometry, Québec, Canada, 223-28.

Günther, O. 1988. Efficient Structures for Geometric Data Management. Lecture Notes in Computer Science No. 337. Springer-Verlag, Berlin.

Günther, O. and E. Wong, 1989. The Arc Tree: An Approximation Scheme to Represent Arbitrary Curved Shapes. In Litwin, W. and H.-J. Schek (eds.), Foundations of Data Organization and Algorithms. Paris, France, Lecture Notes in Computer Science No. 367, Springer-Verlag, 354-70.

Guttman, A. 1984. R-Trees: A Dynamic Index Structure for Spatial Searching. ACM, SIGMOD, 13, 47-57.

Mark, D.M. and M.F. Goodchild 1986. On the Ordering of Two-Dimensional Space: Introduction and Relation to Tesseral Principles. in Diaz, B. and S.B. M. Bell (eds.) Proc. of the Tesseral Workshop No. 2, (Swindon: Natural Environment Research Council), 179-92.

Matsuyama, T., L.V. Hao, and M. Nagao 1984. A File Organization for Geographic Information Systems Based on Spatial Proximity. Computer Vision, Graphics and Image Processing, 26, 303-18.

Molenaar, M. 1989. Single Valued Vector Maps - A Concept in Geographic Information Systems. Geo-Informationssysteme, (2)1, 18-26.

Morton, G.M. 1966. A Computer Oriented Geodetic Data Base, and a New Technique in File Structuring. Unpublished report, IBM, Canada Ltd.

Nievergelt, J., H. Hinterberger and K.C. Sevcik 1984. The Grid File: An Adaptable, Symmetric Multikey File Structure. ACM Trans. on Database Systems, 9(1), 38-71.

Preparata, F. and M. Shamos 1985. Computational Geometry: An Introduction. Springer-Verlag, New York.

Roos, T. 1991. Dynamic Voronoi Diagrams. PhD Thesis, Universität Würzburg, Switzerland.

Rosenberg, J.B. 1985. Geographical Data Structures Compared: A Study of Data Structures Supporting Region Queries. IEEE Trans. on Computer Aided Design, 4(1), 53-67.

Samet, H. 1984. The Quadtree and Related Hierarchical Data Structures. Computing Surveys, 16(2), 187-260.

Sibson, R. 1977. Locally Equiangular Triangulations. Comput. J. 21, 243-45.

Tamminen, M. 1983. Performance Analysis of Cell Based Geometric File Organizations, Computer Vision, Graphics, and Image Processing, 24(2), 160-81.

van Oosterom, P. 1993. Reactive Data Structures for Geographic Information Systems. Oxford University Press.

Worboys, M.F. 1994. Object-Oriented Approaches to Geo-Reference Information. Int. J. Geographic Information Systems, (8)4, 385-99.

Worboys, M.F. and P. Bofakos 1993. A Canonical Model for a Class of Areal Spatial Objects. in Abel, D. and Ooi, B. C. (eds.), Advances in Spatial Databases, Proc. of 3rd International Symposium, SSD'93, Lecture Notes in Computer Science No. 692, Springer-Verlag, 36-52.

Worboys, M.F. 1992. A Generic Model for Planar Geographic Objects. Int. J. Geographic Information Systems, (6)5, 353-72.

Yang, W. and C.M. Gold 1995. Dynamic Spatial Object Condensation Based on the Voronoi Diagram. In Proc. 4th International Symposium of LIESMARS, Wuhan, China, 134-45.

Contact address
Weiping Yang and Christopher Gold
Chaire industrielle en géomatique
Centre de recherche en géomatique
Université Laval
Québec, Qc
Canada G1K 7P4
E-mail: weiping@gmt.ulaval.ca

DECISION-ANALYTIC INTERPRETATION

OF REMOTELY SENSED DATA

B.G.H. Gorte[1], L.C. van der Gaag[2] and F.J.M. van der Wel[3]

[1]International Institute for Aerospace Survey and Earth Sciences (ITC)
Enschede, the Netherlands
[2]Utrecht University, Department of Computer Science
Utrecht, the Netherlands
[3]Utrecht University, Faculty of Geographical Sciences _ Cartography Section
Utrecht, the Netherlands

Abstract
Increasingly, remotely sensed data are used for taking decisions in geographical information systems. Decision making can in principle be based on a classification of such remotely sensed data into nominal information classes. Such a classification, however, typically includes an unknown amount of uncertainty. Moreover, when processing spatial data for decision making, not only the uncertainties inherent in these data but also the objectives and preferences of the decision maker have to be taken into account. This paper proposes exploiting concepts from the mathematical framework of decision analysis for integrating uncertainties and preferences. It aims to solve complex decision problems on the basis of remotely sensed data. The feasibility of the decision-analytic approach to the interpretation of spatial data is demonstrated by means of a case study.

Key words
Remote sensing, GIS, decision analysis, preferences, uncertainty.

1. Introduction

To monitor, analyze and interpret developments in our changing environment, up-to-date spatial data are periodically collected and processed. Increasingly, remote sensing is considered as a valuable source for this purpose. It yields data that can be subjected to further analysis in a geographical information system (GIS) at advantageous average cost. By systematic application of spatial operations and visualization, a GIS is able to generate, on request, derivative data sets contributing to making decisions involving characteristics of spatially-related phenomena of the environment.

Classification of remotely sensed data into qualitative information classes is useful to extract information from the spectral attributes of these data, yielding an insightful representation of the real world. Such a representation can be exploited directly as a thematic map or as part of a time series in a change detection application. Unfortunately, classification generally introduces unknown uncertainty in the information classes assigned to the spectral objects. This uncertainty propagates through the subsequent stages of the decision making process [Lunetta et al., 1991]. The uncertainty can be reduced by using evidence with regard to the real world, usually derived from sources such as domain experts, maps, field work, aerial photographs, or thematic maps from former classifications.

Such evidence can be exploited before, during, and after classification and hence contribute to the accuracy of the final results in various different ways [Strahler, 1980]. Despite all efforts to reduce the uncertainty introduced by classification, it always influences the results. These imperfections may seriously affect the adequacy of using classification results for taking environmental decisions. For example, the commonly used maximum a posteriori probability classification discards useful information that may serve to yield insight in the uncertainties. In this approach to classification, the posterior probabilities that are computed for each spatial object within an information class distinguished during sampling, are used only to select the most likely class. The entire probability distribution for the object, however, reflects highly valuable information about the extent and distribution of uncertainty which could be further utilised in a GIS.

If decisions are to be made on the basis of remotely sensed data, uncertainty tells only part of the story: the objectives to be pursued with interpretation of the data become crucial. In the presence of uncertainty, the best decisions are those that, in view of the objectives, carefully weigh the benefits of correct interpretation of the data on the one hand and the losses due to incorrect interpretation on the other hand. This idea is illustrated by an example dealing with fraud with subsidies assigned to agricultural crops by the European Union. In this example, the main objective is to detect illegal declarations of subsidised crops by taking remotely sensed images from crops on parcels, to avoid waste of public resources. From this objective alone, the number of detected illegal declarations should be maximised. However, unjust implication of fraud is highly unfavourable as it results in extra costs for verification and in loss of face. Therefore, the number of unjust implications should be kept at a minimum. In pursuing both objectives simultaneously, overlooking fraud is considered worse than over-estimating. It now depends on the probabilities computed for the various possible crops for a parcel under consideration whether or not fraud should be implied. Interpretation of remotely sensed data for decision making therefore involves both the extent and distribution of uncertainty introduced by classification and the preferences of the decision maker. These preferences concern the objectives that are being pursued with the interpretation and therefore differ from knowledge about the subject of the interpretation as referred to by [Strahler, 1980]. Both types of knowledge equally contribute to the interpretation, yet at different levels.

Further elaborating on the idea that remotely sensed data can serve as a basis for decision making, the question arises whether or not it is necessary to derive a complete classification before considering viable decisions. In principle, decisions can be taken on the basis of a classification. However, classification contains uncertainty of which the extent and distribution are unknown. By making decision directly on the data, full knowledge about the uncertainties involved can be included, thereby allowing for making better decisions. As decision making does not so much involve classification results as the extent and distribution of uncertainty introduced by classification, deriving a complete classification is no longer required and, in fact, has become obsolete (Figure 1). However, an accurate classification nevertheless serves various purposes beyond decision-making.

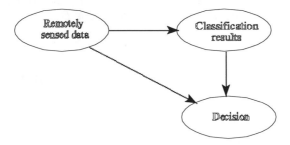

Figure 1 Founding decision making on data.

This paper addresses the interpretation of remotely sensed data in view of the objectives that are to be pursued when exploiting the data for decision making. To this end, various concepts from decision analysis are introduced which allow integration of uncertainties and a decision-maker's preferences. Section 2 expresses the interpretation of remotely sensed data as a decision problem and introduces the mathematics for solving this problem. Section 3 describes the assessment of the various parameters involved in quantification of uncertainties and preferences. A case study will be presented in Section 4, demonstrating the feasibility of the decision-analytic approach.

2. Interpretation of data: a decision problem

Interpretation of remotely sensed data is in essence a decision problem: the problem is to decide upon which decision to take for each spatial object on the basis of available data. The solution to this problem is for each object the decision that is expected to best meet the objectives that are being pursued with the interpretation. The field of Decision analysis provides the mathematical framework for solving complex decision problems such as the data-interpretation problem [Raiffa, 1968, von Winterfeldt & Edwards, 1986, Smith, 1988]. It offers means for structuring decision problems and for computing solutions. In this section, we express the interpretation problem and its solution in decision-analytic terms.

A decision problem involves two types of variable:
- a decision variable is a variable that represents viable decisions or actions that can be taken in the context of the problem at hand;

- a chance variable is a variable that represents the true 'state of the world'; the value of such a variable cannot be selected by the decision maker.

In the data-interpretation problem, there is only one variable of each type: a chance variable C that represents the true information class of a spatial object O and a decision variable D that represents the possible decisions that can be taken with regard to this object. A variable in a decision problem can take its value from among a pre-defined set of values. We assume that

$C1,...,Cn$, n 1, are possible information classes of O. These classes therefore are values for the chance variable C. We further assume that the decision variable D takes its value from among the decisions $D1, ... ,Dm$, m 1.

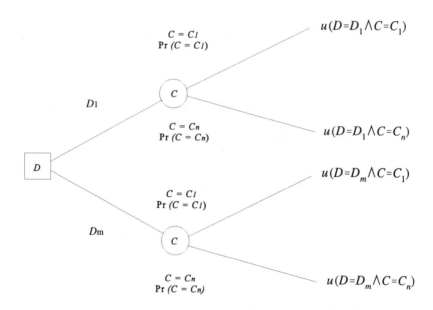

$$u(D=D_1 \wedge C=C_1)$$

$C = C_1$
$\Pr (C = C_1)$

$$u(D=D_1 \wedge C=C_n)$$

$C = C_n$
$\Pr (C = C_n)$

$$u(D=D_m \wedge C=C_1)$$

$C = C_1$
$\Pr (C = C_1)$

$$u(D=D_m \wedge C=C_n)$$

$C = C_n$
$\Pr (C = C_n)$

D_1

D_m

D

Figure 2 A Decision tree for data-intepretation problem.

In a decision problem, there typically is uncertainty regarding the true values of the chance variables involved. In data interpretation, there is uncertainty concerning the true value of the chance variable C since the true information class of O is unknown at the time of interpretation. This uncertainty is expressed as a probability distribution $\Pr(C)$ for the variable C, specifying for every possible information class Ci the probability $\Pr(C = Ci)$ that Ci is the true class of the object. Note that this probability distribution will not be influenced by the various decisions that can be taken.

In addition to uncertainties, a decision problem involves preferences. The desirability of a decision and its consequences, with each other called a scenario, is quantified by means of its utility. In our data-interpretation problem, each combination of a decision $D = Di$ and a true information class $C = Cj$ has associated a utility $u(D=Di \ C=Cj)$. The utility expresses the desirability of the scenario where the decision Di is taken with regard to O while it has Cj as its true information class. Actual utilities associated with the various scenarios depend upon the objectives that are being pursued with the interpretation.

Structuring all aspects of a decision problem can be done with a decision tree. A decision tree is a pictorial, tree-like representation of the problem. The various variables and values of the problem are organised in a (rooted) tree. Each node in the tree models a variable; the edges emerging from a node represent the values of its associated variable. The topological structure of the tree is an explicit representation of all scenarios that can possibly arise from a decision. The root node of the tree represents the initial situation before any decision is taken and each path from the root node to the tip of a terminal edge corresponds with a scenario. Figure 2 shows a tree organising the variables of our object-interpretation problem. To distinguish between the decision and chance variable, the former is depicted as a square box and the latter is shown as a circle. In the tree, the uncertainties concerning the chance variable's values are depicted with the appropriate edges; the utilities are depicted at the tips of the terminal edges of the tree.

Once a decision problem has been structured in a decision tree, the best decision for the problem is easily computed. For this purpose, the tree is evaluated by foldback analysis. Foldback analysis starts the tips of the terminal edges, works its way through all intermediate nodes and edges, and ends at the root of the tree. In foldback analysis, for each viable decision the desirability of taking this decision is computed. The desirability of a decision depends on the values of the chance variables modeling its consequences. However, these values are not known before the decision is taken. The desirability of a decision therefore is computed by weighing the utilities of the various possible scenarios that can arise from taking this decision with the probabilities that these scenarios actually do occur. For each chance variable, the expected utility over its values is computed, which expresses the expected utility of taking the decision corresponding with the incoming edge of the node modeling the chance variable. For each decision variable, the maximum expected utility over its values is computed. In a foldback analysis of the decision tree for the data-interpretation problem, the expected utility $\hat{u}(D = Di)$ for each decision $D = Di$ is computed as:

$$\hat{u}(D = D_i) = \sum_{j=1}^{n} u(D = D_i \wedge C = C_j) \cdot \Pr(C = C_j)$$

The best decision is the decision Dk with the highest expected utility. Computing the best decision with regard to a spatial object O as outlined before will be coined decision-analytic data interpretation. The statistical description of decision analysis provides a general and flexible framework for data interpretation. In fact, the framework also provides for conventional classification by taking for the decision the various possible information classes; the utilities then express the severity of different types of misclassification. As an example, we express the common maximum a posteriori probability classification. The only objective pursued in maximum a posteriori probability classification is to maximise the probability of correct classification: every misclassification is considered equally undesirable. This

objective can be expressed in terms of utilities by taking $u(D=Di \wedge C=Ci) = 1$, for all $i = 1, \ldots, n$, and $u(D=Di \wedge C=Cj) = 0$, for all $i, j = 1, \ldots, n, i \neq j$, where Di is the decision to assign class Ci to the object O.

3. Assessing parameters

For decision-analytic data interpretation, a decision tree to model the interpretation problem is evaluated. This decision tree includes the various uncertainties and preferences involved. The accuracy of the assessment of these quantities directly determines the quality of the decision computed for the problem. This section briefly addresses the assessment of the quantities required for the decision-analytic approach.

3.1 Probability assessment

The uncertainties involved in data-interpretation are expressed as probability distributions over the various information classes distinguished for the spatial object O under consideration. The probabilities in these distributions are computed from remotely sensed data as posterior probabilities given the spectral attributes of these data. Given a vector x of spectral attributes, for each information class $Ci, i = 1, \ldots, n$, the posterior probability $\Pr(C=Ci \mid \mathbf{x})$ is computed using Bayes' formula:

$$\Pr(C=C_i \mid x) = \frac{\Pr(x \mid C=C_i) \cdot \Pr(C=C_i)}{\Pr(x)}$$

where $\Pr(\mathbf{x} \mid C=Ci)$ is the probability that the vector of spectral attributes x occurs in the data given that the true class of the object is Ci. $\Pr(C=Ci)$ is the prior probability that the object has class Ci for its true class and $\Pr(\mathbf{x})$ denotes the probability of the vector x occurring in the data. $\Pr(\mathbf{x})$ is the same for every information class and does not have to be computed independently: $\Pr(\mathbf{x})$ is obtained by normalising the nominators of the right-hand side of the formula over all information classes. The probabilities $\Pr(\mathbf{x} \mid C=Ci)$ and $\Pr(C=Ci)$, however, have to be assessed explicitly for each class Ci.

To assess $\Pr(\mathbf{x} \mid C=Ci)$ for $Ci, i = 1, \ldots, n$, either a parametric or a non-parametric method may be used. The parametric method builds on the assumption that each information class Ci yields a parametric, such as a Gaussian, distribution over the space of spectral attributes. To calculate the required probabilities, for each class the distribution is selected that best fits a set of training data; this distribution is obtained by estimating its parameters from these data, that is, for a Gaussian distribution, its mean and variance-covariance matrix. Alternatively, the k-nearest neighbour method is a non-parametric method for assessing the probabilities $\Pr(\mathbf{x} \mid C=Ci)$. It computes for the vector of spectral attributes $\mathbf{x}$ from the training data a set of neighbouring vectors. These vectors are selected by searching the space of spectral attributes at increasing distance from $\mathbf{x}$ until k vectors are included. From the set of neighbouring vectors thus selected, the number ki of vectors corresponding with information class Ci is determined. Now, if ni is the total number of training data corresponding with class Ci, the probability that the vector **x** corresponds with class Ci is estimated as

$$\Pr(x \mid C = C_i) = \frac{k_i}{n_i \cdot V_k(x)}$$

where $V_k(\mathbf{x})$ is related to the distance at which **x**'s k nearest neighbours are found[Fukunaga & Hummels, 1987]. In practice, the non-parametric method proves to be superior to the parametric method in the sense that it yields probabilities of higher accuracy. Especially in the context of our decision-analytic approach, the non-parametric method therefore is preferred.

The prior probabilities $\Pr(C=C_i)$ for the various information classes C_i generally are more difficult to obtain than the probabilities $\Pr(\mathbf{x} \mid C=C_i)$ as they are independent of the data to be interpreted and therefore require knowledge about the subject of the interpretation for their assessment. Depending on the available knowledge, they may be estimated by different methods with varying degrees of sophistication. The least sophisticated method builds on the assumption that any spatial object is fully characterised by its spectral attributes: this method assigns to each class Ci, $i = 1, \ldots , n$, the prior probability $1/n$. A more involved method assigns a prior probability to each class distinguished in the image based on knowledge about its percentage of coverage. Assessment of the prior probabilities based on local information is the most precise method. To this end, the image is subdivided into segments, for example derived from additional GIS-data. The prior probabilities subsequently are estimated per segment, either on the basis of available knowledge about processes that influence the occurrence of classes in the area under consideration, or extracted from the image data, assuming accurate and representative sampling, by an iterative algorithm [Gorte, 1995].

3.2 Utility assessment

The utilities of a decision problem are derived from the objective which is pursued and express the desirability of the various scenarios that can arise from a viable decision. In most decision problems several different objectives are pursued simultaneously. Therefore, a utility can be a complex combination of quite different commodities, such as monetary gain, status, and time. Decision analysis offers various, more or less formal, methods for performing this task [von Winterfeldt & Edwards, 1986].

The simplest, and least formal, method for utility assessment is to visualise all possible scenarios of a decision problem on a linear scale. The least desirable and the most desirable scenarios are identified and assigned to the ends of the scale. Every other scenario is now positioned on the scale, where the distance between two scenarios is indicative of the difference in desirability between these scenarios. Once all scenarios have been positioned, for each scenario a utility is yielded by projecting its position onto a matching numerical scale. Figure 3 illustrates the basic idea for two scenarios *si* and *sj*.

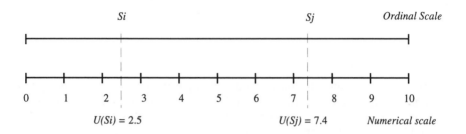

Figure 3 The visualization method for utility assessment.

Instead of first visualising the differences in desirability among scenarios, these differences can be quantified directly, by using a standard reference gamble. A standard reference gamble serves for comparing three scenarios with regard to their desirability. Let *si, sj,* and *sk* be scenarios such that si is less desirable than *sj*, and *sj* in turn is less desirable than *sk*. In assessing utilities for these three scenarios, a probability *p* is found such that scenario *sj* is as desirable as a gamble that yields scenario *sk* with probability *p* and scenario *si* with probability 1 - *p*. Through this probability *p*, the utilities *u(si)*, *u(sj)*, and *u(sk)* have now been assessed to satisfy:

$$u(s_j) = p \cdot u(s_k) + (1-p) \cdot u(s_i)$$

By using the standard reference gamble for appropriate three-tuples of scenarios, a system of equations is obtained from which a set of utilities is computed. The use of a standard reference gamble tends to yield better calibrated utilities than the visual method; the method, however, is more time-consuming. If the utilities of a decision problem are composed of various commodities that are hard to compare, utility assessment can be especially cumbersome. The assessment then often is simplified by decomposing the utilities into their separate commodities. In terms of these separate commodities, marginal utilities are assessed, for example using one of the techniques outlined above. These marginal utilities subsequently are combined to yield overall utilities ([von Winterfeldt & Edwards, 1986]).

4. A case study

The decision-analytic approach to data interpretation has been applied to a case study. Although the situation described in the study in itself is hypothetical, it emerges from a real-life issue. The study concerns fraud with subsidies provided by the European Union to support the cultivation of certain agricultural crops. These subsidies are paid on the basis of declarations submitted by farmers. A fraud detection mechanism can make use of remotely sensed data. For each parcel, the viable decisions to consider on the basis of the data concern approval of the declaration on one hand, and an implication of fraud followed by further investigation on the other hand.

The study area is located around the village of Biddinghuizen in the province of Flevoland, the Netherlands. A Landsat Thematic Mapper image of the area is available (we used spectral bands 3, 4 and 5) from June 1987, as well as crop maps from 1986 and 1987. Seven different land-cover classes are distinguished: grass, wheat, potatoes, sugar beets, peas, beans, and onions. The crop maps, originating from an initial survey that included interviews with farmers, likely contain errors and uncertainties. In our study, we have used the 1986 map to calculate local prior probabilities. In the calculation, crop rotation cycles have been taken into consideration; so, the land-cover classes in successive years are not independent. Part of the 1987 crop map has been used for training sample selection, in combination with a colour composite of the image. From the 1987 map we have subsequently extracted the fields with peas or beans, and considered them as farmers' declarations for subsidy on those two crops.

To investigate viable decisions, various utilities have been assessed. The decision to imply fraud and suggest further investigation is very advantageous if the farmer's declaration specifies peas or beans and there is a different agricultural crop reality: this scenario uncovers an illegal declaration. The scenario is assigned a utility of 10. The decision to not inspect such as field is extremely bad. This scenario is assigned a utility of 0. If a declaration turns out to be legal after further investigation, we have put ourselves (or the farmer) through unnecessary trouble. However, an investigation that turns out superfluous is not so bad as overlooking a false declaration. This scenario therefore is assigned a utility of 3. Avoiding superfluous investigations is more advantageous anyway: we assign a utility of 8. These utilities are summarised in Table 1. Based on these utilities, we have applied our decision-analytic method to the decision for each pixel.

| | inspection | |
crop	yes	no
grass	10	0
wheat	10	0
potato	10	0
sugar beet	10	0
pea	3	8
bean	3	8
onion	10	0

Table 1 Utilities for the detection of illegal farmer declarations.

The result is a binary raster map, indicating the decision per pixel. Subsequently, a majority criterion has been applied to identify the fields that have been indicated for further investigation. These results are shown in Figure 4. Of 81 fields with a declaration of peas or beans, 22 will be inspected.

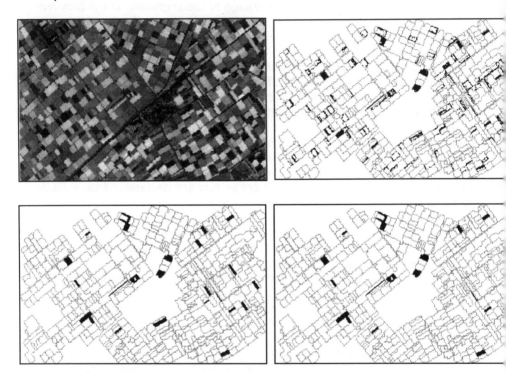

Figure 4 Experimental results. Upper Left: Landsat TM Image (band 4). Upper Right: Pixels with positive 'inspection' decision. Lower Left: Fields to be inspected. Lower Right: Fields to be inspected after modification of utilities.

	inspection	
crop	yes	no
grass	10	6
wheat	10	6
potato	10	6
sugar beet	10	6
pea	0	20
bean	0	20
onion	10	6

Table 2 Modified utilities for the detection of obvious illegal declarations 7.

Now consider a slightly different (perhaps less realistic) situation in which the subsidies paid are rather small and the fraud detection agency is under-staffed. In this situation, farmers generally will be given the benefit of the doubt and only very suspicious looking declarations will be inspected. The utility assigned to the scenarios for this situation are shown in Table 2. After applying our decisionanalytic method to the same data with these new utilities, the number of fields to be investigated has decreased from 22 to 16 as expected.

5. Conclusions

Remotely sensed data are exploited to an increasing extent for decision making. For processing spatial data for this purpose, the objectives and preferences of the decision maker have to be taken into account. In principle, decisions may be taken on the basis of a complete classification of the data at hand. However, as taking the best decision involves the full extent and distribution of the uncertainty in the data, decision making is better founded directly on the data themselves. Decision-analytic interpretation, proposed in this paper, provides such an approach by integrating preferences and uncertainties in a mathematically well-founded way. The aim of the method is to assist a decision maker in taking the best decision and not so much to reconstruct reality, thereby contrasting conventional classification. The decision-analytic approach to the interpretation of spatial data has been illustrated by means of a simple case study. Because of the simplicity of the presented study, it does not serve to fully demonstrate the potential power of the approach. However, it illustrates the issue of customisation: from a single set of spatial data, various results can be obtained tailored to a decision maker's objectives, by interpreting the data with different sets of utility assessments. An interesting issue that remains to addressed is the performance of the decision-analytic approach to data interpretation at a level beyond pixels. The approach is suitable for decision making for spatial objects instead of for individual pixels, as the concepts involved remain the same; however, an image segmentation pre-processing step is required. Applying the approach to spatial objects is expected to benefit from (probabilities of) geometrical and topological properties of objects for decision making. To conclude, we would like to emphasise that our approach is based on a well-known and long-established mathematical framework from decision analysis for solving complex decision problems. The rich field of decision analysis provides a wealth of methods, for example for assessing probabilities and utilities, that can be applied to the problem of interpreting spatial data. Thanks to its flexibility and mathematical well-foundedness, the framework has the potential to become an integral part of geographical information systems.

References

[Lunetta et al., 1991] R.S. Lunetta, R.G. Congalton, L.K. Fenstermaker, J.R. Jensen, K.C. McGwire and L.R. Tinney (1991). Remote sensing and geographic information system data integration: error sources and research issues. Photogrammetric Engineering & Remote Sensing, vol. 57, no. 6, pp. 677 - 687.

[Fukunaga & Hummels, 1987] K. Fukunaga, D.M. Hummels (1987). Bayes error estimation using Parzen and kNN procedures, IEEE Trans. PAMI, Vol. 9, pp. 634 - 643.

[Gorte, 1995] B.G.H. Gorte (1995). Improving spectral image classifications by incorporating context data using likelihood vectors. Procs. IPA 1995, IEE Conf. Publ. no. 410, pp 251 - 255.

[Morrison, 1995] J.L. Morrison (1995). Spatial data quality. In: S.C. Guptill & J.L. Morrison (ed.). Elements of Spatial Data Quality. Pergamon, Oxford, pp. 1 - 12.

[Raiffa, 1968] H.A. Raiffa (1968). Decision Analysis: Introductory Lectures on Choices Under Uncertainty. Addison-Wesley, Reading, Massachusetts.

[Smith, 1988] J.Q. Smith (1988). Decision Analysis. A Bayesian Approach. Chapman and Hall Ltd., London.

[Strahler, 1980] A.H. Strahler (1980). The use of prior probabilities in maximum likelihood classification of remotely sensed data. Remote Sensing of Environment, no. 10, pp. 135 - 163.

[Von Winterfeldt & Edwards, 1986] D. von Winterfeldt and W. Edwards (1986). Decision Analysis and Behavioral Research. Cambridge University Press, New York.

Contact address

B.G.H. Gorte[1], L.C. van der Gaag[2] and F.J.M. van der Wel[3]
[1]International Institute for Aerospace Survey and Earth Sciences (ITC)
P.O. Box 6
7500 AA Enschede
The Netherlands
E-mail : ben@itc.nl
[2]Utrecht University
Department of Computer Science
P.O. Box 80.089
3508 TB Utrecht
The Netherlands
E-mail : linda@cs.ruu.nl
[3]Utrecht University
Faculty of Geographical Sciences _ Cartography Section
P.O. Box 80.115
3508 TC Utrecht
The Netherlands
E-mail : f.vanderwel@frw.ruu.nl

SUPPORTING DIRECTION RELATIONS IN SPATIAL

DATABASE SYSTEMS

Yannis Theodoridis[1], Dimitris Papadias[2] and Emmanuel Stefanakis[1]

[1]Department of Electrical and Computer Engineering
National Technical University of Athens
Athens, Greece
[2]National Center for Geographic Information and Analysis &
Department of Spatial Information Science and Engineering
University of Maine
Maine, USA

Abstract
Direction relations (e.g., *north*, *northeast*) constitute an important class of user queries in Spatial Databases and Geographic Information Systems. However, little work has been done on the formalization and efficient processing of such relations. In this paper we define direction relations between two-dimensional objects and we show how they can be efficiently retrieved in existing DBMSs using alternative indexing methods. We also propose modifications of the traditional techniques that facilitate efficiency for several types of queries. Although for our experiments we use a small set of representative relations motivated by geographic applications, the results of this paper are directly applicable to any type of direction relations that may arise in different application domains.

Key words
Spatial databases, Geographic Information Systems, query processing.

1. Introduction

This paper describes implementations of direction relations in Spatial DBMSs and Geographic Information Systems (GIS). Direction relations describe order in space and constitute an important class of user queries. Despite their importance, direction queries have not been studied extensively in spatial access methods; query processing and optimization techniques have mainly focused on *window* queries [Gutt84], *topological relations* [Papa95b] and *nearest neighbour* queries [Rous95]. The main reason for this, is the lack of well-defined direction relations between actual objects.

In Geography, for example, although the linguistic terms used to describe the direction relation between some pairs of countries are undisputed (everybody will agree that Germany is *north* of Italy) this is not always the case. Consider the query "find all countries *north* of Italy". Should France belong to the result, or is it *northwest* (but not *north*) of Italy? The answer to the above query depends on the definition of direction relations which may vary for each application.

In this paper we define direction relations between two-dimensional objects in

---

Dimitris Papadias is currently with the Artificial Intelligence Research Division, German National Research Center for Information Technology (GMD), D 53754, Germany.

different levels of qualitative resolution to match the application needs. Our work extends previous attempts to formalize direction relations which have concentrated on point objects [Fran92] or Minimum Bounding Rectangles [Peuq87]. Then we show how the relations that we define can be efficiently retrieved in spatial DBMSs using different indexing methods. Essentially we compare the efficiency of indexing methods on multi-dimensional range queries in the context of geographic applications. We also propose modifications of the traditional (spatial) data structures that yield improved performance for certain classes of queries.

The results of this paper are directly applicable to Spatial Databases and GIS where the formalization of spatial relations is crucial for user interfaces and query optimisation strategies. Although we deal with geographic examples, potential applications for direction relations include other domains such as CAD and VLSI design. In these domains, the queries can be very similar, but the linguistic terms used to express them may vary (e.g., *above* instead of *north*).

The paper is organised as follows: in section 2 we define several direction relations between objects in 2D-space and we describe how MBR approximations can be used to retrieve them. Section 3 discusses the retrieval of direction relations using alternative access methods (B-trees, KDB-tress, and R-trees respectively). Section 4 is concerned with extensions of the above indexing techniques that increase performance for some queries. Section 5 compares the different implementations illustrating experimental results, and section 6 concludes with comments on future work.

2. Direction relations and minimum bounding rectangles

Unlike the case of topological relations where there seems to exist a set of widely accepted relations [Egen90], there are no such definitions of direction relations. In order to define direction relations between extended objects we assume an absolute frame of reference and a pair of orthocanonic axes x and y. Our method is *projection based,* that is, direction relations are defined using projection lines perpendicular to the coordinate axes. An alternative approach is based on the *cone-shaped* concept of direction, i.e., direction relations are defined using angular regions between objects. Extensive studies of the two approaches can be found in [Hern94].

For the rest of the paper, p denotes the *primary object* (the object to be located) and q the *reference object* (the object in relation to which the primary object is located). In the following illustrations the reference object is grey. Let p_i be a point of object p, q_j be a point of object q, and X and Y be functions that give the x and y coordinate of a point. In order to define direction relations between objects we use universally and existentially quantified formulas that compare point coordinates. The relation *strong_north* between objects p and q, for instance, denotes that all points of p are *north* of all points of q: strong_north $\equiv \forall p_i \ \forall q_j \ Y(p_i) > Y(q_j)$. The relation *strong_north* can be characterised as low resolution relation because its acceptance area is large. On the other hand, we can define a higher resolution version of *strong_north*, as : strong_bounded_north(p,q) $\equiv \forall p_i \forall q_j\, Y(p_i) > Y(q_j) \wedge \forall p_i \exists q_j\, (Y(p_i) > Y(q_j) \wedge X(p_i) > X(q_j)) \wedge \forall p_i\, \exists q_j\, (Y(p_i) > Y(q_j) \wedge X(p_i) < X(q_j))$. According to this relation, all points of object p must be in the region bounded by the horizontal line that passes from q's northmost point and by the vertical lines that also bound q.

Table 1 illustrates several direction relations between European countries. All of these relations concern the direction *north*, and they are representatives for other relations as well, in the sense that they refer to several levels of qualitative resolution. Depending on the application needs, a large number of direction relations between extended objects can be defined and implemented accordingly. The set of relations can be chosen so that several properties are satisfied: the relations can be pairwise disjoint, provide a complete coverage, or form a relation algebra etc. Notice that the set that we study here does not satisfy any of these properties, since our goal is to show how direction relations of different resolution can be defined and retrieved in spatial data structures. Customised direction relations between extended objects that satisfy the above properties and correspond to certain application needs are straightforward to define by comparing point coordinates.

Although the previous discussion refers to actual two-dimensional objects, usually spatial access methods use approximations to efficiently retrieve candidates that could satisfy a query. In this paper we examine methods based on the traditional approximation of Minimum Bounding Rectangles (MBRs). MBRs are the most commonly used approximations in spatial applications because they need only two points for their representation; in particular, each object q is represented as an ordered pair (q'_l, q'_u) of *representative points* that correspond to the lower left (q'_l) and the upper right point (q'_u) of the MBR q' that covers q. Figure 1 illustrates how a map of Europe can be approximated by MBRs.

 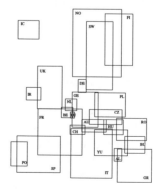

Figure 1 MBR approximations for the map of Europe.

In order to answer the query "find all objects p that satisfy the relation R with respect to an object q" we have to retrieve all MBRs p' that satisfy the relation R' with respect to the MBR q' of object q. Table 1 illustrates the mapping from direction relations R between actual objects to relations R' between MBR representative points. Basically, each direction relation is transformed into a *range query* involving representative points.

Relation R between objects	Example	Definition	Relation R' between representative points
strong_north(p,q) - sn - example: sn(NL,AU)		$\forall p_i \forall q_j\ Y(p_i) > Y(q_j)$	$Y(p'_l) > Y(q'_u)$
weak_north(p,q) - wn - example: wn(BU,GR)		$\exists p_i \forall q_j\ Y(p_i) > Y(q_j) \wedge$ $\forall p_i \exists q_j\ Y(p_i) > Y(q_j) \wedge$ $\exists p_i \exists q_j\ Y(p_i) < Y(q_j)$	$Y(p'_u) > Y(q'_u) \wedge$ $Y(p'_l) > Y(q'_l) \wedge$ $Y(p'_l) < Y(q'_u)$
strong_bounded_north(p,q) - sbn - example: sbn(GE,IT)		$\forall p_i \forall q_j\ Y(p_i) > Y(q_j) \wedge$ $\forall p_i \exists q_j\ (Y(p_i) > Y(q_j) \wedge X(p_i) > X(q_j)) \wedge$ $\forall p_i \exists q_j\ (Y(p_i) > Y(q_j) \wedge X(p_i) < X(q_j))$	$Y(p'_l) > Y(q'_u) \wedge$ $X(p'_l) > X(q'_l) \wedge$ $X(p'_u) < X(q'_u)$
weak_bounded_north(p,q) - wbn - example: wbn(BE,FR)		$\exists p_i \forall q_j\ Y(p_i) > Y(q_j) \wedge$ $\exists p_i \exists q_j\ Y(p_i) < Y(q_j) \wedge$ $\forall p_i \exists q_j\ (Y(p_i) > Y(q_j) \wedge X(p_i) > X(q_j)) \wedge$ $\forall p_i \exists q_j\ (Y(p_i) > Y(q_j) \wedge X(p_i) < X(q_j))$	$Y(p'_u) > Y(q'_u) \wedge$ $Y(p'_l) > Y(q'_l) \wedge$ $Y(p'_l) < Y(q'_u) \wedge$ $X(p'_l) > X(q'_l) \wedge$ $X(p'_u) < X(q'_u)$
strong_north_east(p,q) - sne - example: sne(DE,NL)		$\forall p_i \forall q_j\ (Y(p_i) > Y(q_j) \wedge X(p_i) > X(q_j))$	$Y(p'_l) > Y(q'_u) \wedge$ $X(p'_l) > X(q'_u)$
weak_north_east(p,q) - wne - example: wne(AU,CH)		$\exists p_i \forall q_j\ (Y(p_i) > Y(q_j) \wedge X(p_i) > X(q_j)) \wedge$ $\exists p_i \exists q_j\ Y(p_i) < Y(q_j) \wedge$ $\forall p_i \exists q_j (Y(p_i) > Y(q_j) \wedge X(p_i) > X(q_j))$	$Y(p'_l) > Y(q'_l) \wedge$ $X(p'_l) > X(q'_l) \wedge$ $Y(p'_u) > Y(q'_u) \wedge$ $X(p'_u) < X(q'_u) \wedge$ $Y(p'_l) < Y(q'_u)$

Table 1 Direction relations between objects and mapping to relations between MBRs.

Because MBRs differ from the actual objects they enclose, they are not always adequate to express the relation between the actual objects. For this reason, spatial queries involve the following two-step strategy: First a *filter step* based on MBRs is used to rapidly eliminate objects that could not possibly satisfy the query and select a set of potential candidates. Then during a *refinement step* each candidate is examined (by using computational geometry techniques) and false hits are detected and eliminated. Unlike topological relations, where the refinement step is the rule [Papa95b], the only direction relations of Table 1 that need a refinement step are *weak_bounded_north* and *weak_north_east* [Papa94a]. For the rest of the relations, all retrieved MBRs correspond to objects that satisfy the query.

In [Papa94a] we describe in detail the retrieval of direction relations using MBRs. In this paper we extend our previous results by implementing direction relations in different access methods and comparing the results. Furthermore we propose modifications to the existing data structures, which yield better performance for some relations.

3. Implementation of direction relations

The retrieval of direction relations in existing DBMSs can be accomplished by maintaining traditional indexes (e.g. B-trees), or by incorporating Abstract Data Types (ADTs) with specialised indexes defined by external code (e.g. KDB-trees or R-trees). In the rest of the section we present three alternative indexing techniques.

3.1 Implementation of direction relations using B-trees

The first solution for the retrieval of direction relations includes the maintenance of a set of four alphanumeric indexes, such as B-trees (to be precise, the implementation used in existing systems is the B^+-tree index [Knut73] and, in this paper, we will think of B^+-trees when we use the term B-trees). Each index corresponds to one of the four numbers: p'_{l-x}, p'_{l-y}, p'_{u-x}, p'_{u-y}, where p'_{l-x} stands for the x-coordinate of the lower point of p', p'_{l-y} for the y-coordinate of the lower point and so on. Obviously, some relations imply search on one B-tree while others imply search on more B-trees. For instance, the query "find all objects p that are *strong_north* of object q" is transformed to the constraint $p'_{l-y} > q'_{u-y}$, which is a simple range query in the corresponding B-tree (B-tree for p'_{l-y} in this case). On the other hand, other queries (such as *strong_north_east*) need to search two or more B-trees and, in a second phase, to compute the intersection of the intermediate answer sets. Table 2 presents the constraints and the number of B-trees needed for the retrieval of each direction relation.

In general, the processing of a query of the form "find all objects p that satisfy a given direction relation with respect to object q" using B-trees involves the following steps: (i) select the B-trees to be searched (this procedure involves Table 2), (ii) search each index involved to find the corresponding answer sets, (iii) if multiple B-trees are involved, find the intersection set. Depending on the relation there may be required a refinement step which is not a part of the indexing scheme.

Relation	Constraints on the p'_{l-x}, p'_{l-y}, p'_{u-x}, p'_{u-y} parameters	# of constraints
strong_north(p,q)	$p'_{l-y} > q'_{u-y}$	1
weak_north(p,q)	$(q'_{l-y} < p'_{l-y} < q'_{u-y}) \wedge (p'_{u-y} > q'_{u-y})$	2
strong_bounded_north(p,q)	$(p'_{l-y} > q'_{u-y}) \wedge (q'_{l-x} < p'_{l-x} < q'_{u-x}) \wedge (q'_{l-x} < p'_{u-x} < q'_{u-x})$	3
weak_bounded_north(p,q)	$(q'_{l-y} < p'_{l-y} < q'_{u-y}) \wedge (p'_{u-y} > q'_{u-y}) \wedge (q'_{l-x} < p'_{l-x} < q'_{u-x}) \wedge$ $(q'_{l-x} < p'_{u-x} < q'_{u-x})$	4
strong_north_east(p,q)	$(p'_{l-y} > q'_{u-y}) \wedge (p'_{l-x} > q'_{u-x})$	2
weak_north_east(p,q)	$(q'_{l-y} < p'_{l-y} < q'_{u-y}) \wedge (p'_{u-y} > q'_{u-y}) \wedge (p'_{l-x} > q'_{l-x}) \wedge (p'_{u-x} > q'_{u-x})$	4

Table 2 Constraints for the retrieval of direction relations using B-trees.

As an example, consider the query: "Find the countries p *strong_north_east* of Switzerland (CH)". In this case, two B-trees (for p'_{l-y} and p'_{l-x} parameters) need to be searched. They are illustrated in Figure 2a where each label indicates the appropriate coordinate for the corresponding object p, namely p'_{l-y} and p'_{l-x}. The sets {CZ', LU', BE', PL', UK', NL', IR', DE', SW', NO', FI', IC'} and {SW', CZ', PL', YU', HU', FI', RO', AL', GR', BU'} are the two answer sets. The intersection set {CZ', PL', SW', FI'} contains the object IDs that satisfy the query (illustrated as the dark shaded area in Figure 2b). A refinement step is not needed for the retrieval of *strong_north_east*, that is, all retrieved MBRs correspond to objects that satisfy the query

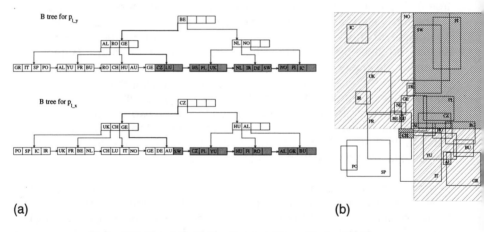

(a)   (b)

Figure 2 Retrieval of relation strong_north_east using B-trees.

The performance of retrieval using B-trees depends significantly on the particular direction relation because the number of B-trees to be searched is equal to the number of constraints that are involved in the definition of the relation. In the rest of the section we describe two other implementations, based on KDB-trees and R-trees respectively, which are more suitable for relations involving a large number of constraints.

3.2 Implementation of direction relations using KDB-trees

The second solution for the retrieval of direction relations includes the maintenance of a group of two point indexes, such as KDB-trees [Robi81]. The two indexes correspond to the representative points p'_l and p'_u. KDB-trees, like B-trees, are height-balanced trees, which consist of leaf and intermediate nodes. Point data are stored in the leaf nodes and intermediate nodes partition the space in *disjoint* regions which entirely contain the corresponding entries. In order to retrieve a direction relation using a KDB-tree we have to specify the region that contains the answer set and recursively search the tree nodes that intersect with this region. Table 3 presents the corresponding range queries for each direction relation, assuming the unit segment [0,1] as global space per axis. Range queries Q_l and Q_u refer to the KDB-trees for lower and upper data coordinates, respectively, and the format is a set of four values that correspond to the four coordinates $(Q_{l-x}, Q_{l-y}, Q_{u-x}, Q_{u-y})$ of Q (Q_l or Q_u).

Relation	Range query Q_l	Range query Q_u
strong_north(p,q)	$(0, q'_{u-y}, 1, 1)$	
weak_north(p,q)	$(0, q'_{l-y}, 1, q'_{u-y})$	$(0, q'_{u-y}, 1, 1)$
strong_bounded_north(p,q)	$(q'_{l-x}, q'_{u-y}, q'_{u-x}, 1)$	$(q'_{l-x}, q'_{u-y}, q'_{u-x}, 1)$
weak_bounded_north(p,q)	$(q'_{l-x}, q'_{l-y}, q'_{u-x}, q'_{u-y})$	$(q'_{l-x}, q'_{u-y}, q'_{u-x}, 1)$
strong_north_east(p,q)	$(q'_{u-x}, q'_{u-y}, 1, 1)$	
weak_north_east(p,q)	$(q'_{l-x}, q'_{l-y}, 1, q'_{u-y})$	$(q'_{u-x}, q'_{u-y}, 1, 1)$

Table 3 Range queries on KDB-trees.

The relations that involve only one of the two representative points (i.e., only one range query, such as *strong_north* and *strong_north_east*) imply search on one KDB-tree while the rest imply search on both KDB-trees. In general, the processing of a query of the form "find all objects p that satisfy a given direction relation with respect to object q" using KDB-trees involves the following steps: (i) depending on the relation to be retrieved, select the KDB-trees involved from the set of two indexes (this procedure involves Table 3), (ii) search each index involved to find the corresponding answer sets, (iii) if both indexes are involved, find the intersection set.

As an example, consider the query: "Find the countries p *strong_north_east* of Switzerland (CH)". The query is transformed to the constraint $(p'_{L\_x} > q'_{u\_x}) \wedge (p'_{L\_y} > q'_{u\_y})$ or, in other words, p'_l should be within the area $(q'_{u-x}, q'_{u-y}, 1, 1)$. In this case, only one KDB-tree (for p'_l data point) needs to be searched. It is illustrated in Figure 3a where each label indicates the low coordinate for the corresponding object p. The intermediate nodes that are selected for propagation are nodes X and Y (1st level), D, F and G (2nd level), i.e., the ones that intersect the shaded area of Figure 3b (these nodes are grey in the tree structure of Figure 3a). Among the leaves of the intermediate nodes D, F and G, the entries that lie within the shaded area and satisfy the range query are: CZ', PL', SW' and FI'.

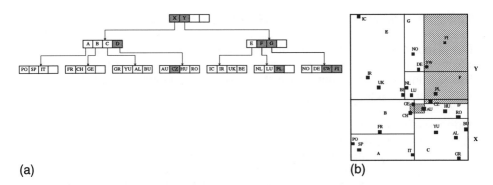

(a)   (b)

Figure 3 Retrieval of relation strong_north_east using KDB-trees.

Intuitively KDB-trees perform efficiently when the range queries Q_l or Q_u are selective (e.g. as in the *strong_north_east* relation). Especially, when only one tree structure needs to be searched, KDB-trees are expected to be a very competitive solution.

3.3 Implementation of direction relations using R-trees

The R-tree data structure [Gutt84] is a height-balanced tree, which consists of intermediate and leaf nodes. The MBRs of the actual data objects are stored in the leaf nodes of the tree. Intermediate nodes are built by grouping rectangles at the lower level. An intermediate node is associated with some rectangle which encloses all rectangles that correspond to lower level nodes. In order to retrieve objects that satisfy a direction relation with respect to a reference object we have to specify the MBRs p' that could enclose such objects (Table 1) and then to search the intermediate nodes P' that contain these MBRs. Table 4 presents the constraints for the intermediate nodes for each direction relation of Table 1 (which are, in general, different from the constraints for the leaf nodes).

The processing of a query of the form "find all objects p that satisfy a given direction relation with respect to object q" using R-trees involves the following steps: (i) starting from the top node, exclude the intermediate nodes P' which could not enclose MBRs that satisfy the direction relation and recursively search the remaining nodes (this procedure involves Table 4), (ii) among the leaf nodes retrieved, select the MBRs that satisfy the relation R' as given in Table 1 for each direction relation.

Figure 4 shows how the MBRs of Figure 1 are grouped and stored in an R-tree. We assume a branching factor of 4, i.e., each intermediate node contains at most four entries. At the lower level MBRs of countries (denoted by two letters) are grouped into seven intermediate nodes A, B, C, D, E, F and G. At the next level, the seven nodes are grouped into two larger nodes X and Y. Consider now the query "Find the countries p *strong_north_east* of Switzerland (CH)". The intermediate nodes that are selected for propagation are nodes X and Y (1st level), B, E, F and G (2nd level) i.e., the ones that have the representative point P_u within the shaded area (these nodes are grey in the tree structure of Figure 4a). Among the leaves of the intermediate nodes B, E, F and G, the ones that satisfy the constraint $(p'_{l-y} > CH'_{u-y}) \wedge (p'_{l-x} > CH'_{u-x})$, are SW', FI', CZ' and PL'.

Relation	Intermediate Nodes P' to be Searched
strong_north(p,q)	$P'_{u-y} > q'_{u-y}$
weak_north(p,q)	$(P'_{u-y} > q'_{u-y}) \wedge (P'_{l-y} < q'_{u-y})$
strong_bounded_north(p,q)	$(P'_{u-y} > q'_{u-y}) \wedge (P'_{l-x} < q'_{u-x}) \wedge (P'_{u-x} > q'_{l-x})$
weak_bounded_north(p,q)	$(P'_{u-y} > q'_{u-y}) \wedge (P'_{l-y} < q'_{u-y}) \wedge (P'_{l-x} < q'_{u-x}) \wedge (P'_{u-x} > q'_{l-x})$
strong_north_east(p,q)	$(P'_{u-y} > q'_{u-y}) \wedge (P'_{u-x} > q'_{u-x})$
weak_north_east(p,q)	$(P'_{u-y} > q'_{u-y}) \wedge (P'_{l-y} < q'_{u-y}) \wedge (P'_{u-x} > q'_{u-x})$

Table 4 Constraints for intermediate nodes of R-trees.

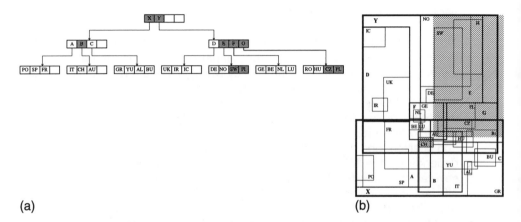

(a)   (b)

Figure 4 Retrieval of relation strong_north_east using R-trees.

Intuitively R-trees perform better than B-trees and KDB-trees in cases where many constraints on both axes are involved in the definition of the direction relation of interest. The implementation of section 5 demonstrates that this is actually the case when four constraints are involved, while for one constraint B-trees have a better performance. For the intermediate relations (two or three constraints) KDB-trees usually win. For these cases modified versions of B-trees and R-trees can have the best overall performance. These modified versions are described in the next section.

4. Extensions

In the case of B-trees, we propose schemes for the maintenance of information regarding the MBR extents in the index; we call the proposed structures *composite B-trees*. As shown in section 3, the number of B-trees to be searched is equal to the number of constraints that compose the query. This inconvenience may be overcome if the B-tree accommodates additional information regarding the MBRs in its *leaf nodes*. That is, a composite key may be maintained instead of a simple numerical value (i.e., the coordinate value of one MBR corner). This composite key consists of some (or even all) coordinate values of the two MBR corners. One of them is the primary component, and based on its value the B-tree is built. The rest of the coordinates are used for the elimination of irrelevant MBRs.

The key of the composite B-tree considered here consists of two values that represent the lower and upper coordinates of each MBR in one of the two axes. This scheme reduces each MBR into two line segments which represent its extents over the two dimensions of the space and are indexed separately. Clearly, the efficient retrieval of direction relations can be achieved when four composite B-trees are maintained: B_{l-x}-tree that keeps p'_{l-x} as primary component and p'_{u-x} as secondary component, and, similarly, B_{u-x}-tree, B_{l-y}-tree and B_{u-y}-tree. Some relations imply the search of only one B-tree (such as *strong_north*), while the rest imply two searches (i.e., one for each axis such as *strong_north_east*).

The processing of a direction relation query involves the four steps described in subsection 3.1. During step 2 the search of each index is based on the primary component of the composite key while, in the leaf nodes, the possible refinement condition for the secondary component is considered to eliminate irrelevant MBRs. The selection of the most effective B-tree by a query processor or optimiser is not always trivial. In general, the selection should take into account: a) the direction relation involved; b) the query window size and position; and c) the distribution of MBR corners over the work space (this depends on the MBRs distribution and size). We implemented and tested several schemes for the selection of the most effective B-tree. The most efficient one proved to be *selection based on the location and size of the query object and the distribution of MBR corners*. This scheme is obtained by maintaining an array (directory) with information about the number of lower and upper coordinates over the work space using a pre-determined resolution.

KDB-trees handle two-dimensional data efficiently when the search procedure involves one of the two-representative points of the MBRs. However, most of the direction relations involve both points and, as a consequence, the intersection of two answer sets should be computed (step 3; Section 3.2). The determination of both answer sets and their intersection can be a highly time-consuming procedure in large databases. We propose the maintenance the opposite MBR corner, along with the corner on which indexing is based, in the leaf nodes of a composite KDB-tree. The additional point will serve the fast elimination of irrelevant MBRs. The efficient retrieval of direction relations can be achieved when two composite KDB-trees are maintained: a KDB_l-tree where indexing is based on the lower left corners of the MBRs; and, a KDB_u-tree where indexing is based on the upper right corners of the MBRs.

R-trees handle two-dimensional data efficiently when the search procedure involves both axes of the work space. However, direction relations such as *strong_north*, involve search on only one axis. In such cases, the information regarding the other axis, which is maintained in the two-dimensional R-tree is useless. Clearly, a one-dimensional R-tree (i.e., segment tree), which is an index of the MBR extents along the axis of interest, would be more efficient because it is more compact (tree nodes accommodate a larger number of entries) and effective (the MBR extents along the other axis do not affect the maintenance of the index). For the retrieval of direction relations that involve both axes, two one-dimensional R-trees are needed to index the MBR extents over each axis separately. Each index provides a set of MBRs, and the intersection of the two sets composes the qualified set. In such cases though, the two-dimensional R-tree is expected to be more efficient.

After the description of the alternative implementations of direction relations in practical systems we compare their efficiency. It is not easy to claim a-priori which solution is the best for each direction relation but it is more sensible to claim that particular structures are more suitable for certain queries.

5. Performance comparison

In the previous sections we have argued that the performance of each structure depends significantly on the particular direction relation. In this section we present several experimental results that justify our argument. In order to experimentally quantify the performance, we created tree structures by inserting 10000 randomly generated MBRs. We tested three *data files* consisting of *small*, *medium*, and *large* MBRs respectively. The search procedure used three *query files* for each data file containing 100 rectangles, also randomly generated, with similar size properties as the data rectangles. We compared classic B-trees, composite B-trees, KDB-trees, composite KDB-trees, 2D R-trees and 1D R-trees (in our experiments we have adopted R*-trees [Beck90] which generally show the highest performance among R-tree variants). The performance of each data structure per relation (for each *data / query size* combination) is illustrated in Figure 5 (e.g., *s/m* means small data objects and medium query objects).

As a first conclusion we notice the way that data and query size affect the performance of the data structures. The performance of B-trees and KDB-trees depends significantly on the query size while the opposite happens for R-trees where the data size is the main factor responsible for the high or low efficiency of the structure. Depending on the relation and the size of the MBRs, the difference of the performance between two data structures can be more than an order of magnitude.

The main results are the following:
- The classic B-tree is the most efficient structure for 1-constraint relations only (e.g., *strong_north*). In this case the classic B-tree outperforms even the (dedicated) composite B-tree because of its higher compactness (node capacity is 126 and 84 keys in classic and composite B-trees respectively). When more constraints (even on the same axis) are involved it is not competitive.

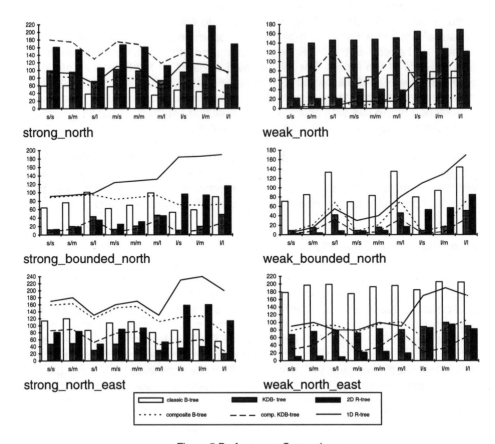

Figure 5 Performance Comparison.

- The composite B-tree is very competitive for large data when more than one constraint is involved (e.g., *weak_north* and *weak_bounded_north* for large data files), because R-trees are unable to index such data efficiently. On the other hand, it is sensitive to query size; large queries are not handled efficiently by composite B-trees.
- The KDB-tree is the most efficient method when two constraints for the same representative point are involved (see *strong_north_east*, *strong_bounded_north*). The composite KDB-tree outperforms classic KDB-tree when both MBR corners are involved in the query.
- The two-dimensional R-tree is the winner when four constraints are involved (see *weak_bounded_north*, *weak_north_east*), assuming that data rectangles are not large. In such cases, R-trees are not able to organise the data efficiently.
- The one-dimensional R-tree is the winner when two constraints along the same axis are involved (see *weak_north* relation), with the exception of large data files because of the reason mentioned above.

From the above results, it is evident that there is not an overall winner but the performance of the structures is affected significantly by the number of constraints involved, the size of the data rectangles and the query size. In a recent paper [Theo95], we describe formulas for the expected performance of spatial access methods in the retrieval of direction (and other) spatial relations based on analytical models presented in [Theo96]. These formulas can be used as guidelines to spatial query processors and optimisers in order to select appropriate indexes for use in answering spatial queries.

6. Conclusion

There has been an increasing interest recently about the representation and processing of spatial relations in Spatial Database Systems (for an extensive discussion see [Papa94b]). This interest has focused on several topics such as Reasoning [Fran92], Consistency Checking Mechanisms [Grig95] and Spatial Query Languages [Papa95a]. In this paper we deal with the retrieval of spatial relations in DBMSs used for geographic, or more generally, spatial, applications. In particular we concentrate on the retrieval of *direction relations* using alternative indexing solutions.

In order to define direction relations between objects we use universally and existentially quantified formulas that compare point coordinates. For the purposes of this paper we define a set of six relations, but a large number of additional ones can be defined in a similar manner. Then we show how these relations can be retrieved in existing DBMSs using B-trees, KDB-trees and R-trees. We also propose extensions of these data structures, called composite B-trees, composite KDB-trees and 1D R-trees, to facilitate efficiency for some types of queries.

For the implementation of the proposed techniques we used data and query files of different sizes. The main conclusion that arises from the experimental comparison of the alternative techniques is that there is not a data structure that performs better in all queries but the performance depends on the following factors:

1. the number and the type of constraints involved in the definition of the direction relations of interest
2. the data size (i.e., the size of the primary MBRs)
3. the query size (i.e., the size of the reference MBRs)

It is possible that in actual systems, two or more of the previous structures can be used incorporated in conjunction with an optimiser that chooses the most suitable one according to the input query.

Future work can be done on the definition of direction relations for use in GIS. Experimental findings from Cognitive and Environmental Psychology can be used as guidelines for the direction relations that people evoke in everyday reasoning. Although so far the psychological results are too vague to be helpful in defining direction relations in actual systems, ongoing research can lead to a set of well defined and psychologically sound relations for geographic interfaces. Progress can also be achieved in specialised data structures that have a better performance for queries involving direction relations or combinations of several types of spatial information (e.g., "find the five closest buildings in the area northeast of the burning factory").

References

[Beck90] Beckmann, N., Kriegel, H.-P., Schneider, R., Seeger, B., "The R˙-tree: An Efficient and Robust Access Method for Points and Rectangles", Proceedings of ACM-SIGMOD Conference, 1990.

[Egen90] Egenhofer, M., Herring, J., "A Mathematical Framework for the Definitions of Topological Relationships", Proceedings of the 4th International Symposium on Spatial Data Handling (SDH), 1990.

[Fran92] Frank, A.U., "Qualitative Spatial Reasoning about Distances and Directions in Geographic Space", Journal of Visual Languages and Computing, Vol. 3, pp. 343-371, 1992.

[Grig95] Grigni, M., Papadias, D., Papadimitriou, C., "Topological Inference", Proceedings of the International Joint Conference of Artificial Intelligence, 1995.

[Gutt84] Guttman, A., "R-trees: a Dynamic Index Structure for Spatial Searching", Proceedings of ACM-SIGMOD Conference, 1984.

[Hern94] Hernandez, D., "Qualitative Representation of Spatial Knowledge", Springer Verlag LNAI, 1994.

[Knut73] Knuth, D., "The Art of Computer Programming, Vol. 3: Sorting and Searching", Addison-Wesley, 1973.

[Papa94a] Papadias, D., Theodoridis, Y., Sellis, T., "The Retrieval of Direction Relations Using R-trees", Proceedings of the 5th Conference on Database and Expert Systems Applications (DEXA), 1994.

[Papa94b] Papadias, D., Sellis, T., " Qualitative Representation of Spatial Knowledge in two-dimensional Space", Very Large Data Bases Journal, Vol. 3(4), pp. 479-516, 1994.

[Papa95a] Papadias, D., Sellis, T., "A Pictorial Query-by-Example Language", Journal of Visual Languages and Computing, Vol. 6(1), pp. 53-72, 1995.

[Papa95b] Papadias, D., Theodoridis, Y., Sellis, T., Egenhofer, M., "Topological Relations in the World of Minimum Bounding Rectangles: a Study with R-trees", Proceedings of ACM-SIGMOD Conference, 1995.

[Peuq87] Peuquet, D., Ci-Xiang, Z., "An Algorithm to Determine the Directional Relationship between Arbitrarily-Shaped Polygons in the Plane", Pattern Recognition, Vol. 20(1), pp. 65-74, 1987.

[Robi81] Robinson, J.T., "The K-D-B-Tree: A Search Structure for Large Multidimensional Dynamic Indexes", Proceedings of ACM-SIGMOD Conference, 1981.

[Rous95] Roussopoulos, N., Kelley, F., Vincent, F., "Nearest Neighbor Queries", Proceedings of ACM-SIGMOD Conference, 1995.

[Theo95] Theodoridis, Y., Papadias, D., "Range Queries Involving Spatial Relations: A Performance Analysis", Proceedings of the 2nd Conference on Spatial Information Theory (COSIT), 1995.

[Theo96] Theodoridis, Y., Sellis, T., "A Model for the Prediction of R-tree Performance", Proceedings of the 15th ACM Symposium on Principles of Database Systems (PODS), 1996.

Contact address

Yannis Theodoridis[1], Dimitris Papadias[2] and Emmanuel Stefanakis[1]
[1]Department of Electrical and Computer Engineering
National Technical University of Athens
Zographou
Athens
Greece15773
E-mail: {theodor, stefanak}@cs.ntua.gr
[2]National Center for Geographic Information and Analysis &
Department of Spatial Information Science and Engineering
University of Maine
ME 04469-5711
USA
E-mail: dimitris@nathan.gmd.de

ON MAP (RE)CLASSIFICATION

Vasilis Delis[1] and Thanasis Hadzilacos[2]

[1,2]Computer Technology Institute
Patras, Greece
[1]Computer Engineering and Informatics Department
University of Patras
Patras, Greece

Abstract

The creation of maps -both from original observation and as derived from other maps-involves the generation of spatial information from spatial data through a process of (re)classification. This is not just an essential geographer's tool: (Re)classification is a crucial cognitive ability and a fundamental GIS operation. In this paper we overview reclassification as used by geographers, and look into its basic mathematical and computational properties. Basic set and relational theory prove sufficient to provide a powerful mathematical formalism for reclassification, which is a special type of equivalence relation. We look into the distinction between spatial and non-spatial classification. We show that relational algebra is not sufficient for the computation of map reclassifications while, on the positive side, we present an efficient generic reclassification algorithm (order of n^2 on the number of map regions).

Key words

Reclassification, map algebra, GIS, layer operations.

1. Introduction

In this work we address map classification as a computational issue of an engineering discipline, but with the view that we can only have good engineering when we have solid theory based on sound mathematical modelling. Our purpose is to respond, with respect to this narrow but important topic, to two demands from GIS research.

First, is the criticism that GIS is application and technology driven (Goodchild 1992) and the corresponding demand that academic research focus on the fundamental research, look into the basic spatial operations and define suitable mathematics (such as Tomlin's map algebra, see (Tomlin 1983, 1990)). Indeed, GIS evolution and advancement was to a certain extent dictated by user needs and restrained by technological limitations. Experience in engineering fields shows that theory almost always comes after the first working system (e.g. it was after the implementation of the first FORTRAN compiler that the theory on compilers was developed). Similarly, the importance of a GIS Theory has already been stressed in the literature (Burrough 1992). Identifying similarities in the functionality of existing GIS and formulating unifying mathematical frameworks is consistent with the desire of scientists to work with general principles. Moreover, a mathematical analysis can always give useful feedback to system developers, in our case by showing that many seemingly different operations are special cases of a fundamental generic one.

Second is the recent but rather strong movement on *open GIS* and *GIS interoperability* (refer to NCGIA initiatives and workshops). At the lowest level, interoperability can only mean common, standardised data format. However, in order to achieve fully interoperable GIS, a "common language" must be used for all the modelling stages of a GIS application (semantic/conceptual, logical, physical). This requirement calls for the specification of fundamental, unifying, cross-GIS operations.

A map being the representation of a classification of geographic phenomena using geometric shapes (regions, lines, points) along with associated properties, the classification (or reclassification, according to GIS terminology) of maps (or layers) is in many aspects the most fundamental class of map operations. It is usually qualitatively discussed in GIS textbooks (see Burrough 1986) or partly referred to in user manuals. In this work we adopt a mathematical/computational point of view.

The main contribution of this paper is a mathematical definition of map reclassification that is sufficiently general to grasp the various and differing ways in which the operation is used in practice; thus it is a contribution towards the theory of GIS. The study of mathematical and computational properties including the distinction between spatial vs. non-spatial reclassification and the proof that relational algebra is not sufficiently powerful to compute spatial reclassifications, as well as the generic algorithm which puts an upper limit on the computational complexity of reclassification, are likewise contributions to the theory of GIS with direct impact on computer systems development.

In Section 2 we review some of the literature, elucidate the term and present a description as well as a categorisation of reclassification from a geographer's point of view. In Section 3 we examine the nature of reclassification operations by outlining some of their most important mathematical and computational properties. In Section 4 we introduce a formal framework for their definition, borrowing concepts from classical set theory and the notion of classes of equivalence. Finally, in Section 5 we present a generic algorithm for the computation of reclassified maps.

2. Reclassification from a geographer's point of view

The generic term *reclassification* refers to the fundamental class of analytical operations that involve the assignment of thematic values to the categories of an existing map, as a function of the initial values, the position, contiguity, size or shape of the individual categories (Berry 1987). This can be visualised as "recolouring", while no new boundaries are allowed. Consider the following simple example where we have a map depicting regions with their associated population and wealth produced attribute values. We can define "per capita wealth", the ratio of wealth over population, as a new attribute and further assign the new values to specific categories, e.g. "low", "medium", "high".

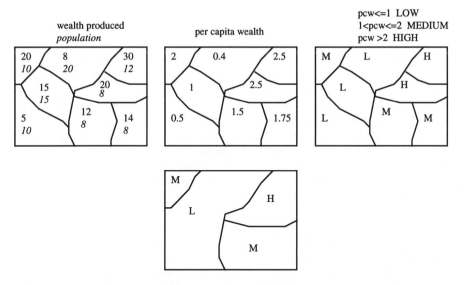

Figure 1 A reclassification example.

The example above represents a general class of operations where the reclassification is based on the non-spatial properties of the features of a map. This class of operations can be further categorised according to the kind of values of the initial attributes combined and that of the result values. Thus, quantitative values (e.g. population and number of cars) can be combined to yield quantitative information (average car ownership). Alternatively, qualitative information (e.g. soil type) can be used to derive quantitative information (the relative suitability for residential development) (Berry 1987).

In both cases reclassification involves the application of a function on the initial set of non-spatial attributes. It can be a "level slicing", which splits a continuous range of values into discrete intervals, or an arithmetic operation (e.g. a map of topographic elevation in feet, can be converted to one in meters, by multiplying each map value by the appropriate conversion factor), a simple masking, or an arbitrarily complex function composition (e.g. a suitability assessment based on altitude, soil-type, land-use, etc.).

Reclassification is a purposeful reading of a map, thus the appearance of the final result (and the potential interpretations) is highly dependent on the class number and intervals used for the reclassification (Burrough 1986, Unwin 1981). The underlying fear is that "a skilled cartographer can manipulate his map like a musician does his instrument, bringing out the quality he wants" (Schultz 1961). Evans (1977) has made a thorough analysis of the problem which resulted in a classification of the class interval systems, very briefly summarised below:
- *exogenous* class intervals are fixed according to threshold values that are relevant to but not derived by the data under study.
- *arbitrary* class intervals are chosen without any predetermined aim.

- *idiographic* class intervals are chosen with respect to specific aspects of the data set.
- *serial* class intervals have limits which are mathematically related.

Reclassification operations can also be based on the spatial properties associated with a map. Position (location, size, shape, orientation) is one of them. Different areas can be reclassified according to their latitudes and longitudes. In addition, size (length or area) and shape can be used as the basis of reclassification. A layer of surface water can be reassigned values to indicate the areal extent of lakes or the length of streams. In the same manner, quantitative considerations of shape can be very promising for many potential applications (e.g. in the classification of digital imagery and the modelling of wildlife habitats). Finally, topology can also form the basis of a reclassification operation (adjacent features with a common property can be grouped and form a new reclassified zone).

Broadly speaking, any aggregation of the existing zones of a map in order to form a new set of zones, can be considered as a reclassification operation. The result of such an operation is a "zoning system", if only contiguous areas are aggregated (electoral districting is a typical example). The non-contiguous case is referred to as a "grouping system". There is a considerably large number of different aggregations that exist for every realistic case. The often extreme variation of the results obtained for different aggregations schemes is the basis of what is called the *Modifiable Areal Unit Problem* (MAUP) (Openshaw 1984).

3. Some mathematical and computational properties of reclassification

In the previous section we drew a distinction between two broad classes of reclassification operations. First, operations which reclassify an existing map in terms of its descriptive information alone and secondly reclassifications based on spatial information such as position and topology. The latter are salient features of spatial entities and therefore, whenever involved, they call for different methods and tools for their manipulation (Tryfona and Hadzilacos 1995). In the general case, the two types of operations described above can be combined into a powerful operation. In the following subsections give an insight into the mathematical and computational properties of reclassification.

3.1 Reclassification = attribute derivation + concatenation

Consider again the reclassification example presented in Figure 1. We can identify three different steps:
1. the attribute *per capita wealth* is calculated as wealth/population.
2. based on the numerical values, the ordinal values 'low', 'medium' and 'high' are assigned.
3. a concatenation is performed among all adjacent regions which have identical values for the attribute per capita wealth; in other words, common borders are eliminated for the above regions.

The first two steps are similar in that values for a new attribute are computed from values of preexisting ones. A function, say f, defined as wealth/population was applied to the original attributes, thus deriving the new attribute *per capita wealth*. Next, a function, say g, defined as

$$\text{'low' for } pcw \leq 1$$

$$g = \text{'medium' for } 1 < pcw \leq 2$$

$$\text{'high' for } pcw > 2$$

was applied to per capita wealth, deriving the final values of 'low', 'medium' or 'high'. In fact, the derivation of new computable attributes based on the initial ones is a very useful operation for GIS (Burrough 1992, Delis et al. 1994). We can further identify different computational subclasses to the operation itself. Attributes can change based on descriptive information alone, or based on descriptive and spatial information but restricted to the geometric feature itself, or on descriptive and spatial information restricted to the geometric feature and a bounded number of neighbours or finally on descriptive and unrestricted spatial information. All the above operations belong to different computational classes and are indicative of the complexity of a generic reclassification operation.

A reclassification operation which consists of just the third step described above is a simple case where we would want to focus on a specific attribute and observe its variation over space (see Figure 2).

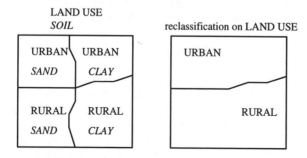

Figure 2 A simple reclassification - the attribute focus case.

However in the general case we wish to identify on a map different classes usually of information that is not already depicted. So this information has to be computed based on the initial attribute values. In these cases we need both attribute derivation and spatial concatenation operations.

3.2 Reclassification = overlay $^{-1}$

In a sense, the reclassification operation is the inverse of the overlay operation and it can be hierarchical or allow for overlap, depending on the way spatial information is arranged into layers. Consider the example described above in Figure 2. The first layer could be the result of an overlay between the two following layers:

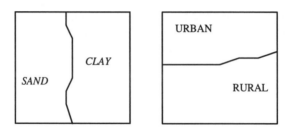

Figure 3 Land-use and soil-type layers.

Retrieving the two initial layers is implemented as an "attribute focus" reclassification operation on each of the two attributes which would consist of the concatenation of adjacent areas with identical values. The same can be generalised for the case of overlay among n layers.

In the aforementioned example, the resulting layers constitute partitions of the initial map's spatial domain (i.e. they consist of non-overlapping regions), which is the case frequently encountered in reality. Nevertheless, there are several occasions where reclassified layers may consist of overlapping classes (e.g. a particular reclassification of a layer consisting of "wolf", "bear" and "mixed" zones of animal habital might yield two overlapping "wolf" and "bear" areas). Despite the fact that any layer allowing overlapping features can be represented by an equivalent hierarchical one, the ability to handle overlapping areas has been recently incorporated to commercial systems, in order to handle a few realistic situations more efficiently.

3.3 Reclassifiability vs. separability

In all the examples presented so far, we implicitly assume that at the final stage of the reclassification the concatenated region acquires the same attribute values as the initial regions. But is this always meaningful? Consider for example a layer containing information about regions and their corresponding population figures and number of cars. If we calculate the ratio "cars per person" and then reclassify on this new attribute, then two adjacent regions with an attribute value of x, would yield a new region with the same value x for this new attribute. However, if we reclassify on the attribute "number of cars", then two regions with exactly the same value should yield a new region with the double value.

This problem is related to the following one. Consider a region and its associated attribute. The value for an arbitrary subset of the region (e.g. a point, a disjoint subset and other) is not always defined, much less known. Some attributes have the property of *separability*, that is, the attribute value of an area is the same for any of its subareas. For example, land use is such an attribute, for a given scale of observation. Researchers often relate this property to the nature of geographic information (Waugh 1974) and its common categorization into *nominal, ordinal, ratio* and *interval* types (see (Laurini and Thompson 1992) for a discussion on the above types).

Thus, nominal attributes, like land use or soil type, are said to retain their value when several regions which share the same value are aggregated to form a bigger one. Analogously, ratio or interval values (like car ownership ratio, population, etc.) are arithmetically transformed (added, averaged, etc.) in the case of region aggregation.

Nevertheless, such a rule could prove wrong, unless we have made certain assumptions for the semantics of a region's attributes. Consider for example a layer consisting of regions with the associated attribute "population", which is of type ratio. Depending on our interpretation of population, values could be added ("population" refers to the actual population figure of a region) or copied ("population" refers to the population of the largest city on the map) in a particular reclassification scenario. We believe that further research and a deeper insight into the nature and the semantics of spatial attributes is needed, in order to answer the questions briefly discussed in this section.

4. Reclassification vs. classes of equivalence

So far, we have explored the range of reclassification operations in GIS from different perspectives and identified some of the associated fundamental properties. In this section we formalise this generic operation, using concepts from set theory. In order to facilitate the reader, we briefly reiterate some of the basic notions (see Lewis and Papadimitriou, 1981).
The *Cartesian product* of two sets A, B, denoted $A{\times}B$, is the set of all ordered pairs (a, b), so that $a \in A$ and $b \in B$.

A *binary relation* R between two sets A, B is a subset of their Cartesian product. For example, the set $\{(i, j) : i, j \in N \text{ and } i{<}j\}$, is the binary relation "less than".

A binary relation $R \subseteq A{\times}A$ is *reflexive*, if $(a, a) \in R$ for every $a \in A$.

A binary relation is *symmetric* if $(b, a) \in R$ whenever $(a, b) \in R$.

Finally, a binary relation R is *transitive,* if when $(a, b) \in R$ and $(b, c) \in R$, then $(a, c) \in R$. A reflexive, symmetric and transitive relation is called *equivalence*. The representation of a relation of equivalence by an undirected graph consists of a set of connected partitions (see Figure 4).

In each partition, every two nodes are connected with an edge. The connected partitions of an equivalence relation are called *classes of equivalence.* In the example presented in Figure 4, the classes of equivalence are the sets $\{a,b\}$, $\{c, d, e, f\}$, $\{g\}$. Generally, we denote as $[a]$ the class of equivalence that contains an element a, given the equivalence relation R, i.e. $[a] = \{b : (a, b) \in R\}$.

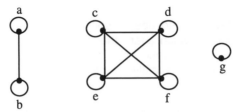

Figure 4 *The representation of an equivalence relation.*

Let R be an equivalence relation on the set A. It is known that the classes of equivalence of R constitute a partitioning of A (i.e. every class has at least one element, no two different classes have an element in common and the union of all the classes is A).

A *chain* on a relation R is a sequence $(a_1, a_2, ... , a_n)$, $n \geq 1$, where $(a_i, a_{i+1}) \in R$, $\forall\ i = 1, ... , n - 1$. A graph is connected if for every pair of nodes a, b, there is a chain from a to b.

The *transitive closure R^** of a binary relation R is defined as:
$$R \cup \{(a, b) : \text{there is a chain in } R \text{ from } a \text{ to } b\}$$
Finally, the *reflexive closure* of a binary relation R is defined as:
$$R \cup \{(a, a) : a \in A\}$$

For example, the reflexive transitive closure of the relation $\{(a, b), (b, d), (c, c)\}$ defined on the set $A = \{a, b, c, d\}$ is the relation:
$$\{(a, a), (b, b), (c, c), (d, d), (a, b), (b, d), (a, d)\}$$

In order to return back to our reclassification considerations, we need a mathematical model of *layer* as defined in the context of GIS and we will borrow the one presented in (Delis et al., 1994).

A layer L is a set of ordered pairs of the form (g, v) where g is a geometric feature (such as region, line, point) belonging to the set G and v is its associated tuple of attribute values belonging to the Cartesian product $A_1 \times A_2 \times ... \times A_n$ of domains. Now consider the binary relation R "adjacency and identical values in L" defined on G.

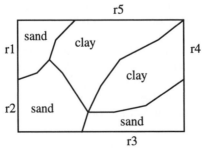

Figure 5 *A layer example.*

For the example presented in Figure 5:

L = {(r1, sand), (r2, sand), (r3, sand), (r4, clay), (r5, clay)}
R = {(r1, r2), (r2, r1), (r2, r3), (r3, r2), (r4, r5), (r5, r4)}

Then the reflexive transitive closure R^* is the set:
R^* ={(r1, r1), (r2, r2), (r3, r3), (r4, r4), (r5, r5), (r1, r2), (r2, r1), (r2, r3), (r3, r2),
(r1, r3), (r3, r1), (r4, r5), (r5, r4)}

We can easily observe that R^* is reflexive, symmetric and transitive, thus a relation of equivalence. The corresponding classes of equivalence are the sets:
{r1, r2, r3}, {r4, r5}

The two above partitions are in fact the resulting classes of a simple reclassification of L according to soil type. Of course in a GIS context, after computing the regions that constitute a new class, a geometric union has to be applied in order to eliminate the common borders.

Although simple, the above example clearly shows the way that classes of equivalence can model any particular reclassification case, as discussed in Section 2. By generalising the relation R (checking for adjacency and for a condition which must be satisfied by a function of the attributes, usually belonging to certain interval classes) we can model more powerful reclassifications (e.g. the "per capita wealth" example). Moreover, this model does not rely on specific geometric figures (points, lines or regions); rather it is based on a generic definition of map layer as a function from a set of spatial entities to an attribute domain. The region case is the most general and frequently encountered in practice. However, lines or points can be incorporated in a straightforward manner. The only modifications required relate to the definition of topological adjacency for points or lines, where appropriate, which can be based on Egenhofer and Herring's (1990) work.

5. Computing reclassifications

So far we have tried to shed light on different aspects of a generic operation so common to geographers and GIS practitioners. In this section we examine reclassification computationally. We answer the question "how difficult is reclassification, or, how difficult can it potentially be?" and we propose a generic algorithm for reclassifying maps.

5.1 Spatial vs. non-spatial reclassification

Adopting a relational approach and assuming that a layer consists of tuples of the form: (geo, a_1, a_2, ..., a_n) where geo is a geometric figure belonging to a geometric type and a_1, a_2, ..., a_n are attribute values belonging to certain domains, we can express a reclassified map as equivalence classes based on a binary relation on a set of spatial entities which satisfy the following generic predicate:

$$R_{spat}(x,y) = R(x.a_1,..,x.a_n,y.a_1,...,y.a_n) \wedge r_\tau(x,y)$$

where x, y are tuples of the above form, R is a condition that must hold among the attribute values of x and y (e.g. equality) and r_τ is the binary topological relation "adjacency", defined in (Egenhofer and Herring 1990).

We pose the question: can we compute reclassifications defined as above using relational algebra? The intuitive answer is yes, since we can calculate new attribute values and then select tuples according to these values. This is true and will work for many cases in practice. However, this computation can not distinguish between classes with the same value lying in different parts of a map. In other words, any region with population density HIGH would be contained in the unique class with value HIGH. Thus, a reclassified map consists of potentially non-contiguous zones. We call this type of operation *non-spatial reclassification*. We contrast this with the more powerful *spatial reclassification* where a region identified as "mountain" at the upper right corner of a map is a different class from a mountain region at the lower left corner of the same map (see Figure 6). Intuitively, the latter is closer to human conceptualisation of space, since we do not treat all the mountains or all the rivers as the same object. Rather, we classify a geographic phenomenon according to location and identify several distinct objects (e.g. Alps, Himalayas, etc.) which all-together constitute the phenomenon in its entirety. We will show that spatial reclassification is out of the realm of relational algebra.

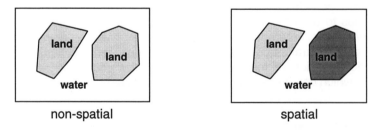

non-spatial   spatial

Figure 6 Spatial vs. non-spatial reclassification.

First, it is easy to see that the resulting regions of non-spatial reclassification are equivalence classes on the predicate R defined above and that spatially reclassified regions form equivalence classes on the predicate R_{spat} (we use the term predicate and relation interchangeably; however this should not cause any confusion since there is a one-to-one correspondence between a relation and the predicate that the elements of its pairs satisfy). The main and essential difference is that by forcing the adjacency condition we distinguish non-contiguous classes. Moreover, we have seen that the core of a reclassification operation is the elimination of common boundaries among bordering regions that share the same value for a subset of their attributes, i.e. we are interested in two distinct relations, equality and adjacency. Equality is transitive, unlike adjacency. Therefore, the transitive closure of adjacency has to be computed in order to reclassify a map (i.e. to define the equivalence classes according to the reclassification criterion). Transitivity in the form of transitive closure of relations has been proved to be a "hard" problem (Ullman 1987), generally not solvable by relational algebra, which is a fact that advocates our claim that spatial reclassification is a computationally more powerful operation, compared to its non-spatial counterpart.

5.2 A Spatial reclassification algorithm

In this section we propose a graph-based generic algorithm for the computation of spatial reclassifications.

A map is represented as an undirected graph $G=(V, E)$ whose vertices $v \in V$ correspond to regions of the map and whose edges $e \in E$ connect two vertices iff their corresponding regions share a common border. According to the theory of *Planar Graphs* (Liu 1968), such a graph is planar (a graph is said to be *planar* if it can be mapped on a plane in such a way that two edges meet one another only at a vertex with which the edges are incident - that is to say, no two edges cross one another). This stems from the fact that maps form partitions of 2-D space, thus they preserve planarity. An important property of planar graphs, known as *Euler's formula*, is that

$$Ne \leq 3^*Nv - 6$$

where Ne, Nv, are the number of edges and vertices, respectively.

Algorithm
Step 1.
> In order to derive new attribute values upon which reclassification is going to be based, we visit each vertex once and perform the necessary calculations.

Step 2.
> Then we visit each edge once and we either keep it if it connects vertices which correspond to adjacent regions with the same value for the specified subset of attributes, or delete it otherwise.

Step 3.
> The result graph G' might not be connected. The connected components of this graph consist of vertices whose corresponding regions constitute the resultant map classes. In order to compute them we use a generic *disjoint set* data structure (Cormen et al. 1990). A disjoint set data structure maintains a collection of disjoint dynamic sets. Each set is identified by a *representative* which is some member of the set. Letting x denote an object, we wish to support the following operations:
> - Make-Set(x) creates a new set whose only member (thus, the representative) is pointed to by x.
> - Union(x, y) unites the sets that contain x and y into a new set that is the union of the two sets, destroying the initial sets. The representative of the new set is any member of their union (usually

one of the former representatives).
- Find-Set(x) returns a pointer to the representative of the (unique) set containing x.

Thus we can determine the connected components of G' using the following simple procedure:

> **for** each vertex v in $V(G')$
> > **do** Make-Set(v)
>
> **for** each edge (v,u) in $E(G')$
> > **do if** Find-Set(v)≠Find-Set(u)
> > > **then** Union(v,u)

Performance Analysis. The 1st step, since we visit each vertex once, takes $O(Nv)$ time. According to Euler's formula, the 2nd step takes also $O(Nv)$, since the number of edges of a planar graph is a linear function on the number of vertices. In the 3rd step we identify two loops. The first takes clearly $O(Nv)$ time while the second takes $O(Ne)*Complexity-of(Union) = O(Nv^2)$ (in the worst union case, all the vertices must be moved to a different set and their pointer to the representative has to be redefined).

Therefore, the worst-case complexity for a spatial reclassification operation of a map with n regions is $O(n^2)$. This result may contradict with our intuition according to which the complexity should be the same with the complexity of computing transitive closure, which is $O(n^3)$ for graphs. It seems that taking advantage of the planarity property we can derive a better result.

6. Conclusions

One of the common charges that GIS research faces is being application and technology-driven (Goodchild 1992), thus lacking deep spatial analytical capabilities (Fotheringham and Rogerson 1993) and a unified theoretical framework. It is true that, like most engineering, GIS technology follows the needs of the market. Even academic research, often driven by available funding rather than fundamental questions, tends to focus on technical issues; thus GIS appear to be an uncoordinated mass of technical information, lacking a core of theory and fundamental principles. This could be an indication of immaturity of our field; more insight and understanding of the nature of geographic data processing is called for.

Classification is an essential part of human cognition and spatial classification is an essential part of geographic understanding and reasoning, but is there something special about spatial classification, from a mathematical or a computational point of view? In this paper we saw that, although any classification is an equivalence relation, spatial reclassification also involves topology, specifically the "adjacency" topological relationship. Two interesting computational conclusions were then derived: that relational algebra is not powerful enough for spatial reclassification (because it involves the transitive closure of adjacency) and that computing reclassification takes no more than n^2 on the number of regions (due to the planarity of the graph that can represent a map).

Acknowledgements

The first author wishes to thank Mr Tom Waugh for an enlightening discussion, during the M.Sc. in GIS course at the University of Edinburgh.

References

Berry, J. K., 1987, Fundamental operations in computer-assisted map analysis, *IJGIS,* vol. 1, no 2, pp. 119-136.

Burrough, P. A., 1986, Principles of Geographical Information Systems for Land Resources Assessment (Oxford: Oxford University Press).

Burrough, P. A., 1992, Development of intelligent geographical information systems, IJGIS, vol. 6, no 1, 1-11.

Cormen, T. H., Leiserson, C. E., Rivest R. L., 1990, Introduction to Algorithms, McGraw Hill.

Delis, V., Hadzilacos, Th., and Tryfona, N., 1994, An Introduction to Layer Algebra, Proceedings of Sixth International Symposium on Spatial Data Handling, Edinburgh.

Egenhofer, M., and Herring, J., 1990, A Mathematical Framework for the Definition of Topological Relationships, Proceedings of the 4th International Symposium on Spatial Data Handling, Zurich, Switzerland, pp. 803-813.

Evans, L. S., 1977, The selection of class intervals, Trans. Inst. Br. Geogrs (N. S.) 2, 98-124.

Fotheringham, A. S., and Rogerson, P. A., 1993, GIS and Spatial Analytical Problems, IJGIS, vol. 7, no 1, pp. 3-19.

Goodchild, M. F., 1992, Geographical Information Science, IJGIS, vol. 6, no 1, pp. 31-45.

Laurini, R., and Thompson, D., 1992, Fundamentals of Spatial Information Systems (Academic Press).

Lewis, H., and Papadimitriou, C., 1981, Elements of the Theory of Computation (Prentice Hall).

Liu, C. L., 1968, Introduction to Combinatorial Mathematics (McGraw Hill).

Openshaw, S., 1984, The Modifiable Areal Unit Problem, Concepts and Techniques in Modern Geography, 38 (Norwich: Geo Books).

Schultz, G. M., 1961, An experiment in selecting value scales for statistical distribution maps, Surv. Map., 21, pp. 224-230.

Tomlin, C. D., 1983, A Map Algebra. In Harvard Computer Graphics Conference (Cambridge, MA).

Tomlin, C. D., 1990, Geographic Information Systems and Cartographic Modelling (Prentice-Hall, Englewood Cliffs, NJ).

Tryfona, N. and Hadzilacos, T., 1995, Geographic Applications Development: Models and Tools for the Conceptual Level, 3rd ACM GIS Workshop CIKM'95, Baltimore, Maryland, USA.

Ullman, J., 1987, Database Theory - Past and Future. In Proceedings of the Sixth ACM Symposium on Principles of Database Systems, pp. 1-10.

Unwin, D., 1981, Introductory Spatial Analysis (Methuen - London and New York).

Waugh, T. C., 1974, Geographical Data in a System Environment, (M.Phil. thesis), Geography Department, University of Edinburgh

Contact address

Vasilis Delis[1] and Thanasis Hadzilacos[2]

[1,2] Computer Technology Institute
Kolokotroni 3
26221, Patras
Greece
Phone: +3061 225073
Fax: +3061 222086
E-mail: {delis, thh}@cti.gr

[1] Computer Engineering and Informatics Department
University of Patras
Patras
Greece

A SPATIAL DATA MODEL FOR NAVIGATION KNOWLEDGE

Christophe Claramunt[1] and Michel Mainguenaud[2]

[1]Swiss Federal Institute of Technology
Lausanne, Switzerland
[2]France Telecom, Institut National des Télécommunications
Evry, France

Abstract
This paper presents a spatial data model suitable for the representation of navigation knowledge. The model analyses a navigation process from a cognitive point of view, with the visual environment necessary for its comprehension. The decomposition is based on the spatial view concept, defined as a logical, dynamic representation of spatial data. The model allows the coexistence of several abstraction levels. A language and operators define and specify the model.

Key words
Navigation knowledge, cognitive collage, spatial collage, spatial view.

1. Introduction

Most current data models describe space in a cartographic, continuous form [e.g. Bur 86, Peu 88]. However, other perceptions of space, narrative or navigational, are also used. This is the case of navigation knowledge, which corresponds to the mental description of a spatial displacement. This perception is called a cognitive map [Kui 78]. Its representation is closer to human thinking than a static map. Cognitive maps, largely studied in psychology and linguistics [Mar 91], inherit the diversity and richness which characterise human thinking but also possess its imperfections. The space represented by a cognitive map has four dimensions (three for space and one for time). Its spatial representation is dynamic (cohabitation of different spatial perceptions) and discontinuous (partial knowledge), and thus less easily identifiable by current spatial data models. The perceived topology is relatively well preserved and essentially linear; notions of distance are, on the other hand, approximated. These properties show the variability of the memory expressiveness when displacement is perceived. In fact, a navigation process comes from a mental memory of our sensory activities. This allows the constitution of a cognitive displacement memory whose quality will vary according to its users and will decrease over time.

The objective of this study is to model a cognitive map representing navigation knowledge (a model of a model). Formalisation of this cognitive representation allows the preservation of these process memories and extends their application potential. Identification of decision criteria during the realization of a navigation process [Gol 95] or of network algorithms will not be developed [Ken 80]. The proposal defines a model and operations that allow decomposition of navigation knowledge and its presentation as a form of dynamic cartography. The model identifies element characteristics which enable the description and presentation of these processes. It uses the concept of spatial view defined as a dynamic and flexible tool for spatial data representation [Cla 94].

Section 2 describes the principles of navigation knowledge; section 3 explains the contribution of the spatial view; section 4 proposes a framework for representing navigation knowledge; manipulation principles of the model are presented in section 5; an application illustrates the model in section 6 and a conclusion is formulated in section 7.

2. Navigation knowledge principles

Navigation knowledge is provided by the application of a mental model which depends on a user perception [Tim 92, Tho 93]. It is the result of an experience if it is known, or it has to be experimented if it acts as a proposed solution. Navigation knowledge is defined by a **procedural knowledge** [Tho 93], the related geographical environment which provides **survey knowledge** [Tho 93] and the **perceptual space** [Che 89]. The procedural knowledge describes the spatial displacement process; the survey knowledge situates it. Perceived elements during a navigation process give visual or symbolic landmarks (e.g. photos or symbols associated with a displacement point). Each form used in a navigation knowledge environment has a specific role: survey knowledge and procedural knowledge own a spatial structure, perceived elements provide poor spatial structure but more meaning. This combination of different expression models gives global coherence to a navigation knowledge representation [Lyn 60]. The language used to describe a navigation process can be textual, gestural or graphic. Language expressions are explicit forms of navigation knowledge. They increase the expressive power of the model (e.g. from the airport, take the highway to the EPFL).

A navigation knowledge representation is based on a **hierarchical** and **incremental** cognitive model [Maa 93]. The hierarchical characteristic allows an incremental ordering of complementary abstraction levels, from strategic to specific according to Kuipers [Kui 78], from basic to secondary according to Maaβ [Maa 93] and from planning to action according to Timpf [Tim 92]. These different abstraction levels used for the description of a navigation knowledge form a **multidimensional** spatial set [Sto 93]. The semantic association of these different abstraction levels is fundamental to ensure the continuity of the cognitive process. The mental association of these multidimensional spaces is expressed by the **cognitive collage** metaphor [Tve 93]. A cognitive collage is defined as the semantic association of different data abstraction levels. A **connection** extends the map collage concept to semantic spatial associations between spatial entities which are members of the represented spaces [Ren 95]. The cognitive collage, when applied to navigation knowledge, makes necessary the identification of a form of cohabitation between different representation models, coordination that few research activities develop [Pau 89]. The goal of the cognitive collage is to associate various representation models designed with different concepts: (1) hierarchical and incremental navigation knowledge representations (2) heterogeneous geographical spaces (several abstraction levels and different spatial structures).

3. Spatial view model

The description of navigation knowledge components is based on the proposal of a spatial view, which is defined as a dynamic and flexible form of spatial data representation [Cla 94]. The spatial view is expressed as an extension of the classic database view as defined within the database context. This allows the definition of an external schema, derived from the logical level, while integrating dynamics by the application of operators [Cla 95]. External schemas provide flexibility for the representation of the spatial models which describe navigation knowledge. The dynamic character allows the identification of relevant data in each represented space. A spatial view is defined by a partially ordered atom set.

A **spatial view atom** is described by a name (Name), one or several spatial queries (Spatial_Query) operating on collections (Collection) and visualization operators (Visualization_Op). It presents a coherent spatial data set at the abstraction and scale levels while providing independence and dynamics (i.e. operators). A spatial query is defined from collections (a collection represents an entity set), and from spatial and non spatial operators which are applied to these collections. Spatial collections describe spatial entities, perceived and symbolic elements.

A **spatial view** is identified by a name (Name). The order attribute allows management of the role of each atom that composes the spatial view (Order). In the context of navigation knowledge, the order attribute distinguishes the spatial view atoms which define the navigation knowledge: the procedural description (Essential attribute, e.g. the trace of a navigation knowledge), geographical spaces (Important attribute, e.g. the network support of a navigation knowledge) and perceived elements (Useful attribute, e.g. the geographical context of a navigation knowledge). This classification defines the spatial view semantics, thus providing a criterion for managing visual conflicts related to spatial view displays and manipulations. Moreover, description of the navigation knowledge elements by a common concept, the spatial view, provides a homogeneous structure for the model. The proposed model is expressed using the notation of complex objects [Adi 87]. This notation allows the extension of classic data types. This is suited to the definition of complex data. The constructors used are aggregation, set and list (respectively noted [ ],{ } and ( )). The definition of spatial view atoms (SVA) and spatial views (SV) follows:

SVA_type = [	Name:	string,
	Collection:	{string} ,
	Spatial_Query:	(string) ,
	Visualization_Op:	{string}]

Order_type = [	Essential:	{SVA_type},
	Important:	{SVA_type},
	Useful:	{SVA_type}]

Elt_Manipulation_Scale_type = [	Min_Limit:	float,
	Scale_Reference:	float,
	Max_Limit:	float]

Manipulation_Scale_type =   {   Elt_Manipulation_Scale_type }

SV_type = [	Name:	string,
	Order:	Order_type,
	Manipulation_Scale:	Manipulation_Scale_type]

The manipulation scale of a spatial view (Manipulation_ Scale) gives the scale at which a user is situated for the manipulation of spatial view atoms [Cla 94]. The manipulation scale is defined by a numerical interval scale (float number type) corresponding to the relevant abstraction level. This interval is centered on a scale reference (Scale_Reference) and two scale boundaries (Min_Limit and Max_Limit). The users's scale position indicates which spatial primitive will be used by a spatial operation [Woo 95]. This scale acts as a filter by allowing the system to choose, for each spatial view atom represented in the spatial view, the appropriate spatial representation for the scale manipulation (e.g displaying a network has different meanings at planning or civil engineering abstraction levels). The manipulation scale of a spatial view is provided if and only if at least one of its spatial view atoms is georeferenced.

Application of the spatial view to navigation knowledge context requires the specification of basic graph theory definitions:

<u>Definition 1</u>: A **graph** is a pair (N, E) where N is a node set and E a subset of the cartesian product N x N. A node is defined by a type (Node_ type).

$N = \{n_1, ..., n_p\}$
$E = \{(n_i , n_j) / n_i \in N, n_j \in N \}$
Node_type = string

<u>Definition 2</u>: A **network** is a graph defined by a name (Name), and identified node and edge sets. A network models a communication system.

Network_type = [Name: string, Nodes: {Node_type}, Edges: {Edge_type}]
Edge_type =   [Name: string, n_i: Node_type, n_j : Node_type]

The following example describes a spatial view which represents a displacement between an origin (e.g. EPFL_Entrance) and a destination (e.g. Arrival) through a network managed by a collection (e.g. Network_Graph). The representation of this displacement is built with a spatial view atom describing its layout (e.g. Track), a spatial view atom representing the network support (e.g. Network) and a spatial view atom situating the spatial environment (e.g. Buildings).

```
SVA: [ Name:                    'Track'
       Collection:              {'Network_graph'}
       Spatial_Query:
              ( 'Select         Path ('EPFL_Entrance','Arrival')
                From            Network_graph' )
       Visualization_Op:        {'bold'} ]

SVA: [ Name:                    'Network'
       Collection:              {'Network_Graph'}
       Spatial_Query:
              ( 'Select         Spatial_Rep
                From            Network_Graph')
       Visualization_Op:        { 'default' }       ]

SVA: [ Name:                    'Buildings'
       Collection:              {'Buildings' }
       Spatial_Query:
              ( ' Select        Spatial_Rep
                From            Buildings'          )
       Visualization_Op:        {'hatch'}           ]
```

A spatial view example (EPFL_Campus) is created from these spatial view atoms (Figure 1). It specifies the respective roles of each spatial view atom (Order attribute). The manipulation scale of this spatial view is an interval centered on the scale of "1/ 10,000".

```
SV: [  Name:               'EPFL_Campus',
       Order:          [      Essential:       {'Track'},
                              Important:       {'Network'},
                              Useful:          {'Buildings'}    ] ,
       Manipulation_Scale:  { [   Min_Limit:       1/25, 000,
                                  Scale_Reference:      1/10, 000,
                                  Max_Limit:       1/5, 000   ]      }           ]
```

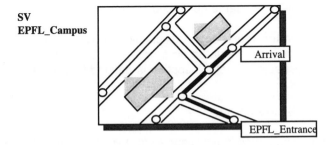

SV
EPFL_Campus

Arrival

EPFL_Entrance

Figure 1 Spatial view example.

4. Navigation knowledge model

The navigation knowledge model is built on a set of spaces represented by different abstraction levels and by semantic connections between these spaces: the spatial collages. Spatial collage semantics are specified by definition 3, and that of the navigation knowledge graph by definition 4.

Definition 3: A **spatial collage** describes a change of space in the description of a navigation process (e.g. from regional to local scale). Each represented space is defined by a distinct spatial view. A spatial collage is identified by a name (Name). The spatial collage translates an oriented semantic relationship between two spatial views identified by their names (SV_from, SV_to) (e.g. a spatial collage between a first spatial view representing a network at a regional scale, and a second spatial view representing the extension of this network at a local scale). The orientation represents the displacement direction.

Definition 4: A **navigation knowledge graph** is an application of the network concept. It is defined by a type (Process_Graph_type). The type specifies the name (Name), the node set (Nodes) and the edge set (Edges) of the navigation knowledge graph. Nodes model spatial views, edges model spatial collages. This graph represents the logical support of navigation knowledge.

```
Collage_type = [          Name:           string ,
                          SV_from:        string ,
                          SV_to:          string ]

Process_Graph_type = [ Name:          string ,
                       Nodes:         {  Node_type  },
                       Edges:         {  Collage_type } ]
```

A navigation knowledge graph is defined using spatial views as nodes and spatial collages as edges. The hierarchical character of navigation knowledge leads to the definition of the incremental displacement element within the same spatial view (at the same abstraction level).

Definition 5 A **section** represents a displacement (without cycle) between an origin and a destination evaluated on a network within the same spatial view (at the same abstraction level). A section is defined by a type (Section_type) and as a network.

```
Section_type = Network_type
```

The semantic spatial collage associating two sections can be spatially materialized by "connecting" the relevant spatial instances of the two sections.

Definition 6: A **connection** is the spatial materialization of the spatial collage of two spatial views. Connection instances are identified by graph functions that give, for each of the two sections, the nodes which allow the connection (end of the section n, origin of the section n+1). There is only one connection for a given spatial collage. A connection is defined by its name (Name), the network names which define the sections (Section_ from_Name and Section_to_Name) and the nodes which allow the connection (Node_End and Node_Origin) by the graph functions (End) and (Origin).

End: Section_type --> Node_type

Origin: Section_type --> Node_type

End is defined by the n_j such as $G^+ (n_j) = ø$; Origin is defined by the n_j such as $G^- (n_j) = ø$ (G^+ and G^- are standard graph functions which identify respectively the nodes which have no successor and no predecessor).

Connection_type = [Name: string ,
                       Section_from_Name: string ,
                       Node_End: Node_type ,
                       Section_to_Name: string ,
                       Node_Origin: Node_ type]

The following example illustrates the concepts of spatial collage and connection (Figure 2). A navigation knowledge graph is defined between two spatial views (Highway and EPFL_Campus) by a spatial collage (EPFL_Arrival) of these two spatial views and by a connection (C2) between the corresponding sections (Highway_to_EPFL and Track) of these two spatial views (respectively Highway and EPFL_Campus). The Highway spatial view represents two spatial view atoms: a section (Highway_ to_EPFL) and a lake that gives a geographical description (Lake). The manipulation scale of the spatial view (Highway) is an interval centered on "1/ 50,000".

SV: [Name: 'Highway',
        Order: [Essential: {'Highway_to_EPFL'} ,
                        Important: { } ,
                        Useful: {'Lake'}   ]   ,
        Manipulation_Scale:
                        { [Min_Limit: 1/75,000 ,
                        Scale_Reference: 1/50,000 ,
                        Max_Limit: 1/25,000]   }]

Collage_type: [Name: 'EPFL_Arrival' ,
                        SV_from: 'Highway',
                        SV_to: 'EPFL_Campus']

Connection_type = [Name: 'C2' ,
                        Section_from_Name: 'Highway_to_EPFL',
                        Node_End: 'Highway_Exit' ,
                        Section_to_Name: 'Track' ,
                        Node_Origin: 'EPFL_Entrance'   ]

A spatial collage of two spatial views does not require a connection if the represented navigation knowledge is defined at the general abstraction level. The spatial collage ensures the semantic association of sections; the connection ensures the spatial materialization of this association. A spatial collage is given by an intentional and declarative process controled by the user, while a connection is constrained by the continuity of the graph. The next definition qualifies connection cases.

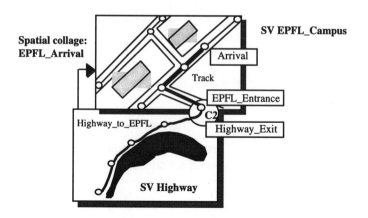

Figure 2 Spatial collage and connection examples

<u>Definition 7</u>: A **convergent connection** (resp. **divergent connection**) is a connection whose the instances (Node_Origin and Node_End) which represent the nodes are spatially merged (resp. non merged) according to geometrical tolerances. Convergent connections ensure the spatial continuity (resp. the spatial discontinuity) of the navigation knowledge graph.

Divergent connections provide flexibility during a preliminary evaluation of a navigation knowledge graph (without a strict spatial continuity constraint, e.g. C2 connexion in the Figure 2). Divergent connections allow default reasoning (i.e. assumption that there is a connection solution whose details will be later identified by the user). A connection can associate two spatial collection instances defined by different spatial representation types but corresponding to the same spatial entity in the real world (e.g. a highway section defined by an area spatial data type in a first spatial view and a highway section defined by a line spatial data type in a second spatial view). A connection can associate two spatial entities situated on non peripheral spatial view positions (semantic continuity but non visual continuity between sections). The connection characterization may be extended, through abstract data types, by the integration of perceived elements such as photographic or symbolic images describing the connection place. These information elements can constitute visual landmarks for a better understanding of a navigation knowledge representation. We introduce the route notion to represent a navigation knowledge displacement:

<u>Definition 8</u>: A **route** represents the ordered vision of a navigation knowledge graph. A route describes the mental representation of a displacement through multidimensional spaces represented by spatial views and associated by spatial collages. A route is defined by a name (Name) and its displacement representation (Navigation_type). This displacement is represented by a spatial view followed by a list of spatial collage - spatial view pairs.

```
Navigation_type =        [ SV_Name:      Node_type ,
                           Sequence:      ( [Collage_Name:    string ,
                                             SV_Name:         Node_type ] ) ]

Route_type =             [ Name:         string,
                           Trace:        Navigation_type]
```

The construction of a route is a function (Trace) of the navigation knowledge graph:

```
        Trace:        Process_Graph_type            -->      Route_type
```


5. Navigation knowledge manipulation

5.1 Definition language

The components of the navigation knowledge model are specified using a classic definition language allowing basic creation and deletion operations. Spatial views and spatial collages allow the design of a navigation knowledge graph (Process_Graph_type). The realization of a spatial view is an explorative process in which the user proceeds by iterations before identifying the manipulation scale which corresponds to the right abstraction level and a coherent order of the spatial view atoms. The route is built according to each user objective and by the definition of a graph order (Route_type). The displacement will be defined by sections specifying its layout in each spatial view and by connections between these sections if allowed by the corresponding abstraction levels. The creation operation set takes the general form of an operator create <component structure of the component>. Conversely, deletion operations take the general form delete <component name>.

5.2 Manipulation language - external level

The data manipulation language relies on classical mechanisms to handle interactions with Process_Graph_type. Two general levels are defined: an identification process and a manipulation (selection and visualisation) process.

Identification operators
Identification operators allow the extraction of a spatial view (SV_Extract) from its name, and extraction of the origin spatial view (SV_Origin) and the last spatial view (SV_End) of a navigation knowledge graph (Process_Graph_type). A SV_Next operator gives the following spatial view from a spatial view of a navigation knowledge graph. Similarly, an operator identifies a spatial collage (Collage_Extract) member of a navigation knowledge graph (Process_Graph_type).

SV_Extract:	Process_Graph_type x string	-->	SV_type
SV_Origin:	Process_Graph_type	-->	SV_type
SV_End:	Process_Graph_type	-->	SV_type
SV_Next:	Process_Graph_type x SV_type	-->	SV_type

SV_Origin defines the spatial view modeled by n_i graph node, such as $\Gamma^-(n_i) = \emptyset$

SV_End defines the spatial view modeled by the n_i graph node, such as $\Gamma^+(n_i) = \emptyset$

Collage_Extract: Process_Graph_type x string   -->   Collage_type

Selection and visualization operators

An operator allows selection of a route by its name (Route_Select_by_Name). Let Database_type denote the type modelling the database.

Route_Select_by_Name: string x Database_type -->   Route_type

Operators are provided to visualize a spatial collage (Collage_Display), a spatial view (SV_Display) and a route (Route_Display). Displayed elements are defined by types (respectively Collage_Display_type, SV_Display_type and Route_Display_type).

Collage_Display:	Collage_type	-->	Collage_Display_type
SV_Display:	SV_type	-->	SV_Display_type
Route_Display:	Route_type	-->	Route_Display_type

5.3 Manipulation language - internal level

A route can be expressed according to several abstraction levels. At each abstraction level there is a specific sequence of spatial views and spatial collages represented by a Process_Graph_type. A spatial view can be decomposed into several spatial views connected by spatial collages; this operation is realized by the application of a development operator (Develop) on a Process_Graph_type. Conversely, a spatial view sequence connected by spatial collages can be grouped within a spatial view; this operation is realized by the application of a group operator (Undevelop) to a Process_Graph_type. These operators allow changes in the abstraction levels of a route representation [Mai 95].

Develop Operator

The signature of the Develop operator is defined as:

Process_Graph_type x Process_Graph_type x Node_type -> Process_Graph_type

The first Process_Graph_type is the graph which models the initial route. The second Process_Graph_type is the graph which specifies a more detailed part of the initial route. Node_type is the node of the graph to be developed (the spatial view whose level of detail will be increased). This node can be obtained, for example, by using the name of the spatial view obtained by a query on a Route_Display_type. The result is a Process_Graph_type (a graph which models a route with a greater level of detail than the initial route).

This operator specification is based on graph properties modeled by Process_Graph_type. These graphs represent routes (without cycles).

Therefore, they have an origin and an extremity. The application of the Develop operator leads to the study of three cases: (a) the developed node is the origin node; (b) the developed node is the extremity; (c) the developed node is neither the origin, nor the extremity (nodes model spatial views, edges model spatial collages).

Therefore, the specification of the Develop operator may be identified by a graph expression:

$G_1 (N_1, E_1) \times G_2 (N_2, E_2) \times N \rightarrow G_3 (N_3, E_3)$.
S_1 and S_2 are the sections modeled by the graphs G_1 and G_2.

Case (a) specifications are the following for the development of the node n_d (Figure 3):

e_{out} is defined as: $e_{out} \in E_1 / \exists\, n_{out} \in N_1 / e_{out} : (n_d, n_{out})$
$N_3 = N_1 - \{n_d\} \cup N_2$
$E_3 = E_1 - \{e_{out}\} \cup E_2 \cup \{ (End (S_2), n_{out}) \}$

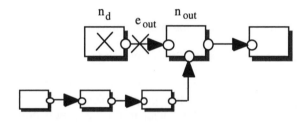

Figure 3 Origin node graph Develop.

Case (b) specifications are the following for the development of the node n_d (Figure 4):

e_{in} is defined as $e_{in} \in E_1 / \exists\, n_{in} \in N_1 / e_{in} : (n_{in}, n_d)$
$N_3 = N_1 - \{n_d\} \cup N_2$
$E_3 = E_1 - \{e_{in}\} \cup E_2 \cup \{ (n_{in}, Origin (S_2)) \}$

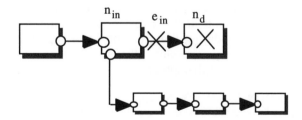

Figure 4 Last node graph Develop.

Case (c) specifications are the following for the development of the node n_d (Figure 5):

$e_{in} = e_{in} \in E_1 / \exists\, n_{in} \in N_1 / e_{in} : (n_{in}, n_d)$
$e_{out} = e_{out} \in E_1 / \exists\, n_{out} \in N_1 / e_{out} : (n_d, n_{out})$
$N_3 = N_1 - \{n_d\} \cup N_2$
$E_3 = E_1 - \{e_{in}, e_{out}\} \cup E_2 \cup \{\, (\, n_{in},\ \text{Origin}\,(S_2)),\ (\text{End}(S_2),\ n_{out})\}$

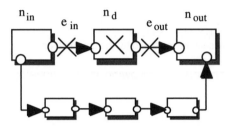

Figure 5 Standard node graph Develop.

Undevelop operator
The signature of the Undevelop operator is defined as:

Process_Graph_type x Process_Graph_type x Node_type -> Process_Graph_type

The first Process_Graph_type is the graph which models the initial route. The second Process_Graph_type is the graph which models the route to be simplified (i.e. a sub-route of the initial route - a graph connected to and resulting from a Develop operator application). Node_type is the node of the graph that represents the abstraction of the deleted sub-route of the initial graph (i.e., the spatial view representing the sub-route). The result is a Process_Graph_type (a graph which models a route with a less precise level of detail than the initial route).

These operator specifications are also derived from graph properties modeled by a Process_Graph_type. As for the Develop operator, these graphs model routes (without cycles). Therefore, they have an origin, and an extremity. There are three cases when applying the Undevelop operator application: (a) the grouping node is placed at the origin node; (b) the grouping node is placed at the extremity node; (c) the grouping node is placed neither at the origin, nor at the extremity. Therefore, the specification of the Undevelop operator is identified by a graph expression:

$G_1\,(N_1, E_1)\ x\ G_2\,(N_2, E_2)\ x\ N \to G_3\,(N_3, E_3)$.
S_1 and S_2 are the sections modeled by the G_1 and G_2 graphs.

Case (a) specifications are the following for the grouping of the node n_u (Figure 6):

e_{out} is defined as: $e_{out} \in E_1 / \exists\, n_{out} \in N_1 / e_{out} : (\text{End}\,(S_2), n_{out})$
$N_3 = N_1 \cup \{n_u\} - N_2$
$E_3 = E_1 - \{e_{out}\} - E_2 \cup \{\, (n_u, n_{out})\, \}$

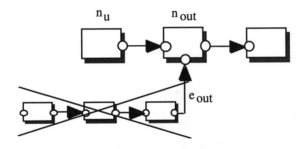

Figure 6 Origin node graph Undevelop.

Case (b) specifications are the following for the grouping of the node n_u (Figure 7):

e_{in} is defined as $\in E_1 / \exists n_{in} \in N_1 / e_{in} : (n_{in}, Origin(S_2))$
$N_3 = N_1 \cup \{n_u\} - N_2$
$E_3 = E_1 - \{e_{in}\} - E_2 \cup \{ (n_{in}, n_u)\}$

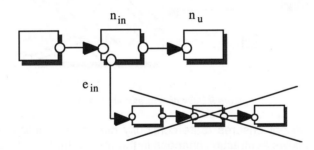

Figure 7 Last node graph Undevelop.

Case (c) specifications are the following for the grouping of the node n_u (Figure 8):

$e_{in} = e_{in} \in E_1 / \exists n_{in} \in N_1 / e_{in} : (n_{in}, Origin (S_2))$
$e_{out} = e_{out} \in E_1 / \exists n_{out} \in N_1 / e_{out} : (End (S_2), n_{out})$
$N_3 = N_1 \cup \{n_u\} - N_2$
$E_3 = E_1 - \{e_{in}, e_{out}\} - E_2 \cup \{ (n_{in}, n_u), (n_u, n_{out})\}$

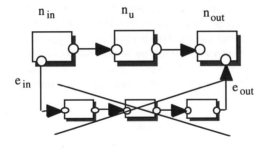

Figure 8 Standard node graph Undevelop.

Example
We introduce a new spatial view, Office, that gives a symbolic representation of the arrival place for this route example, and a new spatial view, EPFL, that gives a symbolic representation. The next figure illustrates the application of the Develop operator applied to the EPFL spatial view (Figure 9).

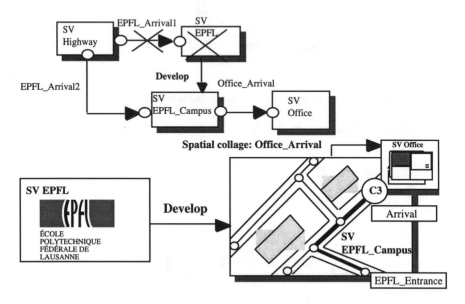

Figure 9 Develop operator application.

The graph concept ensures route continuity during the application of these operators. It allows abstraction changes in the representation of the route and the use of corresponding spatial views for each of these abstraction levels. The model allows representation of partial knowledge within the route description (e.g. a user knows that a route leaves an airport but is unable to describe the airport site displacement).

The route scale is the union of the manipulation scales of the spatial views which compose this route. Considering the diversity of spatial abstractions used in a navigation process, this scale notion gives a global vision of the set of abstraction levels.

The ordered representation of a route allows the application of a textual operator (Textual) that gives a descriptive textual expression using the names of corresponding spatial views. The textual form is provided by an expression From <Initial view name> followed by a sequence of Path to <Name of the next view> applied to the spatial views part of the Sequence attribute of the Navigation_type. The signature of this textual operator is given by the following expression:

Textual: Route_type   -->   string

6. Application

A navigation process example (Geneva_Airport_to_EPFL) begins at the airport (Airport_Site), and continues through a highway (Highway) to the EPFL. At a general abstraction level, a first example of route is:

```
Route:
      [ Name:        'Geneva_Airport_to_EPFL',
        Trace:       [ SV_Name:     'Airport_site',
                       Sequence:    ( [      Collage_Name: 'Airport_exit' ,
                                              SV_Name:       'Highway'   ] ,
                                      [       Collage_Name: 'EPFL_Arrival1' ,
                                              SV_Name:       'EPFL'   ] ) ]    ]
```

The Airport_Site and EPFL spatial views provide significant symbols at this abstraction level. Application of the Textual operator documents the visualization represented by the textual form "From Airport_site Path to Highway Path to EPFL". Figure 10 presents a visualization example of this route (the display arrangement of the spatial views representing the route is an interface level task).

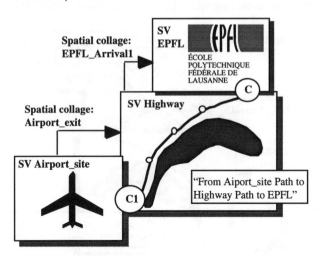

Figure 10 Route example at a general abstraction level.

A second route example illustrates a navigation process at a more precise abstraction level. The objective of this second route example is to specify the presentation of the spatial views which compose this navigation process. The previous EPFL symbolic spatial view is decomposed by the application of a Develop operator applied to the EPFL spatial view.

Route:
      [Name: 'Geneva_Airport_to_EPFL',
        Trace: [SV_Name: 'Airport_site',
                        Sequence: ([Collage_Name: 'Airport_exit' ,
                                    SV_Name: 'Highway'   ],
                             [Collage_Name: 'EPFL_Arrival2' ,
                                    SV_Name: 'EPFL_Campus'],
                           [Collage_Name: 'Office_Arrival' ,
                                    SV_Name: 'Office'   ])]]

The spatial view Office gives a visual signal of the destination place. The textual description is more complete than that of the previous example. Figure 11 illustrates this second route representation example (C1 and C3 are divergent connection examples).

These route representations may be extended by various user experiences or knowledge. The route sequence allows initialization of condition-action events (condition = spatial collage, action = displacement initialisation within a spatial view) necessary to generate a navigation process [Kui 78]. A spatial collage gives a passage condition for continuing navigation through a section. Actions to undertake (e.g. right, left, until, next, [Maa 93]) for the route displacement and associated positional operators [Kui 78] can make use of this model.

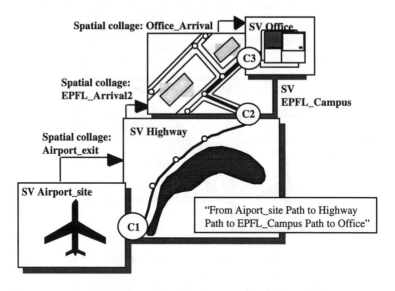

Figure 11 Route example at a more precise abstraction level.

7. Conclusion

The coordinate route representation with multidimensional associated spaces meets the needs of navigation process users. It preserves displacement memory and facilitates its assimilation, favours reusability and ensures updating.

The proposed model allows a displacement action to be situated within its geographical context through complementary abstraction levels that accept partial knowledge. The spatial view gives a representation framework for navigation knowledge. It associates the visualization of a route with multidimensional spaces that allow it to be situated, including significant visual landmarks and textual descriptions. Continuity of route representation is ensured by the graph concept applied to spatial views and spatial collages. Spatial collages are spatially materialized by connections. A route trace in each spatial view spaces is described by a section. Develop and Undevelop operators allow abstraction level changes within the route representation.

This proposal may be applicable to network guidance applications, tourism or professional displacement evaluations. Nautical or aerial navigation are other potentially suitable applications. Further work involves the definition of a manipulation language and the realisation of a demonstrative prototype.

References
[Adi 87] Adiba, M. E., Modeling complex objects for multimedia databases. Entity-Relationship Approach: Ten Years of Experience in Information Modelling, Spaccapietra, S. eds., North-Holland, Dijon, France, November 17-19, 1987, pp. 89-117.
[Bur 86] Burrough, P. A., Principles of Geographical Information Systems for Land Resources Assessment, Oxford: Clarendon Press, United-Kingdom, 1986.
[Che 89] Chen, S., A spherical model for navigation and spatial reasoning. Mapping and Spatial Modelling for Navigation, Pau, L. F. ed., Springer-Verlag, Fano, Denmark, August 21-25, 1989, pp. 59-72.
[Cla 94] Claramunt, C. and Mainguenaud, M., Identification of a definition formalism for a spatial view. Advanced Geographic Data Modelling, Molenaar, M. and de Hoop, S. eds., Netherlands Geodetic Commission, Delft, The Netherlands, September 12-14, 1994, pp. 191-203.
[Cla 95] Claramunt, C. and Mainguenaud, M., Dynamic and flexible vision of a spatial database. Proceedings of the DEXA'95 Int. Workshop on Database and Expert System Applications, Revell, N. and Min Tjoa, A. eds., Omnipress, London, United-Kingdom, September 4-5, 1995, pp. 483-492.
[Gol 95] Golledge, R. G., Path selection and route preference in human navigation: A progress report. Spatial Information Theory: A Theoretical Basis for GIS, Frank, A. U. and Kuhn, W. eds., Springer-Verlag, Semmering, Austria, September 21-23, 1995, pp. 207-222.
[Ken 80] Kennington, J. L. and Helgason, R. V., Algorithms for Network Programming, Wiley, J. ed., New-York, 1980.
[Kui 78] Kuiper, B., Modeling spatial knowledge. Cognitive Science, Vol. 2, 1978, pp. 129-153.
[Lyn 60] Lynch, K., The Image of the City, MIT Press, Massachusetts, 1960, USA.
[Maa 93] Maaß, W., A Cognitive model for the process of multimodal, incremental route descriptions. Theories and Methods of Spatio-temporal Reasoning in Geographic Space, Frank, A. U., Campari, I. and Formentini, U. eds, Springer-Verlag, Elbe Island, Italia, September 19-22, 1993, pp. 1-13.
[Mai 95] Mainguenaud, M., The modelling of the geographic information system network component. International Journal of Geographical Information Systems, v. 9 n. 6, Taylor and Francis, 1995, pp. 575-593.
[Mar 91] Mark, D. M. and Frank, A. U., Cognitive and Linguistic Aspects of Geographic Space, Kluwer Academic Publishers, NATO ASI series, 1991.
[Pau 89] Pau, L. F., Survey, Mapping and Spatial Modelling for Navigation, Pau, L. F. ed., Springer-Verlag, Fano, Denmark, August 21-25, 1989, pp. 1-9.
[Peu 88] Peuquet, D. J., Representations of geographic space: Toward a conceptual synthesis. Annals of the Association of American Geographers, 78 (3), 1988, pp. 375-394.

[Ren 95] Rennison, E. and Strausfeld, L., The Millennium project. Spatial Information Theory: A Theoretical Basis for GIS, Frank, A. U. and Kuhn, W. eds., Springer-Verlag, Semmering, Austria, September 21-23, 1995, pp. 69-91.

[Sto 93] Stonebraker, M., Chen, J., Nathan, N., Paxson, C. and Wu, J., Tioga: Providing data management support for scientific visualization applications. Proceedings of the Very Large Data Base Conference, Dublin, Ireland, August 24-27, 1993.

[Tve 93] Tversky, B., Cognitive maps, cognitive collages, and spatial mental models. Theories and Methods of Spatio-temporal Reasoning in Geographic Space, Frank, A. U., Campari, I. and Formentini, U. eds., Springer-Verlag, Elbe Island, Italia, September 19-22, 1993, pp. 14-24.

[Tho 93] Thorndyke, P. W. and Hayes-Roth, B., Differences in Spatial Knowledge Acquired from Maps and Navigation, Office of Naval Research, Santa Monica, USA, 1993.

[Tim 92] Timpf, S., Volta, G. S., Pollock, D. W. and Egenhofer, M. J., A conceptual model of wayfinding using multiple levels of abstraction. Theories and Methods of Spatio-temporal Reasoning in Geographic Space, Frank, A. U., Campari, I. and Formentini, U. eds, Springer-Verlag, Pisa, Italia, September 1992, pp. 348-362.

[Woo 95] Woodruff, A. et al., Zooming and tunneling in Tioga: Supporting navigation in multidimensional space. Proceedings of Visual Database Systems Conference, Spaccapietra, S. and Jain, R. eds., Chapman et Hall, Lausanne, Switzerland, March 27-29, 1995, pp. 323-332.

Contact address
Christophe Claramunt[1] and Michel Mainguenaud[2]
[1]Swiss Federal Institute of Technology
Rural Engineering Department
Spatial Information Systems
CH1015 Lausanne
Switzerland
Phone: + 41 21 693 57 83
Fax: + 41 21 693 57 90
E-mail: Christophe.Claramunt@dgr.epfl.ch
Url: http://dgrwww.epfl.ch/SIRS/index.html
[2]France Telecom
Institut National des Télécommunications
9 rue Charles Fourrier
F91011 Evry
France
Phone: + 33 1 60 76 47 82
Fax: + 33 1 60 76 47 80
E-mail: Michel.Mainguenaud@int-evry.fr
Url: http://www-inf.int-evry.fr/BasesDeDonnees/Cigales/cigales.html

MULTI-MODAL SPATIAL QUERYING

Max J. Egenhofer

National Center for Geographic Information and Analysis
Department of Spatial Information Science and Engineering and
Department of Computer Science
University of Maine
Orono, USA

Abstract
People who use multiple channels at the same time, communicate more successfully about spatial problems than those who rely exclusively on either voice or pictures. To achieve a similarly successful interaction between a person and a geographic information system (GIS), we use two concurrent communication channels—graphics and speech—to construct a multi-modal spatial query language in which users interact with a geographic database by drawing sketches of the desired configuration, while simultaneously talking about the spatial objects and the spatial relations drawn. Through the combined use of graphics and sketch, more intuitive and more precise specifications of spatial queries are possible. The key to this interaction is the exploitation of complementary or redundant information present in both graphical and verbal descriptions of the same spatial scenes. A multiple-resolution model of spatial relations is used to capture the essential aspects of a sketch and its corresponding verbal description. The model stresses topological properties, such as containment and neighborhood, and considers metrical properties, such as distances and directions, as refinements where necessary. This model enables the retrieval of similar, not only exact, matches between a spatial query and a geographic database. Such new methods of multi-modal spatial querying and spatial similarity retrieval will empower experts as well as novice users to perform spatial searches more easily, ultimately providing new user communities access to spatial databases.

1. Introduction

Today's methods of interacting with geographic databases are largely non-spatial, as they require their users to deal with geographic data primarily through alphanumeric command languages. Currently, spatial querying is done by typing a command in some spatial query language, such as an extended version of SQL (Ingram and Phillips 1987; Herring *et al.* 1988; Egenhofer 1994), or by selecting the same or a similar syntax through a forms interface or from pull-down menus (Egenhofer 1990; Calcinelli and Mainguenaud 1994; Aufaure-Portier 1995). Such spatial querying is a tedious process, because it often requires extensive training in the use of the particular query language. A more serious disadvantage of such textual spatial querying is that it forces users to translate a spatial image they may have in their minds about the situation they are interested in, into a non-spatial language. Graphical user interfaces provide only little improvement for such query languages, because they use the same type of syntax and grammar as the typed languages, and they only release users from remembering the particular syntax (Egenhofer 1992). The problems with communicating a user's request to a spatial database through conventional spatial query languages become most apparent when several users have to work together and have to understand their intentions. Verbal descriptions of spatial situations are frequently ambiguous and may easily lead to misinterpretations, particularly in multi-language working groups. Traditional spatial query languages have serious limitations when geographic concepts are used that are vague, imprecise, little understood, or not standardized. As an example, take the notion of the spatial predicate "cross" whose semantics may vary depending on the context in which it is used, the meaning of the objects the predicate relates to, and the topology and the metric of the particular configuration. These drawbacks make current spatial query languages error-prone and difficult to use.

Sketch-and-Talk is an innovative spatial query language that uses simultaneously graphic and voice input. It is made possible by the advent of pen-based computers and is a response to the increased interest in Mobile Computing (Imielinski and Korth 1995). Such a multi-modal spatial query language will allow users to interact more intuitively with spatial data than traditional GIS query languages and GIS user interfaces do, because it supports more directly human spatial thinking and familiar human interaction techniques through the combination of graphical ("sketch") and voice ("talk") input.

Sketch-and-Talk aims at retrieving from a spatial database those configurations that match a set of constraints specified. In this process, spatial relations play a significant role (Frank 1982; Pizano *et al.* 1989; Egenhofer 1992; Papadias and Sellis 1994) as they often specify the principal constraints about the data to be retrieved. Sketch-and-Talk addresses the particular need for modeling semantics of visual information, which participants at a recent NSF-ARPA workshop identified as essential for the success of Visual Information Management Systems.

"... currently a major bottleneck are, however, the techniques to introduce and manage the semantics in these [visual information] systems. ... The queries require assignment of semantics to data." (Jain and Pentland 1995)

Sketch-and-Talk allows users to choose their favorite interaction mode to compose a spatial query and the query processor integrates the two representations into a canonical form. The redundancy in the voice and graphics mode is critical information for the information system's query processor as it contributes to solving *spatial incompleteness and spatial ambiguities* that may exist in one mode through specifications in the other mode; detecting *spatial contradictions* among the different modes, therefore, saving precious time when trying to process a query that the user did not specify precisely enough; and exploiting *spatial consensus* among the different modes to determine more reliable input.

Sketch-and-Talk is motivated by findings in Human-Computer Interaction where a combined speech and gesture interaction mode for the manipulation of graphic images was preferred over a pure gesture or a pure speech interaction (Hauptmann 1989), as well as results from cognitive science and linguistics, where researchers argued that natural language descriptions of spatial configurations absorb detail, while graphics are generally overspecified (Talmy 1983). For example, a natural-language description like, "the road that enters the park" addresses the salient properties of the configuration—the principal topological concept of getting into without specifying such details as where the road starts or how often it crosses the park's boundary—while it abstracts away other geometric characteristics such as the shape of the objects, their sizes, and the orientation among them. This leaves ample freedom for topological and metric variations and these geometric details remain unspecified when processing such a description as a query against a spatial database. As an alternative to natural-language descriptions, sketches may constrain more precisely a particular spatial situation, but they introduce additional spatial properties that the user did not necessarily intend to specify, such as the objects' shapes, their relative sizes, and directions. When parsing a sketch it becomes impossible to distinguish explicit from implicit spatial constraints. Through the combined use of sketch and voice queries, Sketch-and-Talk balances between underspecifications and overspecifications by giving different priorities to the details of a drawing and a corresponding natural-language description.

This paper discusses the design of Sketch-and-Talk. After a review of related work (Section 2), we give in Section 3 a guided tour through a simple interaction with Sketch-and-Talk. Section 4 presents the model used for integrating verbal and sketched spatial relations, and Section 5 introduces how Sketch-and-Talk queries get processed. Conclusions are presented in Section 6.

2. Related work

Sketch-and-Talk is complementary to other spatial search mechanisms (Kuhn 1992) such as spatial data mining to discover interesting spatial configurations (Koperski and Han 1995), and browsing to explore a spatial data set for selection (Clementini *et al.* 1990). Sketch-and-Talk supports spatial similarity retrieval (Chang and Lee 1991) and complements methods of content-based image retrieval (Hirata and Kato 1992; Faloutsos *et al.* 1994; Flickner *et al.* 1995; Ogle and Stonebraker 1995) and specifying spatial consistency constraints graphically (Pizano *et al.* 1989). Although Sketch-and-Talk employs pen-based interaction with spatial data, it differs from sketch-based CAD languages that employ sketching to construct new configurations (Borning 1986) or update spatial information systems (White 1988; Kuhn 1990), including 3-D drawing systems (Deering 1995). Geometric construction and updating specify a particular configuration and require precise geometric positioning, whereas sketching a spatial query describes a prototypical configuration from which valid query results may deviate by relaxing spatial constraints ("best fit").

Sketching for querying was used in Query by Visual Example (Hirata and Kato 1992) and Query by Image Content (Faloutsos *et al.* 1994), which are targeted for content-based image retrieval. While the interaction mode of these query languages is similar to some of the basics of Sketch-and-Talk—in both cases users draw an approximate spatial configuration of what to retrieve—scope, modalities, and sketch interpretation are considerably different. Sketches for content-based image retrieval assume that the user draws something that matches quite closely the target and that all relations are intended as drawn. Their query processors accommodate primarily metric variations (Del Bimbo *et al.* 1994) and they are very sensitive to variations in sizes, orientations, and shapes. On the other hand, Sketch-and-Talk assumes that the user's sketch and the targets may vary, at times considerably, as long as they match in the most important criteria.

The semantics of spatial relations, essential for the interpretation and processing of a Sketch-and-Talk query, has received attention primarily in the arena of linguistics and artificial intelligence. Clark (1973) suggested a strong correspondence between Perceptual Space, which humans use to perceive the space around them, and Linguistic Space, which is used by language to represent the perceived space. This correspondence has been widely used in subsequent research for eliciting natural language descriptors of scenes. Talmy's (1983) seminal paper on "How Language Structures Space" establishes the link between prototypical spatial configurations and the use of natural language predicates. According to Talmy, at the fine-structural level of conceptual organization, language shows greater affinity with topology than with metric spaces. Work on metaphors (Lakoff and Johnson 1980) and people's conceptualizations of spatial relations (Grimaud 1988; Japkowicz and Wiebe 1991) are fundamental to the study of human cognition of spatial relations. We consider them to be complimentary to our approach and do not explicitly use these ideas of conceptualization in our work, because our formalism is focused on semantics that can be captured from the geometric configuration of spatial relations. Positive results obtained from our earlier work with human subject testing (Mark and Egenhofer 1994b) justify our assumption for taking this approach. Our study also relates to Rosch's (1978) general theory of human categorization whereby prototypical cases are used and objects are specified in terms of their distances from these prototypes. We are, however, dealing with spatial relations and not objects. In this respect, we are building on Herskovits's (1986) work, which used Rosch's method of prototypical categorization and applied it to spatial relations.

3. A sample scenario

The following scenario provides a rough outline of the interaction a user may perform when sketching and simultaneously talking about a query. Sketch-and-Talk uses a touch-sensitive input device—ideally a touch screen with a pen, such as Apple's Newton, or alternatively a tablet with a pen—plus a microphone to record the user's voice (Figure 1).

The user employs a pen to sketch an example of what she wants to find in the database. In this particular case, the user is interested in land parcels with a wooded area on the Penobscot River. While the user starts describing her request verbally (*"I'm interested in a land parcel ..."*), she draws the boundary of a land parcel (Figure 2). Sketch-and-Talk parses that the object drawn is a land parcel and adds the label to the sketch.

Figure 1 Sketch-and-Talk with a pen-based user interface and a microphone.

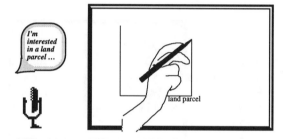

Figure 2 The user draws the outline of a land parcel, while describing that the object of interest is a land parcel.

The user continues the query by specifying that the land parcel should be *"... with some forest on it ..."* and simultaneously drawing part of the forest's boundary to indicate its location with respect to the land parcel (Figure 3). Again, the new object drawn is labeled with its type.

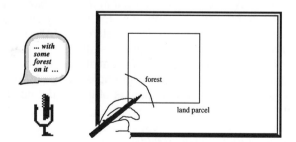

Figure 3 The user adds the boundary of a forest such that it intersects with the land parcel.

Since it is unclear on which side of the line the forest is located, the user gestures with the pen by filling the interior of the forest (Figure 4).

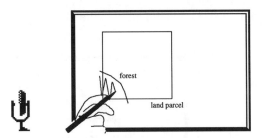

Figure 4 To determine the location of the forest (i.e., on which side of the boundary it lies), the user fills the scribbles into the forest's interior.

Finally the user draws a line that crosses the land parcel, while completing the constraint that the land parcel must be *"... located on the Penobscot River"* (Figure 5).

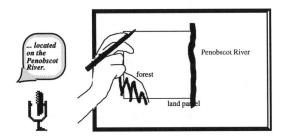

Figure 5 The user adds the location of Penobscot River such that runs along the land parcel, but does not intersect with the forest.

4. Modeling spatial relations for multi-modal querying

A critical component for the success of Sketch-and-Talk is the integration of what the user drew and what she described verbally. This integration must occur at a semantic level such that the meanings of the sketch and the verbal description can be compared. Our approach to this integration is based on the use of a canonical representation for Sketch-and-Talk queries, which is compatible with natural-language spatial relations as well as graphics. Sketch-and-Talk bases its analysis of spatial relations on the premise *Topology Matters, Metric Refines* (Egenhofer and Mark 1995b). The remainder of this section reviews the pieces of the model used as a symbolic representation of spatial relations.

4.1 9-Intersection for topological relations

We base the analysis of spatial relations on the 9-intersection, which is a comprehensive model for binary topological spatial relations. It applies to objects of type area, line, and point (Egenhofer and Herring 1990; Egenhofer and Franzosa 1991) and characterizes the topological relation between two point sets, A and B, by the set intersections of A's interior ($A°$), boundary (∂A), and exterior (A^-) with the interior, boundary and exterior of B (Equation 1). With each of these nine intersections being empty ($\varnothing$) or non-empty ($\neg\varnothing$), the model has 512 possible topological relations between two point sets, some of which cannot be realized. For two simple regions without holes embedded in R^2, the categorization shows eight distinct topological relations. They have been called *disjoint, meet, equal, overlap, inside, contains, covers,* and *coveredBy*. For two simple lines (non-branching, no self-intersections) embedded in R^2, 33 different topological relations can be realized with the 9-intersection, and for a line and a region, 19 different situations are found (Egenhofer and Herring 1990).

$$I(A, B) = \begin{pmatrix} A°\cap B° & A°\cap\partial B & A°\cap B^- \\ \partial A\cap B° & \partial A\cap\partial B & \partial A\cap B^- \\ A^-\cap B° & A^-\cap\partial B & A^-\cap B^- \end{pmatrix}$$ **(1)**

We use the 9-intersection relations as the key for analyzing spatial relations sketched, because it captures topological relations at a coarse level and, therefore, is an appropriate candidate for grouping sketches into classes of similar relations (Figure 6). By mapping the sketched relations onto 9-intersection relations, we capture the most salient features of a sketch in a form that is independent of orientations and sizes. This abstraction is critical to translate a sketched configuration into a database query.

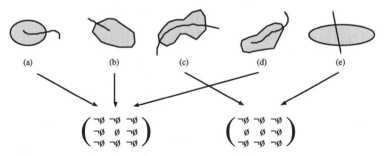

Figure 6 The 9-intersection as a categorization of the spatial relations sketched: (a), (b), and (d) map onto the same 9-intersection, while (c) and (e) map onto a different 9-intersection.

4.2 Component invariants for detailed topological relations

More detailed distinctions about topological relations are possible if further criteria are employed to evaluate the non-empty intersections. In order to establish topological relation equivalence between two regions (i.e., to decide whether or not two pairs of objects have the same topological relations), it is sufficient to describe such invariants for the components (or separations) of the boundary-boundary intersection, since the other intersections can be inferred from them (Egenhofer and Franzosa 1995). The necessary invariants to consider are:

- The *sequence* of components counted along the boundaries (Figure 7a).
- The *dimension* of each component (Figure 7b).
- The *type* of boundary-boundary component intersection—*touching* if the boundary enters and leaves the intersection from the same part, or *crossing* if the boundary enters from a different part than it leaves (Figure 7c).
- The *complement relationship*, i.e., whether a component is a next to a bounded or unbounded exterior (Figure 7d).

Detailed topological relations between two regions are expressed by the *component invariant table* for non-empty boundary-boundary sequences, which lists the sequence of boundary-boundary components and each component's dimension, type, and complement relationship (Egenhofer *et al.* 1994; Egenhofer and Franzosa 1995).

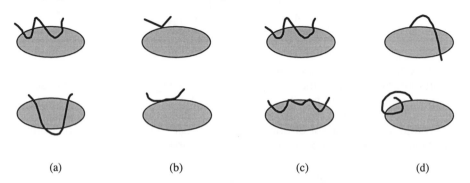

(a)   (b)   (c)   (d)

Figure 7 Pairs of topological relations that distinguish by component invariants: (a) different boundary sequences, (b) different component dimensions, (c) different component types, and (d) different complement relationships.

We use the component invariants as the key for analyzing the intentional complexity of the spatial relations sketched. The component invariants capture complexity of spatial relations. If a user draws a sketch with a high level of complexity, then we assume that this complexity was intended and that it provides the lower bound of what should be retrieved; therefore, a configuration in a spatial database with the same 9-intersection relation, but lower-rated component invariants, would not qualify as a match. On the other hand, a sketch of a low-complexity spatial relation may indicate that more complex configurations under the same 9-intersection category should be considered as well.

4.3 Conceptual neighborhoods for similarity of topological relations

Similarity among topological relations is described in terms of the *conceptual neighborhood graph*, which links those relations that are most similar to each other (Figure 8). It is based on the computational model of determining for each relation those relations with the least number of differences in the 9-intersection matrices (Egenhofer and Al-Taha 1992; Egenhofer and Mark 1995a).

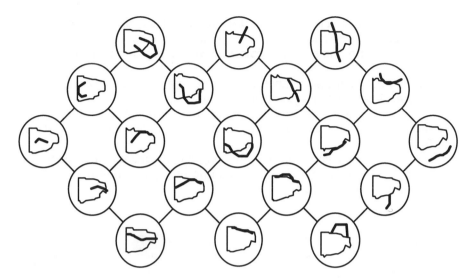

Figure 8 Conceptual neighborhood graph of the 19 line-region relations.

We use the conceptual neighborhoods of 9-intersection relations as the key to the semantics of natural-language spatial relations. In our previous work we found that topology is the primary part of the definition of natural-language spatial relations and that the 9-intersection is an appropriate model for a wide range of natural-language spatial relations (Mark and Egenhofer 1994a), applicable to different natural languages (Mark and Egenhofer 1995). For this investigation we have available a corpus of 1500 drawings for sixty different English-language spatial relations. An analyses of selected sketches found topological agreements in the way subjects referred to spatial relations. Figure 9 shows four examples of natural-language spatial relations and their mappings onto the conceptual neighborhoods of corresponding line-region relations (i.e., the set of highlighted relations).

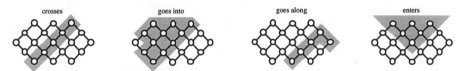

Figure 9 Conceptual neighborhoods of the topological relations for the natural-language terms (a) crosses, (b) goes into, (c) goes along, and (d) enters.

4.4 Quantitative refinements for metric details

Occasionally, topology *per se* is insufficient to characterize the essence of spatial relations. For instance, in order to capture the semantics of the spatial relation between Interstate I-95 and New Hampshire requires the consideration of some metric properties in addition to topological concern—I-95 divides New Hampshire into a very small area to the East of I-95 and a larger piece to the West. We apply such measures about areas, lengths, and directions as refinements of the topological properties. For line-region relations we use the following set of metric properties:

- the ratio by which a line's interior divides a region's interior (Figure 10a);
- how much of a line's interior is inside a region's interior (Figure 10b);
- how much of a line's interior coincides with a region's boundary (Figure 10c);
- the ratio between the distance from a line to and region's boundary, and the line's total length (Figure 10d);
- the ratio between the area made up by an equidistant enlargement (reduction) of the region (also known as a *buffer zone*) and the actual area (Figure 10e); and
- the cardinal direction between the objects, measured in a qualitative orientation scheme (Frank 1992) such as orthogonal half planes with a neutral zone (Figure 10f).

Figure 10 Quantitative refinements of topological relations due to (a) the area ratio, (b) the line-interior ratio, (c) the line-boundary ratio, (d) the line's closeness, (e) the region's closeness, and (f) the orientation.

We use the quantitative refinements of the 9-intersection relations as the key to formalizing detailed geometric constraints about natural-language spatial relations and sketched spatial relations. Some natural-language spatial relations may depend heavily on metric properties. For example, *Northeast* describes a relation where the road is outside of the park, but it is equally important that one object has a particular orientation with respect to the other (Figure 10e). More complex configurations include such natural-language descriptions as *along*, which qualifies for configurations of varying topology in combination with certain metric constraints about the line-boundary ratio (Figure 10 c), the line's closeness (Figure 10d), and the region's closeness (Figure 10e). Qualitative refinements are also critical to interpret sketched configurations that deviate considerably from the prototypical configuration. A particularly short line-closeness measure (Figure 10d), for instance, may indicate that the user would accept as an answer a configuration with a slightly different topology. The qualitative metric properties would indicate which of the configurations targeted through the conceptual neighborhoods would qualify.

5. Processing stages of a sketch-and-talk query

5.1 Recording

The user draws a query and simultaneously talks about what she is drawing. Sketch-and-Talk records graphics, voice, and their temporal concurrencies by time-stamping them at regular intervals (Figure 11). We analyze the users' multi-modal interactions and determine which spatial concepts they prefer to draw and which ones they rather describe verbally.

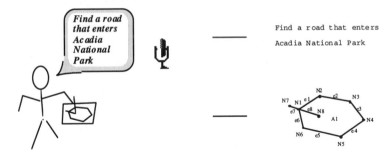

Figure 11 Recording the spoken and sketched queries and translating them into symbolic representations of unstructured text and a topological vector model, respectively.

5.2 Parsing

Sketch-and-Talk parses the graphical and verbal sentences, and translates them into a compatible format (Figure 12). We build a semantic network for the spoken spatial query and translate its natural-language spatial relations into their corresponding conceptual neighborhoods of 9-intersection relations, building on a library of mappings from English-language terms onto 9-intersections and quantitative refinements. For the sketched query, we extract the 9-intersections, components invariants, and quantitative refinements from the topological data model.

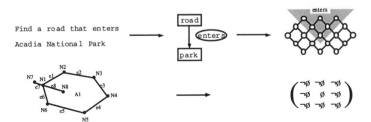

Figure 12 Translating text and graphical query into a compatible representation.

5.3 Integration

Sketch-and-Talk integrates the two representations by cross-referencing in the two sentences the corresponding relations and adding, as available from the verbal description, the object types and their values to the objects sketched. Where necessary, Sketch-and-Talk consolidates the integrated spatial relations resolving incompleteness, ambiguities, and contradictions (Figure 13). We exploit the unifying format of the 9-intersection (and where necessary component invariants and quantitative refinements) to link the query parts expressed in different modes. If the same relations are expressed both verbally and graphically, we use this information to consolidate the query. This includes the completion of otherwise incomplete query parts, and

.the resolution of ambiguities in one representation by using information derived from the other interaction mode. Finally, we exploit consensus by putting higher preferences on those aspects of a spatial query that were well specified both verbally and graphically.

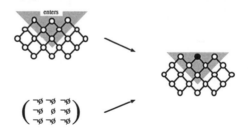

Figure 13 Integrating the spatial relation described verbally and sketched.

5.4 Query processing

Sketch-and-Talk develops a query processing plan, translates it into the database's query language (e.g., SQL-3 MM), and retrieves the scenes that match the query. Retrieval and presentation of the results should be in a prioritized order such that the scenes most similar to the sketch are presented first to the user. We use the integrated representation of spoken and sketched query and iteratively relax the spatial constraints to generate a set of database queries of decreasing specificity. We start with executing the most specific queries first and build for each configuration found a similarity measure to assess how closely it matches the query asked. If the set of answers is small, further relaxations may be necessary. In a post-processing step, the answers are sorted according to the similarity measures so that the best matches are presented first to the user.

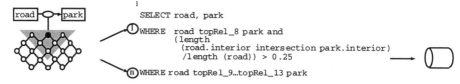

Figure 14 Developing a prioritized query processing plan from the integrated representation and processing these queries against a spatial database.

6. Conclusions

The combination of graphical and voice input enables a new style of spatial querying. The two interaction modes are complementary, and when used together they provide more power than either of them alone. The conceptual design of Sketch-and-Talk will be tested through a prototype implementation, and we will evaluate Sketch-and-Talk with user tests.

Acknowledgments
This work was partially supported by Rome Laboratories under grant number F30602-95-1-0042. Max Egenhofer's research is further supported by grants from the National Science Foundation under grant number No. SES 88-10917 for the NCGIA and grant number IRI-9309230; the Scientific and Environmental Division of the North Atlantic Treaty Organization; Intergraph Corporation; Environmental Systems Research Institute Inc.; Space Imaging, Inc., and by a Massive Digital Data Systems contract sponsored by the Advanced Research and Development Committee of the Community Management Staff and administered by the Office of Research and Development.

References
M.-A. Aufaure-Portier (1995) Definition of a Visual Language for GIS. in: T. Nyerges, D. Mark, R. Laurini, and M. Egenhofer (Ed.), Cognitive Aspects of Human-Computer Interaction for Geographic Information Systems. pp. 163-178, Kluwer, Dordrecht.

A. Borning (1986) Defining Constraints Graphically. in: Human Factors in Computing Systems, CHI `86, Boston, MA, pp. 137-143.

D. Calcinelli and M. Mainguenaud (1994) Cigales, a Visual Query Language for a Geographical Information System: the User Interface. Journal of Visual Languages and Computing 5: 113-132.

C. Chang and S. Lee (1991) Retrieval of Similar Pictures on Pictorial Databases. Pattern Recognition 24(7): 675-680.

H. Clark (1973) Space, Time, Semantics, and the Child. in: T. Moore (Ed.), Cognitive Development and the Acquisition of Language. pp. 27-63, Academic Press, New York, NY.

E. Clementini, A. D'Atri, and P. D. Felice (1990) Browsing in Geographic Databases: An Object-Oriented Approach. in: Workshop on Visual Languages, Skokie, IL, pp. 125-131.

M. Deering (1995) HoloSketch: A Virtual Reality Sketching/Animation Tool. ACM Transactions on Computer-Human Interaction 2(3): 220-238.

A. Del Bimbo, P. Pala, and S. Santini (1994) Visual Image Retrieval by Elastic Deformation of Object Sketches. in: IEEE Simposium on Visual Languages, St. Louis, MO, pp. 216-223.

M. Egenhofer (1990) Interaction with Geographic Information Systems via Spatial Queries. Journal of Visual Languages and Computing 1(4): 389-413.

M. Egenhofer (1992) Why not SQL! International Journal of Geographical Information Systems 6(2): 71-85.

M. Egenhofer (1994) Spatial SQL: A Query and Presentation Language. IEEE Transactions on Knowledge and Data Engineering 6(1): 86-95.

M. Egenhofer and K. Al-Taha (1992) Reasoning About Gradual Changes of Topological Relationships. in: A. Frank, I. Campari, and U. Formentini (Ed.), Theories and Methods of Spatio-Temporal Reasoning in Geographic Space. Lecture Notes in Computer Science 639, pp. 196-219, Springer-Verlag, Pisa.

M. Egenhofer, E. Clementini, and P. di Felice (1994) Evaluating Inconsistencies Among Multiple Representations. in: T. Waugh and R. Healey (Ed.), Sixth International Symposium on Spatial Data Handling, Edinburgh, Scotland, pp. 901-920.

M. Egenhofer and R. Franzosa (1991) Point-Set Topological Spatial Relations. International Journal of Geographical Information Systems 5(2): 161-174.

M. Egenhofer and R. Franzosa (1995) On the Equivalence of Topological Relations. International Journal of Geographical Information Systems 9(2): 133-152.

M. Egenhofer and J. Herring (1990) A Mathematical Framework for the Definition of Topological Relationships. in: K. Brassel and H. Kishimoto (Ed.), Fourth International Symposium on Spatial Data Handling, Zurich, Switzerland, pp. 803-813.

M. Egenhofer and D. Mark (1995a) Modeling Conceptual Neighbourhoods of Topological Line-Region Relations. International Journal of Geographical Information Systems 9(5): 555-565.

M. Egenhofer and D. Mark (1995b) Naive Geography. in: A. Frank and W. Kuhn (Ed.), Spatial Information Theory—A Theoretical Basis for GIS, International Conference COSIT '95, Semmering, Austria. Lecture Notes in Computer Science 988, pp. 1-15, Springer-Verlag, Berlin.

C. Faloutsos, R. Barber, M. Flickner, J. Hafner, W. Niblack, D. Petrovic, and W. Equitz (1994) Efficient and Effective Querying by Image Content. Journal of Intelligent Information Systems 3: 231-262.

M. Flickner, H. Sawhney, W. Niblack, J. Ashley, Q. Huang, B. Dom, M. Gorkani, J. Hafner, D. Lee, D. Petkovic, D. Steele, and P. Yanker (1995) Query by Image and Video Content: The QBIC System. IEEE Computer 28(9): 23-32.

A. Frank (1982) MAPQUERY—Database Query Language for Retrieval of Geometric Data and its Graphical Representation. ACM Computer Graphics 16(3): 199-207.

A. Frank (1992) Qualitative Spatial Reasoning about Distances and Directions in Geographic Space. Journal of Visual Languages and Computing 3(4): 343-371.

M. Grimaud (1988) Toponyms, Prepositions, and Cognitive Maps in English and French. Journal of the American Society of Geolinguistics 14: 54-76.

A. Hauptmann (1989) Speech and Gestures for Graphic Image Manipulation. in: K. Bice and C. Lewis (Ed.), CHI '89, Austin, TX, pp. 241-245.

J. Herring, R. Larsen, and J. Shivakumar (1988) Extensions to the SQL Language to Support Spatial Analysis in a Topological Data Base. in: GIS/LIS '88, San Antonio, TX, pp. 741-750.

A. Herskovits (1986) Language and Spatial Cognition—An Interdisciplinary Study of the Prepositions in English. Cambridge University Press, Cambridge, MA.

K. Hirata and T. Kato (1992) Query by Visual Example—Content-Based Image Retrieval. in: A. Pirotte, C. Delobel, and G. Gottlob (Ed.), Advances in Database Technology—EDBT '92, 3rd International Conference on Extending Database Technology. Lecture Notes in Computer Science 580, pp. 56-71, Springer-Verlag, Vienna, Austria.

T. Imielinski and H. Korth (1995) MOBIDATA Workshop Report. MOBIDATA: An Interactive Journal of Mobile Computing 1(2):
http://www.cs.rutgers.edu/~badri/journal/contents12.html.

K. Ingram and W. Phillips (1987) Geographic Information Processing Using a SQL-Based Query Language. in: N. R. Chrisman (Ed.), AUTO-CARTO 8, Eighth International Symposium on Computer-Assisted Cartography, Baltimore, MD, pp. 326-335.

R. Jain and A. Pentland (1995) Workshop Report: NSF-ARPA Workshop on Visual Information Management Systems. Technical Report available from http://www.virage.com/vir-res/reports/.

N. Japkowicz and J. Wiebe (1991) A System for Translating Locative Prepositions from English into French. in: 29th Annual Meeting of the Association for Computational Linguistics, UC Berkeley, pp. 153-160.

K. Koperski and J. Han (1995) Discovery of Spatial Association Rules in Geographic Information Databases. in: M. Egenhofer and J. Herring (Ed.), Advances in Spatial Databases—4th International Symposium, SSD '95, Portland, ME. Lecture Notes in Computer Science 951, pp. 47-66, Springer-Verlag, Berlin.

W. Kuhn (1990) From Constructing Towards Editing Geometry. in: ACSM-ASPRS Annual Convention, Denver, CO, pp. 153-164.

W. Kuhn (1992) Paradigms of GIS Use. in: D. Cowen (Ed.), Fifth International Symposium on Spatial Data Handling, Charleston, SC, pp. 91-103.

G. Lakoff and M. Johnson (1980) Metaphors We Live By. University of Chicago Press, Chicago, IL.

D. Mark and M. Egenhofer (1994a) Calibrating the Meanings of Spatial Predicates from Natural Language: Line-Region Relations. in: T. Waugh and R. Healey (Ed.), Sixth International Symposium on Spatial Data Handling, Edinburgh, Scotland, pp. 538-553.

D. Mark and M. Egenhofer (1994b) Modeling Spatial Relations Between Lines and Regions: Combining Formal Mathematical Models and Human Subjects Testing. Cartography and Geographic Information Systems 21(3): 195-212.

D. Mark and M. Egenhofer (1995) Topology of Prototypical Spatial Relations Between Lines and Regions in English and Spanish. in: D. Peuquet (Ed.), Autocarto 12, Charlotte, NC, pp. 245-254.

V. Ogle and M. Stonebraker (1995) Chabot: Retrieval from a Relational Database of Images. IEEE Computer 28(9): 40-48.

D. Papadias and T. Sellis (1994) Qualitative Representation of Spatial Knowledge in Two-Dimensional Space. VLDB Journal 3(4): 479-516.

A. Pizano, A. Klinger, and A. Cardenas (1989) Specification of Spatial Integrity Constraints in Pictorial Databases. Computer 22(12): 59-71.

E. Rosch (1978) Principles of Categorization. in: E. Rosch and B. Lloyd (Ed.), Cognition and Categorization. pp. Erlbaum, Hillsdale, NJ.

L. Talmy (1983) How Language Structures Space. in: H. Pick and L. Acredolo (Ed.), Spatial Orientation: Theory, Research, and Application. pp. 225-282, Plenum Press, New York, NY.

R. M. White (1988) Applying Direct Manipulation to Geometric Construction Systems. in: N. Magnenat and D. Thalmann (Ed.), New Trends in Computer Graphics: Computer Graphics International '88. pp. 446-455, Geneva, Switzerland.

Contact address

Max J. Egenhofer
National Center for Geographic Information and Analysis,
Department of Spatial Information Science and Engineering and
Department of Computer Science
University of Maine
Orono, ME 04469-5711
USA
E-mail: max@spatial.maine.edu

AN ONTOLOGY-BASED APPROACH TO SPATIAL

INFORMATION MODELLING

F. Hernández Rodríguez, G. Bravo Aranda and A. Martín Navarro

Departamento de Ingeniería del Diseño E.S.I.
Universidad de Sevilla
Sevilla, Spain

Abstract
This paper presents a scheme for modelling spatial information. This scheme is based on the use of a conceptual framework (metamodel) that describes the basic aspects and concepts underlying the field of spatial information. The modelling task for an application domain is turned into a classification task through the use of the proposed conceptual framework.

Key words
Spatial information modelling, modelling methodology, aspect modelling, reusable ontologies.

1. Introduction

Traditionally, digital description of spatial objects is done by considering their two fundamental components separately; that is, their spatial component (concerning shape and location of objects) and their non-spatial component (non-spatial attributes). As a result two different information models are developed.

Most systems engaged in spatial information management (mainly CAD systems and Geographical Information Systems) basically adopt the above mentioned representation scheme. To describe the geometry of spatial objects, computer aided drafting systems are used, which are characterized by the use of a set of basic low-level graphic primitives for this purpose. For non-spatial attributes of spatial objects Data Base Management Systems are used in which object attributes are described through the basic constructs that they provide, that is, tuples in relations. In both representation schemes, a significant difference exists between the interpretations of descriptions given by people and those given by computers. The meaning that people ascribe to those descriptions is derived from knowledge that is implicit and computers cannot use it. In short, it can be said that the schemes that are used to describe spatial objects in computers are based on very limited concepts from a semantic standpoint.

As work in numerous fields has shown, the representation schemes in use are not suitable for describing the complex nature of spatial information. Hence, new representation schemes have to be found. Any representation scheme for spatial information (to be of general validity) should not attempt to approach the problem of modelling as a whole, that is, in its entire complexity.

The present-day need for computer handling of more and more complex problems has caused new modelling methodologies to emerge. Most of those methodologies emphasize the importance of domain analysis. Particularly, in areas such as knowledge engineering, and more recently in the field of geographical information standards, there is a trend to identify and conceptualize the essential aspects that characterize the objects in different domains as an effective means of supporting the analysis task. The keynote of this approach is to build abstract models for the aspects identified, each of them including its own conceptualization or ontology, and to describe any particular object from the suitable concepts. If such a conceptual framework is developed in a domain-independent way, it becomes a reusable analysis tool for the task of modelling (spatial) information.

2. An outline of the approach proposed

Spatial information presents certain characteristics that make unsuitable the adoption of a modelling scheme that specifies rigidly how the features of interest have to be described. For one thing, the same feature can be modelled differently based on the scale; for the other, different users look at the same feature from different points of view, and the corresponding non-spatial attributes and/or the topological characterization may not coincide.

The observation of the way in which spatial features modelling is usually done can make clear the existence of a set of concepts (the majority of which are implicit) that define a general conceptual framework. If such a conceptual framework is made explicit, it becomes a fundamental support for the development of software applications for handling spatial information.

The aforementioned facts and ideas suggest that spatial information modelling can be done in accordance with the following general scheme:

- First, it is necessary to identify and differentiate the basic aspects incident to the modelling of any spatial feature, that is, the conceptual dimensions required to define them completely.
- Second, for each identified aspect, fundamental concepts have to be established. In other words, a set of concepts or ontology that allows the definition of any feature on the considered dimension is needed.
- Third, it is necessary to define a set of meta-classes from which any spatial feature can be described as a whole. These meta-classes are the results of the superposition of fundamental concepts from the different conceptual dimensions, and represent meaningful associations of those concepts.
- Finally, to model any spatial feature the users of the proposed framework only have to identify the most suitable meta-class for it.

3. Basic aspects of spatial information modelling

The analysis of the peculiarities of spatial information makes feasible to identify the basic aspects inherent in its modelling:

Geometric aspect.- To model this aspect it is necessary to specify a set of basic geometric objects from which to describe, from a geometrical point of view, any spatial object, so that its location and shape can be defined.

Thematic aspect.- Spatial objects also have a thematic component, that is, they have their own (non-spatial) attributes and relations with other objects. To model this aspect it is necessary to specify the attributes and relations of a spatial object that are relevant to the application.

Topological aspect.- Topology is the science of properties and relations of geometric objects that are independent of a particular coordinate system.

Due to the fact that euclidean hypotheses do not hold for geometric information when it is represented in digital form, the use of topologically structured geometric information prevails [1; 2; 3]. Thus, the geometry of a spatial object should be described in terms of basic geometric objects, and two different basic geometric objects cannot be placed at the same location. If geometric information is structured in this way, the operations required to derive spatial relations between objects are significantly simplified.

In accordance with the general principle of topologically structured geometric information, there is no need for the explicit modelling of the topology of spatial objects. However, the benefits from the explicit representation of certain topological concepts (e.g., in order to answer frequent queries) make the adoption of a mixed approach convenient (that is, one that explicitly represents certain topological concepts and also applies the principle of topologically structured geometric information), in spite of the fact that topological relations could be derived. A typical example is the modelling of spatially embeded networks (e.g., road networks, utilities, etc.).

Thus, to model this aspect it is necessary to specify a set of basic topological objects from which to describe, from a topological point of view, any spatial object.

Representational aspect.- Another aspect that has to be considered when modelling spatial objects is that of symbolic representation, as one of the purposes of using digital spatial information is its visualization on a display. To model this aspect it is necessary to analyse the kinds of representations used for describing spatial features on plans and maps, in order to specify a set of basic symbolic objects from which to describe, from a representational point of view, any spatial object.

In short, the modelling of any spatial object can be done by specifying its definition or classification with relation to *four conceptual dimensions* or *basic aspects: geometry, topology, symbology of representation* and *thematic contents.*

In a general sense, spatial objects can also be classified as simple or compound (a compound spatial object is made up of several spatial objects). Simple spatial objects can be classified in one of the following categories [4]:

- *0-dimensional or Point-like*: category reserved for those objects whose dimensions are not considered (i.e., small objects or objects whose dimensions are of no interest, but it is their location).
- *1-dimensional or Linear*: category reserved for those objects where one dimension prevails over the others.
- *2-dimensional or Areal*: category reserved for those objects that represent a surface extension or region.

It is important to point out that the same spatial feature may be modelled using object classes in different categories.

The four basic aspects of spatial objects are illustrated in figure 1 (the notation proposed in [5] has been used in this and the other figures). The modelling of each aspect is the subject of the following sections.

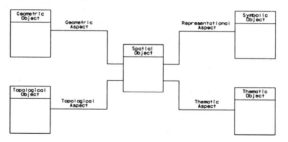

Figure 1 Basic aspects of spatial objects.

4. Geometric aspect

This aspect only deals with defining the location and shape of spatial objects in relation to an appropriate reference system. The different categories in which a spatial object can be classified, as well as the idea that geometric information must be topologically structured, allow the specification of a set of basic geometric objects from which it is possible to model the geometric aspect of any spatial object.

- *Point* (P): specifies a location on the earth's surface. It is used to identify the location of point objects and the locations where more than one feature is placed.
- *Connected chain* (CC): is defined as a sequence of straight chains and/or curves connecting two points. It is used to describe the geometry of linear objects (for which it is important to topologically structure geometric information, in order to analyze their spatial relations with other objects), and the boundary of areal objects.

A *straight chain* is a sequence of non-intersecting straight segments connecting a series of vertices. A *curve* is a locus of points that is defined by a mathematical function.

- *Non-connected chain* (NCC): is defined as a sequence of straight chains and/or curves. It is used to describe the geometry of linear objects for which it is not important to topologically structure geometric information, because their spatial relations are considered of no interest (e.g., contour lines can be modelled as non-connected chains).

From these three basic geometric objects (i.e., point, connected chain, and non-connected chain) it is possible to model the geometric aspect of any spatial object. Thus, the geometric aspect of a point object can be specified by the point identifier whose coordinates correspond to its location. The geometric aspect of a topologically structured linear object can be specified by the identifiers of the connected chains that describe the shape of the linear object. The geometric aspect of a non-topologically structured linear object can be specified by the identifier of the non-connected chain that describes the geometry of the linear object. Finally, the geometric aspect of an areal object can be specified by the identifiers of the connected chains that describe the boundary of the areal object.

5. Topological aspect

The two first previously defined basic geometric objects, point and connected chain, allow the geometric information of spatial objects to be topologically structured. As a result, the spatial relations between them are easily derived and there is no need of explicitly modelling the many different types of such relations that a single spatial object may have (e.g., an areal object such as a province may be adjacent to other provinces, may contain several townships, centres of population, may be crossed by a river, roads, etc.). This diversity of relations would complicate the model's conception if they had to be explicitly described. However, the scheme proposed allows them to be derived exclusively from the geometric modelling of each of the mentioned features [6].

5.1 Explicit topology

There are situations in which it is necessary, or of great use, to explicitly model certain topological relations, in order to exploit their meaning directly. This has been considered particularly the case of spatially embeded networks. With this aim, two basic topological objects are defined:

- *Node* (N): defines a topological connection or end of linear objects for which there is a need to explicitly establish that they are part of a network. A node is specified by an identifier. In addition, the specification should include the topological chains that the node is one end, so that the connectivity between these linear objects can be easily determined.

- *Topological chain* (CT): defines a link between two nodes. A topological chain is specified by an identifier and the identifiers of the connected nodes.

These basic topological objects are used to describe the networks that are part of the model as a single graph. Thus, a node or topological chain may belong to more than one network, but it is defined only once. This means that two different nodes should not exist at the same location, and two different topological chains should not connect the same nodes.

6. Representational aspect

The analysis of plans and maps that describe bi-dimensional spatial information shows that there are basically three types of features to which it is necessary to associate a symbol. As a consequence, to model the representational aspect of any spatial object the following generic symbolic objects are defined:

- *Symbol*: is defined as a figure used in the representation of a spatial object class (point, linear or areal). From a geometric point of view, a symbol is a group of elementary geometric objects (lines, circular arcs, b-splines, etc.), each one of which has certain parameters of representation associated to it (line-type, color and thickness).
- *Point symbolic object* (PS): the symbology associated to a class of point spatial objects is specified through the name of a symbol (it has to be a defined symbol), a scale factor and the symbol's orientation.
- *Linear symbolic object* (LS): the symbology associated to a class of linear spatial objects can be one of two types. If the representation is a line-up of instances of a symbol following the geometry of the linear object (*alignment*), it is necessary to specify the name of a symbol, a scale factor, the orientation of the symbol relative to the linear object, and the spacing between two adjacent symbols over the linear object. If the representation reproduces the lines that define the geometry of the linear object (*sequence*), it is necessary to specify the line-type, color, and thickness.
- *Areal symbolic object* (AS): the symbology associated to a class of areal spatial objects can be one of two types. If the representation is the filling of the areal object with a symbol matrix (*filling*), it is necessary to specify the name of a symbol, a scale factor, row and column orientation, symbol orientation, and the spacing between rows and columns. If the representation is a hatching of the areal object (*hatching*), it is necessary to specify, for each of the different hatch patterns, the line-type, color, thickness, line orientation and spacing between lines.
- *Textual symbolic object* (TS): the symbology associated to toponyms and annotations, since they are chains of characters, is specified through a font name, character height, and, optionally, color and spacing between characters.

7. Thematic aspect

The thematic contents of a spatial object specifies the (non-spatial) attributes associated to the object and its relations with other objects. Each attribute takes its value from a domain, and it is specified through its name and domain. In object oriented data models, it is usual to represent relations between objects as attributes whose domains are object classes; i.e., attributes whose value is a reference to another object. In addition, an attribute may have just one value or a collection of values. So, it can be said that each attribute of a class of objects can be classified in one of the following general categories: *simple attributes*, and *complex attributes*. Figure 2 illustrates the modelling of the four basic aspects of spatial objects.

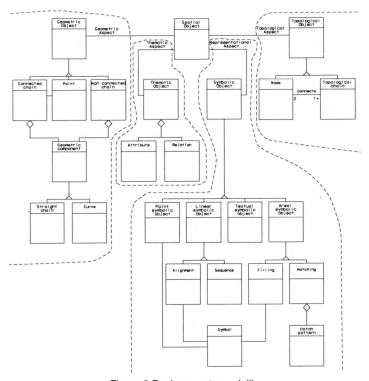

Figure 2 Basic aspects modelling.

8. Generic classes of spatial objects

The aim of this section is to establish a functional definition for each of the generic classes of spatial objects that constitute the modelling set. Making use of this conceptual framework, any user who needs to model spatial information in a domain (cartography, plant lay-out, etc.) would declare each relevant class of spatial objects as a specific case of one of the defined generic classes, and specify its thematic contents and symbology of representation when needed. In the following, those generic classes of spatial objects that may or may not have thematic contents are defined simultaneously.

- *Point entity/Thematic point entity*: generic class of point spatial objects that is used to represent features placed at a location, with which *it is not/it is* associated thematic contents, and there is no interest in making any topological relation explicit.
- *Connecting point/Thematic connecting point*: generic class of point spatial objects that is used to represent features placed at a location, with which *it is not/it is* associated thematic contents and for which topological connection with other spatial objects is made explicit.
- *Toponym*: generic class of point spatial objects that is used to specify the location of an annotation.
- *Line/Thematic line*: generic class of linear spatial objects that is used to represent features *without/with* thematic contents, and for which no topological relation is made explicit, although their geometric information is topologically structured.
- *Isoline/Thematic isoline*: generic class of linear spatial objects that is used to represent features *without/with* thematic contents, and for which no topological relation is made explicit, and their geometric information is not topologically structured.
- *Area boundary*: generic class of linear spatial objects that is used to represent the limits of a region.

The following table details, in terms of basic objects, the aspect modelling for the above defined generic classes. In the table, the identifiers of basic objects instead of their complete names have been used. Braces ({}) around a basic object identifier have to be interpreted as collections of those objects.

Given the definitions provided for the generic classes of point spatial objects (i.e., point entity, thematic point entity, connecting point and thematic connecting point), it is possible to model any point-like feature.

In contrast, the generic classes defined for linear spatial objects do not allow the modelling of any linear feature. It is therefore necessary to define three additional classes:

Generic Classes	Thematic Aspect	Topological Aspect	Geometric Aspect	Representational Aspect
Point entity	-	-	P	PS
Thematic point entity	Yes	-	P	PS
Connecting point	-	N	P	PS
Thematic connecting point	Yes	N	P	PS
Toponym	-	-	P	TS
Line	-	-	{CC}	LS
Thematic line	Yes	-	{CC}	LS
Isoline	-	-	NCC	LS
Thematic isoline	Yes	-	NCC	LS
Area boundary	-	-	{CC}	LS

- *Network line section/Thematic network line section*: generic class of linear spatial objects that is used to represent sections of network lines, and that *does not have/does have* associated thematic contents, and, in addition, is topologically connected with other spatial objects. The geometry of a network line section is specified through the identifiers of the lines/thematic lines that make it up, and its topology through the corresponding topological chain identifier. The description of a network line section class specifies the features modelled as lines or thematic lines that may be part of it.

- *Network line*: generic class of linear spatial objects that is used to model spatial objects that constitute a network and have their own identity (e.g., a river in the river network), and also have associated thematic contents. The geometry and topology of a network line are specified implicitly through the identifiers of the network line sections that make it up. The description of a network line class specifies the features modelled as (thematic) network line sections that may be part of it.
- *Simple network*: generic class of linear spatial objects that is made up of a set of network lines (e.g., a river network, a road network, etc.) and has associated thematic contents. The description of a simple network class specifies the features modelled as network lines that may be part of it.

The generic classes defined thus far only allow the modelling of region boundaries. It is necessary to define additional generic classes for modelling explicitly areal objects:

- *Simple region/Thematic simple region*: generic class of areal spatial objects that is used to model features whose extension can be defined using a single closed sequence of contiguous connected chains, and that *has not/has* associated thematic contents. The extension of a simple region is specified through the identifier of the corresponding area boundary. The description of a simple region class specifies the feature modelled as area boundary that defines its extension.
- *Complex region/Thematic complex region*: generic class of areal spatial objects that is used to model features whose extension cannot be defined using a single, but several closed sequences of contiguous connected chains, and that *has not/has* associated thematic contents. The extension of a complex region is specified through the identifiers of the corresponding area boundaries. The description of a complex region specifies the feature modelled as area boundary that defines its extension.

Finally, it is common to group spatial objects to constitute complex spatial objects. With the aim of modelling complex spatial objects an additional generic class is defined:

- *Composition*: generic class of spatial objects that is used to model features that are made up of a collection, possibly heterogeneous, of simpler features, and that has associated thematic contents. The description of a composition class specifies the features that may be part of it.

Figure 3 summarizes the generic classification of spatial objects. All the classes of spatial objects specified have to be considered generic classes, so that any specific class of spatial objects to be modelled can be described by a suitable one. To illustrate this, province limits in an application could be declared as a specific case of the generic class area boundary.

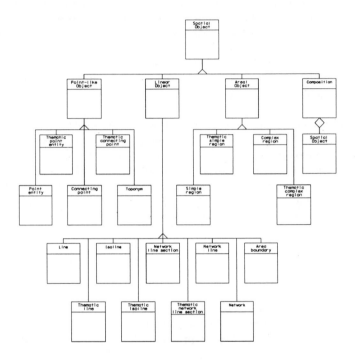

Figure 3 Generic classification of spatial objects.

9. Differences between the proposed scheme and previous ones

The scheme put forward involves a significant change in the way spatial objects have been usually modelled. In this scheme, it is fundamental the identification and definition of a set of high level concepts, that are domain and tool independent. Those concepts have a precise and uniform meaning for both man and computer; that is, both understand the same concepts in the same way, and so a dialogue between man and computer can be naturally had.

In adition, spatial objects are considered as a whole, so that it is not necessary to build several partial models as it is the case when conventional tools are used (e.g., the use of graphic primitives and relational data stuctures to describe, respectively, spatial and non-spatial components of spatial objects).

In short, spatial data modelling becomes a two-stage classification task:

- *Conceptual modelling:* in this stage, domain analysis is done; i.e., spatial object classes that are relevant to the application domain are identified.

For each identified class, basic aspects of interest are defined.
- *Logical modelling*: once the conceptual model is defined, it is necessary to build the logical model. For this purpose, a formal specification language that can be interpreted by computers has to be used (a definition of a formal specification language for the conceptual framework presented can be found in [6]). In this stage two phases are distinguished:
    - The mapping between real-world object classes and generic classes of spatial objects in the metamodel.
    - The specification of a data model using a formal language for this purpose.

Conceptual modelling is an usual activity in the analysis of any application. However, the identification of basic aspects inherent to spatial information and the definition of fundamental concepts for those aspects support and facilitate the modelling task. Logical modelling naturally follows conceptual modelling. In this stage, one just has to classify spatial object classes in the application domain, and does not need to pay attention to low-level data structures, as it is usually the case. An application sample of the scheme proposed to a specific domain, as it is that of water supply networks, can be found in [7].

10. Practical experience

In this section, the experience in the use of the conceptual framework is described. As a research work in course, the practical experience is still limited. At present, only the integration of spatial data models built using the conceptual framework into conventional environments can be reported (the integration in several of the more common environments has been analysed in [6]). In particular, a software tool for the integration into RDBMS has been developed. Several conclusions can be drawn from this development:

- An integrating tool releases the user from specific aspects of the particular environment selected.
- RDBMS's are unsuitable to support spatial data models. Due to the low-level relational constructs, a complex database scheme is necessary to maintain spatial data models.
- To exploit the semantics of the high-level concepts in the conceptual framework it is necessary to implement powerful operations. This implementation is cumbersome when relational operations are the only basis.

The above reasons have motivated a change of orientation towards a different development approach. At present, the development of an object oriented software tool that takes into account the semantics of the conceptual framework has just started. The objective is to build an environment in which the concepts in the conceptual framework and their associated procedural semantics are included.

11. Summary and conclusions

The aim of this section is to summarize the most important proposals and conclusions of this research. A system for spatial information modelling has been presented whose most relevant characteristics are the following:

- The four fundamental aspects incident to the description of spatial features have been identified and modelled: *geometric aspect, thematic aspect, topological aspect,* and *representational aspect.*
- A scheme for structuring geometric information has been adopted that makes it unnecessary to make spatial relations between objects explicit, as these are easily derivable.
- The possibility of explicitly representing the topology of spatially embedded networks has been considered, beacause of its usefulness.
- A generic classification of spatial objects (ontology) has been established which constitutes the *metamodel of spatial information* proposed. Making use of this metamodel, any user who needs to model spatial information in a domain would declare each relevant class of spatial objects as a specific case of one of the defined generic classes.
- Geometric and topological aspects of each generic class of spatial objects have been completely specified. Thus, the users of the framework proposed only have to model the thematic aspect and define the symbolic aspect that characterize their particular view in the application domain.

Such a modelling framework provides great advantages, among which the following two should be highlighted:

Ease of modelling. The modelling task for an application domain is turned into a classification task through the use of a domain-independent ontology. The development of a new application is not started from scratch, but a set of metaclasses is available from which it is posible to identify the most suitable one for each relevant feature. This way, any feature modelled takes on the meaning ascribed to its associated metaclass; which means that all knowledge associated with the metaclass (metainformation and applicable operations) is also valid for the feature. In short, the high level modelling effort is reusable.

Flexibility of information communication. The fact that different interlocutors share a set of high level concepts (ontology) means that they all interpret others information models in the same way. This significantly increases the flexibility of spatial information transfers between systems, as there is no need for a prior agreement on the form of modelling of different features, as long as the model itself is included as part of the transfer. It is clear that this approach makes the development of communication software more complex. However, this is a minor drawback in comparison with the advantages of the fact that just one software development is able to interpret different information models.

References

[1] Herring, J.R. A Fully Integrated Geographic Information System. Proceedings of AUTOCARTO 9, Falls Church, Virginia, pp. 828-37, 1989.

[2] Pullar, D. and M.J. Egenhofer. Towards Formal Definitions of Topological Relations among Spatial Objects. Proceedings of the 3rd International Symposium on Spatial Data Handling. International Geographical Union, Columbus, Ohio, pp. 225-41, 1988.

[3] Egenhofer, M.J. A Formal Definition of Binary Topological Relationships. In Litwin W., Scheck H.J. (eds.) Third International Conference on Foundations of Data Organization and Algorithms (FODO), Paris (Lecture Notes in Computer Science, Vol. 367), Springer-Verlag, New York, pp. 457-72, 1989.

[4] Imhof, E. Positioning Names on Maps. The American Cartographer, Vol.2, No. 2, pp. 128-44, October 1975.

[5] Rumbaugh, J., M. Blaha, W. Premerlani, F. Eddy and W. Lorenson. Object-Oriented Modeling and Design. Prentice-Hall, 1991.

[6] Hernández Rodríguez, F. Modelización de Información Espacial mediante Tecnología Orientada a Objetos. Ph.D. Dissertation, Universidad de Sevilla, 1995 (ISBN: 84-88783-11-6).

[7] Hernández Rodríguez, F., G. Bravo Aranda, and A. Martín Navarro. The Use of Conceptual Frameworks for Spatial Data Transfer. Proceedings of the 2nd Joint European Conference & Exhibition on Geographical Information. Barcelona, Spain, IOS Press, Vol. 1, pp. 258-67, 1996.

Contact address

F. Hernández Rodríguez, G. Bravo Aranda and A. Martín Navarro
Departamento de Ingeniería del Diseño E.S.I.
Universidad de Sevilla
Avda. Reina Mercedes s/n
41012 Sevilla
Spain
Phone: + 34 5 455 69 47
Fax: + 34 5 455 69 41
E-mail: ingedis@esi.us.es

EXPERIENCES WITH METADATA

Sabine Timpf, Martin Raubal and Werner Kuhn

Department of Geoinformation
Technical University Vienna
Vienna, Austria

Abstract
The need to share and integrate spatial data has spurred an interest in metadata. This paper documents the acquisition and modeling of metadata from eleven digital geodata sources in Austria. It shows how the information was modeled according to the proposed CEN standard on metadata, how it was encoded in a database, and what problems were encountered during these processes. The paper concludes with a discussion of recent developments around metadata and of the option to make meta-databases available on the world-wide web.

Key words
Metadata, standards, Open GIS, meta-databases, WWW.

1. Motivation

The topic of metadata has recently received considerable attention [Blott, and Vckovski, 1995; Dorf, and Scholten, 1993; Fisher, 1993; Strobl, 1995]. While many discussions addressed conceptual, architectural, and organizational requirements, practical experience with producing metadata has scarcely been documented. This might be caused by the fact that not a lot of meta-databases exist because metadata is expensive and hard to collect. Also, the possibility of a distribution of geodata to unknown users over the network has only arisen in the last few years due to the technological advancement. This contributed to a shift in attitude towards sharing data collections. More and more users outside the traditional spatial disciplines need spatial data. Data providers need to tell users what they have and what it can be used for. Metadata is destined for this purpose.

Metadata are 'data about data'. For companies working with spatial data, good documentation of the datasets becomes extremely important to make sure that they can still be used after changes of employees, software and hardware [Strobl, 1995]. Metadata can facilitate research on the environment in Europe [Dorf, and Scholten, 1993]. Metadata are also necessary to insure multiple usage of datasets [Frank, 1992]: Spatial data are being collected everywhere and could often be of use to others if they only knew of their existence. Meta-databases are one possible solution to this dilemma.

Metadata describe spatial datasets in a way that one can infer the usability of a specific dataset for a specific task. Some important criteria for the use of a dataset are:
- reference systems and area covered by the dataset
- currency of the dataset
- quality parameters such as positional, thematic and temporal accuracy
- administrative metadata.

The metadata is either gathered during the data collection process itself or at some later time. Generating metadata later requires considerable effort and not all the information might be available.

This paper reports on the experiences that surveying engineering students at the Technical University of Vienna made with collecting and describing metadata and entering them into a database. The metadata was collected from eleven different sources of geodata in Austria, coming from the areas of geodesy, geophysics, hydrology, and geomarketing. The metadata was described using the proposed standard on metadata from CEN TC 287 [CEN, 1995]. The proposed CEN standard defines a minimum set of metadata that should be provided by data suppliers. We used the relational database Microsoft Access 2.0 to implement a metadata repository.

After a short introduction to related metadata work (chapter 2), we introduce the CEN metadata standard (chapter 3) and explain how we applied it to the geodata sources (chapter 4). We review the standard in its present proposed form (chapter 5) and describe how we implemented the meta-database in Microsoft Access 2.0 (chapter 6). Finally, we discuss the issue of standardizing metadata and its possible future (chapter 7).

2. Metadata projects

The first organization to consider data about data was the FGDC (Federal Geographic Data Committee) with its Spatial Data Transfer Standard (SDTS). This started the discussion on metadata and its organization.

In the United States, the national spatial data infrastructure (NSDI) encourages standards and information interchange. The National Geospatial Data Clearinghouse links providers, managers, and users of information in a large network [FGDC, 1994]. In this system, users can search for metadata on the data they need. All federal agencies are required to make their data available to other agencies and to the public.

The Alexandria Digital Library[*] is a library for spatially indexed material. The library will enable users who are distributed over the network to access the information in the space they want. It will also be made available over the network.

---

[*] http://alexandria.sdc.ucsb.edu/

In Europe, the MEGRIN (Multipurpose European Ground-Related Information Network) is an initiative of the Comite Europeen des Responsables de la Cartographie Officielle (CERCO). It provides an information system with metadata on data-sources of the members of CERCO [Salgé, Smith, and Ahonen, 1992]. The information system is currently made available over the network.

3. The proposed metadata standard

The proposed European Standard on metadata (Geographic Information - Data Description - Metadata) has been prepared by the Technical Committee (TC) 287 of the European Committee for Standardization (CEN). Currently CEN TC287 is in the process of soliciting comments on the draft of the metadata-standard.

The standard defines a conceptual schema for metadata, based on two related standards: the proposed standard for quality and the proposed standard for positioning. The main reason for developing this metadata standard is to encourage the widespread use of geographic information. It is explicitly stated that the standard is not concerned with implementation details and therefore the construction of meta-databases. This is also made clear in the choice of EXPRESS as the language for the formal definition of entities. EXPRESS is a language for defining an information model, not a database model.

EXPRESS is the data description language of the Standard for the Exchange of Product Data (STEP), which has been developed by the International Standards Organization [ISO, 1992]. EXPRESS has been selected by CEN TC287 as the standard for the exchange of geoinformation. EXPRESS defines entities and relations in schemas. A schema is a context, which contains several entities and their relations as well as rules for their interaction. For example the context of metadata forms a schema, the context organisation forms a second one. Schemas can be used in other schemas. This avoids the redefinition of entities. For example the schema *organisation* is used in the schema *metadata* for the definition of the organization that manages or supplies the metadata. EXPRESS-G illustrates the EXPRESS definitions. Efforts are made to derive the EXPRESS-G diagram directly from the EXPRESS language.

The proposed standard consists of six parts. Each of the six parts is described below. We give an example from the standard for the verbal description, the EXPRESS-G description, the tabular description, and the EXPRESS description of the item Organisation.

3.1 Introduction to the proposed standard

The first three chapters of the document contain the scope of the project, references to other standards or draft standards, and definitions.

3.2 Verbal description

This most detailed part of the standard contains verbal descriptions of the metadata (see table 1), including EXPRESS-G graphics of each group of metadata (see figure 1).

The groups are:
- Dataset Identification,
- Dataset Overview,
- Dataset Quality Parameters,
- Spatial Reference System,
- Geographic and Temporal Extent,
- Data Definition,
- Classification,
- Administrative Metadata, and
- Metadata Reference.

Organisation and organisation role
Organisation name - the name of the organisation
Abbreviated organisation name - the short name of the organisation
Organisation address - the postal address, telephone, telefax number, electronic mail address of the organisation
Role - the responsibility of the organisation in relation to the dataset, for example, the creator, owner, administrator or distributor of the dataset. An organisation shall have one or more roles
Alternative organisation name - another name of the organisation which is either in the same language or another language
Function of the organization - description of the overall role of the organisation.

Table 1 Verbal description of the metadata.

The metadata items are defined by a name and a short description of what is meant by this name. For example, the name *Role* means the *responsibility of the organisation in relation to the dataset, for example, the creator, owner, administrator or distributor of the dataset. An organisation shall have one or more roles* (see table 1).

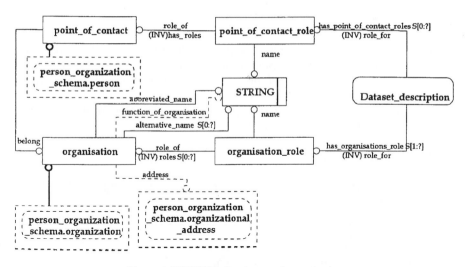

Figure 1 EXPRESS-G schema of organisation.

Administrative Metadata				
Organisation and organisation role		Con-straint	Card-inality	
• Organisation name	The name of the organisation	M	1	string
• Abbreviated organisation name	The short name of the organisation	M	1	string
• Organisation address	The postal address, telephone, telefax number, e-mail address of the organisation	M	1	address
• Role	The responsibility of the organisation in relation to the dataset, for example, the creator, owner, administrator or distributor of the dataset.	M	N	enumeration
• Alternative organisation name	Another name of the organisation which is either in the same language or another language	O	N	string
• Function of the organization	Description of the overall role of the organisation.	O	1	string

Table 2 Tabular description of the metadata.

The EXPRESS-G diagram illustrates the metadata items and their relations graphically. For example, the *role of* the *organisation* can be found in the field *organisation_role*. The EXPRESS-G schema also shows sub- and supertypes (dotted lines). For example the item *organisation address* needs to be represented as defined in the item organization_address in the schema person_organization_schema.

3.3 Annex A: Metadata Table

Annex A is a table with the metadata items including constraints, cardinality, and types (see table 2). Types may be string, numeric, picture, enumeration, address etc. Constraints define if the metadata item is mandatory or optional. Cardinality defines the allowed number of occurrences of an item. For example, the Alternative organisation name is an optional item of type string, which can occur N-times in a dataset.

3.4 Annex B: EXPRESS

In this part a formal description of the metadata items using the EXPRESS language is given (see table 3).

```
ENTITY Organisation
SUBTYPE OF (organization);
        abbreviated_name              :STRING;
        address                       :organizational_address;
        alternative_name              :OPTIONAL SET OF STRING;
        function_of_organisation      :OPTIONAL STRING;
INVERSE
        roles                         :SET OF Organisation_role FOR role_of;
END_ENTITY;
```

Table 3 The formal description in EXPRESS.

In this example the item organisation is defined as an entity in the schema. Its supertype is *organization,* which comes from a different schema. The attributes abbreviated_name, address, alternative_name, and function_of_organisation are defined. The relation *roles* is an inverse of the relation role_of in the entity Organisation_role.

3.5 Annex C: EXPRESS-G

In Annex C, the complete Express-G schema for the standard is given.

3.6 Annex D: Examples

In Annex D, two examples are supplied of how metadata should be described according to the proposed standard. The structure used is the table of metadata from Annex A.

4. The setting

Students in the course "Sources of Geoinformation" at the Technical University of Vienna[*] described eleven digital spatial data sets using the proposed CEN standard on metadata. The goal was to test the usability of this proposed standard for a variety of datasets and to get experience with modeling and describing metadata. Datasets for testing the proposed standard came from four different areas: geodesy, geophysics, hydrology and geomarketing. Students worked with the following databases:

- Coordinate database of Austria
- Parcel database of Austria
- Administrative Boundaries of Austria
- Digital cadastral map
- Terrain database
- Database of the multi-purpose city map Vienna
- Database of the city information system Linz
- Leveling and gravity database
- Fresh water wells database
- Water management database - ground water Vienna
- Geocoding database of Vienna.

The data sources were analyzed in two different ways: The necessary information could partly be extracted from flyers, papers, or brochures that were given out by the collecting organizations. More detailed information was collected by interviewing the responsible people at each organization. The second part of the information gathering process often took considerable time. Data were described by filling in a table taken directly out of the standard (Annex A).

Usually, the contacted persons were very interested in the project and liked to be selected as data providers. However, the students noted that some persons were not very interested in the proposed standard itself. This is, at least partly, due to the fact that the importance of metadata is still underestimated.

When formal descriptions of the datasets had been finished, students entered the metadata into a common database. For this implementation the relational database Microsoft Access 2.0 was used. Our experience with the proposed standard is given in the next chapter, the implementation is critically reviewed in chapter 6.

---

[*] http://www.geoinfo.tuwien.ac.at/Department/Courses/GIQ.html (in german).

5. Comments on the proposed standard

A data description using the proposed CEN standard provides a useful overview over a data set. The students got a generally positive impression of the proposed standard. Major problems were not reported, though some details created obstacles in the processes of modeling or implementing. After describing their datasets, the students prepared individual comments on the proposed standard. As the datasets were very heterogeneous, the results and especially the difficulties encountered were also of a great variety.

5.1 Criteria used

A standard should be clear, comprehensible, consistent, complete, flexible, and simple to use. In other words, it should be easy for the user to apply the standard to a dataset. Only if a standard is easy to use will it be used.
The documentation should be well structured and should give *clear* instructions how to derive metadata from geodata.
It should be *comprehensible*, explaining each step in the definitions, and connecting the different representations.
The various representations of the data description and the use of terminology throughout the document should be *consistent*.
The representations should be *complete* but also complement each other in the description. It is especially important that the examples be complete.
The standard should be *flexible* enough to accommodate different datasets, proveniences, complexities, or cultural differences.
The *ease of use* criterion requires a deep understanding of the process of creating metadata. Ease of use should therefore benefit the most from the experiences made in such a metadata project. It is also the all encompassing criterion in which all other criteria are reflected.
Generally, the standard should clearly point out the purpose of metadata and of meta-databases so that users can see the benefit and be prepared to produce and use meta-databases.

5.2 Application of criteria

- **Structure and clarity**
The description of the proposed metadata standard is not very homogeneous. A verbal description cannot easily give the full information that is detailed in a formal language. This means that the verbal description should be complemented with EXPRESS and EXPRESS-G descriptions. At least, the verbal description should be structured the same way as the EXPRESS description is. The proposed standard describes how metadata should be defined. It does not give instructions for its use. This is a serious impediment to the use of the standard in its current form. This shortcoming may be alleviated with the forthcoming reference framework and other explanatory documents from CEN TC287.

- **Comprehensibility**

The relations of the different representations are partly made clear in the verbal description, where the connections between definitions and the graphical representations are given. The representations should be placed side by side to clarify their equivalence. The use of some expressions or examples obscures rather than helps the proposed standard. For instance, the item Spatial_Reference_of_Metadata presumably means: the location, where the metadata will be available, though this is not evident from the textual definition. It is necessary to know EXPRESS or to consult a manual to understand these expressions and their meaning. Some expressions are not clear, even when provided with examples. For instance, three types of structure primitives are given without further explanations of their meaning. The meaning of these structure primitives needs to be extracted from another standard. Again, this problem may be somewhat reduced once the complete family of standards is available. When technical terms are used, they have to be defined very carefully. For instance, the parameters required when describing the reference ellipsoid are not clearly stated.

- **Consistency**

The proposed standard has some inconsistencies in type definitions and with missing or redundant relationships. For example, a picture is of type image in one annex and of type string in another. Inconsistencies between the EXPRESS and the EXPRESS-G model will not occur, when the EXPRESS-G will be directly derived from the EXPRESS model. All other relations between the representations must be checked for consistency.

- **Completeness**

The representations should be complete in themselves but they should also complement each other. This criterion is not yet satisfied. For instance, some definitions like feature type and attribute type in the introduction are missing and the examples given in Annex D are incomplete. Definitions should be given in the same document and be as concise as possible. Only if they need to be complex, they may be left out and the user referred to other documents. Currently, missing definitions and the need to look them up at different places is a serious impediment to the ease of use of the standard.

- **Flexibility**

Considering the diversity of the data sets described, the proposed standard proved to be very flexible. It was sometimes hard to find out from the document if or why certain items are mandatory or optional. In these cases, the decision should be in favor of the more.flexible solution. For example, the item *support services* is mandatory. But many organizations do not have further services available, so this item should be optional.

- **Ease of use**

From the standard document, it is easy to understand the idea of the standard and its purpose. It is hard to understand the details of applying it to a real dataset. The tricky parts are in the details of the specification language. For example, one needs a better knowledge of EXPRESS to understand all the implications of the formal specifications. Ease of use also requires simple words and ideas. The user of a standard should be incited to work with it. Simpler terms would often have done a better job. For example, the expression *spatial reference of metadata* presumably means simply the location of the meta-dataset.

6. Implementation

People of various disciplines want to use geodata for different applications. To find out if a certain dataset fits their needs, users have to browse the data. One way to do so, is querying and looking through a database. With a meta-database, users will find the desired information within a shorter period of time. This is a major improvement over looking through all kinds of flyers and brochures.

Another advantage of using a database is the possibility to look for specific sets of metadata. This can be done with queries. In SQL the query 'SELECT dataset_title FROM meta-database_Austria WHERE Administrative_Metadata.Abbreviated_organisation_name = BEV' would retrieve the metadata of all datasets that belong to the Austrian Federal Mapping Agency (BEV, Bundesamt für Eich- und Vermessungswesen).

This project used the relational database Microsoft Access 2.0 to implement the sets of metadata. Entities in the form of tables containing various attributes had to be created and relationships with cardinalities between the tables defined. The dataset title was used as key (see figure 2). While implementing the metadata we found a major problem:
EXPRESS uses some concepts of object-orientation, which could not easily be implemented into a relational database. Therefore, the tables do not exactly correspond to entities of the EXPRESS model.

It was also not possible to implement data types like lists or sets with Microsoft Access 2.0. These data types had to be implemented by recursive decomposition to conform to the normalization rules. There is a general problem to describe standards with some sort of object-oriented language, when there are basically only relational databases on the market. Another problem is the impossibility of using two different data types for the same attribute. The proposed standard states that when information is not available, then this shall be described as "no information available". When trying to write this into a column of a table which demands an integer type the user will not succeed. Some students tried to avoid this problem by inserting null values. But null values in databases have to be handled with care and lead to additional problems: what is the value of an undefined value in mathematical expressions, how is an attribute with a null value handled in joins, is there any difference between null values? These questions arise because a null value can be seen either as an unknown or as an arbitrary value [Duerr, and Radermacher, 1990]. These kinds of difficulties have to be overcome before a database like Microsoft Access 2.0 can be used.

The next step of the implementation process will be to connect the meta-database to the World Wide Web (WWW). This should happen as soon as we have revised the metadata and received permission from the data suppliers. Access to the meta-database will be provided through a web-based query form.

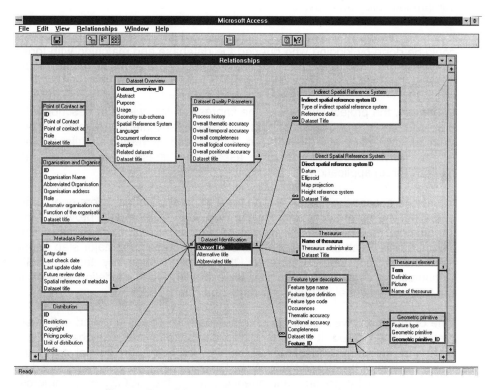

Figure 2 Part of the relationships in Microsoft Access 2.0.

7. Discussion and future perspectives

Metadata is the current approach to supply information to users on the purpose and usability of spatial data collections. Metadata is data about data and helps to decide if the data from specific sources are usable for another purpose than the one they were originally collected for.

Discussions on this topic have led to various definitions of standards and to the creation of some meta-databases. The process of defining metadata is often described abstractly and documented experiences with the collection, definition, and storage of metadata are missing. In this paper we described how a group of students collected and described several metadata sets. We commented on the currently proposed CEN metadata standard and critically reviewed the modeling and implementation issues that arose.

825

The standard was found to be easy to understand, but not very easy to use. This is mainly due to the hidden implications of EXPRESS. It is necessary to know EXPRESS to understand the standard as well as to know how to translate EXPRESS models into implementations. An improved structure of the standard with a better distribution of the different representations of metadata (text, formal language, graphics) would contribute to comprehensibility and ease of use.

It is now necessary to find out if such a meta-database is really useful. This could be done by logging who queried the database and matching with those who used or bought the data. One could also prepare a questionnaire to find out why datasets are used or not used.

In general, the project left us with the impression that we have described the data sources in great detail without necessarily enlightening potential users. With all the metadata available in a database, it remains quite unclear whether anybody can use them to assess the fitness for use of the data sources for an application.

In order to assess the fitness of data for a given application, users need more and different information than the current kinds of metadata provide. Most importantly, they need to know what operations are supported by the data [Kuhn, 1994]. Such an approach to metadata represents a major step ahead from the current state of the art. It has been taken by the OpenGIS Consortium [Doyle, 1995] and has, for example, found modeling support in the form of functional languages [Frank, and Kuhn, 1995].

One of the unresolved issues in any metadata approach is to find out where metadata are located. It could require a database on meta-databases, i.e. a meta-meta-database. To some extent, this problem can be solved in the larger context of the World-Wide Web. Geodata are increasingly provided on WWW servers by government agencies (e.g., the USGS) or commercial providers (e.g., ImageNet), though these activities are still in an experimental stage. The main advantages of distribution via the Web are the low cost of distribution, high currency of information, potential market for geodata to millions of users worldwide and 24 hour access [Vincent, 1995]. When geodata are made available on the World Wide Web, they can serve as distributed repositories of their own metadata.

Acknowledgments
The authors thank the many organizations who provided the metadata and all participants of the course "Sources of Geoinformation" for their contributions: Karin Amesberger, Bernhard Gaupmann, Damir Medak, Hartmuth Schachinger, Franziska Silwester, and Ines Stanzer. Special thanks to Joachim Heinzl and Wilhelm Jachs for helping to integrate the different views from the participants.

References

Blott, Stephen, and Vckovski, Andrej. "Accessing Geographical Metafiles through a database storage system." In Advances in Spatial Databases (SSD'95) in Portland, ME, edited by Egenhofer, Max J., and Herring, John R., Springer Verlag, New York, 117-131, 1995.

CEN. Geographic Information - Data description: Metadata. CEN/TC 287/WG2, 1995.

Dorf, Josh, and Scholten, Henk J. "Meta-Catalog: a necessary requirement for european environmental research." In EGIS'93 in Genoa, Italy, edited by Janjaap Harts, Ottens, Henk F.L., and Scholten, Henk J., EGIS Foundation, Utrecht, 636-643, 1993.

Doyle, Allen. "Software that plays together - a systems integrator's viewpoint." Geo Info Systems (May 1995).

Duerr, M., and Radermacher, K. Einsatz von Datenbanksystemen. Berlin, Heidelberg: Springer-Verlag, 1990.

FGDC. The national geospatial data clearinghouse. FGDC, 1994.

Fisher, P.F. "Conveying Object-Based Meta-Information." In 11th International Conference on Automated Cartography in Minneapolis, MN, ACSM, 113-122, 1993.

Frank, Andrew U. "Acquiring a digital base map - A theoretical investigation into a form of sharing data." URISA Journal 4 (1 1992): 10 - 23.

Frank, Andrew U., and Kuhn, Werner. "Specifying Open GIS with Functional Languages." In Advances in Spatial Databases, ed. Egenhofer, Max J., and Herring, John R. 184-195. LNCS 951. Springer-Verlag, 1995.

ISO. The EXPRESS language reference manual. ISO TC 184, 1992. Draft International Standard ISO/DIS 10303-11.

Kuhn, Werner. "Defining Semantics for Spatial Data Transfers." In 6th International Symposium on Spatial Data Handling in Edinburgh, UK, IGU, 973 - 987, 1994.

Salgé, François, Smith, Neil, and Ahonen, Paula. "Towards harmonized geographical data for Europe; MEGRIN and the needs for research." In 5th International Symposium on Spatial Data Handling in Charleston, SC, IGU Commission on GIS, 294-302, 1992.

Strobl, Josef. "Grundzüge der Metadatenorganisation für GIS." Salzburger Geographische Materialien - Beiträge zur Geographischen Informationswissenschaft (Proceedings AGIT 1995) (22 1995): 275-286.

Vincent, S. "CyberGIS - Providing Geodata via the World Wide Web." GIS Asia Pacific 1 (5 1995): 28-31.

Contact address

Sabine Timpf, Martin Raubal and Werner Kuhn
Department of Geoinformation
Technical University Vienna
E 127.1
Gusshausstr. 27-29
A-1040 Vienna
Austria
Phone: +43 1 58801 - {3791, 3788, 3790}
Fax: +43 1 504 3535
E-mail:{timpf,raubal,kuhn}@geoinfo.tuwien.ac.at
WWW: http://www.geoinfo.tuwien.ac.at/

DRAINAGE QUERIES IN TINS:

FROM LOCAL TO GLOBAL AND BACK AGAIN[*]

Sidi Yu[1], Marc van Kreveld[2] and Jack Snoeyink[1]

[1]Department of Computer Science
University of British Columbia
Vancouver, Canada
[2]Department of Computer Science
Utrecht University
Utrecht, the Netherlands

Abstract
This paper considers the cost of preprocessing a digital terrain model (DTM) represented as a triangulated irregular network (TIN) so that drainage queries — e.g., what is the watershed of a query point, or how much water passes through a point given that rain is falling at a known rate — can be answered by simply evaluating a summary function. Although the worst-case storage and preprocessing costs are high, the experimentally-observed costs are reasonable. In order to compute a compact and consistent summary function, the drainage network needs a rigorous definition. This paper, therefore, also surveys some of the previous definitions, extends them, and establishes a number of properties of drainage networks with a focus on TINs.

1. Introduction

Terrain drainage characteristics provide important information on water resources, possible flood areas, erosion and other natural processes. In natural resource management, for example, the basic management unit is the *watershed*, the area around a stream that drains into the stream. Road building, logging, or other activities carried out in a watershed all have the potential to affect the defining stream. Manual quantification of terrain drainage characteristics is a tedious and time-consuming job. Thus, many researchers have developed spatial analysis algorithms to compute drainage characteristics from digital terrain models (DTMs).

We began this research intending to go one step further. Many of the spatial analysis algorithms on terrains are batch processes whose product is a map depicting drainage characteristics; our aim is to provide this map, but also provide richer structures that could answer interactive queries for watershed areas, regions, or flow rates. We had to start, however, by taking a step back to reconsider definitions of the drainage network and its components, especially for triangulated irregular networks (TINs). Thus, Section 2 begins with a brief critique of definitions and algorithms for drainage in various terrain

[*]Research of the UBC authors was partially supported by an NSERC Research Grant and a B.C. Advanced Systems Institute Fellowship. The last two authors performed part of the research during a workshop sponsored by the a PIONIER grant of the Dutch Organization for Scientific Research (N.W.O.).

models.

Frank et al. [6] point out that formal definitions should be used to define terrain-specific features so that properties of the structure can be established mathematically and contradictory definitions (or at least those in disagreement with other related research) can be avoided. We concur, and propose definitions in Section 2.3 that capture the global behavior of non-inertial flow on a surface. Our definition extends the local definitions of Frank et al. [6] to include accumulation of water. The focus of the definitions and methods is on triangulated irregular networks (TINs).

Using a global definition for the drainage network (with accumulation) is natural but may make it more difficult and time consuming to extract additional drainage information. Section 3 describes how to augment the drainage network with quantitative information like areas of watersheds and flow rates through the network. This information is represented in a compact and consistent manner by summary functions, which are computed for all parts of the drainage network. The summary functions turn global drainage queries into local ones.

Several useful applications of the summary functions cannot be obtained easily from local drainage models. Firstly, user-defined questions on flow rate and basin areas can easily be answered, since these are quantified by the accumulation in the model. Secondly, by thresholding on the flow rate in the drainage network we immediately obtain a multi-scale drainage network, useful for map generalization. Thirdly, by thresholding on basin areas (or volumes), local depressions that could be artifacts of data acquisition or modeling can be removed in a uniform way.

For an n-point TIN, a fiendish (and improbable) worst-case example shows that the size of this structure can be cubic, but preliminary experimental results indicate that the size is more likely to be proportional to n. We then turn to the delineation of the watershed of a point, or basins for whole parts of the drainage network. Section 4 relaxes some of our initial assumptions, indicates possible extensions to the model and at the same time, discusses ideas for further research.

2. Definitions and assumptions for drainage

A terrain is the graph in R^3 of a continuous function $z = f(x, y)$ defined on a closed, compact, connected subset of the xy plane. A TIN is a terrain that consists of triangles.

2.1 Assumptions

When one looks at a terrain, one can intuitively identify a network of watercourses and ridges based on the notions that water flows downhill, tributaries join larger streams, rivers run to the sea, and ridges divide rivers.

Let us be more precise about the first three of these assumptions; we consider the fourth in Section 2.3.

A1 At any point, water follows the steepest descent,

A2 Watercourses can merge, so that the drainage network has the topology of a set of trees, and

A3 Watercourses end only at a local minimum of the terrain.

Precise definitions of watercourse and ridge networks that satisfy these assumptions are rare in the literature, especially when these definitions are applied to a terrain model that is stored in the computer. In hindsight, this should not have been a surprise; from a continuous, two-dimensional terrain, represented in the computer by a discrete set of bits, one wishes to extract connected networks that are one-dimensional but have a discrete (graph) topology. Thus, if the reader will permit some hasty generalizations, the mathematical literature is "too continuous," being concerned with the differential geometry of flows on smooth surfaces. The GIS literature is too focused on computation, often defining the drainage network as the output of some algorithm. And the hydrology literature is willing to introduce uncomputable "infinite regress" to have duality properties for interlocking ridge and channel networks. The issue of interlocking networks is discussed in Section 3.2.

2.2 Earlier definitions in various terrain models

In addition to drainage definitions for *smooth* mathematical surfaces, there are different definitions for the three main forms of digital terrain models (DTMs): gridded DEM (digital elevation model storing elevations at points of a regular grid), digital contours, and TINs (triangulated irregular networks). Drainage has also been defined in terms of other surface networks.

Smooth mathematical surfaces

The study of drainage on *smooth* mathematical surfaces has a long history, including mathematicians such as Cayley [2] and Maxwell [17]. Koenderink and van Doorn [12,13] credit Rothe [27] with the first characterization of ridges and channels in 1915; Koenderink and van Doorn's work expresses this solution in the terminology of modern differential geometry and includes examples of why other attempts at characterizing drainage on smooth surfaces (especially using those using local criteria) are inadequate. Unfortunately, in smooth surfaces, watercourses tend not to join except in the limit, violating A2. Assumption A3 can also be violated; watercourses can end at points that are not local minima. A further complication is that most smooth surfaces and their flows can only be approximated in the computer. Thus, it is debatable whether smooth surfaces are appropriate models of physical terrain.

Contours

Some researchers [8,14,20,32] define the drainage networks in terms of contour lines: A points is on the the watercourse or ridge network if it is a local minimum or maximum on its contour line. Koenderink and van Doorn [12] have shown that this definition can violate A1 — the computed network does not necessarily travel in the direction of steepest descent. This definition also implies that the terrain should be twice differentiable, which again hinders streams from joining (A2). Three of the above-mentioned papers [8,14,32] continue the computation of the drainage network on gridded models.

Grid or raster-based DEM
There is no accepted definition of the drainage network in a grid. Frequently the network is defined as the output of an algorithmic procedure, which may not be completely specified outside of the computer code. We can, as other researchers have done, classify these computations by their locality and by whether they have an explicit surface model.

A number of methods [5,9,10,14,18,21,23,28,31] are based on extracting features from a raster DEM by local operations, using techniques that are common in raster image processing. Local filters are are used to detect potential pit, peak, channel, and ridge cells, then the channel and ridge cells are linked according to some heuristic to form the drainage network. Basin boundaries are formed in a similar fashion [9,18]. Assumptions A1 and A3 are frequently violated, since the global nature of the terrain's drainage network is not adequately reflected. As with image processing, there are inherent ambiguities when combining information from local filters — one research team developed an expert system for this task [26]. Different heuristics will result in different sets of extracted terrain features.

Mark [16] proposed a global computation: sort all the cells in order of decreasing elevation, and initially assign each cell one unit of fluid. Then each cell in order adds its fluid to the lowest of its eight neighbors. In the end, cells receiving more than a threshhold quantity of fluid are declared to be on the drainage network. This method has an implicit surface and approximates a steepest descent; at lower elevations the network may become two-dimensional.

A grid of points with elevations does not give a terrain unless a continuous surface is defined. (One early approach to drainage [23] has no concept of surface.) Approaches that define flow based on 4 or 8 neighbors in the grid [1,9,16,21,24,28] define an implicit surface, which may not have geometric interpretation in which water flows according to steepest descent. If an interpolation method is specified, then the grid can be considered as an explicit surface on which flow can be properly defined. Douglas [5] inserts a virtual grid point at the center of each square for a piecewise linear interpolation; Haralick [7] uses bicubic polynomial surface patches for a continuous and smooth surface. Afterwards, however, flow is still approximated in raster pieces.

We note that both the surface and the flow are discretized in many grid algorithms. The discontinuities in both make it difficult to compute summary functions and lead us to prefer TINs.

TINs
Triangulated irregular networks, or TINs, define a continuous terrain that is not differentiable at the edges and vertices. Flow along triangles and edges

are easy: the steepest descents in a triangle lie on parallel lines; an edge becomes a channel if flow comes from both adjacent triangles. In the literature, the continuous nature of the terrain is often ignored; this happens in one of two ways. First, the flow may be discretized and passed triangle to triangle based on neighbor relations in the direction of steepest slope [22]. This approximates steepest descent (A1). It also calls for small elements, which works against the adaptive nature of TINs. Second, a subset of edges that catch flow (the *confluent* edges or local channels as defined in the next section) may be considered important features and cross-triangle flows may be ignored [6,11,30,33]. Watercourses then stop at places that are not local minima (A3).

Surface networks
Related to drainage networks are the so-called surface networks by Mark [15], Pfaltz [25], Warntz [35], Shinagawa and Kunii [29], and Wolf [39,40]. Also worth mentioning as related work is the computation of terrain data from hydrological information [20,32]; hydrologists use complex models that differentiate between surface flow and subsurface flow, and therefore are also dependent on permeability of the ground, rainfall, and other quantities [19,30,34].

2.3 Definitions for global drainage in TINs

We make one more assumption in order to simplify our definitions:
    A4 At any point, there is a unique direction of steepest descent.
This assumption rules out flat faces and enforces a canonical choice at edges and vertices where there may be more than one descent. It also means that watercourses in our model do not bifurcate. We modify and relax this assumption in Section 4.

General definitions
The following definitions capture infinitesimal, viscous flow on a surface.
- The *trickle path* $\tau(p)$ of a point p on a surface S is the path that begins at p and follows the steepest descent until it reaches a local minimum or the boundary of S.
- The *watershed* of a point p is the set of all points whose trickle paths contain p.
- The *watercourse network* (or *drainage network*) consists of all points whose watersheds have non-zero area (or more mathematically: whose watersheds have two-dimensional Lebesgue measure).
A key observation is that trickle paths cannot cross — if two trickle paths join at some point, then by our assumptions they must continue together along the steepest descent path.
As we shall see next, these definitions capture the local definitions in a TIN by Frank et al. [6,33] and extend them to satisfy our assumptions on global behavior of a drainage network.

In smooth terrains these definitions omit some of the special slope lines that satisfy Rothe's differential equations [12,13] because trickle paths in smooth terrains may merge only in the limit. The expected watercourse at the bottom of a smooth gutter, for example, does not materialize. Rothe's equations do not apply in non-differentiable terrains such as TINs, however.

The relation to TIN-based definitions
Figure 1 illustrates definitions from Frank et al. [6] that are specific to edges of a TIN. We say that an edge *e* of a triangle *receives flow, sends flow,* or does neither, depending on whether the steepest descent direction for the triangle moves flow into, out of, or parallel to the edge. By convention, edges do not include their endpoints.

- An edge *e* of the triangulation is *transfluent* if it is receives flow from one adjacent triangle and sends flow to the other.
- An edge *e* of the triangulation is a *local channel* (or is *cofluent*) if it receives flow from an adjacent triangle and is not transfluent.
- An edge *e* of the triangulation is a *local ridge* (or is *difluent*) if it sends flow to an adjacent triangle and is not transfluent.

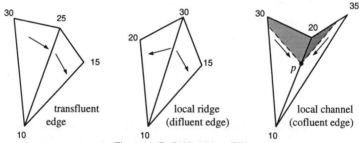

Figure 1 Definitions in a TIN:
Any point p on a local channel has a watershed with non-zero area.

The following definitions apply to general surfaces.
- A *peak* is a local maximum of the surface. In a TIN, a peak is a vertex with higher elevation than its incident edges and triangles.
- A *pit* is a local minimum of the surface. In a TIN, a pit is a vertex with lower elevation than its incident edges and triangles.
- A *basin* is the watershed of a pit.

These definitions imply that all local channels are on the watercourse network and that none of the local ridges are: A point *p* on a local channel collects fluid from the region bounded by steepest ascents in the adjacent triangles, as shown shaded in Figure 1. A point *q* on a ridge is on the trickle path for itself. In fact, the trickle paths from local channels determine the watershed network.

Lemma 2.1   *The watercourse network in a TIN is collection of disjoint (graph-theoretic) trees rooted at pits whose leaves are local channels.*

Proof : The trickle path from any point on the watercourse network is contained in the network. Since a trickle path can only end at a pit, the watercourse network is, what is called in graph theory, a forest of rooted trees. These trees are disjoint because the trickle paths cannot bifurcate.
We have already argued that local channels are contained in the watercourse network, and one can check that points on triangles, transfluent edges, or vertices are on the network only if some point above is also on the network.

Therefore the watercourse network begins at local channels.■

Knowing the topology of the watercourse network, the following definitions are natural. We'll use them throughout the paper.

- A *(watercourse) junction* is a point of the watercourse network that has degree at least 3 in graph terminology, that is, there are at least two separate incoming parts of the watercourse network.
- A *watercourse segment* is a maximal portion of the watercourse network that doesn't contain junctions, leaves, or pits of the watercourse network.
- A *catchment area* of a watercourse segment is the part of the terrain that drains directly into the watercourse segment; it does not enter another watercourse segment first.

Junctions occur, for example, where two local channels have the same downstream vertex or where cross-triangle flow reaches a local channel.

3. Processing a TIN for drainage queries

The definitions from the previous section describe global flow on a TIN surface. There are now a number of queries or operations that we would like to support, including:

1. What is the area of the watershed of point q?
2. What is the area of the projection of the watershed of point q on the xy plane?
3. What is the flow rate through q if water falls at a uniform rate r?
4. Construct the watershed of point q.
5. What is the catchment area (projected area, flow) of a given watercourse segment?
6. What points have watershed areas of size (at most, at least, exactly) A?

With additional processing (after a few more definitions) we can partition a TIN surface into pieces for which simple functions summarize the drainage behavior — in essence turning drainage computations back into local computations.

3.1 Partitioning a TIN based on drainage

To compute the summary functions needed to solve the drainage queries, we need a few more definitions that are tailored for this purpose. The *drain of a point p*, denoted *drain(p)*, is the point where the trickle path $\tau(p)$ first meets the watercourse network. If p is on the network, then *drain(p)* = p.

The set of basins, one for each pit, form a coarse level partition of a TIN. The set of catchment areas for the watercourse segments refines this partition. For a watercourse segment s, its catchment area consists of all points p for which *drain(p)* is on s.

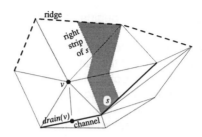

Figure 2 Right strip for segment s.

Define *strips* as a further refinement: For every triangle vertex *v* of the TIN, introduce a node at *drain(v)*. These nodes cut the watercourse segments into yet smaller segments that we call *base segments*. One can easily verify that the nodes can only lie on local channels. Figure 2 gives an illustration. For a base segment *s* of a local channel, define the *right strip* to be the set of points *p* such that *drain(p)* lies on *s* and the trickle path $\tau(p)$ comes from the right, assuming base segments are directed consistent with the flow. Define the *left strip* similarly. Strips, whether left or right, have a simple structure.

Lemma 3.1 *A strip is a region bounded by a base segment from a local channel, a segment of a local ridge, and two steepest descent paths. No vertices of the TIN lie inside a strip.*

Proof : We can establish this lemma for a right strip by induction on the number of triangles that the strip intersects.
Consider the triangle *T* to the right of the base segment *s*. The points of *T* in the strip are contained in a trapezoid (or triangle) formed by extending the endpoints of *s* in the direction of steepest ascent until they hit a triangle edge. Let *s'* denote the segment of this edge that forms the upper boundary of the trapezoid. This really is a trapezoid (or triangle, if an endpoint of *s* is a vertex of *T*): only one of the other two edges of *T* can be hit, since the relative interior of *s* does not contain *drain(v)* for any triangle vertex *v*.
Now, if *s'* lies on a ridge, the trapezoid is the strip. Otherwise, the strip is the trapezoid together with the strip of *s'*. Mathematical induction completes the proof.■
This lemmas has immediate corollaries about the structure of coarser partitions and about summarizing area and flow.

Corollary 3.*2* *Basins, watersheds, and catchment areas are bounded by local ridges and steepest descent paths.*
Proof : All three are composed of left and right strips for watercourse segments.■

Corollary 3.3 *Watershed area, projected area, or steady-state flow rate under uniform rainfall assumptions can be summarized by piecewise quadratic functions on segments of the refined watercourse network.*

Proof : For a query point *q* on the watercourse network, the base segment *s* determines left and right strips that are partially inside the watershed of *q*. Every other strip is either entirely outside of the watershed of *q* or entirely inside and contributes a constant amount of area, projected area, or steady-state flow.

The contribution from the left and right strips with base *s* can be summarized as a quadratic function $at^2 + bt + c$, where t is a parameter indicating the position of *q* on *s*.■

3.2 Ridge networks and duality

Any definition of catchment areas for watercourse segments leads to a natural definition of a ridge network as the (non-watercourse) boundaries of the catchment areas. Thus, the ridge network divides the terrain surface into regions whose flow goes to different watercourse segments. We can exploit the fact that catchment areas are unions of strips, under our definitions, to efficiently compute the ridge network.

Notice that the ridge network is determined by the partition of the watercourse into segments. Taking the finest partition into watercourse segments makes the ridge network consist of all strip boundaries; a coarse partition into major branches of the watercourse means that only a subset of these boundaries lie on the ridge network. We can identify that subset by tracing the trickle paths of junctions of watercourse segments *back upward* to local ridges and tracing up local ridges from passes. Two types of duality are referred to in work defining ridges and courses; both have limited validity.

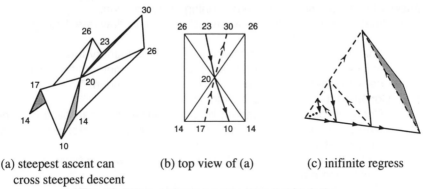

(a) steepest ascent can   (b) top view of (a)   (c) inifinite regress
    cross steepest descent

Figure 3 Ridges and watercourses are not strict duals.

The first is that ridges and watercourses change roles when a terrain is inverted (or when all the elevations are negated). This is true for the local features: Local ridges and local channels, pits and peaks are duals in this sense. The global assumption A3, that water flows to a pit, breaks this duality at vertices, however. We cannot assume that ridges continue to a peak along the lines of steepest ascent without allowing watercourses to flow across ridges. Figure 3(a) shows such an example.

A second is that, in a given terrain, the ridge and watercourse networks interlock — that there is a ridge between two adjacent courses and a course between adjacent ridges. Werner [38] shows that in such a case the ridges and courses can be embedded in dual graphs and that specific properties of one network determine specific properties of the other. Here one must be careful with the meaning of "between." Werner's definition of "between" essentially guarantees that an embedding is possible, and thus his theorems are about graphs and need not reflect a physical situation.

Frank et al. [6] pointed out that this assumption can lead to infinite regress, a fact that was known to Warntz [36], even in a single triangle. Consider the side of a pyramid as having two ridges and a watercourse at the base, as in Figure 3(c). To separate the ridges, one must introduce a course splitting the triangle; one of the pieces now needs a ridge to separate two courses, ad infinitem. Although Frank et al. attributed this to the discrete nature of a TIN, continuous examples can also be given. Dawes and Short [3] argue that the problem occurs when one insists that the ridge and course networks are joined; if they are not joined, however, then watershed boundaries are hard to determine and the word "between" must again be carefully defined.

In our global approach, every division of the watercourse network into segments induces a ridge network. These two networks are not duals for either of the above types of duality, but taken together they form the boundaries of catchment areas for each watercourse segment.

3.3 Answering queries

Using the summary functions or ridge networks described above, we can answer the six queries from the beginning of this section.

1–3. What is the watershed area, projected area, or flow rate through point q?

These queries can be answered from the summary functions attached to the watercourse network at q. (If q is not on the watercourse network, then the answer is "no measurable amount.")

4. Construct the watershed of point q.

Construction queries cannot be summarized, but can be answered in time proportional to the number times the output intersects triangles. Simply extend all strips through the triangles.

5. What is the catchment area (projected area, flow) of a given watercourse segment?

The area can be determined by subtracting the watershed areas for the nodes at the top from that for the node at the bottom of the watercourse segment. The other quantities can be determined analogously.

6. What points have watershed areas of size (at most, at least, exactly) A?

The summary functions can be evaluated to determine where a threshold is reached or exceeded. "At least" or "at most" queries can be answered in time proportional the output size (plus the number of pits in the "at least" case) by starting with each pit or leaf, respectively, and growing the answer until the threshold is reached.

3.4 Data structure size and computation

Of course, one must pay a price for considering a global definition of drainage and again for computing and storing local summaries. We briefly investigate the memory size and computation time requirements for the watercourse network and for augmenting it with local summaries. In a TIN with n vertices, the worst-case size of the network alone can be horrible. In conjunction with a number of computational geometers [4], we have shown that it can require on the order of n^3 memory cells. The upper bound can be proved by showing that a trickle path crosses any edge at most n times; the lower bound is by a construction depicted in figure 4, in which roughly n rivers cross n triangulation edges n times each.

As you can see, this construction is unlikely to occur in physical geography, with the possible exception of when the Egyptians were building the pyramids. In our preliminary experiments on TINs derived from sampling data from St. Mary's Lake, B.C., the network has slightly fewer segments than there are vertices.

No. of vertices	450	1800	4956	11125
Network size	353	1427	4134	10579

The network can be computed in time proportional to its size by simply tracing down from every local channel, marking which local channels we have visited so that we can stop as soon as we encounter a marked edge.

The memory required for local summaries is easy to analyze. For each vertex v of the TIN, $drain(v)$ is a node of the augmented network, which requires a position and three coefficients of a quadratic polynomial.

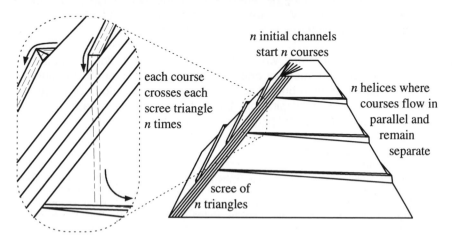

Figure 4 A watercourse network with cubic size.

We compute $drain(v)$ for each vertex v by simply tracing the trickle path $\tau(v)$ through the triangles to $drain(v)$. Several clever data structuring techniques could have been used to trace a several points at once and reduce the worst-case construction time, but our preliminary experiments indicated that this would not be a significant savings over the practical cases that we were attempting. The summary functions for area, projected area, or flow could be computed from areas of trapezoids in strips; we measure the quantities at three points in a watercourse segment and fit a quadratic polynomial to the measurements.

4. Extensions and future research

There are many possible extensions to this model, some of which are vital for its practical application. It can be extended to take into account the fact that water has volume and can collect one clearly should do so in terrains because errors in data or in sampling can create spurious local minima that are shallow pits. Any significant volume of water would collect to fill these and then spill over the lowest pass to rejoin the main network. We may use a threshold on the basin area to determine which local depressions should be filled, or perform smoothing operations. (O'Callaghan and Mark [21] reported 90% removal rate for one pass of a simple binomial smoothing operator.)

As with any elevation-based definitions, flat and near-flat areas must be handled specially. We assign steepest descent directions to triangles in some way that has no cycles. Whether and how to do this will depend on the specific application, desired drainage characteristics, and additional information sources, therefore we decided not to embed these design decisions in our model.

Watershed areas can be used to prune small watercourses; by choosing an appropriate threshold value, generalization of the watercourse to various scales is possible [37]. We are also exploring the effects of terrain generalization on the watercourses and watersheds computed.

Finally, we can incorporate water absorption into the flow model. By employing precipitation models and water absorption models to different areas of the terrain (based on the surface type and vegetation, for instance), we can either fill the small pits into lakes if the influx is greater then the absorption, or leave it as a real pit (sink), if otherwise.

Acknowledgments
The authors are indebted to Facet Decision Systems for their support of this research, David Mark for discussions on drainage networks, and the fellow participants — especially Thomas Roos and Otfried Schwarzkopf — in the Utrecht Workshop on Computational Geometry and GIS, sponsored by a PIONIER grant from the Dutch Organization for Scientific Research (N.W.O.), for discussions on these definitions and on the worst-case size of the query structure [4].

References

[1] L.E. Band. Extraction of channel networks and topographic parameters from digital elevation data. In K. Beven and M.J. Kirkby, editors, Channel Network Hydrology, pages 13-42. John Wiley & Sons, New York, 1993.

[2] A. Cayley. On contour and slope lines. Lond. Edin. Dublin Phil. Mag. and J. of Sci., 18(120):264-268, 1859.

[3] W.R. Dawes and D. Short. The significance of topology for modeling the surface hydrology of fluvial landscapes. Water Res. Res., 30(4):1045-1055, Apr. 1994.

[4] M. de Berg, P. Bose, K. Dobrint, M. de Groot, M. van Kreveld, M. Overmars, T. Roos, J. Snoeyink, and S. Yu. The complexity of rivers in triangulated terrains. To appear, Proc. 8th Canadian Comp. Geom. Conf., 1996.

[5] D.H. Douglas. Experiments to locate ridges and channels to create a new type of digital elevation model. The Canadian Surveyor, 41(3):373-406, Autumn 1986.

[6] A.U. Frank, B. Palmer, and V.B. Robinson. Formal methods for the accurate definition of some fundamental terms in physical geography. In Proc. 2nd Intl. Symp. Spatial Data Handling, pages 585-599, 1986.

[7] R.M. Haralick. Ridges and valleys on digital images. Comp. Vis. Graph. Image Proc., 22:28-38, 1983.

[8] M.F. Hutchinson. Calculation of hydrologically sound digital elevation models. In Proc. 3rd Intl. Symp. Spatial Data Handling, pages 117-133, 1988.

[9] S.K. Jenson. Automated derivation of hydrologic basin characteristics from digital elevation model data. In Proc. AUTO-CARTO 7, pages 301-310. ASP/ACSM, 1985.

[10] S.K. Jenson and J.O. Domingue. Extracting topographic structure from digital elevation data for geographic information system analysis. Photogrammetric Engineering and Remote Sensing, 54(11):1593-1600, Nov. 1988.

[11] N.L. Jones, S.G. Wright, and D.R. Maidment. Watershed delineation with triangle-based terrain models. Journal of Hydraulic Engineering, 116(10):1232-1251, Oct. 1990.

[12] J.J. Koenderink and A.J. van Doorn. Local features of smooth shapes: Ridges and courses. In Geometric Methods in Computer Vision II, volume 2031, pages 2-13. SPIE, 1993.

[13] J.J. Koenderink and A.J. van Doorn. Two-plus-one-dimensional differential geometry. Pat. Recog. Letters, 15:439-443, May 1994.

[14] I.S. Kweon and T. Kanade. Extracting topological terrain features from elevation maps. Comp. Vis. Graph. Image Proc., 59(2):171-182, Mar. 1994.

[15] D.M. Mark. Topological properties of geographic surfaces: Applications in computer cartography. In G. Dutton, editor, First Int'l Adv. Study Symp. on Topological Data Structures for Geo. Info. Sys., volume 5. Harvard, 1978.

[16] D.M. Mark. Automated detection of drainage networks from digital elevation models. Cartographica, 21:168-178, 1984.

[17] J.C. Maxwell. On hills and dales. Lond. Edin. Dublin Phil. Mag. and J. of Sci., 40(269):421-425, 1870.

[18] J.E. McCormack, M.N. Gahegan, S A. Roberts, J. Hogg, and B.S. Hoyle, Feature-based derivation of drainage networks. IJGIS, 7:263-279, 1993.

[19] C. W. Mitchell, Terrain Evaluation: an introductory handbook to the history, principles, and methods of practical terrain assessment. Longman, Harlow, Essex, 1991.

[20] D.G. Morris and R.W. Flavin. A digital terrain model for hydrology. In Proc. 4th Intl. Symp. Spatial Data Handling, pages 250-262, 1990.

[21] J.F. O'Callaghan and D.M. Mark. The extraction of drainage networks from digital elevation data. Comp. Vis. Graph. Image Proc., 28:323-344, 1984.

[22] O. L. Palacios-Velez and B. Cuevas-Renaud. Automated river-course, ridge and basin delineation from digital elevation data. Journal of Hydrology, 86:299-314, 1986.

[23] T.K. Peucker and N. Chrisman. Cartographic data structures. Amer. Cartog., 2(1):55-69, 1975.

[24] T.K. Peucker and D.H. Douglas. Detection of surface-specific points by local parallel processing of discrete terrain elevation data. Comp. Vis. Graph. Image Proc., 4:375-387, 1975.

[25] J.L. Pfaltz. Surface networks. Geog. Anal., 8:77-93, 1976.

[26] J. Qian, R.W. Ehrich, and J.B. Campbell. DNESYS — An expert system for automatic extraction of drainage networks from digital elevation data. IEEE Trans. Geosci. Remote Sens., 28(1):29-45, Jan. 1990.

[27] R. Rothe. Zum problem des talwegs. Sitz. ber. d. Berliner Math. Gesellschaft, 14:51-69, 1915.

[28] W.W. Seemuller. The extraction of ordered vector drainage networks from elevation data. Comp. Vis. Graph. Image Proc., 47:45-58, 1989.

[29] Y. Shinagawa and T.L. Kunii. Constructing a Reeb Graph automatically from cross sections. IEEE CG&A, 11(6):44-51, 1991.

[30] A.T. Silfer, G.J. Kinn, and J.M. Hassett. A geographic information system utilizing the triangulated irregular network as a basis for hydrologic modeling. In Proc. Auto-Carto 8, pages 129-136, 1987.

[31] S. Takahashi, T. Ikeda, Y. Shinagawa, T.L. Kunii, and M. Ueda. Algorithms for extracting correct critical points and constructing topological graphs from discrete geographical elevation data. In F. Post and M.oGbel, editors, Eurographics '95, volume 14, pages C-181 - C-192. Blackwell Publishers, 1995.

[32] L. Tang. Automatic extraction of specific geomorphological elements from contours. In Proc. 5th Intl. Symp. Spatial Data Handling, pages 554-566, 1992.

[33] D.M. Theobald and M.F. Goodchild. Artifacts of TIN-based surface flow modeling. In Proc. GIS/LIS'90, pages 955-964, 1990.

[34] A.K. Turner. The role of 3-D GIS in subsurface characterization for hydrological applications. In J.F. Raper, editor, Three Dimensional Applications in GIS, pages 115-127. Taylor & Francis, Bristol, 1989.

[35] W. Warntz. The topology of a socio-economic terrain and spatial flows. Papers Regional Sci. Assoc., pages 47-61, 1966.

[36] W. Warntz. Stream ordering and contour mapping. J. Hydro., pages 209-227, 1975.

[37] R. Weibel, An adaptive methodology for automated relief generalization. In Proc. Auto-Carto 8, pages 42-49, 1987.

[38] C. Werner. Several duality theorems for interlocking ridge and channel networks. Water Res. Res., 27(12):3237- 3247, Dec. 1991.

[39] G.W. Wolf. Metric surface networks. In Proc. 5th Intl. Symp. Spatial Data Handling, pages 844-856, 1990.

[40] G.W. Wolf. Hydrologic applications of weighted surface networks. In Proc. 5th Intl. Symp. Spatial Data Handling, pages 567-579. IGU Commission on GIS, 1992.

Contact address
Sidi Yu[1], Marc van Kreveld[2] and Jack Snoeyink[1]
[1]Department of Computer Science
University of British Columbia
Vancouver
Canada
E-mail: syu@cs.ubc.ca
E-mail: snoeyink@cs.ubc.ca
[2]Department of Computer Science
Utrecht University
Utrecht
The Netherlands
E-mail: marc@cs.ruu.nl

VARIATIONS ON SWEEP ALGORITHMS:

EFFICIENT COMPUTATION OF EXTENDED VIEWSHEDS AND

CLASS INTERVALS

Marc van Kreveld

Department of Computer Science
Utrecht University
Utrecht, the Netherlands

Abstract
Two novel applications of the plane sweep paradigm are demonstrated, namely, for the computation of extended viewsheds on gridded DEMs and for class interval selection on TIN-based DEMs. In both cases, the efficiency of the plane sweep algorithm is significantly better than a straightforward approach. The algorithms are presented by first giving the plane sweep method as a general approach that requires some ingredients to make it work well. Adaptations to minimize additional storage use are also presented.

1. Introduction

One of the most important paradigms in the design of geometric algorithms is that of sweeping. In a plane sweep algorithm, an imaginary horizontal line traverses the plane from top to bottom, during which some property of the data is computed (a more clear description is given later). Plane sweep algorithms have been used for a variety of geometric problems like map overlay, Voronoi diagrams and hidden surface removal. The plane sweep approach is described in most textbooks on computational geometry [33,36,38]. Implementation of plane sweep algorithms usually is straightforward. In computational geometry, the plane sweep approach has become a standard technique in the design of efficient algorithms. On the other hand, in GIS literature the plane sweep technique is known---in particular for map overlay [4,28,31]---but not yet a standard technique. Certainly, its use hasn't been recognized to its full extent. The purpose of this paper is to give two new applications of the plane sweep method, showing its importance and versatility once more. Both applications address a problem in the important GIS capability of geographical analysis.

The first plane sweep algorithm that will be described solves various problems in viewshed analysis, such as the computation of extended viewsheds [19,20] and visibility indices [22,43]. Given a gridded DEM and a specific pixel on it, we're interested in information like the number of pixels that are visible from the specific pixel (the visibility index), the vertical distance to visibility for the non-visible pixels, and the vertical distance to the local or global horizons. A straightforward algorithm would do these computations on an n by n grid in $O(n^3)$ time. Our variant plane sweep algorithm requires only $O(n^2 \log n)$ time, saving an order of magnitude. The algorithm is based on a half-line rotating around the specified pixel. Rotating sweep algorithms have been described before, but not for viewshed computations, nor for elevation grids. The additional storage required by the algorithm is only $O(n)$. There have been several other papers dealing with viewshed analysis [11,18,30].

The second application of plane sweep involves the computation of class intervals on DEMs. These are needed for displaying isarithmic maps with appropriate isolines. We now assume that the elevation model is a triangulated irregular network, or TIN. The idea applies to grids as well. Since an elevation model represents some surface, the distribution of the elevation values of that surface is modelled best by a density function (or, frequency distribution).

Once the density function is computed, various class interval selection systems may be used, like percentile (quantile) classes, natural breaks, and bounded within-class variance. It is important that classes are chosen based on the whole elevation model, not the underlying data set of point samples, so that percentile classes ensure equal representation of each class on the isarithmic map. For a good discussion on class interval selection, see Evans [15]. Several textbooks also describe the choice of class intervals, sometimes called setting contour levels or indicator thresholds [7,26,45,46].

The algorithm to compute the density function sweeps a horizontal plane vertically through the TIN. For a TIN with n triangles the algorithm requires $O(n \log n)$ time. The sweep algorithm is the first one---to our knowledge---to sweep a horizontal plane through a TIN. The extra storage required by the algorithm is $O(n)$ in the worst case, but in practice it can be reduced considerably.

Section 2 of this paper explains the basic ideas and applications of the plane sweep method. In Section 3 the viewshed algorithm is presented, in Section 4 the approach toward the selection of class intervals is given, and in Section 5 we show that storage requirements of both algorithms can be reduced. Conclusions are given in Section 6.

2. Sweep algorithms

Imagine a set S of objects in the plane about which something must be computed. Let's say a question Q needs to be addressed.

Sometimes the following intuitive approach works: Take a horizontal line above all objects and sweep it downward. During the sweep, make sure that all information relevant to answer Q becomes known when the sweep line passes the objects that give the information. When the sweep line lies below all objects, we have gathered all relevant information and we can output or construct the answer. A more explicit but still general description follows; see also Figure 1.

During the sweep, we nearly always have to maintain the subset of objects that is intersected by the sweep line. Since this subset changes repeatedly, this subset must be maintained in a dynamic data structure. Often a balanced binary search tree suffices. This data structure is called the *status structure*.

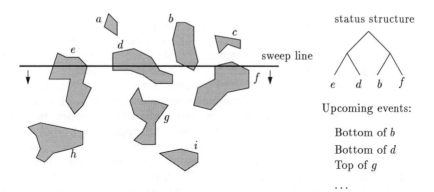

Figure 1 Sketch of the sweep algorithm.

We also have to know somehow when the status changes, and when we can find bits of information to answer the question Q. This happens at certain positions of the sweep line. These positions are called *events* and they are given by a y-coordinate since we assumed the sweep line to be horizontal. Since the sweep line may only move downward, we need to store these events sorted on y-coordinate. The data structure for this purpose is called the *event list*. Often it is simply a priority queue (or a balanced binary tree). The sweep line jumps from event to event by repeatedly removing the one with maximum y-coordinate, until all events have taken place.

There is one more ingredient to sweep line algorithms: the handling of events. This ingredient is the one that varies most from application to application. Two types of event are almost always present: when the sweep line starts intersecting an object of the set S, it must be inserted into the status structure, and when the sweep line has just passed the lowest point of an object of S, it must be removed again. In summary, a sweep line algorithm has the following structure:
1. Initialize the event list.
2. Initialize the status structure.
3. While the event list is not empty do steps 4 to 8.
4. Delete the event with maximum y-coordinate from the event list.
5. If a new object is intersected by the sweep line then insert it in the status structure.
6. If an object stops being intersected by the sweep line then delete it from the status structure.
7. Address the question of interest using the status structure.
8. If necessary, add new events to the event list.

With plane sweep, a two-dimensional geometric problem is solved by using dynamic one-dimensional data structures. More extensive descriptions of plane sweep algorithms can be found in textbooks on computational geometry [33,36,38]. The idea can be generalized to dimension three, where a plane sweeps through space. The status structure now is a dynamic two-dimensional data structure and since these aren't that well-known, three-dimensional sweeping hasn't been applied as often as plane sweep.

Known applications of plane sweep and variants are computation of the Voronoi diagram [21], map overlay [4,31], nearest objects [3,24,25], triangulation [29], hidden surface removal [32,41], two- and three-dimensional point location [8,39,40], separation of point sets [16], rectangle intersection [12], shortest paths [35], median-of-squares statistics [14,42], and many others that don't seem to have immediate applications in GIS. The Geometry Literature Database [23], a database with over 7000 papers related to computational geometry, lists about one hundred papers that employ plane sweep. In most of the two-dimensional applications listed above, the status structure is either a balanced binary tree or a segment tree. Both allow updates and queries to be performed in $O(\log n)$ if the data structure stores $O(n)$ objects. This often results in $O(n \log n)$ time algorithms, plus the time needed to report the answer. The more staightforward algorithms take quadratic time for these problems. For example, one can compute the Voronoi diagram of a set of n sites by computing all Voronoi cells (Thiessen polygons) separately, taking $O(n \log n)$ time per site. In total, this comes down to $O(n^2 \log n)$ time, whereas plane sweep only takes $O(n \log n)$ time.

Some variants of the standard sweep are versions where a line rotates about a point [9,16,34], and the three-dimensional version where a horizontal plane translates through space [37,39]. There are also sweep algorithms where a ' ' "topological" line is used instead of a straight line [13].
Implementation of plane sweep algorithms is not difficult, certainly not if existing code for balanced search trees, sorting, and geometric primitives is used.

A major advantage of sweeping over other methods like divide-and-conquer is that not all objects need be in main memory simultaneously. Objects are needed in the order of their y-coordinates, and only the objects that intersect the sweep line in its current position need be in main memory (in the status structure). There are simple ways to avoid storing the whole event list in main memory [1,2,28]. Divide-and-conquer algorithms usually require all objects to be in main memory at some moment, either during the first divide or during the last conquer step.

Of course, plane sweep is not the appropriate solution to all problems. For certain more difficult tasks, lack of a dynamic geometric data structure prohibits the use of efficient plane sweep. In other cases, plane sweep is just one of several candidates for a problem. Map overlay can also be solved using different techniques, like those based on R-trees [5,44] or on geometric graph traversal [17].

Finally, there are cases where plane sweep is possible but not at all clever. To construct a minimum bounding box of a set of n objects one can use plane sweep and get an $O(n \log n)$ time solution. However, the straightforward inspection of all coordinate values of the objects is much simpler and takes only $O(n)$ time.

3. Viewshed computation

This section presents efficient algorithms for viewshed computations on an n by n grid. We don't just consider standard viewsheds, but also local and global horizon offsets like described by Fisher [19,20]. He notes the use of computing more than just a binary grid of visible/non-visible information, such as, what distance a grid pixel is below the local horizon that obscures it (the *local offset*). In other words, how high a stick must be placed so that it is visible from the viewpoint. Fisher studies the uncertainty aspects of viewshed computation rather than efficiency aspects. A standard algorithm for the visibility information would, however, require $O(n^3)$ time. We show that using sweep, $O(n^2 \log n)$ time suffices, improving the efficiency considerably. Our algorithm is more complex than the straightforward one, but still it is fairly simple to implement. There are other approaches that compute visibility information in DEMs in roughly quadratic time. The ring growing method (described in [22,43]) has the drawback that visibility of a pixel depends not only on the pixels on the line of sight, but also on pixels that---according to the terrain model---cannot block visibility at all. The approximation method that only uses the O(n) lines of sight to the perimeter of the region (described in [22]) doesn't determine visibility to the centers of pixels. Also, it requires a considerable amount of temporary storage space. Our method doesn't have these drawbacks. The sweeping idea can also be applied to visibility computation on a TIN-based terrain model, but this will add to the complexity of the algorithm. The algorithm by Katz et al. [27] is preferred on TINs when the standard viewshed (not extended) is needed.

Suppose a grid G with elevation data is given together with a viewpoint v, which is represented by the center of one of the pixels. The idea of our variant of plane sweep is to rotate a half-line about v a full turn of 2p radians, while computing the visibility or local offset of each pixel when the half-line passes over the pixel center. The status structure stores all pixels intersecting the sweep half-line in the leaves of a balanced binary search tree T, such that the leftmost leaf of T stores the pixel closest to the pixel with the viewpoint, see Figure 2.

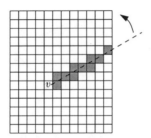

Figure 2 Rotating a half-line over the grid; grey pixels are stored in order of intersection in the status structure.

The tree T is augmented with one real number per leaf and internal node. With each leaf the gradient is stored of the line segment from the center of the pixel with v to the center of the pixel stored in the leaf (the gradient of the line-of-sight LoS). So if the pixel with v has elevation h_v, the pixel in a leaf l has elevation h_l, and the distance between the centers of the pixels is d, we store $\arctan((h_l - h_v)/d)$ in the leaf. In bottom up fashion we define the additional real number stored with internal nodes: it is the maximum of the real numbers stored with its two children. So for each node in T, we have stored the gradient of the ``most obscuring'' pixel in its subtree. One can perform insertions and deletions in $O(\log n)$ time in T, inclusive the updating of the additional real numbers. The method of augmenting data structures, that is, maintaining the additional information in the tree, is described in several textbooks on algorithms (e.g. [10]).

The event list consists of all angles at which the sweep half-line starts to intersect a pixel, at which the sweep half-line stops intersecting a pixel, and at which the sweep half-line goes through a pixel center. So for each pixel exactly three event angles are stored, and in total $O(n^2)$ events will take place. If two angles are the same we let the event of the pixel closer to the viewpoint have preference in the event order.

The sweep algorithm is as follows. Initialize the event list by sorting all the event angles we referred to. Assume for instance we start with a rightward directed half-line (angle 0 with the positive x-axis) and we rotate counterclockwise. Repeatedly take an event---the first one---from the event list and depending on the type of event, insert a pixel in T, delete a pixel from T, or search in T. The latter is done when the event is caused by the center of a pixel, and the algorithm must establish whether it is visible (and if not, its local offset). The pixel is visible if and only if the LoS to it has greater gradient than of all pixels closer to the viewpoint, since only these may obstruct visibility.

We follow the search path to the pixel whose center the sweep half-line has reached, and we consider the nodes in T nearest to the root that are left children of the nodes where the search path turned right. In Figure 3, a search path and the corresponding left children are indicated. Next we determine the maximum of the real numbers stored with these left children (0.25 in Figure 3). This maximum is the maximum gradient of the LoS to a pixel closer to the viewpoint.

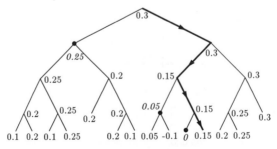

Figure 3 A search path in T.

By comparing the maximum gradient to the gradient of the pixel itself we establish whether the pixel is visible. If not, we use the gradients and the distance of the pixel to the viewpoint to determine the local offset.

Observe that at most $2n$ pixels are stored at any time in T. Since T is balanced, its depth is $O(\log n)$. A search follows one path to a leaf, and the left children of this path. There can only be $O(\log n)$ left children on a path of length $O(\log n)$, and all further computations are trivial. So inserting, deleting, and searching only take $O(\log n)$ time per operation on an n by n grid. Since there are $O(n^2)$ events, and each event is handled in $O(\log n)$ time, we have given an $O(n^2 \log n)$ time algorithm. The initial sorting of events by angle also takes $O(n^2 \log n)$ time. Note that in our model we have chosen to store the gradient to the center of the pixel in the leaves of the tree T. When the LoS passes over a pixel, it usually doesn't pass over the pixel center, which may influence visibility of the pixels beyond it slightly. It depends on the model we assume to hold for the grid entries. We omit this issue here; adaptations to the algorithm can be made to incorporate different grid interpretations.

The algorithm that was given determined visibility of every pixel from the viewpoint, and therefore it can be used to compute the visibility index as well. We simply record how many pixels were visible from v. It is easy to adapt the algorithm for global offsets; we won't discuss this any further. Similarly, one could compute the number of local horizons in each direction, or the number of visible local horizons, with a similar sweep. It comes down to augmenting the status structure T in a different manner.

4. Density function computation and class interval selection

It is well-known that a finite population of interval data can be described by a histogram. For continuous interval data, the density function---or frequency distribution---is the corresponding descriptive statistic. It shows how frequent each elevation occurs in the data. The density function plays an important role in the choice of class intervals. This section studies the computation of the density function of a TIN that models a continuous univariate variable. Similar ideas can be used for some other elevation models, though. Note that it is more appropriate to compute class intervals based on the density function than on the finite set of sampled locations that has been the basis of the data. The presense of auto-correlation in the data will cause that class intervals based on the sample will not classify in a fair way, unless random sampling was used. By interpolating between the data points this problem is overcome (see e.g. [15,26]). A triangulation is one example of an interpolation method.

We begin with a useful observation and a straightforward algorithm.
Consider just one triangle D in 3-space with vertices u,v,w. Assume for simplicity that $h(u) > h(v) > h(w)$, where $h(..)$ denotes the elevation of a vertex. Then the density on the triangle for a given elevation t is $l.\cos(a)$, where l is the length of the intersection of the triangle D with the plane $z=t$, and a is the angle between the normal of D and any horizontal plane. The density is zero for all elevations t with $t > h(u)$ or $t < h(w)$. It is given by a function f_{uv} depending

linearly on t if $h(u) > t > h(v)$, and it is given by a different function f_{vw} depending linearly on t if $h(v) > t > h(w)$. So we have $f_{uv}(t) = a.t + b$, where a and b depend only on the coordinates of u, v, w and thus are fixed. The same holds for f_{vw}, but with different a and b.

For simplicity of exposition we assume that all vertices have different elevations. This restriction can be overcome without problems, but some care must be taken.

Let $v_1, \ldots, v_n$ be the vertices of the TIN, and assume that they are sorted on decreasing elevation.

This holds without loss of generality because we can simply relabel the vertices to enforce $h(v_1) > h(v_2) > \ldots > h(v_n)$. Consider the density for an elevation t, where t is in $(h(v_j), h(v_{j+1}))$. In such an open interval, the density is the sum of a set of linear functions, which is again a function linear in t. We denote the linear function that gives the density over the whole TIN in the interval $(h(v_j), h(v_{j+1}))$ by $F_j(t)$. So the linear functions $F_0, F_1, \ldots, F_n$ form the density function, where each function is only valid in its interval. By default we set $F_0(t) = 0$ and $F_n(t) = 0$ for the intervals $(h(v_1), \infty)$ and $(-\infty, h(v_n))$, because for these elevations the density is zero. We observe:

Proposition: The density function based on a TIN with n vertices is a piecewise linear continuous function with at most $n+1$ pieces (see Figure 4).

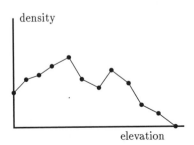

density

elevation

Figure 4 Density function of a TIN.

The density function need not be continuous when there are vertices with the same elevation. The straightforward algorithm to construct the density function on the TIN is the following: Sort the vertices by elevation, and for each interval $(h(v_j), h(v_{j+1}))$, determine the set of linear functions contributing to it. Then add up these linear functions to get one linear piece F_j of the density function. Since we have $O(n)$ vertices, we have $O(n)$ intervals and for each we can easily determine in $O(n)$ time which linear functions contribute. The total time taken by this algorithm is $O(n^2)$.

The efficient computation of the density function is again based on the sweeping approach. We will exploit the fact that the linear function F_j can be obtained easily from the linear function F_{j-1} since the contributing f are for the larger part the same ones.

We compute the summed linear functions F from top to bottom, which comes down to a sweep with a horizontal plane through the TIN. Throughout the sweep we maintain the density function of the current elevation. Using sweeping terminology, every vertex of the TIN gives rise to one event. The event list is a priority queue storing all these $O(n)$ events in order of decreasing elevation. With each event we store a pointer to the vertex of the TIN that will cause the event. The status structure is trivial: it is simply the summed linear function F for the current position of the sweep plane.

The final ingredient to the sweep algorithm is handling the events. When considering how F_{j-1} should be changed to get F_j when the sweep plane passes the vertex with elevation $h(v_j)$, we must examine how the density function changes. The vertex v_j is incident to some triangles, for which it can be the highest vertex, the lowest vertex or the vertex with middle elevation. We update F_{j-1} to get F_j according to the following rules:

1. For all triangles for which v_j is the lowest vertex (lightly shaded in Figure 5), we subtract from F the appropriate linear function (f_1 in Figure 5).
2. For all triangles for which v_j is the highest vertex (darkly shaded), we add to F the appropriate linear function (f_6 and f_7).
3. For all triangles for which v_j is the middle vertex (white in the figure), we subtract the one linear function (f_2 and f_4) and add the other (f_3 and f_5).

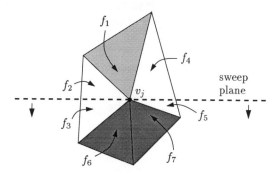

Figure 5 Passing a vertex with the sweep plane.

We don't need to precompute or store the linear functions f on each triangle to update F; the f can be obtained from the coordinates of the vertices on the TIN when the event at vertex v_j is handled. We have fast access to vertex v_j in the TIN; recall that an extra pointer was stored in the event list.

We also evaluate the function F_j at the event.

The sequence of evaluations gives the breakpoints of the (piecewise linear) density function. These breakpoints are computed from right to left in Figure 4 since the sweep goes from high to low elevations.

Considering the efficiency of the algorithm, the initial sorting of the events takes $O(n \log n)$ time for a TIN with n vertices. Extraction of an event takes $O(\log n)$ time; for all events this adds up to $O(n \log n)$ time. Updating the status structure at an event v_j requires time linear in the number of triangles incident to v_j. Summed over all vertices this is linear in n by Euler's formula. The evaluation to determine the breakpoints requires constant time per event.

So in total the sweep algorithm requires O(n log n) time.

Once the density function is computed, the class intervals may be determined. Suppose as an example that the objective is to determine seven classes such that each class occupies an equivalent amount of area on an isoline map. We assume that the isoline map and the TIN have the same domain, otherwise we can clip the TIN with the domain of the isoline map before doing the sweep. The total area of the isoline map is the same as the total area under the density function and is denoted A.

The area under the density function in the elevation interval [a,b] is denoted $A(a,b)$. If $F(t)$ denotes the (piecewise linear) density function, then

$$A(a,b) = \int_a^b F(t)dt.$$

The value of $A(a,b)$ is exactly the area for the class [a,b] on the isoline map. We know the total area A and compute $A/7$, the desired area for each class. We then determine the lowest elevation such that $A/7$ of the area is below that elevation. This operation is easy by scanning over the known density function $F(t)$ from left to right and maintaining the area under $F(t)$ (this is also a kind of sweep). This gives the lowest class boundary. Continuing the scan gives all six boundaries of the seven classes in O(n) time.

In a similar way one can compute a non-fixed number of classes with the property that the within-class variance is less than or equal to a certain threshold, for each class.

Finally, the density function can be used class interval selection by natural breaks in the data: They are the local minima of $F(t)$. We refer to Burrough [7] and Evans [15] for other classification schemes.

5. Reducing the storage requirements

In both of the given sweep algorithms, the event list requires far more storage than the status structure. For the viewshed computations, the status structure required O(n) storage and the event list O(n^2) storage, since we stored all upcoming events in the event list before the sweep started. Similarly, for the density function computation, the status structure required O(1) storage and the event list O(n) storage.

In both algorithms we can bring down the storage requirements of the event list using the same idea. Observe that for a sweep algorithm to work well, we need not have all upcoming events in the event list initially. We only need to be sure that if some event is the next one to be handled (reached by the sweep), then it must be in the event list. The underlying idea is the same as Brown's improvement in storage [6] of the well-known Bentley-Ottmann sweep algorithm [4].

Consider the viewshed algorithm of Section 3. Instead of storing all $O(n^2)$ events, we will only store the events caused by pixels that are not yet encountered, but which have a neighboring pixel that has been encountered. It is easy to see that there are only $O(n)$ of these pixels, and that every pixel is present in the event list before it causes an event. The algorithm has to be adapted slightly: when the event to be handled next is an insertion of a pixel to the status structure, then we must also insert its neighbors in the event list (unless they were already present, or have already been passed by the sweep).

Consider the density function computation of Section 4. The storage needed by the algorithm is $O(n)$ for the event list, but this can be reduced using a simple observation. Every vertex except the local maxima (peaks) have a higher neighbor in the TIN. So we can initialize the event list with the local maxima only. When the event at a vertex v is handled, we insert all lower neighbors of v in the event list. This guarantees that every event is present in the event list when the sweep plane reaches it. The storage required by the algorithm is at most linear in the sum of the number of local maxima and the number of edges in the largest complexity cross-section.

6. Conclusions

This paper presented two new sweep algorithms that are useful in GIS applications. We showed that extended viewsheds can be computed in $O(n^2 \log n)$ time on an n by n grid of elevation data, and that the density function of an elevation model based on a TIN with n vertices can be determined in $O(n \log n)$ time. The density function is the basis of various class interval systems other than the equal-class-width classes. Both algorithms are simple to implement and require only little extra storage. In both cases we reduced the computation time by an order of magnitude, when compared to the more straightforward algorithms. This demonstrates the use and relevance of one of the most important geometric algorithms design methods, namely, that of plane sweep. This type of algorithm has much more promise in geographic information systems than is recognized to date.

References
[1] L. Arge. The Buffer Tree: A new technique for optimal I/O-algorithms. In Proc. WADS, number 955 in Lect. Notes in Comp. Science, pages 334-345.
[2] L. Arge. External-storage data structures for plane-sweep algorithms. Technical Report RS-94-16, BRICS, Aarhus Univ., Denmark, 1994.
[3] F. Bartling and K. Hinrichs. A plane-sweep algorithm for finding a closest pair among convex planar objects. In Proc. 9th Sympos. Theoret. Aspects Comput. Sci., volume 577 of Lecture Notes in Computer Science, pages 221-232. Springer-Verlag, 1992.
[4] J.L. Bentley and T.A. Ottmann. Algorithms for reporting and counting geometric intersections. IEEE Trans. Comput., C-28:643-647, 1979.
[5] T. Brinkhoff, H.-P. Kriegel, and B. Seeger. Efficient processing of spatial joins using R-trees. In Proc. ACM SIGMOD, pages 237-246, 1993.
[6] K.Q. Brown. Comments on "Algorithms for reporting and counting geometric intersections". IEEE Trans. Comput., C-30:147-148, 1981.
[7] P.A. Burrough. Principles of Geographical Information Systems for Land Resources Assessment. Oxford University Press, New York, 1986.

[8] R. Cole. Searching and storing similar lists. J. Algorithms, 7:202-220, 1986.

[9] R. Cole, M. Sharir, and C.K. Yap. On k-hulls and related problems. SIAM J. Comput., 16:61-77, 1987.

[10] T.H. Cormen, C. E. Leiserson, and R.L. Rivest. Introduction to Algorithms. The MIT Press, Cambridge, Mass., 1990.

[11] L. De Floriani, G. Nagy, and E. Puppo. Computing a line-of-sight network on a terrain model. In Proc. 5th Int. Symp. on Spatial Data Handling, pages 672-681, 1992.

[12] H. Edelsbrunner. A new approach to rectangle intersections, Part II. Internat. J. Comput. Math., 13:221-229, 1983.

[13] H. Edelsbrunner and L.J. Guibas. Topologically sweeping an arrangement. J. Comput. Syst. Sci., 38:165-194, 1989. Corrigendum in 42 (1991), 249-251.

[14] H. Edelsbrunner and D.L. Souvaine. Computing median-of-squares regression lines and guided topological sweep. J. Amer. Statist. Assoc., 85:115-119, 1990.

[15] I.S. Evans. The selection of class intervals. Trans. Inst. Br. Geogrs., 2:98-124, 1977.

[16] H. Everett, J.-M. Robert, and M. van Kreveld. An optimal algorithm for the ($\leq k$)-levels, with applications to separation and transversal problems. In Proc. 9th Annu. ACM Sympos. Comput. Geom., pages 38-46, 1993.

[17] U. Finke and K.H. Hinrichs. The quad view data structure - a representatio n for planar subdivisions. In Advances in Spatial Databases (proc. SSD'95), number 951 in Lect. Notes in Comp. Science, pages 29-46, 1995.

[18] P.F. Fisher. Algorithm and implementation uncertainty in viewshed analysis. Int. J. of GIS, 7:331-347, 1993.

[19] P.F. Fisher. Stretching the viewshed. In Proc. 6th Int. Symp. on Spatial Data Handling, pages 725-738, 1994.

[20] P.F. Fisher. Reconsideration of the viewshed function in terrain modelling. Geographical Systems, 3:33-58, 1996.

[21] S.J. Fortune. A sweepline algorithm for Voronoi diagrams. Algorithmica, 2:153-174, 1987.

[22] W.R. Franklin and C.K. Ray, Higher isn't necessarily better: visibility a lgorithms and experiments. In Proc. 6th Int. Symp. on Spatial Data Handling, pages 751-770, 1994.

[23] Geometry Literature Database. http://www.cs.ruu.nl/people/otfried/html/geombib.html.

[24] T. Graf and K. Hinrichs. A plane-sweep algorithm for the all-nearest-neighbors problem for a set of convex planar objects. In Proc. 3rd Workshop Algorithms Data Struct., volume 709 of Lecture Notes in Computer Science, pages 349-360, 1993.

[25] K. Hinrichs, J. Nievergelt, and P. Schorn. A sweep algorithm for the all-nearest-neighbours problem. In Computational Geometry and its Applications, volume 333 of Lecture Notes in Computer Science, pages 43-54. Springer-Verlag, 1988.

[26] E.H. Isaaks and R.M Srivastava. An Introduction to Applied Geostatistics. Oxford University Press, New York, 1989.

[27] M.J. Katz, M. Overmars, and M. Sharir, Efficient output-sensitive hidden surface removal for objects with small union size. Comput. Geom. Theory Appl., 2:223-234, 1992.

[28] H.-P. Kriegel, T. Brinkhoff, and R. Schneider. The combination of spatial a ccess methods and computational geometry in geographic database systems. In Advances in Spatial Databases (proc. SSD'91), number 525 in Lect. Notes in Comp. Science, pages 5-21, 1991.

[29] D.T. Lee and F.P. Preparata. Location of a point in a planar subdivision and its applications. SIAM J. Comput., 6:594-606, 1977.

[30] J. Lee. Analyses of visibility sites on topographic surfaces. Int. J. of GIS, 5:413-430, 1991.

[31] H.G. Mairson and J. Stolfi. Reporting and counting intersections between two sets of line segments. In R. A. Earnshaw, editor, Theoretical Foundations of Computer Graphics and CAD, volume F40 of NATO ASI, pages 307-325. Springer-Verlag, Berlin, West Germany, 1988.

[32] M. McKenna. Worst-case optimal hidden-surface removal. ACM Trans. Graph., 6:19-28, 1987.

[33] K. Mehlhorn. Multi-dimensional Searching and Computational Geometry, volume 3 of Data Structures and Algorithms. Springer-Verlag, Heidelberg, West Germany, 1984.

[34] A. Mirante and N. Weingarten. The radial sweep algorithm for constructing triangulated irregular networks. IEEE Comput. Graph. Appl., pages 11-21, May 1982.

[35] J.S.B. Mitchell. L$_1$ shortest paths among polygonal obstacles in the plane. Algorithmica, 8:55-88, 1992.

[36] J. O'Rourke. Computational Geometry in C. Cambridge University Press, New York, 1994.

[37] M.H. Overmars and C.-K. Yap. New upper bounds in Klee's measure problem. SIAM J. Comput., 20:1034-1045, 1991.
[38] F.P. Preparata and M.I. Shamos. Computational Geometry: An Introduction. Springer-Verlag, New York, NY, 1985.
[39] F.P. Preparata and R. Tamassia. Efficient point location in a convex spatial cell-complex. SIAM J. Comput., 21:267-280, 1992.
[40] N. Sarnak and R.E. Tarjan. Planar point location using persistent search trees. Commun. ACM, 29:669-679, 1986.
[41] M. Sharir and M.H. Overmars. A simple output sensitive hidden surface removal algorithm. ACM Trans. Graph., 11:1-11, 1992.
[42] D.L. Souvaine and J.M. Steele. Time- and space- efficient algorithms for least median of squares regression. J. Amer. Statist. Assoc., 82:794-801, 1987.
[43] Y.A. Teng and L.S. Davies. Visibility analysis on digital terrain models and its parallel implementation. Technical Report CAR-TR-625, Center for Automation Research, University of Maryland, 1992.
[44] P. van Oosterom. An R-tree based map-overlay algorithm. In Proc. EGIS'94, pages 318-327, 1994.
[45] D.F. Watson. Contouring - a Guide to the Analysis and Display of Spatial Data. Pergamon, Oxford, 1992.
[46] R. Webster and M.A. Oliver. Statistical Methods in Soil and Land Resource Survey. Oxford University Press, New York, 1990.

Contact address
Marc van Kreveld
Deptartment of Computer Science,
Utrecht University
Utrecht
The Netherlands
E-mail: marc@cs.ruu.nl

THE INFLUENCE OF CELL SLOPE COMPUTATION

ALGORITHMS ON A COMMON FOREST MANAGEMENT

DECISION

Dr. Robert C. Weih[1] and Dr. James L. Smith[2]

[1]School of Forest Resources
University of Arkansas
Monticello, USA
[2]Canal Forest Resources, Inc. Charlotte
Charlotte, USA

Abstract
While applications of Geographic Information Systems (GIS) have progressed from a descriptive tool to a decision making and modeling tool, an understanding of the errors and variability of the components of a GIS has lagged behind. Slope is one of these components. Eight different previously-used methods for determining cell slope values were compared using elevation data from the USGS Big Stone Gap, Virginia Digital Elevation Model. The 28 pair wise comparisons were all statistically different, but for most practical applications six of the methods produced similar results. In a common natural resource decision model, the effect of changing the method used to determine cell slope values caused up to a 4.5 times difference in the area deemed unsuitable for timber production.

Key words
Slope, DEM, GIS, raster, forestry.

1. Introduction

Slope is the rate of change in altitude between points on a surface. In natural resource management, many analyses rely on slope as an important component of the decision making process, for example, Kessell (1979) used slope as a component in developing a fire management information system for Glacier National Park. The slope of a land surface proved to be important in determining the erodability of soils (Meijerink, 1986). Slope was an important information layer in the application of GIS to estimate peak flow for a watershed (Chieng and Luo, 1993). Katsaridis and Tsigourakos (1993) found slope to be an important layer for land use planning using a GIS. McHarg (1969) considered slope an important factor to consider for ecological planning.

Before the development of Geographic Information Systems (GIS), the most common method for determining the slope of a surface utilized information collected in the field at sample points using some type of angle measuring instrument, or slope was calculated from elevation measurements taken on a topographic map. In a GIS, slope is usually calculated by incorporating elevation data from a Digital Elevation Model (DEM) into one of many mathematical computation algorithms. The calculation of slope, and the errors and assumptions associated with it are invisible to the user of the GIS.

GIS developers don't always document the slope model they use, and different slope algorithms may produce significantly different results from the exact same elevation data.

The objectives of this paper are two-fold. First, to document the statistical and practical differences in slope values for a set of well-known computation methods. Second, to evaluate the effect of changing slope values on a typical forest management decision. In each case, we will utilize actual data similar to that encountered in operational situations.

2. Materials and methods

2.1 Basic data

The area chosen for this study is located in Wise and Lee counties in southwestern Virginia. The selected area consists of ridges with steep slopes and flat valleys, and the elevation ranges from 424 meters (1,391 feet) to 1,079 meters (3,540 feet). The elevations used in this study were from the United States Geological Survey (USGS) Big Stone Gap DEM. The USGS created the DEM in March of 1976 using the Gestalt Photo Mapper II System. No statistical information pertaining to the horizontal and vertical accuracy of this DEM was available. It was assumed that the DEM was a reasonably accurate representation of the Big Stone Gap quadrant surface. The accuracy of the DEM in representing the surface did not affect the comparative aspects of this study, and placed the study in the context of situations that are encountered in an operational sense. Cell size for the DEM was 30 meters, and the reference system was the Universal Transverse Mercator (UTM) projection. Elevation values were recorded to the nearest meter.

Forest management decisions are typically based on administrative or biologic polygons called stands. In order to emulate a real decision-making situation, actual stand polygons were defined using methods common in this region, and digitized in vector format. The polygons used in this study met the following criteria.
- Polygons included in the study do not have a common outside boundary with the USGS 72 quadrangle.
- Polygon elevation values must be available at least 60 meters from the polygon perimeter.

Two hundred and forty polygons with a wide variety of shapes and surface complexities were used in the study (Figure 1). Polygon areas ranged from .6 hectare (1.5 acres) to 371.4 hectares (917.8 acres) with a mean of 28.8 hectares (71.2 acres).

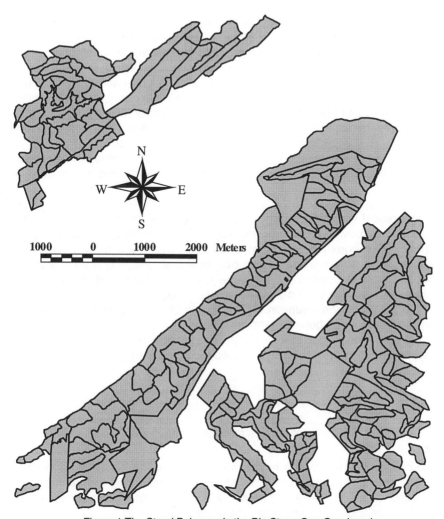

Figure 1 The Stand Polygons in the Big Stone Gap Quadrangle.

A vector-to-raster program written by the author was used to convert the 240 polygons into a raster format. This transformation from vector-to-raster generalizes the original map data into 30 meter cells. This had no effect on the outcome of the study except that the areas for the rasterized polygons were always greater than the area of the polygons in their original vector form, because each individual polygon was rasterized. The vector-to-raster program used Bresham's algorithm and a modified YX scan algorithm discussed by Newman and Sproull (1979). If the cell was in the polygon, or on the border, a one was recorded; otherwise, a zero was recorded for the cell in the raster polygon file. The bounding rectangle for each of the 240 polygons plus 90 meters on each side was used to clip the appropriate elevations from the USGS DEM. This created 240 DEMs that were subunits of the original USGS DEM.

2.2 Slope calculation methods

The slope of a surface is commonly defined as the rate of change in altitude of that surface. A special case of slope, called gradient, is the maximum rate of altitude change. Both slope and gradient computation algorithms are examined in this study, and both are called slope. While this is incorrect terminology in a technical sense, it is an accurate characterization of how these types of algorithms are applied. GIS practitioners only rarely distinguish between slope and gradient in their applications, and software documentation is commonly incomplete or confusing on the differences. In a study such as this one, it was important to examine as many of the potential options as possible. Since there is no universally accepted terminology for these slope/gradient algorithms, they will be referred to as methods one through ten.

The theoretical considerations for slope calculation are briefly discussed in this section to provide a common framework for comparing the different slope methods. The equation $z = f(x, y)$ describes points on a three dimensional terrain surface with z equal to the perpendicular distance from the terrain surface to point P(x, y) lying on a plane referenced by X and Y coordinates. The elevation of the terrain surface at point P(x, y, z) is the perpendicular distance z. When the equation for the terrain surface consisting of points is expressed in a more general form as $F(x, y, z) = c$ and c is a constant set equal to zero then $F(x, y, z) = z - f(x, y)$. The tangent plane at P(x_0, y_0, z_0) can be defined by the normal vector, provided the slope is not zero (Zill, 1985).

$$\vec{N} = \nabla F(x_0, y_0, z_0) = -\frac{\partial f}{\partial x}\hat{i} - \frac{\partial f}{\partial y}\hat{j} + \hat{k} \tag{1}$$

$\hat{i}, \hat{j}, \hat{k}$ = orthogonal vectors directed along the x, y and z axis

Slope can be defined by the magnitude of tilt from the horizontal of a plane perpendicular to the normal vector, which is equivalent to the zenith angle. The slope angle in degrees can be derived from the components of the normal vector and expressed using the following equation.

$$\tan \phi = \sqrt{\left[(\frac{\partial f}{\partial x})^2 + (\frac{\partial f}{\partial y})^2 \right]} \tag{2}$$

For all the methods discussed in this study, slope is derived locally for each cell using elevations from a 3 x 3 cell grid window (neighbors) of the altitude matrix. All the slope methods discussed in this study treat an elevation value of a cell as a point elevation, even though the value represents a 30 x 30 meter area.

The location of the point elevation is at the centroid of the cell. The notation that is used in this study to describe the elevations of the cell neighbors in the slope method equations is shown in Figure 2.

The west-east coordinate direction will be denoted by X, south-north direction by Y and the elevation (altitude) by Z. The equation for calculating slope in degrees shown below, can be derived from equation 2.

$$\tan \phi = \sqrt{\left[\left(\frac{\delta Z}{\delta X} \right)^2 + \left(\frac{\delta Z}{\delta Y} \right)^2 \right]} \qquad (3)$$

Substituting $\left(\frac{\delta Z}{\delta X} \right)$ for $\left(\frac{\partial f}{\partial x} \right)$ and $\left(\frac{\delta Z}{\delta Y} \right)$ for $\left(\frac{\partial f}{\partial y} \right)$

Slope Method One
Slope method one is the most common method for calculating slope in the literature mainly because it is computationally simple and easy to program. Numerous papers have presented this method for calculating slope using a digital elevation model (Fleming and Hoffer,1979; Horn, 1982). The steps involved in determining slope of cell Z_0 using this method are shown below.

$$[\delta Z / \delta X] = [Z_2 - Z_6] / 2\Delta X \qquad (4)$$

$$[\delta Z / \delta Y] = [Z_4 - Z_8] / 2\Delta Y \qquad (5)$$

ΔX is the spacing between points in the horizontal direction
ΔY is the spacing between points in the vertical direction

With substitutions into equation 3, the slope of the cell Z_0 can be determined.

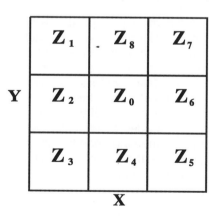

Figure 2 Notation for Elevations used in Computing Slope from an Altitude Matrix Window.

Slope Method Two
Slope method two was used by O'Neill and Mark (1987) to study the frequency distribution of land slope using 18 $7\frac{1}{2}$ minute USGS DEMs. The cell to the north (Y axis) and to the east (X axis) were used to define a triangle as shown in Figure 3. This forms a horizontal right triangle plane with sides

equal to cell width, which is 30 meters for a $7^1/_2$ USGS DEM. There are two steps involved in determining the slope of Z_0 using this method.

$$[\delta Z / \delta X] = [Z_0 - Z_6] / \Delta X \qquad\qquad (6)$$

$$[\delta Z / \delta Y] = [Z_0 - Z_8] / \Delta Y \qquad\qquad (7)$$

With the appropriate substitutions into the equation 3 the slope of cell Z_0 can be determined.

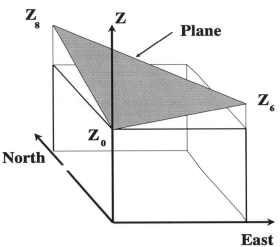

Figure 3 Horizontal Right Triangle Plane used to Calculate Slope.

Slope Method Three
Slope method three of determining slope is used in VIEWIT, a computer program developed by the U. S. Forest Service to determine the terrain visible from a single point or a group of points (Travis et al., 1975). Struve (1977) describes in detail how you can use this method to determine slope. This method is also used in the Map Analysis Package (MAP) (Tomlin, 1980). Slope is determined for Z_0 by calculating the slope from Z_0 to each of its eight neighbors by taking the absolute value of the difference in elevation between Z_0 and one of its neighbors divided by the cell size. The maximum slope of the eight calculated slopes is then assigned to cell Z_0. IDRISI (1990) modifies this method by assigning the maximum absolute slope value of only the neighbors above, below and to either side of Z_0.

Slope Method Four
Slope method four, which is called the plane algorithm by Struve (1977), assigns the maximum slope of the four surrounding right triangle planes that have Z_0 as a common point. The slope for each plane is calculated similarly to Method Two.

Slope Method Five

The eight neighbors of cell Z_0 are used to calculate the maximum slope of two, three dimensional surfaces which are called S and S'. Surface S uses the four nearest neighbors which are Z_2, Z_4, Z_6 and Z_8 to determine the partial derivative for the X and Y directions. The next-nearest neighbors Z_1, Z_3, Z_5 and Z_7 are used to determine the partial derivative for the X' and Y' directions of surface S'. Figure 4 shows the local coordinate system used by this method.

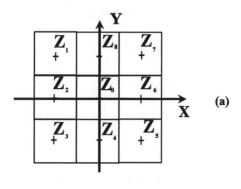

(a)

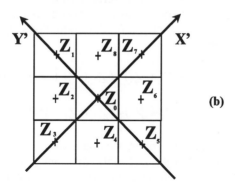

(b)

Figure 4a Coordinate System used by Method Five to Calculate Slope of Cell Z_0
(Struve, 1977) (a) Nearest- Neighbors of Surface S.
Figure 4b Coordinate System used by Method Five to Calculate Slope of Cell Z_0
(Struve, 1977) (b) Next Nearest of Surface S'.

$$[\partial f / \partial x] = [\delta Z / \delta X] = [Z_2 - Z_6] / 2\Delta X \qquad (8)$$

$$[\partial f / \partial y] = [\delta Z / \delta Y] = [Z_4 - Z_8] / 2\Delta Y \qquad (9)$$

$$[\partial f / \partial x'] = [\delta Z / \delta X'] = [Z_3 - Z_7] / 2\sqrt{2}\Delta X \qquad (10)$$

$$[\partial f / \partial y'] = [\delta Z / \delta Y'] = [Z_5 - Z_1] / 2\sqrt{2}\Delta Y \qquad (11)$$

Using the above equations the maximum absolute value derived from the partial derivatives in each direction is substituted into equation 3 to calculate the slope. The partial derivatives do not necessarily have to come from the same surface. Struve (1977) thoroughly describes the benefits of using this method in calculating slope from an altitude matrix.

Slope Method Six
Burrough (1986), Horn (1981, 1982) and Skidmore (1989) proposed a third-order finite difference method for calculating slope. Horn (1981) stated that the method gave good results because it is based on numerical analysis. The method uses a weighting of three central differences.

$$[\delta Z / \delta X] = [(Z_1 + 2Z_2 + Z_3) - (Z_7 + 2Z_6 + Z_5)] / 8\Delta X \qquad \textbf{(12)}$$

$$[\delta Z / \delta Y] = [(Z_1 + 2Z_8 + Z_7) - (Z_3 + 2Z_4 + Z_5)] / 8\Delta Y \qquad \textbf{(13)}$$

By substituting the results of the above equations into equation 3, the slope value is calculated for Z_0.

Slope Method Seven
Sharpnack and Akin (1969) proposed a third-order finite difference model for calculating slope of Z_0 from a 3 x 3 window which is similar to method six. The only difference is a change in the weighting of cell Z_2 and Z_8. Using the Sharpnack and Akin (1969) slope model, equations 12 and 13 can be rewritten as shown below.

$$[\delta Z / \delta X] = [(Z_1 + Z_2 + Z_3) - (Z_7 + Z_6 + Z_5)] / 6\Delta X \qquad \textbf{(14)}$$

$$[\delta Z / \delta Y] = [(Z_1 + Z_8 + Z_7) - (Z_3 + Z_4 + Z_5)] / 6\Delta Y \qquad \textbf{(15)}$$

The third-order finite difference model for calculating slope of Z_0 can have any reasonable weighting scheme, but only this weighting scheme and the method six weighting scheme have been proposed in the literature.

Slope Method Eight
Slope method eight is used in the program TOPO III (Grender, 1976). The program calculates the maximum and average slope for cell Z_0, by determining the slope from Z_0 to the perimeter of a circle surrounding Z_0. The radius of the circle can range from one grid cell to one half the minimum altitude matrix dimension. The radius size successively doubles (30, 60, 120, 240, etc.) in the program. Because of its limited application in the literature, this method will not be included in further analyses nor discussed further in this paper.

Slope Method Nine
This method uses a multiple linear regression model proposed by Travis et al. (1975) and Skidmore (1989). A surface is fitted to a 3 x 3 cell window using least squares. The least squares minimize the difference between the fitted surface and the cell elevations. The regression model is shown below.

$$Z_i = \beta_0 + \beta_1 X_i + \beta_2 Y_i + \varepsilon_i \qquad (16)$$

Assuming that the $E(\varepsilon_i) = 0$, the regression model for the model is:

$$E(Z) = \beta_0 + \beta_1 X_+ \beta_2 Y \qquad (17)$$

Taking the partial derivatives with respect to X and Y and substituting them into equation 3 the slope value for cell Z_0 can be determined.

Substituting $[(\partial E(Z)/\partial X] = \beta_1$ for $(\delta Z/\delta X)$ $\qquad$ (18)

Substituting $[\partial E(Z)/\partial Y] = \beta_2$ for $(\delta Z/\delta Y)$ $\qquad$ (19)

This method uses all nine elevation values to fit the surface and determines the slope of cell Z_0.

Slope Method Ten
Evans (1980), Young (1978) and Skidmore (1989) proposed fitting a full quadratic multiple regression to the nine elevation values of a 3 x 3 window to determine the slope value of Z_0. The regression model is shown below.

$$Z_i = \beta_0 + \beta_1 X_i + \beta_2 Y_i + \beta_3 X_i^2 + \beta_4 Y_i^2 + \beta_5 X_i Y_i + \varepsilon_i \qquad (20)$$

Assuming that $E(\varepsilon_i) = 0$, the regression model is shown below:

$$E(Z) = \beta_0 + \beta_1 X + \beta_2 Y + \beta_3 X^2 + \beta_4 Y^2 + \beta_5 XY \qquad (21)$$

By taking the partial derivatives with respect to X and Y and then substituting them into equation 3 the slope value for cell Z_0 can be derived.

Substituting $[\partial E(Z)/\partial X] = \beta_1 + \beta_3 2X + \beta_5 Y$ for $(\delta Z/\delta X)$ $\qquad$ (22)

Substituting $[\partial E(Z)/\partial Y] = \beta_2 + \beta_4 2Y + \beta_5 X$ for $(\delta Z/\delta Y)$ $\qquad$ (23)

Evans (1980) showed that if the data are arranged in a square grid, the slope value of Z_0 with a local window coordinate of 0, 0 can be determined using the equations shown below. These equations are more computationally efficient than determining the slope value of Z_0 using equations 22 and 23.

$$[\delta Z/\delta X] = [(Z_1 + Z_2 + Z_3) - (Z_7 + Z_6 + Z_5)]/6\Delta X \qquad (24)$$

$$[\delta Z/\delta Y] = [(Z_1 + Z_8 + Z_7) - (Z_3 + Z_4 + Z_5)]/6\Delta Y \qquad (25)$$

Equations 24 and 25 are identical to equations 14 and 15of slope method seven respectively, which is the method proposed by Sharpnack and Akin (1969). Since slope methods 7 and 10 are mathematically equivalent, only the results of slope method 7 will be included in future discussions.

2.3 Comparing slope methods based on cell slope values

A cell by cell slope value comparison for the entire study area was performed for each slope method. Even though the ultimate concern of this study was to evaluate a decision model, an understanding of the effects of slope calculation methods on individual cell slope values was important since polygon slope is based upon an aggregation of individual cell slope values. It should be noted here that this study did not attempt to determine which slope method was "best," but was aimed at determining if there were differences between the methods, and if, so what was their magnitude and direction of that difference. The "best" slope method could not be determined because actual DEM data were used in this study, and thus the "true" slope values were unknown. Simple summary statistics and histograms were used to assess the differences in the eight slope methods, and analysis of variance techniques were used to determine the statistical significance of the differences.

Cell by cell comparisons were examined in previous studies by Evans (1979), Young (1978), Struve (1977) and Skidmore (1989), but this investigation has unique aspects. In this study, commercially available digital elevation data were used in the calculation of cell slope values. Most of the previous studies used elevation data created by mathematical models which did not mimic all the irregularities of the earth's surface or vagaries in DEM data that may affect the calculation of slope. Further, this study examined eight different slope methods for a total of 28 pair wise slope comparisons. Slope method 10 was not investigated because it was the same as slope method 7. Since slope method 8 has only been discussed in one article, and not referenced or found in any other articles, it was not examined in this study. Previous studies usually compared only a small number of the potential slope calculation methods. The majority of the previous studies were trying to determine the "best" slope method, and were not ultimately concerned with characterizing the differences between slope methods.

2.4 Effects of slope method on a decision model

To demonstrate how the slope method used can affect the results of decisions, an example using a real decision model was examined. The model described was developed by the Jefferson National Forest (USFS, 1985) to determine lands suitable for timber production based on three parameters: site index, stand slope and access as shown in Table 1. This is called a Level I suitability classification (USFS, 1985). Level II suitability classification further evaluates Level I suitable stands by considering various other factors, such as stand characteristics and management objectives. Another facet of this decision model is determining the recommended logging method that can be used to harvest the stand, based on the slope of the stand.   Table 2 shows the operability guidelines for determining the appropriate harvesting equipment to use in a the forest stand. These criteria for classifying forest stands into suitable and unsuitable stands for timber production, and separating the suitable forest stands into harvest equipment operability classes were used in this example.

The 240 polygons (forest stands) were randomly assigned site indices between 50 and 80. Site Index (SI) refers to the height of dominant or dominant and codominant trees in even-aged stands at some index age. This allowed some stands to be classified as "unsuitable" because of slope if the site index was less than 60. The possibility of a forest stand being classified as "unsuitable" solely because of site index was eliminated. All stands were considered accessible by roads. The example was designed so only stand slope could alter how a particular stand would be classified. The mean of the individual cell slopes was used to characterize the slope of a stand (polygon). This does not imply that the mean should always be used to characterize the slope of a polygonal area, but it appears to be common.

Stand Status	Site Index	Slope (*percent*)	Access
SUITABLE	> 50	35 <	Roaded/Unroaded
	60	> 35	Roaded
	> 70	> 35	Roaded/Unroaded
UNSUITABLE	40 <	> 0	Roaded/Unroaded
	50	> 35	Roaded/Unroaded
	60	> 35	Unroaded

Table 1 Suitability Guidelines for Forest Service Stands (USDA, 1985).

Slope (*percent*)	Recommended Equipment
0.0 - 35.0%	Ground based skidding
35.0 - 50.0%	Advanced ground based skidding
50.0% +	Cable yarding system

Table 2 Harvest Operability Guidelines (USFS, 1985).

3. Results and discussion

3.1 Cellular comparisons

A cell by cell comparison was performed for the entire study area to evaluate the differences between individual cell values for eight slope calculation methods. A total of 77, 855 cells were compared, and a summary of the results is presented in Table 3. Differences were calculated as row method minus column method, so negative signs indicate that the column method produced larger slope values, and positive signs indicate that the row method produced larger slope estimates.

All possible dual comparisons of the eight slope calculation methods were analyzed using ANOVA techniques. At the 0.05 level, all of the 28 comparisons were statistically different. However, this statistical difference should not be solely relied upon as a indicator of difference. The sample size was very large, which means that virtually any difference will be statistically significant. Further, "statistically" different is not identical to "practically" different, which must be assessed in a somewhat more empirical setting. Finally, there are some characteristics of this data set which could cause some questioning of the validity of these statistical tests (Figure 5). Histograms of the cell slope differences were Normal or Gaussian in appearance (Fig. 5a) for comparisons 1-2, 1-6, 1-7, 1-9, 2-5, 2-6, 2-7, 2-9, 6-7 and 6-9. Comparisons 1-3, 2-3, 3-4, 3-5, 3-6, 3-9, 4-5, 4-6, 4-7 and 4-9 had distributions that appeared to be somewhat asymmetric (Fig. 5b), while 1-4, 1-5, 2-4, 5-6, 5-7, and 5-9 were very skewed (Fig. 5c). The distribution of the differences between methods 7 and 9 was unusual because of its discrete form. Since most parametric statistical tests require the assumption of normality, a shadow of doubt is cast upon some of the statistical differences reported previously.

While all the slope methods were different in a statistical sense, we need to examine the differences in a more practical context since several methods appeared to give similar results according to Table 3. Methods 7 and 9 produced nearly identical slope values, with a mean difference in only the fifth decimal place. Five comparisons (1-6, 1-7, 1-9, 6-7, 6-9) produced mean differences less than 1%, and five others (1-2, 4-5, 2-6, 2-7, 2-9) had mean differences less than about 2%. Thus, the practical differences between these methods may be negligible for some applications. However, the mean difference can be misleading. Even in the methods with mean differences near zero, the individual cell slope values are occasionally +/- 50% (Table 3), probably because of irregularities in the terrain and/or vagaries in the DEM data. Further evidence can be found in Table 4, which contains the eight computed slope values for all 23 cells in polygon 57. Even though the mean slope value for methods 1, 6, 7, 9 are all within 1% for this polygon, the individual slope values varied by several percentage points for many cells, such as cells 3, 6, 13 and 14 (Table 4).

Mean (Min,Max) Variance	Slope Method 2	Slope Method 3	Slope Method 4	Slope Method 5	Slope Method 6	Slope Method 7	Slope Method 9
Slope Method 1	-1.16 (-177.87, 95.28) 62.80	-15.53 (-136.03, 9.82) 98.44	-7.24 (-177.87, 0.00) 31.28	-5.79 (-78.07, 0.00) 20.25	.72 (-37.29, 34.14) 3.84	.92 (-49.73, 45.51) 6.79	.92 (-49.74, 45.53) 6.79
Slope Method 2		-14.37 (-193.19, 35.90) 143.68	-6.08 (-196.62, 0.00) 70.79	-4.63 (-120.44, 83.20) 89.23	1.87 (-71.63, 91.73) 65.23	2.07 (-67.42, 95.00) 67.66	2.07 (-67.40, 94.96) 67.69
Slope Method 3			8.29 (-65.3, 86.67) 77.01	9.74 (-22.76, 14.67) 92.55	16.25 (-7.07, 181.45) 92.40	16.44 (-9.43, 158.08) 91.52	16.45 (-9.40, 169.00) 91.82
Slope Method 4				1.44 (-74.48, 106.51) 56.06	7.95 (-33.70, 127.00) 35.10	8.15 (-46.14, 134.02) 37.93	8.15 (-46.15, 134.00) 37.93
Slope Method 5					6.51 (0.00, 51.38) 17.86	6.71 (0.00, 58.40) 18.87	6.71 (0.00, 58.38) 18.87
Slope Method 6						.20 (-12.44, 11.37) .428	.20 (-12.45, 11.39) .429
Slope Method 7							.00007 (-.05, .05) .00083

N = 77,855   Difference = Row - Column

Table 3 Pairwise Comparisons of Slope Method Differences on a Cell by Cell basis in Units of Slope Percent.

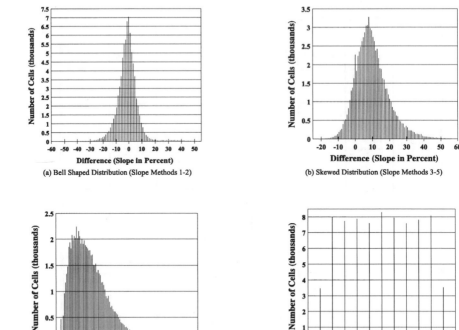

Figure 5 Histograms of Slope Method Comparisons: (a) Bell Shaped Distribution (Slope Methods 1-2); (b) Skewed Distribution (Slope Methods 3-5); (c) Truncated Distribution (Slope Methods 5-6); (d) Uniform Distribution (Slope Methods 7-9).

Some of the results presented in Tables 3 and 4 could be at least partially predicted a priori by examining the particular computational algorithms used by the various methods. For instance, because slope methods 3,4 and 5 compute or select the maximum slope, they should always produce larger slope estimates than methods which average or smooth a set of cell values. The question in these cases becomes how large the differences are, and also, how the various methods which utilize a maximum criteria compare among themselves. Method 3 appeared to be the most different, both from the non-maximum methods (1, 2, 6, 7, and 9) and the other maximum-based methods (4 and 5). Method 5 was more similar to the non-maximum methods than were methods 3 and 4. Users should note that Methods 3, 4 and 5 produce estimates that are generally very different from the other slope calculation methods, which should not be a surprise. However, the magnitude of the differences may surprise many. Clearly, a software user should know if one of these slope calculation methods based on maximum values is being used, and understand the ramifications of that choice.

Cell	Slope Method 1	Slope Method 2	Slope Method 3	Slope Method 4	Slope Method 5	Slope Method 6	Slope Method 7	Slope Method 9
1	44.88	56.67	66.67	56.67	58.63	45.60	45.86	45.90
2	54.42	69.36	90.00	69.36	63.54	53.74	53.52	53.50
3	30.73	44.85	60.00	44.85	31.82	27.70	26.71	26.70
4	48.59	67.08	86.67	67.08	48.59	46.40	45.76	45.80
5	61.87	56.67	83.33	67.08	61.87	58.53	57.42	57.40
6	16.67	30.73	36.67	30.73	18.60	12.98	11.79	11.80
7	35.32	37.27	66.67	40.28	35.77	34.82	34.66	34.70
8	48.07	52.70	80.00	52.70	54.16	48.45	48.59	48.60
9	58.05	59.35	86.67	59.35	65.45	56.18	55.56	55.60
10	18.63	34.32	53.33	34.32	23.89	19.62	20.01	20.00
11	32.19	38.01	66.67	38.01	32.83	32.50	32.61	32.60
12	39.33	49.55	83.33	49.55	46.37	39.64	39.77	39.80
13	53.67	58.31	90.00	60.09	64.03	50.86	49.93	49.90
14	14.24	12.02	43.33	22.36	20.65	17.33	18.40	18.40
15	26.93	39.02	66.67	39.02	36.02	26.65	26.60	26.60
16	41.23	53.75	63.33	54.97	51.11	38.23	37.25	37.30
17	12.13	13.33	26.67	14.91	17.44	10.82	11.18	11.20
18	10.54	27.49	33.33	27.49	18.30	12.43	13.36	13.40
19	13.44	31.45	40.00	31.45	15.63	14.01	14.20	14.20
20	15.09	14.91	16.67	17.95	18.37	13.24	12.67	12.70
21	19.00	21.34	20.00	21.34	24.09	19.44	19.64	19.60
22	21.21	18.86	30.00	23.57	25.98	21.42	21.57	21.60
23	16.67	13.74	26.67	21.34	22.14	17.89	18.39	18.40
Mean	31.87	39.87	57.25	41.06	37.19	31.24	31.11	31.12

Table 4. Cell Slope Values in Percent for Polygon 57 for each Slope Method.

Each of the methods used in this study for calculating cell slope values was affected by the characteristics of the USGS DEM data used, however, some seemed to be more sensitive than others. Since 72 USGS DEM elevation values were recorded in integer values (meters), and the cell size was 30 meters, the smallest cell slope value (not including 0.00%) for the different methods ranges from 0.55% (Method 7) to 3.33% (Method 2) depending on which slope method is used. Also, there were more neighbor cells at the same elevation than would be expected on a real value representation of a surface. This increases the number of cells with a zero slope value which only rarely occurs in most terrain. Some of the slope methods were more sensitive to irregularities (cell elevation blunders) in the elevation data than the other methods. This information can be discerned by analyzing the algorithm each method is based on. Slope method 3 was the most sensitive to large elevation errors while slope method 9 was least sensitive. For example, a 200 meter single cell elevation error (blunders) could cause a 666.6% slope error using method 3, but only a 111.1% slope error using method 9. Slope methods 6, 7 and 9 were the least sensitive to single cell errors because they used all eight neighbors (smoothing effect) to determine the slope of a single cell. Slope methods 1, 2, 3, 4 and 5 were more sensitive since they used fewer elevation values to determine the slope of a single cell. Slope methods 2, 3 and 4 were sensitive to linear (row, column) errors in elevation data (altitude matrix) because they calculated the slope of a cell by using neighbors in only one direction from the cell of concern. The sensitivity of the slope method to data errors must be considered when determining which slope method to use because it is presumptuous to assume that elevation data is error-free.

3.2 Effect on the land suitability decision

While we have demonstrated the effect of slope calculation method on individual values, the context of real interest is the decision-making situation. Based on the previously-described decision criteria, the 240 polygons were categorized into unsuitable, suitable (ground skidding), suitable (advanced ground skidding) and suitable (cable yarding), and the results presented in Table 5.

Membership in the various suitability categories was clearly affected by the slope calculation method used. The number of stands that were classified as unsuitable for timber production ranged from 18 (382.9 hectares) using slope methods 6, 7 and 9 to 59 (1,735.2 hectares) when using slope method 3. This was almost a five times difference in the area of land called unsuitable for timber production based only on the method used to calculate the cell slope values of a stand (polygon). Similar differences exist in the other suitability categories, for instance, suitable, ground skidding area varied from 921 to 4632 hectares using methods 3 and 7. Similarities between methods are also evident. Methods 6, 7 and 9 produced nearly identical suitability decisions, and again, the maximum-based methods (3, 4 and 5) were somewhat similar. Methods 3, 4 and 5 produce conservative suitability decisions because they tend to produce higher slope estimates, and harvesting suitability decreases with increasing slope. Of particular interest here are how stands can change suitability category just by changing the cell slope calculation methods.

Slope Method		Unsuitable	Suitable, Ground skidding	Suitable, Advanced ground skidding	Suitable, Cable yarding
1	Polygons	20	183	32	5
	Hectares*	388.9	4,437.5	817.7	260.3
2	Polygons	24	173	37	6
	Hectares*	518.9	4,230.9	882.9	271.8
3	Polygons	59	49	91	41
	Hectares*	1,735.2	921.0	2,116.4	1,131.9
4	Polygons	38	120	70	12
	Hectares*	1,199.6	2,649.5	1,676.1	379.2
5	Polygons	34	135	57	14
	Hectares*	764.7	3,276.6	1,415.0	448.1
6	Polygons	18	188	29	5
	Hectares*	382.9	4,578.9	682.4	260.3
7	Polygons	18	191	26	5
	Hectares*	382.9	4,632.5	628.7	260.3
9	Polygons	18	191	26	5
	Hectares*	382.9	4,632.5	628.7	260.3

* Based on raster polygon area

Table 5 Classification of Stands into Suitable and Unsuitable Categories.

Of the 240 polygons in which the decision criteria were applied, the suitability/harvest method decision did not change for 137 (57%) polygons regardless of which slope calculation method was used. However, 81 (34%) polygons were in two decision categories, and 22 (9%) were in three decision categories. Clearly, if decisions or recommendations are made based on slope information calculated from DEM's, the algorithm used to compute slope can impact the results.

4. Conclusion

In this study, we have demonstrated the effect of different slope calculation algorithms on individual slope values, and on a decision reached using slope information. Eight different cell slope methods commonly used in the literature were evaluated to determine if they were different, and if so at what magnitude. Cell slope values (77,855) from all 240 stands (polygons) were used in this comparison. All 28 cell pair wise slope comparisons were statically different, but six comparisons were similar for most practical applications. There were five slope groups in which the calculated cell slope values were similar on the average. They were slope method 2, slope method 3, slope method 4, slope method 5 and slope methods 1, 6, 7 and 9. This grouping was based on a pair wise mean difference of less than 1% slope value between the methods. Of the 28 pair wise comparisons only 10 of the slope differences had a normal distribution. The other distributions were skewed or truncated at the zero slope value.

For the eight slope computation methods used in this study, the classification of over 40% of the suitability/harvest stands (polygons) were affected because of differences in the slope values calculated by the slope methods. The effect of just changing the slope method used influenced the results of the suitability/harvest decision model and caused more than a 4 fold difference in land area called unsuitable for timber production.

We cannot make general statements about the effect of slope algorithms, but this is real DEM data, actual stand polygons and real decision making criteria, so the effects demonstrated are real. Clearly, the possibility exists that the results of a decision may be changed just by selecting another slope calculation method, and commonly, that possibility is unknown to the user of software. The real lesson is that the practitioner in GIS should inform themself of what algorithms their software is using, and how they may affect the answer they are seeking.

References

Burrough, P.A. 1986. Principles of Geographic Information Systems for Land Resources Assessment. Oxford University Press, New York. 193pp.

Chieng, S. and J. Luo. 1993. Application of GIS to Peak Flow Estimation. Proceedings of Application of Advanced Information Technologies: Effect Management of Natural Resources. Spokane, Washington. pp. 279-289.

Evans, I.S. 1980. An integrated system of terrain analysis and slope mapping. Z. Geomorph. N. F. 36(12): 274-295.

Fleming, M. D. and R. M. Hoffer. 1979. Machine processing of Landsat MSS Data and DMA Topographic Data for Forest Cover Type Mapping. LARS Technical Report 062879. LARS, Purdue University, West Lafayette, IN.

Grender, G. C. 1976. Topo III: a FORTRAN program for terrain analysis. Computer and Geosciences 2(1): 195-209.

Horn, B. K. P. 1981. Hill shading and the reflectance map. Proceedings of the IEEE 69(1): 14-47.

Horn, B. K. P. 1982. Hill shading and the reflectance map. Geo-Processing 2: 65-146.

IDRISI. 1990. IDRISI, A Grid-Based Geographic Information Analysis System. Clark University, Worcester, MA. 364pp.

Katsaridis, P. and Y. Tsigourakos. 1993. The Use of GIS in Land Use Planning for the Protection of the Delfi Hinterland. Proceedings of Thirteenth Annual ESRI User Conference. (1):321-327.

Kessell, S. R. 1979. Gradient Modelling Resource and Fire Management. Springer-Verlag, New York, NY. 432pp.

Mcharg, I. L. 1969. Design with Nature. Doubleday Natural History Press, Garden City, NY. 197pp.

Meijerink, A. M. J. 1986. A spatial assemblage model for the estimation of gross erosion and sediment yield using remote sensing and geo-database operations. Land evaluation for land-use planning and conservation in sloping areas. International Institute for Land Reclamation and Improvement. pp. 174-195.

Newman, W. M. and R. F. Sproull. 1979. Principles of Computer Graphics (2nd ed.). McGraw-Hill Book Co., New York, NY. 541pp.

O'Neill, M.P. and D. M. Mark. 1987. On the frequency of land slope. Earth Surface Processes and Landforms 12: 127-136.

Sharpnack, D. A. and G. Akin. 1969. An algorithm for computing slope and aspect from elevations. Photo. Eng. Rem. Sens. 35(3): 247-248.

Skidmore, A. K. 1989. A comparison of techniques for calculating gradient and aspect from gridded digital elevation models. International Journal of Geographic Information Systems 3(4):323-334.

Struve. H. 1977. An automated procedure for slope map construction. Technical Report M-77-3. U.

S. Army Waterways Experiment Station, Vicksburg, MS. 98pp.

Tomlin, C. D. 1980. The IBM Personal Computer Version of the Map Analysis Package. Laboratory for Computer Graphics and Spatial Analysis. Harvard Graduate School of Design, Harvard University, Cambridge, MA. 51pp.

Travis, M. R., G. H. Elsner, W. D. Iverson and C. G. Johnson. 1975. VIEWIT: computation of seen areas, slopes and aspect for land-use planning. General Technical Report PSW-11, Pacific   Southwest Forest and Range Experiment Station, Berkeley, CA. 70pp.

USFS. 1985. Jefferson National Forest Land and Resource Management Plan. USDA-Forest Service, Jefferson National Forest, Roanoke, VA. 290pp.

Young, B. 1978. Statistical characterization of altitude matrices by computer. Progress Report (5) DA- ERO-591-73-60040. U.S. Army Corps of Engineers, Fort Belvoir, VA. 29pp.

Zill, D. G. 1985. Calculus With Analytic Geometry. PWS Publishers, Boston, MA. 939pp.

Contact address
[1]Dr. Robert C. Weih and [2]Dr. James L. Smith
School of Forest Resources
Arkansas Forest Resources Center
University of Arkansas
P. O. Box 3468
Monticello, Arkansas 71655
Phone: (501) 460-1248
Fax: (501) 460-1092
E-mail: weih@uamont.edu
[2]Canal Forest Resources, Inc.
9140 Arrow Point Blvd.
Charlotte, North Carolina 28273
Phone: (704) 527-6780
Fax: (704) 527-1245

HANDLING DATA SPATIALLY:

SPATIALIZING USER INTERFACES

Werner Kuhn

Department of Geoinformation
TU Vienna
Vienna, Austria

Abstract
The impact of GIS research extends beyond GIS. Our expertise about space, spatial relations, and spatial knowledge needs to be shared with other domains who can benefit from it, even if their problems are not inherently spatial. This paper discusses the needs for spatial expertise in the field of human-computer interaction. Modern user interface design increasingly employs spatial metaphors to structure objects and operations in understandable and useful ways. Spatialized user interfaces create a virtual space in which users perceive, operate, and move while they solve their application problems. The designers of these virtual spaces need to be informed about human spatial cognition and properties of spaces in order to design "usable" and credible spaces. They also need technological support for implementing spatial data structures or databases and procedures for accessing, navigating, and visualizing them. The paper investigates these needs, proposes a theory to address them, and shows how spatial data handling expertise helps to handle data spatially.

Key words
Spatialization, spatial metaphors, user interfaces, usability, task analysis.

1. Introduction

Over the twelve years of its history documented at Spatial Data Handling conferences, the Spatial Data Handling community has accumulated a large body of knowledge. This expertise covers issues like spatial data integration, spatial databases, algorithms, decision support systems, spatial analysis, digital terrain modeling, visualization and multimedia systems, spatio-temporal GIS, fuzzy processing, uncertainty and error analysis. Its primary purpose has been and will remain to support the effective handling of spatial data in GIS applications.

However, it is time to look beyond the borders of the field and to realize that our collective expertise is not only unique, but also of potential use outside spatial applications. Uniqueness means that we are the experts on spatial data and as such qualified to support any spatial data handling needs. Potential use outside spatial applications means that we should seek other problem domains where our competence can contribute to solutions (Frank and Campari 1993; Frank and Kuhn 1995). This paper shows one such domain and explains its needs for knowledge about space and spatial models: the domain of human-computer interaction. The paper interprets the adjective "spatial" in "Spatial Data Handling" as referring to handling, not to data. It provides an introduction to the *spatial handling* of data, whether these are spatial or not.

More and more user interfaces use spatial concepts, even if the application domains are not spatial. This metaphorical use of space in interaction has been called *spatialization*[1]. It became popular with the desktop metaphor (Smith and others 1982) which presents an operating system in terms of a desktop, documents, folders, and a trash can. Recent examples include the NAVIGATION (Conklin 1987), TOURIST (Fairchild and others 1989), ROOM (Robertson and others 1993), MUSEUM (Travers 1989), MYST (Miller and Miller 1994), and CAVE (Cruz-Neira and others 1992) metaphors. GIS and CAD interfaces often employ spatial metaphors as well (Gould and McGranaghan 1990; Mark and others 1992), but this paper deliberately excludes them from the discussion, in order to make a stronger case for the spatial needs of *non-spatial* domains.

Earlier Spatial Data Handling symposium papers have analyzed the structure and role of spatial metaphors (Kuhn 1990; Kuhn 1992; Mark 1992). Papers at the Conferences on Spatial Information Theory (COSIT) have presented spatialized user interfaces (Chalmers 1993; Dieberger 1995; Erickson 1993; Rennison and Strausfeld 1995). This paper discusses how spatial data handling expertise applies to the spatialization of user interfaces. Section two argues why spatialization of user interfaces is useful. Steps toward a theory of spatialization are described in section three and implementation issues in section four.

2. Why spatialization?

Spatialization maps physical space to abstract computational domains through spatial metaphors. This section argues why users benefit from such metaphorical mappings. It answers the question why spatial metaphors are good for interaction in three steps, explaining briefly what metaphors are, why they serve interaction, and why space is an ideal source domain for them.

2.1 What are metaphors?

Metaphors allow us to understand one thing in terms of another, without thinking the two are the same (Lakoff and Johnson 1980). They are *mappings* from *source* to *target* domains. For example, the metaphor LIFE IS A JOURNEY takes the idea of a journey as a source domain and life as a target. Thus, we

---

[1]While individual interface designs have sometimes been referred to as "spatialized", the term seems to have been used for the first time in the sense of a general design technique at a workshop at INTERCHI'93 [Kuhn and Frank 1993] and subsequently at a tutorial at CHI'96 [Kuhn and Blumenthal 1996]. Given that this design technique is a direct application of what cognitive linguists have called spatialization (in language) [Lakoff and Johnson 1980], it seems appropriate to call it spatialization (of user interfaces). The only other common meaning of the term is closely related: In Virtual Environments, producing sounds that appear to originate from certain positions in space is called *audio spatialization*.

say that *he got off to a good start in life,* but then *lost his sense of direction, stumbled,* and *did not get anywhere,* etc. Lakoff and Johnson argue convincingly that metaphors shape our thinking, talking, and acting in everyday life.

Metaphors are always *partial* mappings. For instance, when we say that an operating system is a metaphorical desktop, we do not expect the documents and folders on it to collect dust. This partial nature is the reason why there is no point in distinguishing metaphors ("A is B") from similes ("A is like B"). The meaning of "is" in a metaphor is always that of a partial correspondence (i.e., "is like, in some respects").

2.2 Why metaphors for user interfaces?

Metaphors have become a key idea in human-computer interaction, permeating the terminology of design theory and practice (Carroll and others 1988; Foley and others 1990; Kuhn 1995; Madsen 1994). Their main role is to make interaction concrete and to help designers and users master complexity. They achieve this by establishing ontologies and affordances.

2.2.1 Metaphors create ontologies
An ontology is the set of objects and operations in a domain, i.e., the universe of discourse for an application. Metaphors create ontologies in applications by projecting structure from the source domain. A useful ontology is one that leads to an appropriate work division and effective ways of solving problems. "A clearly and consciously organized ontology is the basis for the kind of simplicity that makes systems usable" (Winograd and Flores 1986, p.165). Thus, the goal is usability, rather than learnability, though a good ontology is always easier to learn. It makes the experience of using a system coherent, for the novice as well as for the expert. The idea of an ontology makes clear that interface metaphors are a conceptual, not only a presentational device (Kuhn 1991).

2.2.2. Metaphors create affordances
Using information technology is abstract and complex, because it lacks affordances, i.e., it does not tell us how to use it. Affordances are what a thing offers us (Gibson 1986). For instance, a surface affords support to its users. Gibson's theory of affordances is partly based on Gestalt ideas: "things tell us what to do with them" (Koffka 1935, p.353), and has had a strong influence on popular human-computer interaction work (Norman 1988; Norman 1992). Metaphors create affordances for information technology by mapping affordances from source domains. For example, an office document affords filing or throwing away and the desktop metaphor maps these affordances to computer file manipulations.

2.3 Why spatial metaphors?

Three quotations from Cognitive Linguistics, Artificial Intelligence, and Information Science express some key reasons to use *spatial* metaphors in user interfaces:

> "Most of our fundamental concepts are organized in terms of one or more spatialization metaphors." (Lakoff and Johnson 1980, p.17)
>
> "Much of how we think in later life is based on what we learn in early life about the world of space." (Minsky 1985, p.122)
>
> "The fact that space is unimportant to modern information processing systems should mean that it could be placed wholly at the disposition of the user, not that all spatial features . . . be systematically homogenised." (Miller 1968, p.288).

The idea that space plays a key role in our cognitive system has linguistic and other roots. Its early manifestation in linguistics was called *localism* (Anderson 1971; Lyons 1977) and its full development came with *experientialism* in cognitive linguistics (Lakoff and Johnson 1980). Modern psychological research underlines the special role of space from an *evolutionary* perspective:

> "Space is special. We have to survive in a dangerous, three-dimensional spatial environment. Given our species' size (and thus vulnerability to gravity, predation, injury), if we had no ability to represent and reason about space, that survival would be unlikely. However, we can represent space, and in fact it seems that our perceptual system recognizes the importance of space in that many subsystems (e.g., vision, audition, proprioception) provide information useful in forming those representations. (...) Thus, from this perspective, language comprehension is intrinsically spatial." (Glenberg 1993).

Space is fundamental to perception and cognition because it provides a common ground for our senses as well as for our actions. The constant mutual reinforcement of visual, auditory and tactile cues has developed our spatial cognition to an extent unmatched by any other domain. Perception, manipulation, and motion in space are largely subconscious activities, imposing little cognitive load, while offering intuitive inference patterns. Space has a strong inner coherence that proves useful for designing and combining metaphors. For example, we know that opening a door gets us to a room or building, not to a desktop or a road, and that we are likely to encounter other doors and windows behind. Thus, space is not just any domain of experience, but the most important one and as such uniquely qualified as a metaphor source.

3. How spatialization works

Spatialized user interfaces structure application domains through spatial metaphors. In order to understand this process, one needs to look at the structure of spatial source domains and at the mappings to target domains. Since target domains get their structure mainly from the sources, descriptions of target structures tend to be circular.

3.1 The Structure of spatial source domains

3.1.1 Different kinds of spaces

Space is not space. Many different kinds of spaces can serve as source domains for interface metaphors. The notion of space in spatialization should be understood in a broad, but experiential sense. Spatialization is *not* meant to imply a transition from two-dimensional to three-dimensional space. It covers all spaces we can experience, but not the n-dimensional, empty, homogeneous, isotropic, infinite kinds of spaces or those characterized by relativity, space-time curvature, or an uncertainty principle.

A fundamental distinction of spaces is based on their relative size compared to the human body (Kuipers 1978). *Small scale* or table top spaces contain objects smaller than human beings that can be moved and turned around. Direct manipulation provides full information about them. Their objects have sharp boundaries and offer an experiential basis for the Euclidean concept of a straight line (Frank 1996). The objects and phenomena in *large scale* or geographic spaces are larger than human beings. Objects cannot be moved or turned around and observers themselves have to move in order to gain full spatial knowledge. The distinction of small and large scale spaces has been refined in many ways (Mark 1992; Montello 1993; Zubin 1989), leading to three, four and more different kinds of spaces. It is an open question, to what extent user interfaces should cut across scales, given their possibility of arbitrary scalings (e.g., one could pick up a house and turn it around) (Kuhn and Egenhofer 1991).

A theory of space in user interfaces needs to take into account what spatial cognition research has found out about navigation and wayfinding in real spaces (Gluck 1991). Human spatial knowledge of an environment is generally thought to consist of the three levels of *landmark, route, and survey knowledge* (Siegel and White 1975). Landmarks are salient points of reference in the environment; route knowledge contains paths for navigating between the landmarks; and survey knowledge arranges routes into a coherent network. These levels of knowledge are interconnected and humans are believed to progress from landmark to route to survey knowledge when learning an environment: landmarks help to find routes and multiple interconnected routes span a network (Hirtle and Hudson 1991).

Another basic distinction of spaces derives from the two ways in which one thinks of a location: as an attribute of objects in space or as a part of space that has itself attributes. The first is usually called the entity view, the second the field view of space. The *entity* view addresses the question "where is it?" for discrete objects with location, shape, and size. Space is determined by discrete entities and their spatial relations, like bounding, intersecting, and containing. The *field* view asks "what is here?", for continuous properties (say, ozone concentrations or magnetic fields), sampled at locations. Space consists of an infinite set of points with associated attributes, and objects are only derived through classification. The semantics of pointing to a location on screen (or in a virtual environment) are different, depending on whether they are interpreted in an entity or field view.

3.1.2 Spatial image schemata

Mark Johnson suggested that our experiences in interacting with the environment result in the formation of basic conceptual structures that he called *image schemata* (Johnson 1987). Examples are a container schema to represent the idea of containment and a path schema to represent movement in space from a source to a goal along a path. Most image schemata are spatial. They are highly structured, rather than being some kind of pictures, and provide a sort of logic: The container schema includes structural information such as inside-outside, and boundary, as well as logical notions such as either inside or outside, and transitivity of embedded containers. Here are the *spatial* image schemata, extracted from the list in (Johnson 1987, p.126):

Container	*Blockage*	*Path*	*Link*
Center-Periphery	*Cycle*	*Near-Far*	*Scale*
Part-Whole	*Merging-Splitting*	*Full-Empty*	*Matching*
Superimposition	*Contact*	*Surface*	*Object*

Image schemata are used to represent a great variety of objects and ideas, directly in space or metaphorically in abstract domains. Thus, the container schema is used in thinking about human bodies, cups, rooms, lakes, and universes, but also about relationships, states of mind, meetings, and ages. Together with the part-whole schema, it underlies *hierarchies* which are typical for our spatial thinking. Image schemata are central to a cognitive view of spatialization. They make spatialization a mapping of structure, rather than dimensions.

3.1.3. Spatial affordances

Spatialization exploits the natural and cultural affordances of space. These affordances generate potential activities in a target domain. The most important affordances of space are listed here, in an attempt to arrive at a theoretical underpinning for spatialization. The list is not meant to be exhaustive. For instance, it lacks negative affordances, such as getting lost. However, it provides a fairly rich collection of activities afforded by space. The four categories reflect different task situations: a user alone, a user with an individual entity in space, a user with one or more entities, and several users.

Affordances for an individual user
- *move*
- *navigate*
- *reason*
- *represent*

For a user and an individual entity
- *objectify*
- *perceive*
- *recognize*
- *interpret*
- *identify*
- *manipulate*
- *move*
- *place and store*
- *hide,search, and discover*
- *access and enter*

For a user and multiple entities
- *differentiate*
- *organize*
- *rank*
- *relate*
- *associate and aggregate*
- *recognize patterns*

For groups of users
- *encounter*
- *communicate*
- *cooperate*
- *assert rights*

Affordances are related to image schemata. For example, objectifying and perceiving are based on the object schema. Recognizing and identifying are based on spatial relation schemata like location, container, link, near-far, or center-periphery. Motion is based on the path schema, place and store on the surface and container schemata. Organizing and ranking use the order (linear, cyclic, pattern, partial orders), superimposition, container (for categories and hierarchies), and center-periphery schemata. Associating and aggregating rely on the near-far, link, path, and part-whole schemata. Communication takes the link schema and asserting rights (e.g., privacy or ownership) the container, path, and blockage schemata.

Two examples shall demonstrate how spatial affordances structure the tasks and objects of target domains. The first task, moving a file, is taken from the domain of operating systems. Note that it had its metaphorical spatial structure ("move", "file") already before the metaphor was visualized by Xerox and Apple. The second example task, memorizing, is inspired by (Yates 1966) and shows a spatialization without any (external) visualization.

Example 1: Move a file to another directory
- *objectify*: the interface designer created objects for files and directories
- *perceive*: the user can see these objects (documents and folders)
- *search*: look for a specific document
- *recognize*: find the document
- *identify*: select the document
- *move*: pick the document up and drag it to the destination folder
- *place and store*: drop the document into the folder.

Example 2: Memorize a sequence of ideas
- *objectify*: mentally represent ideas as pictures and their context as a museum
- *organize*: order the pictures in the required sequence to produce a story
- *associate and aggregate*: distribute parts of the story to separate rooms
- *place*: hang each picture on a wall
- *access and enter*: visit the museum (sometime later)
- *move*: walk through the rooms
- *perceive*: see the pictures hanging on the walls
- *discover*: find each picture along the way
- *recognize and interpret*: remember what each picture represents.

3.1.4. Prototypical spaces

Most source domains of spatial metaphors are variations of a few common types of spaces from human experience, such as tables, rooms, houses (Bachelard 1957), cities (Lynch 1960), or landscapes (Higuchi 1983). We call these spaces *prototypical*, because each of them is a good example of a category of spaces that humans are familiar with. Each space is characterized by a specific set of experiences and affordances. The differences in human cognition of these spaces need to be taken into account when designing spatialized interfaces.

We list prototypical places, parts and kinds of them, as well as prototypical spatial elements that are not places. The classification is tentative and not the only one possible. Alternative (orthogonal) classifications might distinguish private and public or natural and artificial places. The point is to show the close hierarchical association of spatial elements, suggesting appropriate (and inappropriate) systems and combinations of metaphors.

Prototypical places and some corresponding interface metaphors are:
- *table*
  desktop: (Smith and others 1982), drafting table: (Wilson 1990)
- *house*
  rooms: (Robertson and others 1993), museum: (Travers 1989)
- *city*
  city: (Dieberger and Tromp 1993), mall: (Dieberger 1995)
- *landscape*
  landscape: (Chalmers 1993)

Prototypical subcategories of places:
- tables: desk, dining table, coffee table, conference table
- houses: home, office, library, museum, cafe, hotel, hospital, school, theater
- cities: hometown, village, metropolis, capital
- landscapes: field, mountain, valley, garden, park, forest, swamp

Prototypical parts of places:
- table: table top, leg, drawer, seat
- house: floor, room, cellar, roof, door, wall, garden
- city: block, square, neighborhood, district, suburb, park, market,mall
- landscape: valley, ridge, peak, pit, plane

Additional elements of prototypical spaces:
- links: door, window, gate
- paths: way, road, track, tunnel, river, bridge
- configurations: network, labyrinth.

Prototypical places are populated by living beings and other spatial entities like objects, furniture, or machines. Humans, animals, plants and their parts as well as spatial objects form additional rich sources of spatial metaphors. Finally, there are also some common types of fictive spaces (like the horizon, sky, paradise, hell, or purgatory) that play a role as source domains.

3.2 The structure of metaphorical mappings

A metaphor is more than a flat mapping from source to target. It is usually part of a *hierarchy* with sub- and super-metaphors (Lakoff and Johnson 1980). The hierarchical structure of the sub- and super-mappings relates subcategorizations in the source and target domains. The more appropriate the metaphor, the deeper this hierarchy will run.

3.2.1 Induced hierarchies

An example of hierarchical mappings is the system of subcategorization in the TIME IS MONEY metaphor (Lakoff and Johnson 1980, chapter 2). Mapping the hierarchy of substances (money *is a* valuable commodity *is a* limited resource *is a* substance) generates a generalization hierarchy of conceptualizations of time.

> TIME IS A SUBSTANCE: my time, a lot of, pressure of, flow of
> TIME IS A LIMITED RESOURCE: use up, run out of, have enough of
> TIME IS A VALUABLE COMMODITY: have, give, lose, thank you for
> TIME IS MONEY: spend, invest, budget, cost.

3.2.2. Generic metaphors

At the same level as TIME IS A SUBSTANCE, a range of *generic spatial metaphors* exists. Each of them introduces a similar hierarchy of subcategories. The following examples, from the works of Lakoff and Johnson, provide a rich source of generic spatialization metaphors. They establish top levels of hierarchies from which designers can derive specialized metaphors:

- THE VISUAL FIELD IS A CONTAINER: the screen is *out of* sight, *in* your view
- EVENTS AND ACTIONS ARE OBJECTS: she *takes* this class
- ACTIVITIES ARE SUBSTANCES: she is *in* the midst of *learning* about spatialization
- STATES ARE LOCATIONS OR CONTAINERS: she was *at* her best, he fell *into* a depression
- CHANGING STATE IS MOTION: they *went from* complete harmony *to* war
- CAUSATION IS THE ORIGIN OF A JOURNEY: headaches can *come from* drinking
- MANNERS ARE PATHS: this is *the way* we did it
- PURPOSES ARE DESTINATIONS: you want to *reach* an understanding of spatialization
- THEORIES ARE BUILDINGS: spatialization theory has received new *foundations*

3.2.3. Hierarchies from prototypical spaces

The prototypical spaces with their subcategories and parts provide additional hierarchies for spatialization at a more concrete level. For example, a hierarchy of spatial metaphors could structure a hypertext application as follows:

> THE HYPERTEXT IS SPACE: users enter and leave it, move in it
> THE HYPERTEXT IS LARGE-SCALE SPACE: designers map it, users navigate in it
> THE HYPERTEXT IS A CITY: users remember landmarks, explore a neighborhood
> THE HYPERTEXT IS THE TOWN OF LAKE WOBEGON[2]: users go to the Chatterbox Cafe.

Entering this hypertext through the door of a room or flying among its nodes would violate the hierarchy, since these operations belong to other kinds of spaces. Such violations of hierarchical structure can create incoherent user interfaces. Notorious examples from the desktop metaphor are the violations

---

[2]"The town where all the women are strong, all the men are good looking, and all the children are above average" [Keillor 1985].

of the surface image schema in the clipboard and of the container schema in the trash can (Kuhn and Frank 1991). Conversely, observing the structure of hierarchical metaphor systems is a good approach to achieving systems with a coherent look and feel.

4. Implementing spatialized interfaces

A key reason why the Spatial Data Handling community should know about spatialization is that it has expertise to offer for some of the implementation needs. Spatialized interfaces require efficient technological solutions for spatial data structures, algorithms, storage and access mechanisms. Visualization techniques are needed, ranging from cartographic mapping to issues such as the use of spatialized sound. This section identifies the technologies that shape the current status of spatialization and discusses some key concerns of designers.

4.1 Technologies

Among the technological developments pushing spatialization are the success of spatialized interfaces in large markets (desktop systems, hypertexts); extensions of direct manipulation interfaces to richer spatial domains (information visualization); attempts to master the complexity of multi-user systems through spatial metaphors (virtual environments and CSCW); and explorations of spatial skills in special-purpose computing environments for scientists (e.g., scientific visualization, computer simulations). Thus, the technologies defining the state of the art in spatialization are primarily

- hypertext and hypermedia
- virtual environments
- cooperative work
- information visualization
- scientific visualization.

A look at some of the achievements and issues in these technologies provides an overview of where spatialization currently is and where it could be going in the coming years.

Hypertexts (Dillon and others 1993) put such a strong emphasis on navigation that hardly any other affordances of space are considered. What should the users do once they get "there"? (Laurel 1993). Still, navigation itself poses enough difficulties for designers and the problem of getting lost in hyperspace remains. Part of the problem may be that research on wayfinding has not yet had a strong influence on hypermedia theory and practice. The World-Wide Web has now brought these difficulties to almost everybody's workplace and home (Dieberger and Bolter 1995).

In *Virtual Environments*, the physical environment can become the source of powerful metaphors. This is potentially the richest case of spatialization, where using a system is using space. Most applications, however, are *simulations* of task environments that are already spatial: surgery, flying and driving, vehicle design, landscape and environmental planning (Pratt and others 1995). While these are not really spatializations, they share many challenges with spatialized interfaces. Virtual Environments also offer insights into the creation of convincing, "realistic" spatializations (Fahlén and others 1993).

A relatively unexplored terrain for spatialization are *Computer Supported Cooperative Work* (CSCW) systems. Their need for spatial metaphors arises from the fact that the real world spaces for interaction and cooperation need replacements and extensions. They require a seamless integration of personal, interpersonal, and shared work spaces. A lot of groupware suffers either from vague metaphors lacking detail or from specific metaphors lacking flexibility (Kaplan 1995). "Groupware ... provides an interface to a shared environment" (Ellis and others 1991): as straightforward as this sounds, it is one of the most difficult goals to achieve in interface design.

Information Visualization is motivated by the fact that humans organize their information spatially (on desks, in offices, on bookshelves). MIT's early Spatial Data Management System, SDMS, (Herot and others 1980) and Xerox PARC's Information Visualizer (Robertson and others 1993) were prominent projects in this area. The current transfer of a lot of information from our immediate environments into "net space" creates new opportunities for spatializing information.

Scientific Visualization makes properties and patterns in scientific data perceivable, focusing on the detection of patterns rather than on data access (MacEachren and Taylor 1994). Contrary to a common belief, visualization is often collapsing dimensions (of multidimensional parameter spaces) rather than introducing them. Scientific Visualization has so far emphasized the dimensional mapping of variables and neglected the operations that a user might perform in such spaces. The structural mappings of metaphors make for richer visualizations than a flat mapping of attributes to spatial dimensions.

4.2 Designer needs

What do designers of spatialized interfaces need and how can the Spatial Data Handling community help them? Discussions with designers (Kuhn and Frank 1993) and a look at spatialized interfaces point to some key concerns. Primarily, the lack of a coherent and applicable theory of how humans use space prevents more wide-spread and more successful spatialized interfaces. Without an understanding of the characteristics of space and its uses, spatialized designs will remain abstract, haphazard, and often more confusing than supporting for users. This paper has tried to establish such a theoretical basis in order to support cooperation among spatial disciplines and interface designers. Questions about how to find and assess spatial metaphors and how to implement them can now be asked more precisely and receive some obvious answers.

4.2.1 Finding and assessing spatial metaphors

The basic tenet of our spatialization theory is that a metaphor source needs to be a space of human experience. Thus, designers need to find out what spaces the users have experience with. For a scientist analyzing clusters of phenomena depending on some parameters, a three-dimensional Cartesian space may be appropriate, but for a digital library user probably not. Small scale spaces are often preferable over large scale spaces, as they involve less design effort for visualization and navigation. An application domain with a dominant object structure requires an entity view of space, while others call for a field view, as in a landscape metaphor. The spatial relations expressed by users in their tasks reveal appropriate image schemata. Target activities call for certain spatial affordances which limit the choice of prototypical spaces.

A practical tool for finding metaphors is *protocol analysis*, a technique used in cognitive science and interaction research. Little evidence for its usefulness has been published so far (Davies and Medyckyj-Scott 1995), but design lore points to it and some designers have successfully applied similar techniques (Erickson 1990; Erickson and Salomon 1991; Marcus 1993). The essence of applying protocol analysis is to avoid arbitrary metaphor choices by listening to users, studying their work regulations, observing their behavior, and analyzing think aloud protocols. Potential metaphor sources are certain to appear in these manifestations of how people think and act when they do their work (Kuhn 1995).

Once candidate metaphors have been found, they need to be assessed for their suitability. There are some key properties that metaphors should (+) or should not (-) have in order to support usability:
+   understandable
+   create an ontology with useful tasks
+   suggest an appropriate division of work
+   support intuitive reasoning
-   unfamiliar or complicated source (e.g. publish and subscribe in the desktop metaphor)
-   incomplete mapping to target (e.g., volumes in the desktop metaphor)
-   incoherence with other metaphors (e.g. publish and subscribe)
-   violation of image schemata (e.g., clipboard and trash of desktop).

These criteria apply to spatial as well as non-spatial metaphors, though human experience with space gives them more weight in spatialized interfaces. Users are less likely to adapt to faulty metaphor designs, because they know "how space works". This constraint generally requires to go beyond heuristic criteria and to apply a *formalization* of spatial metaphors (Kuhn and Frank 1991).

4.2.2. Implementing spatial metaphors
The key to successful spatialization is to broaden and open up *communication channels* in order to create a sense of place. Navigation and wayfinding require a great deal of multi-modal input; probably more than a screen can provide. The state of the art in Virtual Environments, on the other hand, is not yet advanced enough. For example, the standard technique for locomotion is to give hand signals with a data glove. This is unnatural, difficult to learn, and gives a strange sensation of motion.

Many implementation problems in existing interfaces are actually underlying problems of metaphor design. Consider, for example, the poor state of zoom and pan operations in many graphics systems (Haunold and Kuhn 1994). The user has to deal with cryptic buttons or even dialogue windows. This is not due to bad implementations, but to incomplete designs of the metaphors. Nevertheless, implementation issues such as the speed of data access and display are also hurting the usability of zoom and pan operations.

5. Conclusions

Space as we experience it daily, from our desktops through the rooms and buildings we live in, to the cities and landscapes of our environment, has essential properties required from source domains of general-purpose interface metaphors. This paper presented a theoretical basis for the spatialization of user interfaces and some guidance to apply it. The main ideas are the proposal to consider spatial affordances and to realize them with metaphors from prototypical spaces.

The paper has shown that there is more than one way to spatialize an interface, but that these ways are not arbitrary. Particular spatial experiences of users are mapped to application domains. This position is captured in the idea of prototypical spaces which helps to find source domains that are familiar and intuitive to the user.
In summary, spatialized user interfaces exploit
* the experience, shared by all users, of living and acting in space
* the affordances offered by spatial environments and artifacts
* the role that spatial layout plays in human memory
* the connection of spatial experience to visual and auditory perception
* the multiple levels of resolution that our cognition imposes on space
* the spatial cues for navigation, decision, and tracing history.
Spatialized user interfaces provide
* a common ground for activities in virtual and real space
* familiar concepts at the user interface
* powerful spatial reasoning
* reduced cognitive load
* coherent operations.
User interface designers have become aware that there are specialists who know how humans deal with space and how computers can model this (Kuhn and Blumenthal 1996). They need this knowledge to design tasks like navigating digital libraries or moving in Virtual Environments (Dieberger and

Bolter 1995). The Spatial Data Handling community can significantly extend its outreach and contribute to the state of the art in human-computer interaction, if it supports the activities in spatialized interface design. This will open up a broader range of applications and secure a stronger basis for the spatial disciplines and markets.

Acknowledgment
The ideas presented here have been influenced by many discussions, primarily with Andrew Frank, David Mark, Max Egenhofer, Sabine Timpf, Andreas Dieberger, and Brad Blumenthal. The work has been supported by the Technical University Vienna, the U.S. National Center for Geographic Information and Analysis (NCGIA), and the ACM Special Interest Group on Computer-Human Interaction (ACM SIGCHI).

References
Anderson, J.M. The Grammar of Case: Towards a Localistic Theory. Cambridge University Press, 1971.

Bachelard, Gaston. La poétique de l'espace. Presses Universitaires de France, 1957.

Carroll, J.M., R.L. Mack, and W.A. Kellogg. "Interface Metaphors and User Interface Design." In Handbook of Human-Computer Interaction, ed. M. Helander. 67-85. Elsevier Science Publishers, 1988.

Chalmers, M. "Using Landscape Metaphors to Represent a Corpus of Documents." In Spatial Information Theory: Theoretical Basis for GIS (COSIT'93), ed. A.U. Frank and I. Campari. Lecture Notes in Computer Science, Vol. 716: 377-390. Springer-Verlag, 1993.

Conklin, J. "Hypertext: An Introduction and Survey." IEEE Computer 20 (9) 1987: 17-41.

Cruz-Neira, C., D. J. Sandin, T. A. DeFanti, R. Kenyon, and J. C. Hart. "The CAVE Audio Visual Experience Automatic Virtual Environment." Communications of the ACM 35 (6) 1992: 64-72.

Davies, Clare and David Medyckyj-Scott. "Feet on the ground: User-GIS Interaction in the Workplace." In Cognitive Aspects of Human-Computer Interaction for Geographic Information Systems, ed. T.L. Nyerges, D.M. Mark, R. Laurini, and M.J. Egenhofer. NATO ASI Series, 123-142. Kluwer, 1995.

Dieberger, Andreas. "Providing Spatial Navigation for the World Wide Web." In COSIT'95: Spatial Information Theory - A Theoretical Basis for GIS, ed. A.U. Frank and W. Kuhn. Lecture Notes in Computer Science, Vol. 988: 93-106. Springer-Verlag, 1995.

Dieberger, Andreas and Jay D. Bolter. "On the Design of Hyper "Spaces"." Communications of the ACM (Special Issue on Hypermedia) 38 (8) 1995: 98.

Dieberger, Andreas and Jolanda G. Tromp. The Information City project - a virtual reality user interface for navigation in information spaces. Georgia Institute of Technology, 1993. http://www.gatech.edu/lcc/idt/Faculty/andreas_dieberger/VRV.html.

Dillon, Andrew, Cliff McKnight, and John Richardson. "Space - the Final Chapter or Why Physical Representations are not Semantic Intentions." In Hypertext - A Psychological Perspective, ed. Cliff McKnight, Andrew Dillon, and John Richardson. Ellis Horwood Series in Interactive Information Systems, 169-191. Ellis Horwood, 1993.

Ellis, C.A., S.J. Gibbs, and G.L. Rein. "Groupware: Some Issues and Experiences." Communications of the ACM 34 (1) 1991: 39-58.

Erickson, Thomas D. "Working with Interface Metaphors." In The Art of Human-Computer Interface Design, ed. B. Laurel. 65-73. Addison-Wesley, 1990.

Erickson, Thomas D. "From Interface to Interplace: The Spatial Environment as a Medium for Interaction." In Spatial Information Theory: Theoretical Basis for GIS (COSIT'93), ed. A.U. Frank and I. Campari. Lecture Notes in Computer Science, Vol. 716: 391-405. Springer-Verlag, 1993.

Erickson, Thomas D. and Gitta Salomon. "Designing a Desktop Information System: Observations and Issues." In Proceedings of ACM CHI'91 Conference on Human Factors in Computing Systems, Information Retrieval, 49-54. 1991.

Fahlén, Lennart E., Charles Grant Brown, Olov Ståhl, and Christer Carlsson. "A Space based model for User Interaction in Shared Synthetic Environments." In Proceedings INTERCHI'93, edited by Steve Ashlund, Kevin Mullet, Austin Henderson, Erik Hollnagel, and Ted White, 43-48, 1993.

Fairchild, K.M., G. Meredith, and A. Wexelblatt. "The Tourist Artificial Reality." In Proceedings Proc. ACM CHI'89, 299-304, 1989.

Foley, J.D., A. van Dam, S. Feiner, and J. Hughes. Computer Graphics: Principles and Practice. 2nd ed., The Systems Programming Series, ed. IBM Editorial Board. Addison-Wesley, 1990.

Frank, Andrew U. "The Prevalence of Objects with Sharp Boundaries in GIS." In Geographic Objects with Indeterminate Boundaries, ed. Peter A. Burrough and Andrew U. Frank. Taylor & Francis, 1996.

Frank, Andrew U. and Irene Campari, ed. Spatial Information Theory: Theoretical Basis for GIS (International Conference COSIT'93). Vol. 716. Lecture Notes in Computer Science. Springer-Verlag, 1993.

Frank, Andrew U. and Werner Kuhn, ed. Spatial Information Theory: A Theoretical Basis for GIS (International Conference COSIT'95). Vol. 988. Lecture Notes in Computer Science. Springer-Verlag, 1995.

Gibson, James J. The Ecological Approach to Visual Perception. Lawrence Erlbaum, 1986.

Glenberg, Arthur. "Comprehension While Missing the Point: More on Minimalism and Models." Psycoloquy (APA) (93.4.31) 1993: reading-inference.13.glenberg.

Gluck, Myke. "Making Sense of Human Wayfinding: Review of Cognitive and Linguistic Knowledge for Personal Navigation with a New Research Direction." In Cognitive and Linguistic Aspects of Geographic Space, ed. D.M. Mark and A.U. Frank. NATO ASI Series, 117-135. Kluwer Academic Publishers, 1991.

Gould, M.D. and M. McGranaghan. "Metaphor in Geographic Information Systems." In Proceedings Fourth International Symposium on Spatial Data Handling, edited by K. Brassel and H. Kishimoto, 433-442, 1990.

Haunold, Peter and Werner Kuhn. "A Keystroke Level Analysis of a Graphics Application: Manual Map Digitizing." In Proceedings CHI'94 Conference on Human Factors in Computing Systems, edited by Beth Adelson, Susan Dumais, and Judith Olson, 337-343, 1994.

Herot, C.F., R. Carling, M. Friedell, and D. Kramlich. "A Prototype Spatial Data Management System." Computer Graphics 14 (3) 1980.

Higuchi, Tadahiko. The Visual and Spatial Structure of Landscapes. Translated by Charles Terry. MIT Press, 1983.

Hirtle, S.C. and J. Hudson. "Acquisition of Spatial Knowledge for Routes." Journal of Environmental Psychology 11 1991: 335-345.

Johnson, Mark. The Body in the Mind: The Bodily Basis of Meaning, Imagination, and Reason. The University of Chicago Press, 1987.

Kaplan, Simon. ""Space" as a Basis for Collaborative Systems." ACM SIGOIS Bulletin, April 1995, 21-22.

Keillor, Garrison. Lake Wobegon Days. Viking, 1985.

Koffka, K. Principles of gestalt psychology. Harcourt-Brace, 1935.

Kuhn, Werner. "Editing Spatial Relations." In Proceedings Fourth International Symposium on Spatial Data Handling, edited by H. Kishimoto and K. Brassel, 423-432, 1990.

Kuhn, Werner. Let metaphors get rid of their WIMP image! NCGIA Research Initiative 13, Report on the Specialist Meeting: User Interfaces for Geographic Information Systems, 1991. NCGIA Technical Report 92-3.

Kuhn, Werner. "Paradigms of GIS Use." In Proceedings 5th Int. Symp. on Spatial Data Handling, edited by IGU Commission on GIS, 91-103, 1992.

Kuhn, Werner. "7±2 Questions and Answers About Metaphors for GIS User Interfaces." In Cognitive Aspects of Human-Computer Interaction for Geographic Information Systems, ed. Timothy L. Nyerges, David M. Mark, Robert Laurini, and Max J. Egenhofer. NATO ASI Series, 113-122. Kluwer Academic Publishers, 1995.

Kuhn, Werner and Brad Blumenthal. Spatialization - Spatial Metaphors for User Interfaces. ACM CHI 96. Tutorial Notes.

Kuhn, Werner and Max J. Egenhofer. "CHI'90 Workshop on Visual Interfaces to Geometry." SIGCHI Bulletin 23 (2) 1991: 46-55.

Kuhn, Werner and Andrew U. Frank. "A Formalization of Metaphors and Image-Schemas in User Interfaces." In Cognitive and Linguistic Aspects of Geographic Space, ed. D.M. Mark and A.U. Frank. NATO ASI Series, 419-434. Kluwer Academic Publishers, 1991.

Kuhn, Werner and Andrew U. Frank. "Workshop on Spatial Metaphors for User Interfaces." In Proceedings INTERCHI'93 Conference on Human Factors in Computing Systems, 220, 1993.

Kuipers, Benjamin. "Modeling Spatial Knowledge." Cognitive Science 2 (2) 1978: 129-154.

Lakoff, George and Mark Johnson. Metaphors We Live By. The University of Chicago Press, 1980.

Laurel, Brenda. Computers as Theatre. Addison-Wesley, 1993.

Lynch, Kevin. The Image of the City. MIT Press, 1960.

Lyons, J. Semantics. Cambridge University Press, 1977.

MacEachren, Alan M. and D. R. Fraser Taylor, ed. Visualization in modern cartography. Modern cartography. Elsevier, 1994.

Madsen, Kim H. "A Guide to Metaphorical Design." Communications of the ACM 37 (12) 1994: 57-62.

Marcus, Aaron. "Human Communications Issues in Advanced User Interfaces." Communications of the ACM 36 (4) 1993: 101-109.

Mark, D. M. "Spatial Metaphors for Human-Computer Interaction." In Proceedings 5th Int. Symp. on Spatial Data Handling, edited by IGU Commission on GIS, 104-112, 1992.

Mark, D.M., A.U. Frank, W. Kuhn, M. McGranaghan, L. Willauer, and M.D. Gould. "User interfaces for geographic information systems: a research agenda." In Proceedings ASPRS/ACSM 1992.

Miller, G.A. "Psychology and Information." American Documentation (July) 1968: 286-289.

Miller, Rand and Robyn Miller. "MYST." Broderbund Software and Cyan, Inc., 1994.

Minsky, M. The Society of Mind. Simon & Schuster, 1985.

Montello, Daniel R. "Scale and Multiple Psychologies of Space." In Spatial Information Theory: Theoretical Basis for GIS, ed. A.U. Frank and I. Campari. Lecture Notes in Computer Science, Vol. 716: 312-321. Springer-Verlag, 1993.

Norman, Donald A. The Design of Everyday Things. Doubleday, 1988.

Norman, Donald A. Turn Signals Are the Facial Expressions of Automobiles. Addison-Wesley, 1992.

Pratt, David R., Michael Zyda, and Kristen Kelleher. "Virtual Environments." IEEE Computer 28 (7) 1995: 17-65.

Rennison, E. and L. Strausfeld. "The Millennium Project: Constructing a Dynamic 3+D Virtual Environment for Exploring Geographically, Temporally, and Categorically Organized Historical Information." In Spatial Information Theory - A Theoretical Basis for GIS (COSIT'95), ed. A.U. Frank and W. Kuhn. Lecture Notes in Computer Science, Vol. 988: 69-91. Springer-Verlag, 1995.

Robertson, George G., Stuart K. Card, and Jock D. Mackinlay. "Information Visualization Using 3D Interactive Animation." Communications of the ACM 36 (4) 1993: 57-71.

Siegel, A. W and S. H White. "The development of spatial representations of large-scale environments." In Advances in child development and behavior: Vol. 10. , ed. H. W. Reese. 9-55. Academic Press: 1975.

Smith, D.C.S., C. Irby, R. Kimball, B. Verplank, and E. Harslam. "Designing the Star User Interface." BYTE, April 1982, 242-282.

Travers, M. "A Visual Representation for Knowledge Structures." In Proceedings Hypertext'89, 147-158, 1989.

Wilson, P.M. "Get your desktop metaphor off my drafting table: User interface design for spatial data handling." In Proceedings Fourth International Symposium on Spatial Data Handling, edited by K. Brassel and H. Kishimoto, 455-462, 1990.

Winograd, T. and F. Flores. Understanding Computers and Cognition - A New Foundation for Design. Ablex, 1986.

Yates, Frances. The Art of Memory. Chicago University Press, 1966.

Zubin, D. "A model for scale differentiation of spatially distributed percepts." In Proceedings Languages of Spatial Relations: Initiative Two Specialist Meeting Report, edited by David M. Mark et al., 14-17, 1989.

Contact address
Werner Kuhn
Department of Geoinformation
TU Vienna
Gusshausstr. 27-29
A-1040 Vienna
Austria
Fax: +43 1 504 3535
kuhn@geoinfo.tuwien.ac.at

INTERACTIVE ANALYSIS OF SPATIAL DATA

USING A DESKTOP GIS

Henning Sten Hansen

National Environmental Research Institute
Ministry of Environment and Energy
Roskilde, Denmark

Abstract
The increasing availability of geographic data and the emergence of true desktop geographic information systems, have created demands for new techniques for spatial data analysis. In this paper I describe how to develop spatial analysis tools for vector data within a commercial desktop GIS product. At the moment Moran's I, Geary's c, the Moran scattergram plus the G_i - and G_i *-statistics are included. The user friendly GUI of ArcView and the dynamic linkages between different windows creates a usefull base for the application development.

Key words
ESDA, spatial association, interactive spatial analysis.

1. Introduction

The most important difference between a geographic information system and a mere computer mapping system is the ability of the GIS to manipulate and analyse spatial data. In order to identify patterns in data we have to go beyond the study of data illustrated in the form of maps. In exploratory data analysis the data are used in an inductive way to gain new insight about patterns and relations within the data, without necessarily having a firm pre-conceived theoretical notion. The first step in exploratory spatial data analysis is to reveal spatial patterns and trends in the data. One of its aims is to identify outliers - cases that are surprising in some way.

However, the manipulation and analysis tools which are usually integrated in commercial vector based GIS products are often limited to simple spatial operations - buffering and overlaying. The possibilities for more advanced forms of spatial analysis and decision making are rather limited and only available in raster based packages e.g. IDRISI and may be the forthcoming ArcView 3. Furthermore, most statistical methods are non-spatial, and standard statistical packages are not designed to handle spatial data. Consequently, there seems to be a growing imbalance between the availability of geographic data in vector format and the limited range of tools for spatial analysis and modeling, and this may result in a failure to make full use of the data being collected.

A lot of mathematical and statistical techniques that seek to identify and quantify spatial relationships have been developed. Spatial autocorrelation tools include for example Moran's I (Moran, 1948) and Geary's c (Geary, 1954). These statistics indicate the degree of spatial association as reflected in the data set as a whole. Getis and Ord (1992) have suggested two statistics to measure the degree of local spatial association for each observation in a data set. These G_i- and G_i*-statistics are particularly useful in the detection of potential non-stationarities, which may occur when the spatial clustering of like values is concentrated in one sub-region of the data. Similarly, Anselin (1993) suggested a device to visualize and assess local instability in spatial associations in the form of a Moran scatterplot.

Openshaw (1991) has suggested to move away from exact inferential statistical methods and regard description as the main purpose in spatial analysis. Thus, the user can identify interesting map patterns without necessarily being able to test any hypothesis relating to the patterns. In fact, exploratory spatial data analysis ranges from simple description to full-blown model-driven statistical inference.

Interactivity between maps, tables and business graphics also gives a good feeling of patterns and relationsships within the data - spatial as well as non-spatial. SPIDER, developed by Haslett et al. (1990) permits geographic data to be examined simultaneously as maps and diagrams in a linked windows environment, and this encourages the user to explore the data from different point of views.

The purpose of the current project has been to develop easy to use ESDA tools using different spatial association statistics, including Moran's I, Geary's c, the Moran scattergram plus the G_i- and G_i*-statistics. The development is based on ArcView 2, which has

- a powerful programming language - Avenue
- a lot of data selection tools
- dynamic linkages between windows
- an easy to use GUI.

2. ESDA tools and GIS

Spatial data analysis can be defined as the statistical study of phenomena that manifest themselves in space. Exploratory spatial data analysis (ESDA) should therefore focus explicitly on the *spatial* aspects of the data. Thus, location, contiguity, distance and interaction become the focus of our attention. Location gives rise to two types of spatial effects: spatial dependence and spatial heterogeneity. The spatial dependence or spatial autocorrelation / association implies that similar values for a variable will tend to occur in nearby locations - leading to spatial clusters, and the smaller the spatial units, the greater the probability that nearby units will be spatially dependent. Spatial heterogeneity pertains to the regional differentiation, which follows from the uniqueness of each location.

The potential value of a GIS lies in its ability to analyse spatial data using spatial analysis techniques, and the power of a GIS as an aid in spatial analysis lies in its georelational data structure - the combination of value information and locational information. Following Openshaw (1990) no easy bridge can be established between GIS and spatial analysis, but the linkage can basically be established in three different ways

• Spatial analysis tools can be fully integrated within the GIS software
• GIS and spatial analysis can be maintained as two separate packages and data can be exchanged between the two systems
• GIS functionality can be embedded in spatial analysis or spatial modelling software.

The strategy to export spatial data from the GIS to standard statistical systems is not an adequate solution, because the nature of spatial data requires specific spatial analytic functions, which are absent in most standard statistical packages. A few spatial analysis packages e.g. SpaceStat (Anselin, 1992) have been developed, but they provides no or poor facilities for mapping or statistical graphics. Furthermore, if spatial analysis and GIS are maintained as separate packages, there are no benefits of linkages between different windows. Therefore, the most promising strategy seems to be a full integration of spatial analysis tools into GIS (Hansen, 1995).

3. Indicators of spatial autocorrelation / association

Spatial autocorrelation is a property that mapped data possesses whenever it exhibits an organized pattern or whenever there is a systematic spatial variation in values across the map (fig. 1). A positive spatial autocorrelation refers to a map pattern where geographic features of similar value tend to cluster on a map, whereas a negative spatial autocorrelation indicates a map pattern in which geographic units of similar values scatter throughout the map. When no statistically significant spatial autocorrelation exists, the pattern of spatial distribution is considered random.

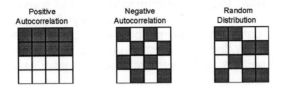

Figure 1 Concepts of spatial autocorrelation

Some methods such as Moran's I and Geary's c summarize a complete spatial distribution into a single number, which indicate the presence or absence of a stable spatial pattern that holds for the whole data set. Moran's I and Geary's c can be calculated according to the following equations

$$I = \frac{n}{\sum_i \sum_j W_{ij}} \frac{\sum_j \sum_i W_{ij} y_i y_j}{\sum_i y_i^2},$$

$$c = \frac{n-1}{2 \sum_i \sum_j W_{i,j}} \frac{\sum_i \sum_j (x_i - x_j)^2}{\sum_i (x_i - \bar{x})^2}$$

where x_i is the value of a variable in region i, and y represents deviations from their mean, i.e. $y_i = (x_i - \bar{x})$. The weighting function W_{ij} is used to assign weigths to every pair of locations in the study area and the spatial autocorrelation depends on these weights as well as the data for the locations. The simplest weighting function for areal data is a set of binary weights that have a value of 1 for areas that share a common boundary and 0 otherwise. These adjacency weights does neglect important spatial elements such as distance between the centers of the polygons.

Instead of interpreting Moran's I and Geary's c, a standardized z-value is obtained by subtracting the expected value for the statistic, and dividing the result by the corresponding standard deviation. The resulting z-value can then be compared to a table of standard normal variates to assess significance. It can be shown that Moran's I is asymptotically normally distributed as n increases. In the current paper we assume Moran's I to be approximately normally distributed. If there are a large number of polygons under consideration then normal approximations are usually accurate and this is used in testing the significance of departures from the null hypothesis. Asymptotic normality cannot be assumed for small sample sizes, and in these situations we use some small-sample corrections as suggested by (Cliff & Ord, 1973). The variance of Moran's I seems to be less affected by the distribution of the sample data than Geary's c (Cliff and Ord, 1973).

As mentioned earlier, these classic methods tend to summarize a complete spatial distribution into a single number. Although this can be useful in the analysis of small data sets, it may not be meaningful in the analysis of spatial association for many thousand spatial units. Due to the degree of non-stationarity in large data sets several regimes of spatial association might be present.

Therefore, Getis and Ord (1992) introduced a technique which allows for a finer classification of the data. The G_i- and G_i*-statistics of Getis and Ord is a measure of spatial association for each individual spatial unit. The G_i-statistics is a measure of clustering of like values around a location, irrespective of the value at that location. Unlike this, the G_i*-statistics includes the value at the location within the measure of clustering.

For each unit, these statistics indicate the degree to which that location is surrounded by high or low values for the variable under consideration. The

G_i*-statistics seems to be more in accordance with our usual interpretation of spatial association. The G_i- and G_i*-statistics may be computed for many different distance bands using the following formulars

$$G_i^{\bullet}(d) = \frac{\sum\limits_{j} w_{ij}(d)x_j}{\sum\limits_{j} x_j},$$

where $w_{ij}(d)$ is a binary symmetric spatial weight matrix with $w_{ij} = 1$ when i and j are within a distance d from each other and zero otherwise. Getis and Ord derive the moments for the G_i and $G_i^{\bullet}$ under the assumption of normality. When d is small, normality is lost, and when d is large enough to encompass the whole area, normality is also lost. Note, that these conditions must be satisfied separately for each locality in order to use the normal approximation. Its significance is assessed by means of a standardized z-value.

The interpretation of the G_i- and G_i*-statistics differs from that of the other measures of spatial association (e.g. Moran's I) in that positive z_i means clustering of high values and negative z_i means clustering of low values. Compared to Moran's I and Geary's c the G_i- and G_i*-statistics are particularly useful in the detection of potential non-stationarities, e.g when the spatial clustering of like values is concentrated in one subregion of the data area.

Following Anselin (1993) Moran's I are (in matrix notation) given by

$$I = \frac{n}{S_0} \frac{y'Wy}{y'y},$$

and if the spatial weights are row-standardized such as the elements in each row sum to 1.0, then $n = S_0$. In this case Moran's I can be expressed as

$$Wy = Iy,$$

where I is equivalent to the regression coefficient in a linear regression of Wy on y. In this case Moran's I can be visualized in form of bivariate scatterplot of Wy against y, and you can easily identify outliers and leverage points. The four quadrants in the Moran scattergram represent different types of spatial association (see fig. 2).

The upper right and lower left quadrants represent positive spatial association - high values are surrounded by high values (upper right) and low values are surrounded by low values (lower left). The upper left and lower right correspond to negative spatial association - high values are surrounded by low values (lower right) and low values are surrounded by high values (upper left). Thus, the Moran Scattergram provides additional information compared to the G_i- and G_i*-statistics.

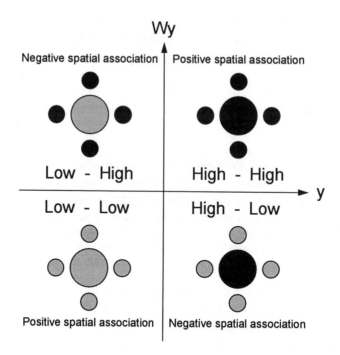

Wy

Negative spatial association | Positive spatial association

Low - High | High - High

_____ y

Low - Low | High - Low

Positive spatial association | Negative spatial association

Figure 2 The interpretation of Moran scattergram.

4. Implementation

The ArcView data model is a so-called georelational data model. Data are stored as coverages or shape files, which are the basic units of vector data storage in ArcView. Moran's I, Geary's c, the Moran scattergram plus the G_i- and G_i*-statistics can be calculated for polygon as well as point themes. Notice, that only the weighting function (contiguity or distance) can be considered "spatial". A vector-topological data model is a very useful basis for the calculation of different spatial statistics (Hansen, 1994a). However, in order to support the shape format, which is a non-topological open format, you have to use another approach. Instead of calculating the weighting matrix for the whole dataset, I use the built-in selection tools (theme-on-theme selections) of ArcView 2

aTheme.SelectByTheme(aTheme, IsWithinDistanceOf, aDistance, New),

which selects all other features within a given distance of the current feature. The described method results in a substantial reduction in processing time compared to the usual method. Using the distance matrix approach the processing time increases as the square of the number of localities, but using the described method the processing time only shows a linear increase. Furthermore, there are in fact no limitations in the number of point or polygon features in the dataset. ArcView 2 contains tools to select adjacent areas as well as areas within a certain distance from a given feature or xy-location.

In the calculation of Moran's I, Geary's c and the Moran scattergram, the adjacent polygons can be found by specifying a selection distance of zero in a theme-on-theme selection. In the current implementation Moran's I, Geary's c and the Moran scattergram are only calculated for polygon features - in accordance with Moran's original ideas. A similar approach was used in the calculation of the G_i- and G_i*-statistics - the features contributing to each G_i^* was found by a theme-on-theme selection with a specified search distance.

Using the description above, we have developed routines to calculate the discussed spatial association statistics. All routines are developed using Avenue, which is a new object-oriented programming language integrated in ArcView 2 (ESRI, 1994 b). The following Avenue code shows how to calculate $\sum_j \sum_i W_{ij} y_i y_j$ in the expression for Moran's I:

```
.....
.....
' Loop through all object Id's in theIdList...
for each objectId in theIdList
    expr = "( [" + theIdField.AsString + "] = " + objectId.AsString + " ) "
    theFTab.Query(expr, theFTab.GetSelection, #VTAB_SELTYPE_NEW)

' Retrieve field value for current Id ...
for each row in theFTab.GetSelection
    Diff1 = theFTab.ReturnValue( theField, row ) - theMean
end

' Select all adjacent polygons ...
theTheme.SelectByTheme(theTheme,#FTAB_RELTYPE_ISWITHINDISTANCEOF,
                    theDistance, #VTAB_SELTYPE_NEW)

' Sum up the Field value for all adjacent polygons ...
for each rec in theFTab.GetSelection
    theId = theFTab.ReturnValue( theIdField , rec )
    if ( theId <> objectId ) then
        Diff2 = theFTab.ReturnValue( theField, rec ) - theMean
        Sum = Sum + ( Diff1 * Diff2 )
    end
end
end
.....
.....
```

ArcView 2 permits several windows to be visible at the same time (fig. 4). The map view is the primary window, showing the spatial distribution of data. Other windows with tables and charts (scatterplots and histograms etc.) are linked dynamically to the map view (ESRI, 1994 a). Thus, selections in one window are simultaneously visible in all the other windows. For example, selections of outliers in a scattergram will highlight the corresponding features in the table and map view windows immediately. Removing spatial outliers and regenerating the chart is a possibility to evaluate the importance of such outliers. Bivariate linear regression can be computed dynamically for different subset of the map by moving a selection rectangle (or circle) over the map view. Due to the dynamic linkages between the different windows in ArcView 2, the exploratory power of the different spatial association statistics is enhanced substantial, e.g.

Figure 3 Calculation of the G_i *-statistics.

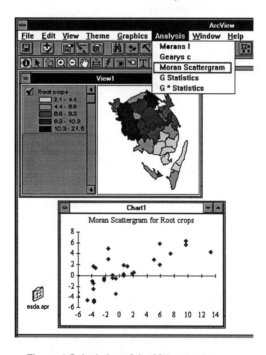

Figure 4 Calculation of the Moran scattergram.

- Moran's I can be calculated for different subsets of the data by moving a selection rectangle or circle over the map.
- Outliers in a Moran scattergram can easily be identified in the map which is dynamic linked to the chart.

5. Example

The user interface of ArcView 2 is object-oriented, and this implies that the menu-, button and toolbars are dependent on the active window (map view, table or chart). The analysis menu is only available when a map view is active. The example is concerned with the regional distribution of root crops in Funen - a Danish island.

With the root crop theme activated you just choose one of the available spatial analysis tools. Choosing Moran's I a z-value of 4.57 is obtained, and this value is appreciably larger than 2.33, the one-sided 99 % significance point of a normal distribution. Thus it is reasonable to reject the null hypothesis of a random distribution of winter root crops in Funen, and we have signifant evidence that the presence of a high percentage of root crops in one municipality has effect on whether the percentage is high in the adjacent municipalities. The G_i*-statistics for root crops in Funen confirms the high value of Moran's I, but identifies furthermore subregions of high and low percentages of root crops.

The Moran scattergram for root crops in the funish municipalities (fig. 4) confirm the conclusion from the calculations above. The point pattern of the scattergram shows, that most of the data are located in the upper right or lower left quadrants. A location in the upper right quadrant means that municipalities with high percentages of root crops are surrounded by other municipalities with high percentages of root crops, and a location in the lower left quadrant means that municipalities with low percentages are surrounded by other municipalities with low percentages of root crops. The scattergram contain a few outliers, which can be easily be identified using the dynamic link to the map view.

6. Conclusions

In the preceding sections I have demonstrated how to implement some measures of spatial autocorrelation and spatial association in an interactive easy-to-use desktop GIS environment. The current paper shows that Moran's I , Geary's c, the Moran scattergram plus G_i- and G_i*-statistics can easily be incorporated in an integrated ESDA-GIS environment at the desktop level. A precondition of the implementation has been the existence of a useful programming language - Avenue - with powerful data selection tools. Together with dynamic linkages between windows, these ESDA tools will offer new possibilities for exploring spatial relationships in data. Any longer ordinary users do not have to possess programming skills in order to use spatial statistical tools. The described tools can be applied through an easy-to-use user interface. Many of the research scientists at the National Environmental Research Institute are skilled ArcView users, and using the described tools they can now involve the spatial dimension of their data more properly. Furthermore, we have experienced that advanced users are able to create their own GIS applications using Avenue, although this programming language is not as simple as for example Visual Basic. The implementation of

ESDA tools into the ArcView 2 environment will be continued with the implementation of process-modelling tools in order to create a powerful ESDA-GIS package at the desktop level.

References

Anselin, L. (1992). SpaceStat: a program for the analysis of spatial data. NCGIA software 92-S-1. Santa Barbara, California.

Anselin, L. and Getis, A. (1992). Statistical analysis and geographic information systems. The Annals of Regional Science, vol. 26, pp. 19 - 33.

Anselin, L. (1993). The Moran scatterplot as an ESDA tool to assess local instability in spatial association. Research paper 9330. Regional Research Inst., West Virginia University.

Cliff, A.D. and Ord, J.K. (1973). Spatial Autocorrelation. Monographs in spatial and environmental systems analysis. Pion Limited, London.

ESRI (1992). ARC/INFO Data Model, Concepts and Key Terms. Environmental Research Institute Inc., Redlands, California.

ESRI (1994 a). Introducing ArcView version 2.0. Environmental Research Institute Inc., Redlands, California.

ESRI (1994 b). Introducing Avenue version 2.0. Environmental Research Institute Inc., Redlands, California.

Geary, R.C. (1954). The contiguity ratio and statistical mapping. The Incorporated statistician, vol. 5, pp. 115 - 145.

Getis, A. and Ord, J.K. (1992). The analysis of spatial association by use of distance statistics. Geographical Analysis, vol. 24, pp. 189 - 206.

Hansen, H.S. (1994 a). Spatial autocorrelation in vector-topological geographical Information Systems. Proceedings of the Fifth European Conference and Exhibition on Geographical Information Systems, Paris, pp. 1252 - 1261.

Hansen, H.S. (1995). Exploratory spatial data analysis in a desktop GIS environment. Proceedings of The Joint European Conference on Geographic Information, The Hague, The Netherlands.

Haslett, J. et al. (1990). SPIDER - an interactive statistical tool for the analysis of spatially distributed data. Int. J. GIS, vol. 4, pp. 285 - 296.

Moran, P.A.P. (1948). The interpretation of statistical maps. Journal of the Royal Statistical Society, Series B, vol. 37, pp. 243- 251.

Openshaw, S. (1990). Spatial analysis and geographical information systems : a review of progress and possibilities. In Geographical Information Systems for Urban and Regional Planning, edited by H.J. Scholten and J.C.H. Stillwell. Kluver Academic Publishers, Netherlands.

Openshaw, S. (1991). Developing appropriate spatial analysis methods for GIS. In Maguire, D. et al. (eds.) Geographical Information Systems: Principles and Applications, vol. 1, Longman, London, pp. 389 - 402.

Contact address

Henning Sten Hansen
National Environmental Research Institute
Ministry of Environment and Energy
Frederiksborgvej 399
DK-4000 Roskilde
Denmark
Phone: +45 46 30 12 00
Fax: +45 46 30 11 14
E-mail: SYHSH@DMU.DK

A DIRECT MANIPULATION INTERFACE FOR

GEOGRAPHICAL INFORMATION PROCESSING

Ming-Hsiang Tsou and Barbara P. Buttenfield

Department of Geography
University of Colorado
Boulder, USA

Abstract
Today the processing and analysis of geographical information is complicated by an increasing volume of information. We present a system to directly manipulate geographical data by using object-oriented approaches and graphic user interface (GUI) design. The study concentrates on vector data and overlay operations. A case study has been conducted by using this system for potential site selection in the Ellington, Connecticut area. The GUI design of the system uses icons to represent the geographical data and their operations. Object-oriented approaches are adopted in establishing a knowledge-based GIS system. This study suggests that the next generation of GIS user interfaces should provide an intelligent agent to assist users to search, query and operate on data. Keywords: direct manipulation, object-oriented approach, graphical user interface.

1. Introduction

With the rapid growth and development of computer technology and data gathering techniques, GIS have become more powerful, tremendous and complicated. A dramatic increase in the volume of geographical information challenges efficient management and analysis of this information. Processing and analysis of spatial data will be more complicated and difficult in the future. GIS operations have become incomprehensible for some users. However, many GIS procedures need to be modified and parameterized by their users, to cope with the complexity of the real world. We need a more intuitive way to help users work with GIS operators and data.

Recently, the graphic user interface (GUI) has emerged throughout GIS software. Case studies show that GUI users complete over one third more work per unit time compared with users of command-line interfaces. (Graham, 1994, p.32) Our approach adopts heuristic methods to improve the interaction between users and systems. However, most GIS software limits GUI commands to display of geographical information rather than on processing. As yet, there is no appropriate GUI designed for the purpose of processing or analysis of geographical data. The reason is that the nature of geographical phenomena is complicated, and the analysis of geographical phenomena is difficult to formalize into fixed functions. Introducing object-oriented approaches for spatial and geographical analysis operations may provide a natural and heuristic design at the conceptual level (Milne, Milton, and Smith, 1993).   Since the processing methods developed by object-oriented approaches can provide reusable and extensible models, future developers of GIS can improve models developed previously by other geographers or scientists. Thus, the processing of geographical information will be more sophisticated and powerful.

Object-oriented concepts originate from object-oriented languages.   The original object language was "Simula" developed in Norway in the 1960s (Taylor, 1992).   The language was designed to simulate real world interactions and resolve complex problems.   Object-oriented methods have been applied in a variety of fields such as business modeling, for example to establish the structure of an organization and the information flows within (Graham, 1994, p.318). In an object-oriented approach, data and operations are combined.   Damage of important data due to unintentional changes by programming mistakes can be avoided.   Spatial data need frequent updates and transformation to many different formats. One might say they need to be protected from external intrusion.   Another advantage of an object-oriented approach is the reuse of object components.   Object-oriented methods encourage people to retain established application models and improve upon them. Thus, geographical knowledge can be accumulated and multiplied.

Early studies of object-oriented methods in GIS applied object-oriented programming to specific applications, such as "object-oriented locational analysis" (Armstrong, 1989). Many geographers advocated the use of object-oriented approaches for spatial data management (Egenhofer and Frank, 1988) and data modeling (Worboys et al 1990). In cartographic research, object orientation has been applied to automate "... knowledge-based symbol selection ... for the visualization of univariate spatial statistical information." (Zhan and Buttenfield, 1995, p.293)

This project adopts a direct manipulation approach to implement iconic processing and analysis. We apply object-orientation to represent spatial data and GIS operations. According to past experience, we limit the project scope for the successful development of a knowledge-based system (Zhan and Buttenfield, 1995).   This research uses a small testbed located in Ellington, Connecticut, and provided with ARC/INFO. We limit the scope of GIS operations to vector data and overlay analysis. This system has been implemented for a case study to locate a residential site. Although the object-oriented geographical processing is not comprehensive in this case study, the complete definition of various spatial relations and GIS operations is the long-term goal.

2. System design and implementation

2.1 Vector data object

In order to transform geographical data to objects, we need to analyze their data structures. Every geographical data object has a specific geometric type such as polygon, line or point and a specific identifying name. For example, public wells are an instance of the "point data" class. Roads are an instance of the "line data" class. And landuse is an instance of the "polygon data" class.   In addition, the complete object also needs geometric and attribute information.   Data objects will be represented as icons indicating their data type (Figure 1).

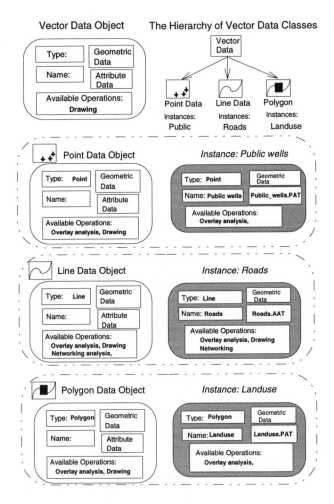

Figure 1 Vector data objects.

The class of "vector objects" is defined as the composite of type, name, geometric data, attribute data and available operations.   There are three subclasses of "vector objects" -- "point objects", "line objects", and "polygon objects". These subclasses inherit a data structure from their superclass "vector objects" and add new operations.   The differences are illustrated in the boxes labeled "type" and "available operations".   Different types of data require different operations.   For example, a "line object" may have a specific "networking analysis" function in its available operations which "point object" and "polygon object" do not have. The advantage of the definition is to generalize the set of all possible GIS functions from hundreds of commands into organized sets of functions appropriate to particular types of data. Object encapsulation allows us to focus on the specific GIS operations associated with a specific type of data without concern for other non-relevant operations. Each data object will have an associated icon to distinguish it in the interface.

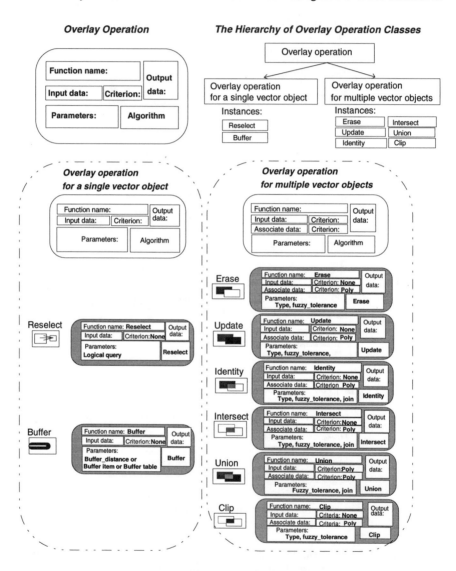

Figure 2 Overlay operation objects.

2.2 Overlay operation objects

It is possible to represent operations as objects. To convert a GIS operation to an object format is more difficult than a data-to-object conversion because an operation is not an item in the same sense as a data item. Operations are composed of algorithms and procedures. In this research, every overlay operation will be represented as an object which consists of specific algorithms, procedures, and associated operation parameters. The elements of an operation object include a function name, the algorithm code and parameters. The object data structure will be filled at execution time by pointers to input and output data objects. Eight operations will be included in this case study. Each has an icon whose image indicates graphically the meaning of the operation. The operations are executed by direct manipulation, by selecting the icon identifying the object.

In Figure 2 "overlay operations" form a class defined as the composite of "function name", "algorithm", "input data and criteria", "output data", and "parameters". Overlay operations" can be divided into two subclasses, "overlay operations for a single vector object" and "overlay operations for multiple vector objects". "Reselect" and "buffer" operations are instances of the first class as these operations are performed on a single data object. The second subclass adds "associate data" into its data structure because this class needs to process two data objects at the same time. Its instances include "erase", "update", "identity", "intersect", "union" and "clip" operations. This Figure also illustrates the hierarchy of object operations on single and multiple vector objects.

2.3 The display of data flow

Major data processing procedures include data input, processing and output. In the interface, icons representing data and operations link between icons represent the data flow (Figure 3). Notice that the geometry contained in the data object (point, line or area) is integrated into the icon image, just as operations icons show simplified representations of the type of encapsulated overlay operator. Two types of data flow identify operators with single and multiple data inputs (data entry paths). Direct manipulation allows users to select data and perform operations by combining data icons with icons of operation objects. Arrows indicate the flow of data from one object to another. Since objects represent both files and operations, the data flow paths indicate data input from a file, processing by a particular operation, and output to another file. Data objects are encapsulated with specific operations, and the interface 'assists' the user by avoiding unlisted (inappropriate) operations. Visualizing the data flow can help users systematically organize their data and develop efficient data flows during a GIS analysis.

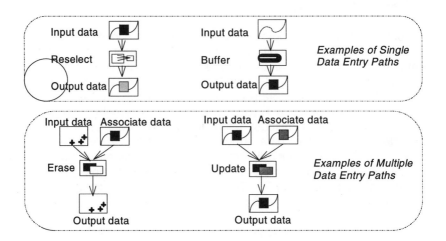

Figure 3 Data flow examples.

2.4 Graphic User Interface design

The direct manipulation design provides an efficient interface for user-system communication. The icons in this system must be simple, easy to operate and to recognize. Following established principles, the GUI includes a control panel and a display window. A GUI control panel has been designed by using the "form menu" provided in ARC/INFO's AML library (Figure 4).

Control panel functions must meet the requirements of system flow control. These functions include selecting and deselecting data objects and submitting operation requests such as "overlay" or "drawing". A data import function lists the available data objects (files) following an external database directory path. If users select an item from the list (a data object), its data file will be automatically imported to the system and appear as an icon in the display window. Four buttons ("select data", "deselect data", "overlay", "drawing" and "data import list") are implemented in the interface. In addition, two system buttons are included, to "restart" and "exit".

The display window consists of a processing box, an operator's box and a map display box. The processing box displays the data flow described above, and provides a visualization of the GIS analysis. The operator's box displays the icons of overlay operations, highlighting those that are encapsulated in data objects whose icons are selected. To execute the overlay, users select an operator and associated parameters. Its icon appears in the processing box; the operation is launched.

The map display box displays the geographical contents of selected data objects (data files) when the "drawing" request is given. The map display of geographical data can help users see the data features in operation results. The display window uses an ARCPLOT canvas. Comprehensive graphic functions in ARCPLOT display icons for data objects, operations and the map features.

The deployment of the interface in Figure 4 illustrates how the control panel and display window appear in a typical situation. A user has selected four items from the data import list (four gray icons at the top of the display window).

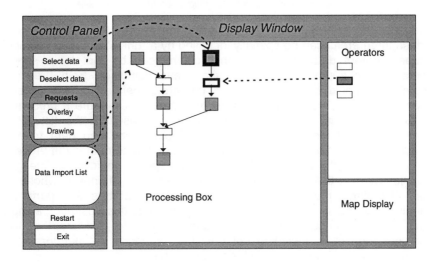

Figure 4 The deployment of the Graphic User interface.

The user has clicked on the "select data" button in the control panel to select one of the four data object icons (gray-filled icon highlighted in black). The user clicked on the "overlay" button to select one of the related operators in the operator's box, and an operation icon appears in the display window (white-filled box highlighted in black). The operation generates a new data object whose icon appears in the processing box (just below the two highlighted icons). This is a single data entry operation. In a previous task, the user has performed a multiple data entry operation, shown by the cluster of icons on the left side of the display window. The control of the system relies almost totally on mouse-driven rather than keyboard input. Using the direct manipulation interface, users can learn and operate the system more easily.

3. A case study

The case study applies the direct manipulation interface to locate a potential housing site. The testbed is "Ellington area". The Ellington database has comprehensive vector geographical data. The criteria for "locating a potential housing site" are listed in Table 1.

1. The site must be more than 100 meters away from wetland areas.
2. The site must be more than 200 meters away from steel tower power lines.
3. The site must be above the 500 year flood plain.
4. The site must be within 200 meters of a major road.
5. The slope of the site must be less than ten percent.
6. The landuse of the site must be "residential urban area".

Table 1. The criteria for locating a potential housing site.

Because the potential site must be distant from any of the first three criteria, these three areas can be spatially identified by a "union" operation combining three different data files. The other criteria can be met by a series of "intersect" operations. After that, the areas combined from the "union" operation should be excluded from the areas defined by the "intersect" operation. The "erase" operation accomplishes this. Figure 5 presents the GIS procedure in this case study. Users can visualize the flow of data processing and identify the interaction of operations and data from their respective icons. In effect, the Figure illustrates the chronology of processing steps to create the final output file. It is a graphical depiction of lineage. To our knowledge, this is the first iconic representation of lineage whose icons are also direct manipulation tools embedding operators that perform the analysis. In theory, one should be able to exchange this illustration along with the object encapsulations to allow another user to replicate the overlay process identically, including not only the sequence of operators but the parameters associated with each operator as well.

4. Discussion

A comprehensive GIS software system usually includes hundreds of commands and functions. Users need to identify which commands are appropriate for their specific data and application. The interface proposed here uses direct manipulation to simplify operations, and provide a visual record of a complex GIS task. If users focus on a specific data object, the system can minimize hundreds of available commands to a smaller number of appropriate operations. In addition, the system can check the parameters of the relevant operations. Users can concentrate on their data and application instead of getting sidetracked in finding the right commands.

The interface developed in this research uses object-oriented approaches to formalize the logical expression: *[Select data objects] ---> [Display the available operations] ---> [Select the operation] -->[Execute the operation].* The iconic approach is more logical than a command-line interface because people can visualize the geographical phenomena "first" and then consider appropriate operations. Therefore, users can execute GIS operations more naturally and intuitively.

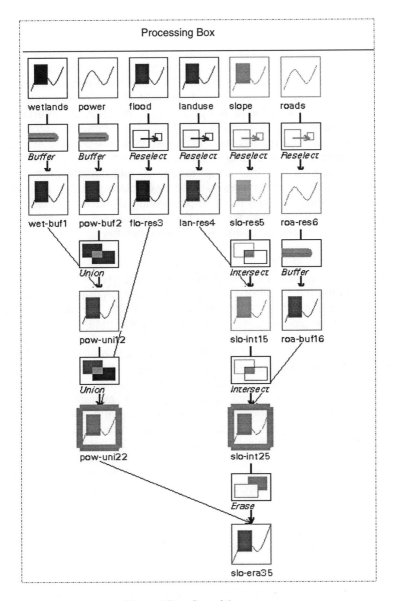

Figure 5 Data flow of the case.

It is hoped that iconic direct manipulation interface design will improve data sharing and communication among geographic researchers and scientists. By saving an image of the icons after a GIS task is complete, researchers can share their models or procedures with other people easily. Exchange of an iconic representation of the model or procedure should not require transfer of source code, AMLs or other programs, but only exchange of the object encapsulations. Recipient users can replicate, revise or tailor the model to meet their specific application.

5. Summary

Recently the GIS community has begun to pay much attention to human-computer interaction (HCI) in the development of GIS user interfaces (Medyckyj-Scott and Hearnshaw, 1993). Vendors have begun to develop GUI to improve the competitiveness of their products. For example, ESRI's ARC/INFO system introduced ARCTOOLs, a menu-driven interface as an easy-and-friendly user interface. MGE, Intergraph's Modular GIS Environment, uses the menus and icon-based interface of Windows. ERDAS' Imagine adopts a graphical user interface. Another example of GIS software which concentrates on the visualization of information processing and analysis is Geolineus, which is developed by GeoDesigns, Inc. (Lanter, 1994). This study extends Lanter's tools, implementing direct manipulation into the icons. All of these systems are intended to make GIS operations easy and self-explanatory for novice users and non-programmers.

Limitations of this study can be discussed from several perspectives. First, the system is specific to ARC/INFO. The encapsulated objects are not directly transferable to other GIS software. The programming capability in AML constrains object definitions and their hierarchy. To get around this, we used a series of global variables to represent the meanings of objects and their relationships. The GUI design is also constrained by the capabilities of ARCPLOT. For example, in a fully direct manipulation environment, functions such as "click and drag" would be implemented. At present, these functions can not be performed in ARCPLOT. Second, from a cognitive perspective, the iconic language has not been empirically evaluated, and may be insufficient for users to understand GIS operators and ancillary information. The authors have begun to design experiments to resolve this issue. Also the classification and hierarchic structure of GIS operations is still preliminary. Third, this study only focuses on vector data objects and overlay operations. Raster data objects and other GIS analysis operations such as networking analysis, hydrological analysis and 3D modeling provide opportunities for further research.

In the past, the development of GIS lacked the application of geographic models that were complicated and diverse. It was difficult to implement geographic models by procedural programming languages. In the future, there are many possible directions for the development of user interface design in GIS. However, the development of a GIS user interface still relies highly on the progress of hardware and software technologies. "GIS represents the interface between geography and an external technology, because developments in computers have been the key enabling factor that has made GIS possible" (Taylor and Johnson, 1995, p.51). The rapid progress of computer science will always have a significant influence on the development of GIS.

Direct manipulation programming capability may be available in the next generation of GIS software. Users should be able to implement their geographical models and spatial theories through object-oriented approaches. The next generation of GIS should provide user interfaces as intelligent agents to assist users to search, query, and operate the system. Interface design should consider comprehensive perspectives, including the system operation level and users' recognition and visualization levels. Future research on GIS system design should focus on the following three areas-- establishing spatial analysis theories and geographical models in GIS by using object-oriented approaches, deeply exploring the meaning of visual language and the cognition of users, and designing a comprehensive and intelligent user interface. This study gives a proof-of-concept for a direct manipulation interface and explores relevant research issues.

Acknowledgments

This paper represents part of NCGIA Research Initiative 8, "Formalizing Cartographic Knowledge", at the National Center for Geographic Information and Analysis, supported by a grant from the U.S. National Science Foundation (SBR-88-10917). Support by NSF and by the University of Colorado is gratefully acknowledged.

References

Armstrong, M. P., Densham, P. J. and Bennett, D.A.,1989, Object oriented locational analysis, Proceedings GIS/LIS '89, Orlando, Florida: 717-726.

Egenhofer, M. J. and Frank, A. U. 1988a. Object-oriented database: database requirements for GIS. in Proceedings of the International Geographical Information Systems Symposium:The Research Agenda, Vol. II: 189-211.

Graham, I., 1994, Object-oriented Methods. Workingham, England: Addison-Wesley Publishing Company (2nd Edition).

Lanter, D. P., 1994, A Lineage Metadata Approach to Removing Redundancy and Propagating Updates in a GIS Database, Cartography and Geographic Information Systems, Vol. 21(2): 91-98.

Medyckyj-Scott, and Hearnshaw, H. M. Editors., 1993, Human Factors in Geographical Information Systems. London: Belhaven Press.

Milne, P. , Milton, S. and Smith, J. L., 1993, Geographical object-oriented databases -- a case study, International Journal of Geographical Information Systems, Vol. 7(1): 39-55.

Taylor, D. A., 1992, Object-Oriented Information Systems: Planning and Implementation. New York: John Wiley & Sons, Inc.

Taylor, P. J. and Johnson, R. J., 1995, GIS and Geography, In: Pickles, John (Ed.) Ground Truth: The Social Implications of Geographic Information Systems, New York: Guilford. Ch. 3: 51-67.

Worboys, M. F., Hearnshaw, H. M., and Maguire, D. J., 1990, Object-oriented data modeling for spatial database. International Journal of Geographical Information Systems, Vol. 4(4): 369-383.

Zhan, F. B. and Buttenfield, B. P., 1995, Knowledge-based Symbol Selection for Statistical Information, International Journal of Geographical Information Systems, Vol. 9(3): 293-315.

Contact address

Ming-Hsiang Tsou and Barbara P. Buttenfield
Department of Geography
University of Colorado
Campus Box 260
Boulder, Colorado 80309-0260
USA
E-mail: tsou@ucsu.colorado.edu, babs@colorado.edu

UNSOLVED PROBLEMS OF SPATIAL REPRESENTATION

Jonathan Raper

Department of Geography, Birkbeck College
University of London
London, UK

Abstract

This paper argues that systems such as GIS developed from spatial information science are based on a very narrow epistemology of spatial representation. On the basis of a brief account of the wider epistemology of spatial representation and a critique of the current representations used in GIS, a new research agenda for GIS is proposed. This agenda for research is divided into problems that can be solved now, those that might be solvable soon and those that currently seem unsolvable.

1. Three objectives

This paper has three not-so-modest objectives: firstly, to explore the epistemological domain of spatial representation; secondly, to examine the epistemological basis on which contemporary spatial information science (SIS) has developed computable spatial representations, and, thirdly to suggest an agenda of 'unsolved problems' for the next generation of research in SIS based on a wider epistemology than the one currently in use. Such large and difficult objectives have been selected for consideration in this paper not in the hope of providing definitive answers but in order to stimulate a debate on the nature of the research agenda for spatial representation. It is argued here that a critical re-evaluation of the epistemology used by SIS suggests that some of its current concerns are only of short term interest and that others are less important when set in the larger context of the whole domain of spatial representation.

2. The epistemological domain of spatial representation

The concern to represent space arises in a large number of disciplines and has a very long history of research. Accordingly, many disciplines have developed their own concepts of space, making the contemporary epistemology of spatial representation multi-paradigmatic in nature. The key differences between the disciplines appear to lie in their metaphysical beliefs about space, the role that space plays in disciplinary knowledge and the way that space is related to time.

Most of the scientific disciplines using the critical rationalist methodology have adopted a 'materialist' metaphysic, that is, a world external to the mind is considered to exist (Musgrave 1993). While the 'real world' in the mind may be modified through the workings of the perceptual apparatus and cognitive filters, the essential external existence of entities and their functional inter-relationship is unquestioned. In this approach space has ontological primacy in the characterisation of the 'real world' since behaviour in the world is based on the laws of physics. By contrast, some disciplines adopting humanist methodologies have argued that a 'non-materialist' metaphysic is appropriate arguing that it is the impressions and experiences in human minds that truly structure and define the external world (Sayer 1984). In this approach space does not have ontological primacy since human reasoning and experience are the key to understanding. Moreover, in the humanistic view experience of occurrences at one place and time do not necessarily hold elsewhere since generalisations about events cannot accommodate individualistic behaviour. Hence, while the methodology of science accepts spatial representation as valid, humanistic approaches may not.

The role that space plays in disciplinary knowledge also distinguishes the different approaches to spatial representation. In physics space is a way to unify matter and energy: the type of spatial representations created depend on whether space is considered to have an independent existence as a universal reference (absolute space), or whether it is considered to be defined by the set of all relations between entities that exist (relative space). In mathematics space is a theoretical frame consisting of an arbitrary number of independent axes: the axes may define the spatial dimensions defined by physics or they may represent the attributes of a dynamic system. In both physics and mathematics space is considered to be continuous although spatial representations may be discrete or continuous (Ray 1991).

In psychology space is considered by some to be a sensory stimulus. The kinesthetic sense detects the movement of the limbs in space while the vestibular sense monitors the influence of gravity on the body. These 'senses' allow an individual to form a representation of position and motion. In the psychology of perception space is considered to be that part of the external world accessible to the senses. The most important perceptual sense for reconstructing the space of the external world is eyesight which is represented in the visual field on the retina (Marr 1982). Some cognitive scientists argue that the sensory apparatus is linked to the mind through image schemas, many of which are spatial (e.g. container, blockage, path, surface, link, near-far, contact, centre-periphery and scale) and that others such as 'object' and 'part-whole' can be used in making spatial representations (Mark and Frank 1991). In linguistics space is important as a factor in lexicalisation: nouns tend to describe stable, bounded entities, i.e. nouns are a form of spatial representation primitive (Givon 1979).

However, in social theory space is considered to be either a resource or a framework for personal interaction. If space is a resource then a spatial representation in social theory might be a right over land (e.g. access, exclusive use) or the tenancy of a building (Hillier and Hanson 1984). If space is a framework for personal interaction then constraints on the mobility of the individual or the cost of mobility may be appropriate spatial representations when expressed as barriers or 'permitted routes'. Social theorists who take a non-materialist metaphysic (e.g. phenomenologists) do not accept the existence of a single universally perceivable external world, but consider space to be an arena within which an individual can shift their 'standpoint' around bringing 'objects of intention' into focus at will (Collinson 1987). By analogy, the dramatic arts recreate space and time from dialogue and staging: this another form of spatial representation whether the audience is co-present (theatre) or not (cinema, radio).

Disciplinary forms of spatial representation also differ in the way in which space and time are related. In physics and in philosophy space and time have been considered to be closely interwoven since Einstein's theories of relativity at the turn of the century. In this formulation a move of any kind even if it is a reverse in three dimensional space is in fact a move forwards through the four dimensional space-time manifold of the evolving universe. This can be represented by a Minkowskian framework of four independent dimensions x,y,z,t (Akhundov 1986). However, in many disciplines space is kept separate from time: in mathematics geometry is the study of static configurations while kinematics is study of moving figures; in most spoken languages nouns are representations of an unchanging entity (space) while verbs are representations of change and behaviour (time); and, in biology organisms are distributed in space while through time they change through evolutionary steps in each generation.

It is clear from this abbreviated account (a full discussion can be found in Raper 1996a) that the epistemology of spatial representation is rich and diverse. However, the integration of the alternative metaphysics, disciplinary foci and spatio-temporal structures is still poor and few interdisciplinary accounts have attempted to make these concepts commensurable. It is argued in the next section that spatial information science (SIS) has adopted a narrow, almost naive subset of these concepts (perhaps by default) and made them computable in the form of GIS and associated systems.

3. Approaches to spatial representation in Spatial Information Science

Defining the contemporary SIS research agenda
Defining the epistemology of spatial representation in SIS has usually been considered uncontroversial: a typical working definition might be that 'spatial representation is concerned with making models of the physical and human environment that are faithful to the real world'. Such a definition might be uncontroversial in SIS since the term 'model' in this context is understood in this research community to be refer to a particular kind of representation, viz. one which is derived from 'conventional' data sources such as official maps, is computable and based upon geometric and alphanumeric data types, incorporates topological relations and deals with a well defined range of spatial and temporal scales. The spatial scales are those larger than a dwelling and smaller than the earth, while the temporal scales tend to focus

on periods longer than a day and shorter than (the last) few hundred years. These scale conventions derive from the traditional concern of the discipline of geography and from the needs of government and commerce. Yet as the discussion above illustrates, this agenda may be too narrow.

It is argued here that the current research agenda for spatial representation in SIS derives from the working definition given above and can be divided into three parts. Firstly, when reduced to its essentials, most current basic research in SIS is limited to work on: handling georeferenced computational geometry; reasoning with topological relations; the design of databases for geometric and alphanumeric data types (such as geographical information systems); and the handling of temporal sequences of georeferenced geometry. Secondly, many more researchers' agenda is to capture data from 'conventional' sources such as published cadastral or topographic maps and use geographic information systems or related software tools to establish inventories, to distribute resources or to study patterns. Thirdly, another group of researchers' agenda is to study how such systems are used on an ergonomic level, in semiotic terms or to make decisions based on the information they provide.

However, there are a number of grounds for being unsatisfied with such 'working' definition of spatial representation and therefore the research agenda derived from it (Raper and Livingstone 1995). It is based on an epistemology of spatial representation that is far too narrow and simply excludes consideration of a number of issues that other disciplines concerned with space regard as crucial. Ironically, it is the wide propagation of geographical information systems (GIS) technology that has (re)opened an interdisciplinary debate about the nature of spatial representation simply because the operational definitions in the technology are manifestly so limited. My own view of the limitations of the representations offered by geographic information systems technology will be described briefly in the following sections, beginning with the more practical issues and then moving to wider contextual problems. This critique will be used to suggest a more comprehensive epistemology from which a new research agenda may be derived.

The restricted dimensionality of spatial representation in GIS
At present most GIS implement a continuous two dimensional geometric form of spatial representation mimicking that used in on maps. However, unlike maps which were designed to be read and reconstructed cognitively, two dimensional geometry must be processed by algorithms. This makes two dimensional geometry a rather limited representation even in the urban environment where topographic elevation provides constraints on interaction and where tunnels, overpasses and multi-story buildings violate the continuity assumption. In the non-urban environment most physical processes operate over surfaces or in true three dimensional domains such as the solid earth, oceans or atmosphere making two dimensional geometry even more limited. Three dimensional GIS that have been developed are still limited representationally with respect to time (Raper 1995).

The spatial representation offered by GIS also assumes that space is absolute, i.e. space is an independently existing frame of reference for entities, in other words, a metric. This concept of space implies that an entity has a distinct identity solely by means of the space that it occupies at a given time. In this form of spatial representation, boundaries partition space into regions which are non-overlapping. Many GIS operationalise this concept of space since it reduces the range of possible topological relations to simple adjacency conditions, or in the case of networks, to connectivity conditions. Clearly, such a design also requires that separate 'layers' be defined to handle the existence of multiple tilings of the 'same' space to avoid violation of the 'non-overlapping' rule. This formulation of spatial representation has been widely implemented since it offers an analogue for the exercise of control over territory in the nation state from land ownership to census recording. However, even the activities which closely mimic absolute space concepts such as land ownership cannot be completely represented in this way. Cadastral systems violate the principles of absolute space through their need to represent the ownership of individuals floors one above another in buildings and rights such as 'easements' where one owner has the right to cross another owners land along a defined path.

The concepts of time employed in spatial representation are also extremely limited. Time is assumed to be absolute (like space) operating as a frame of reference where events partition a single universal timeline. By convention representations of space in two or three dimensions are expected to be realised at an instant in time, ideally ones where instantaneous changes take place. This approach requires the construction of a spatial representation based on a series of 'timeslices' which can be conveniently operationalised as 'layers' in many GIS. Change can then be defined as geometric differences between 'times'. This concept of 'time as separate from space' has been universally adopted by GIS since it makes time external to the main spatial representation. It is also relates closely to the operation of administrative 'updating' procedures. However, compared to the many rich theoretical treatments of space-time this approach is rather limiting.

Hence, continuous two dimensional 'absolute space and time' concepts of spatial representation amount to a compromise strongly influenced by implementational convenience and certain application area priorities. However, many researchers concerned with space find it impossible to operationalise their theories in the limited static, two dimensional domain provided by GIS.

The restricted nature of the spatial representation primitives used in a GIS
Spatial representations used by GIS and associated tools are generally made of geometric primitives. The use of points to make lines and lines to make areas in the vector method is apparently a standard approach to spatial representation, yet few GIS employ precisely the same semantics even to define what combinations of points can form a line (e.g. are self-intersecting lines allowed?). The raster method of spatial representation based on grid cells relies on the assignment of values to the cells as a means of representing spatial configurations. While this is geometrically more

homogenous than the vector approach across the range of implemented GIS, the assignment of values and the implemented functions available to operate on them vary widely among raster GIS. In both raster and vector cases the basic primitives are also treated as uncontroversially isomorphic with selected real world entities to create a spatial representation, for example, a road is treated as a vector line, or a lake is treated as a connected set of grid cells.

Descriptors of the selected real world entities are implemented as alphanumeric characters making up 'attributes' which are generally stored in tables. Such representations also have implicit limitations: there is an assumption that numbers and words can convey the nature of the 'selected real world entities'; adjectival descriptions may not be culturally neutral; categories used in descriptions may be restricted to the nominal, ordinal, interval and ratio scales when other scales exist such as circular (e.g. compass directions) or cyclical types (e.g. days of the year) (Chrisman 1995).

This brief restatement of the representational basis of many GIS and associated tools simultaneously illustrates the importance of the process of establishing a geometric isomorphism between world and GIS and shows the extremely limited expressive power that the process currently gives. Although both vector and raster forms of representation link to certain fundamental concepts (vectors are associated with the cognitive importance of entification, rasters are associated with the visual field), both are limited to a sense of physical extension. There are other modalities in the human senses, notably motion parallax, sound, smell and touch which are critical in the formation of spatial representations cognitively.

Some progress has been made in making these modalities computable, notably in the form of digital video and sound where the representational primitives are the raster frame and the sound sample respectively. Yet these new forms of representation have been adopted in an extremely limited way in GIS: at present video and sound can be associated with just a single vector or raster primitive, and even then only in some systems. Video and sound representational primitives are usually assumed to be spatially and temporally dimensionless as their internal times do not relate to 'world' time but 'playback' time. No GIS can operate on the fields of view or audible ranges of these primitives, neither can they be temporally related to any implemented 'timeline' in the system (if such a thing exists). Neither could a GIS reconcile multiple views of the same entity from different directions or deal with different sounds made simultaneously in different places. Nor can any GIS represent or capture dynamic behaviour of entities 'seen' or 'heard' moving through these new representational primitives. Such issues (even the latter) are though being addressed in multimedia information systems, often in an explicitly spatial and temporal framework (Raper 1996b)

Hence, spatial representation as understood in GIS means blind and deaf raw geometry in a world with no movement. The lack of a concept of 'viewpoint' or 'extent of influence' also makes spatial representations used in GIS into depersonalised 'universal' views of the world. While this is the view that is required by many of the applications that GIS was originally developed for, the fact that other representational perspectives cannot be operationalised illustrates the limits of current designs.

The restricted sense in which a GIS represents the real world
The working definition of spatial representation given above includes the phrase 'faithful to the real world' regarding the models made using GIS. At present the understanding of this phrase seems to be limited to the sense that a GIS renders a geometric facsimile of a paper map source, despite the wider representational scope offered by multimedia information systems and high quality visualisation environments (as cartographers have demonstrated). However, even the arrival of technically much more sophisticated representations should not obscure the fact that what is currently offered is an extremely restricted form of spatial representation. The implication is that GIS can only serve a map-oriented world view when many others can be envisaged in such developing fields as in-car navigation, 'virtual' cities and real-time emergency management systems. GIS are manifestly not 'faithful to the real world' for these users as many physical, psychological, cognitive and linguistic issues are not addressed.

In physical terms GIS offer a two or three dimensional projection of a four dimensional world and insist that all the attributes of stored entities are referenced to the same spatial and temporal granularity (the 'timeslice' model). Such assumptions are rather limiting in cases where (say) environmental theories employ spatially and temporally heterogeneous variables or where information is spatially and temporally incommensurable. GIS also define a physical world which is discrete: this must be so since the representational units are defined by discrete geometric primitives. GIS do not (usually) provide representational forms that handle continuity such as mathematical functions that mirror the continuity of the external world.

Spatial representation must also accommodate psychological studies of perception. The current map-oriented world view assumes that the features marked on a map have a perceptual equivalence to different users. Studies of perception have emphasised the importance of shape, yet few (any?) GIS can store, recognise or recreate shapes (e.g. buildings extruded upwards from their footprints?) as observed from the perspective of an observer in that environment, for example, as might be needed in an in-car navigation system. Psychological studies of wayfinding have demonstrated the importance of landmarks (often street furniture), yet no means of merging multimedia representations into GIS have yet been demonstrated (beyond simple playback of a point-referenced video clip). Neither do GIS offer representational tools for real-time event monitoring where the spatial representation must update itself with multiple inputs of different kinds at various granularities.

In a cognitive sense the design of GIS has been driven by map making priorities. The 'features' that topographic maps capture were largely decided by 18th century army generals and 19th century civil engineers and emphasise physical plant, street furniture, buildings and administrative boundaries. Such 'features' have long lives by comparison with human lifecycles and can easily be approximated by static geometry on 'timeslices'. However, what about spatial representation for: the movement of crowds or flocks; the occurrence of traffic jams; where it rained heavily last week; the noise envelopes for aircraft taking off from an airport over a working day; the delineation of the 'red-light' district or the ghetto; defining where and when I am 'safe' on the street; or, tracking a locust swarm. Spatial representation for these questions cannot be provided by GIS in their current form since they are dynamic or would involve the use of novel representational primitives.

Another question to pose of current GIS-based spatial representation is the relationship between the geometric and linguistic approaches. Language contains many forms of spatial representation in addition to geometry, for example, most languages contain the fundamental distinction between nouns (entities, static) and verbs (motion, change); languages use spatial prepositions to describe positions which are often relative in nature (such as 'between' or 'beyond'); languages also contain asymmetries of meaning between binary contrasts known as 'markedness' (Lakoff 1987)- in a spatial context, the phrase 'how long is the journey' has a different meaning from 'how short is the journey', which is a distinction that may be important in the conversion between natural language and a spatial representation.

Social context of spatial representation
Spatial representation also contains a fundamental assumption about the nature of space, viz. that space is universally understood to be defined as a metric (i.e. space is absolute, and space exists independently of entities). In fact, space is also widely construed as a relation (i.e. space is relative, and space is made of entities). Much contemporary social theory takes the view that space is relation arguing that space is an arena for interaction or a fabric for the conduct of relationships. The implication of this is that 'spatial determinism' (distance explains) is rejected. According to this view creating and storing a record of the locations of entities on the earth's surface and the distance between them (as in a GIS) is, therefore, meaningless in itself (Friedland and Boden 1994).

In contrast to the view that space (in the sense of a metric) has explanatory power, research in social theory has suggested alternative frameworks for explanation. These include approaches based on 'structuralism' (locations of entities are produced by the functioning of social systems) (Gregory 1978), 'structuration' (social structures reproduce across space and time through interactions) (Giddens 1984), 'realism' (social structures generate events and entities in an individualistic non-generalisable fashion) (Sayer 1984), and 'postmodernism' (space is not ordered but a disordered mosaic- place may be destroyed as a meaningful concept by new forms of interaction over communication networks) (Soja 1979). Each of these views regard distance measuring, pattern examination and spatial adjacency/connectivity as meaningless in themselves since they are socially defined. However, the way that social processes are constrained by physical space and time is considered important.

4. The need for a new epistemology for spatial representation

In the light of the shortcomings of GIS-based spatial representation presented above, it is strongly argued here that any new research agenda for GIS should be developed out of the wider epistemology of spatial representation briefly reviewed earlier in this article. For some years the prevailing opinion in the SIS community has been that two dimensional geometry is an acceptable compromise in the goal to represent the external world, and some study it as an end in itself for this reason. However, this agenda for action intentionally ignores the challenges of 'optimising' the current representations or their use, as the aim in this paper is to look to new 'non-incremental' researchable questions.

In the period ahead the new demands likely to be placed on spatial information science seem to be unreachable using the narrow epistemology currently in use. Accordingly, in the following sections preliminary suggestions are made of which may be the most important 'unsolved' problems of spatial representation, divided into three groups of perceived tractability.

Unsolved problems that can be tackled now
Current systems based on geometry and alphanumeric attributes can and should be extended to offer richer spatial representations. For example, representational primitives such as curves and mathematical functions should be brought into GIS and integrated with the ruling topological model to offer tools for reasoning about continuous phenomena; the databases used to store the attributes need to be extended to handle data types which are cyclic and circular; and, GIS need to fully integrate multimedia data types such as video and sound into the architecture of GIS and to develop spatial operators for them.

New systems can be built to handle some spatial representations for which GIS cannot easily be adapted. For example, spatial operations need to be developed for true three dimensional representations (e.g. shortest path in 3D space); the integration of motion (kinematics) into a georeferenced framework, for example, to link inertial navigation with absolute global position-fixing; and, spatial multimedia systems need to be created which merge representations of the different forms of sensory data into systems storing information about georeferenced entities.

There are a range of new techniques that have been developed in computer science, artificial intelligence and cognitive science that can also be used to represent space and process spatial relations. These include: the use of intelligent agents to store and develop knowledge based on application-specific ontologies of space; the application of genetic algorithms to solve spatial optimisation problems or to predict spatial change through time; and, the development of a fuzzy ontology of space that makes it possible to reason about probabilistic relations.

The use of these new techniques in GIS will open up associated questions such as: how to translate from natural language to the geometric/

alphanumeric or multimedia representations that are now in use; what alternative sets of entities or phenomena are recognised as important but not recorded in the new topographic databases being developed by mapping agencies; and, what are the semiotics of these new data types- how are they understood by users of the systems, what are the differences between the encoded intention and the derived meaning? A particular challenge would be to try to capture the perceived constraints on social interaction and mobility existing for different social groups of people and then to try to reproduce them using a new form of spatial representation.

Unsolved problems that may become tractable
There are a number of problems of spatial representation which are not realistically tractable now but may become so in the next decade. These examples are intended to be an illustrative set and should be seen as representative points on the horizon: the implementation of a spatio-temporal framework based on Minkowskian 4D space (x,y,z,t) which permits the handling of temporally heterogeneous variables; the development of real-time application-specific generalisation of information presented in a rich spatial representation to different scales; the development of shape recognition (machine vision) processes which can be integrated into a georeferenced system for navigation; and, the development of topologically realistic spatial representations which can store and analyse holes (tunnels) and handles (flyovers) on otherwise continuous 2.5D representations. Also in this timeframe the explosive development of the Internet may facilitate the development of truly distributed spatial knowledge bases and methods of cooperative work.

Unsolved problems
There are, finally, a class of spatial representation problems that are, and seem likely to remain, unsolved at present. These problems should not though be given up as lost cause since only by continually re-examining them are they ever likely to become tractable. Examples include: the development of spatial representations that are truly accessible to all and which can bear testimony to the experience of the individual; the derivation of spatial intelligence from natural language; and the creation of environments that can recreate dynamic interactions from the scale of room-sized spaces to road networks in cities.

5. Conclusion

This paper argue that the epistemology of spatial information science must now embrace wider questions taken from knowledge of other disciplines that are concerned with the nature of geographic space. To stop at the current stage of development of spatial information science would be to enter a period of consolidation and to become inward looking. This paper therefore offers one view of the directions in which the new departures should be made and hopes to have persuaded sceptics to raise their eyes to more distant horizons!

References

Akhundov, M.D. (1986) Conceptions of space and time. MIT Press, Cambridge, MA.

Chrisman, N.R. (1995) Beyond Stevens: a revised approach to measurement for geographic information. Proc. Autocarto 12, ASPRS/ACSM, pp271-280.

Collinson, D. (1987) Fifty major philosophers: a reference guide. London, Routledge.

Friedland, R. and Boden, D. (1994) NowHere: space, time and modernity. Berkeley, Univ. of California Press.

Giddens, A. (1984) The constitution of society. Oxford, Polity press.

Givon, T. (1979) On understanding grammar. New York, Academic Press.

Gregory, D. (1978) Ideology science and human geography. London, Hutchinson.

Hillier, W. and Hanson, J. (1984) The social logic of space. Cambridge, CUP.

Lakoff, G. (1987) Women, Fire and Dangerous Things: What Categories Reveal About the Mind. University of Chicago Press, Chicago.

Mark, D.M. and Frank, A.U. (1991) Cognitive and Linguisti Aspects of Geographic Space. Kluwer Academic Publishers, Dordrecht.

Marr, D. (1982) Vision. San Francisco, W.H. Freeman.

Musgrave, A. (1993) Common sense, science and scepticism. Cambridge, CUP.

Raper, J.F. (1995) Making GIS multidimensional. Proc. 1st Joint European Conf. on Geographic Information, The Hague, Netherlands, 27-31 March 1995, 1, 232-40.

Raper, J.F. (1996a) Multidimensional geographies. London, Taylor and Francis.

Raper, J.F. (1996b) Towards spatial multimedia. In Craglia, M. and Couclelis, H. (Eds.) Transatlantic GIS. London, Taylor and Francis.

Raper, J.F. and Livingstone, D (1995) Development of a geomorphological data model using object-oriented design. International Journal of Geographical Information Systems 9 (4), 359-83.

Ray, C. (1991) Time, space and philosophy. London, Routledge.

Sayer, A. (1984) Method in social science: a realist approach. London, Hutchinson.

Soja, E.W. (1989) Postmodern geographies: the reassertion of space in critical social theory. London, Verso.

Contact Address

J. Raper
Department of Geography
Birkbeck College
University of London
7-15 Gresse St.
London, W1P 2LL
UK
j.raper@geog.bbk.ac.uk

A MECHANISM FOR OBJECT IDENTIFICATION AND

TRANSFER

IN A HETEROGENEOUS DISTRIBUTED GIS

Bishr Yaser, M.Sc.

Department of Geoinformatics
International Institute for Aerospace Survey and Earth Sciences (ITC)
Enschede, the Netherlands

Abstract
Federating distributed heterogeneous GISs requires a high level of semantics in their underlying databases. Semantics refer to the relationship between objects in the database and their real world entities. Enforcing a certain standard specifications to the federation violates the principal of autonomy. There are three major sources of heterogeneity in a GIS federation: syntactic, schematic, and semantic. The heterogeneity can be solved, partly, during the design phase. However, some of the heterogeneity aspects can only be handled at run-time. The latter is due to the fact that databases population, i.e., instances, are dynamic and are only known at insertion time. It is hence required to develop a high level GISs federation, i.e., fully interoperable, where users can transparently access remote services, and yet they still maintain their autonomy. This could be achieved by devising a mechanism for searching database objects by inferring their semantics rather than syntax. In this paper we present a novel approach to achieve this objective. Identifying objects will be based on their semantics at both the client and the server levels. We introduce four stages for object identification and transfer. The approach is built on the foundation of GIS theory. The semantics are defined onto the well established syntax of geographic objects representation.

Key words
Interoperable heterogeneous GIS, open GIS, syntax and semantic, GIS theory, context hierarchy.

1. Heterogeneity in distributed GISs

[Worboys, M., et al., 1991] classified semantic heterogeneity as generic and contextual. The earlier occurs when different GIS applications are using different generic models of the spatial information. For example one may use a layer-based approach and a second may use an object-based approach. The contextual heterogeneity occurs when the semantics of schemes depend upon the local conditions at particular GISs. For example two spatial databases which store two different objects can have two different meanings and roles, though they refer to the same real world entity, e.g., agricultural fields in environmental database are different from those in a cadastral database. [Spaccapietra, S., et al., 1991] listed 4 classes of heterogeneity or conflicts: semantic conflict, descriptive conflict, data model conflict, and structural conflict. The semantic conflict occurs in the situation where set of objects from two schemes are representing sets of real world entities which are related by a set comparison operators other than equality. Descriptive conflict occurs when two database objects, representing the same real world entity, are described with different sets of properties. Data model conflict is the situation where two schemes are defined with different data models, e.g., relational and object oriented. The situation where two related objects are represented using different data structures is called structural conflict. For example a designer represents a component X of an object O either by creating a new object type X or add it as a property of O.

A relatively similar classification of types of database heterogeneity is presented by [Saltor et al., 1993]. They provided more comprehensive classification of heterogeneity to which we are more inclined, and will be used in this paper. Their classification has three aspects: syntactic, schematic, and semantic. Descriptive and structural conflicts are equivalent to schematic heterogeneity, while the data model conflict is equivalent to syntactic conflict.

1. Syntactic: each database may be implemented in a different DBMS with a different data model, e.g., relational model Vs object oriented model. Moreover, syntactic heterogeneity is also related to the geometric representation of geographic objects, e.g., raster and vector representations.
2. Schematic: where objects in one database are considered as properties or metadata in the other, or object classes of the same real world entity have different aggregation or classification hierarchies in different databases.
3. Semantic: two database objects might have two different meanings in their underlying databases, although they might refer to the same real world entity, giving as a consequence semantic conflict. For example a road network object in a GIS for transportation has different semantics from that in a GIS for topographic mapping.

Object identification and transfer in a heterogeneous distributed database environment must be close enough to users' intuition and conception of real world entities. Users usually capture real world entities and abstract them into a symbolic language in the database. This symbolic language is a mean for representing real world objects in a database, e.g., relational and object oriented[1]. The study of relation between any symbolic language and real world entities, i.e., what the database symbols mean in the real world, is called semantic analysis. The result of semantic analysis is a proper understanding of what exactly a database represents. Users who query remote databases should understand its contents and meaning, i.e., semantics. It is rather difficult to leave users perform semantic analysis whenever they need to retrieve data from a remote database. A proper solution would be allowing users to query remote databases based on their own semantics and leave the system to handle any semantic difference which might arise.

The objective of this paper is to propose a mechanism for object identification and transfer at the semantic level. In section 2 heterogeneity aspects, that

---

[1] We mean here the subjective meaning of language, i.e., a way for representing and describing real world entities in a computer.for example we might use engish language for describing real world things amongst us, we might also use object orinetation for describing the same entities in computer.

could exist in geographic objects, are described. The semantic layers that have to be built on the syntactic representation of objects are presented in section 3. Section 4 is a review on the current development in open geodata interoperability with comments on some of its features The method and a testbed proposals for introducing semantics to the symbolic representation, of real world entities, in a database is introduced in section 5. The paper is concluded in section 6.

2. Heterogeneity in spatial objects

Geographic object representation in a GIS contains both thematic and geometric information. For most applications the thematic aspects of a terrain description and analysis are of prime importance. This means that the querying and processing will be organized and formulated primarily from a thematic perspective. The structuring and formulation of the analysis of the geometric aspects of the data will be secondary [Molenaar, M. 1995]. This statement leads to two important conclusions. Firstly, object identifications in a heterogeneous database environment should not rely on its geometric representation. Secondly, resolving the geometric discrepancies between different GIS has to be solved independently from the thematic ones.

There are two approaches for geographic object representation in GIS: 1) field approach in which the earth surface is presented as spatio-temporal continuum; 2) object-structured planar graph approach in which terrain features are defined by their geometry, shape and position (in addition to thematic descriptors). In addition to the above three types of heterogeneity, mentioned in section 0, there is also spatial heterogeneity which adds more complexity to the problem of federating heterogeneous GIS. [Laurini, R., 1994] listed some discrepancies that could exist in different GIS databases: spatial representation, spatial scale, projection and coordinated systems, and spatio-temporal difference are some examples. Following Molenaar's classification of the types of geometric information we can arrive to a more structured list of possible geometric discrepancies that could exist between heterogeneous GISs. There are three aspects of geometry: topology, size and shape, and position and orientation. Also there are three levels of topology: connectivity of geometric elements, connectivity of objects, and connectivity of geometric elements to objects. Thus, it is not suggested to develop a mechanism for object identification based on its geometry.

In the above section it is demonstrated that developing a mechanism for object identification and transfer, in a heterogeneous federated GIS, based on syntax or geometry is complex and may result an unstable system. It is rather more practical and intuitive to develop a mechanism for object transfer at the semantic level because geometry and syntax are prone to change in a database during a life time of an object while its semantic is more stable.

3. Syntax and semantics in GIS theory

The GIS theory introduced by [Molenaar M., 1993 (a) (b)] [Molenaar M., 1994] and [Molenaar M. et al., 1994 (a) (b)] properly defines the building blocks of spatial objects. The building blocks forms the GIS syntax figure 1. The theory also shows how to use this syntax to build object schemes, i.e., object hierarchies. At the lowest level of the syntactic definition we find the classic data structures, i.e., field and object based approaches. The GIS theory formalizes the topologic relationships amongst objects, uncertainty aspects, and the handling of geometry and topology of fuzzy objects. Finally the theory introduces a consistent framework for object hierarchies, i.e., generalization, aggregation, etc.

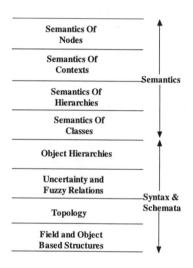

Figure 1 Syntactic and semantic definition.

In this section the emphasis is on which semantic layers that have to be built, the question of how to build them is left to section 5. The proposed theory for semantic data sharing is based on building layers of semantics onto the syntactic definition of geographic objects, figure 1 A node is a collection of interrelated contexts figure 2. Nodes can be divided into sub-nodes. A sub-node can have one or more context. A Context refers to the assumptions underlying the way in which an interoperating agent represents or interprets data. A context is defined by one and only one set of semantic specifications. [Siegel et. al, 91] defines context with respect to the view of an application, i.e., application view. Contexts can be structured in a hierarchical way. Hence, semantic specifications of a lower level context are used as building blocks for those at a higher level, as will explained in section 5.1. A context can have one or more sub-context(s). [Siegel et. al, 91] defines context with respect to the view of an application, i.e., application view. Hence, each context corresponds to one and only one database. A database in turn corresponds to one and only one data model. A data model consists of one or more hierarchy(ies). A hierarchy is formed by one or more object class(s). A class can have one or more instance(s).

Figure 1 shows the semantics that have to be built onto the syntax. Object class hierarchies are considered as syntactic problem while the functional relationships between classes, within and across hierarchies, are considered as semantic problem. This is similar to the association concept between objects [Date C.J., 1995]. Semantics of hierarchies are the second semantic level. The relation between contexts, i.e., different databases, is the third semantic level. Relationship between contexts defines the relationship between different GIS applications. The highest level of semantics is the relationship between nodes.

The intention of current research activities in the filed of interoperable information systems for data sharing and transfer are focused on open system technology. In order to illustrate where our proposal for semantic data sharing fit in these activities, the efforts of open system technology are explained in the next section. The semantics of classes, hierarchies, contexts, and nodes, are explained in section 5.1.

4. Interoperable GIS

Current research trends in interoperable GIS focus on the data transfer at the syntactic level. [Alaam, M., 1994] and [Otoo, J.E., et al., 1994] presented a federated system that provides data identification and transfer at the syntactic level. Object transfer on the syntactic level with pre-defined standards forces a strict level of semantics to the whole federation. [Bordie, M.L., 1992] defines interoperability as the ability of federated information systems to interact effectively to achieve shared goals, e.g., joint activity. Interaction does not only include data transfer but also processes, i.e., application programs. For example a user can access remote systems in order to execute some application programs which are not provided by his local system. Services transfer is a more generic term that will be used in order to refer to both data and processes transfer. Providing services with high transparency can not be achieved by focusing only on the syntactic structure of the data. The Open DataBase Connectivity, ODBC™ [Geiger, K., 1995], and the Open Geodata Interoperability Specifications OGIS™ [David S., 1995], are two solutions to achieve interoperability. The earlier is designed to handle non spatial information systems, while the latter is designed for spatial information systems .

The technical vision of the OGIS project is to enable the application developer to use any geospatial data and any geospatial function available on the net within a single environment and a single work flow. The next section is a review of the ODBC and OGIS. The functions that have to be added to them will also be mentioned.

4.1 Open DataBase connectivity

One of the most prominent and promising technologies which aims to provide connectivity between heterogeneous databases is the Open DataBase Connectivity, ODBC. It is a standard application programming interface (API) for accessing data in both relational and non relational database management systems. Using ODBC API, applications can access data stored in a variety of personal computer, minicomputer, and mainframe DBMSs, even when each DBMS uses a different data storage format and programming interface. ODBC's architecture has four components: **1)** *applications* responsible for interacting with the user via the user interface and for calling ODBC functions to submit SQL statements and retrieve results. **2)** *The driver manager* loads and calls drivers on behalf of applications[2]. **3)** *Drivers* process ODBC function calls, submit SQL requests to specific data sources, and return results to applications. **4)** Data sources consist of sets of data and their associated environments, which include operating systems, DBMS, and networks.

4.2 Open geodata interoperability specifications

OGIS™ has two part specification and conceptual model: **1)** the Open Geodata Model, OGM, which provides a common geodata model for all spatio-temporal data. The model supports both object and field based approaches by two basic components, features and coverage, respectively. **2)** Open Geoprocessing Service, OGS, defines a common consistent set of geoprocessing software interfaces. These interfaces define the behavior of geoprocessing software services which access, interchange, manage, manipulate and present geospatial data specified in OGM.

OGIS Reference Model, ORM, provides a scientific and engineering framework for designing OGM and OGS. There are three parts to the OGIS reference model: **1)** *OGIS Application domain model* which defines the methodology for abstraction from real world entities to database model. **2)** *OGIS Technology Model* which defines the architecture of the OGM and OGS. **3)** *OGIS Standards Model* pertains to the community standards built upon a foundation of existing standards and specifications.

The difference between ODBC and OGIS does not only lie in the type of data they are handling but also in their structure and, in turn, their functional abilities. ODBC is merely a tool that solves DBMSs heterogeneity. It does not include constructs that can handle the schematic and intuitively not the semantic problems. On the other hand the OGM of OGIS includes constructs that can handle the schematic problem, i.e., interoperability at the syntactic level. However, OGIS does not include constructs that allow developers to identify and access services, in remote sites, on the semantic level.

---

[2] There is a set of drivers associated with ODBC that allow users to access several DBMS provided by different vendors.

5. Interoperable GIS - a proposal

Introducing a single canonical model to a federation of heterogeneous GIS is practically not possible. There are two arguments for this impossibility. Firstly, the augmentation of all data models into a single model if not impossible it will add complexity to the global schema which will make it difficult to manage. Secondly, a real world entity can have more than one role in the individual databases and hence different syntax and semantics in the unified data model, which is of course not possible with the available systems. The definition of objects using different aspects of semantics and syntax is always defined with respect to the role the object is playing in a particular context. Different contexts mean different object roles and hence variation in the syntax and semantics. This is where the heterogeneity between databases arises. It is, hence, useful to develop another strategy that enables us to define rules that should be satisfied in order to transform an object in one context to a similar object in another context. The quest is now to clearly define what we need to develop and where it should be.

In section 1 we showed that the GIS theory should be extended in order to accommodate semantic definitions. Semantics are built onto the syntactic definition of geographic objects. Molenaar showed the syntactic similarity between object and field based approaches. This similarity implies that sending a query to a remote site requesting a set of objects should not be affected by whether the data set is field or object based.

GIS theory and OGIS are serving two interrelated purposes: GIS and OGIS are both concerned with defining syntax and semantics of geodata in individual GISs. On the other hand OGIS goes further and defines constructs and protocols for accessing heterogeneous GIS. Two major aspects have to be further developed in order to achieve GIS interoperability on the semantic level:
1. Building semantics on to syntax which is formally considered as an extension to the GIS theory.
2. Building semantic operators, that access object semantics, on top of the OGIS specifications as a complementary component.

In the subsequent section we will only give an explanation of semantics that can be built on top of syntactic definition of objects. [Bishr 1996] introduced the first formal set of constructs and rules for accessing such object hierarchies. The formal rules, i.e., semantic operators, can be built on top of OGIS. Introducing this set of mathematical rules is, however, outside the scope of this paper.

5.1 Hierarchical semantic and schematic specifications

In our approach we propose a hierarchic definition of the semantic specifications of heterogeneous databases as well as the database schemes they describe. The semantic specifications usually describe some database contents. This means that the abstraction level of a schema correspond to the same level of its semantic specifications. This approach avoids the problem of having a single unified data model and/or posing a certain standard.

In section 5 we showed that it is needed to build another layer on top of the comprehensive OGIS. This layer should include constructs (hereafter is called semantic constructs) that can handle database semantics. The semantic constructs should provide developers the capability of ensuring integrity and consistency of GISs' semantics. The rules assume that the semantic specifications of databases are arranged in a hierarchy.

According to [Sheth et. al, 1990] context can be represented in a form of a hierarchy. We can further define at each context level the possible set of object classes that can participate in the corresponding database and the processes and constrains that can be applied. Contexts can also include knowledge about the problem domain, for example in the context of analysis in watershed management, the hydrology expert is considering the logical consequence of applying a certain management practice on the environment. In this tree context definitions are aggregated to higher levels in a way similar to object oriented paradigm.

In his attempt to develop a method for semantical analysis of the meanings of linguistic expressions [Carnap R. 1964] proposed a method called *the method of extension and intension*. The semantic language has two main components:
1. An object symbolic language.
2. A metalanguage which contains translation of the sentences and other expressions in the object symbolic language.

Mapping this to our database domain, the object language is similar to that mentioned in section1, e.g., representation of real world entities in a database using object oriented data modeling. On the other hand the metalanguage is similar to metadata. Metadata are data about the content, quality, condition, and characteristics of data. It contains detailed information about the contents of a database. It is thus our proposal to use metadata to model the relationship between database symbols and real world entities. we call the symbolic language and its corresponding metadata, semantic specifications.

Moreover, human beings normally use their own logic and understanding in order to infer conclusions from expression. This can be simulated (at least to an extend which is enough for our purpose) by using predicate logic in order to infer conclusions from the semantic specifications. Expert system is in fact an implementation of predicate logic. We can hence use this technique in order to inference our semantic system, as will be show in section 5.2.

The semantic specifications (or metadata) will have the same hierarchy as that of the data model it describes. For example we will have a schema for hydrology and one for soil applications. Both are abstraction from agricultural application. There are also metadata for hydrology and another for soil. Both used the metadata of agriculture as their building blocks.

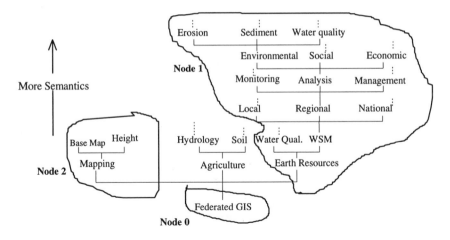

↑

More Semantics

Figure 2 Node context hierarchy.

Each major discipline in geoinformatics has its own vocabulary, e.g., the vocabulary of soil science is different from that of geology. However they all share one thing, that is, they all describe the Earth. In this approach metadata is a recursive definition of specifications. The first stage should include a proper inventory of the spatial applications at the national level. The applications has to be set in a hierarchical form similar to that shown in figure 2. The specifications at the national level of the federation should be global in order to be applicable for all the contexts. We can generally say that at the national level of the syntactic structure of spatial objects is defined. The syntactic structure will be used as building blocks for the next semantic level of contexts and so on. The semantics of each object class in a certain context is unique at the level of object oriented shareable schemes. Thus, the implementation of an object class is unique per context[3]. We call the basic set of class definitions per context, well-known types. This implies that each context has its own well-known types. Because contexts are arranged in a hierarchical tree, well-known types in a specific context can be built from other types defined in another lower context level. Introducing the semantic specification in a hierarchical form has a major advantage, it overcomes the problem of providing a global data model for all applications and disciplines in a federation. Each discipline who share a common vocabulary have their own unified data model.

5.2 Interoperable GISs Testbed

In this section, the architecture and methodology that will be implemented for developing the testbed is explained. The architecture is implemented in a Local Area Network. An FGIS driver component will be added to the global server as well as to each user who wishes to join the federation. The role of the driver is to firstly provide a platform and operating system independent

---

[3] The unique implementation of object classes does not contradict with the concept of polymorphism. The latter is related to the variant response of methods, encapsulated with such classes, according to the source that triggered the method execution.

interface; secondly to provide a transparent access of services from remote GISs. FGDC specifications [FGDC 1994] is used to implement the metadata of the global server and those of the component databases. OGIS specifications are used as the canonical data model for the federation and also are used in the global schema and the component schemes of the members of the federation.

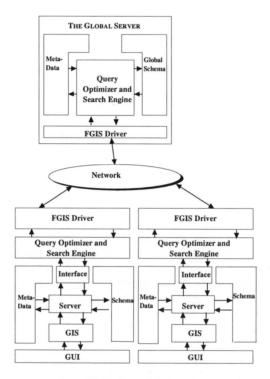

Figure 3 Federated GIS testbed.

At the global server, the context hierarchy, figure 3, is implemented and accessed via an inference engine. This engine is embedded in the query optimizer and search engine. Following our hierarchical approach and the difficulty of implementing a global schema for all possible GIS applications, not all schemes and their metadata are implemented in the global server. Thus the global schema includes a general description of database contents of each organization as well as the semantic relationship between them. The metadata will be at the same level as that of the schema, e.g., transportation, agriculture, earth resources, etc. The query optimizer is enriched with the semantic operators [Bishr 1996].

The graphic user interface, GUI, at the global server will provide non members with limited access to the federation resources. Users at this level will not have the high level of semantic tie provided to the federation members. The data they request will be provided to them either in the global schema specifications or that of the data provider.

At the user's site three components will be added: 1) FGIS driver which is similar to the one at the global server. 2) Query optimizer which will access the schema and the metadata in order to retrieve the required data and/or processes. 3) a sharable hierarchical schema and its supporting hierarchical metadata. The context tree will be further expanded at this level. The schema conforms with the OGIS specifications, and the metadata conforms with FGDC specifications.

This architecture will implement a 4 phases strategy for object identifications: 1) at the global server a process will start to identify the relevant context(s) where the requested objects might reside; 2) identifying the semantic relation between the context of the service requester and the service provider. This is also done at the global server; 3) identifying the relevant classes, i.e., intensions within the context(s). This is done at the component database which owns the data; 4) identifying the set of objects, i.e., extensions. If there exist more than one candidate context that satisfy the query, a cost/benefit model can be applied. However this is outside the scope of this paper.

6. Conclusions

In this paper the aspects of heterogeneity of geographic objects are presented. It is proposed to extend the syntactic specifications of geographic objects with semantic aspects. It is not viable to identify objects using their spatial descriptors, i.e., syntax, or geometry. This is due to the fact that objects in two similar applications can have similar semantics however they can have different syntax. For example transportation data for traffic control, in two different databases, intuitively will have the same semantics but might have different syntax.

It is required to add more constructs to OGIS in order to access objects by their semantics. These constructs will have their operands from the schemes and their corresponding metadata. In order to overcome the problem of a global data model for all possible GIS applications in the federation we propose a hierarchical model for both the data schemes and their corresponding metadata. This will provide a multi-level approach for query optimization.

References

Alaam M., 1994 "management perspective of an infrastructure for GIS interoperability - the delta-x project". Proceedings of ISPRS commission II symposium on System for data processing analysis and presentation, Ottawa, Canada. Mosaad Alaam, and Gordon Plunkett (eds) Vol 30 No. 2. The survey, mapping and remote sensing sector, natural resource Canada.

Bishr Y., Molenaar M., and Radwan M. 1996 "Spatial Heterogeneity Of Federated GIS In A Client/Server Architecture". In Proceedings of the GIS'96 Vancouver Canada. To Be Published.

Bordie M.L., 1992 "The Promise of Distributed Computing And The Challenge Of Legacy Information Systems". In Proceedings of the IFIP WG2.6 Database Semantics Conference on Interoperable Database Systems (DS-5), Lorne, Victoria, Australia, 16-20 November, David K. Hsiao, Erich J. Neuhold, and Ron Sacks-Davis (eds), pp. 1-31.

Date C.J., 1995 "An Introduction to Database Systems", Sixth edition, Addison-Wesley Publishing Company, Inc., 839 pages, ISBN 0-201-54329-X.

David Schell, 1995 "The Open Geodata Interoperability Information Package". Open GIS Consortium, Inc. 35 Main Street, Suite 5 Wayland, MA 01778.

FGDC, Content Standards For Digital Geospatial Metadata, Federal Geographic Data Committee, June 8, 1994. US Geological Survey, 590 National Center, Reston, Virginia 22092.

Kyle Geiger "Inside ODBC", 1995, Microsoft Press, Redmond Washington 98052-6399. 482 pages. ISBN 1-55615-815-7.

Laurini R., 1994 "Sharing Geographic Information In Distributed Databases". Proceedings URISA, pp. 441-454.

Molenaar M., 1995 "An Introduction Into The Theory Of Topologic and hierarchical Object Modeling In Geo-Information Systems". Lecture Notes, Department of Land-Surveying & Remote Sensing, Wageningen Agricultural University, The Netherlands.

Molenaar M., 1993 (a) "Object Hierarchies and Uncertainty in GIS or why is Standardisation so difficult". Geoinformation systems, Vol6, No.3, pp22-28.

Molenaar M., 1993 (b) "Object Hierarchies and Uncertainty in GIS or Why is Standardisation so Difficult?". Geo-Information system, vol. 6, No 3.

Molenaar M., 1994 "A Syntax for the Representation of Fuzzy Spatial Objects". In Molenaar, M. and S. de Hoop (eds), AGDM'94 Spatial data Modeling and Query Languages for 2D and 3D Applications, Publications on Geodesy - New Series, No. 40, pp155-169, Netherlands Geodetic Commission, Delft, 1994.

Molenaar M., and D.E. Richardson, 1994 (a) "Object Hierarchies for Linking Aggregation Levels in GIS". in R. Welch and M. Remillard (eds.), Mapping and GIS, proceedings of ISPRS Comm.IV, Int. Archives of Photogrammetry and Remote Sensing vol. 30 part 4, Athens Georgia, pp610-617.

Molenaar M., and L.L.F. Janssen, 1994 (b) "Terrain Objects, Their Dynamics and Their Monitoring by the Integration of GIS and Remote Sensing". In Ebner, H. et.al. (eds.), Spatial Information from Digital Photogrammetry and Computer Vision, proceedings ISPRS Comm.III, Int Archives of Photogrammetry and Remote Sensing vol. 30 part 3, München, Germany, pp585-591.

Otoo J.E., and Mamhikoff A., 1994 "Delta-X Federated Spatial Information Management System". Proceedings of ISPRS commission II symposium on System for data processing analysis and presentation, Ottawa, Canada. Mosaad Allam, and Gordon Plunkett (eds) Vol 30 No. 2. The survey, mapping and remote sensing sector, natural resource Canada.

Saltor F., Castellanos M.G., and Gracia-solaco M., 1993 "Overcoming Schematic Discrepancies in Interoperable Databases". In Proceedings of the IFIP WG2.6 Database Semantics Conference on Interoperable Database Systems (DS-5), Lorne, Victoria, Australia, 16-20 November, David K. Hsiao, Erich J. Neuhold, and Ron Sacks-Davis (eds), pp. 191-206.

Schell D., "Open GIS Consortium", An information package. 35 Main Street, Suite 5, Wayland, Ma 01778, USA.

Sheth A., and Larson J., 1990 "federated database systems for managing distributed, heterogeneous and autonomous databases". ACM computing surveys, Vol. 22, No. 3.

Siegel M., and Madnick S.E., 1991 "A Metadata Approach to Resolve Semantic Conflict". In Proceedings of the 17th international conference on Very Large Databases, pp. 133-145.

Spaccapietra S. and Parent C., 1991 "Conflicts And Correspondence Assertions In Interoperable Databases". SIGMOD Record, Vol.20, No. 4, pp49-54.

Worboys M.F. and Deen S.M., 1991 "Semantic Heterogeneity in Distributed Geographic Databases". SIGMOD Record, Vol.20, No. 4, pp30-34.

Carnap R., 1964 "Meaning And Necessity, A Study In Semantics And Model Logic". The University of Chicago press, Chicago and London. Library of Congress Catalog Card Number: 556-9132.

Wood J., 1985 "What Is The Link?" Morgan Kaufmann. In Readings in Knowledge Representation.

Contact address
Bishr Yaser, M.Sc.
Department of Geoinformatics
International Institute for Aerospace Survey and Earth Sciences, ITC
P.O. Box 6
7500 AA Enschede
The Netherlands
Phone: +31-534-874 299
Fax: +31-534-874 335
E-mail: yaser@itc.nl

A QUALITATIVE APPROACH TO RECOGNITION OF

MAN-MADE OBJECTS IN LASER-RADAR IMAGES

Erland Jungert

FOA (National Defence Research Establishment)
Linköping, Sweden

Abstract
The usage of sensor data is becoming more and more intensive in many applications and for this reason the amount of data that need be analyzed in real or close to real time is growing rapidly. As a consequence, many of the analysis methods, which traditionally have been used, may no longer be applicable. Neither may parallel methods be useful to all existing types of sensor data applications. For this reason other methods must be developed as well. Possible solutions to these problems can be based on qualitative spatial reasoning. The method introduced here is qualitative and primarily designed for recognition of man-made objects through a matching process involving sensor data from a laser-radar which generates 3D 'images' of geometric type in high resolution.

Key words
Object recognition, spatial reasoning, qualitative spatial reasoning, laser-radar.

1. Introduction

A laser-radar is a sensor that creates images where the information is represented in three dimensions and most often in high resolution, i.e. down to half a meter or less. Therefore, this information can be used for recognition of various types of man-made objects, for instance, different types of vehicles and buildings etc. Traditionally, object recognition has been subject to matching methods that generally are pixel based. Among these can, e.g. a technique based on chamfer matching and distance transforms by Borgefors [1] be mentioned. Another matching approach applied to laser-radar images is suggested by Haala et al. [2].The general problem with these methods is that they are quite slow and require massive computation times which is especially apparent when dealing with large object libraries. This cannot be accepted when real time or near real time solutions are required. Another problem of importance is, that data from most sensors quite often are distorted, in some way, and also include other types of uncertainties. For these reasons, other approaches will be necessary in order to keep down the computation time and to deal with the uncertainties. One way of overcoming these problems is the use of qualitative methods which will allow spatial reasoning and representations that can be useful as cognitive models and as indices to very large spatial data quantities. Jungert [3] is an example of an early qualitative approach to object recognition which, however, was not completely adequate, which will be shown below. A related work, concerned with similar-shape retrieval is proposed by Mehrotra and Gary [4]. In their work features and objects are represented in a normalized co-ordinate system in a way that is not qualitative. Their approach is of interest in that the objects are encoded with respect to scale, rotation and translation in sequences of co-ordinate points. An arbitrary pair (the base pair) is normalized and the remaining vectors are scaled accordingly. As will be demonstrated subsequently, in qualitative approaches, characteristics of concern in the object representation are mainly the same as in the approach by Mehrotra and Gary, i.e. scale, rotation and to a lesser degree, also translation. Special attention must be paid to scale and rotation in all shape describing methods. Especially, if they cannot be made explicitly rotation and scale invariant, so at least, it should be possible to deal efficiently with these aspects. This has in particular been in focus in this work where the object description structure is scale independent and implicitly is rotation invariant. A temporary constraint in this work is, however, the assumption that the objects should not be in motion which makes recognition of the objects somewhat simpler.

In the matching process identified objects are matched against known objects, described in the same terms, and stored in an object library. With this in mind the primary goal is a qualitative approach that can deal with the matching problem. At the same time, the uncertainties of the objects should be handled. Uncertainties of concern are, among other things, of type incomplete or missing information.

2. Laser-radar characteristics

A laser-radar is simply expressed, a laser emitter and a receiver which can send out and receive short laser pulses reflected by a target. The reflected pulses, the waveforms, are received and analyzed. An image generated by a laser-radar is thus more than a conventional image in which the intensity is registered in each pixel but corresponds to a true geometric mapping in 3D, which here is called a geometric image. The laser-radar of concern here is airborne and for this reason three dimensional information is vertically acquired. The so obtained information corresponds mainly to position and elevation. Hence, a laser-radar image corresponds to a terrain elevation model in high resolution. Other information can be determined from the laser-radar waveform as well. For instance, a laser pulse can hit both the top of a tree and the ground beside it since this information is carried by the received waveform and therefore the height of the tree can be determined from the same waveform. A problem is that large areas cannot be covered quickly by the sensor. Nevertheless, laser-radar images may be used in geographical information systems and is also a very good complement to other types of sensors. The laser-radar used here is called Top-eye and is manufactured by SAAB Dynamics. Top-eye uses partial GPS for georeferencing ground data.

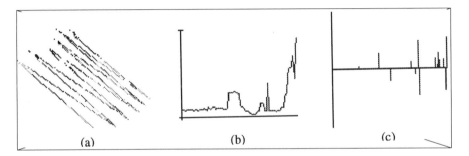

Figure 1 A fraction a laser-radar image (a), a horizontal scan sweep of the image (b) and the Dz peaks of the scan sweep (c).

Since the laser-radar creates information by means of a scanning mechanism the images are generated sequentially corresponding to a zigzag pattern growing in the direction of the flight. A laser-radar image can be seen in figure 1a. In figure 1b is a horizontal view of a singular scan sweep given. The rectangular 'box like' part in the middle corresponds to a cross section of a truck. Trucks are members of a class of man-made objects of interest to recognize in this work. Thus in each single sweep the intervals of the object in focus should be identified. This can be done from the observation that the man-made objects of concern have right angles on both sides of their sweep cross sections. This is in particular of importance since right angles generally are unique to man-made objects in nature. Determination of start and end positions is simple and is done by just looking at the differences in altitude (Dz) in the sweep, see figure 1c. Here a positive peak followed by a zero interval and then by a negative peak determines the cross section of the object. The Dz-sweep is thresholded to eliminate small objects. Thus it is simple to filter out the object interval of interest. However, at this time the orientation of the cross section relative the object is unknown. This orientation can only be determined when more than one sweep have been explored. Once all existing intervals are available for an object a set of points along its contour is available from which an object description be created.

3. The QSD concept

The method described in this paper is designed to automatically recognize objects through a matching process. The first step of the method, that is the step from the simple point set, described in section 2 above, to a intermediate contour description of the object corresponding to a polygon is determined by a variation of Hough transforms, see e.g. Gonzales and Woods [5]. In the next step a sequence of atomic elements corresponding to the given object description is determined. A first outline of such a description was made in [2] which is illustrated in figure 2 where the sequence of angles includes relations from the set $\{<,=,>\}$. This sequential description is similar to that of the projection strings in Symbolic Projection, see Chang and Jungert [6]. It turns out that these angle strings are inadequate as an object description since their qualitative power, with respect to the angles, is not sufficient and for this reason the technique must be further extended. This was done by replacing the angles in the sequences by the *qualitative slope descriptors* (QSD). A QSD is a qualitative angle representation corresponding to a line segment.

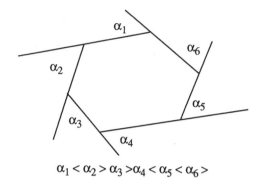

$$\alpha_1 < \alpha_2 > \alpha_3 > \alpha_4 < \alpha_5 < \alpha_6 >$$

Figure 2 A qualitative object description not enough adequate for object matching.

The general representation of a QSD is defined as:

$$(\text{sign}(\Delta x), \text{sign}(\Delta y), <\text{qualitative slope indicator}>)$$

where Δx and Δy normally are used in determining the slope coefficient. The third parameter of the QSD contains the qualitative slope indicator which is equivalent to a qualitative angle or angle interval of the QSD. Hence, the QSD $(+,+,=)$ means an angle in the first quadrant, since both Δx and Δy are positive, and where $\Delta x \approx \Delta y$, i.e. the angle of the QSD is close to $\pi/4$. The angle intervals of the QSDs may be chosen arbitrarily but in this work are $\pi/8$. There are 16 possible QSDs as illustrated in figure 3. The set of qualitative slope indicators is $\{<,=,>,-,|\}$ where '<' means the sector 'less than' the '=' sector in the quadrant determined by Δx and Δy. '>' is defined similarly. '-' is the slope corresponding to one of the 'horizontal' sectors, while '|' means one of the vertical sectors, see figure 3. Thus, a QSD is the carrier of the slope and quadrant information. The qualitative power of the QSDs is obvious and includes the following characteristics:

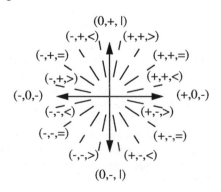

Figure 3 The QSDs and their corresponding sectors.

- A QSD reflects the orientation in space of a line segment but is independent of the coordinates and length of that line segment.
- The structure of the QSD representation allows determination of several parameters of the local vertex of the QSD, e.g. whether it is convex or concave, and whether the angle is acute or obtuse. However, it is at this point not quite clear whether the sector sizes, given in figure 3, are satisfactory or not. They are here all equal but in a practical situation it is perhaps better to define more narrow intervals

The simplest way to determine a QSD is to first determine the slope angle and then look it up in table 1 below.

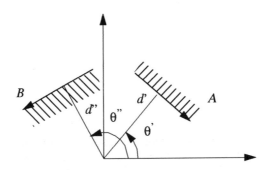

Figure 4 Two object sides, indicated with their inside, where A has a clockwise and B a counter clockwise direction.

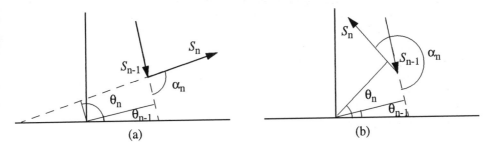

Figure 5 Illustrations of projections for generation of two consecutive QSDs of an object.

4. Global and local QSDs

QSDs are either of global or local type, thus they are either determined and referenced relative to a global or a local co-ordinate system. The former is not useful for a description of a target object that should be matched against a model object in a library since both target and model descriptions must be rotation independent Hence, global QSDs are not useful as object descriptions. Instead local QSDs determined from their precursors should be used, i.e. their orientation should be equivalent to those in figure 2.

It turns out that determination of local QSDs is a simple task that can be accomplished from the line segments of the objects, determined by the Hough-transforms (section 3). This is illustrated in figure 4. As a convention the objects are traversed counter clockwise so that the inside of the object is always to the left, see the arrows in figure 4. From the Hough transforms the distance d and the angle q were determined. As a consequence, the direction of a local QSD is either clockwise or counterclockwise, as illustrated in figure 4 where A is clockwise and B counterclockwise.

Local QSDs are determined by using the nearest preceding object segment as x-axis in a local co-ordinate system. There is, however, no need for a co-ordinate transformation since a can be determined in a much simpler way. This is done by looking at the q-angle for both the current and the proceeding line segments as illustrated in figure 5a and b.

The pairs of sequential line segments can be either clockwise, figure 5a, or counterclockwise or a combinations thereof, figure 5b. It is simple to show that when both segments have equal directions then the angle can be determined from:

$$\alpha_n = \theta_n - \theta_{n-1}$$

For segments where the directions are different a is determined from:

$$\alpha_n = \pi + \theta_n - \theta_{n-1}$$

Eventually, the local QSDs corresponding to a is determined directly from table 1 below.

α_n	QSD	α_n	QSD
$-\pi/16 < \alpha_n < \pi/16$	(+,0,-)	$15\pi/16 < \alpha_n < 17\pi/16$	(-,0,-)
$\pi/16 < \alpha_n < 3\pi/16$	(+,+,<)	$17\pi/16 < \alpha_n < 19\pi/16$	(-,-,<)
$3\pi/16 < \alpha_n < 5\pi/16$	(+,+,=)	$19\pi/16 < \alpha_n < 21\pi/16$	(-,-,=)
$5\pi/16 < \alpha_n < 7\pi/16$	(+,+,>)	$21\pi/16 < \alpha_n < 23\pi/16$	(-,-,>)
$7\pi/16 < \alpha_n < 9\pi/16$	(0,+,l)	$23\pi/16 < \alpha_n < 25\pi/16$	(0,-,l)
$9\pi/16 < \alpha_n < 11\pi/16$	(-,+,<)	$25\pi/16 < \alpha_n < 27\pi/16$	(+,-,<)
$11\pi/16 < \alpha_n < 13\pi/16$	(-,+,=)	$27\pi/16 < \alpha_n < 29\pi/16$	(+,-,=)
$13\pi/16 < \alpha_n < 15\pi/16$	(-,+,>)	$29\pi/16 < \alpha_n < 31\pi/16$	(+,-,>)

Table 1 Look-up table for determination of the QSDs.

5. QSD sequences and their impact on matching

An object can now be described in terms of a set of sequences of QSDs to establish a 3D object description. These sequences, also called strings, correspond to an ordered sequence of QSDs and the binary spatial relations between consecutive QSDs. For this, two types of binary relations have been found useful. The first relation is the angular relation equivalent to those in the string representation in figure 2. The set {<, >, =} contains all possible relations. However, it is not sufficient to just include the angular relations, length relations between consecutive segments are necessary as well, to establish a satisfactory shape description that is scale independent and where the problem of rotation invariance can be mastered. The set of possible length relations is equivalent to that of the angular relations. It has been found necessary to also include information about the length proportions between the segments which here is called the 'ratio'. Normally, the ratio will be presented as n:m, where n and m are integers. The ratio is of particular importance when man-made objects should be distinguished, e.g. a car has generally different proportions than a truck. The ratio can, however, be left out e.g. when close to 1:1. Consequently, a QSD sequence is described in the following terms:

QSD_1 [angular-rel_1,length-rel_1{, $ratio_1$}]
QSD_2 [angular-rel_2, length-rel_2{, $ratio_2$}] ...
QSD_{n-1} [angular-rel_{n-1}, length-rel_{n-1} {, $ratio_{n-1}$ }] QSD_n

The angular and length relations are collectively called segment relations.
The QSD sequences in figure 6 illustrates the method where the length ratio is excluded. The objects in figures 6a and b are equal but differs in orientation, i.e. the object in figure 6a is somewhat rotated counter clockwise compared to figure 6b. The difference in orientation has consequences on the QSD strings. This problem is simple to overcome by shifting either the head or tail QSDs including their succeeding segment relations. This can be illustrated by viewing D_1 and then shift the head of the string right to the end of the string. The resulting string then becomes equal to D_2 in figure 6b. This is also apparent from:

D_1:(+,+,>)[=,<](+,+,>)[<,<](-,+,<)[=,>](-,+,<)[>,>]

Rshift(D_1) = D_2:(+,+,>)[<,<](-,+,<)[=,>](-,+,<)[>,>](+,+,>)[=,<]

Thus Rshift means that the first element in the string is shifted to the end of the string, that is a circular shift to the right. In practice, this means that the object is clockwise rotated. Lshift is the reverse operation to Rshift, in other words, the last element is shifted left in a circular fashion so that it becomes the first in the string. Correspondingly this means a rotation counter clockwise. In this way the problem concerning possible differences in orientation between target and model object is simple to overcome.

The object in figure 6c looks similar to the first two in figure 6 but its QSD-string is different. Two QSDs and their intermediate relations are, however, equal. For this reason there is a partial match between the two objects which is simple to verify.

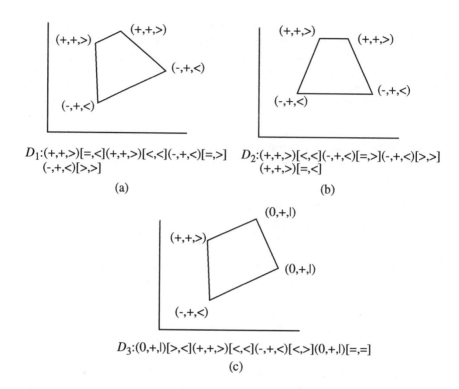

D_1:(+,+,>)[=,<](+,+,>)[<,<](-,+,<)[=,>]
(-,+,<)[>,>]

(a)

D_2:(+,+,>)[<,<](-,+,<)[=,>](-,+,<)[>,>]
(+,+,>)[=,<]

(b)

D_3:(0,+,l)[>,<](+,+,>)[<,<](-,+,<)[<,>](0,+,l)[=,=]

(c)

Figure 6A A simple object and its corresponding QSD sequence (a), the same object somewhat rotated (b) and an object similar to first two (c).

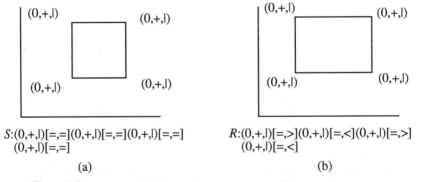

S:(0,+,l)[=,=](0,+,l)[=,=](0,+,l)[=,=]
(0,+,l)[=,=]

(a)

R:(0,+,l)[=,>](0,+,l)[=,<](0,+,l)[=,>]
(0,+,l)[=,<]

(b)

Figure 7 Two rectangular objects with equal QSDs and different segment relations.

In figure 7a a square shaped object is given and in figure 7b a rectangular. The square has a string that is symmetric since all the QSDs are equal including also all segment relations where, if present, the parameters of the ratio would be equal to each other, i.e. 1:1. Thus it is simple to conclude that the string S describes a symmetric object. The symmetry can also be validated from the following string transformation:

$$S:(0,+,l)[=,=](0,+,l)[=,=](0,+,l)[=,=](0,+,l)[=,=] \,¤\, 4(0,+,l)[=,=]$$

The rectangle in figure 7b is similar to the square and the QSDs and the angular relations are all equal. The difference between the objects occur in the length relations. Thus a few obvious observations can be made which will have an impact on the matching process:

- The number of sides of both objects are equal (the number of QSDs are equal).

- The QSDs are all equal (here to (0,+,|), i.e. all angles are approximately right).

- All angular relations are equal (they are all equal to '=').

- The length relations are different.

Besides that figure 7a is a square and figure 7b a rectangle it is clear that in a matching between the two objects a hierarchical process can be obtained. This will be discussed further below.

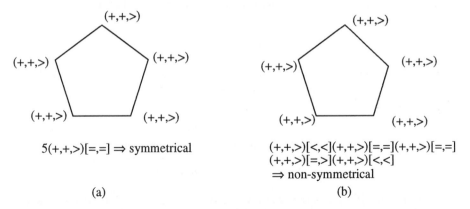

$$5(+,+,>)[=,=] \Rightarrow \text{symmetrical}$$

$$(+,+,>)[<,<](+,+,>)[=,=](+,+,>)[=,=]$$
$$(+,+,>)[=,>](+,+,>)[<,<]$$
$$\Rightarrow \text{non-symmetrical}$$

(a)   (b)

Figure 8 A symmetrical object and its QSD string (a) and a similar non-symmetrical object and its corresponding QSD string (b).

Another case that is similar to that in figure 7 is given in figure 8 where both QSD sequences are equal but differ partly for the segment relations. Hence, two levels of successful matchings can be identified. The first corresponds to sequences of QSDs that are exactly the same. The pentagon in figure 8a corresponds to an object that is symmetrical since all segment relations are equal, but this is not the case in figure 8b. Hence, the latter object is not symmetrical. A substring is, however, common to both objects, i.e.

$$(+,+,>)[=,=](+,+,>)[=,=](+,+,>)$$

Consequently, in this case the two objects are matching on a level indicating that the objects may be less similar than the square and rectangle above since here the strings have a match that is less successful. In other words they match on QSD level but except for the common substring they neither have matching angular relations nor length relations.

6. General aspects on qualitative object matching

The objects, subject to matching, are all represented in three dimensions in the images and, therefore, in order to take full advantage of all available information it is natural to expand the qualitative object descriptions such that they reflect all three dimensions. Thus, the QSD strings will correspond to contour descriptions of the object, projected onto three different planes in space. Except for the vertical projection there are two further possible projection directions, i.e. the planes corresponding to the length- and breadth cross sections. Consider a car described in terms of QSD strings of the long-side projection, the vertical projection and a few of the cross sections. This is normally sufficient for most vehicles; it may not be true for buildings which may require further projection strings onto each plane since the shape of a building quite often is more elaborate. Compared to symbolic projection [6] where the strings correspond to projections from either 2D or 3D into 1D-strings the projections here go from 3D to 2D-strings. An illustration of possible QSD projections for a simple car can be seen in figure 9.

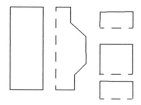

Figure 9 A number of projections of a car that can be transformed into QSD strings, corresponding to the horizontal (left), the longside (middle) and three cross section edges (right).

Since the matching process will involve not only a single string but a set of strings in more than one projection plane it must proceed according to some stepwise strategy one string at a time in a hierarchical fashion. It is, thus, natural to start matching the x-y-plane strings followed by a match of the long-side string and then eventually the cross-sections. That is, from left to right in figure 9. Thus the matching strategy for single strings generally should succeed as follows:

(i)   Determine whether model and target strings have equal number of QSDs. (If 'true' then a complete match is possible otherwise just a partial.)
(ii) Match the QSDs to find if there is either a complete or a partial match.
(iii) Match the segment relations of the string, including the ratio (if available).
(iv) Exclude objects with incorrect size.
(v)   Proceed to next string, if any.

These four steps are simple to accomplish but require more internal reasoning than indicated here which is obvious from step (iii) which is concerned with the angular relations and the length relations separately. The length relation should be given higher priority than the angular since it includes more relevant shape information than the latter which just contains refined slope information. Generally it should also be determined whether:

- the target object is completely represented or not.
- there are any distortions in the target object.

These aspects are further discussed in the next section.

7. Disconnected object segments

Distortions in the images may lead to disconnected object segments. For instance, an object may partly be hidden by a tree or the laser pulse is totally reflected such that no waveform is caught. There are two types that may occur frequently, e.g. a single segment is split into two separate segments or a corner may have disappeared. In the worst case open ended segments from which very little knowledge can be inferred and in which case just a partial match is possible. For the time being, we are here dealing with the first two cases, as illustrated in figure 10.

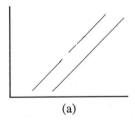
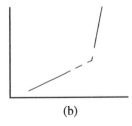

(a)   (b)

Figure 10Two disconnected segments resulting in a single QSD (a) and in two QSDs (b).

Whether two, more or less consecutive, segments with the same slope should be merged or not depends on whether there are reasons to believe that they belong to the same object or nor. Since we are here dealing with man-made objects projected onto a plane that normally correspond to rectangularily shaped objects in the vertical projection there is a line in parallel to the two lines that are subject to a merge. As a consequence, it is simple to verify that the two lines should be represented with a single QSD. This case is illustrated in figure 10a.

Figure 10b illustrates a situation where a corner is missing and for this reason this part of the object is represented with two consecutive QSD. However, some evidence that supports the hypothesis that the two segments belong to the same object must be available as well. This is, for instance, the case when one of the lines is perpendicular to the other and where the latter is in parallel with a third.

8. Discussion

The qualitative matching technique described above is currently under implementation and for that reason no practical results are not yet available. However, the examples presented above illustrates the power in the method. The primary motivation for the proposed qualitative method is the requirement for an efficient matching technique. A secondary motivation is to allow integration of qualitative results in a cognitive model for decision support but that is a long term goal. Furthermore, some experiments in Matlab on data from some recent test flights support the assumption that the method will work sufficiently efficient.

The described object matching method has primarily been developed to deal with object recognition in laser-radar images. Those images are three dimensional geometric images. The QSD approach is for that reason simple to apply since no traditional image processing is required for creation of the QSDs. This does not prevent the method from being applied to conventional images and therefore the method can be considered sensor data or image independent. Generally speaking, qualitative structures that are sensor data independent are also desired in sensor data fusion since they will be able to deal with uncertain information. For this reason, the QSD structure is a candidate for these types of problems as well.

Problems that need be studied further are how symmetries in objects and sub-objects can be determined, to develop a technique for determination of the probability that an object belongs to a certain class and to further study distortions and their consequences on matching.

References
[1] G. Borgefors, "Distance Transformations in Digital Images", Computer Vision, Graphics and Images Processing, vol. 34, 1986, pp 100-107.
[2] N. Haala, M. Cramer and J. Kilian, "Sensor fusion for airborn 3D capture", 2nd international Airborne Remote Sensing Conference, San Fransisco, June 24-27, 1996.
[3] E. Jungert, "Symbolic Spatial Reasoning on Object Shapes for Qualitative Matching", Spatial Information Theory, A theoretical basis for GIS, A. U. Frank, I. Campari (Eds.), Springer Verlag, Berlin Heidelberg, 1993, pp 444-462.
[4] R. Mehrotra and J. E. Gary, "Similar-Shape Retrieval in Shape Data Management", Computer, September 995, pp 57-62.
[5] Gonzales, R. C. and Woods, R. E., Digital Image Processing, Addison-Wesley pub. Co., Readings, Mass. 1992, pp 432-438.
[6] S.-K. Chang and E. Jungert, "Symbolic Projection for Image Information Retrieval and Spatial Reasoning", Academic Press, London, 1996.

Contact address
Erland Jungert
FOA (National Defence Research Establishment)
Box 1165
S-581 11 Linköping
Sweden
Phone: +46 13 31 83 37
Fax: + 46 13 31 81 00
E-mail: jungert@lin.foa.se

SPATIAL DATABASE PROGRAMMING USING SAND[1]

Claudio Esperança[1] and Hanan Samet[2]

[1]Universidade Federal do Rio de Janeiro
Rio de Janeiro, Brazil
[2]Computer Science Department and Center for Automation Research and
Institute for Advanced Computer Studies
University of Maryland
College Park, USA

Abstract
SAND (Spatial and Non-spatial Data) is an interactive environment that enables the
development of spatial database applications. It was designed as a tool for rapid prototyping
of algorithms and query evaluation plans dealing with spatial and non-spatial data. In this
paper we give an overview of SAND's architecture and illustrate how typical spatial and non-
spatial queries can be processed by means of short code fragments.

Key words
Spatial databases, GIS, query optimization.

1. Introduction

The design of spatial database applications involves many stages. The first
stage is choosing a proper development environment which entails the
appraisal of the many existing software packages in which will supply the
basic facilities needed for the task.   The software components most
commonly used in such applications are programs or libraries specialized in
performing operations on spatial data, while non-spatial data is frequently
handled   by database management systems (DBMS). In fact, this
combination has become so common that much effort has been put   into
integrating these components into a single framework. For the most part,
these efforts have concentrated either in adding spatial capabilities to existing
standard DBMS's [1, 2, 15, 21, 23] or in developing new database systems
with the "spatial" aspect in mind [13, 16].

Our own experience with the problem [2, 3] has shown us that the so-called
"extensible" database systems did not provide enough flexibility to experiment
with different data models, spatial indexing structures and query optimization
techniques. In other words, we feel that the integration of spatial and non-
spatial data processing is still an open problem that has to be attacked on a
wide front and at many levels. We would like, however, to be able to use the
result of efforts in that area from an application point-of-view, so that, as new

---

[1] The Support of the National Science Foundation under Grant IRI-92-16970 and of Conselho Nacional
de Desenvolvimento Científico e Tecnológico is gratefully acknowledged.

capabilities are developed, applications that can take advantage of such capabilities can be created or modified with little effort. SAND was created in this spirit. It does not purport to be a full featured spatial database; instead, it is a programming environment dedicated to the development of applications dealing with spatial and non-spatial data.

SAND can also be viewed as a toolkit that is accessed primarily by means of an interpreted language. Thus, all capabilities offered by SAND take the form of commands in that language, and queries are usually expressed by means of short code fragments called scripts. This arrangement allows SAND to emulate different paradigms for query processing. Currently, for instance, SAND offers in its library a set of functions that implement operators of the relational algebra[7]. No query language, however, has been devised for SAND, even though some high-level query languages have been proposed in the past which would be adequate to express the class of queries that SAND proposes to answer [6, 10, 15, 18]. The rest of this paper is organized as follows. Section 2 discusses the overall architecture of the SAND kernel. Section 3 presents query processing with the aid of the SAND library. Section 4 elaborates on SAND's capabilities for spatial data processing. Section 5 concludes the paper.

2. The SAND kernel

SAND consists of a kernel that implements basic objects and functions and a library that is responsible for assembling these into plans for evaluating higher level queries. Presently, the kernel implements atomic objects of common non-spatial types (strings and numbers) as well as a few choice two-dimensional geometric types (points, line segments, polygons, axes-aligned rectangles and regions). These objects are organized in tuples and tables using a relational-like data model. In order to access the functionality of the kernel in a flexible way, we opted to provide an interface to it by means of an interpreted language. We chose *Tcl* [17] for that role, mainly because it offers the benefits of an interpreted language but still allows code written in a high-level compiled language (in our case, C++) to be incorporated via a very simple interface mechanism.

A table, as implemented by SAND, is an abstraction which can better be defined by its functionality. In object-oriented programming, the concept known as *class* is used to refer to software components (*objects*) which share the same functionality. Objects of class *table* are repositories of data that can be handled one tuple at a time by means of a uniform set of functions, or, to use object-oriented terminology, *methods*. The SAND kernel implements a few table varieties (classes) and the SAND library implements a few more. In order to perform operations on any given table, that table must be "opened" -- a process not unlike opening a file. Table varieties implemented by the kernel are opened by using the **sand open** command, while those implemented by the library are opened via a call to a procedure. Either way, the result of such an operation is an open table "handle" which can be used to invoke table methods. Each opened instance of a table contains a memory buffer large enough to contain one tuple of that table, to which we refer here as a *tuple buffer*. A handle of an open table in SAND responds to the following set of methods:

Method	Description
`get`	returns the value of the tuple buffer
`set` *tuple value*	loads the tuple buffer with the given value
`first`	loads the tuple buffer with the first tuple in the table
`next`	loads the tuple buffer with the next tuple in the table
`status`	returns a boolean value that indicates whether the last call of an access method (e.g. `first, next`) was successful

To illustrate this mechanism, consider an example table called `sspring`. It is a relation containing geographical information about an area corresponding to the city of Silver Spring, Maryland. This table stores data provided by the Census Bureau [22] and has the following schema:

Attribute	Type	Description
`name`	char(30)	the name of a street
`type`	char(4)	street type (e.g., road, ave., lane)
`zipleft`	integer	zip code for the left side of the street
`zipright`	integer	zip code for the right side of the street
`line`	line	line segment corresponding to the geographic location of the street or a part thereof

If table `sspring` was part of a standard relational database, then in order to obtain a listing of its contents, we would use the query language supported by the database system (e.g., SQL) to form a statement like:

```
select * from sspring
```

Compare this to the following SAND script:

```
# script to list all tuples of sspring
set handle [sand open sspring]
$handle first
while {[$handle status]} {
    puts [$handle get]
    $handle next
}
$handle close
```

For those not familiar with the syntax of Tcl, a few explanations are in order:
- The pound sign (#) used at the beginning of a line denotes a comment.
- The dollar sign (\$) is used for variable dereferencing. Thus, construct "\$*var*" denotes the value of variable *var*.
- **set** *var value* assigns *value* to variable *var*.
- The construct "[*string*]" used as a value means that *string* is to be evaluated as a command and the result used instead.
- **puts** is one of Tcl's file output commands.

One distinction that should be made at this point is that a table is not necessarily linked to a file or any kind of storage structure. It may perhaps be compared to the concept of *cursor* existing in many implementations of relational databases or, better still, to the concept of *iterator* used in the Volcano system [12]. For instance, the output of SAND's query plan generator is also a table, i.e., opening a query evaluation plan (QEP) is equivalent to starting its execution, with methods *get*, *first* and *next*serving the purpose of accessing the data computed by the plan. The SAND library is discussed in the next section.

The table abstraction serves as a common ground to the various access methods available in SAND. Again using object-oriented terminology, we say that the set of methods presented earlier define the simplest type of table, or a "base" table class. Other, more specialized tables, offer additional capabilities by means of additional methods; these constitute "derived" table classes. For instance, the SAND kernel presently supports direct access tables, buffer tables, B-tree tables [9], PMR-quadtree and region quadtree tables [19]. Each of these are supported by appropriate additional methods:

1. Direct access tables and buffer tables support a method called *goto* which, given a tuple identifier (i.e., a *tid*), loads the tuple buffer with the corresponding tuple.
2. B-tree and hash tables support the *find* method, which locates the tuple whose key is closest to (or equal to, in the case of hash tables) the one presently loaded in the tuple buffer, e.g. via the *set* method.
3. PMR-quadtrees offer methods related to the many varieties of spatial queries: window, closest to feature, incremental nearest, etc.
4. Region quadtrees offer methods for performing operations on raster maps, such as set-theoretic operations, buffer zones, connected component labeling, etc.

The creation and destruction of tables of any of the types described above is performed by commands **sand create** and **sand drop**, respectively. Also, since all of these table types are used for storage, their contents can be altered by the following two methods:

Method	Description
`insert`	the contents of the tuple buffer is added as a new tuple in the table
`delete`	the tuple most recently retrieved from the table is removed

3. SAND library

The most important distinction between a script in SAND and a query expressed in a high-level query language such as SQL is that the latter does not usually describe a series of steps that are required to obtain the answer, whereas the former very often looks like a standard computer program. From the application point-of-view, the power of expression of a high-level query language is certainly advantageous, but not indispensable. A more important issue is how efficiently a query can be answered. In a standard database system, this issue is addressed by providing a query planner and optimizer. In SAND, this same issue is addressed by a library that is responsible for creating efficient plans.

As an example, consider a modification of the previous query so that the listing is restricted to streets corresponding to zipcode 20895. Such a query could be expressed in SQL by:

```
select * from sspring where zipleft=20895 or zipright=20895
```

One way to express this query in SAND is to modify the previous script so that the action of printing a tuple is conditioned to the satisfaction of the predicate. This approach, however, may prove to be inefficient. For instance, it may be the case that `sspring` possesses other access paths that allow the retrieval of tuples by zipcode directly. In a standard database system these factors are weighed by the query optimizer in order to provide the best possible query evaluation plan. In SAND, such a plan is created via a call to a library function. Some of the plan generating functions implemented by the SAND library are listed below:

`selection_plan` *name predicate* generates a plan to retrieve all tuples of table *name* which satisfy the given *predicate*.
`project_plan` *name attr1 attr2 ... attrN* generates a plan to retrieve from table *name* distinct tuples corresponding to the subset of attributes *attr1 attr2 ... attrN*.
`union_plan` *name1 name2* generates a plan to retrieve all distinct tuples from tables *name1* and *name2*.
`semi_join_plan` *name1 name2 predicate* generates a plan to retrieve all tuples of table *name1* such that there exists at least one tuple of *name2* which satisfies *predicate*.

The *predicate* argument used in some of these functions has the form of a common Tcl expression, except that attribute names are replaced by attribute values before its evaluation. The result of calling any of these functions is a table handle which can be used in the same way as the handles returned by the kernel **sand open** command. Thus, the above query may be answered in SAND by using the following script:

```
# script to list all tuples of sspring in the area of
  zipcode 20895
set handle [select_plan sspring
   zipleft==20895||zipright==20895]
$handle first
while {[$handle status]} {
   puts [$handle get]
   $handle next
}
```

The plan generating strategies used in the library vary in complexity from very crude to very sophisticated. Some, such as the ones described above, are modeled after relational algebra operators [7], while others are implementations of classic algorithms such as the index-scan or the hash-join algorithm. Since tables implemented both by the kernel and by the library share a common interface, query plans can be built by drawing on either. In other words, SAND offers table handling capabilities on various levels but does not impose a strict hierarchy among these levels. This arrangement permits applications to use high-level tools when these are available, and to resort to low-level tools either when higher level capabilities are unavailable or, sometimes, in an attempt to extract better performance. This open ended architecture is also valuable in the study of issues in spatial database optimization.

4. Spatial data handling

SAND supports spatial data at various levels. At the attribute level, five 2D geometric data types can be used in table schemas and can be processed with the aid of a set of functions and predicates. SAND also supports two types of spatially organized tables which can be accessed in several ways.

4.1 Spatial features

Attribute Type	Description
`point`	a point in 2D.
`polygon`	a simple (no holes, non-self-intersecting) polygon in 2D.
`rectangle`	an axes-aligned rectangle in 2D.
`region`	an arbitrary axes-aligned polygon in 2D.

The following set of spatial attribute types (also referred to as *spatial features*) are currently implemented in SAND:

The coordinate values of all spatial features are stored as double-precision floating point numbers. Polygons are stored as a list of its endpoints and regions are stored and manipulated with the aid of structures called *vertex lists* [11].

SAND also supports a series of spatial predicates and functions. Most of these are polymorphic, that is, they operate on any type of spatial features. For example:

> **distance** *f1 f2* is a function that returns the Euclidean distance between spatial features *f1* and *f2*.
> **intersects** *f1 f2* is a predicate that returns true if and only if features *f1* and *f2* occupy at least one common point in space.
> **bbox** *f1 f2 ... fN* returns the smallest axes-aligned rectangle enclosing all features *f1 ... fN*.

Since the only primitive data type handled by Tcl is the character string, constants and variables representing spatial features must eventually be converted to and from strings of characters. Thus, for instance, string "point 1.0 2.0" represents a point in 2D with coordinates x and y equal to 1 and 2 respectively. This translation process, however, is relatively costly, and should be avoided in query evaluation plans. For this reason, the SAND kernel provides support for *attribute variables*. These are similar to regular Tcl variables, except that they need not be dereferenced explicitly (and thus converted to strings) when used in functions and predicates implemented by SAND. In particular, the tuple buffer associated with an open table handle can be accessed by means of attribute variables that are created automatically when the table is opened. Thus, for instance, after executing the command

```
set handle [sand open sspring]
```

five attribute variables are created, named **$handle.name, $handle.type, $handle.zipleft** etc.

As an example, consider a query plan for listing all all streets of **sspring** within 1 mile from a given point. One way to obtain such a plan would be to use the **select_plan** library function by specifying a suitable predicate:

```
# script to list all tuples of sspring in the area within
   one mile
# (approx. 0.01 in map units) of point (-77.03, 39.0)
set handle [select_plan sspring {[distance line "point -
    77.03, 39.0"]<=0.01}]
$handle first
while {[$handle status]} {
   puts [$handle get]
   $handle next
}
$handle close
```

Consider now a related query where we wish to obtain a *graphical* output of table **sspring** by drawing all line segments corresponding to attribute **line**, but *emphasizing* those line segments lying within one mile of point (-77.03,

39.0). Clearly, this query requires that all tuples of *sspring* be accessed, and thus it is very likely that a simple scan will prove to be an "optimal" plan. All that is required is to test each line segment for proximity with the given point in order to determine the proper drawing style. This is shown in the following script:

```
# script to draw all tuples of sspring, emphasizing
# those within one mile of point (-77.03, 39.0)
set handle [sand open sspring]
$handle first
while {[$handle status]} {
    if [distance $handle.line "point -77.03, 39.0"]<=0.01}] {
        .r draw_sand $handle.line -style 1
    } else {
        .r draw_sand $handle.line -style 2
    }
    $handle next
}
$handle close
```

In this script, `.r` refers to a graphical output window, **draw_sand** is a SAND command used for drawing and **-style** is an option that modifies drawing parameters such as color or line thickness. The output of this script is shown in Figure 1.

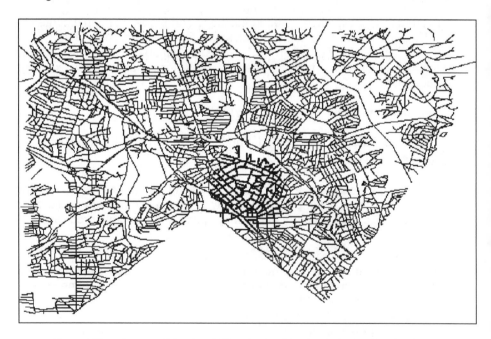

Figure 1 Map of a part of the city of Silver Spring, Maryland, obtained by drawing the value of attribute **line** *of relation* **sspring**.
The highlighted streets (thicker lines) correspond to lines within 1 mile of point (-77.03, 39.0).

This example also illustrates how the application-oriented approach used in SAND can sometimes pay off in terms of a relatively simple implementation. In order to obtain the same results with a full-featured spatial database we would have to use a conventional programming language capable of accessing the query engine (say, embedded SQL). Another alternative would be to use a graphical presentation language like the one proposed by Egenhofer [10].

4.2 Spatially organized tables

In addition to being able to process spatial data at the attribute level, SAND supports two types of spatial access structures implemented as tables: PMR-quadtrees and region quadtrees [19].

A PMR-quadtree table organizes tuples spatially based on one of its attributes whose type must be one of SAND's spatial types, i.e., point, line, rectangle, polygon and region. As with any other table type, tuples in a PMR-quadtree table may be examined by applying methods `first` and `next`, in which case the tuples are retrieved in some arbitrary order. Such a scan order is used in situations where the only requirement is that all tuples be examined exactly once as fast as possible. Additionally, PMR-quadtree tables can be scanned in a number of spatially meaningful orders. Such scan orders are achieved by using special forms of the `first` command. For instance, it is possible to retrieve tuples in order of increasing or decreasing distance from a given feature by using a command called `firstcloseto`. Note that only one command is needed to initiate a scanning order, i.e., one particular form of first. In contrast, the next command performs the function of retrieving successive tuples independently of the scanning order.

The `firstcloseto` command implements incremental spatial ranking as described in [14]. The advantage of its incremental nature is that at any given moment of the scanning process, the algorithm has processed only enough quadtree blocks to determine the position of the current tuple within the scanning order.

It is important to realize the difference between *explicit* scanning orders such as those implemented by sorting a relation or using a B-tree and *implicit* scanning orders such as the one initiated by the `firstcloseto` command. In the first case, the order in which the tuples are retrieved is a consequence of the sequential nature of the table, and therefore is limited to a few variations, e.g., increasing or decreasing value of a B-tree key. In the second case, the ordering is not "pre-computed", that is, the ordering is a function of parameters that define geometric relationships which can be infinitely varied. This is one of the reasons why the processing of spatial data is intrinsically more challenging.

As an example, we will use a PMR-quadtree table defined as an index for attribute `line` of `sspring`, to rewrite the query that returns all tuples within a certain distance of a point. In the script below, this index table, called `sspring.line`, is scanned in such a way as to retrieve the line segments in increasing order of distance from the point of interest. For each qualifying tuple of the index, the corresponding tuple of **sspring** is retrieved by using the **indexfrom** command. The scanning is interrupted if the line segment retrieved from the index is farther than one mile from the point of interest.

```
# Script to list all tuples of sspring in the area
  within one mile
# (approx. 0.01 in map units) of
  point (-77.03, 39.0).
# Uses only the SAND kernel.
set index_handle [sand open sspring.line]
set table_handle [sand open sspring]
set point "point -77.03 39.0"
$index_handle firstcloseto $point
while {[$index_handle status] &&
       [distance $index_handle.line $point]<=0.01} {
   $table_handle indexfrom $index_handle
   puts [$table_handle get]
   $index_handle next
}
$index_handle close
$table_handle close
```

Notice that the above plan could actually have been the plan produced by the library in our previous implementation. This example illustrates how an application is frequently able to obtain the desired results even in the absence of a query plan generator. This characteristic is important for a research tool such as SAND.

Most selection queries based on spatial predicates can be answered efficiently by a examining a PMR-quadtree table using an appropriate scan order. A tuple-by-tuple scan, however, is too poor a mechanism for performing join queries based on spatial predicates -- the so-called *spatial joins*. Spatial join algorithms usually take advantage of the clustering properties particular to the spatial access methods in use. For the type of PMR-quadtree implemented by SAND, for instance, Aref and Samet [4] propose a spatial join algorithm that requires access to the relative order and dimensions of each quadtree block appearing in each   table. The SAND kernel addresses this problem by supporting a special kind of scan iterator for PMR-quadtree tables that implements block-by-block retrieval.

In order to illustrate this concept, consider a query where we want to list all streets within a distance of 1 mile from Georgia Ave. Figure 2 shows a map with the **line** attribute of all tuples that satisfy this query. A possible plan for this query could adopt the following strategy:
1. Perform a select plan to retrieve all tuples of sspring corresponding to Georgia Ave.
2. Create a temporary table tmp to store the line segments of the tuples retrieved by the select plan.
3. Scan the PMR-quadtree sspring.line block-by-block and select those blocks B that lie within one mile of a line segment L in tmp.
4. For each pair B/L, retrieve tuples of sspring whoseline attribute is overlapped by B and test them against L.

5. Print those tuples that qualify.

Figure 2 Example of spatial join. Map of Silver Spring showing line segments corresponding to Georgia Ave (thick line) and all line segments of streets within one mile of Georgia Ave (thin lines).

A SAND script for this strategy is shown below.

```
# Script to list all tuples of sspring  within one
# mile of any other tuple such that name="Georgia".
# Create tmp relation with all lines of Georgia Ave
sand create tmp line line
set table_handle [select_plan sspring name=="Georgia"]
set tmp_handle [sand open tmp]
while {[$table_handle status]} {
    $tmp_handle setfrom $table_handle
    $tmp_handle insert
    $table_handle next
}
$tmp_handle close
$table_handle close

# Scan PMR sspring.line block-wise
set block_handle [sand open sspring.line -blockwise]
while {[$block_handle status]} {
    set block [$block_handle get]
    set tmp_handle [select_plan tmp {[distance line
        $block]<=0.01}]
    while {[$tmp_handle status]} {
        set table_handle [select_plan sspring {[intersects line
            $block]}]
        while {[$table_handle.status]} {
            if [distance $tmp_handle.line
                $table_handle.line]<=0.01 {
                puts [$table_handle get]
            }
```

```
        $table_handle next
      }
      $table_handle close
    }
    $tmp_handle close
  }
  $block_handle close
```

A few notes about the above script:

- In command "**sand create tmp line line**" the first **line** refers to the name and the second to the type of the single attribute in table tmp's schema.

- Command "**$tmp_handle setfrom $table_handle**" copies all attributes with same name from one tuple buffer to the other. In this case, only attribute **line** will be copied.

- Option **-blockwise**, when used for opening a spatially organized table, means that the returned table handle will scan blocks and not tuples. Thus, when the contents of the tuple buffer for the handle is accessed (as in "**$block_handle get**"), a value of type **rectangle** is returned to indicate the position and extent of the block.

This plan demonstrates how the blocks of the PMR-quadtree can be used as bounding boxes (step 3 in the strategy), and thus permit a fast elimination of tuples that cannot satisfy the query. If we analyze this plan more carefully, however, we will notice that the temporary result **tmp** is a simple table, i.e., it has no spatial structure. This forces each block retrieved from the PMR-quadtree to be compared exhaustively with all elements of **tmp**. A possibly better alternative is to use a temporary result which is spatially structured. The idea is to make **tmp** a PMR-quadtree as well. This will allow the usage of a quadtree-based spatial join algorithm (see, for instance, [4]), instead of the tuple-by-tuple nested loop join performed in steps 3 and 4 above. This is a powerful technique, because it prevents the loss of information concerning the spatial proximity of features as multiple steps of a plan are executed.

5. Final considerations

SAND is an on-going project. Most, but not all of the capabilities shown in this paper are implemented. The plan generation and optimization used in the library is still crude and needs to be enhanced. In particular, the heuristics used by the optimizer tend to favor plans using indices whenever possible, which not always results in an optimal plans. Cost estimation is currently not used at all.

In order to improve SAND's plan-generating strategies, we are presently considering a rule-based optimizer in the molds of GRAL [5] and dynamic query optimization as proposed for the Volcano system[8].

In its current state, however, SAND has already been proved of value. For instance, we have built the "Map Browser", an application that uses a simple graphical user interface to answer simple queries. All examples shown in this paper were produced with the aid of the Map Browser. Additionally, SAND has been used for the prototype of an image database [20].

References

[1] D. J. Abel. SIRO-DBMS: A database tool-kit for geographical information systems. International Journal of Geographical Information Systems, 3(2):103--116, April--June 1989.

[2] W. G. Aref and H. Samet. An approach to information management in geographical applications. In Proceedings of the Fourth International Symposium on Spatial Data Handling, volume 2, pages 589--598, Zurich, Switzerland, July 1990.

[3] W. G. Aref and H. Samet. Extending a DBMS with spatial operations. In O. Günther and H. J. Schek, editors, Advances in Spatial Databases - 2nd Symposium, SSD'91, pages 299--318, Berlin, 1991. Springer-Verlag. (also Lecture Notes in Computer Science 525).

[4] W. G. Aref and H. Samet. The spatial filter revisited. In T. C. Waugh and R. G. Healey, editors, Sixth International Symposium on Spatial Data Handling, pages 190--208, Edinburgh, Scotland, September 1994. International Geographical Union Comission on Geographic Information Systems, Association for Geographical Information.

[5] L. Becker and R. H. Güting. Rule-based optimization and query processing in an extensible geometric database system. ACM Transanctions on Software Engineering, 17(2):247--303, June 1992.

[6] R. Berman, M. Stonebraker, and L. Rowe. Geo-quel: A system for the manipulation and display of geographical data. Computer Graphics, 11(2):186--191, 1977.

[7] E. F. Codd. A relational model for large shared data banks. Communications of the ACM, 13(6):377--387, June 1970.

[8] R. K. Cole and G. Graefe. Optimization of dynamic query evaluation plans. In SIGMOD 94, pages 150--160, Minneapolis, Minnesota, May 1994.

[9] D. Comer. The ubiquitous B--tree. ACM Computing Surveys, 11(2):121--137, June 1979.

[10] M. J. Egenhofer. Spatial sql: A query and presentation language. IEEE Transactions on Knowledge and Data Engineering, 6(1):86--95, February 1994.

[11] C. Esperanca and H. Samet. Representing orthogonal multidimensional objects by vertex lists. In C. Arcelli, L. P. Cordella, and G. Sanniti di Baja, editors, Aspects of Visual Form Processing, pages 209--220, Singapore, 1994. World Scientific.

[12] G. Graefe. Volcano--an extensible and parallel query evaluation system. IEEE Transactions on Knowledge and Data Engineering, 6(1):120--135, February 1994.

[13] R. H. Güting. Gral: An extensible relational system for geometric applications. In Proceedings of the 15th International Conference on Very Large Databases, pages 33--44, Amsterdam, August 1989.

[14] G. Hjaltason and H. Samet. Ranking in spatial databases. In M. J. Egenhofer and J. R. Herring, editors, Proceedings of the Fourth Symposium on Spatial Databases, Portland, ME, August 1995.

[15] B. C. Ooi, R. Sacks-Davis, and K. J. McDonell. Extending a DBMS for geographic applications. In Proceedings of the Fifth IEEE International Conference on Data Engineering, pages 590--597, Los Angeles, February 1989.

[16] J. A. Orenstein and F. A. Manola. Spatial data modeling and query processing in PROBE. Technical Report CCA-86-05, Computer Corporation of America, Cambridge, MA, October 1986.

[17] J. K. Ousterhout. Tcl and the Tk Toolkit. Addison-Wesley, April 1994.

[18] N. Roussopoulos, C. Faloutsos, and T. Sellis. An efficient pictorial database system for PSQL. IEEE Transactions on Software Engineering, 14(5):639--650, May 1988.

[19] H. Samet. The Design and Analysis of Spatial Data Structures. Addison-Wesley, Reading, MA, 1990.

[20] H. Samet and A. Soffer. Integrating images into a relational database system. Technical Report CS-TR-3371, University of Maryland, College Park, MD, October 1994.

[21] M. Stonebraker. Inclusion of new types in relational data base systems. In Proceedings of the 2nd International Conference on Data Engineering, pages 262--269, Los Angeles, CA, February 1986.

[22] Economics U.S Department of Commerce and Bureau of the Census Statistics Administration. Tiger/line census files, 1992. Technical documentation, 1993.

[23] A. Wolf. The DASDBS GEO-Kernel: Concepts, experiences, and the second step. In A. Buchmann, O. Günther, T. R. Smith, and Y. F. Wang, editors, Design and Implementation of Large Spatial Databases, Proceedings of the First Symposium SSD'89, pages 67—88. Springer-Verlag, Berlin, 1990. (also Lecture Notes in Computer Science 409).

Contact address
Claudio Esperança[1] and Hanan Samet[2]
[1]Universidade Federal do Rio de Janeiro
COPPE, Programa de Engenharia de Sistemas e Computação
Caixa Postal 68511
Rio de Janeiro, RJ 21945-970
Brazil
E-mail: esperanc@cos.ufrj.br
[2]Computer Science Department and Center for Automation Research and
Institute for Advanced Computer Studies
University of Maryland
College Park
Maryland 20742
USA
E-mail: hjs@umiacs.umd.edu